ROUTLEDGE HANDB
CONTEMPORARY TH

The *Routledge Handbook of Contemporary Thailand* is a timely survey and assessment of the state of contemporary Thailand. While Thailand has changed much in the past decades, this handbook proposes that many of its problems have remained intact or even persistent, particularly problems related to domestic politics. It underlines emerging issues at this critical juncture in the kingdom and focuses on the history, politics, economy, society, culture, religion and international relations of the country.

A multidisciplinary approach, with chapters written by experts on Thailand, this handbook is divided into the following sections.

- History
- Political and economic landscape
- Social development
- International relations

Designed for academics, students, libraries, policymakers and general readers in the field of Asian studies, political science, economics and sociology, this invaluable reference work provides an up-to-date account of Thailand and initiates new discussion for future research activities.

Pavin Chachavalpongpun is Associate Professor at the Center for Southeast Asian Studies, Kyoto University, Japan. He is the chief editor of the online journal *Kyoto Review of Southeast Asia*.

ROUTLEDGE HANDBOOK OF CONTEMPORARY THAILAND

Edited by
Pavin Chachavalpongpun

LONDON AND NEW YORK

First published 2020
by Routledge
2 Park Square, Milton Park, Abingdon, Oxon OX14 4RN

and by Routledge
605 Third Avenue, New York, NY 10017

First issued in paperback 2022

Routledge is an imprint of the Taylor & Francis Group, an informa business

British Library Cataloguing-in-Publication Data
A catalogue record for this book is available from the British Library

Library of Congress Cataloging-in-Publication Data
Names: Chachavalpongpun, Pavin, 1971– editor.
Title: Routledge handbook of contemporary Thailand / edited by Pavin Chachavalpongpun.
Description: Abingdon, Oxon ; New York, NY : Routledge, 2020. | Includes bibliographical references and index.
Identifiers: LCCN 2019032574 (print) | LCCN 2019032575 (ebook) | ISBN 9781138558410 (hardback) | ISBN 9781315151328 (ebook) | ISBN 9781351364881 (adobe pdf) | ISBN 9781351364867 (mobi) | ISBN 9781351364874 (epub)
Subjects: LCSH: Thailand—Politics and government. | Thailand—Social conditions. | Thailand—Economic conditions.
Classification: LCC JQ1745 .R68 2020 (print) | LCC JQ1745 (ebook) | DDC 959.304—dc23
LC record available at https://lccn.loc.gov/2019032574
LC ebook record available at https://lccn.loc.gov/2019032575

ISBN 13: 978-1-03-240104-1 (pbk)
ISBN 13: 978-1-138-55841-0 (hbk)
ISBN 13: 978-1-315-15132-8 (ebk)

DOI: 10.4324/9781315151328

Typeset in Bembo
by Apex CoVantage, LLC

CONTENTS

CONTRIBUTORS

Magnus Andersson is an associate professor in economic geography at the Institute of Urban Research at Malmö University and an advisor on sustainable development for the United Nations Department of Economic and Social Affairs. His research focuses on socio-economic development in contexts with limited access to official statistics and draws primarily on data collected from household surveys and remote sensing data. He holds a PhD in international business from Thammasat University.

Chris Baker has a PhD from University of Cambridge, taught Asian history and politics there, and has lived in Thailand since about 1980. With Pasuk Phongpaichit, he has written widely on the economics, political economy, history and literature of Thailand. In 2017, they jointly won the Fukuoka Grand Prize. He is currently working on the history of Ayutthaya and translation of early Thai literature.

Surachart Bamrungsuk is a professor in the Department of International Relations at Chulalongkorn University, Thailand. He received a BA from the Faculty of Political Science, Chulalongkorn University, an MA from the Department of Government, Cornell University, and an MPhil and PhD from the Department of Political Science, Columbia University. He teaches security and strategic studies in the department, including Thai foreign and security policy. He also heads the Security Studies Project, which produces books and organises conference for Thai security personnel. He has produced a number of publications on strategy, security, terrorism, insurgency, defence policy, democratisation and civil-military relations. He has also edited and written the Security Studies Monograph. His recent books in Thai are *Asymmetric Warfare* (2017), *Militocracy* (2015), *New Security* (2015) and *The Islamic State* (co-author, 2016).

Pongphisoot Busbarat is a lecturer of international relations at the Faculty of Political Science, Chulalongkorn University, Thailand. Prior to this, he was a visiting fellow at ISEAS-Yusof Ishak Institute in Singapore and a research affiliate at Sydney Southeast Asia Centre (SSEAC), University of Sydney. His main research interest is in the role of ideas in international relations and its applications to Southeast Asia, particularly on the issues of Thailand's foreign policy, Southeast Asia's relations with regional powers, and regionalism. He previously held the Dorothy Borg Fellowship in Southeast Asian Studies at Columbia University. He holds a PhD in political science from the Australian National University (ANU) and postgraduate degrees from Columbia University and Cambridge University. He also has practitioner experience as a policy analyst at the Office of the National Security Council of Thailand.

Pavin Chachavalpongpun is Associate Professor at the Center for Southeast Asian Studies, Kyoto University. Pavin teaches Southeast Asian politics at Kyoto University and international relations of Asia at Japan's Doshisha University. He is the chief editor of the online journal *Kyoto Review of Southeast Asia*, where all articles are translated from English into Japanese, Thai, Bahasa Indonesia, Filipino and Vietnamese. He earned his PhD from SOAS, University of London. He is the author of *A Plastic Nation: The Curse of Thainess in Thai-Burmese Relations* (2005) and *Reinventing Thailand: Thaksin and His Foreign Policy* (2010), and is the editor of *"Good Coup" Gone Bad: Thailand's Political Developments since Thaksin's Downfall* (2014) and *Coup, King, Crisis: Time of a Critical Interregnum in Thailand* (2019).

Saksith Chalermpong is Associate Professor in Civil Engineering at Chulalongkorn University where he teaches transport engineering, planning and policy. His research interests include urban transport planning, public and informal transport, and equality issues in transport policy. He has published extensively in the field of transport and has provided consulting services for several government agencies in Thailand, including the Department of Land Transport, Office of Transport Planning and Policy, and Bangkok Mass Transit Authority. He received his bachelor's degree in civil engineering from Chulalongkorn University, his master's degree from MIT, and his doctoral degree from the University of California, Irvine, the latter two in the field of transport.

Paul Chambers serves as a lecturer and special advisor for international affairs at the College of ASEAN Community Studies (CACS), Naresuan University, Phitsanulok, Thailand. He is also a research affiliate at the German Institute of Global Area Studies (GIGA) in Hamburg, the Peace Research Institute Frankfurt (PRIF) in Frankfurt, and the Cambodian Institute for Cooperation and Peace (CICP) in Cambodia. His research interests focus on Southeast Asia in the areas of civil-military relations, democratisation, international politics and the political economy of the Greater Mekong Subregion. He has written numerous books and journal articles about the military, police democratisation and international politics of Southeast Asia. His articles have appeared in *Asian Survey, Critical Asian Studies* and *the Journal of Contemporary Asia*, among others. He has specifically concentrated his research on Thailand, Cambodia, the Philippines, Myanmar and Laos.

Peera Charoenvattananukul is a junior lecturer in international affairs at the Faculty of Political Science, Thammasat University. His research interests primarily focus on ontological security, status-concern theory in international relations, Thai foreign policy and diplomatic history. His publications in Thai are related to the use and misuse of soft power in Thailand, the historical narration of Phibun Songkhram's repetitive resignations during the Second World War and other empirical papers on status-concern theory and Thai foreign policy. His ongoing research project is on the reinterpretation of Thailand's post–Second World War settlements and how the Thai behaviours were driven by the concerns for status.

Michael K. Connors is a lecturer in the Department of International Relations at Xi'an Jiaotong-Liverpool University. He was formerly head at the School of Politics, History and International Relations, University of Nottingham Malaysia and founding director of the Institute of

Asia Pacific Studies (Malaysia). He has taught at La Trobe University, the University of Leeds and Thammasat University. He works in the area of Southeast Asian politics and the international relations of the Asia Pacific. The third edition of his co-authored *The New Global Politics of the Asia Pacific: Conflict and Cooperation in the Asian Century* was published by Routledge in 2018.

Charles David Crumpton received his doctoral degree in public administration and public policy from Portland State University. For 18 years he served as professional city manager, holding senior local government management positions in six US states. As a consultant and researcher in subnational government, he has worked in seven states. He served as director of research for Maryland's judiciary. His research has included subnational governance, justice, child welfare, education, technology innovation in state and local governance, mental health, comparative public administration and policy, and social policy. He currently serves as Senior Research Associate with the Institute for Governmental Service and Research (IGSR) at the University of Maryland, College Park, and teaches research and social policy courses at the University of Maryland, Baltimore. He is also a visiting professor at the College for Local Administration (COLA) at Khon Kaen University in Thailand.

Björn Dressel is an associate professor at the Crawford School of Public Policy at the Australian National University (ANU). His research is concerned with issues of comparative constitutionalism, judicial politics and governance and public sector reform in Asia. He has published in a range of international journals, including *Governance, Administration & Society, International Political Science Review* and *Pacific Review*. He is the editor of *The Judicialisation of Politics in Asia* (Routledge, 2012) and co-editor of *Politics and Constitutions in Southeast Asia* (Routledge, 2016). You can follow him on twitter @BjoernDressel.

Eli Elinoff is a lecturer in the Cultural Anthropology Programme in the School of Social and Cultural Studies at Victoria University of Wellington. He received his PhD in anthropology in 2013 from the University of California, San Diego. His research there examined struggles over land, citizenship and politics in the growing northeastern Thai city of Khon Kaen. Subsequently he took up a postdoctoral fellowship in Asian urbanisms, jointly appointed at the Asia Research Institute and the Department of Sociology at the National University of Singapore. His research publications appear in *South East Asia Research, Political and Legal Anthropology Review, Journal of the Royal Anthropological Institute, Contemporary Southeast Asia, City, Cultural Anthropology Online*, and *New Mandala*. His new research project, funded by the Royal Society of New Zealand's Marsden Grant, explores political ecologies of concrete across Thailand and Southeast Asia.

Anders Engvall is a research fellow at Stockholm School of Economics, where he does research and teaches on the political economy of Thailand and Southeast Asia. He has worked on Thailand's southern conflict for more than a decade, largely in collaboration with Deep South Watch (DSW). His work with DSW has focused on developing a conflict monitoring system and quantitative analysis of the Thai southern conflict. In addition to his research, he has worked as a trade advisor at the Swedish Ministry for Foreign Affairs. He holds a PhD from Stockholm School of Economics.

Federico Ferrara is Associate Professor at the City University of Hong Kong, Department of Asia and International Studies, where he also serves as Associate Head of Department. A graduate of Harvard University, where he was awarded a PhD in political science and government in 2008, Federico's writings on Thai politics and comparative political institutions have been published in a variety of books and academic journals. Most prominently, he is the author of *The Political Development of Modern Thailand*, published in 2015. He is now working on a new, single-authored monograph seeking to improve upon existing explanatory theories of institutional reproduction, institutional decay and institutional change.

Jim Glassman received his PhD in geography from the University of Minnesota in 1999. He is currently Professor in the Department of Geography at the University of British Columbia. His research focuses on the political economy of development, social struggle and geopolitical conflict in East and Southeast Asia, with an emphasis on state theory. He is the author of *Thailand at the Margins: Internationalisation of the State and the Transformation of Labour* (2004), *Bounding the Mekong: the Asian Development Bank, China, and Thailand* (2010) and *Drums of War, Drums of Development: The Development of a Pacific Ruling Class and Industrial Transformation in East and Southeast Asia, 1945–1980* (2018).

Kevin Hewison is Weldon E. Thornton Distinguished Emeritus Professor of Asian Studies at the Department of Asian Studies at the University of North Carolina at Chapel Hill. He has recently held visiting positions with the Institute of China Studies at the University of Malaya, the Centre for Southeast Asian Studies at Kyoto University and the Department of Political Science at the University of Stockholm. He is the editor in chief of the *Journal of Contemporary Asia* and a Fellow of the Academy of Social Sciences in Australia.

Patarapong Intarakumnerd is a professor at National Graduate Institute for Policy Studies (GRIPS) in Tokyo. His research interests include innovation systems in Thailand and newly industrialising countries in Asia, industrial clusters, technological capabilities of latecomer firms, innovation financing and the roles of research technology organisations and universities. He is a regional editor and a member of international editorial boards of several international journals related to innovation management and policies. He has worked as an advisor/consultant for the World Bank, UNESCO, UNCTAD, OECD, JICA, German Development Institute, International Development Research Centre of Canada (IDRC) and the Economic Research Institute of ASEAN and East Asia (ERIA).

Charnvit Kasetsiri is a professor emeritus of Thammasat University, Thailand. He is a prominent historian and Thai studies scholar. He earned his PhD in 1972 from Cornell University. He served as lecturer of history at Thammasat from 1973–2001 and founded, in 2000, the Southeast Asian Studies Programme. He was president of Thammasat University in 1995–1996. He has written approximately 200 articles and a number of publications on Thai and Southeast Asian history. He is a co-author, along with Pavin Chachavalpongpun and Pou Sothirak, of *Preah Vihear: A Guide to the Thai-Cambodian Conflict and Its Solutions* (2013). Charnvit was awarded with a Fukuoka Academic Prize 2012, Japan, and the DCAS 2014 (Distinguished Contributions to Asian Studies) by the Association for Asian Studies, USA.

Arne Kislenko is a history professor at Ryerson University and an instructor at Trinity College, University of Toronto. He teaches modern international relations history, including courses on the world wars, the Cold War, espionage, terrorism, comparative foreign policy and modern Southeast Asia. He has won numerous teaching awards, including the prestigious 3M National Teaching Fellowship, the inaugural Province of Ontario Leadership in Faculty Teaching Award and Ryerson's first President's Award for Teaching. He was also named Ontario's "Best Lecturer" following *TV Ontario's* first "Academic Idol" series. His books include *Culture and Customs of*

Laos (2009) and *Culture and Customs of Thailand* (2004). He currently serves as book review co-editor for the prestigious journal *Intelligence and National Security*.

Vanessa Lamb is a lecturer in the School of Geography at the University of Melbourne with a focus on human-environment geographies and political ecology of Southeast Asia. She completed her dissertation, *Ecologies of Rule and Resistance*, focused on the politics of ecological knowledge and development of the Salween River at York University's Department of Geography in 2014. Her professional experiences include policy analysis and research into the social dimensions of environmental with Oxfam, International Rivers and UN Women. She was also the inaugural Urban Climate Resilience in Southeast Asia (UCRSEA) research fellow at the Munk School of Global Affairs at the University of Toronto and is an affiliated researcher with the York Centre for Asian Research, York University, Toronto, Canada.

Rattana Lao is Programme Officer: Policy and Research at the Asia Foundation. Previously, she worked as a head of the Thai Studies International Programme at Pridi Banomyong International College, Thammasat University, where she taught Thai politics and social economic development policy in Thailand. She was also a post-doctoral fellow at the Faculty of Education, University of Hong Kong. She obtained a doctorate in comparative and international education (political science) from Teachers College, Columbia University. She has published a book, book chapters and research papers on globalisation and higher education reforms. Her book *A Critical Studies of Thailand's Higher Education Reform* was published by Routledge in 2015.

Tomas Larsson is a senior lecturer in the Department of Politics and International Studies at the University of Cambridge and a fellow of St John's College. He has a PhD in government from Cornell University. He is the author of *Land and Loyalty: Security and the Development of Property Rights in Thailand*, published in 2012. In recent years, his research has increasingly focused on religion and politics, and especially on various aspects of state regulation of Buddhism in Southeast Asia, resulting in a number of publications in journals such as *International Political Science Review, Modern Asian Studies* and *Journal of Law and Religion*.

Grichawat Lowatcharin is a lecturer in the College of Local Administration, Khon Kaen University, Thailand. He received a PhD in public affairs from the Harry S. Truman School of Public Affairs, University of Missouri. His research interests include local governance, decentralisation, fiscal federalism, intergovernmental relations and transparency. He is also interested in combining and applying different methodological approaches – from qualitative methods and geographical information system to survey and quasi-experimental designs – to address policy issues. His work appears in *State and Local Government Review, Social Science Asia, Thai Journal of Public Administration*, and *Local Administration Journal*. He also published several policy papers for the University of Missouri Extension and has conducted several service evaluation research studies for local governments in Thailand.

Kanokwan Manorom is Associate Professor and a director of Mekong Subregion Social Research Center (MSSRC) based at the Faculty of Liberal Arts, Ubon Ratchathani University, Thailand. She earned her PhD in rural sociology from the University of Missouri–Columbia, USA. She was granted the Australian Leadership Award (ALA) as a visiting scholar based at the University of Sydney in 2011. She has been working in Thailand and the Mekong region in fields such as political economy and resource governance, livelihoods analysis and development and impact assessment. She had worked on the Mekong Fellowship Project under Water, Land and

Ecosystem Programme (WLE) Greater Mekong during 2015–2017. Since 2014, she has been collaborating with researchers from the University of Wisconsin–Madison and the University of Alberta on importance of local/indigenous knowledge on sustainable resource governance.

Justin Thomas McDaniel's research foci include Lao, Thai, Pali and Sanskrit literature, art and architecture, and manuscript studies. His first book, *Gathering Leaves and Lifting Words*, won the Harry Benda Prize. His second book, *The Lovelorn Ghost and the Magic Monk*, won the Kahin Prize. He has received grants from the National Endowment for the Humanities (NEH), Mellon, Rockefeller, Fulbright, Pacific Rim (PACRIM), Luce, and the Social Science Research Council (SSRC), among others. He has won teaching and advising awards at Harvard University, Ohio University, the University of California and the Ludwig Prize for Teaching at University of Pennsylvania. His recent book is on modern Buddhist architecture, titled *Architects of Buddhist Leisure* (2017). His forthcoming work includes edited books on Thai manuscripts, Buddhist biographies and Buddhist ritual. For his chapter in this volume, Justin co-wrote with **Sara E. Michaels**, who is currently an undergraduate at the University of Pennsylvania studying Buddhism and Asian Studies. She grew up in Bangkok, Thailand, as a practicing Buddhist.

Titipol Phakdeewanich is a dean of the Faculty of Political Science at Ubon Ratchathani University. He completed his PhD in politics and international relations. It has been a key component of his work to highlight the promotion and protection of human rights, the plight and injustice in the lives of the rural poor, and to look towards finding actual solutions to these problems, which can have a tangible positive effect on the lives of the underrepresented as well as the disenfranchised and marginalised groups within Thailand. Under the military rule as a result of the 2014 coup, he has been closely monitored by the junta because of his work on democracy and human rights. However, he remains committed to work to support the promotion of democracy and human rights in his country.

Matthew Phillips is a lecturer in Modern Asian History at the Department of History and Welsh History at Aberystwyth University in the United Kingdom. He completed his PhD at the School of Oriental and African Studies (SOAS) in 2013. His work focuses on the culture of the Cold War in Thailand with a particular focus on propaganda, popular culture and urban space. He has written for the *New York Times* and the *Guardian* newspapers, among other publications, and also previously worked for the BBC World Service.

Pasuk Phongpaichit is a professor of economics at Chulalongkorn University. She has a PhD from the University of Cambridge and has been a visiting professor at universities in Australia, the United States, and Japan. With Chris Baker, she has written widely on the economics, political economy, history and literature of Thailand. In 2017, they jointly won the Fukuoka Grand Prize. Recently, her work has focused on inequality.

Krislert Samphantharak is an associate professor of economics and Associate Dean at the School of Global Policy and Strategy at University of California, San Diego. He also serves as an advisor at Puey Ungphakorn Institute for Economic Research at the Bank of Thailand. His specialisations are in finance and economic development. His current research includes studies on household enterprises and household debt, and he is working on a book manuscript on comparative development of Southeast Asian economies. He received his PhD in economics from the University of Chicago. He previously held visiting positions at the International Monetary Fund, UCLA, University of Tokyo, and Yale-NUS College.

Douglas Sanders is Professor Emeritus, Faculty of Law, University of British Columbia, Canada (1977–2003); Emeritus Visiting LL.M. Professor, Faculty of Law, Chulalongkorn University, Thailand (1999 to 2009); Academic Associate, Institute of Human Rights and Peace Studies, Mahidol University, Thailand. He is a Canadian citizen, resident in Thailand since 2003. His writings are on indigenous and tribal peoples, cultural minorities and LGBTI issues. Notable articles on sexuality issues are "Getting Lesbian and Gay Rights on the International Human Rights Agenda", *Human Rights Quarterly* (1996); "The Unnatural Afterlife of British Colonialism in Asia", *Asian Journal of Comparative Law* (2009); and "What's Law Got to Do with It? Sex and Gender Diversity in East Asia", in *Routledge Handbook of Sexuality Studies in East Asia*, edited by Mackie McLelland (2014). Email: sanders_gwb@yahoo.ca.

Apichat Satitniramai received his PhD in development studies from the University of Wales, United Kingdom, for his thesis *The Rise and Fall of the Technocrats: The Unholy Trinity of Technocrats, Ruling Elites and Private Bankers and the Genesis of the 1997 Economic Crises*. He currently works as Associate Professor at the Faculty of Economics, Thammasat University, Thailand. His most recent book, *Thai State and Economic Reform: From the Origin of Banker's Capitalism to the 1997 Economic Crisis* (2013, in Thai), received the Toyota Thailand Foundation Book Award on Social Sciences. His research interests pertain to political economy and politics of development, particularly in Thailand.

Saichol Sattayanurak was a Thammasat University student during 1973–1976. After several years of teaching at the Department of History, Faculty of Arts, Chulalongkorn University, she moved to the Department of History, Faculty of Humanities, Chiang Mai University. She is currently a professor emeritus of this institution. Most of her published works are about history of ideas, ideologies and discourse on the Thai nation and Thainess. Her works in progress are *Thai Scholars and Contesting Democracy in Thai Society amidst Political Crisis, 2005–2014* and *Methodology and the Body of Knowledge on Human and Culture from the Works of Thai Humanities Scholar, Anthropologists and Political Scientists*. She is currently researching on the value system and regime of emotion of the Thai middle class.

Wolfram Schaffar served as professor for development studies and political science at the University of Vienna between 2010 and 2018. Prior to this position, he worked at the Department of Southeast Asian Studies, University of Bonn, at the Faculty of Political Science, Chulalongkorn University in Bangkok and at the Royal Netherlands Institute of Southeast Asian and Caribbean Studies (KITLV) in Leiden, Netherlands. His fields of interest are state theory of the Global South, social movements, new constitutionalism and democratisation processes, and new authoritarianism, with a regional focus on East and Southeast Asia.

Edoardo Siani is an anthropologist specialising in Buddhism and power in Thailand. He is a researcher at Kyoto University's Center for Southeast Asian Studies, and he taught anthropology at Thammasat University and SOAS, University of London. Edoardo received a PhD in anthropology and sociology from SOAS with a thesis on divination and politics in Bangkok. A long-time resident of Bangkok, he previously worked in education and journalism. He also served as an interpreter for the Thai police.

Aim Sinpeng is a lecturer in the Department of Government and International Relations at the University of Sydney. She is a co-founder of the Sydney Cyber Security Network and a Thailand Coordinator for the Sydney Southeast Asia Centre. Her publications on Thailand include articles in *Pacific Affairs, Contemporary Southeast Asia, Asian Politics & Policy, Journal of Information Technology & Politics* and *Asian Journal of Social Science*. She has recently received multiple research grants to study how social media has shaped contemporary politics in Southeast Asia.

Claudio Sopranzetti is a postdoctoral research fellow at All Souls College, Oxford University and a research associate at the Oxford Programme for the Future of Cities. He is the author of *Red Journeys: Inside the Thai Red Shirts* (2012) and *Owners of the Map: Motorcycle Taxi Drivers, Mobility, and Politics* (2018). He occasionally writes for *Al Jazeera* and a number of international magazines.

David Streckfuss is an honorary fellow with the University of Wisconsin–Madison who has lived in Thailand for more than 20 years. He is interested in legal history, nationalism and ethnic identities. His book *Truth on Trial in Thailand: Defamation, Treason, and Lèse-majesté* was published by Routledge in 2011. He was one of the co-authors of the recent biography, *King Bhumibol: A Life's Work*, published by Editions Didier of Singapore. He also occasionally has pieces published in the *Bangkok Post*, the *New York Times*, the *Wall Street Journal Asia* and *Al Jazeera*.

Khemthong Tonsakulrungruang is a lecturer in constitutional and administrative laws at Chulalongkorn University, Thailand. His research interest includes freedom of expression, constitutional development of Thailand, and Buddhism and laws. He is currently a PhD candidate at the University of Bristol, the United Kingdom.

Peter Warr is John Crawford Professor of Agricultural Economics, Emeritus, at the Australian National University. He studied at the University of Sydney, the London School of Economics and Stanford University, where he received his PhD in economics. His current research is on the relationship between economic policy and poverty incidence in Southeast Asia. He is a fellow of the Academy of Social Sciences in Australia and is a distinguished fellow and past president of the Australian Agricultural and Resource Economics Society. He has been a visiting professor of economics at Thammasat University and at Chulalongkorn University in Bangkok and was for many years the executive director of the ANU's National Thai Studies Centre.

FIGURES

MAPS

TABLES

PART I

The history

1

INTRODUCTION

A timeless Thailand

Pavin Chachavalpongpun

In 1967, Sir Anthony Rumbold, the British ambassador to Bangkok, was leaving his post and preparing to return home. He wrote a ten-page confidential document echoing his thoughts about the country in which he had served during the past two and a half years – Thailand. The document, which the ambassador hoped would be useful for his successor, explained in great detail how he perceived Thailand, particularly during the 1960s at the height of the Cold War. Rumbold talked about the domination of Bangkok, the rigid structure of Thai society and the rules which govern it, as well as the unwillingness of Thais to assume responsibility and the existence of endemic corruption. He went on to discuss Thailand's political characteristics and the Thai attitude towards their king – in this specific context, King Bhumibol Adulyadej (1946–2016). He began this report by saying:

> There is a theory that the Thais are rather easier for Europeans to understand than are other oriental people. I do not believe this theory. It seems to me that Sino/Indian/Malay/Thai ways of thought are so alien to ours that analogies between events in South-East Asia and events in Europe are nearly always misleading, that forecasts based on such analogies are bound to be wrong, that the motives of Asians are impossible for us to estimate with any exactness, and that Thailand and the Thais offer no exception to these precepts. The general level of intelligence of the Thais is rather low, a good deal lower than ours and much lower than that of the Chinese. But there are a few very intelligent and articulate ones and I have often tried to get some of these with whom I believe myself to be on close terms to come clean with me and to describe their national characteristics as they see them themselves and to explain why they behave in this way rather than in that way. The result has never been satisfactory.[1]

Rumbold also wrote:

> Thailand is governed by a benevolent dictatorship without a dictator. It is benevolent in the sense that it does its best according to its lights to promote the welfare of the people. . . . The dictatorship is embodied jointly in the two military leaders, Field Marshal Thanom Kittkachorn, Prime Minister and Minister of Defence, and General Prapass Charusathiana, Minister of Interior and Commander-in-Chief of the Army. . . .

> But Field-Marshal Thanom has the backing of the King and enjoys a greater degree of general popularity and goodwill than does General Prapass.[2]

Rumbold's description of Thai political characteristics still rings true today. Thailand in the 1960s is not much different, in many respects, than Thailand in the first two decades after the millennium. Something in the kingdom has remained intact and in some cases even moved backwards. Certainly, there have been some great developments in the past decades, particularly the economic growth and the opening up of society, the process in which one sees the traditions of Thailand being challenged by new ideas and the emergence of social media, which gives rise to the voice of the younger generation.

While admitting that Thailand has seen changes since the time of Rumbold, this chapter finds that much of what the British ambassador discussed in his report is worth revisiting. The document also brings up the pertinent question of how one defines "contemporary". Where is the beginning of the contemporary period in Thailand? How far can one go back in time and yet define a given period as contemporary in order to explain the present? Since this handbook is about "contemporary Thailand", it is imperative to make clear at this stage the starting point of Thailand in the contemporary period. There is a benefit in so doing. Contributors in this volume were requested to examine their chosen topic and its current situation. But they were also encouraged to go back into the past, particularly the past that has produced persistent implications in the present. In this process, such past can be considered as a part of being contemporary. This allows readers to understand better the topics based on their historical context. The developments or failures of Thailand, like any other country, owe much to its past. To look at contemporary Thailand, topics discussed in the volume cannot be totally divorced from the reality of yesteryear. Hence, "contemporary" in this volume carries a meaning of "continued impacts" on aspects of today's Thai life.

The contemporary life of Thailand

Thailand's contemporary politics might be defined as protracted, convoluted and even violent. For the longest time, many Thais have rested their trust in one key institution – not the democratic institution but rather the monarchy. It is true that the journey of the Thai monarchy on the political road has not always been agreeable. Following the revolution in 1932, which put an end to the centuries-old absolute monarchy, the royal institution has faced numerous challenges, most of which threatened its existence or questioned its anachronism. The first decade after the 1932 revolution witnessed some attempts on the part of the elites to promote democracy, then at its infant stage, while reaching for a compromise with the powerless monarchy. The abdication of King Prajadhipok in 1935 and the mysterious death of King Ananda Mahidol in 1946 further obscured the position of the monarchy in politics. The decline of the monarchy in this period, however, failed to strengthen democracy as a result of internal struggles among different factions within the realm of Thai politics. By the late 1950s, King Bhumibol, having already been on the throne for more than a decade, began a lifelong project of reviving the glorious days of the monarchy, with firm backing of the military represented by Field Marshal Sarit Thanarat. Together, they crafted a version of neo-royalism whereby the monarchy would once again become central to Thai political life. In so doing, not only was the position of the monarchy elevated, but the military itself exploited the king in order to guarantee its interest in the political arena. Defending the monarchy was made equal to defending national security. Hence, the existence of the royal institution in many ways entrenched the prominent role of the military in politics – the role that has solidified over the years until now.

Bhumibol found an ally in the army, Sarit. The process of constructing the neo-royalist ideology commenced and was guided under the newfound alliance between the monarchy and the military. In this process, Bhumibol was reinvented into the utmost venerable king. Thai historian Thongchai Winichakul argues that the embodiment of neo-royalism rests on three important characteristics: "being sacred, popular and democratic".[3] Bhumibol was sacralised and transformed into Dhammaraja. He was down to earth and citizen-centric. But he was also protected by the draconian lèse-majesté law, which forbids anyone from insulting the king, the queen, the heir apparent and the regent.[4] Accordingly, Sarit resurrected old rituals, such as prostration before the king, which was once declared "uncivilised" by King Chulalongkorn (1968–1910). The purpose of reviving old rituals was to elevate Bhumibol to the level of demi-god, hence sacralising the royal institution that would command unconditional allegiance from the people. He boosted his popularity mainly through his royal developmental projects designed to supposedly improve the livelihood of those in marginalised regions. He embarked on endless trips throughout the country, meeting with his subjects, listening to hardship stories and handing out assistance, spiritual or material – introducing the concept of populism long before other politicians. The exhaustion from journeying thousands of miles was portrayed through an image of the hardworking king with "sweat on the tip of his nose". As for his fondness of democracy, Bhumibol's occasional interventions in politics, particularly during the tumultuous periods, earned him the status of a "stabilising force". In intervening in politics bringing an end to political conflict, Bhumibol presented himself as a political alternative – one that was imbued in morality. Thai politics was gradually turned into a realm of ethics, with the monarchy standing on righteousness and politicians on immorality. As Rumbold referred to this, "The god-like position of the King is questioned by nobody".[5]

In this process too, the military openly strolled into the political domain under the pretext of defending the country's key institutions – the nation and the monarchy. The Cold War of ideological conflict provided the perfect arena for the military to deepen its roots in politics. The threat of communism was looming in Thailand's neighbouring states. As Indochina became a stronghold of communists in the region, Thailand became the front-line state, directly encountering the menace to its security. Thai despots during the Cold War successfully made communism so extrinsic to Thai identity, so in the meanwhile, ironically, dictatorship became intrinsic to Thai nationhood. Had communism swept Thailand, then Thainess undoubtedly would have become communist. As the war against communism was brewing, the United States stepped in to play a significant role in the kingdom. The US government, in safeguarding its interests in the region, offered strong support for the monarchy and the military in combating the communist threat. Together, the monarchy, the military and the United States built closely knitted relations, lasting unto the present day. Painting the face of enemies was now a necessary mission as part of legitimising the despotic regime at home.[6] While external enemies were assigned to Indochinese communists, internal enemies were members of the Communist Party of Thailand (CPT). When students at Thammasat University rose against Thai despots, they were accused of being communists and thus dealt with brutally.

The end of the 1970s, traumatised by the massacres of Thammasat students, served as a wake-up call for the monarchy and the military to reassess their position amid public fury with the state's brutality. Neo-royalism was put in doubt, whether the royal institution should be taken as a political alternative. In 1979, Bhumibol initiated another self-reinvention project in order to distance himself from the dark days at Thammasat. He invited the BBC to interview him, a discussion which focused on the essence of neo-royalism. Called "Soul of a Nation", it offered a narrative on the "normal" life of the royal family inside the walls of the palace. While one objective behind this interview was showcasing the royal family as the main pillar holding the Thai nation together against the threat of communism, the other goal was to portray the monarchy as a devoted institution. In this documentary, Bhumibol talked succinctly about his selfless care of his people and his kingdom. The documentary was elaborately staged; in some scenes Bhumibol was sitting on the floor with layers upon layers of blueprints flooding the entire room, signifying yet again a working monarch with a common touch. The emphasis on the people-centric royal institution was accompanied by the restructuring of the network monarchy. And the key figure behind the success of this network was General Prem Tinsulanonda.

The network for royal hegemony

Network monarchy, a term coined by Duncan McCargo, is a useful framework in analysing Thai politics.[7] The best way to understand Thai politics is to look at it as a kind of network. The most powerful political network from the 1970s until at least the arrival of Thaksin Shinawatra as premier had been the network monarchy. It consists of powerful institutions outside the parliamentary structure, namely the military, judiciaries, senior bureaucrats, powerful businesses and royalists. Prem was handpicked by Bhumibol to serve as prime minister for two terms, from 1980 to 1988. At the beginning of the Prem regime, the communists continued to play a role as "the other". But the American withdrawal from the region as a result of its loss in the Vietnam War compelled Thailand to readjust its position vis-à-vis the communist bloc. In search of a new security provider, Thailand established diplomatic relationship with China in 1975 during the Kukrit Pramoj government, four years before US President Jimmy Carter signed a joint communiqué normalising bilateral relations with his Chinese counterpart, Deng Xiaoping. It was another twist in the story of Thai nationhood. Just as Thailand depicted communists as number-one enemies throughout the Cold War, the reconciliation with China raised the question of what really constitutes "otherness" in the world of Thai nationhood. While defining communist Vietnam as a clear and present danger, Thailand defined communist China as a reliable friend. The notion of Thai nationhood was therefore malleable according to the power interests of the Thai political elites.

Domestically, Prem was credited for creating a template of an unelected but relatively effective prime minister during his eight-year tenure.[8] A former army chief, he took advantage of his profound connection with the military in order to sustain political stability at the same time as King Bhumibol continued to firm up his moral and sacred persona. Prem was known as the key engine behind the network monarchy, particularly after he stepped down from the premiership. From 1988, Bhumibol appointed Prem as president of the Privy Council, an advisory body of the king. In the post-Prem government era, the network monarchy remained politically active, working intimately with powerful institutions in placing the monarchy at the top of the political structure. In so doing, the network monarchy eagerly strove to shape the contour of Thai politics in which civilian governments were kept weak and vulnerable. Had they proven to be challenging and strong, they would be removed from power by military coup, hence explaining the frequency of military intervention in politics since 1991, when Prime Minister Chatichai Choonhavan was toppled that way. From 1991 until 2014, there were three military coups – on average, one coup every seven and a half years.

A year after Chatichai was overthrown, in 1992, General Suchinda Kraprayoon, the coup leader, appointed himself as prime minister, irking the middle class in Bangkok so immensely that they took to the streets in protest. The demonstrations ended tragically with military crackdowns on protesters: 52 were killed, many disappeared, hundreds were injured, and 3,500 were arrested. Yet the bloodshed, later known as the Black May incident, failed to cast a shadow over the network monarchy. If anything, the dreadful incident paved the way for the monarchy to re-emphasise its dominant role as the stabilising force. The king summoned the leaders of the two opposing camps, Suchinda and his nemesis, protest leader Major General Chamlong Srimuang. The meeting took place at the royal palace, televised nationwide, with Bhumibol sitting on the sofa with Suchinda and Chamlong below on the floor. The footage of the royal audience on the night of 20 May 1992 lived on to become Thailand's most memorable political event of the 20th century and undeniably the peak of the royal hegemony. But this peak would see its decline a decade later.

In the aftermath of the Prem regime and the Black May, a consortium of intellectuals, scholars, students and civil society was calling loudly for democracy. Prem's military-guided democracy no longer fit the new circumstances. Meanwhile, the method of eliminating elected governments by military coups, manipulated by the network monarchy, as seen in the case of Chatichai, became stridently rejected. The brutal crackdowns on the protesters in the Black May also brought shame on the military so damningly that it considered returning itself to the barracks throughout the 1990s. This decade therefore witnessed serious democratisation which culminated in the writing of the 1997 constitution, dubbed the people's charter designed to advance far-reaching reforms in the country's democratic development. Scholars argue that the 1997 constitution unintentionally paved the way for Thaksin to take Thai politics by storm and indeed consolidate his grip on power – which, ironically, indirectly precipitated the conditions for the 2006 coup that ousted his own government.[9]

The launch of the 1997 constitution coincided with the financial crisis of 1997 that hit the Thai economy devastatingly. After decades of rapid economic growth following the successful transformation of the Thai industry from predominantly agriculture to emerging manufacturing, the financial crisis erupted and gravely impacted economic and social systems. The crisis originated from problems in numerous sectors: finance, real production, government and management. Since the financial liberalisation of the early 1990s, foreign capital had been attracted to Thailand by ballooning stocks, high interest rates and the relatively lower risk in Southeast Asia, due to the Thai baht being pegged to the US dollar. With lower interest rates for offshore loans and a perceived fixed exchange rate, the Thai private sector continued to borrow and cash poured in. This increased the burden of foreign debts. Without effective management and supervision, increasing capital inflows mostly came in the form of short-term loans and went into speculative rather than productive sectors. Excessive private investment, particularly in risky and non-tradable sectors, and property price inflation soon led to a bubble economy.[10] The crisis prompted the Bank of Thailand to float the baht on 2 July 1997. This also led to the collapse of the Chavalit Yongchaiyudh government on 8 November later the same year. His successor, Chuan Leekpai, from the Democrat Party was compelled to apply for a US$17.2 billion loan from the International Monetary Fund (IMF). From 1997 to 2001, the mood for democratisation and the economic vulnerability provided pivotal ingredients for the shift of the political landscape long dominated by royal hegemony.

Bhumibol had long been known as the richest monarch in the world, with an estimated wealth of US$30 billion (in 2014, two years before he passed away).[11] His overwhelming wealth was sustained by the Crown Property Bureau (CPB), as investment arm of the Thai monarchy. Porphant Ouyyanont argues that the CPB was particularly vulnerable because of its dependence on the performance of two private companies in which the CPB was a major stakeholder. Both companies, the Siam Commercial Bank and the Siam Cement Group, were in sectors hit hard by the financial crisis. However, the CPB survived the crisis by making significant changes in its own management and investment policies by promoting vigorous reforms in two affiliated companies. As a result, as Porphant asserts, the CPB emerged with an income significantly higher than its peak pre-crisis level.[12] The real challenge for the network monarchy by this time was clearly not the economic crisis. A bigger threat came in the form of a political challenge from a new breed of politician, Thaksin, who had tremendously conquered electoral politics in Thailand.

Thaksin as game changer

A former police officer, Thaksin reincarnated to become a new prototype of politician with impressive backgrounds in business. A tycoon from the north, Thaksin was different from other political elites; he climbed up the economic and social ladders in competition with the network monarchy. He founded the Thai Rak Thai (TRT) party, translated as "Thais love Thais", signifying his intention to ride on nationalism at a time when the country lost its "economic independence". After years of lacking real democracy and credible political parties which would task themselves to devote for the marginalised, TRT won a landslide election in 2001, convincing many in Thailand of the return of the majority rule. A part of the TRT success rested on its implementation of populist programmes engineered to improve the livelihood of the far-flung regions of Thailand, particularly the north and the northeast regions. These two regions combined produced the majority of eligible voters in the country. Winning these two regions, therefore, augmented a chance to win the majority in the election. The era of Thaksin commenced with great promise, including the empowerment of the electoral process, a fairer contest for political power, greater access to political resources for rural residents and a better management of economic policy that would distribute wealth more proportionally throughout the nation. Soon, Thais began to compare two kinds of populism: one à la Bhumibol and the other à la Thaksin. The comparison between the two players here might be unintentional, but certainly inevitable.

While Bhumibol inculcated national development supposedly in a sustainable manner, Thaksin was assertive in promoting economic productions both from within and outside the Thai economic realm. While Bhumibol propagated his self-sufficiency philosophy, Thaksin encouraged cutthroat consumerism. While in the past Bhumibol handed out "royal pouches" to poor villagers in which they found basic commodities for daily use, Thaksin showered them with high-tech gadgets. Times had changed by the time Thaksin sat comfortably in his position of power. Yet the network monarchy was convinced that the Thaksin phenomenon was a flash in the pan. In early 2003, the Thaksin government managed to repay Thailand's outstanding debts from the previous IMF bailout. In a ceremony celebrating the so-called declaration of economic independence, Thaksin was seen waving a Thai flag, arousing patriotic sentiment among Thais, an act that seemed to mimic a medieval monarch who declared political independence from inimical Burma. In 2005, Thaksin won his second term in office in an even greater landslide, making him the first and only prime minister of Thailand to have served a full four-year term.

Thaksin could have construed his electoral success as a sign of political confidence, yet others might have interpreted it as a political arrogance. His unstoppable electoral victories started to worry the network monarchy. Why so? For a long while, the network monarchy had little interest in participating in electoral politics, and as a result it failed to invest in maintaining its political influence through the electoral channel. Thaksin, riding the wave of democracy, chose to play the game of electoral politics. But it became a game in which the network monarchy felt it had little or no control over the outcome. When the TRT won again in 2005, some individuals identifying themselves as part of the network monarchy voiced its discontentment of the Thaksin government. The anti-Thaksin campaign was thus launched that year, initially accusing Thaksin of committing and condoning corruption. These accusations were not groundless, since Thaksin did make several mistakes caused by his ill judgement and greed. Cronyism became a norm in his government. Amending laws to facilitate his family businesses was another thing. Censoring the opposition and the media was well known. Blurring the line of national interests and personal interests in his conduct of foreign policy was also documented. On top of these, he was called to take responsibility for the iron-fist policy towards the Muslim community in the Deep South and the war on drugs (the latter of which killed more than 2,500 people alone).

But what really invigorated the 2005 anti-Thaksin demonstrations, now under the flagship of the right-wing People's Alliance for Democracy (PAD), was the accusation against Thaksin of being disloyal to the monarchy. Members of the PAD started to wear yellow shirts, since yellow is the colour of King Bhumibol, who was born on a Monday. They also held portraits of the king and the queen to delegitimise the Thaksin government. Exploiting the royal symbols this way, apparently with the green light from the palace, effectively pitted the monarchy against Thaksin. The numbers of protestors grew over time, most of which were the middle class in Bangkok, who might once have demonstrated against Suchinda in 1992. Beneath the psyche of the protestors hid not only their inner drive to defend the revered monarchy but also their own interests, aligned with the monarchy, which they believed were now under attack from the Thaksin government. The rise of the "other" middle class in the once marginalised communities, made possible by the consumerist policy of Thaksin, allowed them to compete with their fellow middle class in Bangkok. Suddenly, the playground for the middle class became crowded. Fearing the mounting competition for political and economic resources, the middle class in Bangkok strove to delegitimise the new money from the rural areas as being "political pawns" of Thaksin. Thus, the war on Thaksin had greater ramifications. To many in Bangkok, it was as much a battle to deny the political and economic rights of the rural residents.

The military returned to the political ring and ultimately resorted to the same old trick in ridding the country of the Thaksin government – a coup on 19 September 2006. On the night of the coup, an image was captured showing the coup-makers, led by General Sondhi Boonyaratglin, having an audience with King Bhumibol and Queen Sirikit, supposedly reporting on the political development. Also in the meeting was Prem, the "CEO" of the network monarchy, suggesting his involvement and manipulation in the move to undermine Thaksin. The coup of 2006, if anything, indicated a heightened anxiety on the part of the network monarchy in regards to new political challenges and the popularised electoral politics. However, the coup occurred not without a total interruption. There emerged an anti-coup group, whose members later chose to wear red shirts to distinguish themselves from the Yellow Shirts. At first, the Red Shirts were careful not to be depicted as hostile towards the monarchy. But the emergence of the Red Shirts, most of which had been core supporters of Thaksin, added another dangerous element to the ongoing conflict. It has further divided the society not only along ideological lines but also along different classes, regions and economic statuses. It gave birth to a long-drawn crisis now known as the colour-coded conflict, between Yellow and Red, which has come to redefine the Thai political landscape.

The network monarchy picked General Surayud Chulanond, another key figure trusted by Bhumibol, to lead the post-Thaksin government. By late 2007, the Surayud government organised an election, being confident that the military had successfully deracinated the political influence of Thaksin and therefore could control the electoral contest. The network monarchy was wrong. Thaksin nominated a veteran politician, Samak Sundaravej, to head a new party, the People's Power Party (PPP), to undisputable triumph. Samak's victory was a slap in the face of the network monarchy. The victory also reaffirmed Thaksin's lasting influence in politics even though he had been in exile overseas. The PAD returned to the streets of Bangkok, finding ways to undermine the Samak government. In mid-2008, his government offered strong support for Cambodia's bid to have the Preah Vihear Temple listed as a UNESCO World Heritage Site.

The Preah Vihear, an ancient Hindu Temple built in the 9th century, has been a thorn in Thai-Cambodian relations for over a century, since both countries claimed ownership of the temple. In 1962 the case was taken to the International Court of Justice, in which the verdict was given in favour of Cambodia. In 2008, the PAD accused the Samak government of giving away a 4.6 km^2 area surrounding the temple to Cambodia in the process of supporting the latter's bid at UNESCO in exchange for the personal interests of Thaksin, who had forged amicable ties with Cambodian Prime Minister Hun Sen. But the stirring up of nationalistic fervour failed to drive Samak from power. In the end, the Constitutional Court intervened and ordered the resignation of Samak on the ground of conflict of interests. Samak was charged for keeping another job while serving as prime minister – a celebrity chef in a televised programme. The intervention from the Constitutional Court marked the new beginning of a serious move of the judiciary in eliminating enemies of the network monarchy of which it had become a part.

Dubbed the "judicial coup", the political interference of the court to protect the interest of the network monarchy has since become increasingly active. Following the forced resignation of Samak, Thaksin nominated his brother-in-law, Somchai Wongsawat, to lead the PPP government. Applying pressure on the government, the PAD seized Suvarnabhumi International Airport and Don Muang Airport for a week in December 2008, costing the Thai economy at least US$100 million a day in lost shipment value and opportunities.[13] As in the case of Samak, Somchai was similarly ousted but not because of protests. The Constitutional Court charged that an executive member of Somchai's party committed electoral fraud and therefore handed down the resolution to dissolve the party. Somchai himself was banned from politics for five years. This departure of the PPP left a vacuum of power, allowing the Democrat Party to form a government from the minority in a backroom deal brokered by the military. But the advent of the Democrat party with Abhisit Vejjajiva as prime minister failed to bring the political stability that the military would have hoped. Growing frustrations in the Red Shirt camp about the unjust verdicts of the court on the two PPP governments, and the perceived illegitimate entry into politics of the Democrat party, drove the Red Shirts to stage months of protest in the heart of Bangkok's financial district. They called for Abhisit to step down and for fresh elections. But in 2010, what they got instead were deadly crackdowns and one of the most heartless incidents in the modern history of Thailand which left corpses littered next to luxury department stores. Almost 100 were killed and 2,500 were injured. Shortly after the crackdowns, the Central World department store was burned down. Immediately, the blame was placed on the Red Shirt camp, giving birth to a new discourse of *pao baan pao muang*, or burning down the home and the city, as a stigma that has tainted the Red Shirts ever since. In April 2011, Princess Chulabhorn openly castigated the Red Shirts for *pao baan pao muang*, a cause for the deterioration of the already ill health of the king. She said, "The last time (the city's burning) was caused by the Burmese. This time, it was more devastating since it was caused by Thais".[14]

The volatility of the political situation mainly stemmed from the start of the long goodbye of the Bhumibol era. Bhumibol had been in and out of the hospital from 2009. A vital pillar of the network monarchy, Bhumibol by then was bedridden, sparking immense angst among supporters of the monarchy regarding the looming loss of royal hegemony. The fact that, since 2006, Bhumibol seemed to have shifted his *modus operandi* from pulling the political strings from behind the scene to playing politics at the forefront, as seen in, for example, the meeting he had with the coup-makers of 2006, prevented him from performing as "force of stability" since he no longer appeared to be neutral and above politics as he liked to portray himself in the past. But Bhumibol was not to be blamed alone. Indeed, the apprehension of the loss of royal hegemony influenced other members of the royal family to directly intervene in politics. On 13 October 2008, Queen Sirikit and her daughter Chulabhorn together attended the controversial funeral of Angkana Radubpanyawut, nicknamed Nong Bo, a Yellow Shirt member who was killed in a protest. At the funeral, Sirikit praised the courage of Nong Bo, who met with an untimely death while participating in the royalist mob led by the PAD against the Somchai government. A large number of Red Shirts were enraged by the obvious partiality of the queen and Chulabhorn. That event now serves as "National Enlightenment Day".[15]

In 2011, Abhisit finally agreed to hold an election. This time, Thaksin nominated his youngest sister, Yingluck, to contest in the polls. The killings of the Red Shirts and the perceived mistreatments against successive Thaksin-backed governments induced his supporters to collectively vote for his new party, Pheu Thai ("For Thais"). The Thaksin faction once again won a landslide election that further irked the network monarchy. Statistically, during the ten years from 2001 to 2011, Thaksin and his proxies had been able to dominate the electoral process. Yet their governments faced ferocious resistance from their enemies, which ended in either military or judicial coups. Yingluck, in the premiership, used her charm to reinforce her position and continued to implement populist policies to confirm loyalty among her supporters. Among many attractive policies was the rice-pledging scheme, which offered Thai farmers the opportunity to pledge and then provide an unlimited supply of rice to the government at a higher price for their crops that they would obtain by selling them at market rates. The ultimate aim was to increase rice prices to safeguard farmers from middlemen.[16] But the scheme was politicised by the enemies of Thaksin and Yingluck; they accused her government of corruption in executing the project.

The Yingluck government also proposed in parliament the blanket amnesty which would have applied to offences committed during the political turmoil after the 2006 coup. This meant that the amnesty could set her brother, Thaksin, free from outstanding charges.[17] The rice-pledging scheme and the amnesty bill provided an excellent opportunity for the anti-Thaksin camp to discredit the government, create a situation of ungovernability, and hence, invite the military to intervene in politics. The Yellow Shirts made a homecoming by setting up the People's Democratic Reform Committee (PDRC), led by Suthep Thuagsuban, from the Democrat Party. The PDRC occupied the financial district of Bangkok for months, pressuring Yingluck to step down. In fact, she did step down and called for a fresh election in February 2014. But the PDRC boycotted the election and harassed voters – and some of their actions were violent. In May 2014, the military decided to stage a coup against the Yingluck government, almost eight years after her brother was removed from power the same way.

The royal succession

The timing of the 2014 coup is worth discussing. King Bhumibol was in a critical medical condition. His wife, Queen Sirikit, although politically active a few years earlier, had been suffering the effects of a stroke since 2012. Royalists worried about the end of the Bhumibol era, yet they became more worried about the coming of the new reign under Vajiralongkorn, the only heir apparent. It is logical to argue that the coup of 2014 was more than just a ploy to purge the Shinawatras and the "Red germs" from politics.[18] The royal succession was imminent. The Crown Prince was not loved. Earlier, members of the network monarchy openly expressed anxiety about Vajiralongkorn being enthroned.[19] The rumours of Thaksin being intimately connected to Vajiralongkorn added to a level of apprehension among the royalists who regarded this political battle as a kind of zero-sum game – either we or they would take it all. The solution was to launch another coup to ensure a smooth and trouble-free royal transition, with the military overseeing the entire process. Vajiralongkorn was willing to debunk the assumption about close ties with Thaksin. He went along with the military in managing his enthronement in a very top-down fashion. Even before Bhumibol's death, Vajiralongkorn commenced the reorganisation of the palace, replacing his father's close confidants with his own trusted men. By the time Bhumibol passed away on 13 October 2016, Vajiralongkorn already had control over palace affairs.

He was crowned on 1 December 2016, although the official ceremony was not until 4 May 2019 (note that the official coronation was a three-day celebration from 4–6 May 2019). During this interval, Vajiralongkorn gleefully made known his political ambition. For example, he intervened in the constitutional drafting process, requesting that provisions related to the monarchy be amended, most prominently to allow him to reside overseas without having to appoint a regent. He has taken Munich as his semi-permanent home, yet he will be able to rule Thailand from his German palace. He restructured the Privy Council, removing some old councillors and filling the positions with military men. He kept Prem as the president of the Privy Council but made him powerless – perhaps as revenge for past insults. He also reformed the Crown Property Bureau. Assets previously registered to the CPB would from June 2018 be held "in the name of His Majesty". In other words, Vajiralongkorn took sole control of the CPB, erasing any ambiguity about the owner of this super-rich organisation.[20] Kevin Hewison argues, "The monarchy now holds more formal power than any king since 1932. The king and the military have an accommodation built around the military's capacity for repression".[21]

Under Bhumibol, political stability was key to the flourishing reign. Politics was predictable. Benefits were shared among major stakeholders. Underpinning Bhumibol's strength was his unsurpassed ability to accumulate moral authority through the invigoration of the neo-royalist ideology. Under Vajiralongkorn, because of his lack of moral authority, the new reign is utilising fear as a mechanism reinforcing his position.[22] There have been cases of those working and then falling out with him being detained and imprisoned. Some have died mysteriously. Others have disappeared. These unexplainable incidents further sully the king's reputation. The ruthless reputation is also juxtaposed with his bizarre lifestyle. Vajiralongkorn has been seen numerous times dressing strangely in Germany, such as wearing a tiny crop top with fake Yakuza-style tattoos, often by the side of his mistress Sineenat Wongvajirapakdi, also known as Koi. The images of them strolling the streets of Munich were published in German media, introducing his mistress long before he wedded Suthida Vajiralongkorn na Ayutthaya, a former air crew member, now the formal queen of Thailand. They married on 1 May 2019, a few days ahead of the grand coronation ceremony. Finally, on 28 July 2019, Sineenat was elevated to become his royal consort, a position last anointed almost a century ago.

In the political domain, the crisis lives on. After almost five years in power, the military government of General Prayuth Chan-ocha hosted an election on 24 March 2019. But nothing is guaranteeing a return of democracy. The junta rewrote the constitution – an exercise that effectively turned back the clock for Thai democracy. The constitution empowers the Senate over the House of Representatives. The senators are all appointed by the junta, guaranteeing the permanent power of the military in politics. The popularity of Thaksin has remained high, but it is not enough to overturn the Thai situation. A rising star has emerged onto the political scene: the new party, Future Forward, with a promising leader, Thanathorn Jungrungruangkit, a self-made billionaire turned politician. With open confrontation with the military on the cards, Thanathorn has become the current target of annihilation. Meanwhile, the force behind the network monarchy, Prem, passed away on 26 May 2019, raising a question of the future trajectory of the most powerful network of Thailand. King Vajiralongkorn has his ways of dealing with his supporters as well as his enemies, and this is not necessarily through the network monarchy. But his ways can be erratic and unduly violent.

Returning to Ambassador Rumbold in his 1967 report to the British Foreign Office, he stated, "All political, economic and social changes of any importance in Thailand are the result of calculations and decisions taken by men in Bangkok and reflect the development of relationships between men or groups of men in Bangkok". In 2019, this situation has changed little. Thailand remains timeless. The Thai destiny is determined by powerful men, mostly in Bangkok (and one in Munich), which does not necessarily generate a positive result for those outside the sphere of power.

The final note

To the best of my ability, I strove to bring together chapters that represent Thailand from various aspects, as widely as possible. For this purpose, I divide this handbook into four parts: the history, the political and economic landscape, the social development and the international relations. In Part I on the history, this book offers the reader some basic background of the modern history of Thailand, the national historiography and the country's early days of economic development. Part II delves into the political and economic life of Thailand, ranging from the issue of democracy, the military, the monarchy, the lèse-majesté law, the Yellow Shirts, the Red Shirts, the police, the judiciary, local governance and public administration, the economy, international trade, industrialisation and transportation. Part III on social development presents topics on national identity, Buddhism, the conflict in the Deep South, classes and races, education, human rights, gender, social media, non-governmental organisations (NGOs) and the environment. In Part IV on international relations, the discussion includes the country's foreign policy, Thai policy with Great Powers, and the Thai role in the Association of Southeast Asian Nations (ASEAN).

I would like to sincerely thank all the contributors who enthusiastically shared their knowledge on their chosen subject while exercising their professionalism from high ground, the result of which saw the completion of this handbook.

Notes

1 "Goodbye Thailand", Sir Anthony Rumbold to Mr. Brown, Foreign Office and Whitehall Distribution, DS 1/6, Thailand 18 July 1967, pp. 1–2.
2 Ibid., p. 8.
3 Thongchai Winichakul, "Toppling Democracy", *Journal of Contemporary Asia*, Vol. 38, No. 1 (February 2008), p. 15.
4 Pavin Chachavalpongpun, "Neo-royalism and the Future of the Thai Monarchy: From Bhumibol to Vajiralongkorn", *Asian Survey*, Vol. 55, No. 6 (2015), p. 1194.
5 "Goodbye Thailand", p. 2.
6 See Pavin Chachavalpongpun, "The Necessity of Enemies in Thailand's Troubled Politics: The Making of Political Otherness", *Asian Survey*, Vol. 51, No. 6 (November/December 2011), pp. 1019–1041.
7 See Duncan McCargo, "Network Monarchy and Legitimacy Crises in Thailand", *The Pacific Review*, Vol. 18, No. 4 (January 2006).
8 Pavin Chachavalpongpun and Joshua Kurlantzick, "Prem Tinsulanonda's Legacy – and the Failures of Thai Politics", *Council on Foreign Relations*, 28 May 2019 <www.cfr.org/blog/prem-tinsulanondas-legacy-and-failures-thai-politics-today?utm_medium=social_share&utm_source=tw> (accessed 29 May 2019).
9 Erik Martinez Kuhonta, "The Paradox of Thailand's 1997 'People's Constitution': Be Careful What You Wish For", *Asian Survey*, Vol. 48, No. 3 (May–June 2008), pp. 373. 373–392.
10 Sauwalak Kittiprapas, "Thailand: The Asian Financial Crisis and Social Changes", in *The Social Impact of the Asia Crisis*, edited by T. Van Hoa (London: Palgrave Macmillan, 2000), pp. 35–36.
11 Jack Linshi, "These Are the 10 Richest Royals in the World", *Time*, 1 June 2015 <http://time.com/3904003/richest-royals/> (accessed 3 June 2019).
12 Porphant Ouyyanont, "The Crown Property Bureau in Thailand and the Crisis of 1997", *Journal of Contemporary Asia*, Vol. 38, No. 1 (2008).
13 "Air Cargo Terminal to Open Soon", *Bangkok Post*, 2 December 2008.
14 On 5 April, Chulabhorn appeared on a TV programme, *Woody Kerd Ma Kui*, saying, "The incident last year, in which we witnessed *pao baan pao muang*, brought a great sorry to His Majesty the King".

See "Nailuang Songtook Hed Pao Baan Pao Muang" ["The King is Saddened by the Arson of Our Home"], *Manager Online*, 5 April 2011 <https://mgronline.com/daily/detail/9540000042546> (accessed 30 May 2019).

15 Pavin Chachavalpongpun, "Princess Chulabhorn's Politics", *New Mandala,* 14 January 2014 <www.newmandala.org/princess-chulabhorns-politics/> (accessed 30 May 2019).

16 See details from the Public Relations Department (PRD) of Thailand <https://hq.prd.go.th/PRTechnicalDM/ewt_news.php?nid=1790>.

17 "Thailand Senate Rejects Controversial Amnesty Bill", *BBC*, 12 November 2013 <www.bbc.com/news/world-asia-24903958> (accessed 30 May 2019).

18 Thongchai Winichakul explained the event at Chulalongkorn Hospital on 30 April 2010 involving the Red Shirts, in his article on the "germs" as the reds' infection of the Thai political body. In "Thongchai Winichakul on the Red 'Germs'", *New Mandala,* 3 May 2010 <www.newmandala.org/thongchai-winichakul-on-the-red-germs/> (accessed 30 May 2019).

19 See this report in WikiLeaks: <https://wikileaks.org/plusd/cables/10BANGKOK192_a.html> (accessed 30 May 2019).

20 "As the Army and Politicians Bicker, Thailand's King Amasses More Power", *The Economist,* 3 January 2019 <www.economist.com/asia/2019/01/03/as-the-army-and-politicians-bicker-thailands-king-amasses-more-power> (accessed 30 May 2019).

21 Kevin Hewison, "The Monarchy and Succession", in *Coup, King, Crisis: Time of a Critical Interregnum in Thailand*, edited by Pavin Chachavalpongpun (Singapore: NUS Press, 2019).

22 See Claudio Sopranzetti, "From Love to Fear: The Rise of King Vajiralongkorn", *Al Jazeera,* 11 April 2017 <www.aljazeera.com/indepth/opinion/2017/04/thailand-junta-king-vajiralongkorn-170411102300288.html> (accessed 30 May 2019). Also see Pavin Chachavalpongpun, "Kingdom of Fear (and Favour)", *New Mandala,* 18 April 2017 <www.newmandala.org/kingdom-fear-favour/> (accessed 31 May 2019).

References

"Goodbye Thailand". 1967. Sir Anthony Rumbold to Mr. Brown, Foreign Office and Whitehall Distribution, DS 1/6, Thailand 18 July.

Hewison, Kevin. 2019. "The Monarchy and Succession". In *Coup, King, Crisis: Time of a Critical Interregnum in Thailand.* Edited by Pavin Chachavalpongpun. Singapore: NUS Press.

Kuhonta, Erik Martinez. 2008. "The Paradox of Thailand's 1997 'People's Constitution': Be Careful What You Wish For". *Asian Survey*, Vol. 48, No. 3 (May–June): 373–392.

McCargo, Duncan. 2006. "Network Monarchy and Legitimacy Crises in Thailand". *The Pacific Review*, Vol. 18, No. 4 (January): 499–519.

Pavin Chachavalpongpun. 2015. "Neo-royalism and the Future of the Thai Monarchy: From Bhumibol to Vajiralongkorn". *Asian Survey*, Vol. 55, No. 6: 1193–1216.

———. 2011. "The Necessity of Enemies in Thailand's Troubled Politics: The Making of Political Otherness". *Asian Survey*, Vol. 51, No. 6 (November/December): 1019–1041.

Porphant Ouyyanont. 2008. "The Crown Property Bureau in Thailand and the Crisis of 1997". *Journal of Contemporary Asia*, Vol. 38, No. 1: 538–540.

Sauwalak Kittiprapas. 2000. "Thailand" The Asian Financial Crisis and Social Changes". In *The Social Impact of the Asia Crisis.* Edited by T. Van Hoa. London: Palgrave Macmillan.

Thongchai Winichakul. 2008. "Toppling Democracy". *Journal of Contemporary Asia*, Vol. 38, No. 1 (February): 11–37.

2

THAILAND IN THE LONGUE DURÉE

Chris Baker and Pasuk Phongpaichit

The present inherits the legacies of the past. Histories written in an era of nation-building seek the roots of the present-day nation. They tend to project the contemporary back into the past. Early 20th-century histories of Thailand pictured 13th-century Sukhothai as a society of Thai rice growers governed by a benevolent king and enjoying a liberal economy – uncannily reminiscent of Thailand in the mid-20th century. This approach diminishes change. In this chapter, we examine five aspects of Thailand's history over roughly a millennium, following time's arrow from past to present. This approach casts historical change in sharper relief. Contemporary Thailand has been shaped by changes in the recent past, particularly in the last century, but with legacies of the deeper past remaining under the surface.

Landscape and resources

Thailand lies in the tropical and subtropical zone and is affected by both the typhoon system of the South China Sea and the Indian Ocean monsoon. High heat and plentiful water generate a rich and varied biomass. Until recently, the country was carpeted with forest – deciduous forest to the north, tropical rainforest further south, and mangroves along the coasts. For humans, these forests were dangerous as they were home to carnivorous animals, poisonous reptiles, disease-carrying insects, and many germs, viruses, and parasites. Malaria and other fevers transmitted by insects ensured the population grew slowly and remained low.

But the forests were also spectacularly fecund, offering food to eat, timber for building, and fibres for dress, while the warm waters of the rivers and coastal seas bred fish in great density. The early hunter-gatherers ate birds, deer, monkeys, crocodiles, rhinos, turtles, frogs, fish, and shellfish, and many different fruits, roots, and seeds. Rice agriculture, which arrived around 2000 BCE, did not replace hunter-gathering but supplemented it.

Until late in the second millennium CE, the impact of humankind on this landscape was limited in extent and intensity. People clustered in settlements in order to keep some distance from the dangerous forest and have allies for defence against predators. On the plains, they favoured sites at river bends which could be converted into moats. On the uplands, they chose mounds ringed by moats to retain water for consumption and agriculture. In the northern hills, the immigrant Tai settled in broad patches of valley with room for cultivation.

Rice and other crops were planted close by the settlements. Much of the diet was still found by everyday forms of hunting and gathering – fish trapped from the waterways, animals hunted from nearby forests, fruits and vegetables found by the wayside, sugar and liquor from a palm. The line from Sukhothai Inscription I of 1292 stating, "There is fish in the water and rice in the fields", is a delightfully understated advert for food security. Early European visitors to Siam were struck by the "superabundance" of food and materials. Jeremias Van Vliet in the 1630s concluded that "Siam is a country that has more than most other countries of everything that the human being needs".[1] Forests also supplied the exotic goods that Siam exported to China.

Engineering works for water management were limited to canals for moating and local distribution of water. The extraction of goods from the forests was small in relation to the stock, except perhaps in the case of elephants and deer. The most highly developed areas of design and manufacture were fish traps, small river craft, and timber housing.

The human impact on the landscape began to change in the 18th century. From 1722, Siam exported increasing amounts of rice to counter a persistent food shortage in China's southwest.[2] People left the cities and escaped bondage to settle in villages and plant rice to meet this demand, creating a "land frontier" that transformed the landscape and the society. From the 1760s this trend was interrupted by the Siam-Burma wars, but it resumed in the early 19th century. Market gardening took off in the lower reaches of the delta and sugar plantations on the fringes. From mid-century, more and more land was cleared and planted with rice to feed the populations in colonised Asia that no longer produced their own food. The frontier spread through the Chao Phraya Plain, from south to north and inner to outer; along the river valleys of the north; down the coastal basins on the east coast of the peninsula; and then, when railways provided access in the 1920s, in the northeast. Under crop diversification policies launched in the 1950s, cultivation of sugar, soyabean, cassava, and maize spread across the rolling uplands of the northeast and lower north and up the lower slopes of the mountain ranges. From the 1980s, plantations of soft fruit (lychee, lamyai) pushed further up these slopes.[3] The low-impact cultivation regime of the Ayutthaya era had been replaced by multiple cropping and high usage of chemicals.

From the mid-18th to the mid-19th centuries, the area under paddy expanded by 50,000 hectares a year. For the following half century, the area under upland crops grew at the same rate. Only at the approach of the millennium did the momentum of the frontier slacken, stop, and slightly reverse.

Higher up the hills, the forests had also diminished. From the late 19th century, colonial timber firms moved into the northern forests where they decimated the high-value timber, especially teak. With the expansion of urban and export demand after 1950, loggers harvested a much wider range of timber, and by the 1970s they were clear-cutting whole hill ranges. The army built roads to facilitate the loggers in order to deny forest sanctuary to communist rebels. From the 1980s, under the cover of "reforestation", government granted concessions for plantations of fast-growing trees for the paper industry. Only after a fatal landslide in 1988 did government revoke all logging concessions, but this had only partial effect.

The upper reaches of the hills were occupied by small numbers of swiddening hill farmers. After an influx of people during World War II and the Indochina War, the area was linked to the international narcotics trade and became the world's major source of opium. From around 1960, government gradually closed down the opium trade and converted the poppy fields to cultivation of coffee, flowers, and temperate fruit and vegetables.[4] By 2000, the forests had been reduced from 80–90 per cent of land area to around 15 per cent. Reforestation schemes had limited results.

Until the early 19th century, the rivers flowed freely, though several bypass canals had been cut to shorten the route from Ayutthaya to the sea. In the early 19th century, several transverse canals were cut through the lower delta to serve as military highways but also attracted agricultural settlement. In the 1880s a grid of canals was dug north of Bangkok to convert over 200,000 hectares of swamp into paddy fields. This Rangsit scheme became the model for many similar schemes in the central plain, northern valleys, and southern coastal basins.[5]

The first major dam, built on the Ping River in 1951 and named after King Bhumibol Adulyadej, was designed to provide hydroelectricity, flood control, and some irrigation. Also in the 1950s, a barrage was constructed across the Chao Phraya River at Chainat to control the rate of water flow in the lower delta.[6] Dam building continued apace until the 1990s, when a wave of protests against displacement brought the momentum to an end. The waterways, which had once been the main highways, were no longer usable for transport except on isolated stretches.

By 2000, the natural abundance of resources from the pre-modern era had disappeared – exploited to underpin the economic growth of the first two centuries of the Bangkok era.

Population and settlement

Until recently, the population of Mainland Southeast Asia was very sparse, perhaps a fifth or a tenth of the density in India, China, and Java. The main reason was disease, particularly the insect-borne parasites that flourish in tropical forests, such as malaria, and the bacteria and viruses which cause epidemics, such as cholera and smallpox. The region's legends are full of epidemics.

The population grew slowly. The prehistoric sites found by archaeology are small and sparse. From the mid-first millennium CE, some 40 towns are known in the lower Chao Phraya Plain, with the largest at Nakhon Pathom measuring 3.7 km by 2.0 km. In the later first millennium more settlements appeared in the upper reaches of the Chao Phraya system. From the 12th to the 14th centuries, many new towns appeared, suggesting a time of modest population growth.

In the early 15th century, warring increased in frequency and intensity. Greater use of elephants extended the range of armies and spurred the ambitions of princes. Forced conscription created massive armies whose size astounded early European visitors. Gunpowder weapons and foreign mercenaries boosted the death rates in battle. Moving armies across the landscape exposed the men to disease, and famine and disease spread to local settlements. Over the 15th and 16th centuries, population must have decreased – perhaps drastically. Around 1600 warfare petered out, partly because people resisted recruitment by revolt, flight, bribery, and taking the robe.[7]

The Dutch who came to Ayutthaya a generation later commented on the sparseness of the population. Jeremias Van Vliet reckoned that "owing to former wars large tracts along the frontiers have grown wild".[8] A half century later, French cleric Nicolas Gervaise recorded that

> The forests of this kingdom are so enormous that they cover over half of its area and so dense that it is almost impossible to cross them . . . there are fearful deserts and vast wildernesses where one only finds wretched little huts, often as much as 7 or 8 leagues distant from one another.[9]

Ayutthaya conducted censuses as part of military recruitment, but the only figure that has survived, recorded by Simon de la Loubère in the 1680s, is "Nineteen Hundred Thousand Souls" (1.9 million).[10] The area covered is not specified but probably included the Chao Phraya Plain below the hills and some adjacent territory. Multiplying the figure up following the current population distribution gives around 3.3 million, but the outer areas were probably sparser than today.

The period 1600–1760 was a time of peace and rising prosperity, and hence probably of population increase. From 1760, a half century of warfare between Siam and Burma again reduced the population by death and flight. Western Siam was severely depopulated and had only begun to recover in the 1830s. Chiang Mai "had become a jungle overgrown by climbing plants . . . where rhinoceroses, elephants, tigers, and bears were living".[11] The upper part of Lanna was not repopulated until the last quarter of the 19th century.

In 1800, the population was probably little different from La Loubère's time, with around two million in core Siam and perhaps three million in the area of modern Thailand. It then grew moderately fast from the "bounce" following war and depopulation and from an aggressive policy to import people including war prisoners from Lao, Khmer, Malay, and Tai/Shan regions as well as refugees from the Mon country. Sir John Bowring reported that the population reached 4.5 to 5 million at mid-century, and Prince Dilok estimated seven million in the early 1900s.[12]

Medicine then removed the restraints. Vaccination was introduced to counter cholera epidemics from the 1900s. A programme to eradicate malaria was begun after World War II and malaria deaths were reduced from 206 per 100,000 in 1949 to 2 per 100,000 in 1987. Improvements in sanitation, nutrition, and medical care reduced infant mortality. Population growth surged to a peak of 3 per cent a year in the 1950s. Over the 20th century, Thailand's population increased almost tenfold to reach 62 million. Under the impact of economic growth and urbanisation, a rapid "demographic transition" of falling birth rate and family size brought the rate below 1 per cent by 2000. The population is now almost static around 70 million, and it will soon begin to decline.

In the Ayutthaya era, the population was concentrated in cities, towns, and their immediate environs. Ayutthaya, with an estimated population of 300,000, contained around a sixth of the total in Siam, and another dozen large cities housed perhaps another third.[13] The agrarian frontier drew the population away to the countryside. By 1950, over four-fifths of the population was rural. The first movements of the land frontier drew people in from the outlying areas, particularly Lao speakers from the east. The later decades reversed the flow, drawing people away to the uplands in the northeast, lower north, and down the peninsula. People were highly mobile, and local populations became very mixed. Some two-thirds of all today's village settlements were founded in the past century.

The urban population grew slowly until 1970 but then accelerated to almost half the total by 2010. Urban growth was at first concentrated heavily in Bangkok, which became a primate city (disproportionately larger than any other in the country). From the 1990s, a dozen provincial centres also expanded rapidly.

Political organisation

Early historians of Southeast Asia were fascinated by the idea of far-flung empires which were a feature of their own time. Nationalist historians were keen to push the modern geo-body back into the past. The resulting works tended to magnify the scale of early political organisation. In practice, technology and environment imposed severe limits. Jungle-clad land was hard and dangerous to traverse. Water was the best medium of transport, but the monsoon climate meant rivers were prone to flooding and drying and the coastal waters were subject to storms.

The early polities were city-states – a central place with a ruler and a penumbra of linked settlements. The Tai term *mueang* (*meng, mang*, etc.) for such configurations became the local word for a polity, later transferred to larger forms.

As a result of increasing population and trade, four coalitions of city-states appeared in the 12th to the 14th centuries: to the north, later known as Lanna; on the middle Mekong River, later known as Lan Xang; in the middle reaches of the Chao Phraya system, called the Northern Cities by Ayutthaya; and Siam around the headwaters of the Gulf of Thailand. Occasionally an imposing and charismatic leader might unite one of these groupings, but such phases rarely survived his death. These were coalitions of city-states, not an expansion of political scale.

Gradually a handful of places emerged as more prominent than others. These places were well sited for trade: Chiang Mai on the route southward from China, Ayutthaya at the outlet to the sea, and Sukhothai/Phitsanulok at the central crossroads. These places invested their trading wealth in armies. These armies raided neighbours for the resources of state-building: people, craftsmen, weapons, sacred objects, and precious materials. These resources were used to build splendid temples and palaces, patronise artistic production, and attract revered monks. Rulers of lesser places honoured the prominence of these emerging capitals out of respect for their military strength and cultural splendour, but there were no administrative links beyond ties of kinship and marriage between ruling families.

From the early 15th century, new technologies transformed warfare. As the first guns came from China, the advantages first accrued to the inland states. But by the mid-15th century the sea had become the main supply route, giving the advantage to port cities on the lower reaches of the rivers. Ayutthaya extended its influence not only over the other city-states of Siam but also over the Northern Cities and ports down the west coast of the peninsula. But the rulers of these other city-states still considered themselves as kings, built capitals in the royal style, and used dynastic titles. In the vocabulary of the time, there were different grades of king, with an *ekkaraja* (or primary king) at the top. The Ayutthaya ruler might call on others to attend at his court, supply troops, or offer a daughter in marriage, but there was no formal system of subordination resembling vassalage.

Trading increased in the 17th century, partly on European demand but more from the exchange between Asian empires ranging from Turkey and Persia in the west to China and Tokugawa Japan in the east. Ayutthaya flourished as a major entrepôt in this Asian trade. The crown took the lion's share of trading revenues through monopolies and taxes, and it used the proceeds to buy goods and services from all over, including the services of administrators from Japan, China, and Turkey. With their help and new techniques, Ayutthaya tightened its control over outlying regions. It extended influence down the peninsula using a Chinese-style tribute system. It established an outpost at Korat on the Isan Plateau, though Lanna and Lan Xang remained beyond its feasible control. It imposed control over subordinate kings by installing a resident envoy to act as Ayutthaya's spy, demanding local rulers spend part of the year in the capital – as in the Tokugawa system – and ultimately replacing some local kings with central appointees.

This gradual integration was disrupted by the Burmese sack of Ayutthaya in 1767. The territory fractured into several sub-states.[14] King Taksin and the Chakri rulers were able to revive trade, establish Bangkok as a successor to Ayutthaya, and rebuild its military hegemony, but they left the provinces under their local rulers for a century. Bangkok concentrated on bringing the city-states of Isan and western Cambodia under a tributary system which provided goods demanded in overseas trade.

The arrival of Western colonialism in the region prompted major changes. First, Siam was obliged to compete with Britain and France to enclose the outer territories within the bounded units known as countries – long familiar in Europe. These boundaries were defined in treaties between 1884 and 1907.[15] Second, with new military technologies from the West – repeating guns (Gatling guns, carbines), a professional standing army, the railway, and the telegraph – Bangkok was able to extend its hegemony over Lanna and parts of Lan Xang and to defeat the dissent this created. Third, the colonial powers introduced new techniques of bureaucratic control, especially the district commissioner system, which Thai royals went to see at work in India and Java and then adapted for local use. Bangkok sidelined local rulers, including the Lanna kings, and imposed a centralised, colonial-style bureaucratic hierarchy.[16] In the years around 1900, Thailand came into being as a territorial entity and a state apparatus. This territorial entity had no historical roots or social coherence. It was pieced together in somewhat haphazard competition with the colonial powers. It included areas where the main language was not Thai but Lao, Khmer, Malay, or a hill language. It included areas such as Lanna, which had never fallen under Siam's control, and much of Isan, where Siam's influence was recent and shallow.

The power of this new state was initially limited by its slim resources of men and money. With economic growth after World War II, state revenues and the number of state personnel increased. From the 1950s, the adoption of responsibility for development, especially economic development, gradually created a closer relationship between the state and more and more people in the nation.

Language and identity

The sparseness of the population in the past, the lack of any forbidding borders (mountains, deserts), and the long coastline meant that mainland Southeast Asia was always open to flows of new people.

The earliest inscriptions found in the lower Chao Phraya Plain and dated to the 6th century are in Khmer or Mon (and Pali-Sanskrit), suggesting these were the local languages. Inscriptions were made in Thai in Lanna from the 13th century.[17] The articles deposited in the crypt of Wat Ratchaburana at Ayutthaya, probably in the 1420s, are inscribed in Thai, Khmer, Mon, and Chinese. The scribe of the Zheng He voyages in the 1420s noted that Ayutthaya already had a settlement of Chinese and another 500 "foreigners".[18]

Thai differs from other languages in the Tai family because it has absorbed so much from Khmer (and Mon), not just loanwords but also structural elements, such as ways to form words and sentences. This hybridised language must have evolved from three language communities living in close proximity over a long period. The fact that some inscriptions were written in Thai and Khmer respectively on two sides of the stele also hints at a bilingual society.

Malay peoples from the archipelago settled on the peninsula, often as refugees from warfare. After Islam took root in the 15th century, traders, professionals, and religious men arrived from the Islamic world. At Ayutthaya from the early 16th century, Portuguese arrived as mercenaries and adventurers, with several settling, falling on hard times, and fading into the local population. Japanese also came as traders and mercenaries, then in larger numbers as refugees from the persecution of Christianity, and also melted into the local population. Persian trader-administrators settled and some converted to Buddhism. Mercenaries also came from Turkey, India, Arabia, the archipelago, and eastern Africa, though little is known of their fate and descent. The Dutch, French, and British kept themselves more separate.[19]

The dynastic conflicts and trading policies in China propelled waves of Chinese emigration to Ayutthaya and to ports around the gulf. By the 17th century, Ayutthaya had a reputation as a city of "forty languages". Gervaise claimed a third of the population of Siam could be counted as "foreigners".[20] A Ceylonese visitor to the city in the 1750s was amazed at the variety of traders seen along a single street: "Pattáni, Moors, Wadiga, Mukkara, men of Delhi, Malacca, and Java, Kávisi, Chinese, Parangis, Hollanders, Sannásis, Yógís, English, French, Castilians, Danes, men from Surat, Ava, and Pegu, representing every race".[21]

In the wars with Burma between the 1760s and 1800s, many people were swept away as prisoners and others fled. The early Bangkok rulers restocked the population by sweeping up people from the Malay, Khmer, Mon, and Lao regions. Lanna rulers swept down Tai peoples from further north. Over the 19th century, more and more people came from China, settling in Bangkok and other urban centres. In the early 20th century, over two-thirds of the population of the capital was reckoned to be non-Thai.[22]

At Ayutthaya, Thai probably became the dominant language after the Phitsanulok dynasty took over the throne of Ayutthaya (with Burmese help) in 1569. Even then, the mainstream language coexisted with others – the royal language or *rachasap* heavily influenced by Khmer; an argot used by monks, larded heavily with words borrowed from Khmer and Pali; and Malay, Persian, and Portuguese used as lingua francas in the trading community.

For the name of the country, early European visitors and mapmakers used terms such as Scierno, probably derived from "Xian", the Chinese word for Ayutthaya, and eventually settled on Siam. The early descriptions of the country by Dutch and French authors call the people Siamese, not Thai. La Loubère also calls the language Siamese. The term "Thai" does not appear in the first indigenous histories, including Van Vliet's compilation from local sources in the 1640, and the earliest known chronicle from 1680. In the chronicles compiled in the late 18th and early 19th centuries, the term "Thai" is used as a synonym of "Ayutthayan", applied to the king, army, soldiers, ambassadors, capital, and dependent towns – not to the people.

From the 17th century, the Europeans' growing sense of superiority provoked greater awareness of local identities. An Ayutthayan decree of 1663 forbidding "carnal relations" with foreigners reveals an emerging definition of Self and Other. The foreigners were "Khaek, Farang, English, Khula, and Malayu", which meant Muslims and Christians. The self was not Thai alone but "Thai, Mon and Lao". Many other peoples present in Siam, such as Khmer, Cham, Japanese, Chinese, and Burmese, seemed to occupy a middle ground. The conception of identity seems analogous to the *mandala*, a conception of space that is not defined by a sharp boundary but by relative closeness to the centre, allowing both variety and variation.

Rulers measured their power as *cakkavattin* emperors by claiming to rule over many different peoples. The eulogy of King Narai stressed how many different peoples "came to shelter under the king's protection". This concept reigned until the late 19th century. During his 1872 visit to India, King Chulalongkorn styled himself as "King of Siam and Sovereign of Laos and Malay".[23]

In an era of colonial domination, Siam had to conform to the European idea of a nation to survive. In the late 19th century, the word *chat* meaning a life or birth was adapted to mean people of the same origin. The term "Thai" was adapted to mean the people of Siam in two ways: first, by claiming that various languages spoken within the country, including Tai dialects, Mon, Khmer and Malay, were in fact variants of Thai; and second, by defining all subjects of the king as Thai.[24] In 1885, King Chulalongkorn said: "The Thai, the Lao, and the Shan all consider themselves peoples of the same race. They all respect me as their supreme sovereign, the protector of their well-being".[25] From around 1902 the country's name in Thai was changed from Siam to *Prathet Thai* or *Ratcha-anajak Thai* (Thai country or Thai kingdom). Over two decades, "Thai" had been transformed from a language to a "nationality". A decade later, King Vajiravudh crafted the phrase "Nation, Religion and King", which neatly fused (and confused) the old concept of royal supremacy and the new language of nationhood. The primary aim of the schooling system was to conform the people with this definition.

From the 1910s, early advocates of democracy tried to craft a non-royal nationalism by positing an alternative source of national unity than loyalty to the king. One theory argued that the original Thai had migrated to Siam from the Altai Mountains in Mongolia and thus were unified by a deep history. More simply, Luang Wichit Watakan suggested that the Thai were mystically united into "one united stream of blood which must never be broken into many streams".[26] This idea was embedded in the national anthem composed in the 1930s: "We are all of the Thai blood-flesh-lineage-race".

In mid-century, when nationalist ideas swept the globe, some Thai leaders hoped that those who did not comply with the national definition – especially the Chinese, the hill peoples in the north, and the Malay Muslims on the southern border – could be converted to Thai-speaking Buddhists by education, persuasion, and legislation. These aspirations were codified in a series of state mandates in 1939, promoted by a new National Culture Commission, and prosecuted by closures of Chinese schools and newspapers and violent repression of dissent in the Muslim south.

Over the second half of the 20th century, a new road network shrunk the space of the nation while universal primary education, radio, and then television made virtually everyone literate in Thai. Various official bodies regularly campaigned to promote "Thainess". In the 1970s, however, the aspiration to absorb everyone into the nation diminished after two million refugees flowed across the borders as a result of wars in Indochina and political chaos in Burma. From the 1980s, an alternative discourse of "diversity" emerged. As China emerged from communism and began its ascent to world power, Chinese began to reclaim their origins and identity. As tourism became the nation's largest export, cultural diversities were recast as economic assets. Local activists pushed back against over-centralisation. In the far south, Malay Muslims actively rebelled against it.

Belief and power

In Southeast Asia from ancient times, supernatural power was attributed to forces in nature, the spirits of the dead, and gods in the heavens. The unseen forces in nature could be predicted by various forms of divination and manipulated by devices including verbal formulas, diagrams, and number magic. Knowledge of the Hindu gods and of the Buddha reached the region before the start of the Common Era. Little is known of early religious practice, but the Hindu gods, the Buddha, and the local spirits seem to have coexisted. The Buddha appeared alongside Hindu gods in early bas-reliefs. Spirit images appear in the iconography of early Buddhist temples. The Indian belief that ascetic practice gives the adept both mastery over the self and over forces in nature was easily married to local forms of supernaturalism. Some schools of Buddhism projected this form of mastery onto the Buddha himself and allowed monks to replicate it. In Siam, this tendency did not become a separate sect but remained essential to mainstream belief and practice.

Between the 12th and the 14th centuries, a revised Buddhist practice (now termed Theravada, the way of the elder) was developed, mainly in Sri Lanka, and spread to Southeast Asia. Under this practice, anyone could become a monk, and the monks and temples depended on the local community for patronage. This practice was immensely popular. Theravada Buddhism became a mass religion deeply embedded in the society. Early Western missionaries first mocked this Buddhism as idolatry, but later they gave grudging respect to its power and made no headway with conversion. Rulers had to negotiate with this powerful popular force. Inscriptions from 13th-century Sukhothai tell stories of warrior princes embracing this new practice and vowing to live and rule by its ethical standards. Sukhothai's rulers boasted of the Buddhist temples in their capital and adopted *Dharmaraja*, meaning a king of Buddhist righteousness, as their dynastic title. These rulers also attempted to exercise some control over the monkhood but met with resistance.

Old scholarship suggests that Sukhothai's Buddhist kingship was superseded at Ayutthaya by a Khmer-derived concept of a *devaraja*, a king who was in some way also a god. Yet the term *devaraja* does not appear in Ayutthaya sources. Rather, a theory of kingship was developed in which a team of 11 Hindu gods (the trinity of Siva-Brahma-Vishnu, plus eight gods of the compass directions) joined together to create a king. This theory was dramatised in the ceremony of royal installation when court Brahmans instilled divine power through a lustration of sacred water. This lustration may have been repeated annually, indicating that the king's divinity was not intrinsic but a gift that needed reaffirmation.

The 17th-century kings who parlayed rising commercial wealth into a new absolutism pushed this theory a bit further. They claimed that their lineage harked back to the kings of Angkor, cast Buddha images decked with crowns and other royal regalia, and elevated the king above other mortals through sumptuous ritual, hiding the royal body, and other forms of mystification. The annual round of royal ceremonial dramatised the king as the military leader, source of prosperity, fount of alms, and medium between this world and supernatural forces in nature and the heavens. The envoys from Bourbon France were greatly impressed: "In the Indies there is no state that is more monarchical than Siam".[27]

From the late 17th century there were reactions against the absolutist kings' monopoly on power and the way it was used. Revolts were led by monks who claimed to have supernatural power through the techniques of self-mastery and magical devices, and who promised that the future Buddha, Maitreya, would shortly arrive and bring in an era of millennial joy. Nobles looked to Buddhism for means to reform the practice of kingship. They promoted two old codes of good Buddhist conduct – the Ten Virtues of Kingship and the Four Principles of Harmony – enjoining the king to abide by high ethical standards and to pay attention to the well-being of his servants and subjects. In this era, two theories were popular to explain kingship. The first was derived from the Akanya Sutta: when the original pristine human society degenerated with conflict, people spontaneously chose the most capable man to be their ruler (sometimes called the Great Elect). By analogy, any subsequent king was the man of greatest merit in the kingdom. In the second theory, the king is a *bodhisatta*, a man accumulating merit through good works across a series of lives in order eventually to become a Buddha. Both theories had no roles for gods and high expectations about the moral qualities of a king.

King Borommakot (1733–1758) responded to these pressures. He promoted Buddhist ritual practice, patronised monks generously, renovated numerous temples, lectured his nobles on living ethically, and passed laws to ban sinful practices. In the nobles' history of the era, Borommakot is not only described as a *bodhisatta*, but he is also credited with the supernatural powers that result from self-mastery.

King Taksin, who re-founded the kingdom after the fall of Ayutthaya in 1767, went further down the same road. Perhaps because he had no claims to royal power, he set out to acquire a personal stock of moral power through Buddhist practices, particularly meditation. He claimed to achieve supernatural powers, and assumed the responsibility to lecture and discipline errant monks, prompting a crisis which contributed to his overthrow in 1782.

King Rama I then completed the process of formulating a new model of Buddhist kingship that Borommakot had begun. He was strict in his own personal practice and generous in his patronage of the monkhood. He passed the first-ever legislation on religious affairs, which banned monks from the magical practices that had figured in revolt, and he encouraged them to become models of good Buddhist principles. He restored the proper division of spheres between crown and monkhood, but he enshrined the right of the monarch to appoint the supreme patriarch at the summit of the monastic hierarchy. The chronicles portrayed him as the king-by-consent from the Akanya Sutta, an upholder of the Ten Principles of Kingship, and as a *bodhisatta*.

From the mid-19th century, as Siam confronted the aggression of the colonial Christian West, the kings became nervous of basing the justification of kingship too heavily on Buddhism. The court gradually grounded kingship on principles that the West could understand, namely dynastic right, effective rule, and a properly royal lifestyle by European standards. This trend towards secularisation of the monarchy continued through the first half of the 20th century, only to be dramatically reversed in the ninth reign.

Conclusion

Examining Thailand's history in the longue durée highlights how dramatically the society has changed in the very recent past.

A resource-rich territory with a sparse population has been transformed by medicine and by resource-hungry economic development. The prevailing landscape of forests has disappeared. The population, once clustered mainly in urban settlements, was first spread across the territory along a moving "land frontier" and is now being gathered back to the cities and towns. City-states were only gradually arranged into coalitions and hierarchies and then dramatically superseded by the nation-state around 1900. With the region open to flows of peoples, many languages prevailed until Thai became not only the dominant language but also a defining feature of society and culture. The religious underpinnings of kingship continue to change over time.

Notes

1 Chris Baker et al., *Van Vliet's Siam* (Chiang Mai: Silkworm Books, 2005), p. 107.
2 Sarasin Viraphol, *Tribute and Profit: Sino-Siamese Trade 1652–1853*, revised edition (Chiang Mai: Silkworm Books, 2004), ch. 5.
3 Pasuk Phongpaichit and Chris Baker, *Thailand: Economy and Politics* (Kuala Lumpur: Oxford University Press, 1995), chs. 1 and 2.
4 Ronald D. Renard, *Opium Reduction in Thailand, 1970–2000: A Thirty-year Journey* (Chiang Mai: Silkworm Books, 2001).
5 Takaya Yoshikazu, *Agricultural Development of a Tropical Delta: A Study of the Chao Phraya Delta* (Kyoto: Center for Southeast Asian Studies, 1987), ch. 4.
6 Steve Van Beek, *The Chao Phya: River in Transition* (Singapore: Oxford University Press, 1995).
7 Chris Baker and Pasuk Phongpaichit, *A History of Ayutthaya: Siam in the Early Modern World* (Cambridge: Cambridge University Press, 2017), ch. 3.
8 Baker et al., *Van Vliet's Siam*, p. 126.
9 Nicolas Gervaise, *The Natural and Political History of the Kingdom of Siam* (Bangkok: White Lotus, 1998), pp. 23, 45.
10 Simon de la Loubère, *A New Historical Relation of the Kingdom of Siam* (London, 1793), p. 11.
11 Volker Grabowsky, "Population Dynamics in Lan Na During the 19th Century", *Journal of the Siam Society*, Vol. 105 (2007), p. 198.
12 Sir John Bowring, *The Kingdom and People of Siam* (Kuala Lumpur: Oxford University Press, 1969), p. 196; Prince Dilok Nabarath, *Siam's Rural Economy under King Chulalongkorn* (Bangkok: White Lotus, 2000), ch. 3.
13 Baker and Pasuk, *A History of Ayutthaya*, pp. 182–187.
14 Nidhi Eoseewong, *Kanmueang thai samai phrajao krung thonburi* [Thai Politics in the Time of the King of Thonburi] (Bangkok: Sinlapa Watthanatham, 1986).
15 Thongchai Winichakul, *Siam Mapped: A History of the Geo-body of a Nation* (Honolulu: University of Hawaii Press, 1993).
16 Tej Bunnag, *The Provincial Administration of Siam, 1892–1915* (Kuala Lumpur: Oxford University Press, 1977); Kullada Kesboonchu Mead, *The Rise and Decline of Thai Absolutism* (London: Routledge-Curzon, 2004).
17 Arlo Griffiths and Christian Lammerts, "Buddhist Epigraphy in Southeast Asia", in *Brill's Encyclopedia of Buddhism,* edited by Jonathan A. Silk (Leiden: Brill, 2015), pp. 988–1009.
18 Ma Huan, *Ying-yai Sheng-lan. "The Overall Survey of the Ocean's Shores"*, tr. J.V.G. Mills (Bangkok: White Lotus, 1997), p. 106.
19 Baker and Pasuk, *A History of Ayutthaya*, ch. 4.

20 Gervaise, *Natural and Political History*, p. 45.
21 P.E. Pieris, *Religious Intercourse between Ceylon and Siam in the Eighteenth Century. I. An Account of King Kirti Sri's Embassy to Siam in Saka 1672* (Bangkok: Siam Observer, 1908), p. 14.
22 Edward Van Roy, *Siamese Melting Pot: Ethnic Minorities in the Making of Bangkok* (Chiang Mai: Silkworm Books, 2017).
23 River Books, *King Chulalongkorn's Journey to India, 1872* (Bangkok: River Books, 2000), p. 8.
24 David Streckfuss, "The Mixed Colonial Legacy in Siam: Origins of Thai Racialist Thought, 1890–1910", in *Autonomous Histories: Particular Truths*, edited by Laurie Sears (Madison: University of Wisconsin, 1993), p. 150.
25 Thongchai, *Siam Mapped*, pp. 101–102.
26 Saichon Sattayanurak, *Chat Thai Lae Khwam Pen Thai Doi Luang Wichit Wathakan* [Thai Nation and Thainess According to Luang Wichit Wathakan] (Bangkok: Sinlapa Watthanatham, 2002), p. 31.
27 Gervaise, *Natural and Political History,* p. 53.

References

Baker, Chris, Dhiravat na Pomberja, Alfons van der Kraan, and David K. Wyatt. 2005. *Van Vliet's Siam*. Chiang Mai: Silkworm Books.

Baker, Chris and Pasuk Phongpaichit. 2017. *A History of Ayutthaya: Siam in the Early Modern World*. Cambridge: Cambridge University Press.

Bowring, Sir John. 1969. *The Kingdom and People of Siam*. Kuala Lumpur: Oxford University Press.

Dilok Nabarath, Prince. 2000. *Siam's Rural Economy under King Chulalongkorn*. Bangkok: *White Lotus*.

Gervaise, Nicolas. 1998. *The Natural and Political History of the Kingdom of Siam*. Bangkok: White Lotus.

Grabowsky, Volker. 2007. "Population Dynamics in Lan Na During the 19th Century". *Journal of the Siam Society*, Vol. 105: 197–244.

Griffiths, Arlo and Christian Lammerts. 2015. "Buddhist Epigraphy in Southeast Asia". In *Brill's Encyclopedia of Buddhism*. Edited by Jonathan A. Silk. Leiden: Brill, pp. 988–1009.

Kullada Kesboonchu Mead. 2004. *The Rise and Decline of Thai Absolutism*. London: RoutledgeCurzon.

Loubère, Simon de la. 1793. *A New Historical Relation of the Kingdom of Siam*. London.

Ma Huan. 1997. *Ying-yai Sheng-lan. "The Overall Survey of the Ocean's Shores"*, tr. J.V.G. Mills. Bangkok: White Lotus.

Nidhi Eoseewong. 1986. *Kanmueang thai samai phrajao krung thonburi* [Thai Politics in the Time of the King of Thonburi]. Bangkok: Sinlapa Watthanatham.

Pasuk Phongpaichit and Chris Baker. 1995. *Thailand: Economy and Politics*. Kuala Lumpur: Oxford University Press.

Pieris, P.E. 1908. *Religious Intercourse between Ceylon and Siam in the Eighteenth Century. I. An Account of King Kirti Sri's Embassy to Siam in Saka 1672*. Bangkok: Siam Observer.

Renard, Ronald D. 2001. *Opium Reduction in Thailand, 1970–2000: A Thirty-year Journey*. Chiang Mai: Silkworm Books.

River Books. 2000. *King Chulalongkorn's Journey to India, 1872*. Bangkok: River Books.

Saichon Sattayanurak. 2002. *Chat thai lae khwam pen thai doi luang wichit wathakan* [Thai Nation and Thainess According to Luang Wichit Wathakan]. Bangkok: Sinlapa Watthanatham.

Sarasin Viraphol. 2014. *Tribute and Profit: Sino-Siamese Trade 1652–1853*, revised edition. Chiang Mai: Silkworm Books.

Streckfuss, David. 1993. "The Mixed Colonial Legacy in Siam: Origins of Thai Racialist Thought, 1890–1910". In *Autonomous Histories: Particular Truths*. Edited by Laurie Sears. Madison: University of Wisconsin, pp. 123–153.

Takaya Yoshikazu. 1987. *Agricultural Development of a Tropical Delta: A Study of the Chao Phraya Delta*. Kyoto: Center for Southeast Asian Studies.

Tej Bunnag. 1977. *The Provincial Administration of Siam, 1892–1915*. Kuala Lumpur: Oxford University Press.

Thongchai Winichakul. 1994. *Siam Mapped: A History of the Geo-body of a Nation*. Honolulu: University of Hawaii Press.

Van Beek, Steve. 1995. *The Chao Phya: River in Transition*. Singapore: Oxford University Press.

Van Roy, Edward. 2017. *Siamese Melting Pot: Ethnic Minorities in the Making of Bangkok*. Chiang Mai: Silkworm Books.

3

THAI HISTORIOGRAPHY

Charnvit Kasetsiri

Tamnan, phongsawadan, prawatisat

Historiography, or the writing of history in Thailand, is one of the oldest tasks and is confined within limited groups of educated, strictly male members of the Buddhist temples and the palace. There were two types of such writings, namely the *tamnan* by Buddhist monks and the *phongsawadan* by the court literati. *Tamnan* means story, legend or myth. The main theme of *tamnan* history clearly revolves around religion. It is the Gautama Buddha who is the moving force in it. Its purpose is to describe the history of Buddhism in connection with a Buddhist kingdom or a certain Buddhist locality: certain temples or important monuments (*cetiya*, stupa, etc.). Kings, kingdoms and to a certain extent the laypeople come into the picture insofar as their actions contribute to promoting Buddhism. Therefore, history in this sense is concerned not only with the past. The past is continuous with the existence of the present and the present is also part of the future. Thus the past, the present and the future are parts of one whole – the history of Buddhism.[1]

The best example of the *tamnan* type of historiography is the *Jinakalamalipakaranam* ("The Sheaf of Garlands of the Epochs of the Conqueror"), written in Chiang Mai, northern Thailand, in 1517. It gives a very clear statement of the *tamnan* concept of time and space. It says that "for therein, the Epoch of the Conqueror means the time as far as the lineal succession of the Dispensation commencing with the time of the aspiration of our Teacher, the Exalted One Gautama; and it is of divers aspects".[2] The history then continues with the career of the Buddha, how he reached enlightenment and how he taught his law, the *dhamma/dharma*. After the passing of the Buddha, the history discusses various world Buddhist councils which took place in India and Sri Lanka. It also emphasises the role of the great Indian King Asoka in the promotion of Buddhism. Afterwards, the history deals with the events which occurred when Buddhism finally travelled to Chiang Mai in today's northern Thailand. The "real history" of the Chiang Mai Kingdom begins at this point. Various kings and other kingdoms are portrayed, with the focus on their roles in supporting the religion. It is usually agreed in these documents that the kingdom under discussion will remain the centre of Buddhist religion until the year 5000 BE (4457 CE) from the time of the Buddha's *nibbana/niravana*.

As for the other type of historiography which is known as the *phongsawadan*, the term derives from the two Pali words *vamsa* (genealogy) and *avatara* (reincarnation); it means the history, chronicles or annals of members of a line, kings and kingdom. The *phongsawadan* historiography became popular within the palace circle sometime in the 17th century, probably during the reign of King Narai (1657–1688) of Ayutthaya Kingdom.

The *phongsawadan* type of palace history was the result of changes within the Thai society and to a lesser extent of the contacts which the Thai ruling elite of Ayutthaya had with foreigners, especially Europeans. By this time, learned men, or literati, were to make their careers at the royal court and were no longer strictly governed by the religious order as in previous days. After a long process of political development, kingship had developed into a powerful autonomous institution, which had increasingly taken the cultural initiative from the religious leadership. Therefore, historians were now men who served the court and belonged to it rather than to the religious order. The *Phraratchaphongsawadan Krung Si Ayutthaya Chabap Luang Prasoet* ("The Luang Prasoet Chronicle of Ayutthaya") of the 17th century was written by a royal astrologer, not a monk. Moreover, the language of the new type of history was Thai, a secular and ethnic language, rather than Pali, a religious and international language of the Buddhist world. This is not to say that Buddhism had lost its influence in Thailand, but rather that religion played a different role from that of earlier days. Learned men were still being educated within the monastic order of Buddhism, but they now served their immediate superior – the king, not the religion.[3]

The *phongsawadan* type of palace history reached its height by the turn of the 19th century to the 20th, especially during the reign of King Chulalongkorn, Rama V (1868–1910). And later on, a new type of writings known as *prawatisat* became more acceptable by the time of King Vajiravudh, Rama VI (1910–1925); it became even more vigorous prior to and after the 1932 coup, a "soft revolution" which abolished absolute monarchy in Siam. A new word was coined, in the 1910s, to look more modern and scientific (i.e. *prawatisat*). The new Thai word comes from Pali-Sanskrit: *paravati* or *puravuttanta* + *sastra* = *prawatisat*, meaning scientific treatise or history, or literally "science of what happens". It is nowadays commonly accepted as an equivalent to the English word "history".

The new writing of *prawatisat* type was born out of modernity, colonialism and the concept of defined boundary nation-states in the middle of the 19th century. During this time, new ideas and modern technologies were brought into Siam from the West. The most important factor which contributed to the new history was the printing press. It is believed that the first printing machine for Thai alphabets was introduced in 1836 by Dr. D.B. Bradley, an American missionary. His main purpose for setting up a publishing house was to propagate Christian doctrines. However, he began to publish old Thai historical works which became widespread among the Thai elite and upper middle class.[4] Moreover, during this period Siam faced the threat of colonialism: the country was caught between the British in India, Burma and Malaya and the French in Indochina. The Chakri kings had to give up their claims over territories bordering Burma, Malaya, Laos and Cambodia in order to remain independent. Therefore, the new boundaries of Siam were shaped to what is seen today as Thailand. In effect, the *prawatisat* history is one of a new nation-state.[5]

There are at least three main figures, of three different generations, who were responsible for the construction of this type of "modern" historiography. They are Prince Damrong (1862–1943), King Vajiravudh, and Luang Wichitwathakan (or Wichit Wichitwathakan; 1898–1962). Interestingly, Damrong is 18 years older than Vajiravudh, who in turn was 18 years older than Wichit.

Prince Damrong

Prince Damrong was one of 82 children of King Mongkut, Rama IV (1851–1868), and also one of the most celebrated statesmen of Siam. He grew up in a period of great change in Thai society, when it faced modernisation and Western colonialism. Damrong was educated in both Western and classical Thai scholarship, thus he was able to bridge these two cultures for the development of a modern Thailand. During the reign of his brother King Chulalongkorn, Damrong rose to prominence in the Thai government, holding many important positions including minister of education (1888–1892) and minister of interior (1892–1915). In spite of his heavy responsibilities of governing the country, Damrong devoted much of his time – almost 30 years from the 1910s to 1943 – to the writing of history, particularly when he was free from administrative works and living overseas. Though maintaining his ministerial post for a few years under his nephew, King Vajiravudh, Damrong was released to pursue his writing habit. After a democratic coup in 1932, the prince was forced into exile in Penang. He left an extensive collection of writings, around 100 books and articles, to which very few scholars could lay claim.

As mentioned before, the works of Prince Damrong revealed a change in the writing of Thai history. The most notable is the usage and selection of sources. It is said that he was very much influenced by the method used by the great German historian Ranke.[6] For example, his account of the foundation of the Kingdom of Ayutthaya introduced a new technique in examining Thai history. Formerly, the *tamnan* and *phongsawadan* schools of thought described the "birth" of the Kingdom of Ayutthaya by associating it with myths and external forces beyond human control. Ayutthaya came into existence, as the old schools see it, as a result of the Buddha's prophecy and/or miraculous acts of certain individuals. However, Damrong explained it by presenting a tangible historical fact. According to him, Ayutthaya was a result of a long period of Thai settlement in the Menam Basin. It became the centre of the Thai when the Khmer began to lose control of the area. Various sources were taken into consideration and a hypothesis was set up to explicate the origin of Ayutthaya. Thus, he rejected the old methods of the *tamnan* and *phongsawadan*.[7]

In 1914 Prince Damrong wrote and established a new concept of historical periodisation in which "the history of Siam is divided into three periods, namely, (1) When Sukhothai was the capital, (2) When Ayutthaya was the capital, and (3) Since Bangkok (Chakri's Ratanakosin) has been the capital".[8] This periodisation became known as *sam krung*, or "three capitals" (Sukhothai, Ayutthaya, Ratanakosin-Chakri-Bangkok). It focuses on the periods when each capital and its kings were considered the centre of what had happened. It was possible that this sequence was influenced by European periodisation of "classical-medieval-modern". Therefore, Sukhothai was "classical", Ayutthaya "medieval" and Ratanakosin/Chakri/Bangkok "modern". Interestingly enough, Damrong even went back beyond classical Sukhothai and added a lengthy elaboration on periods prior to the two ancient capitals (Sukhothai-Ayutthaya). He explained by way of focusing on territories (space) and races (ethnicities). Accordingly, "the territory of which Siam is now made up was originally occupied by people of two races, the Khmers (Khom) and the Lao (not the people of present day country of Laos but of an ethnic called Lawa)".[9] Since the Thais were not the original people of Siam, the prince had to look elsewhere further north in China. In short, the historian prince developed a new hypothesis of the "Thai southward migration" back some thousand years ago. And in their migration movement, the Thais made one major stop. They were able to establish the mighty Kingdom of Nanzhao in Yunnan. Here, from the 6th to the mid-13th centuries, the kingdom lasted for 700 years. Damrong labelled Nanzhao as *muang Thai doem*, literally the original country of the Thais.[10] Nanzhao was defeated by Kublai Khan in 1253, but the link to present-day Thailand continued. Prince Damrong argued that while the Thais were still powerful in their "original home", a great number had already migrated to settle in the valleys of the Salween and the Mekong, south of the present-day Golden Triangle.[11] Sukhothai, Ayutthaya, and Bangkok/Rattanakosin followed, respectively.

In short, the concept of new historiography, as seen from Damrong's works, commenced with the impact of Western penetration. It was influenced by the West, which contributed new methods of using sources and "scientific" analyses. However, many of the old themes of the *phongsawadan* tradition still remained. Thus, although using new methods, Damrong still wrote history in which the monarchy remained the prime driving force in the new nation-state. Thai history became the history of a particular nation, a secular linear forward moving from the past to the present, surrounded by many other powerful countries, especially the Western ones, no longer in isolation.

King Vajiravudh

Besides Damrong, it is imperative to stress the impact of King Vajiravudh and his official nationalist policy. As suggested earlier, the latter part of King Chulalongkorn's reign was crucial for the understanding of the changes within Thai society in connection with the emergence of the new "royal nation-state". Vajiravudh, born in 1881, is one of Chulalongkorn's 97 children. In 1893, at the age of 12, he was one of the first royal children to be sent to schools in Europe. He spent nine years in Great Britain, and through 1902 he went to various schools including the Royal Military Academy Sandhurst and Oxford's Christ Church College (law and history). In this light, being English educated, he was therefore different from his predecessors on the throne. As a prince and later as king, Vajiravudh was known to be rather aloof. He isolated himself from the large and active palace family of his father, his relatives, and from the bureaucracy. In the early years of his reign he often stayed at the newly built Sanam Chan Palace in Nakhon Pathom, some 50 km outside Bangkok. He also surrounded himself with a male entourage, the royal-national paramilitary "Wild Tiger Corps", and kept himself busy in his own peculiar ways.

The 15 years he spent on the throne were caught between the very successful and long rule of his father, whose absolutist reign had lasted for 42 years. As the most prolific writer of all time in Siam, Vajiravudh is now officially remembered as Phra Maha Dhiraratchao, or the "Great Scholar King". Under more than 100 pen names, he wrote travelogues, plays, poetry, songs, articles and sermons, totalling around 200 titles. The king is particularly known for his official nationalistic policy and is often labelled "The Father of Thai Nationalism".[12] The reign of Vajiravudh was a time of change. The first strikes by Bangkok's Chinese merchants and workers took place just before his first coronation on 11 November 1910. The following year, the Celestial Monarchy in Peking came to an abrupt end.[13] The year 1912, in February, witnessed an attempted coup, known as "Kabot R.S. 130", or the 1912 Rebellion, to overthrow Vajiravudh. It happened only a few months after the 13-day extravaganza of his second coronation in November 1911.

Indeed, the 1900s and the 1910s were a very different time. First, with domestic changes, a new, though rather small educated middle class had emerged, a good number of whom were Sino-Thai, who became critical of Siam's absolute monarchy. To them, it had become anachronistic. The time was also marked by the spread of the free press, which claimed to represent *paksiang* (mouth and voice) of the common people. In addition, "print-capitalism" also allowed people to think and imagine their own place in society differently.[14] Second, the rise of nationalism, along with the fall of monarchies in Asia and Europe (the Qing, the Ottomans, the Romanovs, and the Habsburgs), was a reason why Vajiravudh had to consolidate his rule and felt compelled to embark on his own nationalist policy.

Soon after his first coronation, the new king established two organisations: the Wild Tigers Corps (*sue pa*) and the Boy Scouts (*luk sua*, meaning cubs or male child of a tiger), on 1 May and 1 July 1911, respectively. Both were inspired by the British Volunteer Force and Lord Baden-Powell's Boy Scouts. Both were aimed at "instilling the love of the nation among the Thai". Of the two, the Wild Tigers Corps was criticised as *khong len* ("plaything" or "toy") and as an extension of the king's personal bodyguards. It did not fit well with the existing military. The

Wild Tigers, not the Boy Scouts, ceased to exist soon after the end of his reign. However, it was within the circle of the Wild Tigers that the king launched his nationalist programmes. He personally lectured them about *chat* (nation), *satsana* (religion), and *phra maha kasat* (great king), which became the three pillars of the Siamese modern state ideology.[15] Since then, this "Holy Trinity" has frequently been exploited by right-wingers and military regimes.

In terms of historiography and the construction of historical narratives, Vajiravudh employed various strategies to achieve his policy of *pluk chat-pluk chai* ("Waking up the Nation"). Interestingly, it was Vajiravudh who actually coined the new Thai word *prawatisat* (possibly from Pali-Sanskrit: *puravuttanta* = history + *sastra* = scientific treatise) for "history". He may have found it unfitting to use the established Thai word *phongsawadan* in a global context because of its direct translation, such as the reincarnation of a royal family which did not fit with the modern time. The new term *prawatisat* caught on, and by 1917 the word was frequently used for titles of history texts.[16]

Luang Wichitwathakan

Among the post Damrong-Vajiravudh generation of the 1930s commoners, Luang Wichiwathakan was the most outstanding. Wichit was born of a very humble origin, possibly that of a Sino-Thai outside the Bangkok elite circle. He moved his way up by joining the *sangha*. For 12 years from 1906 to 1918, he was ordained, first as a novice and then as a monk. His main education came through this experience, studying Buddhist temple classics and self-educating with modern science and the English language. In 1918, at the age of 20, Wichit left the monkhood and joined the Ministry of Foreign Affairs. He was sent as an assistant secretary to the Thai Legation in Paris, from 1920–1926; he was also assigned to the League of Nations in Geneva and served briefly in London. Upon his return, his first major writing in the Thai language was published in 1931. It is a magnum opus called *prawatsat sakon* ("world history", 12 volumes). A year later, right after the 1932 anti-royalist coup, Wichit acted as a spokesman for the new regime and became intimate with Premier Field Marshal Phibun Songkhram. During his long service in the government, from 1934 until his death in 1962, Wichit held many extremely important posts including director of the Fine Arts Department, cabinet minister and ambassador to various countries. However, his most notable position was as advisor-administrator to the Phibun and Sarit military regimes. He displayed an ability of manipulating, writing and organising cultural activities which were directed towards arousing nationalistic feeling within the country. Similar to Prince Damrong, Luang Wichit was a prolific writer.

Since 1932, successive governments have been trying to create a new kind of legitimacy to replace the absolute monarchy. Military-bureaucratic-oriented nationalism became a means of rallying popular support. The period between 1932 to World War II witnessed the rise of military nationalism in Thailand. Wichit became a prime instigator of these movements. In 1942, he wrote a nationalistic account (in English) condemning the French imperialism, titled *Thailand's Case*. This short essay explained how Thailand unjustly lost its territory in Indochina to France and urged the public to support the Phibun government's irredentist policies.[17] At the heart of Wichit's history, there were strong leaders leading the Thai race. He put aside the role of the monarchs (except Ramkhamhaeng, Naresuan, and Taksin), once predominant in the writings of the *phongsawadan* and of Prince Damrong. Thus Thai history was interpreted according to the purity and glory of the Thai race. Everything was aimed at creating a new nation in which the military and strong leaders took the roles of former kings. His interpretation of Thai history showed uncomfortable similarities to Nazi and fascist writings of the period.

Wichit's history was very popular. It became an ideological weapon of the new ruling elite who sought justifications to rule the country. Furthermore, Wichit had the advantage of exploiting the mass media to disseminate his interpretation of history. As director of the Fine Arts Department, he used this position to propagate his concepts of history through music, plays and songs. These were presented on stage and broadcast on the official Radio of Thailand. His skills in writing and sense of public sentiment contributed to the spread and popularisation of his version of Thai history, no matter how politically biased.[18]

In short, therefore, as far as the *prawatisat* type of official historiography is concerned, one can say that from 1900 to the 1920s, the royal nationalistic version of Damrong-Vajiravudh was vigorously promoted, and by the 1930s the military-bureaucratic version of Wichit and the new commoner elite had become dominant. However, by the 1960s these two were blended together to give birth to a hybrid nationalistic historiography. This outcome was a product of the very long 70-year reign of King Bhumibol, Rama IX (1946–2016), in which the king himself and his special new style of monarchy had become a hegemonic balancing force for Thai society and politics. It can be said that this hybrid version is a product of bureaucrats at various government offices, namely the Ministries of Education and Culture and the Departments of Fine Arts and Public Relations. The new agreeable version of Thai historiography seems to work out successfully.

However, in 1995, Thongchai Winichakul, one of the most prominent historians of Southeast Asia, wrote:

> Historical studies in Thailand have been closely related to the formation of the nation . . ., and until recently the pattern of the past in this elitist craft changed but little. It presented a royal/national chronicle, a historiography modern in character but based upon traditional perceptions of the past and traditional materials. It was a collection of stories by and for the national elite celebrating their successful mission of building and protecting the country despite great difficulties, and promising a prosperous future.

Thongchai went on to say that

> The popular uprising led by the student movement against the military dictatorship in 1973, a political as well as an intellectual revolution, shook this historical paradigm. Historical studies became a centre of intellectual interest for all disciplines as well as an arena of ideological struggles, with dramatic effect. The conventional knowledge of the past was challenged and negated. A new past was needed.[19]

Therefore, it would be misleading to say that development of official Thai historiography, seen from above, has been smooth and uninterrupted from (1) the times of the religious *tamnan*, (2) the royal *phongsawadan*, (3) the ruling elite *prawatisat* and (4) the hybrid royal-official nationalistic historiography constructed under the long, hegemonic reign of King Bhumibol. It would also be misleading to say that there was no need for a new past. But a closer look will tell a different story. Exactly two decades before the "intellectual revolution" in the 1970s, as mentioned above, it is worth looking back to the 1950s historical writings. This was the period of the Cold War when the "free world" clashed with the "Communist bloc" in Thailand and elsewhere in Southeast Asia. This brings us to what is known as a period of Thai "radical discourse". There were numerous left-Socialist-Marxist history writings, historical novels and translations by well-known (and some less well-known) authors like Sri Burapha (1906–1974), Supha Sirimanond (1914–1986), Seni Saowaphong (1918–2014), Assani Pholachan (1918–1987), Udom Sisuwan

(1920–1993), Plueng Wannasri (1922–1996) and Thaweep Woradilok (1928–2005). But the most outstanding of all was a young man named Jit Phumisak (1930–1966).

Jit Phumisak

Jit Phumisak is probably one of the few most original, immensely gifted, brave and committed Thai intellectuals of the 20th century. Although his father was merely a low-level official and his mother a school teacher, Jit's brilliant mind got him an entry to the conservative Chulalongkorn University. His writing career started when he was studying at the Faculty of Arts. In the middle of the 1950s, the Phibun semi-authoritative government began to lose its popularity; to compensate, the government allowed a certain degree of freedom of speech. A large number of Jit's writings came out during this time, including *Chomna Khong Sakdina Thai Nai Patchuban* ("The Real Face of Thai Feudalism Today"), which is a Marxist analysis of Thai history. Jit, together with other writers, were arrested and put in jail when Field Marshal Sarit staged two successive coups in 1957 and 1958. Jit spent six years (1958–1964) in prison for his iconoclastic writing. Despite being strictly banned, his most important writings passed illegally from hand to hand. Most of the time, he was able to smuggle out his articles and have them published under various pen names. At the end of 1964, Jit was released but was threatened and spied on by the regime until he decided to join the guerrilla forces of the Communist Party of Thailand. On 5 May 1966, at the age of 36, he was killed by agents of the military dictatorship.

Jit's creative legacy covered many subjects ranging from history, politics, literature and culture. His writings were very expressive and provocative. As in the cases of Damrong and Wichit, he had good command of the Thai language. However, his Thai was more forceful because he worked from a different social viewpoint. Moreover, he was a poet who had mastered classical Thai and used it accordingly – a quality which Damrong and Wichit did not possess. Jit composed poems and songs for the masses, which are now sung and cited by the Thai left wing and radical students. It should be pointed out that Jit's writings became very popular after the 1973 student uprising. Since then, a period of freedom of speech has flourished in Thailand. His books have been undergone several reprints, and his songs and poems continue to be sung and read. In fact, his name is controversial in Thailand even today.

Jit's approach to Thai history, employing Marxist theory, may not appear radical today. However, in Thailand, where communism was outlawed and Marxist ideas socially ostracised, Jit's writings represent the forbidden fruit and an alternative interpretation of Thai society. Most Thai historians either reject or ignore the question of class structures in Thailand. But Jit backed up his analysis by historical evidence. His knowledge of Thai history and his linguistic command added up the credibility of his interpretation. Thus it is not so much a question of a Marxist approach, which shows in his history, but rather the issues of class exploitation in the form of class differentiation, landownership, corvée labour and oppressive taxation.[20]

In short, the masses were the main interest of Jit in describing history. His books speak out their misery and various means which the ruling class employed for its self-interest. It can be said that Jit's history is a political weapon against the ruling class in Thailand.

Conclusion

After portraying a general picture of Thai historiography from ancient time to the modern period, the chapter should end with today's major trend. During the three decades from the 1960s to the 1980s, history studies had for the first time (though it was well controlled and supervised by various governmental ministries and departments) become firmly established and institutionalised in universities and colleges throughout the country. This was due to socio-economic booms in Thailand and the expansion of education. History departments and history lecturers have been cultivated to the point that there are new generations of historians outside the traditional official domain, unlike those of Damrong's, Vajiravudh's and Wichit's times. Some are direct or indirect outcomes of the unusual time of the 1950s and the 1970s as mentioned earlier. There are at least three different age groups of new historians with new interpretations.

The first group may be called avant-garde historians. They are led by senior academics like Nidhi Aeosriwong of Chiang Mai University. Born in 1940, he became an active promoter of a socio-cultural history breaking away from the traditional ruling elite interpretation.[21] As for Srisak Vallibhotama of Sinlapakorn University, born in 1938, he applies archaeology and anthropology in searching for local history. Chatthip Nartsupha of Chulalongkorn University, born in 1941, introduced political economy into the study of history.[22] In the meantime, Piriya Krairiksh of Thammasat University, born in 1942, interestingly applies art history to understand different parts and to introduce different interpretations of the history of Sukhothai and Ayutthya. Instead of classical or medieval, like historians in the past, it is more of the Thai elite's response to colonialism and survival tactics in the second half of the 19th century.[23]

Then came the second group directly or indirectly connected to the Octobrist generation, the radical youth and student leaders of the mid-1970s. They were mostly male, born in the late 1950s, but it is notable that females had also been an important part of this batch. Leading historians include Winai Phongsripian, Thanet Aphornsuwan, Sunait Chutintaranond, Chalong Suntharavanich, Suwit Therasatwat, Sarasawadi Ongskul, Aroonrat Wichiankeo, Suthachai Yimprasert and Attachak Sattayanurak. The most outstanding are probably Saichol Wannarat (born in 1953), Thongchai Winichakul (born in 1957) and Somsak Jeamteerasakul (born in 1958).[24] Most of these historians are over 60 years old and have officially retired from their universities, but many are still active. They come from many backgrounds and directions and their histories range from revised traditional to liberal and extremely radical ideology. Some of them attempt just to understand Thai society through the study of history. But many see and write history not only to understand the facts about what happens but also as part of their political struggle for a new Thailand.

It is possible that the third group of Thai historians is now in the making. They were born between the 1960s and the 1980s and are now in their forties and fifties; examples are Thamrongsak Petchlertanan (born in 1964), Natthapoll Chaiching (born in 1971), Chanida Chidbundid (born in 1977), Pokpong Chanan (born in 1984) and Warisara Tangkhawanit (born in 1986). They continue from where the second group left off and are inclined to study history from myriad directions, from the "opposite", "below", "subaltern", "post-colonial" and "post-modern", with greater attention paid to alternative issues of gender and LGBTIQ issues.[25]

Notes

1 For *tamnan and phongsawadan* histories, see chapter 1 of my book *The Rise of Ayudhya: A History of Siam in the Fourteenth and Fifteenth Centuries* (Kuala Lumpur: Oxford University Press, 1976). See also Anthony Reid and David Marr, *Perceptions of the Past in Southeast Asia* (Singapore: Heinmann, 1978), pp. 156–170.

2 N.A. Jayawickrama (tr.), *The Sheaf of Garlands of the Epochs of the Conqueror* (London, 1968), p. 2.

3 At this same time, there was a change in Thai education and a new textbook was compiled. It is believed that the 17th-century textbook *Chindamani* was written for use in instruction as well as for the information of foreign diplomats. See D.K. Wyatt, *The Politics of Reform in Thailand* (New Haven: Yale University Press, 1969), pp. 21–22.

4 Amphai Chanchira, *Wiwatthanakan Kanphim Nangsu Nai Prathet Thai* (Bangkok: Wannasin, 1973), pp. 88–91; and Kajorn Sukhapanich, *Kaoraek KhongNnangsuphim Nai Prathet Thai* (Bangkok: n.p., 1966),

pp. 5–11. See also C.J. Reynolds, "The Case of K.S.R. Kulap: A Challenge to Royal Historical Writing in late Nineteenth Century Thailand", *JSS*, Vol. 61, No. II (1973), p. 65; and Aemon Chaloemrak, *Prince Damrong Rajanubhab: His Writings and His Contribution to the National Library*, MA thesis, Chulalongkorn University, 1968, p. 12.

5 See a brief account of Mongkut's reign in A.B. Griswold, *King Mongkut of Siam* (New York: The Asia Society, 1961), ch. 4. For the reform of King Chulalongkorn, see for example Wyatt; F.W. Riggs, *The Modernisation of a Bureaucratic Polity* (Honolulu: East-West Center Press, 1966); W.J. Siffin, *The Thai Bureaucracy: Institutional Change and Development* (Honolulu: East-West Center Press, 1966).

6 Kobkua Suwannathat, "Kansuksa Prawatsat Sakun Damrong Rajanubhab", *Aksonsat Phichan*, Vol. 2, No. VI (1974), pp. 28–44.

7 See *The Rise of Ayudhya*, ch. 4.

8 See translation by O. Frankfurter, "The Story of the Records of Siamese History", in *Miscellaneous Articles Written for the Journal of the Siam Society by His Royal Highness Prince Damrong* (Bangkok: Siam Society, 1962), p. 31.

9 Interestingly, Prince Damrong mistook the word "Lao" for "Lawa". The two are ethnically and linguistically different. Lao belongs to the Tai-Lao family and Lawa is Mon-Khmer. But he went on to say:

> Who were the original Khmers and Lao? Today we only know that the peoples designated under the name of Kha, Khamu, Cambodians, Mons and Meng all speak languages which are of Khmer stock. We may conclude, therefore, that these peoples are descended from the Khmers. As for the original Lao, they are to be identified in the people styled today Lua or Lawa.

See "History of Siam in the Period Antecedent to the Founding of Ayuddhya by King Phra Chao U Thong", in *Miscellaneous Articles Written for the Journal of the Siam Society by His Royal Highness Prince Damrong* (Bangkok: Siam Society, 1962), p. 49.

10 Ibid., p. 61.

11 Ibid., pp. 50–65.

12 The best favourable account of Vajiravudh in English comes from Walter F. Vella, *Chaiyo! King Vajiravudh and the Development of Thai Nationalism* (Honolulu: University of Hawaii Press, 1978). As for his official nationalism, see Benedict Anderson, *Imagined Community: Reflection on the Origin and Spread of Nationalism* (Verso: London, 1991); and Kullada Kesboonchoo, "Official Nationalism under King Chulalongkorn", paper presented at the International Conference on Thai Studies (Canberra, 1987). See also Stephen L.W. Greene, *Absolute Dreams: Thai Government under Rama VI, 1910–1925* (White Lotus: Bangkok, 1999); and a provocative new treatment of the almost-all-male inner-court life of Vajiravudh by Pokpong Chanan, *Nai Nai* (Inner Men) (Bangkok: Matichon, 2013).

13 Anderson, *Imagined Community: Reflection on the Origin and Spread of Nationalism*, ch. 6: Official Nationalism and Imperialism.

14 Ibid.

15 Anderson remarks that Vajiravudh's triad echoes the theme of late Tsarist Russia: autocracy, orthodoxy and nationality, but in reversed order. Anderson, *Imagined Community*, p. 101. Many Thais believe that the king copied the English slogan of "God, King, and Country" and turned it into Thai in a different order.

16 See Krommahun Phitthayalab, *Hokho Prawatisat Phak 1* [Headings of History, Part 1], 1917. The author claimed that he first saw this word used by Ramchitti, one of Vajiravudh's many pen names, published in *Witthayachan*, No. 16, p. 104.

17 However, Wichit ignored the rise of nationalism in Laos and Cambodia, which opposed all forms of imperialism including Thai and French. See Charnvit Kasetsiri, "The First Phibun Government and Its Involvement in World War II", *JSS*, Vol. 62, No. II (1974) pp. 179–228.

18 See a list of his plays in ibid., pp. 9–41. See also Kobkua Suwannathat, "Kankhian Prawatsat Thai Baeb Chatniyom: Picharana Luang Wichitwathakan", *Warasan Thammasat*, Vol. 6, No. I (1969), pp. 149–180.

19 Thongchai Winichakul, "The Changing Landscape of the Past: New Histories in Thailand since 1973", *Journal of Southeast Asian Studies*, Vol. 26, No. 1 (March 1995), pp. 99–120.

20 See a discussion on his writings in Suchart Sawadsri (ed.), *Jit Phumisak*, Association of the Social Sciences (1974). See the best account and translation in English on Jit's life and works by Craig J. Reynolds, *Thai Radical Discourse – The Real Face of Thai Feudalism Today* (Ithaca: Cornell Southeast Asia Program, 1994).

21 For Nidhi Eoseewong, see *Pen and Sail: Literature and History in Early Bangkok*, 2006.

22 See Chatthip Nartsupha, *The Thai Village Economy in the Past*, translated by Chris Baker and Pasuk Phongpaichit, first published 1984. Reprint edition (Chiang Mai: Silkworm Books, 1999), p. 131.
23 For Piriya Krairiksh, see *Charuk Pho Khun Ramkhamhaeng* [Ramkhamhaeng Inscription], revised 2003 and J.S. Chamberlain, ed. *The Ramkhamhaeng Controversy: Collected Papers* (Bangkok: Siam Society, 1991).
24 For Saichol Wannarat, see *Somdet Krom Phraya Damrong Kan Sang Attalak Muang Thai* [Prince Damrong and the Construction of Muang Thai Identity] (2003); and *Kukrit Dap Praditdam Khwam Pen Thai* [Kukrit and the Invention of Thainess], 2 vols. 2007; for Thongchai Winichakul, see his well-known *Siam Mapped: A History of the Geo-Body of a Nation*, 1994; and for Somsak Jeamteerasakul, see his *Prawatisat Thi Phueng Sang* [History Just Invented], 2001.
25 For Thamrongsak Petchlertanan, see *2475 Lae Nung Pi Lang Patiwat* [1932 Revolution and One Year After], 2001; for Natthapoll Chaiching, see *Kho Fan Fai Nai Fan An Lue Chua* [To Dream the Impossible Dream], 2013; for Chanida Chidbundid, see *Kan Sathapana Phra Ratcha Amnat Nam* [The Foundation of the Royal Hegemony], 2007; for Pokpong Chanan, see *Nai Nai* [Inner Men]; for Warisara Tangkawanich, see *Prawatsat Sukhothai Thi Pherng Sang* [Sukhothai History: Newly Invented], 2014.

References

Anderson, Benedict. 1991. *Imagined Community: Reflection on the Origin and Spread of Nationalism*. London: Verso.

Charnvit Kasetsiri. 1976. *The Rise of Ayudhya: A History of Siam in the Fourteenth and Fifteenth Centuries*. Kuala Lumpur: Oxford University Press.

Greene, Stephen L.W. 1999. *Absolute Dreams: Thai Government under Rama VI, 1910–1925*. White Lotus: Bangkok.

Griswold. 1961. *King Mongkut of Siam*. New York: The Asia Society.

Reid, Anthony and David Marr. 1978. *Perceptions of the Past in Southeast Asia*. Singapore: Heinemann.

Riggs, F.W. 1966. *The Modernisation of a Bureaucratic Polity*. Honolulu: East-West Center Press.

Siffin, W.J. 1966. *The Thai Bureaucracy: Institutional Change and Development*. Honolulu: East-West Center Press.

Vella, Walter F. 1978. *Chaiyo! King Vajiravudh and the Development of Thai Nationalism*. Honolulu: University of Hawaii Press.

Wyatt, D.K. 1969. *The Politics of Reform in Thailand*. New Haven: Yale University Press.

4

ECONOMIC DEVELOPMENT OF POST-WAR THAILAND[1]

Peter Warr

At the end of the Second World War, Thailand was one of the world's poorest and economically most backward nations. Its agrarian-based economy had remained stagnant for at least a century and the war itself had caused widespread damage. External observers of the time generally assessed the country's economic potential negatively.[2] In the seven decades following, Thailand has developed to an upper-middle-income, semi-industrialised and technologically advanced economy. But multiple economic and associated social problems remain. This chapter describes the major economic changes that have occurred, analyses the forces driving them, and attempts to identify the principal policy priorities for continued progress.

The next section briefly summarises the performance of the Thai economy over the seven decades following the Second World War, with a focus on aggregate economic growth and structural change. The following sections then address four questions regarding this performance:

- What have been the drivers of Thailand's long-term growth?
- What caused the Asian Financial Crisis, and why did it originate in Thailand?
- How equitably have the benefits of Thailand's economic growth been distributed?
- Is Thailand caught in a "middle-income trap"?

Space limitations prevent full presentation here of the evidence underlying all the analysis summarised in this chapter. I will therefore draw extensively on past published works of myself and colleagues while updating where necessary for the present study.

Economic growth

Aggregate economic performance

Despite its many limitations, the level of real gross domestic product (GDP) per capita, meaning the level of economic output, adjusted for inflation, per member of the population, is widely considered the most useful single measure of economic performance, provided it is supplemented by other relevant indicators. Figure 4.1 summarises real GDP per capita in each year (vertical bars) and its growth rate (solid line) for the period 1951 (the first year for which official GDP data are available) to 2017 (the most recent at the time of writing). Over the six and a

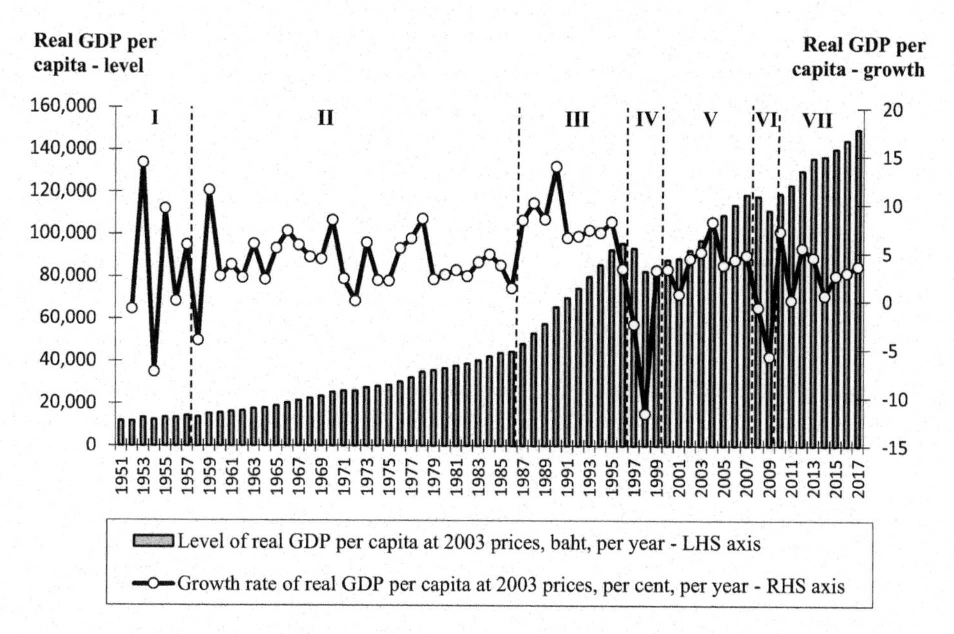

Figure 4.1 Thailand: real GDP per capita and its growth rate, 1951 to 2017

Source: Author's calculations, using data from National Economic and Social Development Board (NESDB), Bangkok, www.nesdb.go.th/ (accessed 15 June 2018).

half decades captured by these data – a little under one human life span – real GDP per person expanded by a factor of about 13, growing at an average annual rate of 4 per cent. Despite this comparatively good long-term outcome, growth was far from uniform over time. The figure identifies seven sub-periods based on the economic outcomes observed.

- Period I: Post-war Recovery (1951 to 1958)

Annual growth of economic output per person fluctuated widely, between −7.4 per cent and +14.3 per cent, averaging 2.5 per cent per annum over this period. The policy priority of this time was not fostering growth but containing price inflation, which had reached almost 100 per cent per annum during the final year of the Second World War. The cause of the hyperinflation had been high levels of spending by the occupying Japanese military regime in Thailand, financed by printing money.[3] The inflation was successfully contained, though at high cost, by restrictive fiscal and monetary policies, including an inefficient multiple exchange rate system, maintained until unification of the exchange rate in 1955.[4]

- Period II: Sustained, Moderate Growth (1959 to 1986)

The average annual growth rate of real GDP per person was 4.3 per cent, compared with an average of just over 2 per cent for all low- and middle-income countries over the same period, according to World Bank data. This was an extended period of moderate growth combined with macroeconomic stability.

- Period III: Economic Boom (1987 to 1996)

Over this critical decade, the Thai economy was the fastest growing in the world, with real GDP per person growing at an average annual rate of 7.3 per cent. In contrast with the pessimism at the end of World War II, by the mid-1990s these negative assessments had been replaced by euphoric descriptions of Thailand as a "Fifth Tiger", following in the footsteps of Korea, Taiwan, Hong Kong, and Singapore. During this boom period, Thailand's economic performance was frequently described by economists and others as an example that developing countries elsewhere might emulate. Its principal economic institutions, including its central bank, the Bank of Thailand, were cited as examples of competent and stable management.[5]

- Period IV: Asian Financial Crisis (1997 to 1999)

The Asian Financial Crisis (AFC) was the collapse of the preceding decade of economic boom. The crisis was a major turning point for Thailand. Mismanagement of exchange rate policy was central to it. Through 1996 and into 1997, growth slowed (in 1996, real GDP growth per person declined to 3.3 per cent) and export performance was especially sluggish. A moderate depreciation of the Thai baht was widely anticipated. This expectation provoked an outflow of mobile financial capital as holders of these funds attempted to avoid a decline in the foreign currency value of their capital. The central bank, the Bank of Thailand (BOT), resisted a depreciation, partly on the grounds that a currency depreciation would increase domestic inflation, but their insistence that there would be no depreciation was widely disbelieved.

Through early 1997 the capital outflow became so large that the foreign exchange reserves of the BOT were exhausted, forcing the government to accept a humiliating bailout by the International Monetary Fund (IMF). The currency was floated on 2 July 1997 and immediately depreciated by 20 per cent. Over the following months the baht/US dollar exchange rate – now market determined – would fall from 25 to over 50, before stabilising at around 40. Thai firms that had borrowed in dollars and were obliged to repay in dollars were quickly bankrupted because their earnings in baht were insufficient to meet their repayment obligations, made much larger in baht terms by the depreciation. Domestic investment declined dramatically. Many financial institutions became insolvent. During 1997 and 1998, real GDP per person fell by a combined 14 per cent.

- Period V: Recovery from the Asian Financial Crisis (2000 to 2007)

Following the AFC, Thailand's rate of economic recovery was moderate, and it has remained so ever since. In a very real sense, Thailand has never fully recovered from the loss of business confidence caused by the AFC. From 2000 onwards, growth of real GDP was positive but below its long-term trend. It was not until 2003 that the level of real GDP per capita again reached its pre-crisis level of 1996. Both private domestic investment and foreign direct investment (FDI) remained sluggish. Despite this gloomy story, it is appropriate to recognise that despite the slower than expected recovery, moderate growth did occur. In 2007 the level of real economic output per person was 20 per cent above its 1996 pre-crisis level and almost 10 times its level of 1951.

- Period VI: Global Financial Crisis (2008 to 2009)

The Global Financial Crisis (GFC) of 2008–2009 originated in the US housing market. It affected Thailand primarily through trade in goods – a contraction in global demand for its manufactured exports – rather than through Asian financial markets, as was the case in the AFC a decade earlier. The effect on Thailand was smaller than the AFC but still significant, and it had political consequences. Unemployment among unskilled and semi-skilled industrial workers, many from the northern and northeastern regions of the country, contributed in part to the political instability from 2008 to 2011 period, culminating in July 2011 with the election of the populist *Pheu Thai* government, led in absentia by the now exiled former Prime Minister Thaksin Shinawatra.

- Period VII: Recovery from the Global Financial Crisis (2010–2017)

Over the eight years from 2010 to 2017, average annual growth of real GDP per person recovered to 3.4 per cent. The first half of this interval was a period of political turbulence. The average rate of GDP growth per person was 3.6 per cent, slightly below the long-term average since 1951 of 4 per cent. A military coup occurred in May 2014, justified by the new government on the grounds that the coup was necessary to establish political stability, implement reforms, promote economic progress, and thereby "restore happiness". Regarding economic progress, over the four years of military government up to the end of 2017 – roughly the maximum duration that a democratically elected government might expect – the average rate of GDP growth per person was just under 3.1 per cent, well below the long-term average growth rate and below the rate achieved over the four years prior to the coup.

Comparison with other East Asian countries

The importance of the AFC for Thailand's economic performance is apparent from Figure 4.1. This point is reinforced by comparing Thailand's economic performance with other East Asian countries. Warr compares pre-crisis economic performance in seven East Asian countries over the pre-crisis decade by indexing the level of real GDP in 1986 to 100 in each country. Over this decade before the AFC, Thailand was the star performer.[6] When this exercise is repeated for the period following the AFC, indexing the 1996 level of real GDP to 100 in each country, Thailand has been the weakest performer. The AFC was a turning point for the Thai economy to an extent not matched by any other country.

Structural change

A universal feature of growing economies is the decline of agriculture both as a share of the value of output, which determines incomes, and as a share of total employment. Correspondingly, the output and employment shares of industry and services both expand. Thailand's experience is typical in this regard. Agriculture's share of GDP contracted from 36 per cent in 1960 to 9 per cent in 2017. Industry's share (including manufacturing and construction) expanded from 19 to 35 per cent, while services expanded from 45 per cent to 56 per cent over the same period.

There are two principal causes for this long-term process of structural change. The first, and best understood, operates on the demand side of the economy. As real incomes rise, consumers allocate an increasing share of their total budgets to non-food items such as manufactured goods and services and a declining share to food. The declining demand for agricultural output as a share of total demand induces a decline in the prices of agricultural products relative to other goods, squeezing resources out of agriculture.

Martin and Warr added a second component to this story, operating on the supply side of the Thai economy, rather than the demand-side phenomena just described. At the economy-wide level, as capital accumulates relative to labour, full employment of both capital and labour requires that economic sectors that are more capital intensive expand at the expense of sectors that are more labour intensive. As capital-intensive sectors expand, they require additional workers, which they obtain by drawing labour from agriculture. In the economics literature, this phenomenon is known as the Rybczynski effect. Martin and Warr provided evidence that in the Thai context, this supply-side phenomenon was a more important cause of agriculture's relative decline than the demand-side effects of a diminishing demand for food relative to other goods.[7]

At the same time as agriculture's output share declines, its employment share also falls, but it typically does so more slowly. This is especially true of Thailand. In 1960 agriculture employed around 65 per cent of the Thai workforce, and by 2017 this share had fallen to 33 per cent. As noted above, its GDP shares in the corresponding years were 36 per cent and 9 per cent, respectively. Based on these data, in 1960 the ratio of agriculture's income share to its employment share was 0.55. It follows that incomes per worker were substantially lower in agriculture than elsewhere, and this fact explains why workers left agriculture. By 2017 these shares were 33 per cent and 9 per cent, respectively, and the ratio of agriculture's output share to its employment share had declined to 0.27.

Output shares are equivalent to income shares. Over time, agricultural workers have become more impoverished relative to those employed elsewhere because the sectoral structure of employment has not kept pace with the rapidly shifting sectoral structure of output. Agricultural workers have exited farming too slowly to maintain their incomes relative to those employed in the expanding and more lucrative sectors of the economy. This simple account explains the lower living standards experienced in rural areas of the country than in urban areas and why poverty incidence has continued to be concentrated in rural areas. Moreover, the disparity has increased over time. Farmers are poor, primarily because there are too many of them. But poor rural people can be mobilised politically. Recognition of this fact was the basis for the political success of the movement led by Thaksin and his successors.[8]

What are the drivers of long-term growth?

Where did Thailand's economic growth come from? In studies of long-term growth, the basic distinction is between growth of the quantities of factors of production employed and growth in their productivity. Empirical attempts to implement this distinction are known as growth accounting, building on the pioneering work of Solow.[9] Warr presented a detailed growth accounting exercise for Thailand, and the discussion in this section draws heavily on its findings.[10]

This kind of analysis rests on the assumption that output is primarily supply constrained – aggregate demand was not the binding constraint on output. That is generally appropriate for long-term analyses but not necessarily for short-term analyses, where aggregate demand may be the binding constraint on growth. The AFC and its aftermath is an example of demand-constrained growth. In growth accounting, detailed data over time are assembled on outputs produced and factor inputs used to produce it. Changes in productivity are then inferred as a residual – the difference between the two. It is important that factor inputs are adjusted for quality. For example, data on labour inputs are adjusted for changes in the quality of the workforce by disaggregating the workforce by the educational characteristics of workers and weighting these components of the workforce using time series wage data for the educational categories concerned. Data on land inputs are similarly adjusted for the changing quality of land inputs by disaggregating by irrigated and non-irrigated land and then re-aggregating these components using data on land prices.

The findings of this analysis for Thailand, covering data for the two decades following 1980, were that factor inputs accounted for 90 per cent of output growth and that productivity growth accounted for the remaining 10 per cent. Drawing on Paul Krugman's metaphor, 90 per cent of the growth was due to perspiration and only 10 per cent due to inspiration. The composition of the factor input growth is informative. Expansion of raw labour inputs – the number of workers – explained 15 per cent of the output growth. The growth of human capital – the skill-adjusted quality of the workforce – accounted for just 5 per cent. Growth of agricultural land (expansion of the cultivated frontier) explained 3 per cent. The principal source of growth, accounting for two-thirds of the output growth, was the quantity of physical capital – machines, buildings, and public infrastructure. In passing, it is important to note the unusually small contribution of human capital growth. Thailand's growth was not based on upgrading the skills of the Thai population. It was based primarily on investment in physical capital.

Expansion of the stock of physical capital comes from investment, derived from three sources: foreign capital inflow, public investment, and domestic private investment. In public discussion, the importance of foreign investment as a contributor to overall capital formation is often exaggerated. Foreign investment has accounted for only 4 per cent of aggregate capital formation in Thailand. Public investment in physical infrastructure accounted for 27 per cent and domestic private investment accounted for 69 per cent of all capital formation. The implication is that to understand Thai economic growth, understanding private investment in physical capital is the key.

Over the long term, financing of private investment in physical capital within Thailand has been primarily through domestic savings. But a significant change occurred during the economic boom of 1987 to 1996. The proportion of total investment that was financed by short-term capital inflows (short-term borrowing from foreigners) increased significantly. This proportion increased from 2 per cent in the decade before the boom (1977 to 1986) to 23 per cent during the boom decade (1987 to 1996). This short-term borrowing, denominated in dollars, helped finance the investment boom of 1987 to 1996, but it also sowed the seeds of the crisis of 1997–1999. As discussed below, the accumulated stock of mobile foreign-owned capital grew to levels exceeding the Bank of Thailand's foreign exchange reserves. If the owners of these funds chose to withdraw them from Thailand, the Bank of Thailand would be unable to defend its fixed exchange rate. This is what happened in July 1997.

The behaviour of aggregate investment helps explain Thailand's economic boom prior to the AFC, the collapse of growth during the AFC, and its sluggish performance ever since.[11] During the pre-crisis boom decade, total investment accounted for an enormous 38.8 per cent of GDP in Thailand. During the crisis period of 1997–1999 it contracted to 24.8 per cent, and in the post-crisis period of 2000–2016 it has averaged 25.2 per cent. Investment collapsed during the crisis and it has never fully recovered. Why? As discussed earlier, private domestic investment is the most important component of total investment and it is driven by business confidence. Thailand has continued to grow slowly since the crisis because business confidence has never fully recovered from the devastation of 1997–1999.

Thailand is not unique in this respect. Figure 4.2 summarises total investment as a share of GDP in the four East Asian countries that were most affected by the AFC: Thailand, Indonesia, Malaysia, and Korea. The contraction of the GDP share of investment in Thailand, beginning in 1997, is very stark. Although Thailand is the extreme case, the same contraction can be seen, to a lesser extent, in each of the other three crisis-affected countries, least so of Korea. The message is that restoration of business confidence, following a crisis like the AFC, is not easily achieved. Each country is different, but caution must be exercised in searching for country-specific causes of events like the loss of business confidence following the AFC. The decline of investor confidence has been region-wide, at least among the countries seriously affected by the crisis. The

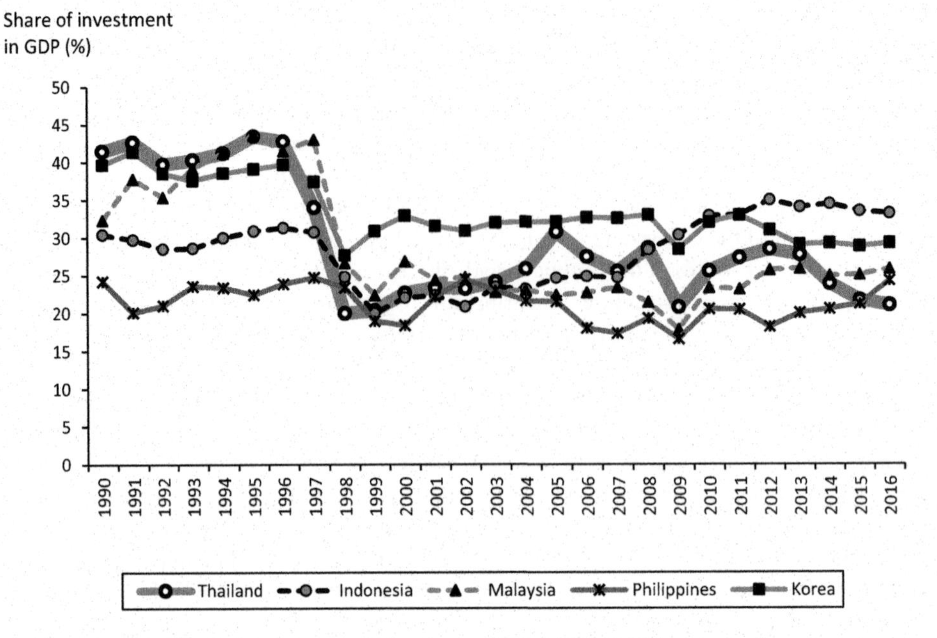

Figure 4.2 Crisis-affected East Asia: investment shares of GDP, 1990 to 2016 (per cent)

Source: Author's calculations, using data from World Bank, *World Development Indicators*. Available at https://datacatalog.worldbank.org/dataset/world-development-indicators (accessed 15 June 2018).

Note: AFC-affected countries are those that were directly impacted by the Asian Financial Crisis of 1997–1999.

AFC showed the possibility that investors could be bankrupted by macroeconomic tsunamis over which they have no control and where they have little or no forewarning.

Why did the Asian Financial Crisis originate in Thailand?

Over the (almost) four decades between 1959 and 1996 (Periods II and III in Figure 4.1), Thailand's real output per capita expanded in every year. Not a single year of negative growth was recorded – a unique achievement among oil-importing developing countries. But this long period of economic success had a downside. It fostered complacency among Thai policymakers. Their prevailing view was that sustained success proved the country was on the right path. The appropriate policy framework was just more of the same. Informed criticism could be safely ignored. During 1997, that over-confidence became very costly.

As noted earlier, in 1996 export growth slowed. Under normal circumstances, an appropriate policy response would have been a moderate currency depreciation. This response was indeed expected, provoking an outflow of mobile financial capital. By converting baht to dollars before a depreciation and then returning to baht after it, holders of mobile funds could achieve a capital gain. The capital outflow was initially small. Perversely, it was interpreted by BOT as the work of malign currency speculators hoping to outwit the bank and profit from depreciation. BOT officials wished to punish this speculative behaviour and to minimise inflation by resisting any such exchange rate adjustment. It was a mistake.

The decade of economic boom prior to the crisis had been fuelled by unprecedented levels of investment. During that decade, this investment had been financed increasingly by inflows of highly mobile, short-term financial capital and not just by domestic savings and long-term inflows of foreign direct investment. During the protracted boom, these short-term inflows had gradually accumulated into a massive sum. Warr showed that by 1994 the accumulated stock of short-term foreign-owned capital exceeded the level of the BOT's international reserves.[12] By 1997 that stock was almost double the level of reserves. These funds are highly mobile and could exit the country almost as readily as they had entered. In 1996 and early 1997, the expectation of a currency depreciation encouraged the conversion of this short-term mobile capital from baht into foreign currency. In this set of circumstances, the widely held expectation of a currency depreciation meant that either an actual depreciation or a currency crisis was inevitable; Thailand's foreign exchange reserves were insufficient to defend the fixed exchange rate in the presence of a sustained capital outflow.

Thailand's fixed exchange rate policy had been in place since 1955. It was initially introduced and subsequently maintained to control domestic inflation, seen by the BOT as its principal task. The fixed exchange rate, combined with an open capital market, meant that the BOT was obliged to convert unlimited quantities of Thai currency into US dollars or other international currencies at this fixed rate upon demand. To do that in the presence of a net capital outflow, the bank had to draw upon its international reserves. The BOT insisted its reserves were adequate to withstand any capital outflow and denied that an exchange rate depreciation was under consideration. In taking this position, the BOT had under-estimated the volume of Thai baht that might be presented for conversion to foreign currency. As the capital outflow gained pace, the BOT literally ran out of reserves. There was then no alternative to a huge depreciation. The belated decision to abandon the fixed exchange rate and float the currency in July 1997 led to a dramatic decline in the commercial value of the Thai baht.

For Thailand and for most of its people, the consequences were ruinous. The domestic economy was in disarray, with levels of output and investment contracting and the incidence of poverty rising sharply. The government was compelled to accept a humiliating IMF bailout package, the financial system was largely bankrupted, and confidence in the country's economic institutions – including the BOT – was shattered. Internationally, far from being an example for others, Thailand was now characterised as the initiator of a "contagion effect" in Asian financial markets, undermining economic and political stability and bringing hardship to millions of people.

The crisis itself also had severe negative economic impacts on Indonesia, Malaysia, and Korea, and to a lesser extent the Philippines. Although Thailand was the first to succumb, and policy failures in Thailand were crucial, as explained earlier, it is superficial to say that events in Thailand "caused" the crisis in other countries. The boom of the previous decade had produced vulnerability to a financial crisis in Thailand, and these conditions went unnoticed by both the Thai authorities and the IMF. But similar conditions of vulnerability had developed in each of the countries that subsequently succumbed to the crisis, and like Thailand, they were all following fixed exchange rate policies.

Thailand's currency crisis provided the trigger for the broader financial effects experienced elsewhere. It caused the expected depreciation of fixed exchange rates in other countries, provoking capital outflows that in vulnerable countries, like Thailand, Indonesia, Malaysia, and Korea, compelled large actual depreciations. The underlying source of this vulnerability – huge stocks of volatile short-term capital combined with misguided attempts to maintain fixed exchange rates – was shared by all the crisis-affected countries, having developed over the preceding decade, but not by the many countries of Asia and elsewhere that did not succumb to the crisis.[13] Given this vulnerability and the commitment to fixed exchange rates, the crisis was a disaster in waiting.

In triggering the expectation of a depreciation, the circumstances of 1996 and early 1997 in Thailand determined the timing of that crisis. But a similar trigger could have arisen in any of the other vulnerable countries at any time. The underlying cause was not this particular trigger but the failure to recognise the developing signs of crisis vulnerability and to act accordingly. In the context of highly mobile financial capital, exchange rate flexibility was essential because if the expectation of a depreciation developed, the attempt to maintain fixed exchange rates would be sure to produce a currency crisis. The central banks of the crisis-affected countries, including Thailand, must bear responsibility for this intellectual and policy failure along with the IMF, whose role is to monitor and advise on exactly these matters.

Within Thailand, the economic damage done by the crisis of 1997–1999 and the hardship that resulted eroded some of the gains from the economic growth that had been achieved during the long period of economic expansion, but it did not erase them. Figure 4.1 shows that at the low point of the crisis, in 1998, the level of GDP per capita was almost 14 per cent lower than it had been in 1996, a level not reached again until 2003. Nevertheless, because of the sustained growth that had preceded the crisis, this reduced level of 1998 was still higher than it had been only five years earlier, in 1993, and was seven times the level of 1951.

The AFC had political consequences within Thailand. As argued earlier, the inept management of the exchange rate crisis of 1997 reflected an over-confidence made possible by the long period of sustained economic success that preceded it. The consequence was not merely widespread suffering but also public doubt regarding the competence of the ruling elite. For many Thai people, it no longer seemed safe to assume that the governing elite knew what was best. Perhaps a different kind of leadership was needed. This shift of public perception contributed to the election in 2001 of the populist *Thai Rak Thai* government led by the businessman Thaksin.

Who benefited from the growth?

Is economic growth really so desirable? If all the gains from growth went to those who were already rich, its social value would surely be dubious. Do the poor benefit? Despite much debate about measurement and conceptual issues, all major studies of poverty incidence and inequality in Thailand agree on some basic points:

- Poverty is concentrated in rural areas, especially in the northeastern and northern regions of the country.
- Absolute poverty has declined dramatically over the last four decades, but inequality has increased.
- The long-term decline in poverty incidence was not confined to the capital, Bangkok, its immediate environs, or to urban areas in general, but it occurred in rural areas as well.
- Large families are more likely to be poor than smaller families.
- Farming families operating small areas of land are more likely to be poor than those operating larger areas.
- Households headed by persons with low levels of education are more likely to be poor than others.

Thailand's official poverty estimates are produced by the government's National Economic and Social Development Board (NESDB) based upon the household incomes and expenditures captured in the National Statistical Office's Socio-economic Survey (SES) of household data, collected periodically since 1962. It is well understood that in all countries, to varying degrees, household survey data of this kind understate true inequality. The very rich and the very poor are both under-represented in their sample coverage.[14] Despite their imperfections, these are the only household level data available covering a long time period. Since 1988 the raw data have been available in electronic form. Table 4.1 summarises these data, focusing on the familiar headcount measure of poverty incidence: the percentage of the population whose household incomes per person fall below the official poverty line, held constant in real purchasing power over time.

The data reveal a massive decline in poverty incidence during the boom decade ending in 1996. Measured poverty incidence declined by an extraordinary 27.9 per cent of the population, an average rate of decline in poverty incidence of 3.5 percentage points per year. That is, each year, on average 3.5 per cent of the population moved from incomes below the poverty line to incomes above it. This was followed by an increase in poverty incidence during the AFC and its aftermath, from 1996 to 2000. Over this four-year interval, poverty incidence increased by 4.5 per cent of the population.

Alternatively, over the eight years ending in 1996 the total number of persons in poverty declined by 11.1 million (from 17.9 million to 6.8 million); over the following four years the number increased by 1.8 million (from 6.8 million to 8.6 million). Thus, according to the official data, measured in terms of absolute numbers of people in poverty, the crisis and its aftermath reversed one-sixth (16 per cent) of the poverty reduction that had occurred during the eight years of economic boom that preceded the crisis.

Table 4.1 Thailand: poverty incidence and Gini coefficient, 1988 to 2016 (income based)

	Poverty incidence (headcount measure, per cent of population)			*Inequality (Gini coefficient)*
	Aggregate	*Rural*	*Urban*	*Aggregate*
1988	44.9	52.9	25.2	0.488
1990	38.2	45.2	21.4	0.515
1992	32.5	40.3	14.1	0.536
1994	25.0	30.7	11.7	0.521
1996	17.0	21.3	7.3	0.513
1998	18.8	23.7	7.5	0.507
2000	21.3	27.0	8.7	0.522
2002	15.5	19.7	6.7	0.508
2004	11.3	14.3	4.9	0.493
2006	9.5	12.0	3.6	0.515
2008	7.2	9.0	2.9	0.499
2010	5.8	7.3	2.6	0.490
2012	4.3	5.5	2.0	0.484
2014	3.5	4.3	1.9	0.465
2016	2.8	3.5	1.5	0.445[a]

Source: Author's calculations using data from National Economic and Social Development Board, www.nesdb.go.th/Default.aspx?tabid=322 (accessed 15 June 2018).

Note: Both poverty incidence and inequality are based on incomes rather than expenditures in these data. Higher values of the Gini coefficient indicate greater inequality.

[a] The Gini coefficient for 2016 is not available. The number shown relates to 2015.

It is notable that this increase in poverty incidence following the AFC was larger in rural areas (5.7 per cent of the rural population) than in urban areas (1.4 per cent of the urban population). Most of Thailand's poor people continue to reside in rural areas. Taking account of the populations of rural and urban areas, in 2016 rural areas accounted for 50 per cent of the total population but 90 per cent of the total number of poor people.

What caused the long-term decline in poverty incidence? It is obvious that over the long term, sustained economic growth is a necessary condition for large-scale poverty alleviation. No amount of redistribution could turn a very poor country into a rich one. Long-term improvements in education have undoubtedly been important, but despite the limitations of the underlying SES data, a reasonably clear statistical picture also emerges on the short-term relationship between poverty reductions and the rate of economic growth. The data are summarised in Figure 4.3, which plots the relationship between changes in poverty incidence, calculated from Table 4.1 and the real rate of growth of GDP over the corresponding periods.

The annual rate of economic growth is negatively correlated with the annual change in poverty incidence. That is, periods of more rapid economic growth were associated with more rapid reductions in the level of poverty incidence. Moderately rapid growth from 1962 to 1981 coincided with steadily declining poverty incidence. Reduced growth in Thailand caused by the world recession in the early to mid-1980s coincided with worsening poverty incidence in the years 1981–1986. Then, Thailand's economic boom of the late 1980s to mid-1990s coincided with dramatically reduced poverty incidence. The contraction of 1997–1998 and the subsequent recession to 2000 led to increased poverty incidence. The recovery since the crisis has been associated with sustained but moderate poverty reduction. It cannot be said that economic growth fails to benefit poor people.

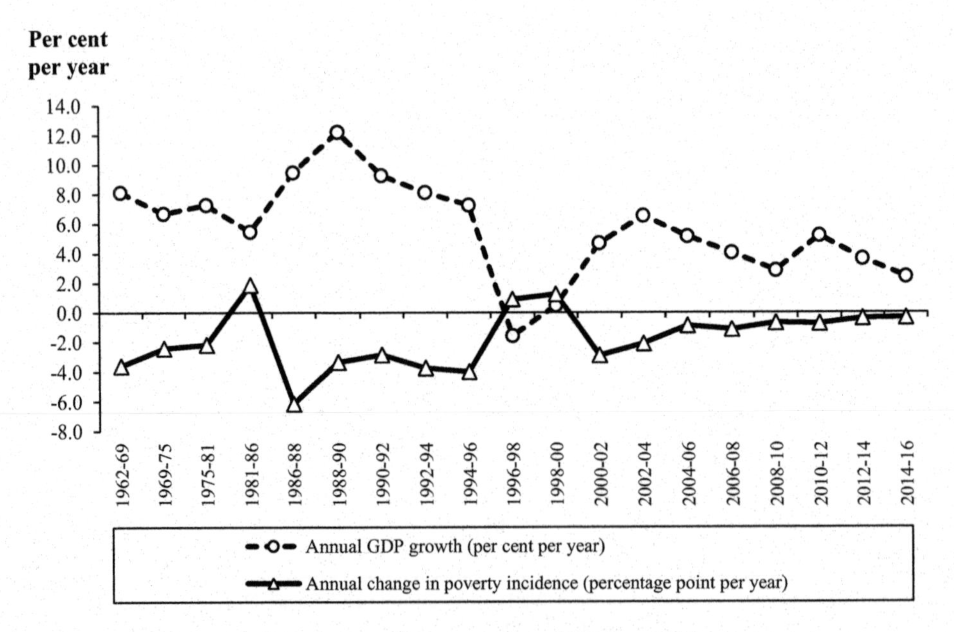

Figure 4.3 Thailand: poverty incidence and economic growth, 1962 to 2016

Source: Author's calculations using poverty data as in table 3 and GDP data at constant prices from World Bank, *World Development Indicators*. Available at https://datacatalog.worldbank.org/dataset/world-development-indicators (accessed 15 September 2017).

The final column of Table 4.1 shows the Gini coefficient of inequality. This index potentially takes values between 0 and 1, with higher values indicating greater inequality. The index for Thailand is high by international standards. It increased sharply during the boom decade of 1986–1996. Combined with the large reduction in absolute poverty which occurred at the same time, this means that during the boom decade the real incomes of the poor increased with economic growth, but the incomes of the rich increased even faster. The measured Gini coefficient has declined since 2000, suggesting a moderate reduction in inequality. This feature of the official data on inequality is disputed, but when combined with the increase in inequality during the boom period it does suggest the hypothesis that rapid economic growth raises inequality while slower growth reduces it.[15]

Finally, Tables 4.2 and 4.3 draw upon the SES household survey data to compare the distribution of household incomes per household member across the population, covering the three decades since 1986. Quintile 1 means the poorest one-fifth of the population; quintile 2 is the next richest one-fifth and so on, up to quintile 5, which is the richest one-fifth of the population. Decile 10 means the richest one-tenth of the population and centile 100 means the richest 1 per cent. In Table 4.3, the real income data shown in Table 4.2 are converted to shares of total income by dividing each real income level by the total of all household incomes. Like the Gini coefficient data shown in Table 4.1, these data suggest an increase in income inequality during the boom decade and a decline since then.

Caught in a "middle-income trap"?

Beneath the macroeconomic events summarised above lies a deeper and longer-term economic process. Between the Second World War and the present, Thailand has achieved a transition from a poor, heavily rural backwater to a middle-income, semi-industrialised and globalised economy. This transition required some elementary market-friendly policy reforms: promoting a stable business environment (not necessarily equivalent to stable politics), open policies with respect to international trade and foreign investment, and public provision of basic physical infrastructure, including roads, ports, reliable electricity supplies, telecommunications, and policing sufficient to protect the physical assets created by business investment. The process was primarily market driven and the central policy imperative was to support the shift from low-productivity agriculture to export-oriented labour-intensive manufacturing and services.

A transition just like this occurred in most of East and Southeast Asia. The pattern was similar in all countries that undertook the basic policy reforms listed above. During this process average real incomes rose significantly, the share of the workforce employed in agriculture contracted, and the incidence of absolute poverty fell. The core of this growth process is expansion of the physical capital stock resting overwhelmingly on private domestic investment. The private financial system facilitates the link between private savings and business investment.

But the process is self-limiting. As labour moves from low-productivity agriculture to more rewarding alternatives elsewhere, wages are eventually driven up. As wages rise, the profitability of labour-intensive development declines. Rising wages lower the return to investment in physical capital, the rate of private investment slackens, and growth slows. The frontier for further expansion of labour-intensive export-oriented development moves to other, lower-wage countries, including Thailand's immediate neighbours to the north, east, and west. Development based on abundant cheap labour can raise a country from low-income to middle-income levels, but it is incapable of raising it from middle-income to higher income levels. Reliance on that mode of development leads to the dreaded but seldom-explained "middle-income trap".

Table 4.2 Real income per person (constant 2015 prices, CPI deflator)

Quintile group	*1986*	*1996*	*2006*	*2017*
Quintile 1 (poorest)	730.6	1,056.0	1,319.6	2,398.8
Quintile 2	1,282.1	1,907.9	2,654.2	4,359.2
Quintile 3	2,022.7	2,987.4	4,235.4	6,530.9
Quintile 4	3,350.4	5,027.8	7,014.2	10,008.9
Quintile 5 (richest)	9,349.3	14,265.7	19,568.4	24,367.1
Population mean	3,347.4	5,050.6	6,958.4	9,533.0
Decile 10 (richest)	13,169.9	20,164.0	27,826.0	33,646.5
Centile 100 (richest)	36,257.1	57,257.9	82,213.5	91,810.7

Source: Author's calculations using data from National Economic and Social Development Board, www.nesdb.go.th/Default.aspx?tabid=322 (accessed 15 June 2018).

Table 4.3 Thailand: income shares (per cent of total income)

Quintile group	*1986*	*1996*	*2006*	*2017*
Quintile 1 (poorest)	4.36	4.18	3.79	5.03
Quintile 2	7.67	7.55	7.63	9.14
Quintile 3	12.09	11.83	12.17	13.70
Quintile 4	20.02	19.91	20.16	21.00
Quintile 5 (richest)	55.87	56.53	56.25	51.12
Total	100	100	100	100
Decile 10 (richest)	39.38	39.95	39.98	35.29
Centile 100 (richest)	10.77	11.34	11.76	9.63

Source: Author's calculations from Table 4.2.

Note: Income share means the total income of the group shown relative to the sum of all household incomes. For example, in 2017 the poorest one-fifth of the population received 5 per cent of all household incomes while the richest one-fifth received 51 per cent. The richest one-tenth of the population received 35 per cent of all incomes and the richest 1 per cent received just under 10 per cent of all incomes.

Escape from the trap requires addressing a key market failure: the undersupply of human capital. Human capital is a crucial input created primarily by investment in education, broadly defined. But unlike physical capital, it does not provide the collateral that can secure repayment of loans. Physical assets may remain in place, but human beings can walk away. Hence, the private financial system does not support investment in human capital. Individual families can and do invest heavily in the education of their own children, but because their resources are limited and because the recipients of the educational investment reap only part of the returns it generates, this is insufficient to resolve the overall underinvestment in human capital.

Increasing the supply of human capital is central to overcoming the middle-income trap. Upgrading the quality of the workforce raises labour productivity and raises the return to physical capital, encouraging greater investment in physical capital as well. In Thailand, as in many other middle-income countries, the problem lies in the quality of education and not just the bare numbers of total school enrolments. The problem is primarily not at the tertiary level but at the primary and secondary levels. This problem is not unique to Thailand and it is not new.[16] Massive public investment and reform of the education curriculum is needed, which in turn requires the raising of sufficient tax revenue to finance it, and combating the self-serving practices of the Ministry of Education and the teachers' unions. Other countries have addressed this problem, but Thailand has not. Until it does, there will be no escape from Thailand's version of the middle-income trap.

Conclusions

This chapter has focused on five aspects of Thailand's economic performance since World War II: the changing rate of growth and its composition; the sources of that growth; the causes and consequences of the Asian Financial Crisis (AFC) of 1997 to 1999, including the reason it originated in Thailand; the distribution among the Thai population of the fruits of long-term growth; and whether Thailand is caught in a middle-income trap.

The AFC was a turning point for Thailand, as it was for several other East Asian countries. During the decade of economic boom preceding the AFC, Thailand's average annual growth rate of real GDP per person was a remarkable 7.3 per cent. Since 2000 the corresponding growth rate has been less than half of that. The immediate cause was a contraction of investment in physical capital. Gross capital formation declined as a proportion of GDP from an average of 38 per cent to 25 per cent over the same two periods. The effect of lower investment was two-fold: it reduced aggregate demand, lowering income in the short run; and it reduced the rate of capital formation, lowering long-run growth prospects.

Following the AFC, a decline in this investment ratio occurred in all the crisis-affected Asian economies, including Indonesia, Malaysia, and South Korea, but the decline in Thailand was especially large. Put simply, after the crisis private investors became less confident about Thailand's prospects and thereby less inclined to invest. An expectation of this kind is self-fulfilling; it reduces investment, which ensures that growth will indeed be lower.

The evidence from Thailand demolishes the notion that economic growth fails to benefit the poor – provided that "benefit" is understood in absolute terms. The story is different if "benefit" is measured in relative terms. The Thai historical experience provides some evidence that very high rates of economic growth have benefited the rich proportionately more than the poor and that lower rates of growth do the opposite.

Finally, it is argued in this chapter that Thailand is now caught in a "middle-income trap" caused by a backward and under-resourced educational system. It is a "trap" of the country's own making, not one imposed by others. Exit from the trap is possible, but it requires a public commitment to overcoming the under-supply of human capital that a market-based economic system inherently produces. This requires raising both the quantity and quality of the country's investment in education, primarily at the primary and secondary levels. It will be costly, requiring an increased level of public revenue. The set of "reforms" ostensibly being implemented by the current Thai government ignores this fundamental problem.

Notes

1 The excellent research assistance of Perada Dulyapiradis, Ramesh Paudel and Moh Agung Widodo is gratefully acknowledged. The author is responsible for all defects.
2 See Sompop Manorungsan, *Economic Development of Thailand, 1850–1950*, Institute of Asian Studies Monograph 42 (Bangkok: Chulalongkorn University, 1989). Also see James C. Ingram, *Economic Change in Thailand: 1850–1970* (Stanford: Stanford University Press, 1971).

3 Bhanupong Nidhiprabha, *Macroeconomic Management of the Thai Economy* (London: Routledge, 2018).
4 W.M. Corden and H.V. Richter, "The Exchange Rate System and the Taxation of Trade", in *Thailand: Social and Economic Studies in Development*, edited by T.H. Silcock (Canberra: Australian National University Press, 1976).
5 Peter Warr and Bhanupong Nidhiprabha, *Thailand's Macroeconomic Miracle: Stable Adjustment and Sustained Growth, 1966 to 1996* (Washington, DC and Kuala Lumpur: World Bank and Oxford University Press, 1996).
6 See Peter Warr, "Thailand: Economic Progress and the Move to Populism", in *Handbook of Emerging Economies*, edited by Robert E. Looney (London: Routledge, 2014).
7 Will Martin and Peter Warr, "Determinants of Agriculture's Relative Decline: Thailand", *Agricultural Economics*, Vol. 11 (1994), pp. 219–235.
8 Kosuke Mizuno and Pasuk Phongpaichit, *Populism in Asia* (Singapore: National University of Singapore Press, 2009).
9 See Robert M. Solow, "Technical Change and Aggregate Production Function", *Review of Economics and Statistics*, Vol. 39 (1957), pp. 312–320.
10 Warr, "Thailand: Economic Progress and the Move to Populism", pp. 416–439.
11 David Vines and Peter Warr, "Thailand's Investment-Driven Boom and Crisis", *Oxford Economic Papers*, Vol. 55 (2003), pp. 440–464.
12 Peter Warr, "What Happened to Thailand?" *The World Economy*, Vol. 22 (1999), pp. 631–650.
13 Peter Warr, "Crisis Vulnerability", *Asia Pacific Economic Literature*, Vol. 16 (2002), pp. 36–47.
14 See F.A. Cowell, *Measuring Inequality*, second edition (Hemel Hempstead: Harvester Wheatsheaf, 1995). Also see J. Houghton and S.R. Khandker, *Handbook on Poverty and* Inequality (Washington, DC: World Bank, 2009).
15 Pasuk Phongpaichit and Chris Baker (eds.), *Unequal Thailand: Aspects of Income, Wealth and Power* (Singapore: NUS Press, 2015).
16 Sirilaksana Khoman, "Education Policy", in Peter Warr (ed.), *The Thai Economy in Transition*, edited by Peter Warr (Cambridge: Cambridge University Press, 1993).

References

Asian Development Bank. (various years). *Key Indicators for Asia and the Pacific*. Manila: Asian Development Bank <www.adb.org/publications/series/key-indicators-for-asia-and-the-pacific> (accessed 15 September 2017).

Bank of Thailand. <www.bot.or.th/english/Pages/BOTDefault.aspx> (accessed 15 September 2017).

Bhanupong Nidhiprabha. 2018. *Macroeconomic Management of the Thai Economy*. London: Routledge.

Chalongphob Sussankarn and Pranee Tinakorn. 1988. *Productivity Growth in Thailand, 1980 to 1995*. Bangkok: Thailand Development Research Institute.

Corden, W.M. and H.V. Richter. 1967. "The Exchange Rate System and the Taxation of Trade". In *Thailand: Social and Economic Studies in Development*. Edited by T.H. Silcock. Canberra: Australian National University Press.

Cowell, F.A. 1995. *Measuring Inequality*, second edition. Hemel Hempstead, UK: Harvester Wheatsheaf.

Houghton, J. and S.R. Khandker. 2009. *Handbook on Poverty and Inequality*. Washington, DC: World Bank.

Ingram, James C. 1971. *Economic Change in Thailand: 1850–1970*. Stanford: Stanford University Press.

Krugman, Paul. 1994. "The Myth of Asia's Miracle". *Foreign Affairs*, Vol. 73: 62–78.

Martin, Will and Peter Warr. 1994. "Determinants of Agriculture's Relative Decline: Thailand". *Agricultural Economics*. Vol. 11: 219–235.

Mizuno, Kosuke and Pasuk Phongpaichit. 2009. *Populism in Asia*. Singapore: National University of Singapore Press.

National Economic and Social Development Board <www.nesdb.go.th/> (accessed 15 June 2018).

Pasuk Phongpaichit and Chris Baker (eds.). 2015. *Unequal Thailand: Aspects of Income, Wealth and Power*. Singapore: NUS Press.

Sirilaksana Khoman. 1993. "Education Policy". In *The Thai Economy in Transition*. Edited by Peter Warr. Cambridge: Cambridge University Press.

Solow, Robert M. 1957. "Technical Change and Aggregate Production Function". *Review of Economics and Statistics*, Vol. 39: 312–320.

Sompop Manorungsan. 1989. *Economic Development of Thailand, 1850–1950*. Institute of Asian Studies Monograph 42. Bangkok: Chulalongkorn University.

Vines, David and Peter Warr. 2003. "Thailand's Investment-driven Boom and Crisis". *Oxford Economic Papers*, Vol. 55: 440–464.

Warr, Peter. 1993. "The Thai Economy". In *The Thai Economy in Transition*. Edited by Peter Warr. Cambridge: Cambridge University Press.

———. 1999. "What Happened to Thailand?" *The World Economy*, Vol. 22: 631–650.

———. 2002. "Crisis Vulnerability". *Asia Pacific Economic Literature*, Vol. 16: 36–47.

———. 2005. "Boom, Bust and Beyond". In *Thailand Beyond the Crisis*. Edited by Peter Warr. London: Routledge.

———. 2014. "Thailand: Economic Progress and the Move to Populism". In *Handbook of Emerging Economies*. Edited by Robert E. Looney. London: Routledge, pp. 416–439.

Warr, Peter and Bhanupong Nidhiprabha. 1996. *Thailand's Macroeconomic Miracle: Stable Adjustment and Sustained Growth, 1966 to 1996*. Washington, DC and Kuala Lumpur: World Bank and Oxford University Press.

World Bank. (various years). *World Development Indicators* <http://data.worldbank.org/indicator> (accessed 15 June 2018).

PART II

The political and economic landscape

5

THE TWO FACES OF DEMOCRACY

Michael K. Connors

Thai democracy, what does your face look like?

Ever since the revolutionaries of the People's Party overthrew the Siamese absolutist monarchy in 1932 and placed the king under a constitution, the conditions for making the sovereignty of the people manifest have been at the heart of political contest.[1] That contest gave rise to "two faces of democracy": the rhetorically native or universal poise struck by adversaries across modern Thai history and which have framed distinctive political communities.[2] However, it is also possible to speak of two-faced democracy (and not in a pejorative way) to recognise varying degrees of syncretism. In what follows, the reader should assume that scare quotes cling to "democracy". Where the reader places them is a political choice. This chapter commences with a discussion of the two faces. They are selectively surveyed from the 1932 revolution up to the recent consolidation of the ideology of Democracy with the King as Head of State (DWKHS). The chapter concludes with that ideology's failure in the early 21st century and the rise of an extraordinary dictatorship.

The two faces of democracy

The idea of there being two faces of democracy, an analytical simplification, in Thailand is conventional in scholarship and popular culture.[3] To take the latter, the title song of rock group Carabao's 1986 album, *Democracy*, asks:

> *Oh Democracy, what does your face look like? . . . May I see, may I see?*

At first, the visage underwhelms: "So that's how you look". But as the void comes into view, Carabao gibes:

> *Oh Oh, there it is. It's huge, huge.*
> *The hole is huge. It's leaking.*

The singer's staccato incantation of the syllables *pra-cha-thi-pa-tai* (democracy), the song's defining feature, is like a musical drill against the then semi-bureaucratic authoritarian regime of General Prem Tinsulanonda. Even so, the singer asks which democratic face will prevail: Will people play "Thai-style democracy" or will it be "universal"? The song is dated (it has a guitar solo), but its teasing question is stuck in a groove.[4]

Here, Thai-style democracy is understood as a conservative nativist logic of what works for "we Thai". It presumes that an evolving monarchy-military-bureaucracy complex steers the people's sovereignty. It takes a statist guardian form even if popularly supported. In its extreme, variants of guided democracy have entailed or promoted prohibition of political opposition, human rights abuse and hyper-nationalism. Despite its reactionary impulses, this face was never a fascist statism where the state is everything and the individual is nothing, even if it mouthed forms of fascist mobilisation in the early 1940s and the 1970s.[5] Quasi-fascist forms have been deployed as mobilisation resources rather than serving as the basis of enduring legitimacy. Within guardian democracy (from outright dictatorship to soft authoritarianism), the monarchy functioned as a moral aristocracy that sponsored conservative developmentalism in Buddhist hue and legitimated hierarchy.[6] And through processes of ritual, it incorporated the once dependent Sino-Thai capitalist class into the ideational structure of a religious-conservative world.[7]

The "universal" face takes a liberal and/or popular form and is characterised by the politics of rights and rule of law constitutionalism. Representative institutions mediate the people's sovereignty. When institutionally emergent (governments from the mid-1940s, mid-1950s, mid-1970s and the 1990s and 2000s), it experienced political instability caused by corruption and coalitional crises or by reactionary opposition. Human rights abuses by elected governments also diminished the face's appeal. Like native guided democracy, this face articulated to monarchy and developmentalism, a combination in part enabled by the affinities between conservative views on the self-reliant subject and the imagined "good citizen" of liberal discourse.[8] Such fusion has meant critics have often mistaken what face they are looking at.

Carabao's image of the two faces chimes with Kasian Tejapira's view of democracy as a floating signifier that entered Siam during the imperial era.[9] Kasian tells us that *prachathipatai* was coined by King Vajiravudh as a Thai equivalent to "republic". Thus, when the ideology of Democracy with the King as Head of State (hereinafter DWKHS) approached hegemonic status in the 1980s and 1990s (mixing the nativist and the universal), the etymologically aware could, as Kasian notes, enjoy the irony of a "Republic with the King as Head of State".[10] The joke suggests that political struggles to signify democracy failed to cover the semantic hole left by the end of absolutism and the rise of "the people". Its incompleteness lies in indeterminacy surrounding sovereignty. Even so, the institutions and practises of the nativist and the universal, separately or combined, were for relatively brief times hegemonic. They were broadly accepted as serving the general good, even if each face served particular interests and opposition remained.

Fundamental questioning on the nature of democracy is a symptom of the precarious nature of Thai political conjunctures. The country's 12 successful coups and 20 constitutions reflect institutional and ideological flux. Only two constitutions have lasted longer than ten years (the December 1932 constitution and the 1978 constitution). The third longest was the lawless interim constitution of 1959 which lasted nine years. Of the 11 permanent constitutions, nine were abrogated after a coup d'état.[11] The predominant political pattern is a "vicious cycle":[12] the perennial return of the coup after a new constitution is promulgated and an election takes place. The resulting constitutional mortuary-generated treatises on democratic failure and what McCargo calls a culture of "permanent constitutionalism"[13] – an obsessive search for a coup-proof mix of institutions to create a "virtuous cycle" of politics. Alone or in combination, the two faces have figured in proposed solutions but to no avail. As soon as a distinctive regime emerged that might have aligned a range of social forces around a common political and economic project, it confronted challengers occupying discrete apparatuses of the state, disruption within or hostile forces outside. Accommodation resulted, leaving the state in a condition of ambivalence and subject to contending forces. The rare moments of directing control over state apparatuses and society have typically been coercive ones (1938–1944, 1957–1973, 1976–1977, 2014–). Despite this gloomy picture, Kasian Tejapira identifies three democratic "power shifts" in the 1930s, the 1970s to the 1990s, and the 2000s widened the circle of participants.[14] These arguably reflect the universal face.

Tutelage, discipline and development

The 1932 revolution was part of a long global moment of national democratic and liberal challenges to monarchy. However awkwardly new principles of governance evolved, after 1932 the idea of sovereignty emanating from the people became a universal fact made real by the unfolding of the French revolution as a global idea. Like the French, the Thai struggled with what that idea meant.

The rise of commercial society, new economic classes and technologies of the state required that the key historic (if unconscious) task of Siam's late 19th-century absolute monarchy was to modernise itself out of existence. As new principles of governance attracted support, the monarchical regime initiated limited self-reform.[15] In the late 19th century the triadic ideology of nation, king and religion linked "the people" to an emerging royal absolutism and its related nationalism.[16] The triad failed on the first count but not before state consolidation was achieved. Siam's aspirant membership of international society in the age of imperialism also drove reform. The kingdom had to demonstrate to established states civilisational capacity to secure its Westphalian sovereignty in the international system.[17] It was especially eager to demonstrate capacity for international membership by adhering to separation of legal institutions from other centres of power (to safeguard foreign capital, property, contracts and citizens). But this was not a simple submission to liberal imperialism, for Siamese elites stamped cultural meaning on new legal concepts to safeguard hierarchy.[18] These self-administered cuts to absolutism were painstaking, marked by the self-regard and half-measures that clings to those of privileged birthright.

Such projects of royalist modernisation with a strong emphasis on the Thai face vied with democratic yearnings. World-travelled and humiliated by their country's circumstance, some of the "promoters" of the 1932 revolution possessed a visceral repulsion of the Siamese everyday. Reminiscing on his time abroad, one promoter observed, "there was equality among inhabitants without status divisions, the deference system and the debasing forms of speech which were so bitter to us in our own land".[19] The revolutionaries wanted to abolish aristocratic privilege and use the state birthed by royalist reforms to modernise Siam. They spoke of royalists as fraudsters and the people as sovereign. One of the revolution's key figures, Pridi Banomyong, was a French influenced constitutionalist who, in Enlightenment fashion, saw bureaucracy as a motor of progress for advancing human rights, freedom, equality and the rule of law.[20]

The first constitution, issued days after the revolution, declared that the highest power belonged to the people and was exercised jointly by the king, the legislature, the people's committee (the executive) and the courts. Six months later, in December 1932, a new constitution reflecting the People's Party's compromise with the old regime now declared that while sovereignty came from the people, it was now to be exercised by the king through the legislature, the executive and the courts. This enduring formula of separation of powers (a liberal-conservative Montesquieuian formula) and deferral to constitutional monarchy replaced the briefly lived doctrine of a revolutionary vanguard vested with people's sovereignty. The change was significant. Monarchists argued in the Constitutional Drafting Committee that the king's position was enabled by an ancient principle of elective kingship and on the practise and symbolisation of Buddhist virtue.[21] The claim was no more theoretically incredible than the myth of a social contract in Western political theory. If 1932 had forced the monarchy to give up its patrimony of the state, the monarchy now sought to shape a new royalist nation-state form. By the late 1950s monarchical discourse had revived and by the 1990s and 2000s it had deepened – even as it articulated to democracy – to the point of sacralisation around the figure of the long-reigning King Bhumibol (1946–2016).[22]

Of either nativist or universal face, those in power often viewed the sovereign people as precocious and in need of tutoring by a self-selecting elite wise to rule.[23] From 1932 to the present the electoral expression of the people's will has mostly been contained. Both the "temporary" constitution issued after the revolution and the "permanent constitution" (December 1932) prescribed an equal proportion of appointed and elected members for ten years. A 1940 constitutional amendment extended this to 20 years dated from June 1932. The various electoral systems since have allowed for appointed members of the national assembly to function as a "state party", the majority of whom were until the 1980s (and again after 2014) typically military and civil service trained. These serve as a buffer against elected members of parliament (MPs). The people, it seems, were in need of permanent mentorship.

No less a figure than Pridi could say, speaking of an economic plan, "We may compare our Siamese people to children. The government will have to urge them forward by means of authority applied directly or indirectly to get them to co-operate in any kind of economic endeavour".[24] The same held for politics. Of course, there was support for the constitutional regime. In some provinces public subscription democracy monuments went up before the iconic democracy monument was built in Bangkok.[25] The formation of a national Constitutional Association also led to the promotion of the constitution as a quasi-sacred object identified with Buddhist virtue, inaugurating the identification of the rule of law with the need for moral instruction in the democratic age.[26] The national education campaigns which proclaimed the people as sovereign only underscored that someone had to tell those yet to identify with the new state what this meant. Since 1932, varieties of democratic moral pedagogy (right or left, royalist or democratic) have been practised to conjure a population that could act with sovereign will. This hierarchical pedagogy defined the state's relation with its subjects and granted effective sovereignty to the dominant definers of the desirable route to national moral and economic development.[27]

Displacement of popular sovereignty has been the characteristic mode of elite democratic politics in Thailand. For example, royalist Seni Pramoj explained his support for the coup against an elected government in 1947 by noting that true democracies were not about numerical majorities but were composed of people "with the greatest knowledge and loyalty to the nation".[28] This aggrandisement of the role of the sovereign legislator is not done by royalists alone; it was present in revolutionaries' desire to constrain the uninstructed electoral will of the people with appointed MPs. And the revolutionary left that became a political force in the late 1960s and 1970s had its own variant of vanguardism by which the people would be led towards their fulfilment.[29]

The will of the people 1

One vanguardist was Field Marshal Plaek Phibunsongkhram (Phibun), who rose up the ranks of the People's Party after participating in suppressing a royalist revolt in 1933. In 1938 he became prime minister. He affected the airs of a fascistic leader in the late 1930s and 1940s only to reinvent himself as a would-be constitutional democrat in the mid-1950s.

In power, Phibun aped fascism with a focus on leaderism, militarist mobilisation and territorial expansion. By the early 1940s the status of the 1932 constitution as a legitimising instrument was diminished.[30] He argued that his strong government was in accord with the people's will (or opinion) and stressed his commitment to the 1932 principles of liberty, equality and fraternity.[31] Phibun's government, influenced by Mussolini's fascist revolution and in line with its striving for a modernising "human revolution", issued a series of proclamations named state preferences (*rathaniyom*) that promoted forms of behaviour that would align with the new regime, including language, dress and work ethics. Phibun saw the preferences as an expression of the people's will.[32] As he put it, in Rousseauian fashion: "In this term (rathaniyom) is included 'public power' which is derived from public opinion. Public opinion brings public power . . . and this enables either reformation, or the suppression of a minority of people who are too stubborn to be reformed".[33] The state preferences were officially compared to royal decrees of the past: "the only difference is that Phrarachaniyom constituted the opinion of the king alone, while Rathaniyom constitutes the opinion of the State formed in conformity with public opinion as a national tradition".[34] Which was to say that the locus of sovereignty and its prerogatives had shifted to the people and was expressed through the state.[35]

In January 1942, Phibun took Thailand into the Second World War on the side of the Axis powers. Pushed from office in late 1944, he subsequently avoided conviction for war crimes. He returned to power in 1948 and navigated competing poles of power while downplaying his erstwhile führerism. In the mid-1950s he self-rehabilitated as a constitutional democrat by mobilising forces for a more competitive party system.[36] After winning a contested election, he was ousted in the coup d'état of 1957.

Which face of democracy did Phibun see in the mirror? His earlier period of guided-democracy hardly conformed to conventional understandings of Thai-style democracy in traditionalist garb (see below), but was an alternative form of Thai nationalist modernisation. His past made him an unlikely late-life devotee of liberal constitutional rule. Yet his foot was on both the nativist and universal faces. His two-faced democratic leanings, a common enough feature of Thai political biography, expressed the state of ambivalence at the core of Thai politics.

The will of the people 2

The coups d'état of 1957 and 1958 began the Saritarian counter-revolution aimed at eviscerating the universal face that 1932 had come to symbolise. Dictatorial rule was the norm until 1973 (with brief and qualified interludes). In this period, classic Thai-style democracy discourse emerged, mixing military rule, royalist moral guardianship and culture.[37] *Pho-khun*, the idea of fatherly kings, was aggrandised by Sarit to legitimate the regime's "despotic paternalism".[38] Some royalists who had ideologised democratic kingship now supported Cold War dictatorship. In return, they gained a space to rebuild the battered position of the palace.

In the early 1960s, radio broadcasts on Thai-style democracy educated the people on the system. One explained that "Thailand at this moment does not have elections and no permanent constitution, but we are a democracy".[39] A political system, one broadcast explained, that gathered the people's opinion and responded to their needs was "government of the people, by the people and for the people".[40] Thais had this, and safeguarding that democracy was the implicit social contract that the king, who headed the system, would reign in abidance with the ten virtues expected of a Dhamma king. In these broadcasts, the idea of public opinion served as a mystical channel by which the sovereign will of the people could be determined, functioning much like Rousseau's general will. Such arguments were not a coherent democratic philosophy but reflected the exigencies of nation-state building during the Cold War and the slow creep of modernisation theory. Indeed, one broadcast entertained that Thailand would one day have a parliamentary democracy once its developing economy generated a sizeable middle class.[41]

Conjuring the will of the people 3: the politico-developmental state

Traditionalist Thai-style democracy was not to last long. In the 1960s, Western modernisation and political development discourse began to shape political thinking in response to the new problems of government, particularly in the powerful Interior Ministry. This new science of democracy drew inspiration from counter-insurgency doctrines in relation to winning the "hearts and minds" of the people.[42] The Interior Ministry educated its officials in the Department of Local Government in the science. Elections in local councils were promoted as learning experiences for villagers; symposiums on party institutionalisation were held to avoid disruptive political mobilisation; and civics education nurtured the formation of a moderate democratic mentality as preparation for democratic enlargement. Although these governmental practices were ostensibly aimed at preparing for the conduct of democracy, they legitimated authoritarian rule: the science judged that existing conditions were unconducive for full democracy.[43] This governmental science provided techniques to shape a citizen able to voluntarily align itself with the regime's stagist aspirations. This superseded traditionalist forms of Thai-style democracy in which the active citizen was neither desired nor imagined.

Despite the turn to science, the mythic triadic national ideology that continued after 1932 remained strong. The hegemonic appeals to national identity around conservative notions of nation, king and religion and the governmental logic of developing democratic citizens would seem to pull in different directions. In actuality, they were unified by the authoritarian purpose of subjecting the people to imagined forms of self-rule. General Praphas Charusathien, interior minister from 1957–1973, understood as much. His book, likely ghost written, *Ideas on Thai Political Development* argued for the fusion of traditional forms of national ideology with the methods of political development applied at local levels first.[44] Higher forms of liberal democracy would come later when a large middle class came into being: this discourse of delayed liberalism, not traditional Thai style democracy, characterised the military state. Others refused to wait.

The 1970s was a most traumatic decade. In October 1973, a student-led constitutionalist movement felled the military dictatorship. The resulting liberal democratic reform embodied in the 1974 constitution was resisted by reactionaries. They viewed liberalisation as a conduit for communism. Rightists associated with the monarchy and military, business and police groupings were alarmed at the radically mobilised nature of politics after 1973. Varied liberal, socialist, labour-democratic, and redistributive farmers' movements attacked decades of injustice.[45] Communist insurgency was growing. The imminent end of the Second Indochina War and the prospect of communist governments in Vietnam, Cambodia and Laos fuelled hysteria. Rightist paramilitary forces intimidated and assassinated both local- and national-level political leaders. Denouement came in the form of the massacre of students at Thammasat University by royalist rightists in October 1976. Dictatorship returned. A new civilian-dictatorship appointed by the king came to power. Its prime minister, Thanin Kraivichien, proposed a 12-year democratic tutelage. He fitfully praised the king as a final checkpoint when disunity and mistrust prevailed. This he saw as an essential function of kingship in DWKHS.[46] Thanin was deposed by the coup of 1977, and there commenced the restoration of order based on limited power-sharing and a more robust national ideology centred on DWKHS.

Towards hegemony: ideological sponginess

The 1980s saw the emergence of a bureaucratic regime subject to liberalising pressures: a "semi-democracy" of hybrid parliamentary-bureaucratic liberal authoritarianism. The two faces became more institutionally recognisable as conservatives and liberals openly contested amid the pursuit of interest by parties, bureaucracy and palace. Conservatives imagined a slow realisation of democracy safeguarded by the moral exemplar of the king and a strong executive, guided by the bureaucratic state. This was no return to the heyday of traditionalism. If nativism was in the mix, it was there to nurture a conservative democracy not unrecognisable to Western corporatist states with their own soil and blood ornamentations. Conversely, liberals wanted to leap into the democratic experiment to advance political development. Ideological and institutional contests ensued.

Institutionally, guardian control over political institutions gradually diminished while retaining veto powers should certain boundaries be transgressed. Reformists seeking parliamentary supremacy pushed against the military, the civil service and the appointed senate. Proponents of a political role for the military were sidelined. Parliament moved to the centre. In 1991, statists fought back and overthrew the elected government of Chatichai Choonhavan, which had transgressed military economic terrain and which the coup-makers claimed threatened DWKHS by its system of "parliamentary dictatorship". The coup-makers attempted to return politics to a semi-democracy.[47] In 1992 a mass democratic movement thwarted this ambition and paid a high price in the May massacre. The military retreated and political liberalisation now accelerated, leading to the adoption of the 1997 "people's constitution".

Ideologically, all governments since the 1980s formally supported DWKHS while differently mixing the nativist and universal elements. Variants of the phrase DWKHS attended royalist resurgence and was registered in constitutions (1949, 1968, 1978 and onwards).[48] In the mid- to late 1970s, when the reputation of the monarchy was soiled, security agencies assessed how to restore it.[49] DWKHS attracted their attention, and over the next decade they remoralised the monarchy around a communitarian democracy attentive to economic injustices. Statists used the ideology to launch a nativist ideological assault to counteract their diminishing political power.

Under General Prem, who served as prime minister from 1980 to 1988, a new policy of national identity was announced that fused traditionalism and moral development.[50] The government declared that DWKHS was integral to national identity. National identity in this account was performed by securing the highest institutions of the country, now a tetrad: "the Nation, Religion, Monarchy and Democracy with the King as Head of State". Democracy, the bloodied corpse of the 1970s, was resurrected by ideologists eager to forge a sustainable ideology that could engage disaffected people. Thus the National Identity Board committed, and the government agreed, to ensuring that people understood royal graciousness and sacrifice, "which are good role models", promote virtue and integrity among the people that aid national development and to "propagate the value of the Democracy with the King as Head of State as well as promote ideology and values that support this system of government".[51] These ideas were widely propagated.

However, rather than a mere official ideology – though arguably it evolved as such – DWKHS would prove to be a hegemonic sponge accommodative to liberal and popular views over the next two decades. Strategic interactions with the monarchy as a form of rights insurance by people, and the king's seeming responsiveness, in part generated the hegemonic position of the monarchy, as did historical narratives of the monarchy's liberality.[52] The notions of virtue and democratic kingship were also promoted by liberal reformists in non-governmental organisations, political parties, universities and the media. The values of freedom and tolerance were at the centre of their civic education projects. Rather than the possessive individualism and liberal neutrality then emergent in the West, this was a kind of liberal-communitarian perfectionism that considered it appropriate that good forms of life and moral ways of being should be promoted that allowed for individual autonomy – and politics was one sphere for this project. For the institutions of liberal democracy to be sustained, liberal forms of good had to be imagined and practiced.[53] Those forms of good tended towards a morally imagined, self-reliant citizen who was neither dependent on the state nor grossly invested in the materialism of the market. On some tendentious readings, the king's actions and speeches on self-reliance could be assimilated to a liberal-communitarian perfectionism.

But it was the Constantian institutional role of the king as a power above power that mostly articulated monarchy to a form of royal liberalism. By the 1990s it was possible to identify a discourse of royal liberalism that not only eulogised the democratic monarchy but which was deployed by liberals in their reform project of good governance, then globally ascendant.[54] As some would have it, the king was the supreme ombudsman, overarching the fractious society that had yet to find a workable political system.[55] All organs of the state, mainstream media and politically oriented elite actors tended to repeat variants of this narrative unreflectively: that processes and institutions were not enough, that embodied transcendence was needed as both safeguard and expression of a national will.

In this two-faced DWKHS, King Bhumibol Adulyadej was imagined as a transcendent power above politics. Advocates could not accommodate the idea that the mortals occupying palace positions could be aspirants for their own power and interests.[56] Indeed, protected by cultish aura and a defamation regime of strict lèse-majesté laws, the palace had now emerged from its weak position in the late 1950s to position itself at the head of a bureaucratic capitalist power bloc by the 1980s and had with some skill adapted to the liberalisation of the 1990s.[57] If Duncan McCargo spoke of "network monarchy" to indicate the mechanisms by which the supreme ombudsman worked and reached into public life, it would be political forces that would determine the direction network monarchy would take.[58]

This politics of royal liberalism partly took shape in the form of the 1997 constitution and the liberal division of power and the anti-statist and anti-majoritarian safeguards it mandated. But the question of sovereignty remained opaque: belonging to the people but exercised by the king. For one of the key constitutional drafters, sovereignty was jointly held by the people and king, but when a crisis occurred it was said to return to the king.[59] This was the practical formula by which it was possible to imagine that Thailand had been a DWKHS ever since 1932.[60]

It was not ideas alone that made possible the march from the statist coup of 1991 to the people's constitution of 1997. The seeming triumph of the liberal world order after the Cold War ended was one factor. Entangled with that order and domestically motivated, there was also public sphere mobilisation against the prevailing state of affairs that were judged incapable of ending the vicious cycle. Political reformists felt swindled when in the 1990s a regime of corrupt money politics and neo-patrimonialism betrayed the democratic martyrs of 1992.[61] That said, the 1990s reform project was largely an elite affair of constitutional crafting and political engineering, however well intentioned. Despite its progressive features – which assumed a conventional relationship of a rights-bearing citizen in relation to the state – it failed to win the electorate to its moral imperatives and its checks and balances on power.

The people have lost the confidence of the governors: it is time to dissolve the people

The billionaire Thaksin Shinawatra and his nationalist party won the 2001 national election on promises of modernisation and welfare measures. In power, an illiberal form of majoritarianism emerged as Thaksin sought political control over formally independent agencies of the state. Thaksin would often repeat the fact of his democratic mandate as a way of questioning the legitimacy of unelected checks and balances on his power that were part of the post-1997 landscape.[62] As Thaksin moved to monopolise the state, he faced opposition from liberals and statists. They questioned his commitment to DWKHS. And with this Thailand's current decade-long conjunctural crisis commenced.

Against Thaksin, a mixed-class movement ostensibly seeking to protect the 1997 constitutional settlement and the monarchy wrapped itself in calls for a royally appointed prime minister and withdrew its loyal opposition by boycotting elections.[63] Collectively they became known as the "Yellow Shirts", identified as such for wearing the colour associated with King Bhumibol. Thaksin was damaged but not defeated. It took the September 2006 coup to remove him. The vicious cycle was now upended. Thaksin and his supporters fought back. They found support among those committed to the principle of democratic elections and parliamentary supremacy. Collectively, they became known as the "Red Shirts", identified as such for wearing "red" – the colour that was initially adopted by the movement against the military's 2007 constitutional referendum – to signal "stop". That constitution provided for an appointed senate and new electoral laws aimed to erode Thaksin's electoral base. Despite this, pro-Thaksin governments were returned through elections in 2007 and 2011. These governments faced belligerent opposition as illegitimate puppets of Thaksin by Yellow Shirts and successor organisations. Both governments were removed, but not by election. Politics had entered into a fully formed enemy/friend logic, leading to intensifying authoritarianism.[64]

From mid- to late 2008, following anti-government protests from Yellow Shirts, judicial intervention removed two pro-Thaksin prime ministers. Military machinations then secured the formation of an anti-Thaksin coalition government. The Red Shirts now congealed as a mass movement with a significant base in the provincial lower-middle classes, seeking to end a government they considered a mask for dictatorship. Some were under the strategic control of pro-Thaksin forces that were focused on his return to power; others adopted a qualified stance given his authoritarian tendencies. Instead they fought for the return of constitutional democracy and to complete the promise of 1932 revolution – the fulfilment of popular sovereignty: "a democratic state with the king as head of state in which the use of law is sacred, equal and just".[65] Thaksin declared himself for the same. In March 2010, Red Shirts occupied central Bangkok in the hundreds and thousands, seeking the government's resignation. As the protest size dwindled over the next two months, political positions hardened. Finally, from April to May scores of protestors were killed by the military under government orders. This violent repression was publicly justified by reference to there being armed elements in the Red Shirt movement. The movement retreated. Its hope was squandered by Thaksin's use of it as bargaining tool with the regime and the government's willingness to punish by association.[66]

In late 2011, Yingluck Shinawatra, Thaksin's sister, took office as elected prime minister. Her government's policies of rice price subsidy to its rural electoral base, attempts at constitutional revision and amnesty legislation, which would allow the return of self-exiled Thaksin, re-activated opposition in the form of a reactionary movement helmed by the People's Democratic Reform Committee (PDRC). The PDRC mobilised hundreds of thousands of people on the streets, calling for a period of sovereign dictatorship to end remnants of the Thaksin regime and to rebuild democracy. After opposition Democrat Party MPs resigned from parliament en masse, the government called elections for February 2014. The PDRC blockaded both candidate registration and on election day the polling stations. The courts declared the election result invalid and later removed Yingluck as prime minister, following which, in May 2014 a coup group seized power.[67]

General Prayuth Chan-ocha, leader of the new junta, the National Council for Peace and Order (NCPO), explained that among the reasons for the coup, "the most important was because we respect the democratic process". Like predecessor guardians, Prayuth spoke the language of delayed sovereignty:

> We understand that we are living in a democratic world, but is Thailand ready in terms of people, form and method? We need to solve many issues; from . . . corruption, and even the starting point of democracy itself – the election. Parliamentary dictatorship has to be removed.[68]

Prayuth would speak about Thai-style democracy, and agencies of the state quickly returned to a time-failed pedagogy to nurture "quality citizens".[69]

After five elected prime ministers (three acting) had been removed from office by the courts or coups between 2006 and 2014, it did indeed seem as if a once hidden gang of governors surmised the people were unworthy of their confidence, as Bertolt Brecht once reported of East Germany's rulers in the wake of the June 1953 uprising in East Berlin.[70] The idea of there being a "deep state" bent on hegemonic preservation now served to cohere a retrospective narrative of democracy lost due to the original sin of a reactionary elite[71] – a powerful idea that nevertheless misses both the contingency of political events and identity, the ambivalence of the Thai state and significant divisions among those designated as deep state agents.[72]

Conclusion

Thailand's conjunctural crisis is defined by the breakdown of two-faced democracy. While DWKHS had hegemonic sting from the 1980s to the 1990s, it cracked in the face of politics of a majority electorate consciously forging in the 2000s a new political settlement. In response a liberal and reactionary revaluation of democracy in explicitly guardian mode (2005–2007) occurred. After 2014, long-form decisionist politics of dictatorship are the fashion. During the crisis the two faces unexpectedly polarised and reformed as distinct political positions. The authoritarian electocrat Thaksin morphed into the symbol of democracy backed by the mainstream of the Red Shirt movement. That movement, peopled by a new generation of activists and followers and a political cadre traceable to the politics of the 1970s, was by its public declarations a liberal constitutionalist movement committed to the electoral return of redistributive policy. Among the dictatorships' supporters or collaborators were many who had once supported the 1970s democracy struggles and the 1990s political reform movement. Erstwhile liberal-communitarians, middle-class supporters of good governance, technocrats and conservatives now supported dictatorship (with insipid criticisms) to the delight of reactionaries. The 'red-yellow' split in the October generation (1970s activists), marked by mutual recriminations of betrayal, underscores the polarisation.[73]

In the years after the coup, the NCPO future-proofed its nativist democratic face. Its 2017 constitution embeds military and bureaucratic control over future elected governments. A range of state agencies will be populated by NCPO appointees who will surveil the untrustworthy people's representatives.[74] Repressive, abusive and nepotistic, the junta will rely on further force should its machinations fail. With force and decisionism replacing a broken hegemony, it is unclear which face of democracy awaits Thailand. Certainly, the successful production of a hegemonic charismatic aura around the late King Bhumibol (October 2014) was a reflection of both propaganda *and* people's negotiation of available politics. It was the pulse of DWKHS. There is little prospect of this in the reign of King Vajiralongkorn, or Rama X, because he does not aspire to the virtues and purpose of monarchy that his father advanced and exploited, nor does a new generation seem to require it. Whether that means a different kind of democratic breakthrough will occur or whether a new geopolitics of autocracy aligns with Thai authoritarianism and sustains a recalibrated two-faced democracy remains unknown.

Notes

1 Arjun Subrahmanyan, "The Unruly Past: History and Historiography of the 1932 Thai Revolution", *Journal of Contemporary Asia* (2018) Online First. DOI: 10.1080/00472336.2018.1556319.
2 "Universal" here refers to nothing other than an idea and practice that is arguably applicable to many places; native is by definition particular. Despite post-colonial provincialisation of the West's experience, the idea of universal democracy referenced to an occidentalist view of a universal West remains strong in Thailand.
3 Nakharin Mektrairat. *Kanbatiwat Siam Pho So 2475* [The Siamese Revolution of 1932] (Bangkok: Amarinwichakan, 1997); Kobkua Suwannathat-Pian, *Kings, Country and Constitutions* (London: Routledge, 2004).
4 The full song and lyrics may be found at Carabao 1986. Prachathipatai. In *Prachathipatai* <http://carabaoinenglish.com/song-translations/by-album> (accessed 1 May 2018).
5 E. Bruce Reynolds, "Phibun Songkhram and Thai Nationalism in the Fascist Era", *European Journal of East Asian Studies*, Vol. 3. No. 1 (2004); Katherine Bowie, *Rituals of National Loyalty* (New York: Columbia University Press, 1997).
6 Kevin Hewison, "The Monarchy and Democratization", in *Political Change in Thailand*, edited by Kevin Hewison (London: Routledge, 1997).
7 Christine Gray, *Thailand: The Soteriological State in the 1970s*, PhD dissertation, University of Chicago, 1986.
8 Danny Unger, "Sufficiency Economy and the Bourgeois Virtues", *Asian Affairs*, Vol. 36, No. 3 (2009); Michael K. Connors, *Democracy and National Identity in Thailand* (Copenhagen: NIAS Press, 2007).
9 Kasian Tejapira, "Signification of Democracy", Paper presented at workshop on "Comparative Perceptions of Democracy and Government", Brisbane, November 1992.
10 Ibid., p. 9. Some authors literally translate DWKHS as "democratic regime with the king as head of state", playing on the negative connotations associated with 'regime' in popular usage.
11 Nakhon Pajonawarapong, Khomun Hetkan Kanmuang Phaitai Ratthamanun Nai Adet, Thuk Chabab Lae Ratthathamnun 2560 [Information on Political Events under all Former Constitutions and the 2017 Constitution] (Bangkok: Hanghun Suan, 2017), pp. 10–11.
12 Chai-Anan Samudavanija, *The Thai Young Turks* (Singapore: Institute of Southeast Asian Studies, 1982).
13 Duncan McCargo, "*Alternative Meanings of Political Reform* in Contemporary Thailand", *The Copenhagen Journal of Asian Studies*, Vol. 13 (1998), p. 5.
14 See Kasian Tejapira, "The Irony of Democratisation and the Decline of Royal Hegemony in Thailand", *Southeast Asian Studies*, Vol. 5, No. 2 (2016).
15 See Kullada Kesboonchoo Mead, *The Rise and Decline of Thai Absolutism* (London: Routledge, 2004).
16 Eiji Murashima, "The Origin of Modern Official State Ideology in Thailand", *Journal of Southeast Asian Studies*, Vol. 19, No. 1 (1988).
17 Thongchai Winichakul. "The Quest for Siwilai", *Journal of Asian Studies*, Vol. 59 (2002); Michael Herzfeld, "The Absent Presence: Discourses of Crypto-Colonialism", *South Atlantic Quarterly*, No. 101 (2002).
18 Tamara Loos, *Subject Siam: Family, Law, and Colonial Modernity in Thailand* (Ithaca: Cornell University Press, 2006).
19 Cited in Yuangrat Wedel with Paul Wedel, *Radical Thought, Thai Mind* (Bangkok: Assumption Business Administration College Bangkok 1987), p. 48.
20 Chris Baker and Pasuk Phongpaichit, *Pridi by Pridi* (Chiang Mai: Silkworm, 2000), pp. xii–xiii.
21 Somchai Preechasilpakul, Khothokthiang Waduay Sathaban Phramahakasat Nai Ongkon Jattham Rattathamanun Khong Thai Dang Dae Pho So 2475–2550 [Debates on the Monarchy in Thai Constitution Making Organs from 1932–2007] (Chiang Mai: D.N. Printing, 2017), pp. 16–18.
22 On the later period see Peter A. Jackson, "Virtual Divinity: A 21st Century Discourse of Thai Royal Influence", in *Saying the Unsayable. Monarchy and Democracy in Thailand*, edited by S. Ivarsson and L. Isager (NIAS Press: Copenhagen, 2010).
23 For the latest critique of this, see Michael Nelson, "Authoritarian Constitution-Making in Thailand, 2015–16", Working Paper Series No. 188. SEARC, City University of Hong Kong, 2016.

24 Cited in Thak Chaloemtiarana (ed.), *Thai Politics: Extracts and Documents 1932–1957* (Bangkok: Social Science Association of Thailand, 1978), p. 168.
25 Whipha Jiraphaphaisan, "Anusaori Ratthathamnun Nai Isan" ["Constitution Monuments in Esaan"], <www.silpa-mag.com/club/article_17795> (accessed 11 July 2018).
26 Puli Fuwongcharoen, "Long Live Ratthathammanoon!: Constitution Worship in Revolutionary Siam", *Modern Asian Studies*, Vol. 52, No. 2 (2018), pp. 624–631.
27 See Connors, *Democracy and National Identity in Thailand,* pp. 21–27 for a critical account of this as a politics of "democrasubjection".
28 Cited in Federico Ferrara, "Democracy in Thailand: Theory and Practice", in *The Routledge Handbook of Southeast Asian Democratization,* edited by William Case (London: Routledge, 2015), p. 353.
29 Thomas Marks, *Making Revolution: The Insurgency of the Communist Party of Thailand in Structural Perspective* (Bangkok: White Lotus, 1994).
30 Puli, "Long Live Ratthathammanoon!: Constitution Worship in Revolutionary Siam", p. 641.
31 Supparat Sangchatkaew, "Wathakam Prachathipatai Nai Sunthoraput Khong Jom Phon P. Phibun Songkhram" ["Democratic Discourse in the Speeches of Field Marshal P. Phibun Songkhram"], *Warasan Mahawithiyalai Silapakon*, Vol. 32, No 2 (2014), p. 116.
32 See Federico Ferrara, *The Political Development in Modern Thailand* (Cambridge: Cambridge University Press, 2015), pp. 113–119.
33 Cited in Yoshifumi Tamada, "Political Implications of Phibun's Cultural Policy 1938–1941", final report submitted to the National Research Council of Thailand, February 1994, p. 21. *Rathaniyom* is typically translated as "state convention".
34 Cited in Anon., *The Centennial of His Royal Highness Prince Wan Waithayakon Krommun Naradhip Bongsprabandh, Thai Great Diplomat and Scholar, 1991, 25 August 1991–25 August 1992* (Bangkok: Office of National Culture, 1991), pp. 31–32.
35 On the appeal to "public opinion" understood as "the people's will", see Saichol Sattayanurak "Matimahachon", in *Kheu Phumjai: Ruam Botkhwaam pheu pen Theraleuk nai Okat Satrajan Chatthip Nartsupha Kasianayu Rachakan* (Festschrift: Articles in Commemoration, on the occasion of the Retirement of Professor Chanthip Nartsupha), edited by Sirilak Sampatchalit and Siriphon Yotkamonsat (Bangkok: Sangsan, 2002).
36 Thak Chaloemitiarana, *Thailand: The Politics of Despotic Paternalism* (Chiang Mai: Silkworm, 2007), pp. 69–71.
37 Classic descriptions are in Thak, *Thailand: The Politics of Despotic Paternalism*, and Kobkua, *Kings, Country and Constitutions.* A reformulation is to be found in Kevin Hewison and Kengkij Kitirianglarp, "Thai-Style Democracy: The Royalist Struggle for Thailand's Politics", in *Saying the Unsayable. Monarchy and Democracy in Thailand*, edited by S. Ivarsson and L. Isager (Copenhagen: NIAS Press, 2010).
38 Thak, *Thailand: The Politics of Despotic Paternalism.*
39 Cited in Connors, *Democracy and National Identity in Thailand*, p. 50.
40 Ibid.
41 Ibid., p. 51.
42 Ibid., p. 71.
43 Ibid., pp. 60–91.
44 Praphas Charusathien, *Khokhit Nai Kanpathana Kanmuang Thai* [*Ideas on Thai Political Development*] (Bangkok: Krom kanpokkhrong, 1973).
45 Tyrell Haberkorn, *Revolution Interrupted: Farmers, Students, Law, and Violence in Northern Thailand* (Madison: University of Wisconsin Press, 2011); David Morell and Chai-Anan Samudavanija, *Political Conflict in Thailand* (Cambridge: Oelgeschlager, Gunn and Hain, 1981).
46 Thanin Kraivichien, *Phramahakasat Thai Nai Rabob* Prachathipatai [The Thai King in the Democracy System] (Bangkok: Ministry of Education, 1976).
47 On the period see Ferrara, *The Political Development of Modern Thailand*, pp. 183–219.
48 See Kasian's "The Irony of Democratization", on the grammatical co-dependence of democracy and kingship after 1978, fn. 3, p. 227.
49 Connors, *Democracy and National Identity in Thailand*, pp. 136–141.
50 Ibid., pp. 141–150.
51 Ibid., p. 143.
52 Nidhi Eoseewong, "The Thai Cultural Constitution", *Kyoto Review of Southeast Asia*, Vol. 3 (March 2003) <http://kyotoreview.cseas.kyoto-u.ac.jp/issue/issue2/article_243_p.html>; Prajak Kongkirati, *Lae Laew Khwam Kleuanwai Prakot; Kanmuang Watthanatham Khong Naksuksa Lae Panyarachon Kon 14 Tula*

[And then the Movement Emerged: Cultural Politics of Thai Students and Intellectuals before the October 14 Uprising] (Bangkok: Thammasat University Press, 2005), pp. 464–519.

53 For details on intellectuals, NGOs and think tanks related to this, see Connors, *Democracy and National Identity in Thailand*, pp. 182–211.

54 Michael K. Connors, "Article of Faith: The Failure of Royal Liberalism in Thailand", *Journal of Contemporary Asia*, Vol. 38, No. 1 (2008).

55 Office of the Parliamentary Ombudsman, "The Ombudsman Institution", Bangkok, 23 April <www.ombudsman.go.th/eng_version/articles_main.asp?id¼100055 2001>.

56 For vivid descriptions of such, see Paul Handley, *The King Never Smiles: A Biography of Thailan's Bhumibol Adulyadej* (New Haven: Yale University Press, 2006).

57 David Streckfuss, *Truth on Trial in Thailand: Defamation, Treason, and Lèse-majesté* (London: Routledge, 2011).

58 Duncan McCargo, "Network Monarchy and Legitimacy Crisis in Thailand", *The Pacific Review*, Vol. 18, No. 4 (2005).

59 Bowonsak Uwanno, *Kotmai Kap Thangleuak Khong Sangkhom Thai* [Law and Solutions for Thai Society] (Bangkok: Samnakphim nitthitham 1994). See discussion in Connors 2008 and Kasian 2016.

60 See Kasian, *The Irony of Democratization*, p. 228.

61 Duncan McCargo (ed.), *Reforming Thai Politics* (Copenhagen: NIAS Press, 2002).

62 Duncan McCargo and Ukrist Pathmanand, *The Thaksinization of Thailand* (Copenhagen: NIAS Press, 2005); Pasuk Phongpaichit and Chris Baker, *Thaksin: The Business of Politics in Thailand* (Chiang Mai: Silkworm, 2004).

63 On the mixed nature of the movement, see Oliver Pye and Wolfram Schaffar, "The 2006 Anti-Thaksin Movement in Thailand: An Analysis", *Journal of Contemporary Asia*, Vol. 38, No. 1 (2008).

64 Michael Connors, "Liberalism, Authoritarianism and the Politics of Decisionism in Thailand", *The Pacific Review*, Vol. 22, No. 3 (2009).

65 United Front for Democracy against Dictatorship (UDD), *Questions and Answers*, p. 7 (undated pamphlet distributed in May 2010 at UDD protests).

66 On this period see Michael Montesano and Pavin Chachavalpongpun (eds.), *Bangkok, May 2010: Perspectives on a Divided Thailand* (Singapore: ISEAS, 2012).

67 Bencharat Sae Chua, "When Democracy Is Questioned: Competing Democratic Principles and Struggles for Democracy in Thailand", in *Political Participation in Asia*, edited by Eva Hansson and Meredith L. Weiss (London: Routledge, 2018).

68 General Prayuth Chan-ocha, "Unofficial translation National Broadcast by Head of the National Council for Peace and Order 6 June 2014" <www.mfa.go.th/main/en/media-center/3756/46368-Unofficial-translation-National-Broadcast-by-Gener.html> (accessed 1 May 2018).

69 See for example, Electoral Commission of Thailand 2017, "Roles of the Thai Citizens in Promoting and Developing Democracy". <www.ect.go.th/ewt/ewt/ect_en/news_page.php?nid=1892&filename=> (accessed 1 May 2018).

70 The subheading obviously derives from Brecht's famous poem "The Solution". Brecht's twist on sovereignty has been used before in relation to Thailand, see Clifford Noonan, "Thai Crisis Could End in Street Clashes or New Coup", *Irish Times*, 20 April 2010.

71 Eugénie Mérieau, "Thailand's Deep State, Royal Power and the Constitutional Court (1997–2015)", *Journal of Contemporary Asia*, Vol. 46, No. 3 (2016).

72 See Connors, "Liberalism, Authoritarianism and the Politics of Decisionism in Thailand", 2009 and Ukrist Pathmanand and Michael K. Connors, "Thailand's Public Secret: Military Wealth and the Thai State", *Journal of Contemporary Asia*, forthcoming.

73 Kanokrat Lertchoosakul, *The Rise of the Octoberists in Contemporary Thailand* (New Haven, CT: Yale Southeast Asia Studies, 2016).

74 See Veerayooth Kanchoochat, "Reign-seeking and the Rise of the Unelected in Thailand", *Journal of Contemporary Asia*, Vol. 46, No. 3 (2016).

References

Anon. 1991. *The Centennial of His Royal Highness Prince Wan Waithayakon Krommun Naradhip Bongsprabandh, Thai Great Diplomat and Scholar, 1991, 25 August 1991–25 August 1992.* Bangkok: Office of National Culture.

Anon. 1965. *Prachathipatai Baeb Thai Lae Khokhit Kap Ratthammanun* [Thai-style Democracy and Ideas on the Constitution]. Bangkok: Samnakphim Chokchaitewet.

Baker, C. and Pasuk Phongpaichit. 2000. *Pridi By Pridi*. Chiang Mai: Silkworm.

Bencharat Sae Chua. 2018. "When Democracy Is Questioned: Competing Democratic Principles and Struggles for Democracy in Thailand". In *Political Participation in Asia*. Edited by Eva Hansson and Meredith L. Weiss. Routledge: London.

Bowie, Katherine. 1997. *Rituals of National Loyalty*. New York: Columbia University Press.

Bowonsak Uwanno. 1994. *Kotmai Kap Thangleuak Khong Sangkhom Thai* [Law and Solutions for Thai Society]. Bangkok: Samnakphim Nitthitham.

Carabao. 1986. "Prachathipatai". In *Prachathipatai* (Album) <http://carabaoinenglish.com/song-translations/by-album> (accessed 1 May 2018).

Chai-Anan Samudavanija. 1982. *The Thai Young Turks*. Singapore: Institute of Southeast Asian Studies.

———. 1993. *Panha Kanphathana Thang Kanmeuang* [Problems of Political Development]. Bangkok: Samnakphim Julalongkon Mahawithayalai.

Connors, M.K. 2007. *Democracy and National Identity in Thailand*. Copenhagen: NIAS Press.

———. 2008. "Article of Faith: The Failure of Royal Liberalism in Thailand". *Journal of Contemporary Asia*, Vol. 38, No. 1: 143–165.

———. 2009. "Liberalism, Authoritarianism and the Politics of Decisionism in Thailand". *The Pacific Review*, Vol. 22, No. 3: 355–373.

Dressel, B. 2010. "Judicialisation of Politics or Politicization of the Judiciary? Considerations from Recent Events in Thailand". *The Pacific Review*, Vol. 23, No. 5: 671–691.

Electoral Commission of Thailand. 2017. "Roles of the Thai Citizens in Promoting and Developing Democracy". <www.ect.go.th/ewt/ewt/ect_en/news_page.php?nid=1892&filename=> (accessed 1 May 2018).

Ferrara, Federico. 2015. *The Political Development of Modern Thailand*. Cambridge: Cambridge University Press.

———. 2015. "Democracy in Thailand: Theory and Practice". In *The Routledge Handbook of Southeast Asian Democratization*. Edited by William Case. London: Routledge.

General Prayuth Chan-o-cha. 2014. "Unofficial Translation National Broadcast by Head of the National Council for Peace and Order 6 June 2014" <www.mfa.go.th/main/en/media-center/3756/46368-Unofficial-translation-National-Broadcast-by-Gener.html>.

Gray, G. 1986. *Thailand: The Soteriological State in the 1970s*. PhD dissertation, University of Chicago.

Haberkorn, T. 2011. *Revolution Interrupted: Farmers, Students, Law, and Violence in Northern Thailand*. Madison: University of Wisconsin Press.

Herzfeld, M. 2002. "The Absent Presence: Discourses of Crypto-Colonialism". *South Atlantic Quarterly*, Vol. 101: 899–926.

Hewison. Kevin. 1993. "Of Regimes, State and Pluralities: Thai politics Enters the 1990s". In *Southeast Asia in the 1990s*. Edited by K. Hewison, R. Robison and G. Rodan. Melbourne: Allen and Unwin.

———. 1997. "The Monarchy and Democratization". In *Political Change in Thailand*. Edited by Kevin Hewison. London: Routledge.

Hewison, Kevin and Kengkij Kitirianglarp. 2010. "Thai-Style Democracy: The Royalist Struggle for Thailand's Politics". In *Saying the Unsayable. Monarchy and Democracy in Thailand*. Edited by S. Ivarsson and L. Isager. Copenhagen: NIAS Press.

Jackson, P.A. 2010. "Virtual Divinity: A 21st Century Discourse of Thai Royal Influence". In *Saying the Unsayable. Monarchy and Democracy in Thailand*. Edited by S. Ivarsson and L. Isager. Copenhagen: NIAS Press.

Kanokrat Lertchoosakul. 2016. *The Rise of the Octoberists in Contemporary Thailand*. New Haven, CT: Yale Southeast Asia Studies.

Kasian Tejapira. 1992. "Signification of Democracy". Paper presented at workshop on "Comparative Perceptions of Democracy and Government", Brisbane, November 1992.

———. 2016. "The Irony of Democratization and the Decline of Royal Hegemony in Thailand". *Southeast Asian Studies*. Vol. 5, No. 2: 219–237.

Kobkua Suwannathat-Pian 2004. *Kings, Country and Constitutions*. London: Routledge.

Krom kanpokkhrong. 1964. *Lakkan Lae Hetphon Prakop Khrongkan Phathana Phon-Lameuang RabopPprachathipatai* [Principles and Rationale of the Project to Develop Democratic Citizens]. Bangkok: Interior Ministry.

Kullada Kesboonchoo Mead. 2004. *The Rise and Decline of Thai Absolutism*. London: Routledge.

Loos, Tamara. 2006. *Subject Siam: Family, Law, and Colonial Modernity in Thailand*. Ithaca: Cornell University Press.

Marks, Thomas. 1994. *Making Revolution: The Insurgency of the Communist Party of Thailand in Structural Perspective*. Bangkok: White Lotus.

McCargo, D. 1998. "Alternative Meanings of Political Reform in Contemporary Thailand". *The Copenhagen Journal of Asian Studies*, Vol. 13: 5–30.

———. (ed.). 2002. *Reforming Thai Politics*. Copenhagen: NIAS Press.

———. 2005. "Network Monarchy and Legitimacy Crisis in Thailand". *The Pacific Review* Vol 18, No, 4: 499–519.

McCargo, Duncan and Ukrist Pathmanand. 2005. *The Thaksinization of Thailand*. Copenhagen: NIAS Press.

Mérieau, Eugénie. 2016. "Thailand's Deep State, Royal Power and the Constitutional Court (1997–2015)". *Journal of Contemporary Asia*. Vol. 46, No 3: 445–466.

Morell, D. and Chai-Anan Samudavanija. 1981. *Political Conflict in Thailand*. Cambridge: Oelgeschlager, Gunn and Hain.

Murashima, Eiji. 1988. "The Origin of Modern Official State Ideology in Thailand". *Journal of Southeast Asian Studies*, Vol. 19, No. 1: 80–96.

Nakharin Mektrairat. 1997. Kanbatiwat Siam Pho So 2475 [The Siamese Revolution of 1932]. Bangkok: Amarinwichakan.

Nakhon Pajonawarapong. 2017. *Khomun Hetkan Kanmuang Phaitai Ratthamanun Nai Adet, Thuk Chabab Lae Ratthathamnun 2560* [Information on Political Events under all Former Constitutions and the 2017 Constitution]. Bangkok: Hanghun Suan.

Nattapoll Chaiching. 2010. "The Monarchy and the Royalist Movement in Modern Thai Politics, 1932–1957". In *Saying the Unsayable. Monarchy and Democracy in Thailand*. Edited by S. Ivarsson and L. Isager. Copenhagen: NIAS Press.

Nelson, M. 2016. "Authoritarian Constitution-Making in Thailand, 2015–16". Working Paper Series No. 188. SEARC, City University of Hong Kong.

Office of the Parliamentary Ombudsman. 2001. "The Ombudsman Institution". Bangkok, 23 April <www.ombudsman.go.th/eng_version/articles_main.asp?id¼100055> (accessed 1 May 2018).

Pasuk Phongpaichit and Chris Baker. 2004. *Thaksin: The Business of Politics in Thailand*. Chiang Mai: Silkworm.

Prajak Kongkirati. 2011. *Lae Laew Khwam Kleuanwai Prakot; Kanmuang Watthanatham Khong Naksuksa Lae Panyarachon Kon 14 Tula* [And then the Movement Emerged: Cultural Politics of Thai Students and Intellectuals before the October 14 Uprising]. Bangkok: Thammasat University Press.

Praphas Charusathien. 1973. *Khokhit Nai Kanpathana Kanmuang Thai* [Ideas on Thai Political Development]. Bangkok: Krom Kanpokkhrong.

Puli Fuwongcharoen. 2018. "Long Live Ratthathammanoon!: Constitution Worship in Revolutionary Siam Modern". *Asian Studies*, Vol. 52, No 2: 609–644.

Pye, O. and W. Schaffar. 2008. "The 2006 Anti-Thaksin Movement in Thailand: an Analysis". *Journal of Contemporary Asia*, Vol. 38. No. 1: 38–61.

Reynolds, Bruce E. 2004. "Phibun Songkhram and Thai Nationalism in the Fascist Era". *European Journal of East Asian Studies*, Vol. 3. No. 1: 99–134.

Saichol Sattayanurak. 2002. "Matimahachon". In *Kheu Phumjai: Ruam Botkhwaam Pheu Pen Theraleuk Nai Okat Satrajan Chatthip Nartsupha Kasianayu Rachakan* [Festschrift: Articles in Commemoration, on the occasion of the retirement of Professor Chanthip Nartsupha]. Edited by Sirilak Sampatchalit and Siriphon Yotkamonsat. Bangkok: Sangsan.

Somchai Preechasilpakul. 2017. *Khothokthiang Waduay Sathaban Phramahakasat Nai Ongkon Jattham Rattathamanun Khong Thai Dang Dae Pho So 2475–2550* [Debates on the Monarchy in Thai Constitution Making Organs from 1932–2007]. Chiang Mai: D.N Printing.

Subrahmanyan, Arjun. 2018. "The Unruly Past: History and Historiography of the 1932 Thai Revolution", *Journal of Contemporary Asia*, Online First. DOI: 10.1080/00472336.2018.1556319

Streckfuss, David. 2011. *Truth on Trial in Thailand: Defamation, Treason, and Lèse-majesté*. London: Routledge.

Supparat Sangchatkaew. 2014. "Watthakam Prachathipatai Nai Sunsap Khong Jomphon Po Phibunsongkhram" [Democratic Discourse in the Speeches of Field Marshal Phibun Songkhram]. *Warasan Mahawithayalai Sinlapakon*, Vol 34. No. 2: 109–130.

Tamada, Yoshifumi. 1994. "Political Implications of Phibun's Cultural Policy 1938–1941". Final Report Submitted to the National Research Council of Thailand, February.

Thak Chaloemtiarana. (ed.) 1978. *Thai Politics: Extracts and Documents 1932–1957*. Bangkok: Social Science Association of Thailand.

———. 2007. *Thailand: The Politics of Despotic Paternalism*. Chiangmai: Silkworm.
Thanin Kraivichien. 1976. *Phramahakasat Thai Nai Rabob Prachathipatai* [The Thai King in the Democracy System]. Bangkok: Ministry of Education.
Thongchai Winichakul. 2002. "The Quest for Siwilai". *Journal of Asian Studies*, Vol. 59, No, 3: 528–549.
Ukrist Pathmanand and Michael. K. Connors. (forthcoming). "Thailand's Public Secret: Military Wealth and the Thai State". *Journal of Contemporary Asia*.
Unger, Daniel. 2009. "Sufficiency Economy and the Bourgeois Virtues". *Asian Affairs*, Vol. 36, No. 3: 139–156.
United Front for Democracy against Dictatorship (UDD). n.d. *Questions and Answers*.
Veerayooth Kanchoochat. 2016. "Reign-seeking and the Rise of the Unelected in Thailand". *Journal of Contemporary Asia*, Vol. 46. No. 3: 486–503.
Whipha Jiraphaphaisan. 2018. "Anusaori Ratthathamnun Nai Isan" [Constitution Monuments in Esaan] <www.silpa-mag.com/club/article_17795> (accessed 11 July 2018).
Yuangrat Wedel with Paul Wedel. 1987. *Radical Thought, Thai Mind*. Bangkok: Assumption Business Administration College.

6

THE LOGIC OF THAILAND'S ROYALIST COUPS D'ÉTAT

Federico Ferrara

The good people of Thailand might just as well have muted their television sets when General Prayuth Chan-ocha appeared live on the air on the afternoon of 22 May 2014, flanked by the leaders of all major branches of the state's security apparatus. The army commander in chief read a curt statement announcing that the "National Council for Peace and Order" (NCPO) had seized administrative power, terminating the country's latest three-year spell of electoral democracy. The themes referenced in the statement had already featured in the public rationales provided for most of the ten successful coups staged during the reign of the late King Bhumibol Adulyadej (1946–2016). The junta's first pronouncement referenced the dangers associated with a further escalation in the political violence that had broken out intermittently over the previous six months, as royalist street protesters had sought to overthrow an embattled elected government they had already prevented, in concert with royalist courts and "independent" agencies, from maintaining order and holding new elections.[1] The standoff having already cost the lives of some 28 people, the NCPO pledged to restore a semblance of normality, uphold the monarchy, enact unspecified "reforms", and "make [. . .] love of unity spring from the nation's people as it did before" (*hai* [. . .] *prachachon nai chat koet khwam rak khwam samakkhi chen diao kap huang thi phan ma*). Days later, General Prayuth himself refused to say when or on what terms an election would be organised, reiterating that his priority was to heal the country's deep political divisions.[2]

So far, so coup d'état. Indeed, while the NCPO's mouthpieces could not resist trotting out the old royalist cliché about Thailand being somehow "unique" and, therefore, presumably exempt from the standards to which less "exceptional" countries might be held – there was nothing terribly unique about this coup's rationalisation, which relied on pretexts few military juntas around the world have missed a chance to highlight over the past century.[3] Even so, Thailand's history of royalist military coups – coups staged in order to expand royal prerogatives or otherwise entrench the country's "monarchy-centred hierarchical political order" – has at least two outstanding features.[4] The first is that the central theme emphasised in each of these coups – a generalised abhorrence for the disunity and the disorder associated with democratic rule – is at the heart of the country's official ideology: a hierarchical "royal nationalism" (*rachachatniyom*) that was formulated during the country's state-building process at the turn of the 20th century, re-tooled and rendered culturally hegemonic by the royalist dictatorship of Field Marshal Sarit Thanarat (1958–1963), revised and propagandised to saturation levels since the late 1970s, and increasingly challenged in the new millennium as a combined result of social change and the rise of elected politicians.[5] The second is that royalists have at times taken it upon themselves to make sure that the prophecy of their guiding ideology – the axiom that democracy leads to anarchy and chaos – would come to be fulfilled whenever the Thai public shrugged off their warnings and rejected their tutelage.

Indeed, all royalist rationalisations of the last coup studiously omit the fact that the military had helped engineer the conditions of disorder and ungovernability cited by the NCPO to justify the suspension of electoral democracy. Whether or not Suthep Thaugsuban spoke the truth when he boasted that he had conspired with General Prayuth since 2010 to overthrow the system, the behavior of the armed forces in the intervening years had left little doubt as to whose side their chiefs had taken.[6] In April and May 2010, when Suthep was deputy prime minister in a royalist government formed thanks to a series of controversial judicial rulings, the armed forces did not hesitate to open fire on the thousands of Red Shirt protesters who had taken to the streets to demand an early election. Less than four years later, the military did nothing to prevent Suthep's "People's Committee to Change Thailand into an Absolute Democracy with the King as Head of State" (PCAD) from violently disrupting an early general election called by a duly elected government on 2 February 2014. Aside from providing covert material support to the PCAD, high-ranking military officers repeatedly warned the government not to disperse the protesters, going so far as to deploy troops to checkpoints and bunkers set up around the rally sites in order to deter any attempt to reclaim the areas.[7] It is also telling that, upon seizing power, General Prayuth sought to give the impression that the NCPO had intervened to stop an "escalation" of violence. This was a deliberate misrepresentation, as the violence had in fact dropped off sharply since its peak in February. At the end of the day, the NCPO did not step in to prevent disorder from spiraling out of control – on the contrary, it acted once it became clear that the disorder military officers had orchestrated with the PCAD had failed to bring down a popular government that had proven uncharacteristically resilient to the onslaught of royalists in the streets, the courts, and the bureaucracy.

The long-standing tendency of Thailand's royalist establishment to cite the threat of chaos – or when that is not enough, to engineer the chaos themselves – in order to make the suspension of electoral democracy appear necessary to "maintaining" or "restoring" order calls to mind an analogy drawn by the late historian and sociologist Charles Tilly between the methods of states and organised crime. "If protection rackets represent organised crime at its smoothest", Tilly wrote, "then war risking and state making – quintessential protection rackets with the advantage of legitimacy – qualify as our largest examples of organised crime".[8] Indeed, while authoritarian regimes especially almost invariably base their claim to power on their capacity to protect the population from internal or external dangers, Tilly reminds us that it is often the same actors who take it upon themselves to "simulate, stimulate, or even fabricate" such threats.[9] In turn, "to the extent that the threats against which a given government protects its citizens are imaginary or are consequences of its own activities", Tilly concluded, "the government has organised a protection racket". Thailand, to be sure, is *not* the only country where the workings of the state have at times resembled those of a protection racket. In Thailand, however, the phenomenon takes special significance for its sheer ubiquity in the succession of 13 coups and 20 constitutions the country has experienced since 1932, as well as for the practice's close connection to the nation's official, state-sanctioned ideology. By investigating the logic of Thailand's royalist military coups – the governing principles undergirding the series of events regularly entailed by a royally sanctioned coup and their affinity with those involved in the running of a protection racket – this chapter hopes to shed light on the longevity of Thailand's royalist political order and on the state of crisis into which the arrangement has been thrust by the de-stabilising political conflict that has engulfed the nation since the turn of the century.

Coups, crackdowns, and constitutions

Thailand's royalist political order is the product of a loose, decades-old alliance between the palace, the military, the bureaucracy, and the "old-money" Sino-Thai families identified with the country's largest business conglomerates. The reason why the arrangement in question is best described as a "political order" is that while the coalition has not always controlled the country's *government*, it has maintained its effective control of the *state* in spite of rather extreme levels of regime instability. The alliance itself originates in the pair of military coups staged by Field Marshal Sarit Thanarat in 1957 and 1958, which put an end to the hostility that had marred the relationship between royalists and high-ranking commoners in the armed forces and the civil service since members of latter, organised in the People's Party (Khana Ratsadon), had abolished the absolute monarchy on 24 June 1932. While the 1957 coup had left in place the semi-democratic regime that operated under an amended version of the 1932 constitution, Sarit's assumption of absolute powers in 1958 was followed by the complete rollback of the 1932 revolution, complete with a revival of the hierarchical "royal nationalism" that had been crafted in the days of the absolute monarchy in an attempt to identify the nation with the king. In this endeavour, Sarit's government benefited from the fact that the monarchy's popularity had experienced an improbable resurgence since King Bhumibol had ascended the throne in 1946 following the tragic death of his brother, King Ananda Mahidol.

The announcements issued by Field Marshal Sarit's "Revolutionary Council" (*Khana Patiwat*) on the evening of 20 October 1958 blamed the coup on the actions of communists, said to have infiltrated Thailand's political, social, and economic institutions in an "attempt to assert their influence over the spirit of the Thai people".[10] While acknowledging that the Thai people had fought for basic rights and freedoms, the junta explained that "some individuals" had taken advantage of the protections afforded by the constitution to destroy the unity of the nation. The constitution being utterly powerless to prevent "the ultimate disintegration of the country", the problem could only be rectified by means of a "revolution" staged with the goal of providing the government with more adequate instruments to confront the nation's enemies.[11]

King Bhumibol is described as not having minded "one bit" the establishment of a military dictatorship that did away with freedom of expression and "Western" procedures of selection and accountability.[12] Royalist thinkers also concurred with Sarit's foreign minister, Dr. Thanat Khoman, that the political instability the country had experienced since 1932 had been caused by the "transplantation of alien institutions" without "proper regard to the circumstances that prevail in our homeland" or "the nature and characteristics of our own people".[13] For the royalists who eulogised Field Marshal Sarit as a new "Thai-style leader" in the mold of ancient kings, the constitutional regime's vicissitudes over the previous quarter century served to validate arguments about the incongruousness of "Western" institutions and the dangers such institutions presented to the country's unity and stability. None saw fit to recall that much of the instability was the result of activities undertaken by royalists themselves – to say nothing of the military officers with whom they were now allied – who had done their utmost to sabotage each of the major efforts made since 1932 to move the country in the direction of greater democracy.[14]

Having set out to overthrow every regime that had made any effort to lead the country in the direction of greater democracy since 1932, royalists finally met in Field Marshal Sarit's lawless military dictatorship a system of government they could support wholeheartedly. The royalist counterrevolution was finally complete, as the state was now in the hands of military men committed to the vision of a hierarchical society led by an all-powerful government, whose legitimacy was derived not from popular mandate but from the endorsement of a re-sacralised monarchy the people were duty-bound to revere. The palace, its ideologues, and its armed auxiliaries went on to exact a heavy price for "protecting" the nation from threats that were either largely imaginary or the result of their own activities, as royalists pointed to the instability they had fomented since 1932 – and the contrived threat of communism they had used against their enemies throughout that time – as the reason why ordinary people must renounce all claims to democratic rights and freedoms for the sake of "order" and "unity".

As previewed in this chapter's introduction, investigating the tendency of Thailand's royalist establishment to run the state as a protection racket yields valuable insights into important aspects of the country's political development. This section addresses one such aspect: namely, the longevity of Thailand's royalist political order despite its ruling class's fragmentation and the country's attendant regime instability. In a fairly recent comparative study of "authoritarian Leviathans" in Southeast Asia, Slater characterised the relatively strong states and the durable non-democratic regimes established in Malaysia and Singapore as the product of a "protection pact" that joined those countries' urban middle classes in a cohesive coalition with state, communal, and economic elites, the purpose of which was the pooling of resources controlled by each component of alliance in order to defeat a revolutionary threat from below, or indeed to prevent any such movements from gathering strength in the first place.[15] In Thailand, conversely, the weakness of the state and the instability of successive authoritarian regimes are ascribed to the fact that the less cohesive alliance formed under Sarit between the monarchy, the military, and wealthy business families took the form of a "provision pact" – that is to say, an agreement to share resources extracted from the state. Under the terms of the alliance, big business families and the urban middle class were never asked to submit to the kind of direct taxation that could fund the formation of a stronger state, while the monarchy and the military exploited their influence over the state to extract resources that could have otherwise been used to maximise state capacity in order to increase their own wealth and prestige. On repeated occasions, moreover, the monarchy, the capitalist bourgeoisie, and the urban middle class have pursued their ambitions of power at the expense of the military, causing the country to oscillate between forms of democracy, non-democracy, and pseudo-democracy.

The first major episode of this kind may be ascribed to a chain of events set in motion following Sarit's death in 1963. The ensuing years, in particular, had seen the regime's internal balance of power shift in favour of King Bhumibol, thanks in part to the fact that Field Marshal Thanom Kittikachorn (as prime minister) and Field Marshal Praphas Charusathien (as commander in chief of the army) – to say nothing of their heir apparent, Colonel Narong Kittikachorn – were almost as crooked as Sarit but had none of his stature and charisma. Having recently asserted himself as the kingdom's most powerful political figure, King Bhumibol pressured Thanom and Praphas to introduce a semi-democratic constitution in 1968 and hold minimally competitive multi-party elections in 1969. The king would soon have occasion to regret his actions upon witnessing the difficulty with which the government dealt with civil society's growing mobilisation, an increasingly critical press, and a newly elected legislature.

As ever, the royally endorsed self-coup staged by Thanom and Praphas in November 1971 was justified on the basis of contrived threats to the monarchy and the nation, whose protection was said to require a military regime wielding unaccountable, absolute powers.[16] By the early 1970s, however, socio-economic transformations brought about in part by developmental policies originally designed to bolster military rule had compromised the viability of a regime of this kind, increasing the availability of farmers, workers, students, and middle-class citizens for mobilisation in opposition to authoritarianism. While it took time for the opposition to muster the strength, things escalated rather rapidly once the new student movement organised the first protests in defiance of martial law in late 1972. Things would come to a head less than a year later, when student-led demonstrations staged between 6 and 13 October 1973 brought hundreds of thousands of middle-class and working-class citizens out to the streets to demand a "real" constitution. After the brutal military crackdown staged on 14 October 1973 failed to quell the demonstrations, despite claiming the lives of at least 77 protesters, King Bhumibol pressured the "Three Tyrants" into resigning and leaving the country. A royally appointed government headed by Privy Councilor Sanya Thammasak was tasked with overseeing the drafting of a new constitution, promulgated in late 1974, and leading the country to free general elections in January 1975.

The events of 1973 exposed the Achilles' heel of the alliance formed by the palace, the military, the civilian bureaucracy, and the families controlling Thailand's largest business conglomerates. Each authoritarian regime established in the period following Sarit's death in 1963 has been brought down by the activation of some or all of the following sources of vulnerability: (1) the urban middle class's willingness to support demands for democracy in its quest for greater inclusion/empowerment; (2) the tendency on the part of economic elites (urban and, later, rural) to support elections as a source of increased wealth and power; and (3) the readiness on the part of the monarchy to turn against military leaders whenever its popularity was threatened by episodes of state violence that met with the disapproval of the wealthy and the middle class. Then again, the events that followed the removal of the "Three Tyrants" also illustrate the fact that Thailand's royalists have never lacked the motivation or the unity of purpose to subvert democratic institutions whenever their interests and prerogatives came to be threatened by the intensifying mobilisation of students, workers, and farmers, or by the growing assertiveness of elected governments. Even as the palace and the successors ofThanom and Praphas in the armed forces sought credit for setting in motion the country's transition to democracy, therefore, they acted quickly to limit the damage. In the months that followed 14 October 1973, the palace and the military responded to the country's democratisation by funding, training, indoctrinating, and promoting groups whose violent actions had the effect of reminding the country of just how badly it still needed the protection of its self-appointed guardians, even at the cost of sacrificing some of its hard-fought freedoms. The target constituency for this campaign was the urban middle class, whose support for the student-led demonstrations of 1973 had caused the crackdown of 14 October to backfire on the Three Tyrants.

Between 1973 and 1976, the urban middle class was to be made an offer it could not refuse. King Bhumibol took a leading role in the effort made by royalists to transform the insecurity felt by urban middle class citizens in a context characterised by a deteriorating economy and escalating social unrest into a veritable hysteria over largely imaginary threats to the nation's existence. The king peppered his speeches with references to "dangers from all sides, from within and from without",[17] conspiracies to "obliterate our country from the world map",[18] and the potential for political conflicts to spiral into a "total free for all".[19] The song "We Fight" (*Rao Su*), composed by the king at the end of 1975, spoke of those who "want to destroy" the country and "threats of annihilation" against which the Thai people should prepare to fight to the death. Radio stations controlled by the army added fuel to the fire, broadcasting hateful right-wing songs and rabid speeches that incited violence against leftists, students, and other "scum".[20]

The rest of the work was done by three paramilitary organisations mobilised by officials in the Internal Security Operations Command (ISOC) and the Border Patrol Police (BPP): the Village Scouts (*Luk Suea Chao Ban*), the Red Gaurs (*Krathing Daeng*), and *Nawaphon*, whose ranks swelled to as many as two million civilian members.[21] Between 1974 and 1976, these groups waged a campaign of bombings and assassinations that claimed the lives of some 15 student leaders, at least 46 provincial activists, and leftist politicians and scores of their supporters. The wave of state-sponsored violence culminated in the gruesome massacre of dozens of students committed by royalist vigilante groups at Thammasat University on 6 October 1976. On that same evening, a new military junta, the National Administrative Reform Council (NARC), seized power and abrogated the constitution. Thousands of arrests followed the instalment of a new government led by jurist and palace favourite Thanin Kraivichien – by some accounts "the most repressive in Thai history" – who speculated it would take his administration 12 years to make the Thai people "ready" for democracy.[22] Ultimately, however, it took Thanin's government far less time to alienate most of its original backers. Given King Bhumibol's continuing support of the government, it fell upon a faction of the military led by General Kriangsak Chamanan to act on the rampant dissatisfaction in the civil service, the business community, and the middle class and remove Thanin on 20 October 1977.

The semi-democratic regime established with the constitution promulgated in late 1978 made important concessions to the aspirations of inclusion/empowerment of the urban bourgeoisie while exploiting its uneasiness with the prospect of the entire electorate's inclusion in order to justify the authority vested in non-elected officials and institutions.[23] While the country was allowed to transition back to a fully elected government in 1988 without the need for a popular uprising, it is a testament to the success of the efforts that the administration of General Prem Tinsulanonda (1980–1988) had made throughout the 1980s to "re-ideologicise" the state and the population[24] – efforts founded on the monarchy's revival and "re-mystification"[25] – that the military's forceful removal of elected Prime Minister Chatichai Chunhavan in 1991 required none of the extreme measures that had been necessary in preparation for the 1976 coup. All that was needed this time was for royalists to discredit Chatichai's government – whose real crime had been to threaten the prerogatives of unelected institutions in the pursuit of an ambitious agenda focused on economic development – as a "buffet cabinet", a "parliamentary dictatorship", and a threat to the monarchy. Then again, the arguments employed to justify the government's removal had largely failed to convince the press, the business community, and the urban middle class that the nation faced threats severe enough to warrant a more generalised rollback of their own rights and freedoms. As a result, the attempt made in the months that followed the 1991 coup to restore the semi-democratic regime that had operated in the 1980s set in train a series of events culminating with a new round of mass demonstrations and another showdown between the forces of democracy and the military. As in 1973, the unpopularity of the botched military crackdown carried out on 18–20 May 1992, causing the death of several dozen protesters, induced King Bhumibol to intervene in order to ease General Suchinda Kraprayoon out of office and appoint a new interim government tasked with overseeing yet another democratic transition.

Once again, the popularity of King Bhumibol and the staying power of Thailand's royalist establishment ensured that the democratic transition undertaken in 1992 did not threaten the interests and prerogatives of unelected institutions nearly as severely as the democratic regime established following the 1973 mass uprising. Indeed, the apparent willingness on the part of the so-called network monarchy[26] to coexist with the institutions of electoral democracy derived from the ease with which royalists managed to domesticate the governments led by Democrat Party leader Chuan Leekpai (1992–1995; 1997–2001), as well as the ease with which royalists discretely rid themselves of the less servile Prime Ministers Banharn Silpa-archa (1995–1996) and Chavalit Yongchaiyudh (1996–1997). By 1997, royalists had grown so confident in their capacity to command the deference of elected governments that they spearheaded the design and promulgation of a new constitution whose primary purpose was, from their standpoint, to empower the national executive vis-à-vis the parliamentary factions commanded by increasingly assertive provincial politicians, whose alleged "corruption" had already featured as one of the main excuses adduced in support of the 1991 military coup.

The making of a crisis of legitimacy

The establishment of a strong state and a durable regime certainly do not feature among the accomplishments of King Bhumibol's 70-year reign. Thailand's royalists, however, have been rather more successful in defending their effective control of the state and its coffers, most commonly against forces seeking to exercise powers theoretically vested in them by the constitution. Elected governments that failed to bend to the will of the royalist establishment were removed in military coups invariably justified on the basis of contrived threats to the nation and the monarchy. And, in instances where simply *referencing* such threats would likely not have sufficed for the people of Thailand to surrender their individual freedoms, royalists in the palace, the military, and the bureaucracy did not hesitate to create conditions of ungovernability designed to impress upon key segments of the population – above all the urban middle class – that they could ill afford to support democratic regimes not subject to the tutelage of the royalist establishment. While this modus operandi is no doubt more reminiscent of a "protection racket" than a "protection pact", it worked just as well to guarantee the survival of Thailand's royalist order through the vicissitudes of the country's formal constitutional structure.

Thailand's history of royalist coups d'état also helps to illuminate a second aspect of the country's political development – namely, the state of crisis in which the country's royalist order now finds itself, its secular decline being at the core of the political strife the nation has experienced over the past 15 years. In a well-run protection racket, the racketeer must be willing to make an example of whoever dares to reject an offer of protection. At the same time, a successful racketeer is one whose verbal cautions generally suffice for victims to accept an offer of protection. Punishing recalcitrant victims, in fact, not only increases the costs of operating the racket, but also raises the risks of incurring a backlash by undermining the benign image racketeers often seek to project in their communities. This is roughly the predicament in which Thailand's royalist order has become ensnared since the beginning of the century. During this time, the fact that royalists have found it necessary to inflict on the country increasing levels of chaos and economic damage in order to pave the way for the military and the judiciary to overthrow elected governments – only to have to do so all over again just a few years later, and ultimately be forced to unleash levels of repression not seen since the dictatorships of Sarit Thanarat (1958–1963) and Thanin Kraivichien (1976–1977) – shows that something is broken in the royalist political order established under the aegis of King Bhumibol and Field Marshal Sarit.

Once again, this is not the place to revisit in any detail the rise and fall of Thaksin Shinawatra. The basic facts are that the billionaire telecommunications tycoon rose to power in 2001, at the head of the most formidable electoral organisation in Thailand's history, by taking advantage of the devastation that the 1997 Asian Financial Crisis had wrought on the country's economy and politics. He subsequently built on that initial success, aided by new constitutional provisions that had been designed to strengthen the executive and empower national over local politicians, through decisive leadership and popular economic politics that stimulated demand and investment in the provinces. In 2005, Thaksin earned himself a second term as prime minister in general elections dominated by his party *Thai Rak Thai*. Famously, Thaksin's first term in office was marred by the government's prosecution of illiberal, deadly initiatives – such as the 2003 "war on drugs" and the 2004 response to the re-emergence of a dormant insurgency in the Muslim Deep South. Still, the reasons why royalists turned on Thaksin, whom they had initially supported, were others. Like Chatichai before him, Thaksin wanted to govern the country as per his constitutional authority. What is worse, Thaksin's overwhelming popular support made him, in the eyes of royalists, not only difficult to control or virtually impossible to defeat through conventional means but also a threat to the ageing king's lock on the people's hearts and minds, the authority royalists have long arrogated to divine the nation's true will, and the fate of the royalist network of power and patronage beyond the looming royal succession.

Whereas the eventual failure of the royalist coup staged in 1991 had in part been caused in the inability of royalists to impress upon Bangkok's middle and upper-middle class that the prevailing circumstances warranted a generalised rollback in democratic rights and freedoms, royalists were a great deal more scrupulous in their preparations to unseat Thaksin. The public component of the anti-Thaksin struggle had essentially two prongs. The first was an effort to discredit Thaksin with the urban middle class, significant parts of which had voted for Thai Rak Thai in 2001 and 2005. Having accused Thaksin of corruption and disloyalty to the king, royalists successfully tapped into middle-class anxieties over ongoing social change,[27] arguing that Thaksin's "populist" policies were nothing but a scheme to manipulate ignorant provincial masses using funds generated by defrauding hard-working middle-class citizens.[28] The second prong of the royalist offensive was, as per usual, an effort to make the country ungovernable, and thereby to weaken the public's confidence in the administration and its faith in the democratic process. In this endeavour, royalists were only partially successful, for while a series of well-attended demonstrations were staged in late 2005 and early 2006 by a movement that eventually coalesced in the People's Alliance for Democracy (PAD), the protests did not come close to threatening Thaksin's position. More damaging on this count was the failure of the snap election scheduled by the government for 2 April 2006, which was boycotted by the opposition and promptly annulled by the Constitutional Court. The failed election went some way towards diminishing the government's standing and power, particularly given the limitations placed by the constitution on "caretaker" administrations, as well as the press's exposure of the underhanded attempts made by Thaksin's allies to induce minor parties to join the contest. When the time came on the evening of 19 September 2006, the military coup was executed rather effortlessly. King Bhumibol's conspicuous endorsement helped ensure that military officers and ordinary citizens still loyal to the elected government accepted the fait accompli.

In retrospect, it is not hard to see why royalists tend to regard the 2006 coup as a "wasted coup". By the time the junta bowed out 15 months later, little of what threatened their control the state had been dealt with conclusively. In the intervening time, to be sure, royalists had placed Thaksin under investigation on charges that would later lead to a two-year prison sentence and the forfeiture of much of his wealth, while Thaksin and 110 of his allies were disqualified from holding elected office by a judicial decision taken concurrently with the dissolution of Thai Rak Thai. In addition, the military managed to ram through a new constitution that reduced the autonomy and power of elected officials for the benefit of courts and "independent" agencies dominated by staunch royalists. For all that, however, Thaksin's supporters won the general elections of 23 December 2007 handily; worse, the coalition government installed in early 2008 seemed intent on pursuing amnesties and constitutional amendments that threatened quickly to undo each of the measures described above. Mindful of the risks of staging another military coup, royalists pinned their hopes on the judiciary. Still, the judicial route to the government's removal, having been initiated immediately after the election, was to take time; not unlike an actual coup, moreover, it required a context in which the judiciary's actions would be accepted as the way out of a crisis of some kind.

The responsibility for contriving the pretext to remove the newly installed, Thaksin-backed government fell upon the PAD. Compared to its previous, 2006 incarnation, "Yellow Shirt" protesters were far fewer in number but a great deal more rabid in their rhetoric and violent in their actions, aiming at the complete breakdown in the elected government's capacity to run the country. After months of rather lacklustre demonstrations, the PAD placed the Government House under indefinite occupation in late August 2008 and started engaging the police in a series of violent skirmishes culminating with the deadly clashes of 7 October 2008. The move that effectively checkmated the administration, however, was made in late November, when the PAD proceeded to seize and shutter both of Bangkok's international airports. As the standoff took its daily toll on Thailand's economy and the government's authority, the stage was set for the planned "judicial coup". The Constitutional Court, which had already removed Prime Minister Samak Sundaravej less than three months earlier, issued a series of rulings on 2 December 2008 that stripped Prime Minister Somchai Wongsawat of his post, dissolved three more political parties supporting the government, and disqualified in excess of a hundred politicians from elected office for a period of five years. As in the Samak decision, the Constitutional Court's president at the time later admitted that the rulings were based not on the merits of the case but on the court's desire to restore "order" in society.[29]

The sequence of events that eventually ushered Democrat Party leader Abhisit Vejjajiva into the prime minister's office in mid-December 2008 provides a compelling illustration of the manner in which Thailand's royalists have appropriated the methods of a protection racket. As in the mid-1970s, privileged constituencies that have long grounded their claim to power in the imperative to safeguard "order" and "unity" had little hesitation to foment disorder and disunity when the voters selected the wrong government. Indeed, it was only through the funding of business elites, the inaction of the military, and the public backing of personalities in the Democrat Party and the palace that the PAD was able to deny the elected government's capacity to govern. Once the disorder had succeeded in highlighting, in the eyes of enough voters, the futility of insisting that the election results be allowed to stand, the Constitutional Court stepped in, in the name of restoring "order", by selectively enforcing laws introduced under military rule for the benefit of those who had engineered the disorder. In the aftermath of the court decision, business elites, the military, and the palace mustered the financial resources, the coercive power, and the moral authority, respectively, required to put together a new government that featured some of the old one's most corrupt and opportunistic politicians.[30] Undertaken in the name of fighting corruption and upholding "good government", the series of machinations initiated with the 2006 coup had in fact served a different agenda: make democracy so dysfunctional as to both legitimise and materially enable the tutelage offered by the palace, the military, the bureaucracy, and the courts.

There was only one problem: in their haste to remove Thaksin and his "nominees", royalists had let the "invisible hand" (*mue thi mong mai hen*) through which they had previously been known to shape the course of events slip its cloak of invisibility – their extra-constitutional role having been exercised too conspicuously to go unnoticed, especially given the electorate's increased politicisation. While the actions taken by royalists to undermine the Thaksin-aligned administration succeeded in creating the conditions that rendered a coup – "military" or "judicial" – acceptable in the eyes of the urban middle class, it is hard to overstate the magnitude of the damage these machinations have done the legitimacy of Thailand's royalist order. Indeed, the "judicial coup" of 2008 arguably did more to outrage, politicise, and polarise Thaksin's supporters than the military's seizure of power two years earlier. In short order, Abhisit's government was confronted with a surge in public expressions of resentment and disgust for the monarchy, whose image had been tainted by the institution's association with coup-makers, activist judges, and the PAD. At the same time, the street movement that had been founded by Thaksin's loyalists after the coup – the "Red Shirts" of the United Front for Democracy Against Dictatorship (UDD) – truly came into its own as Abhisit became prime minister, asserting itself as the primary vehicle for the anger and frustration that much of the electorate felt, especially in the country's north and northeast, over the undoing of successive election results. In April 2009, and then again beginning in March 2010, the UDD staged large, disruptive, and at times violent protests in the heart of Bangkok. By 2010, the stated purpose of the demonstrations went well beyond propitiating Abhisit's resignations, as the Red Shirt leaders openly described the showdown as the "final battle" against the rule of unelected officials and institutions.

The distasteful measures taken by Abhisit's government in order to neutralise the inevitable backlash only made matters worse. Its censorship of the media and its unprecedented number of arrests for the "crime" of lèse-majesté invited even more scrutiny and criticism of the monarchy.[31] Still more counterproductive were the military crackdowns staged in April and May 2010 in order to disperse the Red Shirts' rallies, which resulted in the deaths of nearly a hundred people – most of them unarmed protesters picked off by military snipers firing from elevated positions. Perhaps worst of all for Abhisit and the forces backing him, the government had only managed to secure the urban middle-class support for the decisive push against the Red Shirts by fabricating evidence that implicated Thaksin, the UDD, and others whose supported demands for new elections in a tenebrous anti-monarchy conspiracy. Once again, the stratagem may have served its purpose in the short run; beyond that, however, it helped to turn the monarchy into a focal point for the frustration of many among those who had seen their votes rubbished and their friends murdered in the king's name. When the time finally came for the people of Thailand to choose their own government, royalists were made to pay a heavy price for their machinations. On 3 July 2011, Abhisit was trounced by Yingluck Shinawatra, whose party Pheu Thai won a comfortable majority in the House of Representatives, outdistancing the Democrat Party by over four million votes.

The proportions of the defeat counseled royalists to bide their time, if only to strike back even more ferociously a little over two years later. In late October 2013, some of the most prominent politicians in the Democrat Party used the pretext of fighting against the amnesty legislation introduced by the government to resign from parliament and set up a new street movement – alternatively referred to as the People's Democratic Reform Committee (PDRC) or, in a more literal translation of its Thai-language name, the "People's Committee to Change Thailand into an Absolute Democracy with the King as Head of State" (PCAD) – with the stated mission of dismantling electoral democracy. Indeed, while the purpose of the "people's revolution" launched by PCAD in late 2013 was ultimately the same as the PAD's in 2008 – to engineer the conditions of paralysis, chaos, and violence necessary for the military, the courts, and "independent" agencies to remove the administration – the scale of its operations was now far larger, involving the repeated mobilisation of tens of thousands of supporters and the logistically daunting occupation of multiple locations across the sprawling city of Bangkok. This time the airports remained open, but the PCAD took the extraordinary step – unprecedented in Thailand's history – to violently disrupt the general elections that had been called by the government on 2 February 2014 and boycotted by the Democrat Party. Right on cue, the Constitutional Court stepped in to annul the election, leaving the government in place in a limited, caretaker capacity, while the Election Commission dragged its feet on choosing the date for a new election. The government's utter powerlessness was cemented by a further court ruling that effectively prohibited the government from doing anything to prevent or stop the occupation of roads and government buildings, effectively stripping away its legal authority to counter the PCAD's intentional disruption of public order.[32] Once again, when General Prayuth Chan-ocha took charge of the country's administration on 22 May 2014, following a failed attempt to strong-arm the government into submitting to the PCAD's demands, "order" was restored in accordance with the designs of those responsible for the disorder. As always, the price royalists extracted from the people of Thailand – in exchange for protecting them against the unrest caused by royalists themselves – was to be measured in political freedoms, few of which survived the advent of the "National Council for Peace and Order" (NCPO).

Dreams of a "permanent coup"

The coup staged on 22 May 2014 was the last in a succession of "royalist" coups d'état that took place in Thailand during the 70-year reign of King Bhumibol – among the ten successful coups staged during this time (in 1947, 1951, 1957, 1958, 1971, 1976, 1977, 1991, 2006, and 2014), only those staged in 1951 and 1977 do not qualify as "royalist" coups. Viewed against the backdrop of Thailand's history of royalist coups d'état, the events of 2014 may be said to have been characterised by the same "logic" ascribed to all previous instances, as the military's eventual seizure of power came at the end of a series of manoeuvres made by royalists in the palace, the military, the bureaucracy, the press, and civil society to pressure the public into submitting to the tutelage of unelected officials and institutions. When observers speak of the unusual nature of the 2014 coup, therefore, they generally refer to the measures taken by the NCPO *after* seizing power. Perhaps most notably, the 2014 coup has shown in restrictions to civil and political rights not seen in the wake of military coups dating back to 1976. Having banned all dissent, the NCPO prosecuted a campaign of censorship and arbitrary detentions against all organised oppositions. Deposed Prime Minister Yingluck and a few of her cabinet ministers were subjected to show trials resulting in heavy prison sentences while a host of opposition activists were brought before military courts on charges of sedition and lèse-majesté. All of these actions were made "legal" by the provisions of an interim constitution that gave General Prayuth, like Field Marshal Sarit before him, the power to take any actions he saw fit, irrespective of any existing laws, to enforce "order" and "unity". Unwilling to renounce its absolute powers, the NCPO saw to it that the "permanent" constitution promulgated in 2017 kept the provisions concerned in place until a new government is chosen.[33]

Beyond the transitory period, the 2017 constitution's main purpose is to institutionalise military coups and entrench the rule of unelected institutions. Regular legislative elections are contemplated, but future governments are set to be chosen through a process dominated by unelected members of parliament. Worse still, future governments will be held to the pursuit of a predetermined 20-year agenda bequeathed to them by the NCPO, under penalty of expulsion. As others have pointed out, the new constitution provides for something of a "permanent coup", in that it empowers unelected officials to remove governments deemed insufficiently royalist, overly assertive, or unduly popular "without resorting to deploying the tanks".[34] The NCPO's ideologues and supporters have spoken openly about what they hope to accomplish at the end of the 20-year period – namely, to re-establish what some of them referred to as the "submission culture" in place of the aspirations of empowerment that have recently found their expression in the people's demands for greater "democracy".[35]

It is no coincidence that one has to go as far back as 1976–1977 in order to find a regime as oppressive. After all, not since the mid-1970s has Thailand's royalist order experienced a crisis of faith as deep or as generalised as the current one. Now as then, the crisis is rooted in the public's failure readily to embrace the tutelage of unelected institutions – the escalating severity of the measures royalists have taken over the past dozen years but a reflection of their fading legitimacy and moral authority. After over four years of military dictatorship, the people of Thailand show no signs of being on the verge of rebelling against the NCPO. The junta's position was further bolstered after the generals successfully navigated the dreaded royal succession, having reached a modus vivendi that allowed King Vajiralongkorn Bodindradebayavarangkun to arrogate constitutional prerogatives no king has exercised in Thailand since the early 1950s in exchange for his backing of the NCPO's current rule and future plans. Still, it is notable that it has taken so long for the NCPO to schedule new elections or relax the extreme levels of repression that have been in place since 2014. In fact, one must go even further back in Thailand's history – back to the regime established by Field Marshal Sarit in 1958 – in order to find a military dictatorship as long-lived as the one currently in place.

It almost goes without saying that it is hardly a sign of confidence or strength that the NCPO has clung to absolute powers as long as it has. Indeed, the fact that the repression unleashed over the past four years has not weakened the support enjoyed by the opposition – and electoral democracy more generally – enough for the NCPO to allow the formation of a civilian government demonstrates that the methods of racketeers and tyrants are of little use in overcoming what is, at its core, a crisis of legitimacy. Of course, the NCPO may not have much of a choice on the matter, as its efforts to re-educate provincial masses to embrace their own submission are complicated, not only by the constraints that the information age places on its ability to crowd out alternative ideas but also by the Thai public's increased political awareness and sophistication. Having previously failed to extinguish popular demands for greater democracy, even at the height of the monarchy's popularity, it is doubtful that "royal nationalism" will ever again stand a chance of doing so in the future. And in the continuing absence of alternative bases of legitimacy, the denial of democracy requires royalists to insist on current levels of repression at the risk of further antagonising constituencies that are already inclined to disbelieve.

In this context, the realisation of the NCPO's vision of a "permanent coup" is anything but a sure thing. The main problem with it – and very likely the reason for the NCPO's own reluctance to put it to the test – is that once the military hands over the reins of power to a civilian government *not* entrusted with the absolute powers it currently exercises, it is virtually guaranteed that the opposition will resume its struggle for democracy in parliament and in the streets. While the NCPO did not neglect to stack the constitutional amendment process in favour of unelected officials conceivably least susceptible to the pressure generated by the opposition, recent history suggests that the mobilisation of several tens of thousands of people rotating in and out of demonstrations stretching on over a period of weeks may suffice for the opposition to prevent an unelected, royalist government from functioning well enough to prosecute its programmatic objectives. If and when that happens, Thailand's royalist elites will be placed before a familiar choice: accede to the opposition's demands for democratic reform or rely on the military to roll out the tanks, violently subdue the opposition, and restore the extreme levels of repression typically imposed in the aftermath of military coups, pending the promulgation of yet another non-democratic, royalist constitution.

The bottom line is that the increasingly low propensity of ordinary people to consider themselves in need of the protection and tutelage offered by their self-described superiors ensures that the denial of democracy will remain a source of instability for the foreseeable future. On the one hand, the silent, "permanent" coup envisioned by the constitution, which involves a return to a weak but stable government – one completely beholden to unelected officials and institutions – presiding over a newly quiescent, submissive population is likely to remain something of a pipe dream. Simply put, a government that is by design too weak and unassertive to throw off the tutelage of unelected institutions is not especially likely to appear strong enough, cohesive enough, or decisive enough to deter the opposition from pressing demands for reform, thereby rendering the denial of democracy contingent upon the military's readiness periodically (if not continually) to use its firepower on a recalcitrant civilian population. On the other hand, an argument can be made that the survival of Thailand's royalist order requires instability, for it is only in the presence of instability – real or perceived – that the coalition now backing the NCPO can be held together in support of a non-democratic regime. Alas, as long as royalist elites remain implacably opposed to democratisation, and as long as they can count on Bangkok's middle and upper-middle class to support the effective disenfranchisement of the majority of the population, Thailand's political development will most likely continue to be defined by a grim, unceasing succession of military coups, civilian massacres, and disposable constitutions.

Notes

1. National Council for Peace and Order (NCPO), *Prakat Khana Raksa Khwam Sa-ngop Haeng Chat Chabap Thi 1/2557 Rueang Kan Khuap Khum Amnat Kan Pokkhrong Prathet*, 22 May 2014.
2. "Phon Ek Prayuth Hua Na Kho So Cho Mai Top Nang Nayok Chua Khrao Eng Rue Mai Lueaktang Yang Mai Kamnot", *Matichon*, 26 May 2014.
3. Kho.So.Cho, "Chi Kan Ang Sitthi Prachathippatai Bang Klum Song Phon Koet Khwam Sun Sia", *MCOT*, 25 May 2014 <www.mcot.net/site/content?id=53819b7fbe04703b878b4569> (accessed 2 July 2014).
4. See Thitinan Pongsudhirak, "Thailand's Stalemate and Uneasy Accommodation", *Bangkok Post*, 15 February 2013.
5. Thongchai Winichakul, "Prawatsat Thai Baeb Rachachatniyom", *Sinlapa Watthanatham*, Vol. 23 (2001), pp. 56–65.
6. "Suthep in Talks with Prayuth 'Since 2010': Thaksin Regime Target in Secret Talks", *Bangkok Post*, 23 June 2014.
7. Jason Szep and Amy Sawitta Lefevre, "Powerful Forces Revealed Behind Thai Protest Movement", *Reuters*, 13 December 2013.
8. Charles Tilly, "War Making and State Making as Organised Crime", in *Bringing the State Back In*, edited by Peter Evans, Dietrich Rueschemeyer and Theda Skocpol (Cambridge: Cambridge University Press, 1985), p. 169.
9. Ibid., p. 171.
10. Revolutionary Council, "Prakat Khong Khana Patiwat Chabap Thi 2", *Royal Thai Gazette (Special Issue)*, Vol. 75, No. 81 (1958), p. 5.
11. Revolutionary Council, "Prakat Khong Khana Patiwat Chabap Thi 4", *Royal Thai Gazette (Special Issue)*, Vol. 75, No. 81 (1958), p. 12.
12. Kobkua Suwannathat-Pian, *Kings, Country, and Constitutions: Thailand's Political Development 1932–2000* (London: RoutledgeCurzon, 2003), p. 158.
13. Thak Chaloemtiarana, *Thailand: The Politics of Despotic Paternalism* (Chiang Mai: Silkworm Books, 2007), p. 100.
14. Federico Ferrara, *The Political Development of Modern Thailand* (Cambridge: Cambridge University Press, 2015), chs. 3–4.
15. See Dan Slater, *Ordering Power: Contentious Politics and Authoritarian Leviathans in Southeast Asia* (Cambridge: Cambridge University Press, 2010).
16. Thak, *Thailand: The Politics of Despotic Paternalism*, p. 228.
17. Bhumibol Adulyadej, "Royal Speech [. . .] on the Occasion of the Royal Birthday Anniversary at the Dusidalai Hall, Chitralada Villa, Dusit Palace, on Wednesday, 4 December 1974", in *Royal Speech [sic] Given to the Audience of Well Wishers on the Occasion of the Royal Birthday Anniversary at the Dusidalai Hall, Chitralada Villa, Dusit Palace, on 4 December 1974, 1975, 1976, 1977, 1978* (Bangkok: Amarin, 1979 [1974]), p. 36.
18. Bhumibol Adulyadej, "Royal Speech [. . .] on the Occasion of the Royal Birthday Anniversary at the Dusidalai Hall, Chitralada Villa, Dusit Palace, on Thursday, 4 December 1975". In *Royal Speech [sic] Given to the Audience of Well Wishers on the Occasion of the Royal Birthday Anniversary at the Dusidalai Hall, Chitralada Villa, Dusit Palace, on 4 December 1974, 1975, 1976, 1977, 1978* (Bangkok: Amarin 1979 [1975]), p. 24.
19. Bhumibol Adulyadej, "Royal Speech [. . .] on the Occasion of the Royal Birthday Anniversary at the Dusidalai Hall" (1974), p. 10.
20. Somsak Jeamteerasakul, *Prawatsat Thi Phoeng Sang: Ruam Bot Khwam Kiao Kap 14 Tula Lae 6 Tula* (Bangkok: Samnak Phim 6 Tula Ram Luek, 2001), pp. 115–148.
21. Prajak Kongkirati, "Counter-Movements in Democratic Transition: Thai Right-Wing Movements after the 1973 Popular Uprising", *Asian Review*, No. 19 (2006), p. 17.
22. Charles F. Keyes, *Thailand: Buddhist Kingdom as Modern Nation-State* (Boulder, CO: Westview Press, 1987), p. 100.
23. Chai-anan Samudavanija, "Thailand: A Stable Semi-Democracy", in *Democracy in Developing Countries*, edited by Larry Diamond, Juan Linz, and Seymour Martin Lipset (London: Lynne Rienner Publishers, 1989), pp. 334–335. Also see, Benedict Anderson, "Murder and Progress in Modern Siam", *New Left Review*, Vol. 181 (1990), pp. 33–48.
24. Michael K. Connors, *Democracy and National Identity in Thailand* (Copenhagen: NIAS Press, 2007), p. 130.

25 Roger Kershaw, *Monarchy in South-East Asia: The Faces of Tradition in Transition* (London: Routledge, 2001), pp. 139–140, 200. Also see Connors, *Democracy and National Identity in Thailand*, p. 145.
26 See Duncan McCargo, "Network Monarchy and Legitimacy Crises in Thailand", *Pacific Review*, Vol. 18 (1995), pp. 499–519.
27 Michael J. Montesano, "Four Thai Pathologies, Late 2009", in *Legitimacy Crisis in Thailand*, edited by Marc Askew (Chiang Mai: Silkworm Books, 2010), p. 280. Also see Nidhi Eoseewong, *Kan Mueang Khong Suea Daeng* (Bangkok: Open Books, 2010), pp. 132–137.
28 Pasuk Phongpaichit and Chris Baker, *Thaksin* (Chiang Mai: Silkworm Books, 2009), pp. 253, 264–266.
29 "Wasan Soipisut Kham to Kham San Ratthathammanun Phalat", *Matichon*, 16 March 2016.
30 Chang Noi, "When the Beauty of Democracy Is Not So Beautiful", *The Nation*, 22 December 2008.
31 "Discussing Lèse-Majesté Law", *Prachatai*, 28 June 2011 <https://prachatai.com/english/node/2623> (accessed 30 August 2018).
32 "Court Strips Gov't of Various Emergency Powers", *Khaosod English*, 19 February 2014 <www.khaosodenglish.com/politics/2014/02/19/1392808934/> (accessed on 30 August 2018).
33 Wasamon Audjarint, "Some Article 44 Orders to Become Permanent Laws", *The Nation*, 11 January 2018.
34 Pithaya Pookaman, "Thailand's Junta Invents a New Coup D'état", *Asia Sentinel*, 2 August 2018 <www.asiasentinel.com/opinion/thailand-junta-invents-new-coup-detat/> (accessed on 25 August 2018).
35 "NCPO May Pick Trusted Allies", *The Nation*, 25 July 2014.

References

———. 2011. "Discussing Lèse Majesté Law". *Prachatai*, 28 June <https://prachatai.com/english/node/2623>.
———. 2013. "Wasan Soipisut Kham To Kham San Ratthathammanun Phalat". *Matichon*, 16 March.
———. 2014. "Court Strips Gov't of Various Emergency Powers". *Khaosod English*, 19 February <www.khaosodenglish.com/politics/2014/02/19/1392808934/>.
———. 2014. "Kho.So.Cho. Chi Kan Ang Sitthi Prachathippatai Bang Klum Song Phon Koet Khwam Sun Sia". *MCOT*, 25 May <www.mcot.net/site/content?id=53819b7fbe04703b878b4569>.
———. 2014. "Phon Ek Prayuth Hua Na Kho So Cho. Mai Top Nang Nayok Chua Khrao Eng Rue Mai Lueaktang Yang Mai Kamnot". *Matichon*, 26 May.
———. 2014. "Suthep in Talks with Prayuth 'Since 2010': Thaksin Regime Target in Secret Talks". *Bangkok Post*, 23 June.
———. 2014. "NCPO May Pick Trusted Allies". *The Nation*, 25 July.
Anderson, Benedict. 1990. "Murder and Progress in Modern Siam". *New Left Review*, Vol. 181: 33–48.
Bhumibol Adulyadej. 1979[1974]. "Royal Speech [...] on the Occasion of the Royal Birthday Anniversary at the Dusidalai Hall, Chitralada Villa, Dusit Palace, on Wednesday, 4 December 1974". In *Royal Speech [sic] Given to the Audience of Well Wishers on the Occasion of the Royal Birthday Anniversary at the Dusidalai Hall, Chitralada Villa, Dusit Palace, on 4 December 1974, 1975, 1976, 1977, 1978*. Bangkok: Amarin.
Bhumibol Adulyadej. 1979[1975]. "Royal Speech [...] on the Occasion of the Royal Birthday Anniversary at the Dusidalai Hall, Chitralada Villa, Dusit Palace, on Thursday, 4 December 1975". In *Royal Speech [sic] Given to the Audience of Well Wishers on the Occasion of the Royal Birthday Anniversary at the Dusidalai Hall, Chitralada Villa, Dusit Palace, on 4 December 1974, 1975, 1976, 1977, 1978*. Bangkok: Amarin.
Chai-anan Samudavanija. 1989: "Thailand: A Stable Semi-Democracy". In *Democracy in Developing Countries*. Edited by Larry Diamond, Juan Linz, and Seymour Martin Lipset. London: Lynne Rienner Publishers.
Chang Noi. 2008. "When the Beauty of Democracy Is Not So Beautiful". *The Nation*, 22 December.
Connors, Michael K. 2007[2003]. *Democracy and National Identity in Thailand*. Copenhagen: NIAS Press.
Ferrara, Federico. 2015. *The Political Development of Modern Thailand*. Cambridge: Cambridge University Press.
Handley, Paul M. 2006. *The King Never Smiles: A Biography of Thailand's Bhumibol Adulyadej*. New Haven: Yale University Press.
Kershaw, Roger. 2001. *Monarchy in South-East Asia: The Faces of Tradition in Transition*. London: Routledge.
Keyes, Charles F. 1987. *Thailand: Buddhist Kingdom as Modern Nation-State*. Boulder, CO: Westview Press.
Kobkua Suwannathat-Pian. 2003. *Kings, Country, and Constitutions: Thailand's Political Development 1932–2000*. London: RoutledgeCurzon.
McCargo, Duncan. 1995. "Network Monarchy and Legitimacy Crises in Thailand". *Pacific Review*, Vol. 18: 499–519.

Montesano, Michael J. 2010. "Four Thai Pathologies, Late 2009". In *Legitimacy Crisis in Thailand*. Edited by Marc Askew. Chiang Mai: Silkworm Books.

National Council for Peace and Order (NCPO). 2014. "Prakat Khana Raksa Khwam Sangop Haeng Chat Chabap Thi 1/2557 Rueang Kan Khuap Khum Amnat Kan Pokkhrong Prathet". 22 May.

Nidhi Eoseewong. 2010. *Kan Mueang Khong Suea Daeng*. Bangkok: Open Books.

Pasuk Phongpaichit and Chris Baker. 2009. *Thaksin*. Chiang Mai: Silkworm Books.

Pithaya Pookaman. 2018. "Thailand's Junta Invents a New Coup D'état". *Asia Sentinel*, 2 August. <www.asiasentinel.com/opinion/thailand-junta-invents-new-coup-detat/>.

Prajak Kongkirati. 2006. "Counter-Movements in Democratic Transition: Thai Right-Wing Movements after the 1973 Popular Uprising". *Asian Review*, Vol. 19: 1–33.

Revolutionary Council. 1958a. "Prakat Khong Khana Patiwat Chabap Thi 2". *Royal Thai Gazette (Special Issue)*, Vol. 75, No. 81: 5–6.

———. 1958b. "Prakat Khong Khana Patiwat Chabap Thi 4". *Royal Thai Gazette (Special Issue)*, Vol. 75, No. 81: 10–16.

Slater, Dan. 2010. *Ordering Power: Contentious Politics and Authoritarian Leviathans in Southeast Asia*. Cambridge: Cambridge University Press.

Somsak Jeamteerasakul. 2001. *Prawatsat Thi Phoeng Sang: Ruam Bot Khwam Kiao Kap 14 Tula Lae 6 Tula*. Bangkok: Samnak Phim 6 Tula Ram Luek.

Szep, Jason and Amy Sawitta Lefevre. 2013. "Powerful Forces Revealed Behind Thai Protest Movement". *Reuters*, 13 December.

Thak Chaloemtiarana. 2007[1979]. *Thailand: The Politics of Despotic Paternalism*. Chiang Mai: Silkworm Books.

Thitinan Pongsudhirak. 2013. "Thailand's Stalemate and Uneasy Accommodation". *Bangkok Post*, 15 February.

Thongchai Winichakul. 2001. "Prawatsat Thai Baeb Rachachatniyom". *Sinlapa Watthanatham*, Vol. 23: 56–65.

Tilly, Charles. 1985. "War Making and State Making as Organised Crime". In *Bringing the State Back In*. Edited by Peter Evans, Dietrich Rueschemeyer, and Theda Skocpol. Cambridge: Cambridge University Press.

Wasamon Audjarint. 2018. "Some Article 44 Orders to Become Permanent Laws". *The Nation*, 11 January.

7

THE DEVELOPMENT OF THE HYBRID REGIME

The military and authoritarian persistence in Thai politics

Surachart Bamrungsuk

This chapter aims to study the post-transition regime in Thai politics since the political transition in the country has never reached the state of democratic consolidation. That is to say, democracy is not fully adopted as "the only game in town" because the new regime, many times, is in between liberalism and authoritarianism, or it represents the middle categories of what is usually referred to as "hybrid regime". The in-between regime type allows the elites and the military to maintain their prerogatives in the post-transition politics. As a result, this kind of regime becomes an ideal type for them. With this direction, democratic transition is always controlled in order to produce the system of "partly free" politics. Indeed, the discussion of post-transition regime suggests that we need to think in terms of "degree" of the so-called Thai democratic regime and also why it happens that way in this country. Therefore, this chapter will not study the military intervention in Thai politics as such, since there are many studies on this subject. But it will put major emphasis on the questions of what kind of regime will emerge after the end of direct military involvement in politics. The other side of the problem is the question of what civil-military relations in the post-authoritarian regime will be.

The Thai hybrid regimes

The military is a prominent actor in contemporary Thai politics, and it is able to maintain its influence and dominance in politics due to its knowledge and monopoly and its institutional cohesion.[1] Moreover, as a Thai political scientist mentioned, the military was the formidable political force since 1932 and coups, not elections, became the norm of changing government in the country.[2] The coups do not always lead to the authoritarian regimes, but to hybrid ones.

The purpose of the hybrid regime is to pave the way for the existence of the authoritarian regime in the post-transition era. Although its existence differs from before, it means that it is not totally annihilated. Interestingly, its existence also bases on the mechanism in the democratic regime as the mechanism/instruments of the former authoritarian regime could no longer respond to the political change/the change in the political landscape. Especially the military, which is the main mechanism of the authoritarian regime, faces so much pressure that it could not maintain their role. As such, it has to withdraw from politics and the political arena in order to preserve the military institution from that pressure.

According to Thai politics, there are hybrid regimes in four cases: the Phibun Songkhram era (1955–1957), the Sarit Thanarat era (1957–1958), the Thanom Kittikachorn era (1968–1971), and the Prem Tinsulanonda era (1980–1988).

The first wave of the hybrid regime (the Phibun era)

The 1947 coup

After the 1947 coup, the military government tried to the utmost to control Thai politics. This coup was not mainly a result of the conflict among the leaders of the People's Party (or the Khana Ratsadon), who succeeded in the revolution that changed the system of government in 1932. Rather this coup reflected the changing political landscape in Thailand after the end of the Second World War.

Meanwhile, the country at that time also confronted the fluctuating domestic politics coupled with the economic conditions in the post-war period. The situation was critical and finally ended up in a coup on 8 November 1947. This coup brought an end to the democratic regime.

However, as a matter of fact, this coup aimed to overthrow the People's Party (Khana Ratsadon), and simultaneously it aimed to destroy the Free Thai (Seri Thai) Group which gained more of a political role from the anti-Japanese movement during the World War II because this group was a major political base of the civilian wing of the People's Party. Therefore, the 1947 coup is the revival of the old/former establishments with the military as the prime supporter.

Hence, this coup is the starting point of the establishment of political power of the authoritarian regime in the post–World War II era, and the position of the government became stronger because all political parties were constitutionally banned.[3] At the same time, it marked the end of the People's Party era in Thai politics.

The external factor

The external factor played an important role in the 1947 coup as it occurred during the Cold War period. Therefore, for the Western government, the threat in the contemporary period was no longer concern with the "anti-Axis" but rather with the communist threat.

Later, Phibun's decision to participate in the Korean War in 1950 indicated the government policy's orientation in anti-communism. At the same time, Phibun successfully transformed his position from "a pro-Axis" to "a pro-Western" position.

The experiment of the first hybrid regime

While the military government tried to strengthen the authoritarian regime, Phibun visited many countries, especially in Europe and the United States. After returning to Thailand, the government began to adjust its policy towards more democratic characteristics.

The new image creation of Phibun to be "a supporter of democracy" led to the political atmosphere consistent with the government's effort to push forward the democratic trend, for example:

1 The government allowed political activists to express or give political speeches.
2 The government arranged for regular official interviews and press conferences.
3 The government announced the upcoming election and promulgated the Political Party Act of 1955, which came into effect on 26 September 1955.

The opportunity that allowed the politics to be partly free seemed to be "the experiment of the semi-authoritarianism". As mentioned earlier, once the military leader decided to open up the political regime through the permission of the political movements and the establishment of political parties, the political atmosphere had the chance to move forward to democracy in the near future.

Since 1956, the student movements tried to gather students from various universities to establish a "student union". From then on, the students played more active roles in politics, especially in the protest against the government due to the "dirty election" of March 1957. Although the government could control the situation, the dissent among the military leaders circled around the political game and each side prepared to stage a coup. Finally, the political victory belonged to the pre-emptive side.

Ultimately, the experiment of the first wave of the hybrid regime came to an end with the coup on 16 September 1957.

The second wave of the hybrid regime (the Sarit era)

After the military officers, led by Sarit, the chief army commander, staged a coup against the Phibun government on 16 September 1957, they established the civilian "puppet government" with Pote Sarasin, secretary-general of SEATO (the Southeast Asia Treaty Organisation), appointed as prime minister. Meanwhile, Lieutenant General Thanom Kittikachorn was appointed as minister of defense and Lieutenant General Praphas Charusathien as minister of interior. Soon after the seizure of power, Sarit revived the establishment of "Supreme Command Headquarters" and also served as the supreme commander.

Sarit's positions strengthened the military power. At the same time, the military also employed the parliamentary mechanisms as the instruments to maintain its power in the political arena. Not long after the coup, the general election was held on 15 December 1957, in a time that the military was very powerful in politics.

The elected government in 1957

The election successfully held after the coup indicated the political competition under military control. On the other hand, Sarit also had strong political support from the Sahaphum Party.[4] The victory in this election led to the formation of the hybrid government, as it was the power succession of the military through the election. Thanom became the prime minister and minister of defense and Praphas became deputy prime minister in January 1958.

The policy orientation of this government was clearly on anti-communism and close relationship with the Western power, the United States. However, the government was not as strong as expected due to the political instability and intra-party conflict, and Thanom was unable to control the politics.[5] In response, the military decided to launch a coup on 20 October 1958.

The second hybrid regime came to an end and it clearly indicated that the military leaders could not bear the hybrid regime. They realised that this type of regime did not really give the opportunity for the military to control the politics, and in the meantime the military leaders could not tolerate any issues/problems in the parliament, as they considered these situations to be disorder.

The 1958 coup

The coup led by Sarit claimed in its statement that this decision was based on "the consent of the majority people who deeply concern with the situation of the country . . . especially the communism that threaten the country severely. . . . it is inevitable to launch a coup".[6] In consequence, the military regime was established because the hybrid regime could no longer be responsible for the interests of the military leaders. Although the hybrid regime portrayed the positive image as it originated from the election, it was unable to fully control and stabilise the politics. In addition, Sarit never had confidence that the elected regime was capable of sustaining the military regime.

Furthermore, the military leaders came up with the new idea to take full control of the Thai politics. The concept of despotic paternalism was applied to politics: the prime minister would serve as "the father" of the society and the people would be the "children" who were governed and expected to obey.[7] The dissidents had to be captured and eradicated. The media needed to be under tight control.[8]

The authoritarian regime under the Sarit premiership was based on four discourses: the communist resistance, the societal development, the Thai-style governmental regime, and the support of the Western power. However, he did not enjoy his power for long. He died on 8 December 1963 and was succeeded by General Thanom, who later was promoted to field marshal in January 1964.[9]

The new constitution was promulgated in June 1968, and in October Thanom established the "Saha Pracha Thai Party" (United Thai People's Party) in which he himself was the leader together with some other military leaders. Their intention was to succeed the military regime in the form of elected government.

The third wave of the hybrid regime (the Thanom era)

The elected government of 1969

As the latest election was held on December 1957 and followed by the 1958 coup, the civilian regime was a missing link in Thai politics for a long time until the next election in February 1969. In the meantime, in order to make this constitution permanent and durable, Thanom asked a monk for the best timing for the king to grant the new constitution. The election resulted in the victory of the Saha Pracha Thai Party, which gained 76 seats. While the Democrat Party gained 57 seats, the independent contenders gained 71 seats and some seats from the small party. In order to form the government, Thanom convinced some independent contenders to join and finally got 115 seats in total. Once again, the hybrid regime was established in Thai politics. This regime was the third hybrid regime, and it was confronted with many challenges, as the two previous regimes stayed in power for only a short time.

The victory from the recent election placed Thanom as prime minister again, with Praphas serving as the deputy prime minister and minister of interior. The formation of the coalition government caused the military leaders to rely on the support from the other politicians. Therefore, the existence of this hybrid regime reflected the compromise among these military factions, or it was the interest of conformity of these factions under the condition of the elected government.

Hence, the survival of this government depended on two urgent issues: the capability to compromise and harmonise the interests of the various military factions and the stability of the politics. If these two issues occurred, the hybrid regime was no longer an effective means to help maintain the existence of the "semi-military" regime.

The 1971 coup

Eventually, what the military leaders feared took place. These military leaders began to have different viewpoints on the administration of Saha Pracha Thai Party. Although this appeared to be a minor problem, it indicated the dissent inside these factions.

On the other hand, the international political situation began to change when the United States normalised diplomatic relationship with the People's Republic of China (PRC) and recognised its status at the United Nations. This circumstance made the government in Thailand deeply concerned that the communist insurgency, which started on 8 August 1965 when the Communist Party of Thailand (CPT) launched a guerrilla war against the Thai government, would be out of control.

The hybrid regime under Thanom reached its end on 17 November 1971 when he staged a coup against his own government. At last, Thai politics returned to the military regime again. The 1971 coup brought the military government under the leadership of Thanom back to Thai politics again. However, this government was smashed by many challenges – for instance, the internal conflict of the military regime, the division inside the military, the development of the students' and the middle-class movements with anti-military perception, and the first energy crisis.

The demand for democracy in 1973

All the problems coupled with the division among the military leaders and the dissatisfaction of the people led to the "14 October 1973 incident", which was a major popular uprising in Thai politics. Although the military force was used to suppress the demand for democracy, in the end the military government was defeated. On the night of 15 October 1973, Thanom, Praphas, and Narong fled the country.

The transition to democracy returned to Thai society once again, and even though military power in politics was destroyed, the military was still a strong institution. On the other hand, this transition occurred at the time of the shift in the Second Indochina War due to the US military withdrawal. The decision of the great power brought about some concerns among the elites and the conservatives, especially the communist victory in Indochina or "the domino theory".

The situation in Thailand reached its turning point when the dominoes in Indochina fell in 1975 accompanied by the fear of the elites, the military leaders, and the conservatives. They firmly believed that the democratic regime was a weakness and that might lead to the defeat to the communist threat as it had happened in Indochina. Hence, the military regime would serve as a security guarantee for the country and would be a strong mechanism to counter the communist expansion.[10] The other important issue to be realised was that Thailand was not only confronted with the changing circumstance in Indochina but also the expansion of internal movements from various groups. For the elites, it was time to cease these political movements. The response to these situations was the decision to turn the politics back to the authoritarian regime through the military coup on 6 October 1976. Furthermore, the student movement was suppressed with the military measures to signal the strength to fight with the communists. These measures were under the "killing communists is not sinful" discourse. Again, the new military government rejected the election.

A failed coup

The 1976 coup was a crucial expectation for the right-wing groups who believed that "violent measures" were the most effective ways to counter communism. However, these measures produced the opposite results. The ultra-right policies and the violent measures were the impulse for many people to decide to head for the rural areas and join the "armed insurrection" with the CPT.

On the other hand, the United States also altered its policy orientation away from the endorsement of the authoritarian regime as an alliance in the anti-communist campaign. The new policy orientation came with the change of the leader at the White House. President Jimmy Carter clearly indicated that "human rights" would be a major ground for his foreign policy.[11] However, the military government in Thailand did not understand the signal and still portrayed itself as a strong anti-communist government. It did not conform to the superpower's policy because the anti-communism was not the core of US policy. The US human rights policy had strong impacts on the military governments in Latin America and led to the end of the authoritarian regimes in that region, including the authoritarian regime in Bangkok.

At the domestic level, the ultra-right civilian government was also confronted with the conflict with the military that led to two attempted coups in 1977. The military decided to launch a third one on 20 October, which was successful. The 20 October 1977 coup led to the end of the civilian authoritarian regime.[12] The situation in 1977 was a reflection that the military still had much power and played a major role in Thai politics. Remarkably, the military group that later launched a coup in 1977 had a different perception and policy orientation. They did not adhere to the extreme anti-communist campaign, as the expansion of the insurgency in the rural areas indicated the failure of this policy orientation. In their perspective, the ultra-right policy would be a decisive factor for Thailand to lose the communist insurgency and would be "the fourth domino". Therefore, if they wanted to change the policies, they would need to change the government. Finally, this military group decided to overthrow the civilian authoritarian government.

The fourth wave of the hybrid regime (the Kriangsak and Prem era)

The hybrid regime of 1979

The 1977 coup resulted in the military government under General Kriangsak Chamanan leadership. This government point of view about the constitution totally differed from the former military governments. During the Sarit era, any measures were applied to delay the constitutional drafting in order to maintain the power of the military government. For example, the Interim Constitution of Thailand of 1959 did not specify when the constitution needed to be completed, so the drafting process lasted for 7 years 11 days (from 9 February 1961 to 22 February 1968). It is the longest constitution drafting process in the political history of Thailand. However, the Kriangsak government did not stay in power for long, as it promulgated the new constitution on 22 December 1978. This military government had a different standpoint from the Sarit-Thanom model, which stayed in power and ruled the country for a long period. The general election was held on 22 April 1979.

Interestingly, from the coup in 1977 to the promulgation of the constitution in 1979 was quite a short period for the government to bring the politics back to normal. This decision was a response to domestic politics and international environment and impacted the government directly. The decision to hold a general election not only helped support the image of Kriangsak in the international arena but also to depressurise domestic political conflict.[13] Moreover, it helped lessen domestic pressure, especially after the implementation of the ultra-right policy which relied on the resistance and suppression of communism. The idea of lessening pressure (Latin American political proverb calls this type of situation "decompression") was the policy orientation that gained wide acceptance.

For the military, this policy brought about the strategic adaptation in the anti-communist warfare. Previously, the military held to a "use of force" strategy and strongly believed that suppression was the most effective instrument to defeat the communist insurgency. Unfortunately the outcome was undesirable, as more suppression led to more expansion of communism. This situation caused military strategists to doubt whether military measures were an effective means for the Thai government to gain victory in the communist insurgency and whether military measures would lead to the end of the country, as in South Vietnam.

The military began to accept that military measures would never defeat this kind of warfare. The failure of the United States in the Vietnam War was a good lesson that superior firing power was not the factor that leads to victory. So they began to adjust policy orientation towards a more political strategy. In their perspective, they admitted that the authoritarian regime was the condition for the defeat, as General Chavalit Yongchaiyudh's phrase "communism overcomes authoritarianism and democracy overcomes communism". As mentioned earlier, the newly elected government in 1979 came with the new political ideas/concepts. Interestingly, they won the election because it was held not long after the successful coup. As a result, the election led to "the transformation of the military government" and the hybrid regime was back to Thai politics again.

The military-led political transition was very important to determine the future of Thai politics because in this transition the military power was not diminished and the military was able to preserve its influence in the new political regime. In comparison with the political transition in Latin America, the outcome stemming from this condition was called the "Protected Democracy". The transition, in this case, would not lead to a "Full Democracy" but rather a "Half- or Semi-Democracy". This type of democracy, on the other hand, was the return of the hybrid regime after the fall of the hybrid regime under the Thanom premiership in 1971.

The hybrid constitution

The return of the hybrid regime in 1979 came with the effort to institutionalise this regime through legal instruments, for example:

1. The number of the members of the senators amounts to three-quarters of the total number of members of parliament (MPs). There were 225 senators out of the total 301 MPs. The senators strengthened the power of the government as they were appointed by the government, and the government could use the senators to balance the member of the House of Representatives directly. The government no longer had to concern itself much about the role of the lower house.
2. The constitution stated that the president of the Senate was the president of parliament who would serve as the recipient of the royal command to appoint the prime minister. Therefore, the president of the Senate would endorse one who was suitable to be the prime minister. On the other hand, the prime minister, Kriangsak, appointed the senators. It implied that one who was appointed by the prime minister would endorse that prime minister to resume the power. This constitution mechanism was established to bring back the military leader to be the government leader.
3. The constitution granted the government to appoint the senators. As a result, the total number of senators were three-quarters of the members in the house of representative, which indicated the power succession of the former military regime.
4. The constitution was promulgated, but there were as many as 12 provisions. It clearly stated that some articles would not be implemented in the first four years but applied the content

of the provisions instead. In this case, the constitution was not fully implemented and the country was regulated by the provisions.

5 The provisions gave the opportunity for the military officers and the civil servants to serve as ministers or political officials. They could serve as both government officers and political officials at the same time. In the case of the military officers, the constitution allowed them to be ministers and commanders concurrently. The consequence of these provisions was the overlapping power between political officials and the opportunity for the government officers to play a major role in controlling politics directly. Eventually, this constitution revived and strengthened "the bureaucratic polity" in Thai politics.

6 According to the constitution, the prime minister was not necessary from the election. This specification gave a chance for Kriangsak to resume to the prime minister position, without running for the election. In other words, the constitution enabled "the middleman prime minister". Under this circumstance, the political process, via the election, was not the mechanism to select the prime minister.

7 The provision of the constitution increases the power of the Senate. The important amendments/bills were the considerations of both the House and the Senate, for example, the government statement, the annual budget bill, and the general debate. The empowerment of the Senate would transform this institution to be the "political shield" for the government when faced with checks and balances from the House of Representatives.

8 The provisions also stated that the electoral contenders were not necessary to be members of any political parties and that they could run for the election independently. As such, the constitution was not to promote and strengthen the political parties, and simultaneously did not aim to develop the political party system of the country. In consequence, the contenders for the premiership would not be from the political parties.

9 The constitution did not cover political party law which aimed not to strengthen the political party system. It also allowed independent contenders to run for election which helped the military leaders to form the government easily. The members of the House of Representatives were able to vote independently as they were not under the party.

10 The government had to make the policy statement to the parliament. However, it did not require a vote of no confidence. It was just the statement and it indicated that there was no check from the parliament during the government formation process. At the same time, the constitution gave the opportunity for the military leaders to form the government easily after the election and they would not face resistance from the political groups in parliament.

The political assurance mechanism

It was obvious that the constitution was drafted to assist the junta's power succession after the election. Even though it was a minority government, the military leaders managed to survive in parliament, and at the same time, it put forth many efforts to launch new policies in order to gain political legitimacy and popular support. These policies were:

1 The adjustment of foreign policy away from the early Cold War style and did not adhere to the political ideology, for example, the establishment of diplomatic relations with China and the strengthening relationship between Thailand and the neighbouring countries of Indochina after the dominoes fell in 1975. These actions portrayed the new orientation in foreign affairs of the country as it lessened the anti-communist sentiment and did not regard socialist ideology to be the essential factor in determining diplomatic relations anymore.

2 The promotion of the reconciliation policy. The military leader recognised that the ultra-right policy of the government in 1976 led to the expansion of the communist insurgency in the rural areas and this policy also created domestic conflict. The Kriangsak government, consequently, made a crucial decision to approve an amnesty bill for the 1976 incident student leaders, along with other people involved in the bloody incident. This bill also provided a chance to reintegrate the students joining the communist party back into the society. The amnesty bill was the beginning of the end of the communist insurgency in Thailand and was the key to national reconciliation.

Therefore, the important foundation of the hybrid regime was the support from the military. Evidently, the military was the core of the military regime as well as the core of the hybrid regime. Unfortunately, when the government was confronted with the energy crisis (the Thanom government) and popular resistance, the Young Turks decided not to support the government. They put so much pressure on the government and asked Kriangsak to resign in order to prevent another mass uprising.[14] Eventually, Kriangsak announced his resignation at parliament on 29 February 1980. It was widely acknowledged that his resignation was the "pressure" from the Young Turks themselves. These officers turned to support Prem for the premiership under the same political order. At last, the military continued to be the decisive factor in the hybrid regime. After the military officers' successful pressure to oust the Kriangsak government, following with the rise to power of Prem, the mechanisms of the hybrid regime stemming from the 1978 constitution continued to function.

The Prem government had the military officers and civilian bureaucrats who were appointed to be the senators as the political supporters. At the same time, some military officers joined the cabinet, especially Colonel Chamlong Srimuang, member of the Young Turks clique, who served as secretary to the prime minister. Although there was a change in government leader, the orientation of the hybrid regime stayed intact. In other words, Prem inherited the hybrid regime from Kriangsak and it was an ideal succession as it involved a change of the head of the military government without staging a coup.[15] Interestingly, this indicated the role of the military leaders behind the politics who applied so much pressure to lead to the replacement of the prime minister.

The semi-democratic government of 1980

The political transition due to the 1979 election occurred under the condition that the military was so powerful in politics. This condition made it inevitable to consider the role of the military leaders. When Prem stepped into power, he was also the commander in chief of the army and the minister of defense.

Prem did not want his government to be represented as the "military regime", so he appointed the retired civilian bureaucrats to his cabinet. Moreover, he also invited major political parties to join the government. Prem was able to control the parliament as he had strong political support from both Houses. He did not establish his own party as in the case of Phibun, Sarit, and Thanom. There was no need for the "military-supported party" when Prem chose to serve the interests of all parties in the government. The political parties were included to form the coalition government in which the military leader as the head of the government and the prime minister was the middleman, not the formation of a coalition government under the military party leadership. This circumstance allowed Prem to be independent from the political parties. At the same time, the military leaders were able to arbitrarily invite the political parties to join the government. This was the factor that created stability in parliament for a long period.

The characteristics of the government

The hybrid regime was considered through the status of the prime minister. Prem not only was the prime minister but also the minister of defense and the commander in chief of the army. Additionally, before his retirement in September 1980, the Defense Council approved to extend his retirement for a year. The resistance arose, however, according to the "special support" Prem possessed. The Democrat Party, which strongly opposed that extension, made a statement to accept the decision because it had just had "new information" on that matter. In this case, this government was similar to the other military regimes in that the military leaders held many positions both in the government and the bureaucracy. This factor strengthened the hybrid regime to control the military.

The hybrid regime of the Prem government indicated that the stability of the regime was under the condition of the military leader's capability to control parliament. Although there were two coups, the Prem government was able to survive. During 1–3 April 1984, the middle-ranking officers staged a coup. This coup was later called "the Young Turks Rebellion".[16] The coup happened again on 9 September 1975, this time staged by the retired officers in the Young Turks clique. This coup was later dubbed "the retired officers rebellion".

After the defeat of the Young Turks clique and the sharp decline of its influence in the 1980s, the class 5 graduates of the Chulachomklao Royal Military Academy began to rise.[17] This momentum also shifted the support in the military. These circumstances displayed the special characteristics of the hybrid regime under Prem, that the military control was a factor contributing to the survival of the government. The Prem government had a unique characteristic that could not be found in the civilian government.

The new security strategy

One of the major successes of the Prem government was the termination of the domestic communist insurgency. This success originated from the strategic adjustment of the government in the struggle with the communist party. In reality, the strategic adjustment had begun in the Kriangsak government since it preferred political measures to military ones to fight communism.[18] And when Prem rose to power, he obtained this strategy as his policy orientation to counter communist insurgency. Therefore, on 23 April 1980, the Order of the Office of the Prime Minister No. 66/2523 (simply called Order 66/2523 or Order 66/23) was issued to adjust the counter-communist orientation. Two years later, another order, the Order of the Office of the Prime Minister No. 65/2525, was issued to emphasise the new strategy of the government.

These new strategies were in line with change at the regional level, as conflict arose between the Communist Party of Thailand and the Communist Party of Vietnam due to the Vietnamese invasion of Cambodia in early 1979. Soon after that, the intensity of the communist insurgency in 33 provinces where the CPT operated gradually decreased; 25,884 people decided to end their struggle and left the jungle. During 1982–1983, the communist insurgency in Thailand reached its final point.

The stability of the hybrid regime not only stemmed from stabilising domestic politics but also the end of the communist insurgency. Even though the government was confronted with border security due to Vietnamese occupation of Cambodia, this issue did not affect Thai security dreadfully, as it was still under control. Moreover, the capability to preserve the stability to respond to this threat was a major concern for the Thai military.[19] As a result, the Prem government received massive support as it was the government that created stability, both in the political and security realms. The government was managed to not only control the military but also the political parties and parliament.

The hybrid regime of the Prem government was terminated. Undeniably, Prem served as the prime minister for eight years and five months, and his regime was the successful model of the longest hybrid regime in Thai politics.[20] The end of the hybrid regime paved the way for General Chatichai Choonhavan, the head of the Chart Thai Party (Thai Nation Party), to be the new prime minister on 9 August 1988. Thai politics reverted to the elected regime once again.

Back to the future: the new wave of the hybrid regime

After the end of the hybrid regime under the Prem premiership, Thai politics was brought back to "elected prime minister" model in 1988. However, the political conflicts persisted and the military was still so powerful in politics. It was widely accepted that no matter how the politics changed, the influence of the military still remained. As a result, the conflict between the Chatichai government and the military ended with a coup staged by the National Peace Keeping Council (NPKC) on 23 February 1991.[21]

The 1992 political crisis

Although there was an effort to establish the hybrid regime within Thai politics again through the establishment of the military party, the Seri Tham Party (Liberal Integrity Party), so that the military leaders would resume the government, the election result on 22 March 1992 was not as expected.[22] The winning party claimed that it was unable to form the government with the elected prime minister. Therefore, General Suchinda Kraprayoon, a class 5 graduate and the commander in chief of the army, was invited to be the prime minister, despite his announcement earlier that he would not accept this position. He gave the reason for doing so through his famous phrase "dishonest for the country".

This decision caused the widespread government resistance and led to popular protest. The government dispatched some troops to disperse the protest during 17–19 May 1992 but it was a political defeat. The May 1992 crisis led to the establishment of a new government accompanied by the changing perspective inside the military. Every faction seemed to accept that it was time to bring Thai politics back on the democratic path. Meanwhile, the political orientation at the international level in the post–Cold War era moved in that direction, or it was "the third wave of democracy" during that time and the politics in Thailand followed that momentum.[23] In this circumstance, the political transition was expected to strengthen the democratic regime in the near future. The democratic trend was blooming so much that no one thought the coups would return to Thai politics.

The turmoil

Thai politics moved along the democratic path together with the 1997 constitution drafting process.[24] This constitution, a significant turning point of the country's constitutional and political history, was accepted as "the people's constitution" and was an expectation of democratisation in Thailand. The politics confronted many challenges due to the occurrence of the newcomer political party, the Thai Rak Thai Party (Thais Love Thais, or TRT). This party was headed by Police Lieutenant Colonel Thaksin Shinawatra and won the landslide election on 6 January 2001. The Thaksin government had new policy orientations and strategies. The TRT's so-called populist policies that were the important linkage between the government and the rural masses and the administrative style of the new prime minister concerned the traditional elites, the conservatives, and also the bureaucrats, especially the military, who distrusted this party's intention and began to perceive the TRT as a threat.

Therefore, when the political conflict emerged, the right wing decided to establish the People's Alliance for Democracy (PAD), symbolised by the "Yellow Shirts", to counter and protest against the government.[25] The massive PAD-led street protests led to political deadlock and paved the way for military intervention in politics. On 19 September 2006, a coup led by General Sonthi Boonyaratglin successfully overthrew the civilian elected government. The coup was the first since the military's defeat in the May 1992 incident. The coup returned to Thai politics only 14 years after 1992 and contributed significantly to the "military era". Thai politics was opposed to the global political trend which admitted democracy as the main political value.

The country was back to elections soon afterward, however, it was estimated that Thai politics would be unstable as the military was still a stronghold in politics and the elected civilian government was contemporary. Besides, when the new round of political conflict emerged in 2013, the conservatives contributing to the 2006 coup organised the "People's Committee for Absolute Democracy with the King as Head of State" (PCAD), also known as the "People's Democratic Reform Committee (PDRC)", aiming to abolish Thaksin's influence from politics. This clique clearly expressed the "anti-democratic standpoint" and "supported the military".

The massive protests paved the way for another coup — the second in a decade – on 22 May 2014 by the National Council for Peace and Order (NCPO). Unbelievably, there were two coups within eight years and they indicated that the political transitions since 1992 were failure. The political transition in 1992 was terminated by the 2006 coup and the coup repeated in 2014. Some scholar suggested calling these coups "the twin coups" as they were staged by the same group within the army and they had the same purpose: to eradicate the influence of Thaksin and the pro-Thaksin party from Thai politics.[26] Furthermore, these coups not only signal failures of political transitions but also indicated that the military was still the major actor in politics. At the same time, the military's role in politics, which was extended after the last two coups, is a crucial obstacle in political transition.

The election without democracy

The NCPO government organised an election on 24 March 2019. It was an election without democracy, as are hybrid regimes in many countries. The election became the instrument to create political legitimacy for the authoritarian regime. The election without democracy also indicated that the political competition would be under the order and regulation of the military. In other words, the military leaders would exploit the political rules and regulations to serve their political advantage. The current military leaders expect to bring Thai politics in the near future back to the hybrid regime through various mechanisms. For example:

1 The creation of the undemocratic constitution, which implies the hybrid regime after the election.
2 The enactment of political rules and regulations that limit/restrict activities of opposition parties.
3 Under a provisional clause of the current constitution, it states that during the initial period, all 250 members of the first Senate will be appointed by the king upon the advice of the NCPO.[27] They will be important actors to nominate and select the prime minister in the future.

4 The approval of the 20-year National Strategy will be the political restriction for the future elected government.[28] The new government will be unable to carry its own policies and strategies, but has to follow the 20-year plan inevitably. On the other hand, the NCPO could maintain its role for the next 20 years.
5 The amendment of internal security legislation and the expansion of power and responsibilities the Internal Security Operations Command (ISOC), the anti-communism tool during the Cold War period. As a result, the ISOC will be like the "Ministry of the Military" overlapping in the administrative structure. The regime's decision indicated the expansion of military power, which is usually extended at wartime when its missions may overlap civilian work.[29]
6 The establishment of the military regime party is necessary in order to succeed the military regime. This move is inevitable, as the military is essential to maintaining the hybrid regime and the military leaders' role in politics in the future.
7 The use of political propaganda to promote the image of the military leaders to assure that these leaders are the best choices. The psychological operations are applied to represent that the military leaders are more effective than the civilian politician.
8 The existence of the undemocratic regime must itself rely on "anti-democracy discourse". It is obvious that the NCPO government tries to promote new policies, for example, the civil state policy integrating Thai-style nationalism and the government's political and economic policies. It is seen as the countermeasure of the populist policy and democracy. On the other hand, the military government makes an effort to represent its effectiveness to controlling political conflict in the society, regardless of freedom and liberty.
9 The expansion of the "anti-politics ideology" comprising the popular distaste of political parties and politicians.[30] The idea aims to seek support for the military leader to resume the premiership. The military leader is the best choice amid the antagonism towards the political parties and the politicians. The military leader will be the "middleman" prime minister under the hybrid regime.
10 The characteristics of the political transition in the future still witness the role and the influence of the military in Thai politics. After the election, the military will definitely be "a major political actor" and the political transition under the guidance of the military will lead to the "protected democracy" as the model of Latin American politics.

As the aforementioned elements suggest, Thai politics in the post-NCPO era reflects the intention of the military leaders to bring the politics back to the hybrid regime. Nevertheless, the political attentiveness soars higher and higher, in parallel with the NCPO's failure in many aspects. Under this condition, the NCPO's expectation to establish the hybrid regime after the election is uncertain, and it is unclear what the characteristics of the regime will be. It is worth mentioning that, historically, the hybrid regime was designed by Kriangsak and successfully implemented by General Prem.

Conclusion

Either in theory or practice, it is clear that the hybrid regime is designed to halt the political transition process in the "grey zone" in order to set up the "in-between politics". That is, the new regime will not be either the full democratic government or the absolute authoritarian government. In fact, this kind of regime leads to authoritarian persistence in politics, and it is rather difficult to achieve democratic consolidation as "democracy is not the only game in town" in Thailand.

In the Thai context, the hybrid regime is the most desirable model for the military leaders who want to stay in power in the post-transition politics. The pattern of regime helps provide legitimacy and build political image for them. As always, the hybrid regime looks better than the military government. Besides, the military regime is the "uncivilised political product" that cannot draw any support, either domestically or internationally. As a result, the hybrid regime is the rational choice for the military leaders in this case.

Besides, most of the political transitions in Thailand are initiated by the military, which allows the military leaders to maintain their prerogative in the post-transition politics. Therefore, the Thai military can maintain its role in some way or another. Although there were military regime breakdowns in 1973 and 1992, it does not mean that the process of military withdrawal from politics has been accomplished. Interestingly, the military as a political institution has survived. The concept of military return to the barracks is problematic in this country. Then it is very difficult and complicated to apply the concept of civilian control to Thai politics.

For instance, the 1955 transition (the Phibun Era), the 1957 transition (the Sarit Era), and the 1968 transition (the Thanom Era) indicate the major role of the military in these processes. It is true that the political process in these periods help maintain military power in politics. But at the same time, it decreases political pressure by allowing elections or political competition under military guidance. It is the kind of democracy which is protected by the military, as we have seen in Latin America. The condition of "protected democracy" gives an opportunity for the military to control the game. In other words, it allows the military to control politics. And a new government which is the product of military-controlled transition has to depend on military support. Therefore, either one of the military leaders or an individual selected by these leaders can become the prime minister. And the military has to accept the election since it is a legitimate tool to bring the military to power. As a result, the hybrid regime in Thai politics is a transition from one military regime to the other. Moreover, it provides a political space for the military in post-transition politics.

In order to guarantee the regime survival, the military has to create political as well as legal mechanisms, such as the constitution and the organic law, to promote the military role in politics. Moreover, there are two additional conditions: (1) the military has to develop the capability to control political conflicts in the parliament and (2) the military has to develop the capability to control political conflicts or factional conflicts inside the armed forces. Without these two capabilities, the hybrid regime will come to an end, as we have seen with evidence from the Phibun, Sarit, and Thanom eras.

Therefore, the fall of the Thai hybrid regime does not stem from economic conditions or from international pressures. Rather, it depends on the military leadership to control the army and the parliament. Without them, the regime will break down and end up with another military coup. That is the reason why the Prem hybrid regime could survive. The Prem regime was the most effective semi-democratic government since he had developed his leadership to control the military and the parliament simultaneously. His regime lasted for eight years, from 1980 to 1988, long enough to be labeled as "the golden age of the half-democratic regime". It is clear that the military could be in power in a parliamentary system, not in a military government system. Eventually, this pattern indicates that the Thai hybrid is the military regime in a parliament.

For the future, the 2014 coup-makers tried to establish a hybrid regime through the 2017 constitution with the high expectations that this constitution framework will facilitate the military to control the politics effectively. Besides the constitution, the coup government designs the ISOC to strengthen the military role in the post-transition regime proclaiming that the military will bring peace, order, and stability back to the country. And these are military missions, not political. The current military leaders hope they can turn Thai politics back to the golden age of the hybrid regime once again.

Notes

1 Chai-Anan Samudavanija and Sukhumbhand Paribatra, "Thailand: Liberalization without Democracy", in *Driven by Growth: Political Change in the Asia-Pacific Region*, edited by James W. Morley (New York: M.E. Sharpe, Inc.), p. 126.
2 Suchit Bunbongkarn, *The Military in Thai Politics, 1981–86* (Singapore: Institute of Southeast Asian Studies, 1987), p. 2.
3 Kobkua Suwannathat-Pian, *Kings, Country and Constitutions: Thailand's Political Development 1932–2000* (New York: Routledge, 2013), p. 46.
4 Sukit Nimmanahemin was a leader of the Sahaphum Party, while Sanguan Chantharasakha, Sarit's half-brother, was a secretary-general.
5 Frank C. Darling, "Marshal Sarit and Absolutist Rule in Thailand", *Pacific Affairs*, Vol. 33, No. 4 (December 1960), p. 349.
6 Prachuap Amphasawet, *Radical Change: The History of Thai Politics from June 24, 1932-October 14, 1973.* [Phlik Phaendin: Prawat KanmueangThai 24 Mithunayon 2475-14 Tulakhom2516] (Bangkok: Sukkhaphap Chai, 2000), p. 457.
7 Thak, *Thailand: The Politics of Despotic Paternalism* (Chiang Mai: Silkworm Books, 2007), p. 116.
8 Darling, "Marshal Sarit and Absolutist Rule in Thailand", p. 351.
9 Donald E. Nuechterlein, "Thailand after Sarit", *Asian Survey*, Vol. 4, No. 5 (May 1964), p. 846.
10 See the communist expansion and the government's countermeasure in Phinyo Watcharathet, Colonel, *The Prevention and the Suppression of Communism in Thailand* [Kan Phatthana Kan Pongkan Lae Prappram Khommionit Nai Prathet Thai], Thesis, Royal Thai Army War College, 1968 [in Thai], Counter-Revolutionary Warfare Committee, Department of Joint and Combined Operations, US Army Command and General Staff College, *A Case Study of Thailand's Counterinsurgency Operations, 1965–1982*; Kanok Wongtrangan, *Change and Persistence in Thai Counter-Insurgency Policy*, ISIS Occasional Paper, Institute of Security & International Studies, Chulalongkorn University, 1983; Saiyud Kerdphol, *The Struggle for Thailand: Counter-Insurgency, 1965–1985* (Bangkok: S. Research Center Co., 1986).
11 Clair Apodaca and Michael Stohl, "United States Human Rights Policy and Foreign Assistance", *International Studies Quarterly*, Vol. 43, No. 1 (1 March 1999), pp. 185–198.
12 The coup on 20 October 1977 was headed by General Kriangsak Chamanan, a supreme commander. See Surachart Bamrungsuk, *From Dominance to Power Sharing: The Military and Politics in Thailand, 1973–1992*, PhD dissertation, the Graduate School of Arts and Science, Columbia University, 1999, p. 14.
13 Surachart, *From Dominance to Power Sharing: The Military and Politics in Thailand, 1973–1992,* p. 16.
14 Ibid., pp. 106–107.
15 See this power transition in *Prem the Stateman* (Bangkok: The General Prem Tinsulanonda Foundation, 1995), pp. 107–112.
16 See the details of this rebellion in Larry A. Niksch, "Thailand in 1981: The Prem Government Feels the Heat", *Asian Survey*, Vol. 22, No. 2 (February 1982), pp. 191–199.
17 See Chai-Anan Samudavanija, Kusuma Snitwongse and Suchit Bunbongkarn, *From Armed Suppression to Political Offensive: Attitudinal Transformation of Thai Military Officers since 1976* (Bangkok: Institute of Security and International Studies, Faculty of Political Science, Chulalongkorn University, 1990).
18 The Royal Thai Army, *The Army in Forty Years* (Bangkok: O.S. Printing House, 1995), pp. 123–127 [in Thai].
19 Surachart Bamrungsuk, *Thailand's Security Policy Since the Invasion of Kamphuchea*, Essays on Strategy and Diplomacy, No. 10 (Clearmont: Keck Center for International Strategic Studies, Clearmont McKenna College, 1988).
20 See the end of Prem's government in Yos Santasombat, "The End of Premocracy in Thailand", *Southeast Asian Affairs* (1989), pp. 317–335.
21 More on this issue, see David Murray, *The Coup d'etat in Thailand, 23 February, 1991: Just Another Coup?* Occasional Paper, Indian Ocean Centre for Peace Studies, University of Western Australia, 1991.
22 For more details about the 1992 elections see Surin Maisrikrod, *Thailand's Two General Elections in 1992: Democracy Sustained* (Singapore: Institute of Southeast Asian, 1992).
23 Daniel E. King, "The Thai Parliamentary Elections of 1992: Return to Democracy in an Atypical Year", *Asian Survey*, Vol. 32, No. 12 (December 1992), pp. 1109–1123.
24 For more detail about the 1997 constitution drafting process please see Björn Dressel, "Thailand's Elusive Quest for a Workable Constitution, 1997–2007", *Contemporary Southeast Asia*, Vol. 31, No. 2 (2009), pp. 296–325.

25 For more details, see Oliver Pye and Wolfram Schaffar, "The 2006 Anti-Thaksin Movement in Thailand: An Analysis", *Journal of Contemporary Asia*, Vol. 38, No. 1 (2007), pp. 38–61.

26 Chris Baker, "The 2014 Thai Coup and Some Roots of Authoritarianism", *Journal of Contemporary Asia*, Vol. 46, No. 3 (2016), p. 389.

27 "Constitution of the Kingdom of Thailand", Unofficial translation from the Office of the Council of State website www.krisdika.go.th/wps/wcm/connect/d230f08040ee034ca306af7292cbe309/CONSTITUTION+OF+THE+KINGDOM+OF+THAILAND+%28B.E.+2560+%282017%29%29.pdf?MOD=AJPERES&CACHEID=d230f08040ee034ca306af7292cbe309 (accessed 12 June 2018).

28 Chatrudee Theparat, "Cabinet to Review 20-Year Strategy in May" <www.bangkokpost.com/news/politics/ 1443747/cabinet-to-review-20-year-strategy-in-may> (accessed 10 June 2018).

29 My interview with the press, Nattaya Chetchotiros, "Abhisit, Surachart at Odds on Security Law" <www.bangkokpost.com/news/general/1366467/abhisit-surachart-at-odds-on-security-law> (accessed 10 June 2018).

30 The anti-politics ideology originated from Latin American politics. For more details, see Brian Loveman and Thomas M. Davies (eds.), *The Politics of Antipolitics: The Military in Latin America* (Lincoln: University of Nebraska Press, 1978); Alain Rouquie, *The Military and the State in Latin America* (Berkeley: University of California Press, 1987); David Pion-Berlin (ed.), *Civil-Military Relations in Latin America: New Analytical Perspectives* (Chapel Hill: The University of North Carolina Press, 2001); Alfred Stepan, *Rethinking Military Politics: Brazil and the Southern Cone* (Princeton: Princeton University Press, 1988).

References

Brownlee, Jason. 20017. *Authoritarianism in an Age of Democratization*. Cambridge: Cambridge University Press.

Ezrow, Natasha M. and Erica Frantz. 2011. "Hybrid Dictatorships". In *Dictators and Dictatorships: Understanding Authoritarian Regimes and Their Leaders*. London: Continuum International Publishing Group.

Huntington, Samuel P. 1991. *The Third Wave: Democratization in the Late Twentieth Century*. London: University of Oklahoma Press.

Kobkua Suwannathat-Pian. 2013. *Kings, Country and Constitutions: Thailand's Political Development 1932–2000*. New York: Routledge.

Levitsky, Steven and Lucan A Way. 2010. *Competitive Authoritarianism: Hybrid Regimes after the Cold War*. Cambridge: Cambridge University Press.

Przeworski, Adam. 1991. *Democracy and the Market: Political and Economic Reforms in Eastern Europe and Latin America*. New York: Cambridge University Press.

Suchit Bunbongkarn. 1987. *The Military in Thai Politics, 1981–86*. Singapore: Institute of Southeast Asian Studies.

Surachart Bamrungsuk. 1988. *United States Foreign Policy and Thai Military Rule, 1947–1977*. Bangkok: Duang Kamol.

Thak Chaloemtiarana. 2007. *Thailand: The Politics of Despotic Paternalism*. Ithaca: Cornell Southeast Asia Program.

Zagorski, Paul W. 2009. *Comparative Politics: Continuity and Breakdown in the Contemporary World*. New York: Routledge.

8

SECURING AN ALTERNATIVE ARMY

The evolution of the Royal Thai Police[1]

Paul Chambers

With approximately 230,000 officials – slightly less than the number of army personnel – the Royal Thai Police (and its paramilitaries) is nevertheless an overlooked security actor in a country long dominated by monarchy and military, with the latter as junior partner.[2] Overall, the mission of the police has been to maintain and enforce law and order, engage in counter-insurgency efforts alongside the military, and protect the royal family. Though the police might be seen as Thailand's fourth armed force (after the army, navy, and air force), the fact is that the police are only slightly smaller than the army.[3] As a result, police have often acted as a sort of alternative army in competing with the military for power. However, despite a long-standing power rivalry, except when the military has experienced internal crises, it has always prevailed against the police. This chapter examines the evolution of Thailand's police as security providers for ruling regimes in which they have also been competitive with or controlled by the military.

Yet how has the Royal Thai Police evolved across time? What has been the character of police-military relations? What is the state of police influence in Thailand's post-2014 era? This chapter first argues that the police, as a massive, quasi-independent security institution, have demonstrated a capacity to behave as an alternative army which has tended to fiercely resist military domination. As a result, police-military rivalry has often degenerated into military attempts to control the police. Second, the chapter contends that the monarchy has generally favoured the army over the police, while the police have more often found their patrons in powerful elected civilians.

Evolution of the police as an alternative army

Until the mid-1800s, the Royal Siamese Armed Forces were charged with national defense as well as basic law enforcement. When soldiers went off to war, informal police units were created to keep order in their stead. Such police remained under the official control of the army. Under King Mongkut, reforms in law enforcement were enacted, with a rudimentary police force created in 1860. Mongkut's successor, King Chulalongkorn, continued the restructuring, and in 1897 a Police Patrol Unit and Royal Siamese Provincial Gendarmerie Department were separately created. The former was specifically tasked with urban security, while the latter was charged with safeguarding the kingdom's frontiers and railways. In 1915, King Vajiravudh merged the Police Patrol Unit and Gendarmerie into the Gendarmerie and Patrol Department

(GPD), also combining the two units' academies into one school.[4] By the end of the 1920s, the police had become the security tool of the Interior Ministry, just as the Royal Siamese Armed Forces were the mechanism of the Defense Ministry.

Following the 1932 overthrow of monarchical absolutism, royalists in the police were extirpated and it was renamed the Thai National Police Department (TNPD), though the unit remained under the Ministry of Interior. From 1933 until 1944, with the Thai military dominating the Khana Ratsadon regime, the police became the armed forces' junior partner in terms of administration. The period 1938–1944 especially saw the expansion of a military-police state, though it was personally dominated by Field Marshal Phibun Songkhram. Army General Adul Decharat served as police chief (and deputy prime minister) from 1936 until 1945. He also personalised his control over the police in the manner of a "Gestapo chief", was a clandestine leader of the insurgent Seri Thai, and in 1944 withdrew police backing for Phibun's martial law regime.[5]

Thailand's brief democracy in 1946–1947 witnessed attempts at police and army restructuring. By 1946, Adul had moved from the police to become the first ex-police chief to serve as army commander (also agreeing to become Pridi Banomyong's next appointed prime minister). Yet Adul's affinity among police and weak support within the army enabled mid-ranking military personnel to successfully carry out a 1947 coup, which allowed for what the coup leaders thought would be the resurrection of a military-dominant regime (in which police were co-opted) across the country.

However, the next nine years witnessed a period of competition for power between the army and the police. Police clout became personalised under another army general, Phao Sriyanond. Married to the daughter of Army Commander General Phin Choonhavan, Phao rapidly advanced to the rank of deputy police commander in 1948 and then police chief (and deputy interior minister) in 1951. With US financial assistance and sponsorship, Phao's TNPD grew into a formidable security apparatus, eventually rivaling the army itself.[6] As Phao's power grew, police finances skyrocketed to a point comparable to army funding. Phao also became the preferred conduit of US assistance to Thailand. In 1951, the US Central Intelligence Agency (CIA) organised a new and specialised paramilitary police called the Gendarmerie Patrol Force (GPF), under the Provincial Police (PP) of the TNPD. In 1955, the GPF was integrated with the recently formed Border Defense Police to become the Border Patrol Police (BPP). The BPP's mission included assisting the army to guarantee border security, intelligence, counter-insurgency, anti-smuggling, and anti-banditry, and giving security training to border villagers. Civic action programmes were a principal duty of the BPP. Designed to gain the trust of local peoples, the BPP built and operated numerous structures in rural areas.[7] Phao and the CIA also created a Police Aerial Reinforcement Unit (PARU), responsible for supporting special airborne missions. In 1954, the Ministry of Interior established the Volunteer Defense Corps (VDC), a civilian militia under the authority of the GPF but based upon the 1911–1925 Wild Tiger Corps. The VDC was trained, armed, and informally overseen by the BPP.

A 1957 coup led by Army Commander Sarit forced Phao and Prime Minister Phibun into exile, but it also served to cement army control over the police, as Sarit's loyal army officers assumed control over leading police positions. Junta leader Sarit ordered the transfer of police weaponry to the army, personally served as police chief from 1959 until 1963, permitted only three police officials to serve in the new 120-member appointed Senate, and diminished the size of the police budget relative to the army.[8] Suspicious of the BPP and overall US backing for the TNPD, he saw to it that most BPP commanders became ex-army officers, which gave the army informal influence over the BPP, with the BPP becoming "virtually independent of PP operational control".[9] Only pressure from the United States and the palace prevented Sarit Thanarat from disbanding the BPP altogether.[10]

Following Sarit's death in 1963, the new police chief, Army General Prasert Ruchirawong, sought to remain outside the shadow of the military regime. Prasert ostensibly answered to General Praphas Charusatien, who simultaneously served as deputy prime minister, interior minister, and army commander and dominated the Communist Suppression Operations Command (CSOC). The latter organisation, formed in 1965, merged civilians and the police under an army-led counter-insurgency structure. CSOC was renamed ISOC (Internal Security Operations Command) in 1974. Prasert, serving as police chief until 1972, managed to insulate the TNPD from army control with buoyant budgetary allocations, though receiving less funds than the army. By 1970, Praphas and Prasert had become bitter bureaucratic foes. The 1971 army-supported coup was partly meant to weaken Prasert so that Praphas and the army could solidify control over the police.[11] Praphas undoubtedly thought he had secured such supremacy when he replaced Prasert as police chief in 1972. However, rivalry persisted between Praphas and Prasert's ally, Deputy Police Chief General Prachuab Suntrangkul, in cooperation with the new army chief, General Krit Sivara. Krit and Prachuab refused to support the junta in 1973[12] and it was forced from power.[13] Thereupon, the two army generals respectively served as army and police chiefs from 1973 until 1975, backing Thailand's limited electoral democracy.[14] However, Prachuab's and Vitoon's 1975 resignations and Krit's sudden death in April 1976 strengthened army factions supportive of another coup, which ultimately occurred on 6 October 1976, hours after BPP officials and vigilantes engaged in the bloody massacre of student protestors at Thammasat University.[15] The palace-endorsed putsch that evening produced a junta possessing only one police official.

Nevertheless, the TNPD did not become completely overshadowed by the armed forces. This was owed to the 1978 Police Service Regulation Act. Meant to enhance police efficiency and autonomy, the act brought the police under the control of a newly created Police Commission (PC), which was dominated by the Interior Ministry and the police director general. Meanwhile, King Bhumibol and the Princess Mother remained committed patrons of the BPP (which by extension gave royal legitimacy to the TNPD), and they closely backed the BPP's tasks of combating narcotics, providing civic action, and supporting royal projects along Thailand's borderlands. In 1978 at the behest of the king, the TNPD created the Office of the Royal Court Security Police, with palace confidant Police General Vasit Dejkunchorn becoming its first commander. Such royal favouritism "arouse[d] military rivalry and suspicion".[16] In 1981, following a bloody coup attempt, Prime Minister General Prem Tinsulanonda purged several senior security officials, including Police Chief General Monchai. Under Prem, the army sought greater control over the police. In addition to ISOC, a new Capital Security Command (CSC), inaugurated in 1976, was upgraded in 1981 to give the army commander (and First Army chief) direct police powers over Bangkok and nearby areas.[17] Nevertheless, simply the fact that the police remained a bureaucracy separate from military control enabled it to continue enjoying substantial independence. Police General Narong Mahanon, a close Prem associate who headed the TNPD from 1982 until 1988, succeeded in personalising his own influence over the police during that time. As prime minister from 1980 until 1988, Prem tabled proposals to place police under the Office of the Prime Minister and "demilitarise the police hierarchy".[18] However, amid strong opposition from the Ministry of Interior and the TNPD, the reforms did not become law.[19]

Following the 1991 coup, junta leaders General Sunthorn Kongsompong and General Suchinda Kraprayoon attempted once and for all to establish army control over the police, eliminating TNPD autonomy. Their plan was to remove the police from under the Ministry of Interior and formally subordinate it directly under the Ministry of Defense and the Armed Forces Supreme Headquarters. However, the plan was dropped following shrill resistance to this proposal from law enforcement officials.[20] Months later, in May 1992, the junta was forced from power and the army lost its control over the CSC, thus depriving it of authority over Bangkok police.

Under Thailand's elected governments of 1992 until 1997, there were renewed efforts to reform the police. The Chuan Leekpai government (1992–1995) enacted the 1993 Decree to Restructure the Police Department. However, this decree was only meant to resolve a "conflict of interest within the coalition government" so that each coalition partner could place their favoured police officials in key positions.[21] Meanwhile, the successor Banharn and Chavalit governments were not in office long enough to have sufficient time to reform the police.

Only after 1997 did efforts commence towards more durable police reforms. Following the 1997 Asian Financial Crisis, popular pressure grew to reform the civil service, much of which was considered to be corrupt, overly bureaucratic, and partisan. Reforming the police was a specific priority. In 1998, the civilian Chuan government successfully enacted a Royal Decree to Establish the Royal Thai Police. The decree took the TNPD from under the Ministry of Interior and made it an independent governmental agency under the direct supervision of the prime minister. It was renamed the Royal Thai Police (RTP). Police commanders were henceforth called commissioner-generals rather than director-generals. The reform diminished the levels of bureaucracy between the prime minister and police, gave the prime minister closer monitoring authority over the police, and diminished the clout of the Interior Ministry.

Thailand's police after the advent of Thaksin

Following the 2001 election of Thaksin Shinawatra, the new prime minister's power over the police intensified dramatically. The first former policeman to become prime minister, Thaksin was intent on personalising control over the police and morphing it into a political agent of his prime ministerial authority. He attempted to accomplish this objective using three strategies.

First, as a former police lieutenant colonel himself, Thaksin identified with the police and sought to promote police interests. Thus, he gave the police a larger role in security affairs and enhanced the police portion of the national budget. In 2001, he diminished the army's role over security in Thailand's Deep South borderlands, placing the Royal Thai Police in charge. As part of this policy shift, Thaksin dissolved the Southern Border Provinces Administrative Centre, which had been created by the army in 1981. Thaksin also used the police to spearhead highly popular law-and-order campaigns such as the 2003 "war on drugs" and crackdowns on organised crime. As police roles expanded, so did their funding. As shown in Appendix 8.1, Thaksin dramatically expanded the police budget from 42 billion baht (US$1.33 billion) in 2001 to 51 billion baht (US$1.62 billion) in 2006. From 2001 until 2006, the number of ex-police cabinet ministers had increased from one to four.

Second, he appointed personal loyalists to leading police positions based upon police academy, kinship, or other related linkages. Several of his appointees, such as Police Generals Surasit Sangkapong, Suchart Muankaeo, Wongkot Maneerin, and Jumpol Manmai, had graduated with Thaksin in Police Academy class 26. Thaksin's wife Pojaman was from a leading police family, and Thaksin saw to it that her brother, Police General Priewpan Damapong, rapidly ascended the police hierarchy. Other police appointees, such as General Chitchai Wannasathit, were simply close to the Shinawatra family.[22]

Third, Thaksin enshrined new laws which granted greater prime ministerial power over the police and potentially gave the police more security powers. Thus, in 2004, he implemented a new Royal Thai Police Act. The law integrated all previous police regulations together and placed the police solidly under the control of the prime minister who alone was responsible for executing the act. The act also gave police new responsibilities for assisting in national development projects, tasks which had heretofore been almost entirely assigned to the military. In addition, the act gave more power for the prime minister to influence police reshuffles, select the police commissioner-general, and manipulate her/his tenure. Never before had a prime minister possessed such a high degree of legal power over police reshuffles.[23] Another law enacted by Thaksin which enhanced prime ministerial power but also could increase police power itself was the Emergency Decree on Public Administration in a State of Emergency, implemented in 2005. The law, which regulates the maintenance of order, places the prime minister in charge of state-led security efforts, with approval also needed from the Council of Ministers, parliament and the Constitutional Court. Under the decree, the prime minister can choose to enforce a state of emergency through the use of police officials instead of military officials.[24]

With Thaksin and the Thai Rak Thai Party boosting the image, tasks, and funding of police, most police personnel became his avid supporters. Thaksin seemed to have transformed the police into a powerful guardian of democratising Thailand. But such was not to be. Had the insurgency in Thailand's Deep South not become more intense in 2004, then Thaksin's opponents, especially in the army, would likely have found it difficult to blame him for security problems. Such blame contributed in 2004–2005 to growing irritation at Thaksin in the palace and the resurrection of army control in Deep South security policy. The post-2004 army leadership was independent of Thaksin and willing to overthrow the government. With the monarch's endorsement, it eventually carried out a 2006 coup.

Recognising the extent to which Thaksin had successfully dominated and strengthened the police, the 2006–2008 junta immediately set out to weaken it and keep it under military control. This was difficult at first: the junta was unable to dislodge Police General Kowit Wattana (Thaksin's police commissioner-general since 2004) from power. Kowit (pre-cadet class 6, Police Academy class 22), a rumored favourite of then Crown Prince Vajiralongkorn, thus continued on in this post until his 2007 retirement while also becoming a deputy junta leader. Nevertheless, in 2006, the Surayud Chulanond government appointed a Police Reform Committee containing opponents of Thaksin and chaired by arch-royalist Police General Vasit Dejkunchorn. This committee's recommendations included placing civilians in greater control of police affairs, decentralising the police and the police budget to metropolitan and provincial bureaus (with police supervised by governors), and transferring several functions out of the police's authority.[25] A related recommendation was for police to be placed under the control of army regional commanders and ISOC regional commands.[26] However, amid enormous police resistance and the fact that Surayud's government only lasted until February 2008, none of these reform proposals was ever passed.

After an elected pro-Thaksin government took office in February 2008, Prime Minister Samak Sundaravej and Interior Minister Kowit Wattana sidelined police commissioner-general and former junta deputy leader, Police General Sereepisut Taemeeyaves, eventually replacing him with Police General Patcharawat Wongsuwan,[27] brother of ex-army chief (2003–2004) General Prawit Wongsuwan, who was seen as close to Pojaman Shinawatra and other pro-Thaksin politicians while also acceptable to the military. However, Patcharawat's appointment brought with it a lingering influence by the Wongsuwan faction.

Under the anti-Thaksin Abhisit Vejjajiva government (2008–2011), Vasit was again appointed to vet police reform, but this effort floundered again amid renewed pressure from powerful police. Only a Decree to Restructure the Royal Thai Police (2009) was implemented. This law expanded police divisions and bureaus, including one devoted to crowd control and another, the Technology Crime Suppression Division, tasked to deter computer-related lèse-majesté activities.

Abhisit's Democrat Party was no friend of Patcharawat, who they held responsible for the deaths and injuries of Yellow Shirt demonstrators during October 2008. When Patcharawat retired as police commissioner-general in September 2009, six eligible candidates could succeed him: Police Generals Priewpan Shinawatra, Wongkot Maneerin, Jumpol Manmai, Watcharapol Prasamrajkit, Pratheep Tanprasert and Wichean Photposri. The Democrats rejected the first three as being too close to Thaksin. Watcharapol was also rejected since he was a confidant of Patcharawat. Pratheep was favoured by Queen Sirikit and the Yellow Shirt movement. Wichean had previously commanded the Royal Court Security Police, graduated in pre-cadet academy class 12 (same as General Prayuth Chan-ocha) and Police Academy class 28 and was considered close to the Democrats. Abhisit tried to use his prime ministerial control over the police to ensure the selection of his favourite, Pratheep. However the committee voted to select Jumpol, who was close to then Crown Prince Vajiralongkorn.[28] Thus, Abhisit maintained Pratheep as the acting police commissioner-general until Pratheep's retirement in 2010, the same year that Jumpol retired. With Jumpol out of the picture, Abhisit successfully appointed Wichean to succeed Pratheep.

The 2011 landslide election of Pheu Thai's Yingluck Shinawatra resurrected the influence of pro-Thaksin senior police officials while enhancing the clout of police once again. Yingluck did nothing to change her brother's 2004 Police Reform Act. With Pheu Thai dominating the government, Wichean was transferred to make way for Thaksin's brother-in-law, Police General Priewpan Damapong to become police commissioner-general. Like Thaksin, he was a class 10 graduate of the pre-cadet academy. When Priewpan himself retired in 2012, Police General Adul Saengsingkaew succeeded him. Adul, Police Academy class 29, had long been an ally of the Democrats but was favoured by Thaksin because when pro-Thaksin Red Shirts rallied in Pattaya in 2009, Adul refused to repress them for fear of bloodshed. Adul was also proactive in promoting a more active role for police in a renewed war on drugs and in the Deep South counter-insurgency.[29]

Yingluck supported the police in other ways as well. Under her administration, the police budget grew from 76.1 billion baht in 2011 to 86.7 billion baht in 2014 (see Appendix 8.1). Yingluck also gave the police a US$200 million "emergency" fund, making it more financially competitive with the military.[30] Moreover, she maintained four ex-policemen in each of her cabinets with two taking the slots of deputy prime minister. Meanwhile, Pongsapat Pongcharoen, deputy police commissioner, was chosen by Thaksin to contest (unsuccessfully) the Bangkok gubernatorial election in 2013. Furthermore, the Yingluck government enhanced police power rather than that of the armed forces regarding domestic security operations. Indeed, when defending itself against anti-Yingluck demonstrators, the government used police rather than soldiers. In November 2012, the government applied the Internal Security Act against anti-Thaksin demonstrators but selected the police to carry out this task, not the army. Similarly, during the 2013–2014 People's Democratic Reform Committee (PDRC) demonstrations against the Yingluck government – a time during which the army refused to help keep order in Bangkok – Yingluck called upon Adul's police to enforce first the Internal Security Act and later the Emergency Decree to bring the demonstrations under control, retake occupied government buildings, and guard Yingluck's residence.[31] Yet Adul, who could have faced charges from the anti-Yingluck judiciary or military opposition (or coup) if he had used force, decided not to break up the demonstrations, which were partly guarded by plain-clothes military officials. In the end, the protests continued unabated until the 2014 coup, which overthrew the Pheu Thai government.

Thailand's police since the 2014 coup

Following the putsch, a principal objective of the newly established National Council for Peace and Order (NCPO) was to repeat the 2006–2008 junta's attempt to solidify military control over the police, considered an overly independent security service which was a bastion of pro-Thaksin sentiment. As in previous Thai juntas, the fact that only one police official (the police commissioner) became part of the NCPO directorate was already a signal that the police were to have diminished powers under the junta. Indeed, the NCPO announced shortly after the coup that it was "chief among the priorities" to rein in the police once and for all.[32] The strategies used for capping police power were threefold: first, personalising control over the police; second, enshrining laws over the police; and third, continuing to offer police a role in security operations as well as enhanced funding, though under the control of the army.

In terms of the first strategy, the deputy junta leader, retired General Prawit Wongsuwan, was determined to personalise control over the police by appointing police to senior positions who were loyal to him, part of the police faction belonging to his brother Patcharawat or close to PDRC leader Suthep Thaugsuban. The personalisation of junta power over the police involved a post-coup purge of its senior officers in the days after 22 May 2014. To begin, Police Commissioner-General Adul was replaced by Police General Watcharapol Prasamrajkit, a confidante of Patcharawat (though in the same Police Academy class 29 as Adul). Watcharapol served as acting police commissioner-general until his retirement at the end of September 2014. Second, Police General Tarit Piengdit of the Department of Special Investigations was replaced by Police General Chatchawal Suksomjit, who was trusted by the junta. Third, Metropolitan Police Bureau (MPB) Chief Police General Kamronwit Thoopkrachang was replaced by Police General Chakthip Chaijinda, who possessed close ties to the PDRC's Suthep but was also a friend of Prawit and Prayuth since they had cooperated in repressing the 2010 Red Shirt demonstrations. Third, Provincial Police Region 1 Chief Nares Nanthachote, considered close to Yingluck, was replaced by Police General Srivara Ransibrahmanakul, another confidante of Patcharawat. Fourth, Provincial Police Region 2 Chief Kawee Supanan (also considered close to Yingluck) was replaced by Police General Sanit Mahataworn who, like Chakthip, was close to Suthep. Meanwhile, the chiefs of Provincial Police Regions 4, 5 and 7 were replaced by Detnarong Suthichanbancha, Wanchai Thanadkit, and Somboon Huabbangyang, respectively, all three of whom possessed close connections with Suthep and Patcharawat.[33]

In August 2014, the retiring Watcharapol was succeeded by Police General Somyot Poonpanmuang as police chief until the latter's own retirement in 2015.[34] Formerly a deputy police chief and known as a pro-Abhisit staunch junta supporter and avid lèse-majesté enforcer, Somyot had directed the violent repression of anti-junta demonstrations shortly after the coup. He was a close friend of Patcharawat and Preecha Chan-ocha, brother of Prayuth.[35] Upon Somyot's retirement in 2015, Chakthip Chaijinda was selected to succeed him. Chakthip, a confidante of the Wongsuwan brothers, is a member of pre-cadet academy class 20. Class 20 is the same class as the leading army officers who were appointed in the 1 October 2018 military reshuffle, including newly posted army commander, General Apirat Kongsompong, who retires in 2020. Chakthip too could remain until his retirement in 2020. However, it has been speculated that Prayuth and Prawit might transfer him out early from the police commissioner-general position to make way for one of two pro-Prawit police deputy chiefs (Srivara Ransibrahmanakul or Chalermkiat Siworakhan) to succeed him.[36]

A second strategy which the junta has used to try to control the police was the enshrinement of laws over them. First, under Decree 88/2557 (2014), the NCPO modified the 2004 National Police Act, diminishing the power of elected officials on the Public Policy Board to choose police chiefs and conferring it to bureaucrats, especially military officials. Under the Police Policy Board's new structure, the prime minister now acted as board chairperson while a deputy prime minister chosen by the prime minister was to be deputy board chairperson. Also, the permanent minister for defense (who had never been a board member) now sat on the board alongside the permanent interior and justice ministers as well as the director of the Budget Bureau while the justice and interior ministers were removed. Meanwhile, where there had previously been four specialists on the board, there were now two, and they were to be selected by the Senate, the members of which were themselves selected by the junta. Only the incumbent national police commissioner-general could nominate a candidate to the board for consideration.[37] Previously, only the prime minister had this power. Ultimately, the new eight-member Public Policy Board includes a military member, three other bureaucrats likely to side with the military, two "experts" chosen by the pro-junta Senate and only two possible elected civilians.

Moreover, under Decree No. 89, only "senior police" previously serving as deputy national police chiefs or police inspectors-general would be qualified for the nomination to be national police commissioners-general.[38] By the time civilian rule returns to Thailand, any "senior police" would be the ones which the junta leadership has thoroughly vetted.

Following the criteria of Decrees No. 88 and 89, the retiring Police Commissioner-General Watcharapol was given the power to nominate his successor (unlike before the coup when the decision was made by a Police Commission). Thereupon, the nomination was sent to a six-member Police Policy Board dominated by the NCPO and military. Both Somyot and then Chakthip were selected as police commissioners-general in this manner. It is probable that the military will continue to influence police reshuffles given that these decrees will most likely persist in one form or another even after the return to civilian rule.

Another law which the NCPO used to boost its influence was Decree 13/2559 (2016), which bestows police powers under the Criminal Code and Criminal Procedures Code upon military officials, or military-directed paramilitaries, from the rank of sub-lieutenant and higher. Such powers include the authority to summon, arrest, and detain people suspected of illegalities without judicial due process of law on 27 different forms of crimes from robbery to loan sharking to prostitution.[39] The decree basically deputises soldiers as police, with the exception that it gives greater legal impunity to army officials as opposed to police. Indeed, the new decree even allows military officials to arrest police as part of law enforcement procedure. The official rationale for the decree is to enable military officers to assist police in countering crime. Yet with the law granting military (especially army) officers combat and policing powers, it legally allows them to prevail over police as the leading security officials in Thailand.

Yet another NCPO legal modification affecting police is with regard to the Internal Security Operations Command (ISOC). In 2017, NCPO chief Prayuth used the all-powerful section 44 of the 2014 constitution to amend the ISOC security law. The amendment grants ISOC veto authority over all bureaucrats at the provincial and regional levels, including over police. The modification fundamentally generates an army-dominant hierarchy over all other bureaucracies to help resolve any vaguely defined security predicaments in favour of the army.[40]

Meanwhile, the NCPO early on announced that it would enact a wide-ranging set of reforms pertaining to the police. Nevertheless, one mid-level police opined in 2017 that the junta was "just paying lip-service [since] they only care about staying in power and don't really care about reforming the force".[41] In mid-2017, a committee headed by Prayuth mentor, army General Boonsang Niempradit, was established to vet potential reforms. However, at the committee's conclusion in 2018, it decided not to decentralise the police, instead focusing on subtle changes in police authority while offering salary increases to police personnel. The committee was later criticised for being too friendly towards the police. By May 2018, a new police committee was recommending downsizing the police force and transferring "non-police missions to local administrative bodies".[42] Such a move would decrease the police budget and diminish its bureaucratic size. By the time the junta dissolved itself in July 2019, it had failed to pass any police reform bill. Yet such a failure might have been useful for the NCPO – though controlling the police was a junta goal, completely weakening the police could have alienated police officials from ever supporting the junta. As General Boonsrang reportedly said, "The boss [then General Prawit Wongsuwan who had been overseeing the Royal Thai Police] didn't want to pursue [reform] due to vested interests."[43]

The third strategy of the NCPO has been to offer police enhanced security roles while guaranteeing them continuing high levels of funding. First, the high-level role which the police enjoyed under the previous government in countering narcotics, tackling migration, and assisting the army in the Deep South counter-insurgency has persisted since the 2014 coup. For example, the NCPO has welcomed the use of forensic police against the Deep South insurrection.[44] Meanwhile, between 2014 and 2018, the police budget grew from 86.7 billion baht to 107 billion baht (see Appendix 8.1).

Ultimately, the NCPO sought to use a carrot-and-stick approach to control the police, including personalised power over reshuffles, laws to control the police, and continuing growth of police budgets. The mindset which the junta wanted to instill in police was perhaps best voiced by a police cadet in 2018: "The military gives the orders and uses force, while the police stick to law enforcement".[45] Indeed, since 2014 the police have enforced laws made by the military (junta) to perpetuate arch-royalist armed forces dominion across the country.

Conclusion

In 2014, the Royal Thai Police experienced a reign of terror. At least seven senior police officials were imprisoned while another died suspiciously. The purge of police involved allegations of lesè-majesté relating to the then crown prince's divorce from his third wife. It enabled Police Commissioner-General Somyot, charged with the investigation, to prove his arch-royalist hatchet-man credentials. Likewise, the NCPO contributed soldiers to the witch-hunt, demonstrating loyalty to the throne. By thoroughly supporting a palace purge of the police, the junta enthusiastically legitimised its preference for clamping down on the RTP. Though the purge occurred for personal royalist intentions, it also helped to satisfy the military that there would be no change in the monarchy/military power partnership across Thailand. After all, in 2016, at the beginning of King Vajiralongkorn's reign, the new sovereign depended upon the junta to help consolidate his hold on power. He thus acquiesced in its domination of the RTP, with the NCPO avidly supporting punishments for police who had lost royal favour.[46]

This chapter has argued that Thailand's police, historically Thailand's second leading security institution after the army, has itself tended to act as an alternative army and has often remained insulated from military influence. Such police resistance has led to enhanced police-military competition. From its inception in 1860, the police have been either controlled (e.g. pre-1932, 1957–1963, 2006–2008) or competitive (e.g. 1951–1957, 2001–2006, 2011–2014). Since the 2014 putsch, a stauncher effort has been made to legally cement junta and army power over the RTP. Meanwhile, monarchy has generally favoured the military over the police, while the police have more often found their patrons in powerful elected civilians. Given that it has almost always been coup-making military officials who have championed palace influence, while the police have traditionally been bureaucratic foes of the armed forces, the monarchy has tended to throw more support to the military. Police units (e.g. the BPP) and personages with palace connections have been few in number, though they have helped to maintain some police independence from the military. The military coups of 1957, 1958, 1971, 1991, 2006, and 2014 strengthened the monarchy and arch-royalist military while weakening the police. Powerful police-politicians, including Prasert Ruchirawong and Thaksin, tended to be less zealously arch-royalist. Since 2001, the police, instead of military officials, have tended to favour elected civilian governments rather than military rule. After all, Thaksin was an elected ex-policeman who appeared to support the police over the military regarding finance, political interests, and policy areas. Thus, most police favoured pro-Thaksin governments – elected governments – time and again. Since the 2014 coup, the military has sought to enshrine its supremacy over police as a method of sustaining the junta's control. As a result, for the near future, such military domination will persevere. Only if the palace or another strong, elected civilian prime minister decides to endorse greater police power will Thailand ever see the police again become an alternative army in competition with the army itself. However, with the military still strong, such a scenario might only provoke yet another coup.

Notes

1 The author would like to acknowledge the assistance of Napisa Waitoolkiat, Katsuyuki Takahashi, Poowin Bunyawejchewin, and Nithi Nuangjamnong.

2 Interpol, *Thailand: Royal Thai Police*, 2018 <www.interpol.int/Member-countries/Asia-South-Pacific/Thailand> (accessed 16 March 2018).

3 The far majority of police pre-cadets (like pre-cadets of the army, navy, and air force) attend Thailand's Armed Forces Academies Preparatory School before attending the Police Academy.

4 Eric Haanstad, "A Brief History of the Thai Police", in *Knights of the Realm: Thailand's Military and Police, Then and Now*, edited by Paul Chambers (Bangkok: White Lotus Press, 2013), p. 461.

5 Jayanta Kumar Ray, *Portraits of Thai Politics* (New Delhi: Orient Longman, 1972), p. 75.

6 Thak Chaloemtiarana, *Thailand: The Politics of Despotic Paternalism* (Chiangmai: Silkworm, 2007), p. 56.

7 See Sinae Hyun, "Mae Fah Luang: Thailand's Princess Mother and the Border Patrol Police during the Cold War", *Journal of Southeast Asian Studies*, Vol. 48 (October 2017), pp. 262–282.

8 Thak, *Thailand: The Politics of Despotic Paternalism*, p. 85.

9 United States, "Reorganisation of the Thai Armed Forces Security Center, Royal Thai Government, 22 October 1972, Report Number, 2222036672, Confidential (Declassified) <https://ia801306.us.archive.org/10/items/ThailandIntelligenceServicesINSCOMDossierZF400117W/Thailand-Intelligence-Services-ZF400117W.pdf> (accessed 10 April 2018), p. 124.

10 The BPP and PARU later engaged in CIA-sponsored anti-Communist military operations in Laos.

11 Suchit Bunbongkarn, "Political Institutions and Processes", in *Government and Politics in Thailand*, edited by Somsakdi Xuto (New York: Oxford University Press, 1987), p. 50.

12 Arch-royalist Police General Monchai Pongponchuen and his deputy Narong Mahanon ordered attacks on students. Both later served as police chief.

13 Kullada Kesboonchoo Mead, "The Cold War and Thai Democratisation", in *Southeast Asia and the Cold War*, edited by Albert Lau (London: Routledge, 2012), p. 230.

14 Army General Vitoon Yasawasdi, Commander of Thailand's CIA-supported elite ranger force in Laos also refused to support Thanom in 1973. He became Prachuab's deputy police chief that year.

15 The queen's favourite Monchai again participated in the anti-student violence. Two months later he was appointed police chief.

16 United States, *Thailand: Challenges to Political Stability*, National Foreign Assessment Center, Central Intelligence Agency, September, 1981 <www.cia.gov/library/readingroom/docs/CIA-RDP03T02547R000100240001-5.pdf> (accessed 4 February 2018), p. 1, footnote 1.

17 Suchit Bunbongkarn, "The Military in Thai Politics", in *Government and Politics in Thailand*, edited by Somsakdi Xuto (Singapore: ISEAS, 1987), p. 60.

18 Arisa Ratanapinsiri, "A History of Police Reform in Thailand", in *Knights of the Realm: Thailand's Military and Police, Then and Now*, edited by Paul Chambers (Bangkok: White Lotus Press, 2013), p. 502.

19 However, Prem's administration in 1983 established Special Operations Unit "Naresuan 261" under the BPP and PARU. Among 261's responsibilities are guarding royal family members when they travel. Following the accession of King Maha Vajiralongkorn, 261 has expanded.

20 United States Embassy Cable, "The New Push for Thai Police Reform, *Wikileaks*, 20 December 2012, 06BANGKOK7501_a <https://wikileaks.org/plusd/cables/06BANGKOK7501_a.html> (accessed 3 February 2010).

21 Arisa, "A History of Police Reform in Thailand", p. 504.

22 Duncan McCargo and Ukrist Pathmanand, *The Thaksinization of Thailand* (Copenhagen: NIAS, 2005), pp. 226–232.

23 Arisa, "A History of Police Reform in Thailand", pp. 513–514.

24 Section 5, Government of Thailand, Emergency Decree on Public Administration in a State of Emergency, 16 July 2005 <www.krisdika.go.th/wps/wcm/connect/4c7e12804152eac69414d593530f59d5/EMERGENCY+DECREE+ON+PUBLIC+ADMINISTRATION+IN+EMERGENCY+SITUATION%2C+B.E.+2548+%282005%29.pdf?MOD=AJPERES&CACHEID=4c7e12804152eac69414d593530f59d5> (accessed 2 March 2017).

25 Arisa, "A History of Police Reform in Thailand", p. 519.

26 This was the proposal of anti-Thaksin Police General Watcharapol Prasamrajkit. See United States, 2006.

27 Patcharawat graduated in pre-cadet class 9 and Police Academy class 25 (the same as Thaksin's ex-interior minister, Police Captain Purachai Piumsumbon, and ex-deputy prime minister, Police General Chidchai Vanasatidya).

28 United States Embassy Cable, "Thailand: Ambassador's Meeting with the Opposition Pheu Thai Party", *Wikileaks*, 21 September 2009, 09BANGKOK2402_a <https://wikileaks.org/plusd/cables/09BANGKOK2402_a.html> (accessed 2 April 2017).

29 Avudh Panananda, "PM out on a Limb with Adul Appointment". *The Nation*, 17 July 2012 <www.nationmultimedia.com/politics/PM-out-on-a-limb-with-Adul-appointment-30186305.html> (accessed 10 March 2018).

30 Paul Chambers, "Yingluck's Affinity with the Police and Thailand's Divided Security Sector", *China Policy Institute*, 2 February 2014 <https://cpianalysis.org/2014/02/25/yinglucks-affinity-with-the-police-and-thailands-divided-security-sector/> (accessed 4 February 2014).

31 *Phuket Gazette Thailand News*, "Army-police Rift Widens; Yingluck Ponders Martial Law; Rice Farmers Join PDRC; Rhino Horn for Hanoi Bagged", 21 January 2014 <https://thethaiger.com/thailand-news/Phuket-Gazette-Thailand-News-Armypolice-rift-widens-Yingluck-ponders-martial-law-Rice-farmers-join-PDRC-Rhino-horn-Hanoi-bagged> (accessed 30 October 2014).

32 Wassayos Ngamkha, "Regime 'Lacks Will' to Reform Police Force". *Bangkok Post*, 19 May 2016 <www.bangkokpost.com/news/politics/978817/regime-lacks-will-to-reform-police-force> (accessed 2 May 2017).

33 *ISN Hotnews*, "Police Shake-up Cuts Ties to Thaksin, Special Report: Junta Moves in Officers It Can Trust", 29 May 2014 <http://en.isnhotnews.com/?p=45676> (accessed 5 March 2017).

34 Somyot graduated from Police Academy class 31 and pre-cadet class 15, the same as then NCPO member General Paiboon Kamchaya.

35 "Staunch Junta Supporter Appointed Chief of Thai Police", *Khaosod*, 20 August 2014 <www.khaosodenglish.com/life/2014/08/20/1408523886/> (accessed 5 March 2017).

36 Sakda Samerphop, "Prayut Tipped to Put Competence over Friendship in Cabinet Reshuffle", *The Nation*, 9 November 2017 <www.nationmultimedia.com/detail/politics/30331163> (accessed 1 September 2016).

37 Decree 88/2557, 2014. "Amendments to the Laws on National Police".

38 Decree 89/2557, 2014. "Rules for Appointment of Police Officers".

39 Decree 13/2559, 2016. "The Prevention and Suppression of Certain Offenses that are a Danger to Public Order or Undermine the Social and Economic System of the Country".

40 Kasamakorn Chanwanpen, "Sinister Motive Seen in Move to Empower ISOC", *Bangkok Post*, 24 November 2017 <www.nationmultimedia.com/detail/politics/30332385> (accessed 2 January 2018).

41 *Prachatai*, "Decentralisation Key to Police Reform: Police Watch", 2 February 2017 <https://prachatai.com/english/node/6896> (accessed 3 April 2018).

42 "Police Reform Hits a Wall", *Bangkok Post*, 28 April 2018 <www.pressreader.com/thailand/bangkok-post/20180428/281831464345980> (accessed 2 May 2018).

43 Veera Prateepchaikul, "Not a Scintilla of Change in Police Force", *Bangkok Post*, 5 August 2019 <https://www.bangkokpost.com/opinion/opinion/1724635/not-a-scintilla-of-change-in-police-force> (accessed 13 August 2019).

44 *Thai PBS*, "Security Beefed up in Deep South Following Spate of Attacks, 30 April 2018 <http://englishnews.thaipbs.or.th/security-beefed-deep-south-following-spate-attacks/> (accessed 5 May 2018).

45 "Reformers Design Game Plan in Police Reform Rejig", *Bangkok Post*, 23 April 2018 <www.pressreader.com/thailand/bangkok post/20180423/281483571983619> (accessed 5 May 2018).

46 In early 2017, reports surfaced that ex-Police General Jumpol Manmai was detained in a prison inside a royal palace. He remains incognito.

References

Arisa Ratanapinsiri. 2013. "A History of Police Reform in Thailand". In *Knights of the Realm: Thailand's Military and Police, Then and Now*. Edited by Paul Chambers. Bangkok: White Lotus Press.

Avudh Panananda. 2012. "PM out on a Limb with Adul Appointment". The Nation, 17 July < www.nationmultimedia.com/politics/PM-out-on-a-limb-with-Adul-appointment-30186305.html> (accessed 17 March 2017).

Bangkok Post. 2018. "Police Reform Hits a Wall". 28 April < www.pressreader.com/thailand/bangkok-post/20180428/281831464345980> (accessed 30 April 2018).

———. 2018. "Reformers Design Game Plan in Police Reform Rejig". 23 April <www.pressreader.com/thailand/bangkok post/20180423/281483571983619>.

Chambers, Paul. 2014. Yingluck's Affinity with the Police and Thailand's Divided Security Sector". *China Policy Institute*, 2 February <https://cpianalysis.org/2014/02/25/yinglucks-affinity-with-the-police-and-thailands-divided-security-sector/>.

Decree 88/2557. 2014. "Amendments to the Laws on National Police".

Decree 89/2557. 2014. "Rules for Appointment of Police Officers".

Decree 13/2559. 2016. "The Prevention and Suppression of Certain Offenses that are a Danger to Public Order or Undermine the Social and Economic System of the Country".

Government of Thailand. 2005. "Emergency Decree on Public Administration in a State of Emergency". 16 July <www.krisdika.go.th/wps/wcm/connect/4c7e12804152eac69414d593530f59d5/EMERGENCY+DECREE+ON+PUBLIC+ADMINISTRATION+IN+EMERGENCY+SITUATION%2C+B.E.+2548+%282005%29.pdf?MOD=AJPERES&CACHEID=4c7e12804152eac69414d593530f59d5>.

Haanstad, Eric. 2013. "A Brief History of the Thai Police". In *Knights of the Realm: Thailand's Military and Police, Then and Now*. Edited by Paul Chambers. Bangkok: White Lotus Press.

Hyun, Sinae. 2017. "Mae Fah Luang: Thailand's Princess Mother and the Border Patrol Police during the Cold War". *Journal of Southeast Asian Studies*, Vol. 48 (October): 262–282.

Interpol. 2018. "Thailand: Royal Thai Police" <www.interpol.int/Member-countries/Asia-South-Pacific/Thailand>.

ISN Hotnews. 2014. "Police Shake-up Cuts Ties to Thaksin, Special Report: Junta Moves in Officers it can Trust". 29 May <http://en.isnhotnews.com/?p=45676>.

Kasamakorn Chanwanpen. 2017. "Sinister Motive Seen in Move to Empower ISOC". *Bangkok Post*, 24 November <www.nationmultimedia.com/detail/politics/30332385>.

Khaosod. 2014. "Staunch Junta Supporter Appointed Chief of Thai Police". 20 August <www.khaosodenglish.com/life/2014/08/20/1408523886/>.

McCargo, Duncan and Ukrist Pathmanand. 2005. *The Thaksinization of Thailand*. Copenhagen: NIAS.

Mead, Kullada Kesboonchoo. 2012. "The Cold War and Thai Democratisation". In *Southeast Asia and the Cold War*. Edited by Albert Lau. London: Routledge.

Phuket Gazette Thailand News. 2014. "Army-police Rift Widens; Yingluck Ponders Martial Law; Rice Farmers Join PDRC; Rhino Horn for Hanoi Bagged". 21 January <https://thethaiger.com/thailand-news/Phuket-Gazette-Thailand-News-Armypolice-rift-widens-Yingluck-ponders-martial-law-Rice-farmers-join-PDRC-Rhino-horn-Hanoi-bagged>.

Prachatai. 2017. "Decentralisation Key to Police Reform: Police Watch". 2 February <https://prachatai.com/english/node/6896>.

Ray, Jayanta Kumar. 1972. *Portraits of Thai Politics*. New Delhi: Orient Longman.

Sakda Samerphop. 2017. "Prayuth Tipped to Put Competence over Friendship in Cabinet Reshuffle". *The Nation*, 9 November <www.nationmultimedia.com/detail/politics/30331163>.

Suchit Bunbongkarn. 1987. "Political Institutions and Processes". In *Government and Politics in Thailand*. Edited by Somsakdi Xuto. New York: Oxford University Press, 1987.

———. 1987a. *The Military in Thai Politics*. Singapore: ISEAS.

Thai PBS. 2018. "Security Beefed up in Deep South Following Spate of Attacks". 30 April. <http://englishnews.thaipbs.or.th/security-beefed-deep-south-following-spate-attacks/>.

Thak Chaloemtiarana. 2007. *Thailand: The Politics of Despotic Paternalism*. Chiangmai: Silkworm.

United States Embassy Cable. 2009. "Thailand: Ambassador's Meeting with the Opposition Pheu Thai Party". *Wikileaks*, 21 September, 09BANGKOK2402_a <https://wikileaks.org/plusd/cables/09BANGKOK2402_a.html>.

———. 2006. "The New Push for Thai Police Reform". *Wikileaks*, 20 December, 06BANGKOK7501_a <https://wikileaks.org/plusd/cables/06BANGKOK7501_a.html>.

United States. 1972. "Reorganisation of the Thai Armed Forces Security Center, Royal Thai Government". 22 October, Report Number, 2222036672, Confidential (Declassified) <https://ia801306.us.archive.org/10/items/ThailandIntelligenceServicesINSCOMDossierZF400117W/Thailand-Intelligence-Services-ZF400117W.pdf>.

———. 1981. "Thailand: Challenges to Political Stability". National Foreign Assessment Center, Central Intelligence Agency, September <www.cia.gov/library/readingroom/docs/CIA-RDP03T02547R000100240001-5.pdf>.

Wassayos Ngamkha. 2016. "Regime 'Lacks Will' to Reform Police Force". *Bangkok Post*. 19 May <www.bangkokpost.com/news/politics/978817/regime-lacks-will-to-reform-police-force>.

Appendix 8.1

Table 8.1 Thai police budgeting (1957–2018) in Thai baht

Year	*Police Budget*	*MOD Bd*	*Security Bd*	*National Budget*	*%PlcBd to Nat'l Bd*	*%PlcBd to MOD Bd*	*%Plc Bd to Sec Bd*
1957	200,725,181	758,463,193	959,188,374	5,069,990,082	3.96%	26.5%	20.9%
1958	NA	NA	NA	NA	0.0%	0.0%	0.0%
1959	NA	NA	NA	NA	0.0%	0.0%	0.0%
1960	540,000,000	1,360,000,000	1,900,000,000	7,700,000,000	7.01%	39.7%	28.4%
1961	428,200,000	1,080,800,000	1,509,000,000	6,660,000,000	6.43%	39.6%	28.4%
1962	605,170,000	1,500,500,000	2,105,670,000	8,800,000,000	6.88%	40.3%	28.7%
1963	610,170,000	1,621,800,000	2,231,970,000	10,380,000,000	5.88%	37.6%	27.3%
1964	639,400,000	1,760,700,000	2,400,100,000	11,430,000,000	5.59%	36.3%	26.6%
1965	729,780,000	1,922,310,000	2,652,090,000	12,420,000,000	5.88%	38.0%	27.5%
1966	790,280,000	2,147,087,000	2,937,367,000	14,440,000,000	5.47%	36.8%	26.9%
1967	900,000,000	2,518,000,000	3,418,000,000	18,480,000,000	4.87%	35.7%	26.3%
1968	1,080,000,000	3,264,000,000	4,344,000,000	21,262,000,000	5.08%	33.1%	24.9%
1969	1,267,000,000	3,772,350,000	5,039,350,000	23,960,000,000	5.29%	33.6%	25.1%
1970	1,434,979,700	4,645,938,000	6,080,917,700	27,299,889,100	5.26%	30.9%	23.6%
1971	1,469,498,400	5,068,141,000	6,537,639,400	28,645,000,000	5.13%	29.0%	22.5%
1972	1,694,250,000	5,268,100,000	6,962,350,000	29,000,000,000	5.84%	32.2%	24.3%
1973	1,900,000,000	5,754,872,000	7,654,872,000	31,600,000,000	6.01%	33.0%	24.8%
1974	2,009,543,000	6,318,000,000	8,327,543,000	36,000,000,000	5.58%	31.8%	24.1%
1975	2,426,750,000	7,710,000,000	10,136,750,000	48,000,000,000	5.06%	31.5%	23.9%
1976	2,845,285,260	9,823,119,500	12,668,404,760	62,650,000,000	4.54%	29.0%	22.5%
1977	3,365,690,400	12,319,270,000	15,684,960,400	68,000,000,000	4.95%	27.3%	21.5%
1978	3,931,100,261	15,213,784,422	19,144,884,683	81,000,000,000	4.85%	25.8%	20.5%
1979	4,381,702,000	17,877,886,000	22,259,588,000	92,000,000,000	4.76%	24.5%	19.7%
1980	5,288,645,000	20,307,213,700	25,595,858,700	109,000,000,000	4.85%	26.0%	20.7%
1981	6,336,630,000	26,215,097,800	32,551,727,800	140,000,000,000	4.53%	24.2%	19.5%
1982	NA	NA	NA	NA	0.0%	0.0%	0.0%
1983	8,296,658,500	33,055,400,000	41,352,058,500	177,000,000,000	4.69%	25.1%	20.1%
1984	8,907,104,550	35,926,668,000	44,833,772,550	192,000,000,000	4.64%	24.8%	19.9%
1985	9,405,856,400	39,331,939,000	48,737,795,400	213,000,000,000	4.42%	23.9%	19.3%
1986	9,283,113,300	39,266,220,000	48,549,333,300	218,000,000,000	4.26%	23.6%	19.1%
1987	9,330,907,740	39,155,502,000	48,486,409,740	227,500,000,000	4.10%	23.8%	19.2%
1988	9,998,059,200	41,170,734,950	51,168,794,150	243,500,000,000	4.11%	24.3%	19.5%
[illegible]	[illegible]	[illegible]	[illegible]	[illegible]	[illegible]	[illegible]	[illegible]

1990	12,547,607,000	52,632,502,500	65,180,109,500	335,000,000,000	3.75%	23.8%	19.3%
1991	16,797,954,200	60,575,221,500	77,373,175,700	387,500,000,000	4.33%	27.7%	21.7%
1992	19,450,824,700	69,272,982,400	88,723,807,100	460,400,000,000	4.22%	28.1%	21.9%
1993	24,875,090,000	78,625,342,500	103,500,432,500	560,000,000,000	4.44%	31.6%	24.0%
1994	28,109,399,400	85,423,916,700	113,533,316,100	625,000,000,000	4.50%	32.9%	24.8%
1995	30,535,948,000	91,638,768,200	122,174,716,200	715,000,000,000	4.27%	33.3%	25.0%
1996	36,701,808,300	100,603,034,800	137,304,843,100	843,200,000,000	4.35%	36.5%	26.7%
1997	40,129,473,700	108,573,602,400	148,703,076,100	984,000,000,000	4.08%	37.0%	27.0%
1998	41,782,911,053	97,766,348,000	139,549,259,053	923,000,000,000	4.53%	42.7%	29.9%
1999	38,111,829,100	77,066,937,000	115,178,766,100	825,000,000,000	4.62%	49.5%	33.1%
2000	42,822,339,200	77,194,873,500	120,017,212,700	860,915,829,100	4.97%	55.5%	35.7%
2001	42,554,490,800	77,210,552,300	119,765,043,100	910,000,000,000	4.68%	55.1%	35.5%
2002	40,447,614,600	78,584,183,600	119,031,798,200	1,023,000,000,000	3.95%	51.5%	34.0%
2003	41,457,410,700	79,923,271,800	121,380,682,500	999,900,000,000	4.15%	51.9%	34.2%
2004	44,201,537,000	78,551,324,500	122,752,861,500	1,028,000,000,000	4.30%	56.3%	36.0%
2005	44,018,428,800	81,241,389,900	125,259,818,700	1,200,000,000,000	3.67%	54.2%	35.1%
2006	51,843,287,800	85,936,118,000	137,779,405,800	1,360,000,000,000	3.81%	60.3%	37.6%
2007	58,668,132,100	115,024,014,800	173,692,146,900	1,466,200,000,000	4.00%	51.0%	33.8%
2008	62,510,661,700	143,518,901,100	206,029,562,800	1,660,000,000,000	3.77%	43.6%	30.3%
2009	69,882,918,200	170,157,393,800	240,040,312,000	1,835,000,000,000	3.81%	41.1%	29.1%
2010	66,594,572,400	154,072,478,600	220,667,051,000	1,700,000,000,000	3.92%	43.2%	30.2%
2011	74,190,555,300	168,501,828,300	242,692,383,600	2,070,000,000,000	3.58%	44.0%	30.6%
2012	77,759,750,700	168,667,373,500	246,427,124,200	2,380,000,000,000	3.27%	46.1%	31.6%
2013	83,758,831,500	180,491,535,700	264,250,367,200	2,400,000,000,000	3.49%	46.4%	31.7%
2014	86,768,728,300	183,819,972,100	270,588,700,400	2,525,000,000,000	3.44%	47.2%	32.1%
2015	90,488,393,600	192,949,090,200	283,437,483,800	2,575,000,000,000	3.51%	46.9%	31.9%
2016	103,479,291,900	206,461,310,900	309,940,602,800	2,720,000,000,000	3.80%	50.1%	33.4%
2017	102,767,491,600	210,777,461,400	313,544,953,000	2,733,000,000,000	3.76%	48.8%	32.8%
2018	107,066,321,000	222,400,000,000	329,466,321,000	2,900,000,000,000	3.70%	48.1%	32.5%
2016	103,479,291,900	206,461,310,900	309,940,602,800	2,720,000,000,000	3.80%	50.1%	33.4%
2017	102,767,491,600	210,777,461,400	313,544,953,000	2,733,000,000,000	3.76%	48.8%	32.8%
2018	107,066,321,000	222,400,000,000	329,466,321,000	2,900,000,000,000	3.70%	48.1%	32.5%

Source: Author's calculations based upon raw Thai budgetary data.

9

THE MONARCHY AND SUCCESSION[1]

Kevin Hewison

Thailand's contemporary monarchy is largely an invention of the second half of the 20th century. In 1932 the absolute monarchy was overthrown, stripped of its most significant powers and became a constitutional monarchy. From late 1934 until late 1951, with no king residing permanently in Thailand, the monarchy's authority was greatly diminished.

By the end of the 20th century, however, the monarchy had regained and expanded its wealth, political authority and social prestige. Its political and economic resurgence saw King Bhumibol Adulyadej preside over the making and breaking of governments. Incessant propaganda from the late 1950s meant the king's prestige was at a level not seen even in the last decades of the absolute monarchy. This return to economic and political prominence is due to the efforts of King Bhumibol, who reigned from 1946 until 2016 with loyal military dictators, aged princes, royal courtiers and Sino-Thai tycoons working to promote the monarchy and advance its political model of Thai-style democracy (TSD). This chapter examines the monarchy's economic and political restoration in the 20th century and examines the succession of his son, Vajiralongkorn, following Bhumibol's death in 2016.

Political restoration: People's Party versus royalists

When the People's Party seized power on 24 June 1932, there was almost no opposition, not least because several senior princes were detained and King Prajadhipok was golfing in Hua Hin, several hours from Bangkok. Once Prajadhipok realised that his forces were unlikely to restore absolutism, he first agreed to negotiate with the new regime but was soon in dispute with it. Meanwhile, princes and royalists began organising to oppose the constitutional regime.[2] In an initial act of defiance, Prajadhipok demanded and received a constitution rather less radical than some in the People's Party had demanded. The king's followers, scheming on his behalf and exploiting rifts within the People's Party, worked to destabilise the new regime.[3] The most significant efforts were to sideline Pridi Banomyong, regarded as the principal People's Party ideologue. A series of covert actions were launched against the new regime, eventually resulting in an attempted coup d'état in 1933 by a pro-royalist military faction, led by Prince Boworadej.[4]

While that uprising was seen off, much of the period from 1932 to 1946 was a constant political and ideological struggle between royalists and the People's Party regime. For its part, the regime exiled, jailed and generally repressed its royalist opponents while engaging in a political contest with the royal family over its constitutional position and wealth. The princes were constitutionally prevented from engaging in politics and most were purged from government and military positions.[5] Yet royalist plotting continued, including several attempted assassinations. As a result of this struggle, the military faction of the People's Party under Field Marshal Plaek Phibun Songkhram (Phibun) became dominant. In 1935, from self-imposed exile in England, Prajadhipok abdicated, replaced by the nine-year-old King Ananda Mahidol, who resided in Switzerland. This reduced the monarchy's ideological challenge to the regime, which also arrested a large group of royalists, including high-ranked princes. It seemed that the royalist challenge had finally been defeated.[6] Phibun's government, however, aligned with the Japanese during the Pacific War, allowing the royalists a way back to prominence through their involvement with the Free Thai Movement and support for the Allies.[7]

The anti-Japanese Free Thai movement included royalists but was dominated by Pridi's faction. Ironically, Pridi was regent for the absent king when the Free Thai contacted the Allies. With the military sidelined, Pridi's Free Thai faction gained the upper hand in the post-war governments, rehabilitating royalists as wartime allies and opponents of Phibun's military group. Pridi backed a royalist party to lead government. However, when hostilities ended, the royalists and Pridi fell out when an election loomed. Parties supporting Pridi won a clear majority, leaving the royalists in opposition. A new opportunity for a royalist political comeback came with the gunshot death of King Ananda on 9 June 1946, who had returned for a visit to Thailand in December 1945. While the king's death has never been satisfactorily explained, the royalist Democrat Party seized the opportunity and its supporters accused Pridi of orchestrating an assassination.[8]

Royalist attacks on Pridi forced him to flee abroad. After returning and becoming involved in failed coup attempts, he was exiled in China and France until his death. It was a disgruntled military that took advantage of political instability to overthrow the last pro-Pridi elected government in 1947. In a brief alliance with royalists, it backed the 1947 charter and the 1949 constitution that returned considerable authority to the monarchy. Yet, in seeking a leader to restore a measure of stability, the coup group invited Phibun to return as premier. He re-established his control over the military and allied with the United States to limit the royalist political revival.[9] Phibun soon pushed the royalists aside, scrapped the 1949 constitution and restored the 1932 charter. The political price paid for curbing royalist power was the political dominance of the military.

Bhumibol, military and Thai-style democracy

Following his brother's death, 18-year-old Bhumibol came to the throne, only to immediately depart for Switzerland. In his stead, the regent, Prince Dhani Nivat, supported the 1947 coup and the royalist charters.[10] As mentioned, however, the political crisis caused by Ananda's death saw the return of Phibun, who maintained a deep distrust of the royalists. He disliked the increased royalist role in government, which saw royals being appointed to parliament and the regent attending cabinet meetings.[11]

As Bhumibol sailed back to Thailand in 1951, Phibun moved against royalists, firmly establishing his grip on government. Not surprisingly, the relationship between Phibun and the young king was rocky. Indeed, advised by the regent and senior princes, the king's first political move was to oppose the 1951 coup and the change back to the 1932 charter, threatening to abdicate. However, he soon backed down, acknowledging Phibun's predominance.[12]

The relationship between the palace and Phibun remained strained until 1957, when Phibun was ousted by pro-royalist and anti-Pridi army leader Field Marshal Sarit Thanarat in yet another coup d'état.[13] The coup was a breath of fresh air for the besieged royals. Sarit's dictatorship, from 1958 to 1962, welded the monarchy and military in a partnership that re-established the monarchy's ideological position and political role. The king was not a passive partner. For example, the United States believed he played an active role in promoting the 1957 coup. Certainly, he did all that Sarit would have wanted, publicly endorsing the coup, dissolving parliament and approving Sarit's interim legislature, packed with military officers.[14] The re-establishment of the monarchy as a significant political institution also permitted the Crown Property Bureau (CPB) to expand its businesses and the crown's wealth (see below).

Bhumibol appreciated Sarit's efforts, viewing him as both loyal and a trusted father figure. In one public address, cheering Sarit, he declared, "This is an expression of thanks for his administration of our country which has brought happiness and content[edness] to everybody".[15] Well might the king have cheered Sarit, for the new alliance of a royalist military and the monarchy made these the defining institutions for Thailand's politics.

Sarit used the monarchy to legitimise his rule. The king began to travel the country and, with the queen, internationally, mainly to countries of the Western anti-communist alliance. The United States, where the king had been born, featured prominently, with the Thai-US alliance intensifying under Sarit and the military regime that followed. The United States provided enormous economic and military aid, underpinning Thailand's own counter-insurgency efforts.[16] The royal family was an enthusiastic participant in these anti-communist efforts, visiting allies and soldiers, dressed in military uniforms and strongly supporting the military's counter-insurgency.[17]

An important outcome of the military-monarchy partnership was the development of notions of "democracy" and "constitutional monarchy" that limited electoral politics. In distinguishing his regime from previous anti-monarchy governments, Sarit needed a new ideological cement for a society that was to be ruled by the military. This was TSD. Later referred to as "democracy with the king as head of state" (*prachathipatai an mi phramahakasat song pen pramuk*), this ideological position became Bhumibol's preferred political arrangement.[18]

Sarit's conservative redefinition of politics meant a regime that was "harsh, repressive, despotic, and inflexible".[19] Sarit viewed his political system as embodying "Thainess", rejecting Western notions of democratic government considered ill-suited to Thailand having been prematurely transplanted with insufficient preparation of the citizenry.[20] In this view, Thai society was amenable to strong leadership through an authority figure who could unify the country, upholding notions of unity (*samakhitham*) based in moral principles. It was Sarit who was the strong leader, but following his passing, it was the king who came to be viewed as the essential and moral leader. As a conservative conception of how society and politics should be arranged, social hierarchy was emphasised, social mobilisation limited and traditional institutions like the monarchy kept strong.[21] The nation was considered a patriarchal family and the unity of this family-nation determined a notion of unelected "representation" by a father-leader who would visit his children to learn of their problems and their needs before helping them. Such "representation" was considered "democratic" even without elections or constitutions.[22]

TSD makes the military and monarchy the custodians of Thainess and allocates them the role of "protecting" Thais from the bad politicians and corrupt parliamentary politics claimed to promote political instability. In this view, the military coup, approved by the monarch, was considered a legitimate mechanism for changing (elected) governments lacking good and moral leadership.[23] The moral leadership of the king was also bolstered by notions of virtuous Buddhist kingship, revived by royalist scholars, who argued that the king had an unquestionable right to rule due to his accumulation of great merit.[24]

From this perspective, the monarch's political role is to watch over government assuring the best interests of all are met. As the father of the family-nation, a benevolent and moral leader, the king protects his people. This places the monarchy at the centre of politics with "good" political leaders defined by the respect and loyalty they display for the king, defending his honour and position. TSD considers the monarch indispensable to the peace, prosperity and stability of the Thai nation-state.[25]

These ideological notions were incubated under a military rule that extended from 1958 to 1973, coinciding with pro-American anti-communism and the counter-insurgency war against the Communist Party of Thailand (CPT). During this period, Bhumibol exhibited no particular concern for democracy or constitutionalism and remained close to the regime, providing it with the legitimation for its public policies and the repression of opposition.[26] In 1973, however, students and academics demanded a constitution. The regime faltered, tottered and then fell in October 1973 after troops began shooting student demonstrators. It was the king who, whether by chance or design, provided belated sanctuary for the embattled students and ordered the military leadership into exile. Needing to disentangle the monarchy from a set of political arrangements that had come to be hated, Bhumibol's reaction was constructed as acts conforming with the ideology of a virtuous paternalist protecting "his people". The king then cemented his political centrality over a disgraced military by appointing a trusted royal adviser and law professor as prime minister. Again, the king was protecting "his people". In addition, he was positioning himself as a champion and arbiter of the new democracy.[27]

Within three years, however, the king's efforts to establish his political paramountcy had failed as electoral politics became, from the palace's perspective, chaotic and dangerously leftist. That Communist regimes were taking power in Cambodia, Vietnam and Laos caused the palace to move further to the political right. Bhumibol once again aligned himself with a resurgent military by supporting the 6 October 1976 coup. That coup followed months of destabilisation perpetrated mainly by rightists and culminated in a mock hanging by leftist students that some media outlets claimed was of a crown prince look-alike. Accusing the students of lèse-majesté, royalist thugs and police descended on Thammasat University, egged on by the military, ending three years of electoral politics. The bloody attack on students was conducted in the monarchy's name.[28] The victims of the royalist-inspired attack were damned as "disloyal". Just days after the bloodshed, Crown Prince Vajiralongkorn distributed awards to the paramilitary personnel involved while the king immediately and warmly welcomed the coup.[29] The new prime minister, the rightist judge Thanin Kraivichien, was the king's choice.

Stridently anti-communist, Thanin's rightist regime was also ultra-royalist and strengthened the lèse-majesté law to protect the monarchy from criticism of its role in bringing an end to electoral politics. The regime's political repression drove thousands of students, workers and farmers to the CPT's jungle bases. As a result, the CPT expanded its insurgency. So blunt was the regime's repression that even the military accused Thanin of unnecessarily stoking political opposition, causing splits within the military and expanding anti-monarchism.[30] Another coup in late 1977 threw Thanin out, replacing him with General Kriangsak Chamanan. The king was unimpressed, immediately appointing Thanin to his Privy Council. It wasn't long before Prem Tinsulanonda, Bhumibol's preferred military general, was able to push Kriangsak from office and replace him in early 1980.

General Prem was an ardent royalist and his slavish loyalty was publicly displayed in ways that indicated he was, in line with TSD, a "good" leader who placed the royal family in an exalted position. Owing his position to the palace, Prem had hundreds of royal projects funded by state funds and arranged numerous events that promoted the king, queen and royal family monarchy as defining of Thai identity.[31] When a group of "Young Turks" rose against Prem in 1981, the royal family fled Bangkok and joined Prem, with both the king and the queen making broadcasts supporting their prime minister. This effectively ended the putsch by making it a coup against the monarchy. Indeed, the putschists were accused of lèse-majesté.[32] Prem remained in power, his position consolidated by the fact that he was unequivocally the palace's preferred premier. At the same time, his grip over the military leadership and middle-level officers was strengthened.

Prem's tenure as prime minister until 1988, was a period where TSD was most determinedly fostered and the role of the king elevated to levels not seen since 1932. The alliance of a "good" leader and a moral monarch was portrayed as indispensable to the peace, prosperity and stability that was threatened by the disloyal and "evil" politicians.[33] Prem also elevated royalists in leadership positions in the military, state enterprises and the bureaucracy. Indeed, displaying one's loyalty became a prerequisite for promotion. It was in this period that the "network monarchy" or "deep state" was embedded as a means for the king to have his will expressed in government policy. McCargo defined the network monarchy as an illiberal mode of governance, "a form of semi-monarchical rule" where

> the monarch was the ultimate arbiter of political decisions in times of crisis; . . . the primary source of national legitimacy; . . . a didactic commentator on national issues, helping to set the national agenda, especially through his annual birthday speeches; the monarch intervened actively in political developments, largely by working through proxies such as privy councillors and trusted military figures. . . . At heart, network governance of this kind relied on placing the right people (mainly, the right men) in the right jobs.[34]

Likewise, Eugénie Mérieau identifies senior bureaucrats, military, police and judiciary as comprising a deep state. As well as working in the "visible state", this "anti-democratic alliance" operates a "shadow set of activities" that works in the interests of the monarchy.[35]

Such characterisations describe how informal and formal political networks developed under the Prem regime and persisted throughout the 20th century. It was a set of persons and arrangements that emerged to "manage" politics without those people and arrangements having to be a particularly visible element of the hurly-burly of politics. Importantly, the king's networks ensured that the "right" people were placed in strategic positions, in the military, judiciary, universities and at the top of the ministries. Such networking provided Bhumibol with unprecedented loyalty within the civil and military bureaucracies, ensuring the tone of governance was royalist and the national agenda reflected the king's political preferences. No administration could govern without due consideration of the king's wishes or biases. More broadly, as the king's prestige grew to levels unimaginable just a few decades earlier, business leaders, academics, media figures and others clustered close to the king and the royal family, seeking mutual benefits, be they political, economic or social status related.[36]

The king's political preferences, interests and biases came to define many aspects governance. When there was a military coup in 1991, the king approved. Significantly it was a royalist, Anand Punyarachun, who became the new military regime's civilian face and prime minister. When the military, intent on maintaining its control, violently put down protesters in May 1992, the king – kept informed of the military's plans – belatedly intervened and Anand again became appointed prime minister.[37] In his intervention, many considered the king had "solved a crisis" as, in McCargo's words, "the ultimate arbiter". In fact, Bhumibol had helped pave the way for the 1991 coup and only intervened in 1992 when the military was already "defeated". He had exhibited little enthusiasm for either constitutionalism or elections.[38] Despite this, the king's political stocks went even higher as his role was carefully crafted and managed as a "moral authority".[39]

By the mid-1990s, the king was in a powerful position. The throne was secure, well-funded and politically influential. Palace public relations emphasised that the king was indispensable for Thailand's future, couched in terms of TSD.

Thaksin, anti-monarchism and military coups

One result of the 1992 events was a new constitution, managed in its development by royalists and designed to permit electoral politics while meeting the basic requirements of TSD; the charter was designed with checks and balances sufficient to ensure the election of "good people". It also aimed to establish strong governments rather than the shaky coalitions that had led to political instability, vote-buying and the dominance of provincial political "godfathers". In practice, the 1997 constitution did all of this. Businessman Thaksin Shinawatra led his party to a huge victory in early 2001 and an even bigger landslide in 2005. A prickly relationship with the palace soon developed. Like all Thai politicians, Thaksin professed loyalty to the throne, but his wealth, policies, politics, style and popularity came together in ways that royalists considered posed a challenge for the monarchy and its networks.

The desire to increase his family's wealth is often seen to have been Thaksin's primary motivation for entering politics and his alleged cronyism and corruption were widely criticised.[40] The notion that Thaksin privileged family and cronies challenged the status of CPB as Thailand's largest conglomerate. Politically, Thaksin's aggressive reorganisation of the civil and military bureaucracies also posed a threat to the position and influence of the network monarchy. Arguably, however, the most significant contest was for the hearts and minds of the masses. With the king long portrayed as the champion of the downtrodden and his development projects symbolic of his connection to the rural masses, Thaksin's formidable electoral appeal among the poorer classes was threatening for royalists. As General Prem was reported to have explained this perception when speaking with the US ambassador in early 2006. Prem stated that Thaksin's electoral victories "had gone to his head", making him believe that "he's number one". That was a grave mistake, for according to Prem, referring to the king, "We already have a number one".[41]

By the time, thousands of People's Alliance for Democracy (PAD) demonstrators had come out in opposition to Thaksin, claiming loyalty to the king, wearing shirts in his yellow birth colour and calling for the king to throw Thaksin out. In the end, with Prem playing an active role, Thaksin was ousted by a military coup in September 2006. The king and queen immediately met with the junta leaders to offer support for the coup, and within a short time, General Surayud Chulanond was made prime minister, having hastily stepped down from the king's Privy Council. The 2006 coup unleashed a political conflict that pitted a coalition of Thaksin supporters, anti-coup and anti-military activists (known as Red Shirts) against PAD's conservatives, royalists and the military (Yellow Shirts) who opposed Thaksin and rejected electoral politics in favour of versions of TSD.

The 2006 coup unleashed more political contestation pitting Red Shirts against Yellow Shirts. Importantly, the monarchy was involved. Red Shirts identified Prem as a covert leader of their opponents while the royal family demonstrated support for the Yellow Shirts.[42] Indeed, many royalists felt they were supported, as they considered they battled for king and country. For anti-monarchists, the throne was part of an "traditional elite" that had cooperated to prevent the emergence of genuine democratic politics, made a minority hugely wealthy and had even urged the military to repress and murder their political opponents. Two military crackdowns on Red Shirt protesters in 2010 with more than 100 deaths and thousands of injuries appeared to confirm such views and resulted in a surge in anti-monarchism.[43]

With Bhumibol incapacitated and mostly hospitalised from September 2009, royalists felt threatened by rising anti-monarchism and political conflict intensified. It was another coup in May 2014 that marked a concerted military push to crush anti-monarchism. The post-2014 military junta's repression has been extensive. While the junta has released few statistics, thousands have been summoned to report to the military, been arrested or taken before military courts.[44]

Most significantly, the numbers accused of sedition and lèse-majesté have spiked substantially, probably exceeding 200, while dozens of anti-monarchists have been forced to flee Thailand.[45]

The monarchy was also chided by anti-monarchists for its own great wealth. Its economic power had increased exponentially during Bhumibol's reign, coinciding with the rise of Sino-Thai conglomerates that came to control large parts of the economy. The families controlling the conglomerates began closely associating themselves with the royal household from the 1970s, making donations to the royal family, sponsoring royal events and supporting royal foundations. By linking to the palace, capitalists once seen as alien were "Thai-ified" and their businesses prospered. During political conflict, anti-monarchists considered he rich and powerful were allied against them and that, far from matching the palace's rhetoric, the monarchy was not protecting all its people. Rather, the monarchy was protecting the wealth of a minority and its own treasure.

Economic restoration

The economic gains made by the monarchy under Bhumibol are as significant as the political gains. At the end of the Second World War, the royal family had lost much of the assets it had controlled until 1932. The rebuilding of that economic base is as important a story as the intertwined restoration of the monarchy's political power.

Prior to the 1932 revolution, the Privy Purse Bureau (PPB) was one of the major investors in Thailand's economic development, investing in numerous ventures and owning large plots of land and urban commercial developments. It invested independently with Sino-Thai capitalists and with foreign investors. As an absolute monarchy, there was a lack of separation between the king's personal spending and expenses and those of the PPB, leading to considerable dissatisfaction and criticism. Such arrangements changed soon after the People's Party seized power, becoming a source of conflict between Prajadhipok and the new regime.

In 1933, the PPB was placed under the Ministry of Palace, which was downgraded to an office under the Office of the Prime Minister, along with the PPB. With these demotions in status, the People's Party also abolished allowances paid to members of the royal family. The PPB's budget allocation was reduced to just 400,000 baht a year, or about 5 per cent of its pre-1932 allocation, with the king, princes and other royalists correctly viewing these as an effort to weaken the royalist faction. This threat to the king's wealth had him moving funds and property from the Privy Purse to his own accounts.[46] As a result, in 1934, the government took more control of the PPB and subjected royal property to taxation. Prajadhipok objected and threatened to abdicate, resulting in a compromise that separated the king's personal property from that of the crown, with only the former being subject to taxation.[47] Before long, Prajadhipok did abdicate. In 1936, the government promulgated the Crown Property Act (1936), which made the minister of finance head of crown property, with the CPB established in 1937.[48]

Effectively, the monarchy lost control of the CPB. This was this situation that faced Bhumibol when he came to the throne in 1946. However, the political changes that resulted from his brother's death and the 1947 coup saw the palace regain control of the CPB. In 1948, a new law on the CPB allowed the king to appoint the CPB's director-general, confirmed its tax-free status and affirmed that no crown property could be sold without the king's assent. While the minister of finance continued to chair the board, this "restriction" was largely for public relations effect. Freed of the previous restrictions and with trusted royalists at its helm, the CPB was able to expand its investments and provide handsome returns to a palace that had considered itself down-at-heel in 1946.[49] Its major assets were the Siam Commercial Bank, the Siam Cement Company and its stock of prime commercial land in Bangkok and vast provincial estates.

By the beginning of the 1970s, the CPB held the largest assets of any business in Thailand, was the country's biggest landowner and reportedly employed 500 people.[50] The CPB's annual income was "presented to His Majesty to be used for Royal Household expenses, entertainment for private and state visitors, donations for charitable purposes as well as for support of palace dependents and pensions for members of the royal family".[51] Those funds were further enhanced by the king's personal wealth and investments which, like those of the CPB, expanded considerably during the boom associated with the Vietnam War spending and massive US aid. With only a slight dip in the 1970s, Thailand's economic boom of almost 30 years until 1997 made the CPB and the royal family fabulously wealthy. By that time, the CPB had become a sprawling conglomerate with assets far exceeding those of other domestic businesses. The CPB held investments in almost 400 companies spanning property, manufacturing, finance and insurance, media, hotels and construction.[52]

While the 1997–1998 economic crisis caused considerable losses for the CPB, with state support and a restructuring, its growth had another spurt.[53] By 2011, the CPB had about 1,200 staff managing its land and corporate assets. It held some 6,560 hectares of land (with 1,328 hectares in Bangkok), operated more than 40,000 rental contracts (with 17,000 in Bangkok) and held corporate investments of almost US$7 billion.[54] Forbes annual ranking of the world's wealthiest monarchs had Thailand's monarchy at the top of its list, estimating its wealth at more than US$30 billion.[55] In 2010, the CPB estimated its Bangkok assets at about US$40 billion;[56] by 2016 the CPB was probably worth US$60–70 billion.[57]

As noted earlier, in addition to losing control of the PPB/CPB in the aftermath of 1932, the monarchy had its state allocations substantially reduced. However, during Bhumibol's reign, this situation changed. While information is opaque and because of the dampening impact of the lèse-majesté law, reports of the cost to taxpayers of maintaining the royal household are rare. However, as conflict deepened in the 2000s, analysis of official Budget Bureau papers began to shed some light on this topic. In 2005, it was calculated that the cost was about 5.5 billion baht (17 billion baht in 2015), making the maintenance of Thailand's monarchy one of the most expensive for taxpayers in the world.[58]

By all measures, under Bhumibol, the CPB and the royal family's economic position was vastly improved during his reign. So wealthy were the royals that they dwarfed all of Thailand's wealthy Sino-Thai tycoons. Crown wealth has grown with these tycoons but has outstripped them, not least because of the support of the state and of state regulations that granted the CPB huge tax benefits. All of this is somewhat ironic given that a persistent image of Bhumibol is that of a frugal man.[59] Related, and of huge propaganda importance, he has long promoted the idea of "sufficiency economy", which urges "moderation, prudence, and social immunity, . . . [using] knowledge and virtue as guidelines in living", portrayed as a fundamental element of true Thai culture.[60]

Succession

When Bhumibol died on 13 October 2016, he left his nominated heir apparent, Crown Prince Vajiralongkorn, with a remarkable legacy: the monarchy was politically stronger and wealthier than at any time since 1932. Yet there was trepidation regarding succession. Not only is it difficult to move beyond a 70-year reign, but the prince was also seen to lack the personal qualities that marked his father's reign and made it so successful. Claims that the succession was contested also abounded.[61] The anxiety was eased somewhat for royalists by the 2014 coup and a military junta determined to manage a smooth succession and crush anti-monarchism.

It is likely that rising anti-monarchism was a motivation for the 2014 military coup. At the same time, the fact that Bhumibol was on his deathbed was a matter of great concern for royalists.

When a monarch who has reigned for decades dies, a political and ideological chasm is left to be filled. In Thailand, some commentators suggested that even more than this, there was a "succession crisis" and a battle over who would be the next monarch. In fact, this crisis was seen to underpin post-2005 political conflict and two military coups. However, when Bhumibol died, Vajiralongkorn, crown prince since 1972, ascended the throne, suggesting that talk of a crisis was overblown.

Even if there was no succession crisis, Vajiralongkorn's path to throne was punctuated by a series of personal crises that undoubtedly caused difficulties for the monarchy and its image. Over several decades, rumour and scandal dogged Vajiralongkorn. However, Bhumibol's 1972 decision to make Vajiralongkorn crown prince and heir apparent was never rescinded, meaning that succession was to follow the 1924 law on succession.

Over time, Vajiralongkorn's idiosyncratic behaviour repeatedly undermined his public position and despite repressive laws and politics, rumour dogged his life as crown prince.[62] At times accused of bullying, gangsterism and serial womanising in the 1980s and 1990s, Vajiralongkorn responded through the palace's public relations machine.[63] By the 2000s, however, his crises had been "normalised" and he seemed confident to live with the gossip and crises. Indeed, as he took more of his ailing father's royal duties, sharing some with his sisters, Thailand's contentious politics made the management of succession somewhat easier, not least because royalists came to consider the monarchy under threat and requiring protection.[64] In fact, Vajiralongkorn's succession was a long process that eased him into his role over about a decade. Over that time, as a kind of "deputy" king while Bhumibol was hospitalised, Vajiralongkorn reorganised the palace for his reign. He and his sister, Princess Sirindhorn, became the public faces of the monarchy for ceremonial purposes, with the prince holding precedence.

The 2014 coup saw the junta throw its repressive resources behind the prince. It used lèse-majesté, computer crimes and sedition laws to stifle criticism of both the monarchy and the junta. The junta also arranged showcase events for the prince, showing him as a fit and loyal son. The junta and its leader, General Prayuth Chan-ocha, looking to Thailand's authoritarian past for its political model, have tended to mimic the Sarit regime in its repressiveness and its monarchism. The junta has put in place sets of rules, laws and a constitution all aimed at maintaining military political dominance for many more years. Such an arrangement, with ultra-royalists leading the military and the civil administration, is essentially a re-establishment of TSD as the only acceptable political system. Military-dominated politics would appear to suit a monarch trained as a military officer and who cannot immediately meet the expectations associated with his father. The military-monarchy partnership may well continue for years to come.

Importantly, as Vajiralongkorn and the junta prepared for succession, the regime supported the prince in a reorganisation of his personal life by ousting his third wife Srirasmi in a nasty and public separation. Vajiralongkorn also moved against people alleged to have misused their relationship with him for personal gain. In one case, a five-year jail term for lèse-majesté involved a sister of the former princess accused of selling overpriced food products to the palace. This purge and separation also included a grand chamberlain of the Royal Household Bureau and left Srirasmi's relatives and friends destroyed, jailed or dead.[65]

Vajiralongkorn's "housecleaning", marked by incarcerations and unexplained deaths of former associates, went even deeper. Dozens of palace officials were stripped of royal decorations and publicly abused. The language used in these announcements indicated the depth of Vajiralongkorn's displeasure. One deputy commander of a security unit to the royal family was accused of "gravely evil behaviour" and "disobedience . . . and exploiting ties to the Royal Family". He was declared a threat to the security of the monarchy.[66] He had unwavering support from the junta in such purges, with Prime Minister Prayuth signing all orders while reaffirming his promise to eliminate critics of the monarchy.

Following his father's passing, Vajiralongkorn reorganised the Privy Council and the Crown Property Bureau, filling them with loyalists and military figures, some of them plucked from the ranks of the ruling junta. This remaking of two of the palace's most significant agencies was not simply a change of faces on succession. It amounted to yet another purge of those Vajiralongkorn considered opponents or as persons who had allegedly misused their positions. In a blizzard of sackings and expulsions, with announcements couched in furious language, some of his father's favourites were retired.[67] Others were favoured, decorated and promoted. The result of such changes was said to have created a kingdom of fear and favour.[68]

Vajiralongkorn also worked with the junta to allocate more control of palace affairs to the king. To be sure, Bhumibol had done much on this score, but Vajiralongkorn was intent on further gains. An early and symbolic move was the junta's 2016 amendment of the Sangha Act, which governs official Buddhism. The amendment gave the king the power to appoint a new supreme patriarch, which Vajiralongkorn did, bypassing the rightful candidate for a palace- and junta-preferred monk. In 2018, the Sangha Act was again amended, giving the king control over the entire Sangha Supreme Council.[69] Soon after, the junta's parliament met in secret session to move five state agencies responsible for the palace's security and management to the king's control, undoing arrangements since the 1930s for government control over palace agencies.[70] Displaying particular concern for policies and legislation reflecting on the status and power of the monarchy, Vajiralongkorn also demanded changes to the junta's 2017 constitution.[71] The junta acquiesced, again ordering the legislature into secret session to make the required changes.

Most importantly, in 2017 Vajiralongkorn had the junta make changes to the act governing the CPB. That the Ministry of Finance and its minister continued to have nominal roles in managing the CPB seemed to irk Vajiralongkorn. Repealing the 1936 act, the new arrangement gave the king total control over the bureau. The new act meant the monarchy returned to arrangements of the PPB era, amalgamating the previously separated categories of crown and personal wealth. All management of crown property was now officially at the king's discretion. Significantly, all crown property is now owned by the king.[72] Vajiralongkorn also removed the long-time CPB Director-General Chirayu Issarangkul and replaced him with his personal secretary, Air Chief Marshal Sathitpong Sukwimol.[73] By these moves, not only was the work of the 1932 revolution further undone, at least for crown property, but the king gained full personal control of Thailand's largest business empire.

Conclusion

King Bhumibol came to a throne in 1946 that was deprived of the vast political power and great wealth that it had enjoyed prior to the 1932 revolution. His 70 years on the throne witnessed a remarkable restoration of both political authority and wealth. Much of this renaissance was based on an idea at the heart of TSD that monarchy was a surer representative of merit and "good" leadership than that provided by elected politicians.

Bhumibol favoured "behind the scenes" manipulation in establishing his preferred polity, operating through trusted mediators and political manipulators referred to as the deep state or network monarchy. Now that Vajiralongkorn has taken the throne, he has already demonstrated a congruence with his father's political views, seeking to further increase the political and economic significance of the monarchy. His disdain for the 1932 arrangements has been made clear, and he has had the military junta's support in rolling back constraints on the monarchy.

The monarch now holds more formal power than any king since 1932. The king and the military have an accommodation built around the military's capacity for repression. At the time of writing, the junta was scheming to maintain a military-backed control of Thailand's politics well into the future. For those who hoped for a more open and democratic Thailand, this is indeed a foreboding political alliance.

Notes

1 The author is grateful for comments on an earlier draft provided by Chris Baker, Andrew Brown and Pavin Chachavalpongpun. They are not responsible for any errors that remain.
2 Nattapoll Chaiching, "The Monarchy and the Royalist Movement in Modern Thai Politics, 1932–1957", in *Saying the Unsayable. Monarchy and Democracy in Thailand*, edited by Søren Ivarsson and Lotte Isager (Copenhagen: NIAS Press, 2010), pp. 147–178.
3 Virginia Thompson, *Thailand. The New Siam* (New York: Paragon Books, 1967), pp. 70–79.
4 Federico Ferrara, *The Political Development of Modern Thailand* (Cambridge: Cambridge University Press, 2015), pp. 75–99.
5 Thompson, *Thailand. The New Siam*, p. 64.
6 Nattapoll, "The Monarchy and the Royalist Movement in Modern Thai Politics, 1932–1957", pp. 158–163.
7 Sorasak Ngamcachonkulkid, *Free Thai. The New History of the Seri Thai Movement* (Bangkok: Institute of Asian Studies, Chulalongkorn University, 2010), ch. 2.
8 Daniel Fineman, *A Special Relationship. The United States and Military Government in Thailand, 1947–1958* (Honolulu: University of Hawaii Press, 1997), ch. 1. On King Ananda's death, see Paul M. Handley, *The King Never Smiles. A Biography of Thailand's Bhumibol Adulyadej* (New Haven: Yale University Press, 2006), ch. 4; and Andrew MacGregor Marshall, "Thailand's Saddest Secret", *Zen Journalist Blog*, 7 March 2013 <www.zenjournalist.com/2013/03/thailands-saddest-secret/> (accessed 13 July 2018).
9 Fineman, *A Special Relationship. The United States and Military Government in Thailand, 1947–1958*, parts 2 and 3.
10 Nattapoll, "The Monarchy and the Royalist Movement in Modern Thai Politics, 1932–1957", p. 166.
11 Ibid., p. 168.
12 Office of the Historian, *Foreign Relations of the United States, 1951, Asia and the Pacific*, Vol. VI, Part 2, Washington, DC: US Government Printing Office, 1977, pp. 1638–1643. The young king came under the influence of the conservative Prince Dhani and Prince Rangsit (see Nicholas Grossman (ed.), *King Bhumibol Adulyadej. A Life's Work* (Singapore: Editions Didier Millet, 2011), p. 41). Rangsit had been arrested for his alleged involvement in royalist plots and was jailed for almost five years. He was rehabilitated by the royalist government in 1946.
13 Nattapoll, "The Monarchy and the Royalist Movement in Modern Thai Politics, 1932–1957", p. 170; and Central Intelligence Agency, "Probable Developments in Thailand", National Intelligence Estimate Number 62–57, 18 June 1957 <www.cia.gov/library/readingroom/document/cia-rdp98-00979r000400300001-4> (accessed 15 June 2017).
14 Central Intelligence Agency, "NSC Briefing", 21 September 1957 <https://thaipoliticalprisoners.wordpress.com/2017/02/15/the-interventionist-king/sarit-coup_21-sept-57/> (accessed 14 February 2017).
15 "The King and Sarit", *Siam Rath Weekly Review*, 2 February 1961, p. 4.
16 Robert J. Muscat, *Thailand and the United States* (New York: Columbia University Press, 1990).
17 Handley, *The King Never Smiles. A Biography of Thailand's Bhumibol Adulyadej*, chs. 8–9; and Sinae Hyun, "Mae Fah Luang: Thailand's Princess Mother and the Border Patrol Police during the Cold War", *Journal of Southeast Asian Studies*, Vol. 48, No. 2 (2017), pp. 262–282.
18 Chalermkiat Piu-nual, *Prachatippatai Baep Thai Kuam Kit Tang Kanmuang Kong Thaharn Thai (2519–2529)* [Thai-style Democracy: The Political Ideas of the Thai Military (1976–1986)] (Bangkok: Thammasat University Press, 1990). For a focused discussion of TSD, as well as the references below, see Kevin Hewison and Kengkij Kitirianglarp, "Thai-Style Democracy: The Royalist Struggle for Thailand's Politics", in *Saying the Unsayable: Monarchy and Democracy in Thailand*, edited by Søren Ivarsson and Lotte Isager (Copenhagen: NIAS Press, 2010), pp. 179–202.
19 Thak Chaloemtiarana, *Thailand: The Politics of Despotic Paternalism* (Bangkok: Thai Khadi Research Institute, 1979), p. 10.
20 Michael K. Connors, *Democracy and National Identity in Thailand* (London: RoutledgeCurzon, 2003), p. 48; Saichol Sattayanurak, *Kukrit Lae Kan Sang Kuam Pen Thai 2: Yuk Jom Phon Sarit Tung Tossawat 2530* [Kukrit and the Construction of Thainess, Volume 2: From Sarit's era to 1997] (Bangkok: Matichon, 2007), pp. 31–32, 69, and Thak, p. 100.

21 Thak, *Thailand: The Politics of Despotic Paternalism*, pp. 104–105.
22 Ibid., pp. xiii, 101–106. The similarities between Sarit's views and those expressed by King Bhumibol over the next 70 years are remarkable. Sarit appears to have had a deep and lasting impact on the king's ideas and ideology.
23 Saichol, *Kukrit Lae Kan Sang Kuam Pen Thai 2: Yuk Jom Phon Sarit Tung Tossawat 2530*, pp. 32–34, 54.
24 Kobkua Suwannathat-Pian, *Kings, Country and Constitutions* (London: RoutledgeCurzon, 2003), p. 21.
25 Kriangsak Chetpattanawanich, *Prachatippatai Baep Thai Chak Yuk Ratchakhru Tung Yuk Chomphon Sarit Thanarat* [Thai-style Democracy from the Rajakhru Era to Sarit Thanarat's Era] (Bangkok: The Foundation for the Promotion of Social Sciences and Humanities Textbooks Project, 2007); and Saichol, pp. 40–47, 61.
26 Indeed, as well as supporting particular regimes, Bhumibol spoke in praise of dictatorship. See for example Bhumibol Adulyadej, *Collection of Royal Addresses and Speeches During the State and Official Visits of Their Majesties the King and Queen to Foreign Countries 1959–1967 (B.E. 2502–2510)* (Bangkok: n.p., 1974), p. 52.
27 A useful account of this period is David Morell and Chai-anan Samudavanija, *Political Conflict in Thailand* (Cambridge: Oelgeschlager, Gunn & Hain, 1981).
28 The role of the military and the royal family is discussed in Thomas A. Marks, "The Status of Monarchy in Thailand", *Issues & Studies* (November 1977), pp. 55–62.
29 On the king's attitude to the coup, see Emily Willard, "Declassified U.S. Documents Help Fill Void Left by Thailand's Silence on 38th Anniversary of Thammasat University Massacre", Unredacted: The National Security Archive Blog, 10 October 2014 <https://unredacted.com/2014/10/10/declassified-u-s-documents-help-fill-void-left-by-thailands-silence-on-38th-anniversary-of-thammasat-university-massacre/> (accessed 18 July 2018). British documents confirm the king's position. See Andrew MacGregor Marshall, "British Cable on Royal Involvement in Thai Politics, November 1976", *Zenjournalist Blog*, 13 March 2013 <www.zenjournalist.com/2013/03/british-cable-on-royal-involvement-in-thai-politics-november-1976/> (accessed 7 July 2018).
30 Frank C. Darling, "Thailand in 1977: The Search for Stability and Progress", *Asian Survey*, Vol. 18, No. 2 (1978), pp. 153–163; and Thomas A. Marks, "The Thai Monarchy under Siege", *Asia Quarterly*, No. 2 (1978), pp. 109–141.
31 The flow of state funds to royal projects is substantial but remains deliberately opaque. For an insight, yet still failing to provide detailed budget figures, see Royal Project Foundation, *The Peach and the Poppy* (Chiang Mai: Highland Research and Development Institute, 2007), pp. 155–157.
32 See Chai-anan Samudavanija, *The Thai Young Turks* (Singapore: ISEAS, 1982), including texts of broadcasts and statements by the Young Turks, Prem, the king and queen and the military high command (pp. 84–95).
33 For more on Bhumibol and politics, see Kevin Hewison, "The Monarchy and Democratisation", in *Political Change in Thailand*, edited by Kevin Hewison (London: Routledge, 1997), pp. 58–74.
34 Duncan McCargo, "Network Monarchy and Legitimacy Crises in Thailand", *The Pacific Review*, Vol. 18, No. 4 (2005), p. 501.
35 Eugénie Mérieau, "Thailand's Deep State, Royal Power and the Constitutional Court (1997–2015)", *Journal of Contemporary Asia*, No. 3 (2016), p. 446.
36 Serhat Ünaldi, *Working Towards the Monarchy* (Honolulu: University of Hawaii Press, 2016), Part I.
37 Grossman, *King Bhumibol Adulyadej. A Life's Work*, pp. 154–155.
38 Handley, *The King Never Smiles. A Biography of Thailand's Bhumibol Adulyadej*, pp. 338–357; and Hewison, "The Monarchy and Democratisation", pp. 58–74.
39 Grossman, *King Bhumibol Adulyadej. A Life's Work*, pp. 148–167. Essentially providing the palace's account of the 1992 events, this book portrays the demonstrators as "infiltrated" by "violent elements" and a threat to the king's palace (pp. 152–153) but acknowledges his deliberate silence as soldiers and police gunned down protesters (p. 154). In the end, though, it is the king who is credited with having resolved the crisis.
40 Pasuk Phongpaichit and Chris Baker, *Thaksin* (Chiang Mai: Silkworm Books, 2009).
41 Wikileaks, "Prem on Thaksin", 6 July 2006 <https://wikileaks.org/plusd/cables/06BANGKOK3997_a.html> (accessed 12 May 2017). From late 2001, the king had publicly criticised Thaksin and his government (see "HM Warns of Catastrophe", *The Nation*, 6 December 2001, p. 1). See also Joshua Kurlantzick, "The Mixed Legacy of King Bhumibol Adulyadej", *Council on Foreign Relations Expert Brief*, 13 October 2016 <www.cfr.org/expert-brief/mixed-legacy-king-bhumibol-adulyadej> (accessed 29 June 2018).

42 See Anon., "Anti-Royalism in Thailand since 2006: Ideological Shifts and Resistance", *Journal of Contemporary Asia*, Vol. 48, No. 3 (2018), pp. 363–394.

43 Some blamed the monarchy for supporting the Yellow Shirts and military while others considered the monarchy was issuing orders to the military. See Serhat Ünaldi, "Working Towards the Monarchy and its Discontents: Anti-royal Graffiti in Downtown Bangkok", *Journal of Contemporary Asia*, Vol. 44, No. 3 (2014), pp. 377–403.

44 The iLaw website keeps some records. See "Latest statistics" <https://freedom.ilaw.or.th/en/content/latest-statistic> (accessed 23 July 2018).

45 Thai Lawyers for Human Rights, *Collapsed Rule of Law: The Consequences of Four Years under the National Council for Peace and Order for Human Rights and Thai Society* (Bangkok; TLHR, n.d. [2018]), pp. 17–32 <www.tlhr2014.com/th/?wpfb_dl=100> (accessed 28 July 2018).

46 Grossman, *King Bhumibol Adulyadej. A Life's Work*, p. 287; and Supot Chaengrew, "Khadi Yeud Phrarajasap Phrabat Somdet Phrapokklaew" [The Case of Appropriating the Assets of King Prajadhipok], *Sinlapawathanatham* (June 2002), pp. 70–74.

47 Federico Ferrara, "The Legend of King Prajadhipok: Tall Tales and Stubborn Facts on the Seventh Reign in Siam", *Journal of Southeast Asian Studies*, Vol. 43, No. 1 (2012), p. 23.

48 Grossman, *King Bhumibol Adulyadej. A Life's Work*, pp. 287–288. This arrangement coincided with a scandal over the sale and purchase of crown property (see Thompson, *Thailand. The New Siam*, pp. 93–95).

49 Porphant Ouyyanont, "The Crown Property Bureau in Thailand and the Crisis of 1997", *Journal of Contemporary Asia*, Vol. 38, No. 1 (2008), pp. 170–171.

50 For an illuminating account of some of the crown's property dealing, see Ünaldi, "Working Towards the Monarchy and its Discontents: Anti-royal Graffiti in Downtown Bangkok", part II.

51 "The Biggest Estate. The Crown Property Bureau – A Major Investor", *Investor*, February 1971, pp. 124–125.

52 Porphant, "The Crown Property Bureau in Thailand and the Crisis of 1997", p. 174.

53 Somluck Srimalee, "CPB Offers Land Deal to Lift Stake", *The Nation*, 10 April 2004, p. B1.

54 Grossman, *King Bhumibol Adulyadej. A Life's Work*, pp. 283, 295, 298.

55 For example, "The World's Richest Royals", *Forbes*, 29 April 2011 <www.forbes.com/sites/investopedia/2011/04/29/the-worlds-richest-royals/#408425f8739f> (accessed 19 July 2018).

56 Grossman, *King Bhumibol Adulyadej. A Life's Work*, pp. 283–284.

57 Author's calculations based on data for Thailand's 50 wealthiest families, 2010–2017, and estimates of property price increases in central Bangkok.

58 For examples of calculations, see David Streckfuss, "A Comparison of Modern Monarchies", Freedom Against Censorship Thailand Blog, 17 March 2012 <https://facthai.wordpress.com/2012/03/17/a-comparison-of-modern-monarchies-dr-david-streckfuss/> (accessed 29 July 2018); Thomas Fuller, "With King in Declining Health, Future of Monarchy in Thailand Is Uncertain", *The New York Times*, 20 September 2015; and Pavin Chachavalpongpun, "A Very Wealthy Monarch Grows Wealthier", *The Japan Times*, 9 September 2017.

59 Matthew Phillips, "How a Thai King Made Wealth Seem Sacred", *The New York Times*, 24 October 2017.

60 Chaipattana Foundation, "Philosophy of Sufficiency Economy", Chaipattana Foundation Website, n.d. <www.chaipat.or.th/eng/concepts-theories/sufficiency-economy-new-theory.html> (accessed 29 July 2018).

61 The most prominent proponent of a succession crisis has been Andrew MacGregor Marshall in his *A Kingdom in Crisis* (London: Zed Books, 2015). Underpinning much of this analysis was a 2010 Wikileaks cable by US Ambassador Eric John that, while acknowledging that Vajiralongkorn would come to the throne, outlined concerns held by Prem and another senior privy councilor. See Wikileaks, "Ambassador Engages Privy Council Chair Prem, Other 'Establishment' Figures on Year Ahead", 25 January 2010 <https://wikileaks.org/plusd/cables/10BANGKOK192_a.html> (accessed 15 May 2014).

62 For more on Vajiralongkorn's life, see Handley's *The King Never Smiles* and Andrew MacGregor Marshall, 2012. *Thailand's Moment of Truth, #Thaistory*, Version 4.0 <https://drive.google.com/file/d/0B5815z2tpnFXWlhqNzhBNFIxdW8/view> (accessed 30 June 2017).

63 Paul Handley, "Prince Hits Back", *Far Eastern Economic Review*, 14 January 1993, p. 13.

64 Thongchai Winichakul, *Thailand's Hyper-royalism: Its Past Success and Present Predicament* (Singapore: ISEAS, Trends in Southeast Asia, TRS 7/16, 2016).

65 Pavin Chachavapongpun, "Thai Prince Cleans House with an Eye on Throne", *Straits Times*, 8 December 2014.

66 See *Ratchakit Chanubeksa* [Royal Gazette], 132 [290], 1, 9 November 2015 <www.ratchakitcha.soc.go.th/DATA/PDF/2558/E/290/1.PDF> (accessed 12 June 2017).

67 "Palace Announces Promotions, Demotions and Expulsions", *Khaosod*, 14 March 2017 <www.khaosodenglish.com/politics/2017/03/14/palace-announces-promotions-demotions-expulsions/> (accessed 31 July 2018).

68 Pavin Chachavapongpun, "Kingdom of Fear (and Favour)", *New Mandala*, 18 April 2017 <www.newmandala.org/kingdom-fear-favour/> (accessed 30 July 2018). The use of lèse-majesté has moderated under Vajiralongkorn. It is, however, impossible to conclude whether this is his approach to the law or whether the junta's repression has successfully stamped out anti-monarchist discourse.

69 "New 'Monk Act' Allows King to Appoint Sangha Members", *Khaosod*, 5 July 2018 <www.khaosodenglish.com/politics/2018/07/05/new-monk-act-allows-king-to-appoint-sangha-members/> (accessed 20 July 2018).

70 "King Granted Direct Control over Palace Agencies", *Khaosod*, 2 May 2017 <www.khaosodenglish.com/politics/2017/05/02/king-granted-direct-control-palace-agencies/> (accessed 20 July 2018).

71 Duncan McCargo, "Thailand in 2017: Politics on Hold", *Asian Survey*, Vol. 58, No. 1, p. 186.

72 At the time of writing an explanation of the changes was at <www.crownproperty.or.th/ข่าวสารและสาระน่ารู้/เกร็ดความรู้/คำชี้แจง-การเปลี่ยนชื่อผู้ถือหุ้นจากสำนักงานทรัพย์สินส่วนพระมหากษัตริย์-เป็นพระปรมาภิไธย> (accessed 30 July 2018).

73 "King Appoints Sathitpong", *Bangkok Post*, 29 January 2017, p. 3.

References

Anonymous. 2018. "Anti-Royalism in Thailand Since 2006: Ideological Shifts and Resistance". *Journal of Contemporary Asia*, Vol. 48, No. 3: 363–394.

Bhumibol Adulyadej. 1974. *Collection of Royal Addresses and Speeches During the State and Official Visits of Their Majesties the King and Queen to Foreign Countries 1959–1967 (B.E. 2502–2510)*. Bangkok: no publisher.

Central Intelligence Agency. 1957. "Probable Developments in Thailand". National Intelligence Estimate Number 62–57, 18 June <www.cia.gov/library/readingroom/document/cia-rdp98-00979r000400300001-4> (accessed 15 June 2017).

———. 1957. "NSC Briefing". 21 September <https://thaipoliticalprisoners.wordpress.com/2017/02/15/the-interventionist-king/sarit-coup_21-sept-57/> (accessed 14 February 2017).

Chai-anan Samudavanija. 1982. *The Thai Young Turks*, Singapore: ISEAS.

Chaipattana Foundation. n.d. "Philosophy of Sufficiency Economy". Chaipattana Foundation website, < www.chaipat.or.th/eng/concepts-theories/sufficiency-economy-new-theory.html> (accessed 29 July 2018).

Chalermkiat Piu-nual. 1990. *Prachatippatai Baep Thai Kuam Kit Tang Kanmuang Kong Thaharn Thai (2519–2529)* [Thai-style Democracy: The Political Ideas of the Thai Military (1976–1986)]. Bangkok: Thammasat University Press.

Darling, Frank C. 1978. "Thailand in 1977: The Search for Stability and Progress". *Asian Survey*, Vol. 18, No. 2: 153–163.

Ferrara, Federico. 2012. "The Legend of King Prajadhipok: Tall Tales and Stubborn Facts on the Seventh Reign in Siam". *Journal of Southeast Asian Studies*, Vol. 43, No. 1: 4–31.

———. 2015. *The Political Development of Modern Thailand*. Cambridge: Cambridge University Press.

Fineman, Daniel. 1997. *A Special Relationship. The United States and Military Government in Thailand, 1947–1958*. Honolulu: University of Hawaii Press.

Fuller, Thomas. 2015. "With King in Declining Health, Future of Monarchy in Thailand Is Uncertain". *The New York Times*, 20 September.

Grossman, Nicholas (ed.). 2011. *King Bhumibol Adulyadej. A Life's Work*. Singapore: Editions Didier Millet.

Handley, Paul. 1993. "Prince Hits Back". *Far Eastern Economic Review*, 14 January, p. 13.

———. 2006. *The King Never Smiles: A Biography of Thailand's Bhumibol Adulyadej*. New Haven: Yale University Press.

Hewison, Kevin. 1997. "The Monarchy and Democratisation". In *Political Change in Thailand*. Edited by Kevin Hewison. London: Routledge, pp. 58–74.

Hewison, Kevin and Kengkij Kitirianglarp. 2010. "Thai-Style Democracy: The Royalist Struggle for Thailand's Politics". In *Saying the Unsayable. Monarchy and Democracy in Thailand*. Edited by Søren Ivarsson and Lotte Isager. Copenhagen: NIAS Press, pp. 179–202.

Hyun, Sinae. 2017. "Mae Fah Luang: Thailand's Princess Mother and the Border Patrol Police during the Cold War". *Journal of Southeast Asian Studies*, Vol. 48, No. 2: 262–282.

iLaw. 2018. "Latest Statistics" <https://freedom.ilaw.or.th/en/content/latest-statistic> (accessed 23 July 2018).

Kobkua Suwannathat-Pian. 2003. *Kings, Country and Constitutions*. London: RoutledgeCurzon.

Kriangsak Chetpattanawanich. 2007. *Prachatippatai Baep Thai Chak Yuk Ratchakhru Tung Yuk Chomphon Sarit Thanarat* [Thai-style Democracy from the Rajakhru Era to Sarit Thanarat's era]. Bangkok: Foundation for the Promotion of Social Sciences and Humanities Textbooks Project.

Kurlantzick, Joshua. 2016. "The Mixed Legacy of King Bhumibol Adulyadej". *Council on Foreign Relations Expert Brief.* 13 October <www.cfr.org/expert-brief/mixed-legacy-king-bhumibol-adulyadej> (accessed 29 June 2018).

Marks, Thomas A. 1977. "The Status of Monarchy in Thailand". *Issues & Studies*, November: 51–70.

———. 1978. "The Thai Monarchy Under Siege". *Asia Quarterly*, No. 2: 109–141.

Marshall, Andrew MacGregor. 2012. *Thailand's Moment of Truth, #Thaistory*, Version 4.0 <https://drive.google.com/file/d/0B5815z2tpnFXWlhqNzhBNFIxdW8/view> (accessed 30 June 2017).

———. 2013. "Thailand's Saddest Secret". *Zenjournalist Blog*, 7 March. <www.zenjournalist.com/2013/03/thailands-saddest-secret/> (accessed 13 July 2018).

———. 2013. "British Cable on Royal Involvement in Thai Politics, November 1976". *Zenjournalist Blog*, 13 March. <www.zenjournalist.com/2013/03/british-cable-on-royal-involvement-in-thai-politics-november-1976/> (accessed 7 July 2018).

———. 2015. *A Kingdom in Crisis*. London: Zed Books.

McCargo, Duncan. 2005. "Network monarchy and legitimacy crises in Thailand". *The Pacific Review*, Vol. 18, No. 4: 499–519.

———. 2018. "Thailand in 2017. Politics on Hold", *Asian Survey*, Vol. 58, No. 1: 181–187.

Mérieau, Eugénie. 2016. "Thailand's Deep State, Royal Power and the Constitutional Court (1997–2015)". *Journal of Contemporary Asia*, Vol. No. 3: 445–466.

Morell, David and Chai-anan Samudavanija. 1981. *Political Conflict in Thailand*. Cambridge: Oelgeschlager, Gunn & Hain.

Muscat, Robert J. 1990. *Thailand and the United States*. New York: Columbia University Press.

Nattapoll Chaiching. 2010. "The Monarchy and the Royalist Movement in Modern Thai Politics, 1932–1957". In *Saying the Unsayable. Monarchy and Democracy in Thailand*. Edited by Søren Ivarsson and Lotte Isager. Copenhagen: NIAS Press, pp. 147–178.

Office of the Historian. 1977. *Foreign Relations of the United States, 1951, Asia and the Pacific, Volume VI, Part 2*. Washington, DC: United States Government Printing Office.

Pasuk Phongpaichit and Chris Baker. 2009. *Thaksin*. Chiang Mai: Silkworm Books.

Pavin Chachavapongpun. 2014. "Thai Prince Cleans House with an Eye on Throne". *Straits Times*, 8 December.

———. 2017. "Kingdom of Fear (and Favour)". *New Mandala*, 18 April <www.newmandala.org/kingdom-fear-favour/> (accessed 30 July 2018).

———. 2017. "A Very Wealthy Monarch Grows Wealthier". *The Japan Times*, 9 September.

Phillips, Matthew. 2017. "How a Thai King Made Wealth Seem Sacred". *The New York Times*, 24 October.

Porphant Ouyyanont. 2008. "The Crown Property Bureau in Thailand and the Crisis of 1997". *Journal of Contemporary Asia*, Vol. 38, No. 1: 166–189.

Royal Project Foundation. 2007. *The Peach and the Poppy*. Chiang Mai: Highland Research and Development Institute.

Somluck Srimalee. 2004. "CPB Offers Land Deal to Lift Stake". *The Nation*, 10 April: p. B1.

Sorasak Ngamcachonkulkid. 2010. *Free Thai. The New History of the Seri Thai Movement*. Bangkok: Institute of Asian Studies, Chulalongkorn University.

Streckfuss, David. 2012. "A Comparison of Modern Monarchies". *Freedom Against Censorship Thailand Blog*, 17 March <https://facthai.wordpress.com/2012/03/17/a-comparison-of-modern-monarchies-dr-david-streckfuss/> (accessed 29 July 2018).

Supot Chaengrew. 2002. "*Khadi Yeud Phrarajasap Phrabat Somdet Phrapokklaew*" [The Case of Appropriating the Assets of King Prajadhipok]. *Sinlapawathanatham*. June: 63–80.

Thai Lawyers for Human Rights. n.d. [2018]. *Collapsed Rule of Law: The Consequences of Four Years under the National Council for Peace and Order for Human Rights and Thai Society*. Bangkok; TLHR, pp. 17–32 <www.tlhr2014.com/th/?wpfb_dl=100> (accessed 28 July 2018).

Thak Thak Chaloematiarana. 1979. *Thailand: The Politics of Despotic Paternalism*. Bangkok: Thai Khadi Research Institute.

———. 2007. *Thailand: The Politics of Despotic Paternalism*. Chiang Mai: Silkworm Books.

Virginia Thompson. 1967. *Thailand. The New Siam*. New York: Paragon Books.

Thongchai Winichakul. 2016. *Thailand's Hyper-royalism: Its Past Success and Present Predicament*. Singapore: ISEAS, Trends in Southeast Asia, TRS 7/16.

Ünaldi, Serhat. 2014. "Working Towards the Monarchy and Its Discontents: Anti-royal Graffiti in Downtown Bangkok". *Journal of Contemporary Asia*, Vol. 44, No. 3: 377–403.

———. 2016. *Working Towards the Monarchy*. Honolulu: University of Hawaii Press.

Wikileaks. 2006. "Prem on Thaksin". 6 July <https://wikileaks.org/plusd/cables/06BANGKOK3997_a.html> (accessed 12 May 2017).

———. 2010. "Ambassador Engages Privy Council Chair Prem, Other 'Establishment' Figures on Year Ahead". 25 January <https://wikileaks.org/plusd/cables/10BANGKOK192_a.html> (accessed 15 May 2014).

Willard, Emily. 2014. "Declassified U.S. Documents Help Fill Void Left by Thailand's Silence on 38th Anniversary of Thammasat University Massacre". Unredacted: The National Security Archive Blog, 10 October <https://unredacted.com/2014/10/10/declassified-u-s-documents-help-fill-void-left-by-thailands-silence-on-38th-anniversary-of-thammasat-university-massacre/> (accessed 18 July 2018).

10

LÈSE-MAJESTÉ WITHIN THAILAND'S REGIME OF INTIMIDATION

David Streckfuss

Over the last 250 years, the law of lèse-majesté has always been a contentious crime that has allowed monarchs to exercise special legal powers to protect themselves and their institutions. In most modern monarchies, though, these laws have been repealed (Japan), heavy restrictions have been put on their use, or while the law may still remain on the books, some monarchies are loath to use the law for fear of engendering opposition to the institution itself.

In Thailand, the elite and many commoners in Thailand believe that survival of the institution is essential, as if its ever-chaotic politics would tear the country apart without it. Many believe it is the core of Thai identity, and without it, or with one that acted like other modern monarchies (which stay out of politics), there would simply be no Thailand. Thai governments have consistently claimed an exception to human rights norms at the United Nations Human Rights Council when it has come to lèse-majesté, arguing that the institution and its relationship to the people is unique, hence necessitating the law.

It is impossible to say how many Thais respect and love their monarch and the institution because the lèse-majesté law makes it illegal not to be a devoted monarchist. In comparison to other monarchies, Thailand's lèse-majesté law is the most draconian in the world by a long shot. Its 3- to 15-year punishment – per count – makes Thailand quite exceptional indeed.

In the past dozen years, there have been four distinct trends and a curious note related to lèse-majesté prosecutions in Thailand. The first trend is that since 2008, more and more defendants facing near certain conviction have argued that their right to freedom of expression should extend to commentary on the monarchy and have pleaded their innocence. The second trend is that there has been a groundswell of criticism of the monarchy among some segments in Thai society that are critical of the monarchy. Although not a "movement", holders of this sentiment were emboldened after 2010 and became increasingly vocal – a sentiment that has largely been silenced by the junta after the 2014 coup. The third trend is that the boundaries of what is deemed lèse-majesté have widened while at the same time social media has expanded the opportunities to run afoul of the law. The fourth trend, contrasted with the view of many observers, is that the actual number of lèse-majesté cases prosecuted has dropped somewhat since the military coup in 2014. And finally, a curious note: there have not been any lèse-majesté arrests since 13 October 2017 to the time of this writing (mid-August 2019) – an unprecedented 22-month period of no new cases of the crime, a phenomenon not seen since the early 2000s.

Trend 1: refusing guilt – the case of Somyot Pruksakasemsuk

Since the use of the lèse-majesté law began in earnest during the 1960s, most defendants capitulate to the near certain prospect of conviction and "plead guilty" to the crime, in hopes of landing a lighter sentence, quite often cut in half for the contrite heart. There have been, through the decades, a handful of cases where defendants argued their innocence and prevailed.[1] The first of these was a case in 1957 where a defendant, who had claimed that the king then had killed his brother to take the throne, argued that he had the right to state such an opinion and so did not violate the law. The court did not agree and sentenced him to three years in prison, but it cut the sentence by one year as the defendant admitted he had uttered the defamatory words. The beginning of a new trend started in 2008 when a defendant, who had declared similar sentiments in public, refused to declare herself guilty. She was given 18 years in prison for her defiance and started a modest trend.[2]

A 2011 case of similar defiance was that of Somyot Pruksakasemsuk, who was charged with lèse-majesté for material published in the magazine *Voice of Thaksin*, of which he was the editor. He was originally sentenced to 11 years, but he appealed the lower court's decision and in the end the Supreme Court reduced his sentence to seven years. However, as someone found guilty of a national security law, he was ineligible for the various kinds of reductions on sentence available to other kinds of prisoners.

His arrest surprised him, and he felt the law used against him was "much more violent" than it needed to be – putting him on par with murderers. He felt resentment, bitterness and hopelessness during for the first three months of his imprisonment. He had gotten a small cut which went untreated and became much more serious. He was denied bail 16 times and was seriously considering suicide.

He began to feel some hope after he learned that he was not just forgotten, but on the contrary he had strong support, both domestically and internationally. Staying in prison now had value. He said he felt he suffered discrimination as a lèse-majesté prisoner – the conditions under which he lived were highly controlled, such as who could visit him. Only his registered family was able to leave him with money to buy things in prison. At first, he wanted to do education work in prison, but this was refused as he could not work outside his block. For two years he was allowed to work in the library, but then that privilege was refused and he was restricted to his cell. He took the opportunity to study political science at Sukhothai Thammathirat Open University, pursuing a degree in prison that allowed him, through his textbooks, to finish a four-year degree.

Somyot said that the lèse-majesté law is a tool used to "suppress people and make them feel threatened". The military government has made "Thai society even more divided than before" and created more resentment. To address the problem of the lèse-majesté law, he said, there first needs to be a democratic government which will end the arrests and the need for people to exercise self-censorship. Once military rule is repudiated and democracy returned, the law can be changed. "It is not the duty of just activists", he argued, "but all citizens to know their duty and not accept dictatorship".[3] Somyot is not alone in fighting for justice and affirming his right to speak out. Since 2008, at least 81 charges of lèse-majesté have been received by the Appeals and Supreme Courts for adjudication, indicating substantial resistance to pleading guilty.[4]

Trend 2: widespread sentiment critical of the monarchy – its rise and fall?

Does the skyrocketing of lèse-majesté cases in the last dozen years indicate opposition to the institution itself as it is constituted in Thailand? According to a recent (and brave) study, it does, and it is quite unlike earlier anti-royalist movements harboring republican sentiments.[5]

An anonymous writer points out that a number of scholars have described the threat that the Red Shirt movement posed to the royalist narrative, how the elite tried to reassert its hegemonic control, and how the Red Shirts envisioned a different kind of democracy in Thailand. Anonymous argues that despite assiduous claims made by the Thai elite and military about the universal and timeless respect and love for the monarchy, a distinct and undeniably critical sentiment towards the institution has emerged, especially since the military crackdown on Red Shirt protesters in 2010. After their leaders were rounded up and jailed, an untamed and angry populace turned its ire on the institution, an "anti-royalism" which "strived for liberal politics in which freedom of expression, basic equality among people and principles of equal representation were upheld".[6] However, facing the harsh lèse-majesté law, this "anti-royalism hid beneath metaphorical ambiguity, humour, vulgarity, and absurdist parody" expressed in apparently non-political ways.[7]

Statistics bear out this writer's point: beyond the 94 new cases of lèse-majesté between 2010 and 2013, there were 243 new cases of lèse-majesté taken up by the prosecution department where the identity of the offender was unknown – an average of more than 60 such cases per year. The debate among Red Shirts about whether to "whisper" or "shout" one's resistance to the monarchy came to an end, the anonymous writer maintains, after the 2014 coup, driving critics to veil their criticisms through "particular non-political themes and signs".[8]

Trend 3: widening scope of lèse-majesté when boundaries of "the public" shifts

The range of what might be deemed as lèse-majesté in Thailand has always been broad, but in the last decade it has seized more and more of the public sphere. It has covered ancient kings, royal dogs, and mere "likes" or sharing on Facebook. The clearest new trend, though, is not so much what is said but where it is taking place – the Internet, a new kind of public space.

- A human rights lawyer faces 50 years in prison for lèse-majesté when he allegedly "imported content deemed defamatory to the monarchy".[9]
- The Ministry of Digital Economy and Society threatens to prosecute anyone who befriends three notorious critics of the regime on Facebook.[10]
- A group of administrators of a satirical Facebook page, "We Love General Prayuth", are charged with sedition, which carries up to seven years' imprisonment; two are additionally charged with lèse-majesté.[11]
- A person who made ten posts deemed defamatory of the monarchy on a fake Facebook page he created to get back at a friend is given a record 70-year sentence.[12]
- A 59-year-old Thai traditional herbalist is given a 20-year sentence for having in a corner of his website a small banner linked to anti-monarchy activity.[13]
- A man is found guilty of "attempted lèse-majesté" for having unsent defamatory draft messages in his email.[14]
- A student who shares a translated biography of Thailand's new king is given a five-year sentence.[15]
- A woman who says "uh-huh" in response to a comment deemed as defamatory to the monarchy in a Facebook chat is threatened with lèse-majesté herself for not informing the authorities of the crime.[16]

There are a number of factors – many of them quite new that became subject to lèse-majesté accusations and convictions. The anonymous writer contends that "the government's surveillance has penetrated formerly semi-private and private arenas. Personal online chats, conversations in taxis, or graffiti on restroom doors no longer escape the eyes of those in power and charges and harsh sentences have resulted". Moreover, the military regime made it illegal not to report lèse-majesté crime. A BBC report said that in failing to "condemn" remarks that may be deemed as a violation, a person can be accused of complicity in the crime.

Prior to the appearance of pervasive social media in Thailand, most lèse-majesté charges derived from what had been taped (by the police, typically) or printed. The "public" in these cases was an audience at a seminar or protest or, in the case of print media, the public at large. Critics of the monarchy had to resort to a coded language or to ambiguous metaphors, aware that what they voiced or wrote in public could be scrutinised by society as a whole.

With social media, the delineations separating "public" and "private" are much less clear. Is what someone might say to "friends" on Facebook a public or private act? What if what was written to just "friends" was then shared with the public? To what extent? With the exception of perhaps a handful of cases, the vast majority of lèse-majesté cases in the past decade in Thailand were in some form shared on social media.

Trend 4: a reduction of lèse-majesté cases and adjudications

There has been an assumption among many scholars and activists that the number of lèse-majesté cases has, naturally, increased under the military junta that seized power from a democratically elected government in May of 2014. For instance, *Time* magazine argued that the lèse-majesté law was "increasingly being used to crush dissent".[17] The International Federation for Human Rights (FIDH) reported that lèse-majesté measures have "reached alarming levels" since the coup.[18] The anonymous writer asserts that the lèse-majesté law "has become the junta's primary strategy to silence its opposition".[19] Metta Wongwat claims that the military junta "holds the record for prosecuting the largest number of Article 112 cases in the past decade".[20]

These assertions are largely correct, for in many ways, as discussed earlier, the law remains a potent weapon with an increasing reach and severe sentencing. But despite the temptation to think that the problem of lèse-majesté abuse became suddenly worse under the military junta, it needs to be said that use of the law has represented a serious challenge to freedom in Thailand for more than a decade.

It has been difficult to get exact figures because of the "normal" course of lèse-majesté cases as many have been tried in military courts and information is only partially available in many cases. Also, as the anonymous writer notes, the actual number of cases has to be provisional as they "could be higher but the authorities often conduct cases in secret".[21]

The organisation iLaw has compiled the most comprehensive list of national security cases since the coup, allowing a rather exact comparison between the four years from 2010 to 2013 and then from 22 May 2014 to mid-May 2018.[22] Where the data was complete, the current situation of lèse-majesté cases is depressingly impressive, as shown in Table 10.1.

These numbers are distressing, but they are hardly unique as lèse-majesté cases are not new to Thailand. According to the statistics of the Office of the Attorney General, the yearly average of total cases considered by the Department of Prosecution between 2010 to 2013 was 23.5, involving 136 individuals. The average number of cases actually prosecuted from 22 May 2014 to 21 May 2018 was 17 cases per year, totaling 94 individuals. According to iLaw's statistics, the average number of cases and persons involved has gone down under the military regime.

The comparison is not exact, but there is something to note between the statistics between the two periods, as shown in Tables 10.1 and 10.3. To take one item, two persons of prosecuted

Table 10.1 Disposition of lèse-majesté cases in Thailand, May 2014–May 2018

	Cases	*Per cent*	*Persons*
Prosecuted	29	43%	40
Refused bail and pending	12	18%	23
Granted bail and pending	14	21%	15
Released after being held for 84 days	6	9%	6
Prosecution decided not to prosecute	1	1%	2
Other	6	9%	8
Total	**68**	**100%**	**94**

Table 10.2 Comparison of number of lèse-majesté cases and persons charged in Thailand, 2010–2013 and from 22 May 2014 to 21 May 2018

Year	*New Cases*	*No. of Persons*	*Period*	*New Cases*	*No. of Persons*
2010	29	44	22 May 2014–21 May 2015	34	47
2011	33	38	22 May 2015–21 May 2016	14	18
2012	18	26	22 May 2016–21 May 2017	17	19
2013	*14*	*28*	22 May 2017–21 May 2018	*3*	*10*
TOTAL:	**94**	**136**	**TOTAL:**	**68**	**94**
YEARLY AVERAGE:	**23.5**	**34**	**YEARLY AVERAGE:**	**17**	**23.5**

Table 10.3 Disposition of lèse-majesté cases in Thailand, 2010–2013, in terms of prosecution

	Cases	*Per cent*	*Persons*
Prosecuted	73	62%	88
Prosecution decided not to prosecute	10	8%	15
Other	9	8%	15
Pending	26	22%	*51*
Total	**118**	**100%**	**169**

cases escaped a verdict of guilty since the coup, while 15 persons achieved the same between 2010 and 2013. It was safer during a democratic period, certainly, with only a single case where defendants escaped punishment.[23]

But these statistics show a distortion not present previously – the use of military courts to try lèse-majesté. As I have argued previously, defamation represents an entirely different class of crimes. Judges must decide whether what was said (or done) would cause the one defamed to be looked down on or disparaged by (implicitly) "society". Cases are "proven" by prosecutors attesting to the defamatory character of what was said, and defendants and their lawyers provide testimony from others who say the words were not defamatory. This is very difficult to assess, even more so for military court judges who presumably have little experience with this type of case.[24]

Accordingly, we see a significant difference between how military and civilian courts adjudicate these cases, especially in terms of sentencing. Since the coup, 58 per cent of lèse-majesté cases have landed in military courts – the vast majority of these in the Bangkok Military Court – while 42 per cent have been handled by civilian courts. The rights of defendants are generally more restricted in military courts – and of course, military court decisions are final. What is most surprising between the military and civilian court decisions is the severity of the former. Of the 29 cases fully adjudicated – 65.5 per cent were made by the military courts – in the preliminary sentencing, the civilian courts' average sentence was 6.4 years, whereas the military courts' average sentence was 14.3 years. In the actual sentencing, which is usually halved with a confession of guilt, the average of civilian court sentences was 2.9 years vs. 8.6 years for the military courts. The median final sentence for the civilian courts was 2.5 years while the median of the military courts was 5 years. The average time of imprisonment in sentences for lèse-majesté cases adjudicated by the military courts is roughly similar to the average in the period prior to the coup.[25] It is minimally positive that the average of years sentenced for lèse-majesté in the last four years that were tried in criminal courts has dropped.

Moreover, while each type of court refused bail in nearly equal numbers, the average amount of time in jail was 1 year for civilian courts and 2.7 years for the military courts. The military courts were more lenient was in granting bail and in deciding not to prosecute in one case involving two persons. Only in one case did a civilian court in Sakon Nakhorn find a defendant not guilty.

The number of lèse-majesté cases in the first year after the coup was very high, at 35. But in the two following years, the annual number of cases dropped to 14 and 17 cases, respectively. That number dropped even further to just two in the fourth year after the coup, resulting in an annual average in four years of 17 cases. The reality is that under democratic governments in Thailand, the lèse-majesté law was the prime weapon the Bangkok elite and its allies used to clip the leaders of the majority opposition. But at the same time, this military regime has made up a whole set of new "legal" options to suppress those supporting democracy. Though the sentences for lèse-majesté have increased significantly through military courts after the coup, what has to be taken into account is the "legal" context created by the military that has become part and parcel of its imprint on Thai politics.

In a gathering discussing academic freedom in January of 2018, a prominent Thai scholar asked participants to look at the larger context of which diminishment of academic freedom is just one part of what this academic called a "regime of intimidation". Many academics have fled the country while others have been summoned by the military for "attitude adjustment". Some foreign scholars have not been allowed to enter the country and were sent home without knowing why they were banned, making it impossible to make a legal appeal. Many academics notice certain books disappearing from bookshops, and some online reviews have disappeared. Organisers of academic seminars are routinely threatened with legal action, and many seminars have been banned altogether. Some scholars have stopped publishing for the time being, and some publishers have put off launching new books, fearful they might fall afoul of the junta's encompassing decrees. On the whole, university leaders have shown little interest in protecting their own teachers or students in the name of academic freedom. In an effort to save their futures within Thai academia, younger scholars especially have become reluctant to publish anything that may be deemed as controversial. This senior Thai scholar concludes, "These events have affected all of us, perhaps in ways we are reluctant to admit. The end-result of this 'regime of intimidation' is self-censorship" (Name withheld).[26]

It is not just subtle degradations of academic freedom but a whole array of other not so subtle measures that have kept the opposition quiet. There are decrees issued by the National Council on Peace and Order (NCPO), such as the ban on gathering of more than five people or not reporting to the NCPO when ordered. Then there are new "laws" passed by the NCPO's hand-selected legislative body. Or there are old, restrictive laws that have been revived after decades of disuse, such as Article 116 of the Revised Criminal Code on sedition. All told, these laws, along with the lèse-majesté law, have led to 198 cases of the arrests of 708 Thais over the past four years – an annual average of 177. Details are as follows:

- Lèse-majesté (Article 112): 68 cases, 94 persons.
- Sedition (Article 116): 32 cases, 91 persons.
- NCPO ban on gathering of more than five people: 49 cases, 421 persons.
- Creating unrest surrounding constitutional referendum: 10 cases, 41 persons.
- Order to report to the NCPO: 14 cases, 14 persons.
- Other charges that are due to political motives: 15 cases, 27 persons.
- Smaller political infractions: 10 cases, 20 persons.

So while the lèse-majesté law has continued to be a weapon used against opposition, there has been a whole new array of laws that has joined it to provide what sometimes seems a total silencing of Thai society. The lèse-majesté law has been vital to the military junta's hold on power, but it is just one piece of a legal strategy used by the military to dismantle Thai civil society.[27] The chilling effect has been phenomenal in silencing what remains of civil society.

A curious note

Finally, we arrive at a curious note that recently occurred. A reliable source, who wishes to remain anonymous, affirmed that in a meeting, King Vajiralongkorn Bodindradebayavarangkun (Rama X) asked about three areas of concern: (1) How can the monarchy remain popular? (2) How can the monarchy be made sustainable? (3) Is democracy possible in Thailand?

In asking these questions, this source said the monarch was impressively well-versed in history and politics and displayed a remarkable knowledge of events in general. This source said the king asked about lèse-majesté and what could be done to address its excesses and put it more in line with other lèse-majesté laws in the world. The result, according to this source, is that the king ordered the prosecution department to initiate no new cases of lèse-majesté. The source said that the king issued an order that called on prosecutors to stop lèse-majesté prosecutions. The source might just be right, as there have not been any new lèse-majesté cases since 13 October 2017, thus inaugurating a period of no new cases of lèse-majesté in more than the last 22 months.[28] Moreover, there are other encouraging signs that might show the monarch's different approach to what are perceived as violations against the institution. As part of celebration of the king's coronation ceremony in 2019, a number of high-profile lèse-majesté convicts were granted a royal pardon.[29]

So could this be a rather dramatic change in lèse-majesté accusations and bringing them to trial? With the special royal order, it appears at this point, mid-August of 2019, that maybe it really has had some effect. Cases that are really outside of coverage of the lèse-majesté law have been dropped.[30] Defendants who might have faced decades in jail have been freed, at least temporarily, after little more than a year.[31] A serial offender of lèse-majesté, Sulak Sivaraksa, successfully petitioned the new king and gained relief.[32]

Conclusion: some optimism about reorienting lèse-majesté in Thailand?

It has proven impossible for any government in Thailand – democratic or dictatorial – to alter the lèse-majesté law, as the mere suggestion of any change has prompted accusations of threatening "Thainess".[33] An effort to raise 30,000 signatures during the last democratically elected government (2011–2014) to revise the law was ignored by the parliament; after the coup it provided a handy list of suspicious people for the military to monitor (and sometimes arrest).

Given the contentious nature of the law, no one but the monarch could put on the brakes. If it is true that new prosecutions have ceased due to the king's recognition that use of the law actually endangers the long-term prospects of the institution itself, then it would be similar to the actions of King Wilhelm II of Germany who, when faced with hundreds of lèse-majesté cases every year, sometimes for the most paltry of reasons, ordered prosecutions of lèse-majesté cases to stop in 1904. By 1907, at least according to the international press at the time, lèse-majesté disappeared entirely from Germany.[34]

In a time when opposition is quickly and efficiently silenced, there is no power that can challenge the military dictatorship other than the monarchy, owning to its symbolic status and actual wealth. For decades, the military was assured that its power – its many seizures of power – would never be countered by a compliant monarchy. In return, the military appointed itself the institution's protector. Even one of the reasons behind the 2014 coup was so that the military could ensure a smooth succession. The junta has ensured that the potency of the lèse-majesté law remains high by throwing cases into the military court and initiating many lèse-majesté cases itself.

But what if this law, which has very much been the linchpin holding this ideological structure in place, were suddenly denied the military and the Bangkok elite? It might upset the decades-long arrangement between the military and the monarchy. Diminished use of the law would deprive the military of an important weapon and in turn would empower democratic forces in Thailand. The only two questions are: Is this merely a consolidation of one elite institution (the monarchy) against another (the military)? Or does this herald a fundamental shift in the relationship between the monarchy and the people, all at the expense of the military?

Notes

1 David Streckfuss, *Truth on Trial in Thailand: Defamation, Treason, and Lèse-Majesté* (London and New York: Routledge, 2011), pp. 187–223.

2 Ibid., p. 87.

3 Phone interview with Somyot Pruksakasemsuk, 25 May and 3 June 2018; "Embattled Lèse-majesté Convict Somyot Released", *Prachatai*, 30 April 2018 <https://prachatai.com/english/node/7729> (accessed 14 Aug. 2019).

4 Office of the Judiciary, *Rai-Ngan Statiti Khadi San Thua Ratcha-Anajak Prajam Pi 2553* [Annual Judicial Statistics, 2553–2556 (2010–2013)].

5 The new study is Anonymous, "Anti-Royalism in Thailand since 2006: Ideological Shifts and Resistance", *Journal of Contemporary Asia*, Vol. 48, No. 3 (2018), pp. 363–394. Some observers have touched on historical threads of anti-monarchism in Thailand, such as Patrick Jory, "Republicanism in Thai History", in *A Sarong for Clio: Essays on the Intellectual and Cultural History of Thailand*, edited by Maurizio Peleggi (Ithaca: Cornell Southeast Asia Program Publications, 2015), pp. 97–117; and Federico Ferrara, "Unfinished Business: The Contagion of Conflict over a Century of Thai Political Development", in *"Good Coup" Gone Bad: Thailand's Political Development since Thaksin's Downfall*, edited by Pavin Chachavalpongpun (Singapore: ISEAS, 2014), pp. 17–46. Others have looked at its modern incarnations, such as Thongchai Winichakul, *Thailand's Hyper-royalism: Its Past Success and Present Predicament*

(Singapore: ISEAS, Trends in Southeast Asia 7, 2016); and Serhat Ünaldi, *Working towards the Monarchy: The Politics of Space in Downtown Bangkok* (Honolulu: University of Hawaii Press, 2016).

6 Anonymous, "Anti-Royalism in Thailand since 2006", p. 367.

7 Ibid., p. 366.

8 Ibid.

9 "Court Accepts Charges against Lawyer Facing 50 Years in Jail for Lèse-majesté", *Prachatai*, 26 July 2017 <https://prachatai.com/english/node/7292> (accessed 14 Aug. 2019).

10 Metta Wongwat, "A Decade of Article 112 Cases", *Prachatai*, 15 November 2017 <https://prachatai.com/english/node/7466> (accessed 14 Aug. 2019).

11 "Sedition Law Political Tool of the Junta: iLaw", *Prachatai*, 30 August 2017 <https://prachatai.com/english/node/7350> (accessed 14 Aug. 2019).

12 "Military Court Breaks Record with 35-year Jail Term for Lèse-majesté", *Prachatai*, 6 September 2017 <https://prachatai.com/english/node/7194> (accessed 14 Aug. 2019).

13 "Military Court Sends Elderly Man to Jail for Almost 20 Years for Lèse-majesté", *Prachatai*, 9 August 2017 <https://prachatai.com/english/node/7319> (accessed 14 Aug. 2019).

14 Thaweeporn Kummetha, "Thai Man Found Guilty of Attempted Lèse-majesté", *Prachatai*, 12 December 2013 <https://prachatai.com/english/node/3790> (accessed 14 Aug. 2019).

15 "Lèse-majesté under King Rama X: Six Observations", *Prachatai*, 12 December 2016 <https://prachatai.com/english/node/6763> (accessed 14 Aug. 2019).

16 "Police to Activists: Take Ja New's Mom as Example", *Prachatai*, 9 May 2016 <https://prachatai.com/english/node/6129> (accessed 14 Aug. 2019); Human Rights Watch, "Thailand: Junta Arrests Activist's Mother", 6 May 2016 <www.hrw.org/news/2016/05/06/thailand-junta-arrests-activists-mother> (accessed 14 Aug. 2019).

17 Charlie Campbell, "Thailand's Leader Promised to Restore Democracy, Instead He's Tightening His Grip", *Time*, 21 June 2018 <http://time.com/5318235/thailand-prayuth-chan-ocha/> (accessed 14 Aug. 2019).

18 FIDH, "And Counting Lèse-majesté Imprisonment under Thailand's Military Junta", Paris: International Federation for Human Rights, February 2016 <www.fidh.org/IMG/pdf/fidh_thailand_report_lese_majeste.pdf> (accessed 14 Aug. 2019).

19 Anonymous, "Anti-Royalism in Thailand since 2006", p. 364.

20 Metta Wongwat, "A Decade of Article 112 Cases", *Prachatai*, 15 November 2015 <https://prachatai.com/english/node/7466> (accessed 14 Aug. 2019).

21 Anonymous, "Anti-Royalism in Thailand since 2006", p. 391.

22 iLaw, see the first section of "Rai Choe Phu Thuk Tang Khoha Jak Thang Kanmuang Lamg Kanratthaprahan" [List of Names of those Charged Politically after the Coup] of the "Rai-ngan Kantang Khoha Thang Kanmuang Lang Ratthaprahan 2557" [Report of Political Charges after the Coup of 2014] <https://freedom.ilaw.or.th/politically-charged> (accessed 14 Aug. 2019); "Sathiti Khadi 112 Thi Na Sonjai Tangtae Pi 2557–2560" [Interesting Statistics in Cases of 112 from 2014 to 2017] <https://freedom.ilaw.or.th/Thailand-Lèse-majesté-Statistics-Until-2014-2018%20> (accessed 27 May 2018). The following analysis of lèse-majesté cases is entirely drawn from this set of statistics.

23 These figures are drawn from the Statistics of the Office of the Attorney General's Annual Reports of 2010–2013. Another set of statistics comes from the Office of the Judiciary, which records counts rather than cases. According to the Annual Statistical Yearbooks of the Office of the Judiciary from 2010–2013, a total of 748 charges of lèse-majesté were received (an average of 187 per year) and 350 charged were adjudicated (an average of 87.5 per year). In those same years, 25 charges were received by the Appeals Court and another 10 were adjudicated by the Supreme Court. It was during these same four years that there were even four charges of lèse-majesté considered by the Juvenile Court, with one adjudicated. These statistics, which do not list number or cases or individuals, show a rather sharp disparity from the statistics of the Offices of the Attorney General. *Rai-ngan Statiti Khadi San Thua Ratcha-Anajak Prajam Pi 2553–2557 [Annual Judicial Statistics, Thailand 2010–2014]* (four separate volumes): (2010: pp. 8, 29, 68, 124, and 139); (2011: pp. 8, 29, 84, 144, 160, 193); 2012 (pp. 9, 31, 119, 180, 193, 261); 2013 [pp. 31, 45, 117, 178, 191, 259].

24 Streckfuss, *Truth on Trial*, pp. 37, 60.

25 David Streckfuss, "European Constitutional Monarchies as Democracies Versus the Kingdom of Thailand as Theocracy", Unpublished manuscript, 2013.

26 Name Withheld. 2018. The private gathering was convened at an undisclosed foreign ambassadorial residence in Bangkok in January.

27 Amnesty International (AI), "Thailand: Civil Society under Attack as Authorities Criminalise Dissent", 9 February 2017 <https://prachatai.com/english/node/6914> (accessed 14 Aug. 2019).

28 The source (and his conversation with the new king) was later publicly revealed as Sulak Sivaraksa. See Political Prisoners in Thailand website, 17 Oct. 2018, "On the lese majeste regime" <https://thaipoliticalprisoners.wordpress.com/tag/asia-times-online/> (accessed 14 Aug. 2019). While the apparent stop to new lèse-majesté cases is encouraging, there is, however, some reason to believe that more new cases of lèse-majesté been have initiated, apart from what iLaw lists. See "Tom Dundee Faces Fourth Lèse-majesté Charge", *Prachatai*, 13 February 2018 <https://prachatai.com/english/node/7624> (accessed 14 Aug. 2019). There is some irony, at the same time, if the new king has actually put brakes on the law, as there was a flurry of cases when the law was used in 2014 and 2015 against "a large number of people" close to the then crown prince. See Metta, "A Decade of Article 112 Cases".

29 "Jailed red-shirt singer released under royal pardon", *Bangkok Post,* 17 July 2019 <https://www.bangkokpost.com/thailand/politics/1713940/jailed-red-shirt-singer-released-under-royal-pardon> (accessed on 14 Aug. 2019); "Activist 'Pai Dao Din' to be released on Friday by the King's Pardon", *Prachatai,* 9 May 2019 <https://prachatai.com/english/node/8038> (accessed on 14 Aug. 2019).

30 "2 Suspects not Guilty of Lèse-majesté for Defaming Princess Sirindhorn", *Prachatai* <https://prachatai.com/english/node/7643> (accessed 14 Aug. 2019).

31 "Embattled Lèse-majesté Lawyer Released after 16 Months in Jail", *Prachatai*, 17 August 2018 <https://prachatai.com/english/node/7802> (accessed 14 Aug. 2019).

32 Sulak Sivaraksa, "Lessons from the latest political lawsuit of Sulak Sivaraksa", *Prachatai*, 29 January 2018 <https://prachatai.com/english/node/7592> (accessed 14 Aug. 2019).

33 The threat was real. There were efforts by the newly elected government in 2011 to actually revise the lèse-majesté law, changing the punishment from 3 to 15 years to up to 7 years, which was the sentence, ironically, under the absolute monarchy. See "'คอป.' เสนอรัฐสภาแก้ 'ม.112' หนักสุดคุกไม่เกิน 7 ปี", *Prachatai*, 30 December 2011 <https://prachatai.com/journal/2011/12/38535> (accessed 14 Aug. 2019).

34 David Streckfuss, "The Intricacies of Lèse-majesté: A Comparative Study of Imperial Germany and Modern Thailand" in *Saying the Unsayable: Monarchy and Democracy in Thailand*, edited by Søren Ivarsson and Lotte Isager (NIAS Press: Nordic Institute of Asian Studies, 2010), p. 119.

References

Amnesty International (AI). 2017. "Thailand: Civil Society under Attack as Authorities Criminalise Dissent". 9 February <https://prachatai.com/english/node/6914> (accessed 14 Aug. 2019).

Anonymous. 2018. "Anti-Royalism in Thailand Since 2006: Ideological Shifts and Resistance". *Journal of Contemporary Asia*, Vol. 48, No. 3, pp. 363–394.

Bangkok Post. 2019. "Jailed Red-Shirt Singer Released under Royal Pardon", *Bangkok Post*, 17 July < https://www.bangkokpost.com/thailand/politics/1713940/jailed-red-shirt-singer-released-under-royal-pardon> (accessed 14 Aug. 2019).

Campbell, Charlie. 2018. "Thailand's Leader Promised to Restore Democracy, Instead He's Tightening His Grip". *Time*, 21 June <http://time.com/5318235/thailand-prayuth-chan-ocha/> (accessed 14 Aug. 2019).

Ferrara, Federico. 2014. "Unfinished Business: The Contagion of Conflict over a Century of Thai Political Development". In *"Good Coup" Gone Bad: Thailand's Political Development since Thaksin's Downfall.* Edited by Pavin Chachavalpongpun. Singapore: ISEAS, pp. 17–46.

FIDH. 2016. "And Counting Lèse-majesté Imprisonment under Thailand's Military Junta". Paris: International Federation for Human Rights February <www.fidh.org/IMG/pdf/fidh_thailand_report_lese_majeste.pdf> (accessed 14 Aug. 2019).

Human Rights Watch. 2016. "Thailand: Junta Arrests Activist's Mother". 6 May <www.hrw.org/news/2016/05/06/thailand-junta-arrests-activists-mother> (accessed 14 Aug. 2019).

iLaw. 2014. "Rai-ngan Kantang Khoha Thang Kanmuang Lang Ratthaprahan 2557" ["Report of Political Charges after the Coup of 2014"] <https://freedom.ilaw.or.th/politically-charged> (accessed 14 Aug. 2019).

———. 2018. "Sathiti Khadi 112 Thi Na Sonjai Tangtae Pi 2557–2560" ["Interesting Statistics in Cases of 112 from 2014 to 2017"] <https://freedom.ilaw.or.th/Thailand-Lèse-majesté-Statistics-Until-2014-2018%20> (accessed 27 May 2018).

Jory, Patrick. 2015. "Republicanism in Thai History". In *A Sarong for Clio: Essays on the Intellectual and Cultural History of Thailand.* Edited by Maurizio Peleggi. Ithaca: Cornell Southeast Asia Program Publications, pp. 97–117.

Metta Wongwat. 2017. "A Decade of Article 112 Cases". *Prachatai*, 15 November <https://prachatai.com/english/node/7466> (accessed 14 Aug. 2019).

Offices of the Attorney General. *Rai-ngan Statiti Khadi San Thua Ratcha-Anajak Prajam Pi 2553–2557* [Annual Judicial Statistics, Thailand 2010–14].

Office of the Judiciary. *Rai-ngan Statiti Khadi San Thua Ratcha-Anajak Prajam Pi 2553* [Annual Judicial Statistics, Thailand 2010–13].

Political Prisoners in Thailand [website]. 2018. "On the lese majeste regime". 17 Oct. < https://thaipoliticalprisoners.wordpress.com/tag/asia-times-online/> (accessed 14 Aug. 2019).

Prachatai. 2019. "Activist 'Pai Dao Din' to be released on Friday by the King's Pardon". 9 May 2019 < https://prachatai.com/english/node/8038> (accessed 14 Aug. 2019).

———. 2017. "Court Accepts Charges against Lawyer Facing 50 Years in Jail for Lèse-majesté". 26 July <https://prachatai.com/english/node/7292> (accessed 14 Aug. 2019).

———. 2018. "Embattled Lèse-majesté Convict Somyot Released". 30 April <https://prachatai.com/english/node/7729> (accessed 14 Aug. 2019).

———. 2018. "Embattled Lèse-majesté Lawyer Released after 16 Months in Jail". 17 August <https://prachatai.com/english/node/7802> (accessed 14 Aug. 2019).

———. 2011. "'คอป.' เสนอรัฐสภาแก้ 'ม.112' หนักสุดคุกไม่เกิน 7 ปี". 30 December <https://prachatai.com/journal/2011/12/38535> (accessed 14 Aug. 2019).

———. 2016. "Lèse-majesté under King Rama X: Six Observations". 12 December <https://prachatai.com/english/node/6763> (accessed 14 Aug. 2019).

———. 2017. "Military Court Breaks Record with 35-year Jail Term for Lèse-majesté". 6 September <https://prachatai.com/english/node/7194> (accessed 14 Aug. 2019).

———. 2017. "Military Court Sends Elderly Man to Jail for Almost 20 Years for Lèse-majesté". 9 August <https://prachatai.com/english/node/7319> (accessed 14 Aug. 2019).

———. 2016. "Police to Activists: Take Ja New's Mom as Example". 9 May <https://prachatai.com/english/node/6129> (accessed 14 Aug. 2019).

———. 2017. "Sedition Law Political Tool of the Junta: iLaw". 30 August <https://prachatai.com/english/node/7350> (accessed 14 Aug. 2019).

———. 2018. "Tom Dundee Faces Fourth Lèse-majesté Charge". 13 February <https://prachatai.com/english/node/7624> (accessed 14 Aug. 2019).

———. 2018. "2 Suspects not Guilty of Lèse-majesté for Defaming Princess Sirindhorn". 22 February <https://prachatai.com/english/node/7643> (accessed 14 Aug. 2019).

Somyot Pruksakasemsuk, phone interview with the author, 25 May and 3 June 2018.

Streckfuss, David. 2013. "European Constitutional Monarchies as Democracies versus the Kingdom of Thailand as Theocracy". Unpublished manuscript.

———. 2010. "The Intricacies of Lèse-majesté: A Comparative Study of Imperial Germany and Modern Thailand". In *Saying the Unsayable: Monarchy and Democracy in Thailand*. Edited by Søren Ivarsson and Lotte Isager. NIAS Press: Nordic Institute of Asian Studies, pp. 105–144.

———. 2011. *Truth on Trial in Thailand: Defamation, Treason, and Lèse-Majesté*. London and New York: Routledge.

Sulak Sivaraksa. 2018. "Lessons from the Latest Political Lawsuit of Sulak Sivaraksa". *Prachatai*. 29 January <https://prachatai.com/english/node/7592> (accessed 14 Aug. 2019).

Thaweeporn Kummetha. 2013. "Thai Man Found Guilty of Attempted Lèse-majesté". *Prachatai*. 12 December <https://prachatai.com/english/node/3790> (accessed 14 Aug. 2019).

Thongchai Winichakul. 2016. *Thailand's Hyper-royalism: Its Past Success and Present Predicament*. Singapore: ISEAS, Trends in Southeast Asia 7.

Ünaldi, Serhat. 2016. *Working towards the Monarchy: The Politics of Space in Downtown Bangkok*. Honolulu: University of Hawaii Press.

11

FROM THE YELLOW SHIRTS TO THE WHISTLE REBELS

Comparative analysis of the People's Alliance for Democracy (PAD) and the People's Democratic Reform Committee (PDRC)

Aim Sinpeng

Who were the Yellow Shirts? Were the Yellow Shirts who helped topple Thaksin Shinawatra the same as the ones that contributed to Yingluck's downfall? How were they mobilised and why? What did the Yellow Shirts want, and how have they mattered to contemporary Thai politics? The chapter charts the origin of the Yellow Shirt movement centred on the People's Alliance for Democracy (PAD) and its evolution to the "whistle rebels" – the People's Democratic Reform Committee (PDRC) – between early 2000s and 2014. These broad conservative forces were largely urban middle class with strong elements of royalism, nationalism, Buddhism and anti-majoritarianism. When focusing on the leadership structure, movement organisation and mobilisation strategies of the PAD and the PDRC, I argue that these broad opposition movements were best understood as composing independent factions rather than like-minded coalition partners. With differing backgrounds, motives and aims, each faction united only temporarily in order to oust the incumbent but soon disintegrated as faction leaders squabbled for power in the post-coup environment. The most successful factions were the ones with the greatest resources: the most ability to organise and mobilise popular support for their cause. The PAD and the PDRC were ideologically similar but different in organisational leadership and structure. Both movements tapped into the same discontent among the politically and socially conservative populace, which had grown to despise the authoritarian-style majoritarianism that the Thaksin-aligned parties had come to represent. Looking forward, as opposition coalitions like the PAD and the PDRC would come and go and the leadership was recycled, the conservative grassroots support would remain because their grievances were not met.

What lies beneath the Yellow Shirts?

What was the origin of the Yellow Shirts? There are two answers to this question: one is simple; the other is complex. The Yellow Shirts came about when Sonthi Limthongkul, a former media mogul in staunch opposition to the Thaksin government, started wearing yellow shirts with a caption *Rao Ja Su Peu Nai Luang* ("we will fight for the king") on his popular talk show, *Muang Thai Rai Supda* (Thailand Weekly) in late 2005. As his talk show grew in popularity, so too did the number of people wearing yellow shirts as he took his show on a road tour, after Thaksin had cancelled it from a free-to-air channel. Criticising Thaksin turned out to be lucrative for Sonthi. It made him personally famous and made his media business a lot of money. Eventually, as various opposition groups loosely amalgamated to form the People's Alliance for Democracy (PAD), the yellow shirts were adopted as the opposition symbol against Thaksin Shinawatra and his government. The colour yellow, symbolising the much-revered King Bhumibol Adulyadej (the royal flag colour), was cleverly used to unite the otherwise fragmented, ideologically diverse and eclectic groups under one roof. The PAD leadership reflected on their decision to frame the opposition as a royalist movement: "Manipulating the issue of the monarchy was an effective tool for mobilisation. A broad-based movement like the PAD needed a common ground. . . . That's why we had to use the monarchy".[1] So why wear yellow shirts? From the opposition movement standpoint, it was strategic and pragmatic to brand themselves as fighting for the monarchy. A more complicated answer to this same question of who the Yellow Shirts were would require us to think beyond the colour of the shirts – to examine what lies beneath what appears to be, at first glance, a "pro-monarchy movement". This chapter examines what lies beneath the yellow-shirted political phenomenon. It argues for a more nuanced analysis of the two key opposition movements – the PAD and the PDRC – which were understood broadly as the Yellow Shirts in contemporary Thai politics. Strategically bound together through the royalist frame, the leadership and organisation structure of both opposition movements were neither unequivocally pro-monarchy nor united. Instead, various anti-government factions came together out of convenience to form these two opposition alliances.

The strategic choice to unite under the pro-monarchy banner by the PAD was both a blessing and a curse to the opposition movement. Royalism, as a mobilisation frame, helped to recruit supporters and garner financial donations – both were critical to demonstrating the movement's popularity and orchestrating mass-scale protests. However, the problem with the core factions inside of the PAD was that none of them were particularly royalists.[2] Their main objectives for opposing the government had nothing to do with monarchical loyalism. The focal point for both the PAD and the PDRC-led protests was corruption and executive abuse of power, but protecting the monarchy was used to rally mass support. As "reluctant royalists" they were strategically forced to adopt the royalist frame in order to appear united so that they could emerge as a credible force opposing the seemingly more united Thai Rak Thai–led coalition government. Royalism sat like a wafer-thin roof of a multi-storied house on shaky foundations, with each of the units on each floor fighting with one another. It was a matter of time before the house came tumbling down as leaders of each unit fought over the control of the house. The house was then reassembled, with some new and some renovated (re-branded) parts and a new faction controlling the top floor. Again, under the PDRC, the house used royalism as its roof, which although still effective acted merely as a glue to bind feuding and unruly parts together. As the Yingluck government was toppled, the various factions inside the PDRC began to fall apart; some went to work in the new military-backed government and others were sidelined. Although one could say the Yellow Shirt political mobilisation – headed by the PAD and the PDRC – achieved its first objective of removing the elected governments of the Shinawatras, their subsequent focus on fighting for their own vested interests meant collectively it remained unclear what the movements achieved for the masses who supported them. Ultimately, these elite-driven opposition movements of the PAD and PDRC may have achieved limited aims for the few select groups but little overall to the rest of their supporters.

What did the Yellow Shirt supporters want? Their concept of democracy was intertwined with being righteous and moral because they saw themselves as being the leaders and organisers in society who could bring order and governance. They felt strongly that they were "the enlightened" who saw through Thaksin's many colours and recognised the "evil" in him.[3] Unlike Thaksin' supporters, who sold their votes and were duped by him, the Yellow Shirts could tell the difference between right and wrong. The most potent and powerful discourse constructed by the Yellow Shirts was one about the "gullible, stupid voters". To legitimise the opposition movement and counteract the fact that Thaksin was massively successful in elections, they argued that rural, uneducated voters sold their votes in exchange for money and/or gifts. They also believed in the monarchy as the utmost important institution in Thailand. Their reverence towards King Bhumibol and their affinity with and emotional attachment to the monarchical institution helped to propel their opposition to Thaksin.

This chapter provides an overall background for the origin and the evolution of the Yellow Shirt political phenomenon in contemporary Thai politics. It focuses on analysing the leadership, organisational structure and mobilising strategies of the PAD and the PDRC, both comparing and contrasting their similarities and differences. Although elite focused in its scope, the chapter will touch upon explaining motivations and issues of the opposition support bases. It will then conclude by reflecting on the future of opposition politics in Thailand after the 2014 coup.

The PAD and the PDRC in comparative perspective

The Thaksin government saw a flurry of protest activities during its administration. Between 2001 and 2006, there were more than 1,850 protest events reported in the media.[4] More than 60 per cent of these protests were organised by non-governmental organisations (NGOs). The biggest and most continuous hike in the number of protests reported was between mid-2005 and the end of the first quarter in 2006. Much of the opposition movement was directly targeting particular sets of government policies, such as bureaucratic reforms, opposition media harassment and plans to privatise state enterprises. The growing lack of effective constraints on the executive – particularly Thaksin – was the most contentious issue for the opposition. This "policy-based corruption" (*corruption cheung nayobai*), defined as a large-scale corruption committed by those at the highest political positions through national policy, allowed the ruling party elites to abuse power for personal gains. Policy-based corruption was rife within Thai Rak Thai (TRT), according to critics, because many of the TRT cabinet members were elite businessmen who themselves had monopolies or semi-monopolies in their industry.[5] By being at the highest level of government, their businesses continued to enjoy far greater monopolies due the fact that they themselves were policymakers. After he came to power, Thaksin's own businesses in telecommunications – Shin Corporation, AIS, and Jasmine – posted impressive profits.[6] The most explosive example of policy-based corruption was the sale of Shin Corporation to Singapore-based Temasek Holdings in 2006 – the biggest stock market trade in Thailand's history. The biggest point of contention was the fact that the sale was completed exactly two days after the passage of a new telecommunications bill that allowed increases in the foreign ownership of Thai companies from 25 to 50 per cent. The deal was, moreover, exempted from capital gains tax because prior to the sale Thaksin had conveniently amended the law regarding foreign investment in the telecommunications sector. This sale outraged so many people that it became one of the biggest rallying points for the opposition movement, prompting hundreds of thousands to join protests on the streets.

At the heart of the social movement against Thaksin has been the People's Alliance for Democracy (PAD), or the Yellow Shirts. This was a movement that brought together a broad range of groups whose interests were adversely affected by the Thaksin government. Despite the diversity of the groups that allied themselves under the rubric of PAD, the nature of the PAD core was exemplified by their five top leaders: (1) Sonthi Limthongkul, (2) Chamlong

Srimuang, (3) Pipob Thongchai, (4) Somkiat Pongpaibul and (5) Somsak Kosaisuk.[7] Indeed, the pre-coup PAD anti-Thaksin rallies were supported largely by networks of NGOs, state enterprises and trade/labour unions. Somsak, for instance, was the leader of the State Enterprise Workers' Relations Confederation, which represented over 200,000 workers. Similarly, Pipob was a highly respected NGO leader and the head of the Campaign for Popular Democracy, which drew support from a large network of NGOs all over the country. Drawing on the networks of the core leaders themselves and other non-NGO anti-Thaksin groups, such as the 40 Senators Group, university academics, Luang Ta Mahabua Students, high-ranking civil servants, students and opposition parties,[8] the PAD came together to form an alliance in February of 2006, just months before the September coup. The PAD had two levels of networks: (1) hardcore supporters, some with a vertical relationship to the leadership and (2) transient supporters, with horizontal relationships to the leadership. While all the opposition groups formed networks and relationships with one another under the umbrella of the PAD, the type of relations and degrees of separation from the PAD's core leadership differed. Moreover, over the course of the PAD's three-stage mobilisation, the structure of the PAD network was far from static. Indeed over time, while some of the groups remained as core supporters, others left and rejoined. The dynamic nature of the PAD structure had serious consequences to the size of its membership, which waxed and waned over time.

Originally the name the People's Alliance for Democracy[9] was adopted following the suggestion of Suriyasai Katasila, a PAD secretariat.[10] The "people" represented the various people from all walks of life who were bent on fighting for the country as they envisioned it. Using the word "people" portrayed the idea of "mass" in this movement and that the PAD was truly a popular mobilisation. "Alliance" meant two things. First, an alliance represented the fundamental commonality among all PAD supporters, despite their own backgrounds or origins. Various groups and interests allied together to fight as one against the Thaksin regime. Second, the alliance illustrated the structure of the PAD as a web of networks that came together to form the movement. This was why the PAD had multiple leaders to represent the various broad interests that came together to defy the Thaksin government. "Democracy" for the PAD meant there was a constitutional monarchy in Thailand, whereby the monarch was above the constitution, and the Thai democracy must protect the very foundation of the state, which was nation, religion and king. Also the people must be enlightened enough to elect the "right" and "good" leader.

The PAD self-identified as a "largely middle-class urbanised" movement.[11] Chaiwat Sinsuwong, one of PAD top leaders and former Palang Tham MP concurred, "The majority of PAD supporters are middle class, although we have some lower class folks too. Many had followed Sonthi's TV show 'Thailand Weekly' and joined the protests on their own. But people who attended rallies often had money".[12] Existing empirical studies on the socioeconomic and demographic backgrounds of Thailand's political divides have shown that the Yellow Shirts are generally better off than their Red Shirt adversaries. According to a 2,200-person survey across five provinces in Thailand in 2012, "yellow-leaning respondents" enjoyed greater economic well-being, income and educational attainment than their Red Shirt counterparts, but the latter cannot be considered as poor.[13] The Red Shirts, who will not be discussed in much detail here, were more likely to have unstable jobs, less income and a lower level of educational attainment. An earlier survey by Thailand Development Research Institute, which polled 4,097 respondents nationwide, concluded that Bangkokians were more likely to endorse the yellow camp and had higher levels of income than the Red Shirt supporters, but that poor people generally did not side with any colour.[14] Naruemon and McCargo's interview of 400 Red Shirts during the May 2010 protests also finds that they were "not well-off but not poor": 42 per cent had monthly incomes of more than 10,000 baht (US$334); 33 per cent had secondary or vocational education; and many owned an average of 6 acres of land and were generally in their forties and fifties.[15] Supporters of the United Front for Democracy against Dictatorship (UDD) were economically insecure, which stood in stark contrast to their Yellow Shirt adversaries who were "more solidly middle class".[16]

The core leadership of the PAD was divided into three parts. The first part was the First Generation leadership, which comprised five leaders drawn from a variety of groups. Sonthi Limthongkul represented the ASTV/Manager portion of the PAD, which was among the largest. The membership of the ASTV was able to mobilise thousands of protesters at every major rally. In fact, more of Sonthi's followers subscribed to ASTV or were regular consumers of Manager Media. The second group was under the de facto leadership of Chamlong Srimuang, which apart from drawing the devout members of Santi Asoke (who numbered in the thousands as well) also elicited support from those who liked Chamlong. Chamlong was a very popular Bangkok governor in the 1990s and he was among the top leaders of the Black May uprising. Somkiat Pongpaibul and Pipob Thongchai were representatives of the NGO sector. While numerous in terms of the number of organisations that joined the PAD, the absolute numbers of supporters that these NGOs brought in to the movement were not large. For instance, the Network of 30 Organisations against Corruption had 30 members, each with its own members, but they may have collectively sent only one or two of the Network's representatives to the PAD protest rally at any particular time. The overall support of NGOs inside the PAD was thus less visible if one went to the rallies. The same was true of the labour unions, under the leadership of Somsak Kosaisuk. Somsak was a chairman of the State Enterprise Workers' Relations Confederation, which represented at least 200,000 workers. He was asked to join the leadership rank and file because of the labour portion of the PAD movement that he represented, but the number of workers that actually attended PAD rallies was not a true representation of their overall support. Again, a few leaders of the labour groups were sent in, while the majority stayed behind as they were unable to miss work to join long rallies.

Of note was the fact that all of the PAD's first generation leaders had significant experience in mass mobilisation and in organising protest activities. Those representing NGOs and the labour section of the PAD, in particular, had a number of decades of experience in "street politics". They had the know-how of organising protests, which came in particularly handy when conducting protect strategies. They understood to a great extent the logistics of protest activities, knowing what worked and what might not. Prior to the PAD being officially established, each of the core leaders had in fact mounted their own protest against the Thaksin government. For instance, Sonthi's Thailand Weekly Mobile drew thousands of supporters to his weekly shows. Somsak himself staged a number of protests against the privatisation of state enterprises between 2003 and 2005. The PAD movement, in some respect, is a conglomeration of opposition movements against Thaksin into one major, unified movement. The second part of the core PAD leadership was composed of the PAD coordinators. They were responsible for the day-to-day running of the PAD movement. The coordinators were also the link between the core leaders and other groups inside the PAD. Suriyasai Katasila, Chachawal Chartsuthichai, Panthep Pongpuapan and Prapan Koonmee, for example, were key coordinators of the PAD. They often partook in the decision-making process along with the first generation of leaders, particularly when it came to press releases and media relations.[17] Following the victory of the Palang Prachachon Party (PPP), a reincarnation of the TRT, in the 2007 election, the PAD was re-mobilised. This phase of the mobilisation was notable for the stronger and more public collaboration between the PPP and the Democrat Party, which had kept its official distance to the movement prior to the 2006 coup. The PAD-Democrat alliance ran from top to bottom. At the leadership level, a number of Democrat MPs frequented PAD protest sites and many were regular speakers on

PAD stages. Notable top Democrat Party executives, such as Prapan Koonmee and Khunying Kalaya Sophonpanich, became an active part of the PAD and helped to finance it, indicating the coalition approval at the highest level. Indeed some frequent PAD protesters revealed they have seen Abhisit visiting the rally site, offering "support" to the PAD leaders backstage.[18] The biggest contribution the Democrat Party could make to the PAD movement was to provide mass support. In fact, Democrat Party leaders admitted mobilising their masses to PAD rallies, particularly to the infamous 193-day protest. While figures varied, according to party estimates, the Democrat Party forces accounted for at least 50 per cent of total PAD masses.[19] "Democrat Party members, mostly southerners, mobilised the mass to join PAD rallies".[20] Democrat southern MPs, particularly Suthep Thaugsuban, were instrumental in bringing in the masses from party supporters from its stronghold. Pusadee Thamtai, Democrat MP and former party executive, argued:

> Officially the [Democrat] Party does not have a specific policy to endorse the PAD. Abhisit told us not to go on PAD stage and let the mass movement take its own course. Unofficially, however, if you [Democrat MPs] want to attend PAD rallies you do it on your own terms.[21]

The end result of this strong PAD-Democrat alliance was not only sheer mass, but it contributed significantly in bringing down both Samak and Somchai governments and allowing the Democrats to form a government coalition. "It's a victory for the PAD . . . we rid of the Somchai government . . . Abhisit can bring us the new kind of politics with policy innovations and good people to govern the country".[22]

In what their leaders termed "the last war", the movement engaged in the longest, most violent anti-government rally to date.[23] The 193-day protest began soon after then Prime Minister Samak Sundaravej of the PPP announced he would seek to amend the 2007 constitution, which the PAD saw as a "national crisis", and that the Thaksin regime was very much alive and remained a threat to Thailand's constitutional monarchy.[24] While not engaging directly with the Red Shirts' *Phrai* versus *Ammat* discourse,[25] the PAD believed democracy in Thailand would not work like in Western countries because politicians were corrupt and many citizens sell their votes. For the PAD, elections were essentially steeped in "money politics".[26] After a long drawn-out rally that included raiding the Government House and occupying the country's main airport, among other actions, the PAD declared "victory" in December 2008 when the PPP was dissolved by the court, paving the way for the Democrat Party to cobble together a coalition and ascend to power. The honeymoon period between the PAD and the Democrats was over, however, soon after Abhisit came to power. The Democrat Party was, in PAD's view, reneging on the promises they made when they fought together against the Thaksin regime. Resentment began to build up as the PAD felt they was not getting their share of what they wanted even though they were responsible for Abhisit coming to power. The straw that broke the camel's back, which became the key issue of the third wave of the PAD protest, was the one involving the Thai-Cambodia territorial dispute. The more radical wing of the PAD, which included the ultra-nationalist and ultra-royalist groups,[27] began to hold anti-government rallies during the Abhisit administration, and the PAD masses dwindled significantly. One by one, Democrat MPs who used to vocally and proactively support the PAD began to distance themselves from the movement.[28]

In the 2011 election, the PAD's most notable action was the "Vote No" campaign, in which the PAD called on voters to check "none of the above". Although the "Vote No" campaign was a resounding failure, with less than 3 per cent of the constituency vote, it created a massive rift between the Democrats and the PAD, as well as the latter's breakaway party, New Politics Party. The fallout between the PAD and the Democrat Party did make a difference in some constituencies, as the "Vote No" campaign took away votes that could have gone to the Democrats and could have meant a victory for the Democrat Party in a tight race.[29] The PAD suffered additional setbacks internally, including dwindling financial support, leadership breakup,[30] poor coordination and fatigue. Losing Democrat Party support was severely detrimental to PAD mass appeal.[31] As the PAD movement died down, the People's Democratic Reform Committee (PDRC) emerged in 2013 to replace the former following two years of low-level discontent against Yingluck and what her opponents believed to be nepotistic politics. In fact, when the PAD leadership fell into disarray towards the end of 2010, there was an opposition vacuum in desperate need of a leadership. But subsequent anti-Thaksin groups that had emerged between 2010 and 2013 were small, and their leaders lacked the charisma and skill to mobilise the masses and unite a number of disparate opposition groups. Much like coalitional politics, groups such as the "multi-coloured shirts" led by Tul Sithisomwongsa or the Pitak Siam Organisation (Protect Siam) led by General Boonlert Kaewprasit (aka Seh Ai) could neither galvanise the opposition support base nor provide coherent and unifying goal necessary to churn out people into the streets. Both Sonthi Limthongkul (former PAD leader) and Suthep Thaugsuban (future PDRC leader) were asked to lead the opposition, but they refused on the basis that the time was not right.[32] It was not until late 2013 that the PDRC would officially come together. Ironically, Suthep the de facto leader of the PDRC, and his Democrat Party would be the last to join but the most powerful unit of the entire movement.

The key trigger for the large-scale anti-government mobilisation and the subsequent establishment of the PDRC came mid-2013 following the government's attempt to pass an amnesty bill seen by opponents as a ploy to vindicate Thaksin, who had been in a self-imposed exile since the 2006 coup. More than 100,000 protesters poured onto the streets of Bangkok, some of which occupied key government agencies, Government House and telecommunication centres despite the bill being thrown out by the Senate.[33] Demands for Yingluck's resignation did not subside, and to break the deadlock Yingluck dissolved parliament at the end of 2013. All opposition MPs by that time had resigned and most would boycott the subsequent election. The PDRC launched the campaign "Reform before Election", outlining their demands to put an indefinite end to electoral democracy before the country could be rid of "corrupt politicians". As Suthep declared:

> Thailand needs a political system that comes from the people – system that chooses moral and knowledgeable candidates to represent the people. The system must also ensure honest elections, prevent vote buying and select only good people to govern. People need to be empowered to keep politicians accountable.... The PDRC demands that a people's assembly be formed to change laws and implement these reforms.[34]

The PDRC would soon stage six-month-long protests and rallies throughout the Bangkok area sabotaging the February 2014 election, which Yingluck hoped would break the political deadlock. When election day finally rolled around, nine provinces in southern Thailand – the Democrats' heartland – had no voting at all, while the overall turnout was 47 per cent, the lowest in decades and a far cry from the 75 per cent turnout in the previous two elections. The No Vote movement was believed to have succeeded in keeping ten million Thais at home on election day, combined with the unusually high number of invalid and Vote No ballots, the PDRC declared its anti-election campaign a victory and was quickly followed by a military putsch in May.

The PDRC and the PAD shared some important similarities and differences. First, key networks of the PDRC were all part of the PAD networks but under different names and organisations. The dissolution of the PAD had produced four large networks led by four former PAD leaders: (1) the Student and People Network Thailand Reform, (2) Santi Asoke and the Dharma Warriors, (3) the Green Group and (4) the People's Networks from 77 Provinces. The Student and People Network Thailand Reform (Koh Poh Toh – STR), known locally as the "Urupong protesters", united some of the smaller opposition groups together and built momentum for what would become the PDRC. The STR was the newly reformed PAD with a close connection to Sonthi and his ASTV networks. The STR's top leaders were Nithithorn Lamleua, PAD's legal advisor, and Uthai Yodmanee, Ramkamhaeng University's student leader, who had protested against Thaksin with the PAD. Both leaders revealed a very close relationship to former PAD leaders, particularly Suriyasai and Sonthi, and had borrowed money from PAD figures and utilised their kitchen and mobile toilets to care for staff and protesters who had gathered at Urupong junction, their main protest stage.[35] All former PAD leaders appeared multiple times on their stage and publicly endorsed them as opposed to the Democrat-led protests.[36] Chamlong and his Santi Asoke crew were already staging their protests at the Makawan Rangsan Bridge, although the numbers were small. Somkiat Pongpaibul, who led the People's Network from 77 Provinces, joined forces with the STR quite early on, followed closely by Suriyasai's Green Group. The STR had their first protest on 10 October 2013, with the mass largely composing university students, former PAD core supporters and local residents in the Urupong area. For the next six weeks, various former PAD networks with all their leaders officially joined forces, with the STR with Sonthi's ASTV media networks helping to popularise the Urupong protests, resulting in greater popular support.[37] The STR was very clear why it was against Yingluck and why it wanted to end the current form of electoral democracy. In its open letter to foreign dignitaries and journalists, the group argued strongly that elections did not equal democracy. In the case of the Thaksin regime, it was "election plus bribed politicians plus corrupted elements in [the] bureaucracy, armed forces, the lack of check and balance and corrupted mass media" that led to the demise of Thailand's so-called democracy.[38] Hence they demanded reforms before election, although they were less clear about the specific of reforms required.

Second, the PDRC represented a much greater role played by the military in street politics. During the anti-Thaksin mobilisation prior to the 2006 coup, there was not a notable role played publicly by former military personnel or police in the streets. Following the PAD's breakup, Pitak Siam, which was led by a former army general, saw an opportunity to rally civil servants, military officers and police, many of whom were supportive of the PAD. But Pitak Siam never materialised and eventually evolved into the "People's Army to Uproot the Thaksin System" (PAUTS), which featured seven leaders, most of whom were former generals.[39] They were the early movers of the anti-Yingluck mobilisation, with their first protests at Lumpini Park in April 2013 before officially joining force with the STR in October and subsequently the PDRC a month later. Their support base was largely drawn from the former PAD as many supporters wore yellow shirts and banners used during previous PAD protests.

Third, the PDRC was led by the Democrats and highly personalised around its leader, Suthep Thaugsuban. The Democrats took a back seat during PAD mobilisation, but seeing no way to compete with the Thaksin-aligned parties following the latter's subsequent electoral victories in 2007 and 2011, they entered street politics full time. One of the biggest contributors to the PDRC's expansion in mass appeal was social media, particularly Facebook. When Suthep, an opposition MP from the Democrat Party announced that he, along with eight other MPs, would walk away from formal politics to start a grassroots movement, his Facebook popularity exploded.[40] Within weeks, Suthep went from a rather unpopular politician to a beloved "uncle" or *kamnan*[41] – championing the cause of many Thais disenchanted with Yingluck. Suthep's

Facebook profile was akin to that of a rock star: he garnered more than a 5,000 per cent increase in likes, with each post eliciting more than 100,000 interactions from net users. Existing research on the relationship between Facebook and mobilisation of the PDRC has shown that social media engagement has both created new groups of supporters who were previously disenfranchised or politically inactive to join the movement, particularly among the lower socio-economic positions.[42] Suthep had been coordinating with former PAD leaders at the Urupong and Makawan Bridge stages, with Suriyasai and former Democrat MP Thavorn Saneneam acting as coordinators between the STR and the Democrat-led protests. By late November, the STR recognised the strength and expansive reach of Suthep and the Democrat machineries which helped grow and amass significant support. On 23 November, all the leaders appeared together on the Urupong stage, officially uniting their forces as one. On 29 November, the PDRC was established with Suthep chosen as the top leader.

In sum, the opposition movements of 2005–2006 and 2013–2014 exemplified by the leadership of the PAD and the PDRC represented two key grievances in Thai politics. The first was the discontent towards majoritarian form of democracy, particularly regarding the executive abuse of power. The two key triggers that really mobilised both movements were not about Thaksin and his family per se, but rather about what they did. The controversial sale of Shin Corporation and Yingluck's attempt to pass an amnesty bill represented, for the Yellow Shirts and the whistle rebels, the ultimate act of unconstrained executives who used their political office for personal gain. The second was about the economic insecurities of the urban middle class, whose political and economic dominance in society were challenged by the new and upwardly mobile lower-middle class, many of whom were Thaksin supporters. Royalism and nationalism were deployed by both the PAD and the PDRC to galvanise and unite variously large and often ideologically disparate groups of protesters together in order to mount a credible challenge to the incumbent and signal to the military popular support for extraconstitutional intervention. The PAD and the PDRC were similar in their networks of support bases and key actors but differed in the extent of the grassroots support and the hierarchy of movement leadership. The PDRC was much more direct in its support for military putsch and temporary secession of electoral democracy. It was also much larger and youthful in popular support, thanks to social media, but its alliances were looser than the PAD. In the post-2014 coup environment, this temporary union of opposition forces under the PDRC banner cracked and broke away, with the Suthep-led faction maintaining much of what used to be the movement. It was unclear, however, how much of what the PDRC demanded that it really achieved for its supporters, other than the coup. With grievances potentially unresolved, it is likely that the "Yellow Shirted political phenomenon" will continue following the next election under a different name and colour.

Notes

1 PAD leader (personal communication, 19 December 2012). Interviewee chose to remain anonymous.
2 Michael Connors, "Article of Faith: The Failure of Royal Liberalism in Thailand", *Journal of Contemporary Asia*, Vol. 38, No. 1 (2008), pp. 143–165.
3 Sonthi Limthongkul and Sarocha Pornudomsak, *Muang Thai Rai Sabda San Jon* [Thailand weekly mobile] (Bangkok: Ban Pra Athit Press, 2006).
4 Aim Sinpeng, *The Power of Political Movement and the Collapse of Democracy in Thailand*, Doctoral dissertation submitted to the University of British Columbia (2013).
5 For a detailed discussion, see Duncan McCargo and Ukrist Pathmanand, *The Thaksinization of Thailand* (Copenhagen: NIAS Press, 2005).
6 From 2001 to 2004, Shin Corporation's net profit jumped 260 per cent while AIS increased its profit by 85 per cent for the same period.

7 Terdtham Songthai, *Kha Kue Nakrop Prachachon Ku Chat* [We are the People's Warriors to Save the Nation] (Bangkok: GPP Publication, 2012), p. 99.
8 ASTV Online, "Thap Phanthamit Ku Chat Yan Khluean Phon Khao Lan Phrarup 14.00 Nor [PAD Forces to Save the Nation Confirmed Mobolisation at 2pm]", *ASTV*, 11 February 2006 <www.manager.co.th/Home/ViewNews.aspx?NewsID=9490000019100> (accessed 18 June 2014).
9 Suriyasai Katasila, *Panthamit Prachachon Prachathipathai* [The People's Alliance for Democracy] (Bangkok: Openbooks, 2006).
10 Suriyasai Katasila, speech at the Leadership School, Kanchanaburi province, 15 January 2009.
11 Interviews with various PAD leaders confirmed this statement.
12 Chaiwat Sinsuwong (personal communication, 10 July 2011).
13 Ammar Siamwalla and Somchai Jitsuchon, "The Socio-economic Bases of the Red/Yellow Divide: A Statistical Analysis", in *Bangkok, May 2010: Perspectives on Divided Thailand*, edited by Michael Montesano, Pavin Chachavalpongpun, and Aekapol Chongvilaivan (Singapore: Institute of Southeast Asian Studies, 2012), 64–71.
14 Wanwipang Manachotpong, "Mong Lueang Daeng Phan Khua Khwamkhit Thang Kanmueang" [The Shades of Red and Yellow: Evidence from Survey Data], *Thammasat Economic Journal*, Vol. 32, No. 3 (2014), pp. 31–68.
15 Duncan McCargo and Naruemon Thabchumpon, "Urbanised Villagers in the 2010 Thai Red Shirt Protests: Not just Poor Farmers", *Asian Survey*, Vol. 51, No. 6 (2011), pp. 993–1018, 1002.
16 McCargo and Naruemon, "Urbanised Villagers", p. 1001.
17 They connected the PAD to the outside world, were the mouthpiece of the PAD and supported and realised the goals of the leaders. They were crucial to the mobilisation efforts of the PAD, particularly for long rallies.
18 Based on interviews with various PAD demonstrators during the 193-day rally and the 2011 round.
19 Witaya Kaewparadai (personal communication, 12 July 2011).
20 Thawil Paison (personal communication, 12 July 2011).
21 Putsadee Thamtai (personal communication, 12 July 2011).
22 Sonthi Limthongkul, "Guarding the Motherland", *ASTV*, 23 November 2008.
23 Ibid.
24 Suriyasai Katasila, "The Big Demonstrations", *PAD Press Conference*, Bangkok <www.manager.co.th/Politics/ViewNews.aspx?NewsID=9510000059495> (accessed 18 June 2014).
25 For more discussion on this see Pitch Pongsawat, *The Politics of Phrai* (Bangkok: Openbooks, 2011). Also Nithi Aeosriwong, *Reading Thai Politics 3: The Politics of the Red Shirts* (Bangkok: Openbooks, 2010).
26 People's Alliance for Democracy, "New Politics", *Speech at PAD Rally Outside Government House*, Bangkok, 8 November 2008.
27 For example, the Thai Patriot Network and Santi Asoke.
28 Kasit Piromya took part in the PAD's occupation of Suvarnabhumi Airport. He gave an interview to the Telegraph on 21 December 2008, saying that the airport protests were "fun with good food".
29 A Democrat incumbent from Bangkok revealed, "Before the coup I was attending PAD rallies everyday because the majority of people in my constituency went. . . . I went to garner votes . . . now the Vote No campaign really hurt the Democrat Party. . . . I was even kicked out from some houses" (personal communication, 14 July 2011).
30 Somsak Kosaisuk, one of the PAD core leaders, left PAD in early 2011. He became the head of the New Politics Party (NPP), a new party formed after the PAD general meeting in 2009. The NPP and the PAD were not officially affiliated.
31 Michael H. Nelson, "Vote No! The PAD's Decline from Powerful Movement to Political Sect?" Paper presented at the conference on "Five Years after the Military Coup: Thailand's Political Developments since Thaksin's Downfall" (ISEAS: Singapore, 19 September 2011).
32 Rungmanee Meksophon, *Suu Mai Toy* [We will Fight to the End] (Bangkok: Baan Pra Athit Press, 2014), p. 207.
33 "Yon Roi Jed Sip Ha Wan Koh Poh Poh Soh Gon Shut Down Muang Krung [A Look Back at the 75-day before the PDRC's Bangkok Shutdown], *Thairath*, 12 January 2013 <www.thairath.co.th/content/395551> (accessed 24 January 2017).
34 Kamnan Suthep Taugsuban, *The Power of Change* (Bangkok: LIPS Publication, 2014), pp. 243–244.
35 Nithithorn Lamleoa, *Rongtao Pah Bai Kub Jai Tueng Tueng* [A Pair of Sneakers and a Big Heart] (Bangkok: Baan Pra Athit Press, 2014), pp. 34–35.
36 Ibid., p. 212.

37 "Koh Poh Toh Gonghtubtham Kruekai Jedsipjed Joh Ok Thalang Khao Yok Radab [SNT, the Dharma Warriors, the 77-Province Network, Declared They Would Escalate Their Mobilisation], *Daily News*, 9 November 2013 <www.dailynews.co.th/politics/193627> (accessed 24 January 2017).
38 Lamleoa, *Rongtao Pah Bai*, p. 202.
39 "Kong Thap Prachachon Khon Rabob Thaksin Ped Tua Jed Gane Num" [The People's Army to Uproot the Thaksin System Announced their 7 Leaders], *Thai PBS*, 4 August 2013 <https://news.thaipbs.or.th/content/187923> (accessed 24 January 2014).
40 "9 Soh Soh Pochopo Nam Doi 'Su Thep' Yuen Bai La-Ok Thi Sapha Kokoto Phrom Chat Lueaktang Som" [Nine MPs Led by 'Suthep' Handed in Resignation Notice at the Parliament. The Election Commission Gets Ready for a By-election]. *Prachatai*, 12 November 2013 <www.prachatai.com/journal/2013/11/49748> (accessed 15 December 2018).
41 *Kamnan* is a governing official in the sub-district level (tambon). Suthep rebranded himself from a politician (a national level official) to uncle or *kamnan* (used as a term of endearment) to increase proximity to the populace and demonstrated his willingness to form and lead a "grassroots movement".
42 Aim Sinpeng, "Participatory Inequality in Online and Offline Political Engagement in Thailand", *Pacific Affairs*, Vol. 90, No. 2 (2017), pp. 253–274.

References

Aim Sinpeng. 2017. "Participatory Inequality in Online and Offline Political Engagement in Thailand". *Pacific Affairs*, Vol. 90, No. 2: 253–274.

Ammar Siamwalla and Somchai Jitsuchon. 2012 "The Socio-economic Bases of the Red/Yellow Divide: A Statistical Analysis". In *Bangkok, May 2010: Perspectives on Divided Thailand*. Edited by Michael Montesano, Pavin Chachavalpongpun, and Aekapol Chongvilaivan. Singapore: Institute of Southeast Asian Studies.

Connors, Michael. 2008. "Article of Faith: The Failure of Royal Liberalism in Thailand". *Journal of Contemporary Asia*, Vol. 38, No. 1: 143–165.

Nithi Aeosriwong. 2010. *Reading Thai Politics 3: The Politics of the Red Shirts*. Bangkok: Openbooks.

McCargo, Duncan and Ukrist Pathmanand. 2005. *The Thaksinization of Thailand*. Copenhagen: NIAS Press.

Pitch Pongsawat. 2011. *The Politics of Phrai* Bangkok: Openbooks.

12

MASS POLITICS AND THE RED SHIRTS[1]

Claudio Sopranzetti

In the years between the second election of Thaksin Shinawatra in 2005 and the 2014 military coup that removed his sister Yingluck from power in 2014, Thailand lived through an unprecedented period of popular social movements and street protest. Even though this era of mass politics is almost hard to remember after five years of Prayuth Chan-ocha's military government,[2] in that decade millions of people participated in street protests, organising meetings, political schools, and rallies across the country, from the highlands of northern Thailand to the sleepy Chinese towns of southern Thailand.

This period of unprecedented political awakenings and popular direct actions, whether under yellow or red shirts, has generated a sea of analysis, articles, and books and will continue to do so once historians will begin to write about it. It would be well beyond my scope here to attempt drafting a preliminary history of street protests over that decade, let alone over the previous few which laid the groundwork for those social movements. In this chapter, instead, I focus on the Red Shirts and particularly on the five years between the 19 September 2006 military coup and the beginning of their largest political mobilisation in 2010. While many scholars have focused on their protests and violent dispersal in 2010,[3] I want here to leave out the 2010 mobilisation and shed light on what often remains unexplored in the study of social movements, namely the genesis of their social composition and political imaginary.

The first baby steps

On 19 September 2006, the military staged the first coup since 1991 and removed from power the prime minister, Thaksin, who had been re-elected in 2005 with the largest majority in Thai history. Many in Bangkok celebrated the military takeover, seeing Thaksin as the impersonation of an immoral authoritarian leader who obtained power through corruption and vote-buying. Many others around the country disagreed but at first did not voice their dissent.

On 1 November 2006, one and a half months after the military coup, a lonely protester dramatically showed his frustration with the new political order. Nuamthong Praiwal, a taxi driver who had driven his car into a military tank at Royal Plaza the day after the coup, hanged himself under a pedestrian flyover on the Vibhavadi-Rangsit Highway, leaving a note opposing the military intervention. A few weeks later, three anti-coup groups composed largely of university students, radicals, and Thaksin supporters staged small protests of a few dozen people – lone screams into the silence of militarised Bangkok. Soon, however, their voices gained volume. By 10 December, Constitution Day, a crowd of a few thousand people protested in Sanam Luang, the enormous grounds in front of the Royal Palace. It was not, however, until 15 June 2007 that the United Front for Democracy against Dictatorship (UDD) was created.

From the central office in the Imperial World working-class shopping mall along Lad Phrao Road and a myriad of meeting places across the country, the UDD organisers feverishly set up small protests and sit-ins. On 2 September, the Red Shirts clashed with the opposing Yellow Shirts that had driven Thaksin out of office and laid the ground for the military coup. After the violent confrontation, the UDD halted its rallies, waiting to hear the results of the December 2007 elections, the first since the coup. The vote was easily won by Samak Sundaravej, a proxy of Thaksin, himself known for his authoritarian tendencies. Other state forces, however, were determined to exclude Thaksin and his allies from re-entering electoral politics. Almost immediately, the Yellow Shirts revived protests and a series of complaints were submitted to the Constitutional Court which, in September of 2008, forced Samak to resign because of a conflict of interest, based on his participation in a TV cooking show while he was prime minister. The premiership was given to Somchai Wongsawat, an uncharismatic bureaucrat and Thaksin's brother-in-law. His rise to power was encountered with extensive mobilisation among the Yellow Shirts, who saw in him a puppet guided by their arch-enemy.[4] On 6 October 2008, they blocked the parliament and Government House, demanding the resignation of the democratically elected yet (in their eyes) illegitimate government. On 25 November, a few thousand Yellow Shirt supporters marched towards Suvarnabhumi International Airport with the declared objective of preventing Somchai from returning to Thailand after an international forum. A convoy of hundreds of Yellow Shirts blocked the two ends of the road in front of the airport's terminal building and the main road to the airport. Boosted by the army's rejection of the government request to intervene and clear the airport, on 27 November the Yellow Shirt protesters moved to occupy Don Muang, the other international airport, effectively arresting air-based transportation in and through Bangkok. After protesters had blocked the nation, their bureaucratic allies proceeded to remove its elected government. On 2 December 2008, the Constitutional Court dissolved all three parties of the government coalition on questionable charges of electoral fraud.[5] The next day the Yellow Shirts protesters, satisfied with the outcomes of their actions, left the airports.

This charade – as Jim Glassman has said, "provinces elect, Bangkok overturns"[6] – escalated a few days after the creation of a new government headed by Abhisit Vejjajiva, the leader of the opposing Democrat party, thanks to the defection of 22 MPs from Samak's party. Red Shirts immediately took to the streets again, demanding the resignation of a government that had never received a popular majority but was the result of a game of military and judicial pressures, elite alliances, and cabinet reshufflings. Until then the movement had mostly coalesced around a wide and vague agenda of returning to electoral power and a motley crowd of groups, symbols, and demands. Nowhere was their heterogeneity more evident than on the bodies of the thousands of protesters who protested the new government. On their heads, Maoist hats emblazoned with red stars; on their chests, shirts with the face of the former prime minister and capitalist tycoon, Thaksin; and on the soles of their flip-flops, the face of Prem Tinsulanonda, the most powerful member of the Privy Council who advises the king and was rumoured to be behind the 2006 coup.

The protesters had articulated quite clearly their desires to change established relations of power, put an end to the military and bureaucratic interference in democratic politics, and rebalance the political-economic and legal inequality that divides Bangkok from the countryside (the rich and powerful from the poor and helpless). Yet their visions diverged and lost clarity as to how these changes should be brought about. Indeed this vagueness was fundamental in allowing collective action among a wide range of factions and actors who had often been in open conflict during the governments of Thaksin and, even once unified in the Red Shirts, operated under diverging political trajectories. While functional, this heterogeneity posed a challenge for observers as well as for the movement itself, which struggled to keep its different currents unified and to present itself to the press in a consistent and coherent way.[7] Up until this point, the multiplicity of actors, motivations, and objectives that came together under the umbrella of the UDD had clustered and dispersed in the streets of Bangkok. During this period of mobilisation, three streams started to emerge.[8]

The Red Shirts' composition

The first stream, which I call the Thaksinites, was largely composed of Thaksin Shinawatra's acolytes, former allies, and voters of the Thai Rak Thai party. Its main exponents were former party members who were banned from electoral politics following the 2006 coup. The Thaksinites' objectives, with significant internal variations, revolved around the erasure of the political changes brought by the military coup, the return of Thaksin to Thailand, and the revocation of the 2007 constitution drafted by the military junta. The second stream, the democracy activists, gathered people who had often opposed the Thaksin government, its policies, and authoritarianism, yet considered unacceptable any political change brought about by military, judicial, or bureaucratic interventions. Their goal was the establishment of liberal democracy in Thailand, with a system of checks and balances and direct control over military interventions in the political sphere. Its main exponents tended to be established personalities in street politics, former student activists, and community organisers who saw the Red Shirts as a new phase in the long and incomplete history of Thai democratic struggles. The third stream was composed of radicals. Their numbers were difficult to estimate given the potential legal consequences of voicing their opinions in public. The radicals' agenda coalesced around a call for the restructuring of political-economic relations between traditional elites and citizens and the withdrawal of the monarchy and the military from any active role in politics, putting them at risk of arrest under the increasingly used lèse-majesté law. In their vision, the Red Shirts needed to move beyond their connections to the ousted prime minister and traditional representative democracy and become a truly revolutionary movement.

These three streams, however, were currents rather than clearly established and discrete groups or organisations. During the protest they often mixed into one another, condensed, and then parted again. Most protesters voiced an idiosyncratic and at times contradictory mixture of demands and objectives. Similarly, the protest leaders and most visible political actors sat at the confluence of these streams, often riding multiple currents at the same time or drifting towards a different stream as the protest evolved. Nonetheless, by the end of 2008, the UDD began to work towards the creation of a unified "social imaginary".[9] This process was not the premise of their political mobilisation but rather emerged through it. The adoption of the colour red – a reference to the national flag – was one of the first signs of this new process of unification. The Thai flag is composed of three colours: the centre horizontal stripe is blue, which represents the monarchy; banded on each side by white, which stands for Buddhism; and with red on the top and bottom, symbolising the nation. Claiming to be the expression of the people, the Red Shirts began to reformulate the dissatisfaction expressed by millions of people around the country into a collective identity, an interpretation of contemporary Thai society, and a definition of their enemies and demands.

The Red Shirts' social imaginary

On 30 December 2008, Nattawut Saikua, one of the protest leaders, delivered a historic speech at the Government House which began to solidify this imaginary. Standing in front of thousands of Red Shirts protesters, Nattawut stared into the crowd. "We're denied many things", he began, echoing what Id, Yai, and Adun had told me:

> We're denied justice; respect in the way governmental bodies treat us; accurate and direct reporting about us in the media. We're denied the chance to declare our fight openly, to declare openly and directly, with our clarity and sincerity, what it is that we are fighting for. For sure, we don't have connections.

He paused as the crowd cheered.

> What's most important for us all to remember, brothers and sisters, is that we are the salt of the earth. We are the people with no privileges. We were born on the land. We grew up on the land. Each step that we take is on this same land. We stand, with our two feet planted here, so far away from the sky.

Nattawut looked up. "Tilting our heads upwards, we gaze at the sky, and we realise how far away that sky is. Standing on this land, we only have to look down to realise that we are worth no more than a handful of dirt". A deep silence descended on the crowd.

> But I believe in the power of the Red Shirts. I believe our numbers are growing day-by-day, minute-by-minute. Even though we stand on this land, and we speak out from our place in the dirt, our voice will rise to the sky. Of this I have no doubt. I have no doubt.

He repeated as the crowd roared.

> The voice we're making now – our cries and shouts – is the voice of people who are worth only a handful of dirt. But it is the voice of the people who were born and grew up on this land, and it will rise to the level of the sky. Of this I have no doubt. No doubt. We, the Red Shirts, want to say to the land and sky that we too have heart and soul. We, the Red Shirts, want to remind the land and sky that we too are the Thai people. We, the Red Shirts, want to ask the land and sky whether we have been condemned to seek, by ourselves, a rightful place to plant our feet here.

The crowd roared and cheered. "No matter what happens, we already have the greatest thing in our lives: the democratic spirit. For this great spirit, for the greatness of all of you, the only thing I can do is this". He concluded by kneeling down on stage and bowing to the crowd.

This speech, with its focus on the inequality in Thai society, its coded reference to monarchic elites (the sky), its denunciation of the protesters' mistreatment by bureaucratic forces (governmental bodies) and of the reduction of the Red Shirts' demands to worthless dirt resonate with what the drivers had been repeating, and became the template for their political mobilisation. Yet it did not ossify the Red Shirts' political discourse. Over the course of the following year, in fact, the movement's identity, opposition, and relations continued to evolve and remain open to transformations of their discourses, agendas, and tactics. On 26 March 2009, the Red Shirts set up a permanent protest camp in front of the Government House and demanded Abhisit's resignation. On 8 April, a crowd of more than 100,000 joined the camp and rallied there and at the adjacent Royal Plaza, while parallel rallies were held in a dozen provincial centres. Overnight, mobility through the transportation hub of the Victory Monument was brought to a halt by a crowd of taxi and motorcycle taxi drivers.

On 11 April, a group of Red Shirts broke into the ASEAN (Association of Southeast Asian Nations) summit being held in Pattaya, effectively bringing the meeting to an end and forcing Thai and foreign heads of state to flee. On 13 April, 10,000 military troops were moved into Bangkok to clear the streets and re-establish usual urban flows. In the first serious clash between state forces and Red Shirts around the Victory Monument, at least 70 people were injured and the army seemed to have won the confrontation. The protest at Government House dispersed, and many observers thought that would be the end of the Red Shirts.

The movement, on the contrary, began to reorganise its forces. The Red Shirts rebuilt their local branches, extended their presence in rural Thailand, and trained their members. More than 450 Red Shirts' schools opened all around the country. These schools – a tactic developed by student activists in the 1970s – were central to the elaboration of the movement's demands and rhetoric. It was during this period, in fact, that two of the central tenets of the Red Shirts movement emerged: the discourse of double standards, which Id used in our train ride to talk about the exclusions of villagers from legal, economic, and political rights; and their self-identification as commoners, or *phrai*. The terminology of double standards had entered the Thai political discourse in 2001 when it was used, in English, to criticise a court decision to acquit the newly elected Prime Minister Thaksin Shinawatra of proven accusations of failing to declare the full extent of his assets when entering politics.[10] Few would have guessed that this term would become so central to a movement that began by protesting Thaksin's removal by the military coup. After the 2009 crackdown, the Red Shirts adopted the term in its Thai version, *song mattrathan*, to point out the difference in treatment between the violent repression of their protest and the complete lack of repercussions and persecutions, legal and military, when the opposing Yellow Shirts blocked Bangkok's airports.[11] Soon, however, this formula was used not just to refer to judicial bias but also to describe larger political economic inequality, regional disparities, and exclusion in Thai society, on par with the word *phrai*.

The opposition between *phrai* (commoners, serf) and *ammat* (aristocracy) dominated feudal relations in Siamese society until the 1892 administrative reform that created the modern Thai bureaucracy. Red Shirts resuscitated this distinction to define themselves as well as their enemies. They conceptualised their political struggles as a fight between *phrai*, represented by their supporters, and the *ammat*, composed of the military, bureaucratic, and monarchic elites as well as the governing party led by the Oxford-educated Prime Minister Abhisit Vejjajiva. As Thongchai Winichakul has argued,

> the UDD discourse of their struggles as the "*phrai*" against the "*ammat*" reveals as much as belies the configuration of class and hierarchy in Thai context. Many Thais and foreign reporters translate the word "*phrai*" as serf, or bonded subject in the Thai feudal society. The pro-government scholars argue correctly that such a feudal social order no longer exists. But the "*phrai*" in the Reds discourse does not mean the historical bonded subjects. "*Phrai*" and its opposite, "*ammat*" (the noble, the lords) in the UDD discourse targets the oppression and injustice due to social class and hierarchy such as the one in Thai political culture. The struggle of the Reds is a class war in this sense of the revolt of the downtrodden rural folks against the privileged social and political class, the "*ammat*".[12]

This war, however, was predicated upon a very specific notion of class. When talking about *phrai*, the Red Shirts were not referring to themselves only as the poor, the peasantry, or the working class but rather as a new category, one defined by exclusion from services, such as education or access to credit, consumption, and fair legal and political treatment. In this sense, the language of *phrai* resonated deeply with the drivers' experiences, struggles, and everyday lives, both in the city and in the village: invisible infrastructure excluded from enjoying the product of their work; migrants stuck between rural backwardness and urban poverty; entrepreneurs torn between a new freedom and new forms of exclusion and exploitation; formalised service workers who had fallen back into the clutches of people of influence.

Conservative commentators responded to these claims by dismissing them with sarcasm. In an article in the pro-Yellow Shirts *Bangkok Post*, Voranai Vanijaka joked that "*phrai* need not be poor, they say. *Phrai* can have money, they insist. *Phrai* are only *phrai* in that they don't have the power, they say".[13] Although intended at dismissing the Red Shirts' discourse, this observation actually provided one of the most perceptive analyses of the movement to appear in Thai mainstream media. When seen in this light, a movement of *phrai* led by a billionaire was not a contradiction, rather it revealed the extent of the oppression and exclusion perpetrated by the *ammat*. No matter how rich you were, the Red Shirts seemed to say, those people would try to disenfranchise and oppress you, whether you were the majority of the population or a democratically elected prime minister.

This discursive shift had a double effect: first, it transformed a rarely used derogatory term (*phrai*) into a term of self-representation and a source of pride. As Isan anthropologist Bunthawat Weemoktanondha argued, the Red Shirts

> [were] breaking a cultural taboo by using this word so openly to describe themselves without feeling ashamed of being *phrai*. It is well known that this word indicates class discrimination. [The word] "*phrai*" is so sensitive that its use to describe a person could lead to serious consequences, even physical attack. This word is not used frequently because it means the speaker is calling a person low-class, ignorant, stupid.[14]

The Red Shirts were now reclaiming it and wearing it as badge of honour and a challenge to equating the oppression and unfair treatment they experienced with that against *phrai* in feudal Siam. Second, it bound the movement together around a shared sense of injustice and unfairness that equated very different groups – from rural poor to economic elites, from former politicians to rural organisers, from urban working classes to radical intellectuals. Their unity was funded on submission to the same structural relations of exclusion to the *ammat*, an equally generic term that stood for everything and everybody who oppressed the Thai population and constrained its full democratic potential – from the military to the government, from the Privy Council to the palace. As Weng Tojirakarn, one of the key leaders of the Red Shirts explained to me:

> *Phrai* simply explains everything. We are *phrai*. Nobody wants to be treated like a dog. Everybody must be treated equally as a human being. For the *phrai*, they only fight to let society accept that they are human. This society dehumanises people, so that is why the majority of the people now understand what the Red Shirts are fighting for. It is only the *ammat* that is fighting against this.

Scholars of social movements have argued that "social movements have three dimensions: (a) Identity, the definition which the social movement actor gives himself [*sic*]; (b) Opposition, the definition of his [*sic*] adversary; and (c) Totality, the stakes over which the movement and its adversary are in conflict".[15] The social imaginary that the Red Shirts developed between 2009 and 2010 accomplished all three objectives: they defined an identity for themselves as *phrai*, construed their adversary as *ammat*, and traced the stakes of the conflict as the end of this system of oppression and exclusion, condensed around a discourse of *song mattrathan* (double standards). Strong from their newly defined worldview, on 12 March 2010 the Red Shirts leaders called on supporters to descend over Bangkok. Thousands and thousands from villages all over the country started to move towards the city and took control of the city. The rest is history.

Notes

1 Parts of this chapter where previously published in the author's monograph *Owners of the Map* (Sopranzetti 2017b).

2 Claudio Sopranzetti, "Thailand's Relapse: The Implications of the May 2014 Coup", *The Journal of Asian Studies*, Vol. 75, No. 2 (2016), pp. 299–316. doi:10.1017/S0021911816000462.

3 Pinkaew Leungaramsri. 2010. *Becoming Red = Kamnoet Lae Phattanakan Sua Deang Nai Chiang Mai*, First Edition (Phim Khrang), p. 2553; Ji Giles Ungpakorn, "Class Struggle between the Coloured T-shirts in Thailand", *Journal of Asia Pacific Studies*, Vol. 1, No. 1 (2009), 76–100; Jim Glassman, *Bounding the Mekong: the Asian Development Bank, China, and Thailand* (Honolulu: University of Hawaii Press, 2010a); Pavin Chachavalpongpun, "Thai Prince Cleans House with an Eye on Throne", *Straits Times*, 8 December 2014; Jim Taylor, "Larger Than Life: 'Central World' and Its Demise and Rebirth – Red Shirts and the Creation of the Urban Cultural Myth in Thailand", *Asia Research Institute Working Paper Series*, Vol. 150 (2011), pp. 1–13; Nick Nostitz, "The Red Shirts from Anti-Coup Protesters to Social Mass Movement", in *"Good Coup" Gone Bad: Thailand's Political Development since Thaksin's Downfall*, edited by Pavin Chachavalpongpun, (Singapore: Institute of Southeast Asian Studies, 2014); Sonthisamphan Kittiphong, *Red Why = Daeng – Thammai: Sangkhom Thai Panha Lae Kanma Khong Khon Sua Deang*, First Edition (Krung Thep: Samnakphim Openbooks, 2553); Kasian Tejapira *Songkhram Rawang Si*, first edition, 2 Vols (Krung Thep: Samnakphim Openbooks, 2553); Claudio Sopranzetti, "Burning Red Desires: Isan Migrants and the Politics of Desire in Contemporary Thailand", *South East Asia Research*, Vol. 20, No. 3 (2012), pp. 361–379; Claudio Sopranzetti, "Framed by Freedom: Emancipation and Oppression in post-Fordist Thailand", *Cultural Anthropology*, Vol. 32, No. 1 (2017a), pp. 68–92; Felicity Aulino, Eli Elinoff, Claudio Sopranzetti, and Ben Tausig, "Introduction: The Wheel of Crisis in Thailand". *Fieldsights – Hot Spots, Cultural Anthropology Online* 23 (2014); Sopranzetti 2017b, 2014; Naruemon Thabchumpon and Duncan McCargo, "Urbanised Villagers in the 2010 Thai Redshirt Protests Not Just Poor Farmers?" *Asian Survey*, Vol. 51, No. 6 (2011), pp. 993–1018.

4 See Nick Nostitz. *Red vs. Yellow*, Vol. 2 (Bangkok: White Lotus Press, 2009).

5 See John Funston, *Divided over Thaksin: Thailand's Coup and Problematic Transition* (Chiang Mai: Silkworm Books, 2009).

6 Glassman (2010b).

7 The multiplicity of representations was particularly evident in international media – especially BBC and CNN – which covered the Red Shirts alternatively as a peasant movement, a rented mob under the control of the media Tycoon Thaksin Shinawatra, or a socialist uprising, depending on their sources in Thailand.

8 This distinction is a gross over-simplification of a segmented multiplicity of forces. I divided them up for the purpose of analytical clarity. I do so with the awareness that this is just one of many possible artificial sectionings of a movement that has retained a fluid and multidimensional nature.

9 Nancy Abelmann, *Echoes of the Past, Epics of Dissent: A South Korean Social Movement* (Berkeley: University of California Press 1996), p. 6.

10 Immediately after his election as prime minister in 2001, Thaksin was charged with illegally concealing the full amount of his assets while deputy prime minister in 1997. The Thai Constitutional Court, in a ruling which seemed to rely more on Thaksin's vast electoral support than on firm legal ground, dismissed these accusations without explaining in detail the reasons behind their decision. As a result the National Counter Corruption Commission (NCCC) pointed to a "double standard" in the ruling when compared to similar cases against other public officeholders.

11 Interestingly, however, the legal origins of this phrase, *song mattrathan*, evidence of the centrality of the relationship between authority and influence to inequality Thai society, is the dynamic I explored in chapter 5 (Sopranzetti 2017b). As Chiang Noi argued,

> In Thailand there is a close connection between power and illegality, between social status and defiance of the law. Often, laws seem to exist precisely to allow certain people the very special privilege of being able to flout them. [. . .] The growing political significance of the judiciary over the last few years has given this trend a new twist. While the judicial system may not in fact perform in the service of equity, the justification for the importance of the judiciary is that there really is a rule of law that applies to all. [. . .] The idea of equity under the law is now very prominent in public debate, yet the political structure still in place is designed precisely to preserve privileges by the evasion or manipulation of the law.
>
> (Chiang Noi 2010)

12 Thongchai Winichakul, "Coming to Terms with the West: Intellectual Strategies of Bifurcation and Post-Westernism in Siam", in *The Ambiguous Allure of the West: Traces of the Colonial in Thailand*, edited by Rachel Harrison and Peter Jackson (Ithaca: Southeast Asia Program Publications, Cornell University, 2010).

13 Voranai Vanijaka, "Ammat and Phrai: The Facebook War", *Bangkok Post*, 15 May 2011.

14 He made this declaration to Inter Press Service News (Marwaan Macan-Markar, "Anti-Gov't Protesters Use Cultural Taboo as Weapon", *Inter Press Service*, 18 April 2010).

15 John A. Hannigan, "Alain Touraine, Manuel Castells and Social Movement Theory a Critical Appraisal", *The Sociological Quarterly*, Vol. 26, No. 4 (1985), p. 445.

References

Abelmann, Nancy. 1996. *Echoes of the Past, Epics of Dissent: A South Korean Social Movement*. Berkeley: University of California Press.

Aulino, Felicity, Eli Elinoff, Claudio Sopranzetti, and Ben Tausig. 2014. "Introduction: The Wheel of Crisis in Thailand". *Fieldsights – Hot Spots, Cultural Anthropology Online* 23.

Funston, John. 2009. *Divided over Thaksin: Thailand's Coup and Problematic Transition*. Chiang Mai: Silkworm Books.

Glassman, Jim. 2010a. "From Reds to Red Shirts: Political Evolution and Devolution in Thailand". *Environment and Planning A*, Vol. 42: 765–770.

———. 2010b. "The Provinces Elect Governments, Bangkok Overthrows Them: Urbanity, Class and Post-democracy in Thailand". *Urban Studies*, Vol. 47, No. 4: 1301–1323.

Hannigan, John A. 1985. "Alain Touraine, Manuel Castells and Social Movement Theory a Critical Appraisal". *The Sociological Quarterly*, Vol. 26, No. 4: 435–454.

Ji, Ungpakorn. 2009. "Class Struggle between the Coloured T-shirts in Thailand". *Journal of Asia Pacific Studies*, Vol. 1, No. 1: 76–100.

Kasīan, Tēchaphīra. 2553. *Songkhrām Rawāng Sī*, First Edition, 2 Vols. Krung Thēp: Samnakphim Openbooks.

Kittiphong, Sonthisamphan. 2553. *Red Why = Daeng – Thammai: Sangkhom Thai Panha Lae Kanma Khong Khon Sua Deang*, First Edition. Krung Thep: Samnakphim Openbooks.

Macan-Markar, Marwaan 2010. "Anti-Gov't Protesters Use Cultural Taboo as Weapon". *Inter Press Service*, 18 April. < http://www.ipsnews.net/2010/04/thailand-anti-govrsquot-protesters-use-cultural-taboo-as-weapon/>

Naruemon Thabchumpon and Duncan McCargo. 2011. "Urbanised Villagers in the 2010 Thai Redshirt Protests Not Just Poor Farmers?" *Asian Survey*, Vol. 51, No. 6: 993–1018.

Nostitz, Nick. 2009. *Red Vs. Yellow*, Vol. 2. Bangkok: White Lotus Press.

———. 2014. "The Red Shirts from Anti-Coup Protesters to Social Mass Movement". In *"Good Coup" Gone Bad: Thailand's Political Development since Thaksin's Downfall*. Edited by Pavin Chachavalpongpun. Singapore: Institute of Southeast Asian Studies.

Pavin Chachavalpongpun. 2014. *"Good Coup" Gone Bad: Thailand's Political Developments since Thaksin's Downfall*. Singapore: Institute of Southeast Asian Studies.

Pinkaew Leungaramsri. 2010. *Becoming Red = Kannoet Lae Phatthanakan Sue Deang Nai Chiang Mai*, First Edition, Phim Khrang.

Sopranzetti, Claudio. 2012. "Burning Red Desires: Isan Migrants and the Politics of Desire in Contemporary Thailand". *South East Asia Research*, Vol. 20, No. 3: 361–379.

———. 2014. "Political Legitimacy in Thailand". Fieldsights: Hot Spots, Cultural Anthropology Online, September 23. < http://www.culanth.org/fieldsights/578-political-legitimacy-in-thailand>.

———. 2016. "Thailand's Relapse: The Implications of the May 2014 Coup". *The Journal of Asian Studies*, Vol. 75, No. 2: 299–316. doi:10.1017/S0021911816000462.

———. 2017a. "Framed by Freedom: Emancipation and Oppression in post-Fordist Thailand". *Cultural Anthropology*, Vol. 32, No. 1: 68–92.

———. 2017b. *Owners of the Map: Motorcycle Taxi Drivers, Mobility, and Politics in Bangkok*. Berkeley: University of California Press.

Taylor, Jim. 2011. "Larger Than Life: "Central World" and its Demise and Rebirth – Red Shirts and the Creation of the Urban Cultural Myth in Thailand". *Asia Research Institute Working Paper Series*, Vol. 150: 1–13.

Thongchai Winichakul. 2010. "Red Germs". 15 May. < http://asiapacific.anu.edu.au/newmandala/2010/05/03/thongchai-winichakul-on-the-red-germs/> (accessed date 20 May 2010).

Voranai Vanijaka. 2011. "Ammat and Phrai: The Facebook War". *Bangkok Post*, 15 May.

13

JUDICIARY AND JUDICIALISATION IN THAILAND

Björn Dressel and Khemthong Tonsakulrungruang

In the last 20 years, judges have become central players in the Thai political landscape. Since the far-reaching constitutional reform and the liberalisation of politics in the 1990s, they have upheld the rule of law by protecting human rights, enforcing environmental standards, and checking on executive abuses. However, they have also dissolved political parties, toppled governments, and actively shaped public policies.[1] Since the military coups in 2006 and 2014 and amid deepening political polarisation, they have also been seen to support the illiberal military-led regime that the coups put in place, specifically by draconian application of lèse-majesté and computer crime laws and more generally by validating the legality of the junta-backed regime.[2] One result, however, has been harsh criticism of the role of the court and even protests – a phenomenon never before seen in Thailand.[3]

Such developments mark a considerable change for a traditionally conservative institution which traces its roots to revivalism and westernisation under King Chulalongkorn (Rama V, 1868–1910) and his son, Prince Ratchaburi (1874–1920), whom official Thai historiography has termed the "Father of Modern Thai Law and the Judicial System".[4] Under Chulalongkorn, the system of customary justice was replaced by professional courts and predictable procedures.[5] The judicial structures and practices he decreed have been basically unchanged, even though the 1932 revolution transformed Thailand's political and administrative systems by replacing absolute with constitutional monarchy.

This is not to say that there has been no formal institutional change. Most notably, the 1997 "People's Constitution" not only gave the judicial branch more independence but also created two new institutions, the Constitutional Court (CC) and the Administrative Court (AC), as part of a broad effort to build a comprehensive system for judicial review of legislative and administrative actions.[6] These two courts and the previously established Court of Justice (COJ) have since 1997 assumed such a central role in the political sphere that observers have been led to speculate that in Thailand, politics is becoming judicialised.[7]

Yet while these courts occasionally have acted in unison against elected politicians, particularly during the 2006–2008 political crisis, the three courts also behave quite differently; if there is a politicisation trend, it is hardly uniform or clear-cut. For instance, although the CC is regularly accused of political bias and failure to adhere to professional standards, this is less true of the AC, and may be even less true for the COJ. This in turn raises important questions for any nuanced understanding of the judiciary and the judicialisation trend in Thailand – as well as for debates on judicial politics in general within and beyond Thailand.

To discuss these questions, this chapter will outline the basic structure of the Thai judicial system before asking how the obvious divergences in court behavior may relate to judicialisation. Although the judiciary may in general be understood as a royalist conservative institution, differences in how political cases may be adjudicated may have to do with differences in how judges are recruited, the jurisdiction of each court, and the degree to which judges in each are still embedded in monarchy-aligned elite networks.

The Thai judicial system

Until 1997, there was only the Court of Justice. Constitutional disputes were settled by the Constitutional Council and administrative disputes by the Council of State.[8] Because this model was perceived as having failed adequately to check the government's exercise of power, those drafting the 1997 constitution also introduced the Administrative Court and the Constitutional Court.[9] Since 1997, these three courts have been independent of each other: none reviews a decision of any other, and judicial independence is guaranteed both internally and externally because each court has its own administration, recruitment, and appellate functions. However, while none is superior to the others, if there is a jurisdictional conflict, it can be referred to a special tribunal.[10]

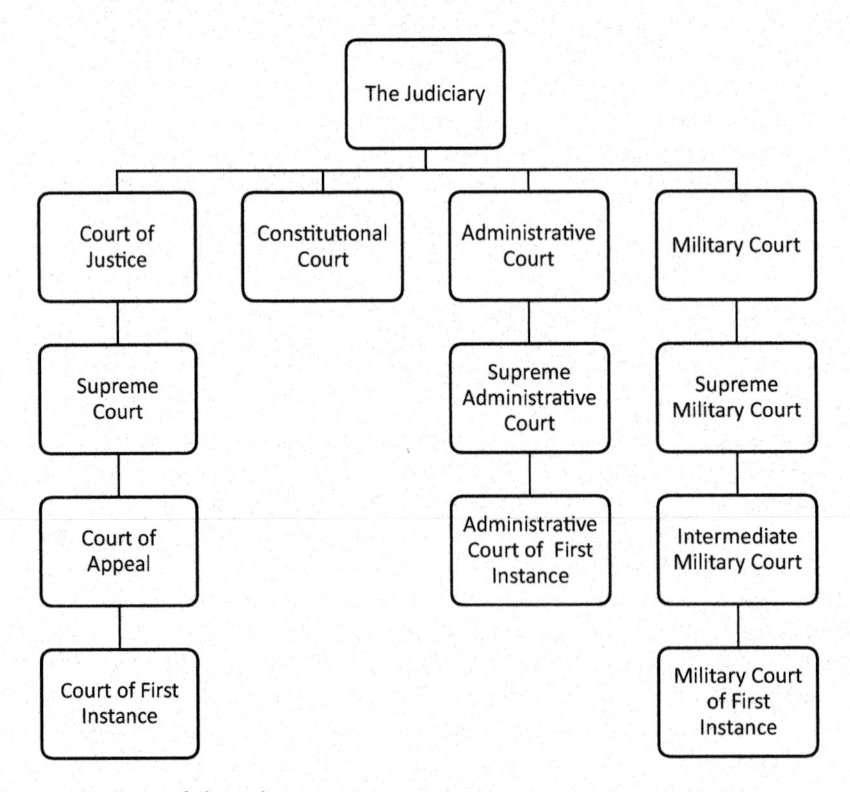

Figure 13.1 Structure of Thai judiciary

Court of Justice

In 1892, in response to threats from Western colonial powers, King Chulalongkorn founded the Court of Justice to showcase to Western nations that Siam also had a civilised justice system.[11] It eventually grew to over 4,700 judges.[12] Unless a statute states otherwise, any legal dispute can be brought before the Court of Justice.

The COJ is structured in three layers: the Courts of First Instance, the Appeal Courts, and the Supreme Court. Over time, specialised court divisions were created within the COJ for tax, bankruptcy, intellectual property, international trade, labour, juvenile and family matters, and, in 2016, corruption.[13] The Thai Supreme Court has 150 judges, though only if a case is unusually important will there be deliberation en banc.[14] The Supreme Court has one special division: the Corruption Division for Political Office Holders (Corruption Division). When politicians or high-ranking bureaucrats are charged with corruption, nine Supreme Court judges are appointed ad hoc to hear the case. An accused retains the right to appeal his case to another ad hoc panel of nine Supreme Court judges.[15] Over the last decade this division has been responsible for many high-profile decisions that have confiscated assets and imprisoned high-profile politicians.

Judges are career officers who enter public service via examination. A graduate from a certified law school is qualified to take the entrance examination at age 25. There are three options for entry: a general round, a small round (for those with an LL.M.), and a special examination (for those holding a foreign law degree earned in two years of study abroad).[16] This method of judicial appointment has been criticised on several accounts because examination is about remembering legal precedent – for instance, it encourages rote learning rather than critical thinking. The small round privileges a small pool of wealthy students. Finally, 25 is thought to be too young to understand how the law actually works.[17] In fact, in 2015, the Constitution Drafting Committee proposed raising the minimum age to 30; opposition was fierce enough that the proposal was dropped.[18] Once admitted, unless convicted of corruption or other serious misconduct, a judge may stay on the bench until age 70. Promotion is based strictly on seniority, not merit. The system may be designed to avoid lobbying or cronyism, but it certainly inhibits creativity.[19]

There is little independent oversight of the COJ. The Judicial Commission, chaired by the president of the Supreme Court, consists of six Supreme Court judges, four judges from Appeal Courts, two judges from Courts of First Instance, and two lay members appointed by the Senate.[20] Young judges have attempted to get equal representation (four from each court) but with no success so far.[21] In 2015, when the constitution drafters also sought to increase a number of lay members, the COJ rejected the proposal as an attack on its independence.[22] As a result, the court is largely isolated from parliamentary oversight.

Although the COJ is in general respected, its reputation has been questioned because of its role in political conflicts. In the legal battle against the Shinawatra family, the Corruption Division has delivered decisions that imprisoned former Prime Minister Thaksin Shinawatra and his men and confiscated a portion of his wealth.[23] Thaksin and his sister Yingluck were forced into exile to avoid jail terms decreed by a court process they claimed lacked independence – one their supporters believed was biased by political consideration. Another contentious issue has been the court's willingness to uphold the legality of the military coups d'état in 2006 and 2014 on the basis that a junta had successfully toppled the government; the COJ has done so by following a precedent dating back to 1953.[24] Lastly, the COJ shows deep concern about national security and appears to feel that it must protect the country from Thaksin. Thus it has given harsh sentences to Thaksin supporters for computer crimes, terrorism, lèse-majesté, and sedition.[25] Often these crimes are simply a pretext for political harassment and intimidation.[26] Meanwhile, it treats Thaksin's enemies leniently.[27] Hence, for a significant portion of the Thai public, the court's impartial image has been severely eroded.

Constitutional Court

Founded in 1998, the CC was initially opposed by the COJ on the grounds that it would thrust the judiciary into politics – as it certainly has.[28] While on the one hand, the CC has acted to defend private rights – gender equality, occupational rights, and fair trial – its successes in those areas have been overshadowed by the controversial political cases that have come before it.[29]

Since its founding the CC has seen drastic changes in its composition and jurisdiction. From 15 judges, it was downsized to nine, both professional and lay. Three professional judges are nominated from the Supreme Court and two from the Supreme Administrative Court, which is a component of the AC. A special commission adds to the CC bench one law expert, one political science expert, and two career bureaucrats.[30] A judge has one non-renewable nine-year term or may retire earlier at 70. Its unique configuration resembles that of watchdog agencies like the National Counter-Corruption Commission, the Election Commission of Thailand, or the National Human Rights Commission, with which it is sometimes compared. Its remit covers four broad categories: (1) the constitutionality of legal provisions, (2) jurisdictional disputes between constitutional bodies, (3) disqualification of public office holders, and (4) safeguarding the constitution and democratic values. The 2017 constitution has assigned it more duties, notably to review the ethics of politicians.[31]

In its first eight years, the CC was perceived as deferential to the government, reinforcing fears by the public that it had been captured by the Thaksin administration – a perception largely driven by the controversial acquittal of Thaksin for failing to disclose his assets.[32] But it became increasingly aggressive after a galvanising 2006 royal speech by King Bhumibol Adulyadej (Rama IX) and the subsequent coup. The CC dissolved two major political parties affiliated with Thaksin and temporarily banned hundreds of party executives from running for election.[33] It also blocked a series of constitutional amendments, citing protection of the minority.[34] Moreover, it upheld the lèse-majesté law and ruled that the PDRC occupation in 2013–2014 of inner Bangkok was peaceful and constitutional while at the same time endorsing heavy-handed actions against Red Shirt protesters.[35]

Thus, over the last decade, the increasingly assertive CC has regularly intervened in politics and kept pressure on the government. Considering the empirical evidence, its opinions in high-profile cases have often raised reasonable doubts about its independence.[36]

Administrative Court

The AC was established in 1999. It consists of nine courts of first instance and the Supreme Administrative Court and hears disputes about the legality of rules, orders, and administrative contracts, as well as physical acts by administrative agencies.[37] The Administrative Procedure Act under which it operates is part of a package with the Government Tort and the Government Information Acts.

The AC is heavily influenced by the continental legal tradition. It takes a practical inquisitorial approach in order to make equitable decisions related to government relations with private individuals. A consultant judge is appointed to investigate and present his or her opinion to the bench; the system is designed to provide internal checks and balances.[38] Statutes of limitations can be waived if a case may affect the public interest.[39]

Judges are appointed to the AC from a diversity of backgrounds in law and related disciplines (e.g. political science, public administration). Candidates must be over 35 or 45 to be eligible to take an examination for either an AC Court of First Instance or its Supreme Court, respectively. As with the CC, a judge need not have a law degree – they may also be experts in political science, economics, social sciences, or public administration – but they must have work experience such as mid- to high-ranking civil servants, academics, judges, or lawyers.[40] With only 210 judges, the full AC is comparatively small.[41]

Since the inception of the AC, its performance has been mixed. It has actively promoted constitutional rights for marginalised people, environmental groups, and people with disabilities, for instance when upholding the signing of bilateral trade agreements, eviction of indigenous people, and contentious irrigation projects in the Mekong River.[42] In a few cases, it clashed directly with the Thaksin government in its rulings on the plan to privatise utilities – a central Thaksin policy[43] – but it also refused to revoke the privatisation of the national petroleum company.[44] In general, the AC tries to strike a delicate balance between being progressive and assertive but also professional and nonpolitical.[45] Time and time again, the AC has emphasised the principle of proportionality. Its decisions are not necessarily winner take all. Usually they discuss at length the need to uphold legal certainty but balance that with the interests of the individual. The court may, for example, uphold an agency action because revocation would place an undue burden on the state but indemnify the plaintiff for the loss of rights and properties.

However, when times are more politically contentious, its behaviour, too, has been erratic. For instance, the AC issued an injunction preventing a by-election in the controversial 2006 election that was later invalidated by the CC.[46] In 2008, it ordered the government not to sign a treaty with Cambodia.[47] In both cases, it drew heavy criticism for acting ultra vires – beyond the scope of its usual jurisdiction – without explicit authorisation.[48] In the Cambodia-Thai treaty case, the president of the Supreme Administrative Court was accused of tampering with the outcome because he reassigned the case to an ad hoc panel;[49] the president refused to cooperate with the investigation by the National Anti-Corruption Commission (NACC).

Military Court

The Military Court is not usually considered in discussions of the Thai judiciary because it is a unit of the Ministry of Defence rather than an independent court. It is also principally concerned with criminal cases involving military personnel. Yet for the last ten years its profile has been rising; martial law may, for instance, assign it jurisdiction over crimes committed by civilians[50] – often critics of the military regime. This has raised serious concerns for the defendants because commissioned officers sit with military judges, and martial law allows no right to appeal.[51]

Towards the judicialisation of politics? Judicial patterns, performance, and drivers

For more than ten years, judges in Thailand have been making broad statements about social and political issues as well as legal matters. This has profound implications for governance. Consider that the CC has destroyed the country's largest and most successful political party, forced the resignation of three prime ministers, and kept pressure on popular former Prime Minister Thaksin to remain in exile. And other courts are following the CC's lead: between 2010 and 2016, the Supreme Administrative Court and the Supreme Court handed down decisions that interfered directly with government policy, often because of powers given them by the 2007 constitution.[52]

This seems a clear demonstration of the judicialisation of politics – a process that occurs when courts, especially constitutional or supreme courts, come to dominate the making of public policy, which is normally the prerogative of legislatures and executives.[53] More narrowly, however, it also reflects a growing reliance in Thailand on courts and judges to deal with "mega politics" – core political controversies and deep moral dilemmas related to such purely political areas as the prerogatives of the executive branch, electoral politics, and regime change.[54]

Though the judicialisation process in Thailand has been well documented,[55] there are different explanations of what is driving it. For some, judicialisation is the direct outcome of the exalted status given to Thai judges based on traits like being a "good person" (*khon dí*), a "proper gentlemen" (*phudí*), a "learned and wise person" (*phuru*), and "a loyal servant to the king" (*phupakdí*). This constructed legal identity has propelled judges to go beyond their limited formal role in Thailand's civil law system,[56] at times encouraged by leading academics who have described the judiciary as the only institution independent enough to adjudicate political conflicts in a polarised political environment.[57]

Others have instead linked the growing role of the judiciary to its traditional closeness with the Thai monarchy and the fact that it has traditionally been embedded in powerful conservative elite networks – often described as "network monarchy" or the "deep state".[58] Highlighting their traditional close relationship with King Bhumibol, who honoured judges more than any other state officials, these authors point in particular to the royal speech in 2006 that asked the judiciary to address the growing political crisis – prompting all three courts to not only act in concert but in particular prompting the CC to become much more aggressive.[59] Cases in point are the CC's decisions against political actors associated with Thaksin (the "Red Shirt" faction, as it is known). This stance has in turn prompted public concern about whether the CC is truly independent – a concern that has found support in recent empirical studies.[60]

Whatever is driving the process, it is clear that the counter-majoritarian role of courts in Thailand, particularly in political cases, has become problematic. Yet it is also true that the involvement of the three major courts in political matters differs considerably. This suggests that the judicialisation trend may be more nuanced than current literature would suggest. The divergence is also reflected in public opinion polls. Although all courts are ranked more favourably than other public institutions, the public ranks the COJ as the most trusted and the CC as the least trusted.[61]

How can this puzzling outcome be explained? Here we suggest attention to three intertwined factors: institutional, agential, and structural.

Institutional design

The differences between the courts in powers, scope, and access are crucial. For instance, the CC was created as a specialised Kelsenian court of centralised judicial review with auxiliary powers related to elections, review of statutes, disqualification of political office holders, asset disclosure, party dissolution, and foreign affairs – all distinctly political matters. The COJ and the AC deal with more mundane matters, except for the Corruption Division, which hears cases against politicians.

Institutional design also affects issues of internal accountability and thus personal conduct. Precisely because their independence is constitutionally guaranteed, for judges accountability largely concerns adherence to personal ethics and is thus internally enforced. Yet this "self-enforced model" may work better in a larger, more institutionalised body like the COJ than in the small, committee-like CC. The COJ is a court staffed with career judges at different levels of the internal judicial hierarchy; the CC operates more similar to other constitutional watchdog agencies such as the NACC, the National Human Rights Commission, or the Ombudsman; like them, it has neither hierarchy nor established precedent. Like these independent bodies, the CC operates on a more personal basis – and it can be arbitrary.[62]

There is also evidence that internal enforcement works better in the COJ and the AC than in the CC for several reasons. For instance, because the COJ is much older, there are more regulations, guidelines, and an internal organisational culture to hold its members accountable. A decision can be appealed to a higher court, and deviation from a generally accepted precedent must be convincingly justified. Judges implicated in bribery or other unethical conduct are disciplined or even dismissed. The culture in the AC is similar, perhaps because AC judges with a bureaucratic background are used to the idea of disciplinary action. But when CC judges have been found to be provide benefits to relatives (e.g. employing their children, granting them scholarships, leaking exam results),[63] no action has been taken. The fact that the president of the AC resigned after a month-long battle when he was found to have misused the office budget demonstrates the differences in the AC and the CC cultures in terms of judicial integrity.[64]

Agential factors

Leadership of and recruitment to the bench also support differences in how judges behave. For instance, unlike the COJ and the AC, leadership of the CC has not been consistent due to ever-changing rules and the fact that it was dissolved during the 2006 coup. CC presidents have averaged barely two years in office.

Perhaps even more important is the nomination process itself. The nine-member CC bench was originally designed to be nonpolitical, but despite an elaborate process for nominating its judges, the reality has been very different. Perhaps because the criteria are vague and subjective, and perhaps because a third of its members are noncareer judges from academe and the bureaucracy, recruitment to the CC has become more politicised and less transparent than recruitment to the COJ. Even whether candidates recruited actually have constitutional expertise is somewhat doubtful.[65] Some CC judges (e.g. Jaran Pakdeethanakul, Nakarin Mektrairat, and Taweekiat Meenakanit) publicly had an anti-Thaksin stance before being nominated, reinforcing the impression that CC verdicts in high-profile political cases are based on ideology rather than law.[66]

The dynamics of the COJ and AC are different. Not only have their leaders had much longer terms, but judges for both are generally recruited from the career judiciary and are well-known experts. In fact, although AC members need not have previously been judges, as with the COJ, recruitment to the AC is less politicised, and both the AC and the COJ are large enough to prevent judges with strong ideological positions from exerting much influence. However, the selection of judges for the COJ Supreme Court Corruption Division is somewhat similar to that of the CC: nine judges constitute the division, case by case. Thus, the Supreme Court, if necessary, may apply political criteria in appointing judges to each ad hoc panel.

Socio-structural dimension

Finally, differences in their political roles have invited a variety of efforts to embed the courts in royal and political networks. For instance, while judges in general tend to be closely linked to monarchical and state-centred networks, the COJ and the CC have been particularly active in participating in bureaucratic and military training courses – something the AC stopped offering in 2014. Considering that appointment to the CC is more political than appointment to the other two courts, it is probably harder for the CC to maintain its independence from these formal and informal political networks – as is demonstrated in its highly biased pattern of decisions against Thaksin-affiliated governments.[67]

Clearly, then, differences in recruitment, internal culture, and the degree to which these courts are embedded in Thailand's sociopolitical structures have led to very different internal cultures, which in turn have influenced the extent to which judges consider political rather than legal factors in making decisions. That is also why the trajectory of the judicialisation of politics in Thailand has been very erratic.

Conclusion

On the one hand, against the background of rapid modernisation and a growing legal profession, Thailand's judicial institutions have adapted and professionalised in response to Thailand's changing sociopolitical landscape. On the other, the judiciary as an institution is a secluded, conservative institution still deeply embedded in Thailand's traditional power structures, having a close and long-standing affiliation with the monarchy and related elite networks. Because any change would be slow without firm political will or external pressures for major political reform, struggles over judicial professionalisation and independence – particularly in times of heightened political tension – are likely to continue.

To be sure, none of this is necessarily unique to Thailand. Many judicial institutions in the region struggle with issues of integrity, independence, and enforcement of the rule of law in a context of persistent informal influences and clientelist political structures.[68] What has set the Thai judiciary apart from its Asian counterparts, however, is its unprecedented willingness to engage in political issues – unprecedented not only in Thailand but also anywhere else in the region. Not surprisingly then, it has often been cited as a prime example for the judicialisation of politics in Southeast Asia.[69]

But even in Thailand the trend has hardly unfolded uniformly. Mitigated by institutional, agential, and structural factors, its major courts – particular in high-profile cases – have behaved rather differently. The question that demands exploration now, therefore, is what differences in their internal dynamics and judicial cultures have directed them along different paths. As yet there have been few comprehensive studies of Thailand's major courts,[70] nor (perhaps with the exception of the CC) have there been systematic empirical analyses of decisions and their legal and political effects.[71]

Equally absent has been debate about the appropriate balance between judicial independence and external oversight. The latter is of particular importance. Since the 1997 constitution, "judicial empowerment" has also meant that those in power – the junta, constitution drafters, and those opposed to Thaksin – have used the judiciary for their own ends, thus exploiting flaws in the Thai constitutional design. With the 2017 constitution accelerating this trend, there are few indications that the judicialisation of politics in Thailand will subside – in fact, the associated risk of the politicisation of the judiciary is starkly evident, particularly for the Constitutional Court.

The Thai case is thus a warning to drafters, in the region and beyond, who put their hopes in empowerment of the judiciary to safeguard democracy. Like any other institution, judiciaries may be captured and (mis)used for narrow political interests by authoritarian or populist leaders (see Cambodia, Venezuela, Turkey, and the Philippines). As what has been happening in Thailand shows clearly, institutional safeguards can only go so far. Ultimately much depends on the motivations of judges themselves. And while Thailand's judiciary has benefited from its traditional closeness to the monarchy and its high public standing, it needs to rethink its long-tern survival in light of the surge in questions about its independence. In fact, the very same institutions that have guaranteed the court's insulation are now also driving the erosion of liberal political order generally. Much is at stake here – and the path the Thai judiciary charts will certainly reveal much about whether the rule of law can gain ground in Thailand and the whole diverse region.[72]

Notes

1 Khemthong Tonsakulrungruang, "Entrenching the Minority: The Constitutional Court in Thailand's Political Conflict", *Washington International Law Journal* (April 2017).

2 David Streckfuss, *Truth on Trial in Thailand: Defamation, Treason, and Lèse-Majesté* (Abingdon and New York: Routledge, 2011); Tyrell Haberkorn, "Martial Law and the Criminalisation of Thought in Thailand", *The Asia Pacific Journal*, Vol. 12 (2014), pp. 1–7.

3 See "อภิชาต-พันธ์ศักดิ์-รังสิมันต์-ปิยบุตร: อภิปราย องค์กรตุลาการในสถานการณ์พิเศษ" ["Apichart-Pansak-Rangsiman-Piyabutr Discuss Judiciary under Extraordinary Circumstance"], *Prachatai*, 22 February 2016 <https://prachatai.com/journal/2016/02/64214>.

4 Pawat Satayanurug and Nattaporn Nakornin, "Courts in Thailand: Progressive Development as the Country's Pillar of Justice", in *Asian Courts in Context*, edited by Jiunn-rong Yeh and Wen-chen Chang (Cambridge: Cambridge University Press, 2014), p. 408. But some argue otherwise. See Somchai Preechasilpakul in Duncan McCargo, "Reading on Thai Justice: A Review Essay", *Asian Studies Review*, Vol. 29 (2015), pp. 30–32.

5 T. Loos, *Subject Siam. Familiy, Law, and Colonial Modernity in Thailand* (Ithaca: Cornell University Press, 2006), pp. 41–70.

6 J.R. Klein, "The Battle for the Rule of Law in Thailand: The Constitutional Court of Thailand", in *The Constitutional Court of Thailand. The Provisions and the Working of the Court*, edited by Amara Raksasataya and J.R. Klein (Bangkok: Constitution for the People Society, 2003), pp. 34–90.

7 B. Dressel, "Judicialisation of Politics or Politicisation of the Judiciary? Considerations from Recent Events in Thailand", *The Pacific Review*, Vol. 23 (2010), pp. 671–691.

8 Banjerd Singkaneti, *Das Thäilandische Verfassungstribunal im Vergleich mit der Deutschen Verfassungsgerichtsbarkeit* [The Thai Constitutional Tribunal in Comparison to the German Constitutional Court System] (Frankfurt: Lang Verlag, 1998).

9 Borwornsak Uwanno, กฎหมายมหาชน เล่ม *2* การแบ่งแยกกฎหมายมหาชน-เอกชน และพัฒนาการกฎหมายมหาชนในประเทศไทย [Public Law Book II Separation of Public and Private Law and Development in Thailand], third edition (Winyuchon, 2004), pp. 338–374.

10 See the Jurisdictional Dispute Act B.E. 2542 (1999).

11 <www.coj.go.th/home/history.html>.

12 See the Office of Judiciary Annual Report Fiscal Year 2017 (2016).

13 Pawat Satayanurug and Nattaporn Nakornin, "Courts in Thailand: Progressive Development as the Country's Pillar of Justice", in *Asian Courts in Context*, edited by Jiunn-rong Yeh and Wen-chen Chang (Cambridge: Cambridge University Press, 2014), pp. 413–414; <www.aljazeera.com/news/2016/10/thailand-introduces-anti-corruption-court-161002162235702.html>.

14 Civil Procedure Code, Section 140.

15 See, Organic Act on Criminal Procedure for Political Office Holders B.E. 2560 [2017], Section 60–64.

16 Judicial Service Act B.E. 2543 (2000), Section 28.

17 *Bangkok Post*, 17 January 2016.

18 <www.bangkokbiznews.com/blog/detail/622277>.

19 *Bangkok Post*, 17 January 2016.

20 Judicial Service Act, Section 26.

21 Atukkit Sawangsuk, "ผู้พิพากษาทวงสิทธิเลือกตั้ง" [Judges Demand Right to Vote], *Khao Sod*, 7 September 2017 <www.khaosod.co.th/politics/news_501435>.

22 "บวรศักดิ์วอนคณะผู้พิพากษาอย่าเข้าใจผิด" [Borwornsak Plead Judges Do Not Misunderstand], *Post Today*, 13 July 2015 <www.posttoday.com/politic/news/375937>.

23 Duncan McCargo, "Competing Notion of Judicialisation in Thailand", *Contemporary Southeast Asia*, Vol. 36 (2014), pp. 417, 431–432.

24 Piyabutr Saengkanokkul, "ศาลรัฐประหาร ตุลาการ ระบอบเผด็จการ และนิติรัฐประหาร" [Coup Court: Judiciary, Dictatorship, and Judicial Coup] (Bangkok: Same Sky Books, 2017).

25 See case statistics at iLaw <https://freedom.ilaw.or.th/en> (accessed 27 July 2018).

26 D. Streckfuss, "Freedom and Silencing under the Neo-absolutist Monarchy Regime in Thailand, 2006–2011", in *"Good Coup" Gone Bad: Thailand's Political Developments since Thaksin's Downfall,* edited by Pavin Chachavalpongpun (Singapore: ISEAS, 2014), pp. 109–138.

27 Witness examination in the 2008 airport occupation case has been postponed to mid-2019.

28 A. Harding and P. Leyland, *Constitutional System of Thailand. A Contextual Analysis* (Oxford and Portland, OR: Hart Publishing, 2011), pp. 161–162.

29 Ibid. Also see McCargo, pp. 417–441. And Kla Samudavanija, "ขอบเขตอำนาจหน้าที่ศาลรัฐธรรมนูญเพื่อส่งเสริมการปกครองในระบอบประชาธิปไตยและคุ้มครองสิทธิเสรีภาพของประชาชน" [Scope of Power and Duty of the Constitutional Court for Promotion of Democratic Rule and Protection of Civil Rights and Liberties] (Bangkok: King Prajadhipok's Institute, 2016).

30 Thai Constitution B.E. 2560 (2017), Section 200–03.

31 Khemthong Tonsakulrungruang, "The Anti-Majoritarian Constitutional Court of Thailand", in *Constitutional Courts in Asia*, edited by Jiunn-rong Yeh and Wen0chen Chang (Cambridge: Cambridge University, 2018).

32 James Klein, "The Battle for Rule of Law in Thailand: The Constitutional Court of Thailand", in *The Constitutional Court of Thailand: The Provisions and the Working of the Court,* edited by Amara Raksasataya and James R. Klein (Bangkok: Constitution for the People Society, 2003), pp. 74–76.

33 Khemthong Tonsakulrungruang. "Thailand: An Abuse of Judicial Review", in *Judicial Review of Elections in Asia,* edited by P.J. Yap (Abingdon and New York: Routledge, 2016), pp. 173–192.

34 Khemthong Tonsakulrungruang, "Entrenching the Minority: the Constitutional Court in Thailand's Political Conflict", *Pacific Rim Law and Policy Journal,* Vol. 2 (2017), pp. 247–267.

35 Const Ct Decision 28–29/2555 (2012), Order 45/2557 (2014), cf. Decision 9/2553 (2010) and 10–11/2553 (2010).

36 B. Dressel and Khemthong Tonsakulrungruang, "Coloured Judgement? The Work of the Thai Constitutional Court, 1998–2016", *Journal of Contemporary Asia*, Early print, 13 June 2018.

37 Admin Ct Act, Section 9.

38 Admin Ct Act, Sections 57–58.

39 Admin Ct Act, Section 52.

40 Admin Ct Act, Sections 13 and 18.

41 The Administrative Court Annual Report, 2016.

42 Marginalised group's nationality case Sup Admin Ct Decision O. 117/2548 (2005); Environment Group, Cent Admin Ct Decision 1352/2553 (2010); disability case, Sup Admin Ct Decision O. 142/2547 (2004); bi-lateral agreement Sub. Admin. Ct. Order 178/2550 (2007); eviction, Sup Admin Ct 4/2561 (2018); and Mekong construction Sub Admin Ct Order 8/2557 (2014).

43 Dominic Nardi, "Thai Institutions: Judiciary", *New Mandala,* 12 July 2010 <www.newmandala.org/thai-institutions-judiciary/>.

44 Viparat Jantraprap, "Thai Court Rejects Bid to Delist Energy Giant PTT", *Reuters,* 14 December 2007 <www.reuters.com/article/thailand-ptt-ruling-idUSBKK7700920071214>.

45 Harding and Leyland, *Constitutional System of Thailand. A Contextual Analysis,* p. 208.

46 McCargo, "Competing Notion of Judicialisation in Thailand", pp. 417, 426.

47 Sub Admin Ct Order 547/2551 (2008).

48 See Borwornsak Uwanno, "แฉเอกสารลับที่สุดปราสาทพระวิหาร พ.ศ. 2505–2551" [Confidential Documents on Preah Vihear Temple], *Matichon,* 2008.

49 Prasong Lertrattanawisut, "หลังดองคดีเปลี่ยนองค์คณะเขาพระวิหารนานกว่าปี ปปช เพิ่งมีมติสอบ 5 ตุลาการศาลสูง" [After Years of Inaction, NACC Will Investigate 5 Supreme Administrative Court Judges], *Prasong,* 7 February 2012 <www.prasong.in.th/%E0%B8%81%E0%B8%8E%E0%B8%AB%E0%B8%A1%E0%B8%B2%E0%B8%A2/5-10/>.

50 The Martial Law Act, B.E. 2457 (1914), Section 7 and 7bis.

51 The Military Court Act B.E. 2498 (1955), Section 61. Also see, "Review of the Situations in 2015: Justice Made to Order, Freedom Still Out of Stock", *iLAW,* 23 December 2015 <https://freedom.ilaw.or.th/en/report%E0%B9%81%E0%B8%AD%E0%B8%9A%E0%B8%AD%E0%B9%89%E0%B8%B2%E0%B8%87/review-situations-2015-justice-made-order-freedom-still-out-stock>.

52 Khemthong, "The Anti-Majoritarian Constitutional Court of Thailand".

53 R. Sieder, L. Schjolden, and A. Angell, *The Judicialisation of Politics in Latin America* (New York and Houndsmills: Palgrave Macmillan, 2005). Also see N.C. Tate and T. Vallinder, *The Global Expansion of Judicial Power* (New York: New York University Press, 1995).

54 R. Hirschl, "The Judicialisation of Mega-Politics and the Rise of Political Courts", *Annual Review of Political Science*, Vol. 11 (2008), pp. 93–118.
55 B. Dressel, "Courts and Judicialisation in Southeast Asia", in *Routledge Handbook of Southeast Asian Democratisation,* edited by William Case (Abingdon and New York: Routledge, 2015), pp. 268–282.
56 Kitpatchara Somanawat. "Constructing the Identity of the Thai Judge: Virtue, Statu, and Power", *Asian Journal of Law and Society*, Vol. 5 (2018), pp. 91–110.
57 Thirayuth Boonmee, "ตุลาการภิวัฒน์" [Tulakanpiwat], *Judicial Review* (Bangkok: Winyuchon, 2006).
58 D. McCargo, "Network Monarchy and Legitimacy Crisis in Thailand", *The Pacific Review*, Vol. 18 (2005), pp. 499–518; E. Mérieau, "Thailand's Deep State, Royal Power and the Constitutional Court (1997–2015)", *Journal of Contemporary Asia*, Vol. 46 (2016), pp. 445–466.
59 Khemthong, "Thailand: An Abuse of Judicial Review", pp. 173–192.
60 See, Dressel and Khemthong, "Coloured Judgement?"
61 See, King Prajadhipok's Institute, "ผลการสำรวจความคิดเห็นของประชาชนเกียวกับ ความพึงพอใจการให้บริการสาธารณะและการทำงานของหน่วยงานต่างๆ ประจำปี 2560" [Public Survey on Satisfaction on Public Services and Agencies, 2017] (King Prajadhipok, 2017).
62 Khemthong Tonsakulrungruang, "Thailand's Forgotten Key", *New Mandala*, 2014 <www.newmandala.org/thailands-forgotten-key/>.
63 Khemthong Tonsakulrungruang, "The Anti-Majoritarian Constitutional Court of Thailand", in *Constitutional Courts in Asia*, edited by A.H.Y. Chen and A. Harding (Cambridge: Cambridge University, 2018).
64 '"หัสวุฒิ' น้อมรับพระบรมราชโองการ ปลดพ้น 'ปธ.ศาลปกครองสูงสุด' เชือแจงคดี 'ยกฉัตร-รถหลวง' ต่อ ก.ศป. ได้" [Hassawut Accepts Resignation from Supreme Administrative Court Presidency], *Thai Publica*, 12 January 2016 <https://thaipublica.org/2016/01/haswut-11-01-2559/>.
65 Somchai Preechasinlapakun, "ศาลรัฐธรรมนูญทีไร้ผู้เชียวชาญรัฐธรรมนูญ" [The Constitutional Court without Constitutional Experts], *The 101 World*, 11 June 2018 <www.the101.world/constitutional-court-without-constitution-expert/>.
66 See Dressel and Khemthong, "Coloured Judgement?"
67 Ibid.
68 M. Curley, B. Dressel, and S. McCarthy, "Competing Visions of the Rule of Law in Southeast Asia: Power, Rhetoric and Governance", *Asian Studies Review*, Vol. 42 (2018), pp. 192–209.
69 B. Dressel, *The Judicialisation of Politics in Asia* (Abingdon and New York: Routledge, 2012).
70 A.M. Johnson, "The Judicialisation of Politics: An Examination of the Administrative Court of Thailand", *Department of Poilitical Science* (DeKalb: Northern Illinois University, 2016). Also see D.M. Engel, *Code and Custom in a Thai Provincial Court: The Interaction of Formal and Informal Systems of Justice* (Tucson: University of Arizona Press, 1978). Also D. McCargo, *Indicting Legalism: The Politics of Justice in Thailand* (Ithaca: Cornell University Press), forthcoming.
71 Panthip Pruksacholavit and N. Garoupa, "Patterns of Judicial Behavior in the Thai Constitutional Court, 2008–2014: An Empirical Approach", *Asian Pacific Law Review*, Vol. 24 (2016), pp. 16–35. See Dressel and Khemthong, "Coloured Judgement?" Also McCargo, *Indicating Legalism.*
72 B. Dressel and M. Bünte, "Constitutional Politics in Southeast Asia: From Contestation to Constitutionalism?" *Contemporary Southeast Asia*, Vol. 36 (2014), pp. 1–22. Also, M. Curley, B. Dressel, and S. McCarthy, "Competing Visions of the Rule of Law in Southeast Asia: Power, Rhetoric and Governance", *Asian Studies Review*, Vol. 42 (2018), pp. 192–209.

References

Banjerd Singkaneti. 1998. *Das Thäilandische Verfassungstribunal im Vergleich mit der Deutschen Verfassungsgerichtsbarkeit [The Thai Constitutional Tribunal in Comparison to the German Constitutional Court System]*, Frankfurt am Main, Berlin, Bern, New York and Wien: Lang Verlag.

Curley, M, B. Dressel and S. McCarthy. 2018. "Competing Visions of the Rule of Law in Southeast Asia: Power, Rhetoric and Governance". *Asian Studies Review*, Vol. 42: 192–209.

Dressel, B. 2010. "Judicialisation of Politics or Politicisation of the Judiciary? Considerations from Recent Events in Thailand". *The Pacific Review*, Vol. 23: 671–691.

———. 2012. *The Judicialisation of Politics in Asia.* Abingdon and New York: Routledge.

———. 2015. "Courts and Judicialisation in Southeast Asia". In *Routledge Handbook of Southeast Asian Democratisation.* Edited by William Case. Abingdon and New York: Routledge, pp. 268–282.

Dressel, B. and M. Bünte. 2014. "Constitutional Politics in Southeast Asia: From Contestation to Constitutionalism?" *Contemporary Southeast Asia*, Vol. 36: 1–22.

Dressel, B. and Khemthong Tonsakulrungruang. 2019. "Coloured Judgement? The Work of the Thai Constitutional Court, 1998–2016". *Journal of Contemporary Asia*, Vol. 49:1, 1–23.

Engel, D.M. 1978. *Code and Custom in a Thai Provincial Court: The Interaction of Formal and Informal Systems of Justice*. Tucson: University of Arizona Press.

Haberkorn, T. 2014. "Martial Law and the Criminalisation of Thought in Thailand". *The Asia Pacific Journal*, Vol. 12: 1–7.

Harding, A. and P. Leyland. 2011 *Constitutional System of Thailand. A Contextual Analysis*. Oxford and Portland, OR: Hart Publishing.

Hirschl, R. 2008a. "The Judicialisation of Mega-Politics and the Rise of Political Courts". *Annual Review of Political Science*, Vol. 11: 93–118.

Johnson, A.M. 2016. "The Judicialisation of Politics: An Examination of the Administrative Court of Thailand". *Department of Political Science*. de Kalb: Northern Illinois University.

Khemthong Tonsakulrungruang. 2014. "Thailand's Forgotten Key". *New Mandala*. < www.newmandala.org/thailands-forgotten-key/ > (accessed 15 August 2019).

———. 2016. "Thailand: An Abuse of Judicial Review". In *Judicial Review of Elections in Asia*. Edited by P.J. Yap. Abingdon and New York: Routledge, pp. 173–192.

———. 2017a. "Entrenching the Minority: The Constitutional Court in Thailand's Political Conflict". *Pacific Rim Law and Policy Journal*, Vol. 2: 247–267.

———. 2018. "The Anti-Majoritarian Constitutional Court of Thailand". In *Constitutional Courts in Asia*. Edited by A.H.Y. Chen and A. Harding. Cambridge: Cambridge University.

Kitpatchara Somanawat. 2018. "Constructing the Identity of the Thai Judge: Virtue, Status, and Power". *Asian Journal of Law and Society*, Vol. 5: 91–110.

Kla Samudavanija. 2016. "ขอบเขตอำนาจหน้าที่ศาลรัฐธรรมนูญเพื่อส่งเสริมการปกครองในระบอบประชาธิปไตยและคุ้มครองสิทธิเสรีภาพของประชาชน" [Scope of Power and Duty of the Constitutional Court for Promotion of Democratic Rule and Protection of Civil Rights and Liberties]. Bangkok: King Prajadhipok's Institute.

Klein, J.R. 2003. "The Battle for the Rule of Law in Thailand: The Constitutional Court of Thailand". In *The Constitutional Court of Thailand. The Provisions and the Working of the Court*. Edited by Amara Raksasataya and J.R. Klein. Bangkok: Constitution for the People Society, pp. 34–90.

Loos, T. 2006. *Subject Siam. Family, Law, and Colonial Modernity in Thailand*. Ithaca: Cornell University Press.

McCargo, D. 2005. "Network Monarchy and Legitimacy Crisis in Thailand". *The Pacific Review*, Vol. 18: 499–518.

———. 2014. "Competing Notions of Judicialisation in Thailand". *Journal of Contemporary Southeast Asia*, Vol. 36: 417–441.

———. (forthcoming) *Indicting Legalism: The Politics of Justice in Thailand*. Ithaca: Cornell University Press.

Mérieau, E. 2016. "Thailand's Deep State, Royal Power and the Constitutional Court (1997–2015)". *Journal of Contemporary Asia*, Vol. 46: 445–466.

Panthip Pruksacholavit and N. Garoupa. 2016. "Patterns of Judicial Behavior in the Thai Constitutional Court, 2008–2014: An Empirical Approach". *Asian Pacific Law Review*, Vol. 24: 16–35.

Pawat Satayanurug and Nattaporn Nakornin. 2014. "Courts in Thailand: Progressive Development as the Country's Pillar of Justice". In *Asian Courts in Context*. Edited by Yeh Jiunn-rong and Chang Wen-chen. Cambridge: Cambridge University Press, pp. 407–446.

Piyabutr Saengkanokkul. 2017. ศาลรัฐประหาร ตุลาการ ระบอบเผด็จการ และนิติรัฐประหาร [Coup Court: Judiciary, Dictatorship, and Judicial Coup]. Bangkok: Same Sky Books.

Sieder, R., L. Schjolden and A. Angell. 2005 *The Judicialisation of Politics in Latin America*. New York and Houndsmills: Palgrave Macmillan.

Streckfuss, D. 2011. *Truth on Trial in Thailand: Defamation, Treason, and Lèse-majesté*. Abingdon and New York: Routledge.

———. 2014. "Freedom and Silencing under the Neo-absolutist Monarchy Regime in Thailand, 2006–2011". In *"Good Coup" Gone Bad: Thailand's Political Developments since Thaksin's Downfall*. Edited by Pavin Chachavalpongpun. Singapore: ISEAS, pp. 109–138.

Tate, N.C. and T. Vallinder. 1995. *The Global Expansion of Judicial Power*. New York: New York University Press.

Thirayuth Boonmee. 2006. "ตุลาการภิวัฒน์" ["Tulakanpiwat"]. *Judicial Review*. Bangkok: Winyuchon.

14

LOCAL GOVERNMENT AND INTERGOVERNMENTAL RELATIONS IN THAILAND

Grichawat Lowatcharin and Charles David Crumpton

The roots of the local government and decentralisation movements in Thailand can be traced back more than a hundred years.[1] Despite several attempts to nurture and strengthen the concept and practice of local self-government in the 1930s and 1940s after the democratic revolution, political instability hindered its institutionalisation and growth. From the 1980s to the early 1990s, Thailand experienced rapid economic development that led not only to infrastructure and telecommunication improvements but also to socio-cultural and political developments, particularly political awareness and calls for self-governance nationwide.[2] Beginning in the early 1990s, Thailand experienced two decades of movement towards more governmental decentralisation. While local economic development was flourishing, there were a number of challenges impeding the movement towards more local governance. The most notable among these barriers has involved attempts by central government bureaucrats to maintain and protect their power and interests.[3] Relations between the central government and local governments in Thailand involve fluctuation due to political instability caused by a series of coups over the past 60 years. The most recent coup on 22 May 2014 has posed new threats to already vulnerable local governments. Local government in Thailand faces a seemingly perennial disruptive world because politics at the national level and structural reforms imposed by the national government dramatically affect public service provision and governance at the local level.[4] This chapter considers the context and challenges associated with local government in Thailand in the following terms. First, it provides an overview of public administration and local administration in Thailand. Second, it describes the evolution of decentralisation and the current state of local government. Third, it discusses alternative analytic frames and recurring issues associated with Thai local governance. Finally, it articulates problems impeding the progress of and addresses existential threats to local government under the junta rule.

Overview of public administration and local administration in Thailand

Thailand is a constitutional monarchy in which the monarch serves as head of state and a prime minister is the head of government. The revolution in 1932 brought an end to absolute monarchy under the Chakri dynasty regime and marked the beginning of a constitutional monarchy that includes representation of the populace. Since the end of the absolute monarchy the political system has been almost continuously unstable, with military coups and military rule via juntas alternating with periods of democratic institution building. The instability in national government is reflected in the number of constitutional changes since 1932: 20 charters and constitutions to date.[5]

The roots of the Thai administrative state can be found from the Ayutthya period of the monarchy to the reign of King Rama IV, Mongkut, of the Bangkok period. During this period, two dominant administrative structures emerged that continue to dominate in 21st-century Thailand: the vast bureaucratic domains of the defense and interior ministries.[6] The transition to modern Thai public administration began in the early 19th century when the visionary King Rama V, Chulalongkorn, directed a series of administrative reforms, stripping power from tributary states and introducing a centralised government structure across functional areas of administration. With the establishment of this public administrative cadre organised according to historic Thai hierarchical principles, a bureaucratic elite emerged that has become a continuing powerful political, economic, and social influence in Thailand. In this light, the 1932 revolution in part represented a transition from rule by a king to rule by a bureaucratic elite.[7]

For over 40 years – the time of the transition from absolute to constitutional monarchy until the early 1970s – Thai public administration was dominated by the Thai military bureaucratic elite. During this period, administrative "reform" involved increasing administrative centralisation under the ultimate supervision of the military elite in Bangkok. The idea that government officials could be accountable to citizens did not exist in Thai public administration. In 1973, the dominance of the Thai military bureaucratic elite was challenged by a mass uprising which overthrew the military government. A transformation began from control by the bureaucratic elites to that by elected politicians. At least theoretically, democracy and national elections increasingly became norms.[8]

Thailand's modern local administration system can be seen as dating back to the Local Government Act of 1914. For the past century this system has remained fairly consistent.[9] During the first 40-year period of the constitutional monarchy, and more or less continuing for another 20 years, administration of local affairs followed the centralisation theme of the absolute monarchy under the control of the Ministry of Interior (MOI). It could be seen as symbolic of the Thai model of administrative centralisation wherein policies, budgets, and personnel decisions were made in Bangkok, and implementation was affected through MOI provincial and district offices. Wongpreedee and Mahakanjana describe the centralised MOI administrative dominance in the following terms:

> The MOI . . . was the very symbol of a centralised administrative system. Appointed provincial governors, apart from being the most senior executive officials of the MOI in each of Thailand's 75 provinces, also presided over most of the branch offices and agencies of other ministries located in the province. In addition, most other ministries and departments devolved power to the provincial governors to supervise and control their field officials in the provinces. Moreover, governors and other MOI bureaucrats also held *ex officio* positions in local government, which enabled them to control these bodies.
>
> (p. 56)[10]

With the National Administrative Organisation Act of 1991, the current structure of Thailand's public administration emerged. According to the terms of this law, Thai government is divided into three levels: central, provincial, and local. The central administration refers to 20 function-based ministries and their subordinate departments and agencies headquartered in

Bangkok. The provincial level, sometimes referred to as the regional level, refers to administrative offices in Thailand's 76 provinces and 878 districts. The 1991 law should not be seen as a move towards government decentralisation. Rather, provincial administration is based on the concept of deconcentration, meaning that while central ministries delegate some power to peripheral offices in the provinces and districts, ultimate control and direction resides in Bangkok ministerial headquarters.[11] The provincial offices are headed by section chiefs who are appointed by their respective ministries. The MOI appoints provincial governors and district chiefs who, by title, serve as heads of the provinces and districts. However, in practice, provincial governors and district chiefs are not chief executive officers in their provinces and districts. Rather, their roles involve coordination, supervisory, and ceremonial duties under the control of their bureaucratic masters in Bangkok.[12]

Prior to 1999, according to provisions of the National Administrative Organisation Act of 1991, local government comprised the provincial administrative organisation (PAO), municipalities, sanitary districts, the Bangkok Metropolitan Administration, and the City of Pattaya.[13] Despite the fact that Thailand was never a colony of a European imperial power, this pattern of central-provincial-local relations was modelled on the British colonial administrative approach. Bangkok essentially treated jurisdictions outside of the capital as colonies. This was intended to maintain control over provincial Thailand. Despite strong opposition from MOI, only since the 1990s have Thai governments shown (tenuous) support for decentralisation.[14]

The 60-year arc of Thai local government after 1932 can be described in the following discouraging terms of capacity to govern: weak financial management, insufficient resources, inefficient planning and service delivery, deficient public infrastructure, inadequate revenue resources, poor mobilisation of existing revenues, lack of technical capabilities and personnel, and unclear responsibilities.[15]

The 1997 economic crisis, the pressures of globalisation, and service demands from Thailand's burgeoning middle class have underlined a need for administrative reform geared towards improving government performance and increasing local responsibility and control. In response to these pressures, a decentralisation policy was included as a focus in Thailand's 1997–2001 National Economic and Social Development Plan.[16]

History of Thai decentralisation

The beginning of decentralisation movements in Thailand can be traced back a hundred years to when King Chulalongkorn established the Bangkok Sanitation Districts in 1897 and the Tha Chalom Sanitation District in 1905. After the 1932 revolution, the newly established democratic government attempted to nurture and strengthen the concepts and practices of local self-government by introducing several laws from the 1930s to the 1950s. These included the Municipality Act of 1933, Provincial Council Act of 1938, and the Sub-district Administrative Organisation (SAO) Act of 1956. However, political instability and authoritarian regimes in the late 1950s and 1960s hindered the establishment of the concept of local self-government. This was seen after the 1972 coup, when the military junta government abolished all SAOs.[17]

From the 1980s to the early 1990s, as Thailand experienced rapid economic development, the nation saw substantial infrastructure and telecommunication improvements. It also experienced socio-cultural and political development, including political awareness and interest in self-governance. The 1990s saw a set of phenomena that would mark a tremendous change in the course of the political history, and particularly the decentralisation, of Thailand.[18] After the Black May incident in 1992, where popular protest against the junta resulted in a bloody military crackdown in Bangkok, decentralisation was focused upon as one of the most important measures to move Thailand towards democracy. At least five political parties proposed the decentralisation of governmental power as a major policy during the election of September 1992. Although many of the decentralisation policies proposed by the political parties were not implemented, a forceful trend to strengthen local self-government persisted after the general election.[19]

During the years 1994, 1997 and 1999, three important decentralisation-oriented laws were enacted: the Sub-district Council and Sub-district Administrative Organisation Act of 1994; the Constitution of the Kingdom of Thailand in 1997; and the Decentralisation Plan and Process Act of 1999.[20]

The Sub-district Council and Sub-district Administrative Organisation Act 1994 provided that any Sub-district Council, wherein a group of centrally appointed officers and headmen met and supervised public services of a designated rural jurisdiction, must be, if ready, transformed into a SAO. The law essentially reintroduced SAOs after they were abolished in 1972.[21] This would be a type of local government in which an elected administrator and legislators possess both political and administrative power. By the end of 1997 this law led to the incorporation of over 6,000 SAOs nationwide.[22]

The 1997 constitution stipulated that local administrative organisations (LAOs) must have the freedom in their policy processes, government administration, human resources management, and fiscal and finance management, and must possess authority to provide local public services. The 1997 constitution was a milestone in accelerating the movement towards decentralisation and enhancing the importance of local governments.[23] The Decentralisation Plan and Process Act of 1999 specified detailed functions of and public services to be transferred to local governments, and also stipulated that 35 per cent of total expenditure of the national government in each fiscal year must be allocated to LAOs by 2006.[24]

Twenty years later, increases in development are evident throughout Thailand, especially in terms of infrastructure and the basic quality of life in rural jurisdictions.[25] As the Thai people witnessed tangible changes in their local conditions, they have focused more attention on and have become more aware of the importance of local politics and local self-governance. Decentralisation has brought about a number of positive outcomes, including the increase of citizen participation, needs-based service delivery, accountability, and political education.[26]

In terms of more specific characteristics of local government that appeared after the 1997 constitution and the Decentralisation Act of 1999, the nation experienced the establishment of locally elected bodies in municipalities, sub-district administrative organisations, and PAOs. As a result, there are now more than 7,000 of these subnational units of government. For the most part, they are supported by revenue transfers from the central government. The National Decentralisation Committee determined that there were 245 government activities that should be transferred to the local level. This would require coordination of 11 central government ministries and 50 of their constituent departments with thousands of local governments. Eighty-seven specific activities related to public infrastructure previously administered by 17 central government departments were also identified. There were also 103 social services administered by 26 departments identified to be moved to local governments. The education system was also identified for gradual decentralisation.[27]

Specific forms of local governments were assigned particular responsibilities. For example, PAOs were directed to formulate provincial development plans and provide support to other local governments. They were also supposed to allocate funding to them while providing encouragement for inter-local cooperation. City municipalities were assigned a wider range of functions. These included law and order, irrigation and transport infrastructure, waste management, public health, firefighting, water management, and some welfare services. SAOs were assigned responsibility for law and order, irrigation and transport infrastructure maintenance, waste management, public health, and social service and environmental activities. With local governments assuming more control for personnel actions, under the decentralisation plan the MOI was to no longer have sole control over this important administrative area.[28]

Within the decentralised framework described above, Thailand can be seen as having adopted a two-tier local government system. The lower tier is responsible for basic public services on a sub-district-wide jurisdictional scale, while the upper tier is responsible for public services on a province-wide jurisdictional scale. Local administrative organisations (LAOs) are classified into two major forms: ordinary and special. The ordinary form comprises three types of LAOs: the PAOs, municipalities, and SAOs. PAOs are upper-tier local government units. Municipalities and SAOs are lower-tier local government units – the former for urban areas and the latter for rural areas. The special form refers to Bangkok Metropolitan Administration (BMA) and Pattaya City. BMA enjoys a status of both upper- and lower-tier providing public services in the 50 districts of the capital city. Pattaya's status is comparable to that of a municipality.

Reframing the intent of the Decentralisation Plan and Process Act, we can first view it in terms of administrative decentralisation, with a number of administrative responsibilities handed off to the subnational units of government.[29] LAOs were assigned roles and responsibilities for the provision of public services that fall into six categories: (1) infrastructure; (2) quality of life; (3) order and security of communities and society; (4) planning, investment promotion, and commerce and tourism; (5) natural resources and environmental protection; and (6) arts and culture, traditions, and local wisdom. The functions of LAOs can also be assessed in terms of two categories: statutory and discretionary. LAO statutory (or mandatory) functions include maintenance of law and order; provision of public transport; provision of sanitary services (water supply, waste disposal, sewage and drainage); provision of fire engines; prevention and control of communicable diseases; provision of slaughterhouses; provision of public health services; provision of welfare for mothers and children; provision and maintenance of public recreation space and facilities; and provision of primary education. LAO discretionary functions include provision of marketplaces, ports and ferry services; provision of crematoriums; provision and maintenance of hospitals; provision of public utilities; provision and maintenance of parks, zoos, and recreation areas as well as sport facilities; provision of vocational training; promotion of citizens' occupation; improvement of slum dwellings; and maintaining government enterprises.[30]

The intent of the Decentralisation Plan and Process Act can also be reframed in terms of financial decentralisation. There are four major sources of revenue for the LAOs: locally collected taxes, centrally collected taxes, shared taxes, and grants. Locally collected taxes include land and building taxes, local maintenance tax, and signboard tax. In the 2018 fiscal year, locally collected taxes accounted for 15 per cent of the total local revenues. Centrally collected taxes refer to taxes that entirely accrue to LAOs but are collected by national agencies. Shared taxes refer to taxes that the national government collects, but portions are allotted to LAOs. In the 2018 fiscal year, centrally collected taxes and shared taxes accounted for 32 per cent and 16 per cent, respectively, of total local government revenues. Grants are intergovernmental transfers, supervised and distributed by the National Decentralisation Committee (NDC), which is chaired by the prime minister. Grants are classified into two types: general and specific. General grants are those allocated to LAOs to spend on any projects related to their functions. Specific grants are those allocated to LAOs with specific instructions mandated by the national government. While the allocation of general grants appears to be substantially systematic in accordance with a formula based on demographic and socioeconomic indicators, the distribution of specific grants is ambiguous and discretionary. In the 2018 fiscal year, grants made up the greatest portion of local revenues – 36 per cent of the total. This means LAOs are highly dependent upon the national government. This dependency notably challenges the reality of fiscal decentralisation. Figure 14.1 graphically represents LAO revenues by type for the period 1997–2018.

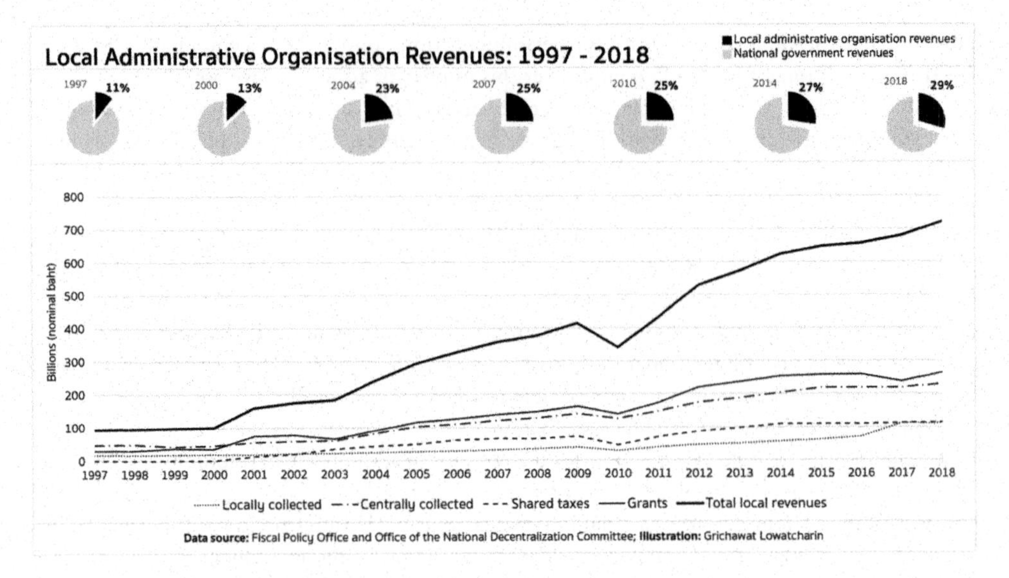

Figure 14.1 Local administrative organisation revenues, 1997–2018

Alternative analytic frames and recurring issues in Thai local governance

Intergovernmental relations

Due to political instability over the past 80 years, relations between central and local governments in Thailand are best described as a tug-of-war between centralisation and decentralisation, or actually between authoritarianism and democracy.[31] During the period from the early 1990s when Thailand reintroduced decentralisation as a key mechanism for providing basic public services nationwide, and while local development was flourishing, challenges to a move to more local governance remained.[32] The MOI/DOLA central and provincial administration continued to cast a shadow over the operation of the LAOs. The central administrative structure continued to exert supervision and control both directly and indirectly.[33] The coup of 2014 has worsened this state of affairs in that the military government has ordered a halt to local elections and the latest constitution downplays the role of local government. Confusion or ambiguity in relations between the central government's administrative agencies and local governments allows ministries and their agencies to control and intervene in local governance in both formal and informal manners.[34]

Thailand's public administration, particularly in relation to local administration, can be assessed based on the concept of fragmented centralism. This means that functions of public service provision are dispersed to government agencies under different ministries and coordination among these agencies is generally poor. While MOI and the Department of Local Administration (DOLA) have lead responsibility for coordinating local public service provision, many other Thai central government agencies have functional responsibilities at the local level. MOI/DOLA have taken advantage of this situation to demonstrate its indispensability in providing necessary implementation coordination at the local level.[35]

Efficiency and effectiveness

Researchers have found that Thai local governments (rural ones in particular) lack capacity for efficient provision of local public goods and services.[36] The inefficiency of LAOs has provided rationale for central administrations to show reluctance in handing over local government responsibilities and specific functions. Most important among these agencies are the Ministry of Interior, Ministry of Public Health, and Ministry of Education. They have sought to retain their power based on the allegation that local officers lack knowledge and experience, are incapable, and tend to be corrupt. In the case of the Ministry of Education, not only have central officials demonstrated hesitation in relinquishing administrative control to the local level, but teachers have also protested against transferring of administrative power, functions, and resources to LAOs.[37] Similar to central government officials, they allege that (1) the quality of education would be lower under supervision of the LAOs, (2) their morale and professional advancement would be abridged, and (3) local administrators and officials lack sufficient knowledge and capability to handle educational policy and functions – responsibilities which the central government has handled for more than a century.[38] Further considering education policy and administration, in recent years several central Thai government cabinets have de-emphasised decentralisation.[39]

To address the inefficiency and ineffectiveness of LAOs, the national government has promoted two approaches in an attempt to promote cost-effective administration at the local level. The first approach is inter-local cooperation, a scheme that aims to promote collaboration among LAOs and the pursuit of economies of scale for the provision of public services among adjacent jurisdictions. However, this has rarely taken place. Not only are LAOs in general not enthusiastic to work with adjacent jurisdictions, but the hierarchical structure and red tape of the central and regional bureaucracies impede the organisation and implementation of inter-local cooperation initiatives.[40]

The second approach is local consolidation, merger, or amalgamation of local jurisdictions. Local consolidation is a common approach employed in countries around the world based on the premise that it helps improve efficiency in public service provision to local residents.[41] Since the 2014 coup, the national government has attempted to implement nationwide compulsory municipal consolidation in the hope of increasing administrative efficiency. The National Reform Committee proposed a draft Local Administrative Organisations Code that would force SAOs and municipalities that have populations of fewer than 7,000 people or revenues of less than 20 million baht to consolidate with adjacent jurisdictions. Sixty per cent of SAOs and municipalities would be affected by this requirement. Due to opposition from local officials, the draft code was not enacted. To date, there only have been two cases of voluntary local consolidation in Thailand. The first was Wang Nam Yen Town Municipality in Sa Kaeo Province, which resulted from the consolidation between an SAO and a sub-district municipality in 2010. The second one is Wang Nuea Town Municipality in Lampang Province, which was a product of the consolidation of an SAO with a sub-district municipality in 2017. Thus far the observed effects of the first case have been mixed, while the second is too recent to offer meaningful evidence. The emerging Thai experience may be consistent with the international experience. A review of literature has found mixed effects from municipal consolidation in federal and unitary countries.[42] The success of consolidation is highly context dependent, and consolidation does not necessary lead to expected outcomes: the benefits of consolidation are often exaggerated. In some cases, consolidation has not improved service provision efficiency. In other cases, consolidation led to increases in costs due to the introduction of new public services. In the case of the

United States, consolidation has been resisted for reasons that have been in evidence in Thailand thus far: local officials and the local electorates that put them in office are reluctant to relinquish political control to new consolidated or amalgamated jurisdictions.[43]

Public finance and fiscal autonomy

The fiscal autonomy of local governments in Thailand is very tenuous. Local tax sources and amounts are limited and, as a result, LAOs have to depend heavily on the national government's decisions regarding how much tax revenues to share with and grant to local jurisdictions.[44] The Decentralisation Plan and Process Act of 1999 stipulated that 35 per cent of the total revenue of the national government in each fiscal year must be allocated to local governments by the year 2006. In 2007, an amendment to the act abolished this time frame for full implementation of the 35 per cent allocation, claiming that necessary transfers of authority and functions had not proceeded as planned.[45] Because local sources of revenues are limited, the national government has employed two measures to stimulate local revenues: shifting some tax bases from the national government to local authorities and increasing local grants.[46] However, rules determining the allocation of grants have varied from year to year and from one regime to another, thus making it difficult to describe or predict a pattern of specific grant allocation.[47] Complicated formulas and rules for specific grants have impacted the effectiveness of this approach. Furthermore, in a study of inter-governmental fiscal relationships in northeast Thailand offers evidence that grants from the national government to LAOs have been driven more by politics than considerations of need and inter-local equalisation of fiscal capacity.[48]

A more recent study offers confirmatory evidence regarding recurring problems associated with local public finance in Thailand.[49] These researchers found that the way that local revenue is structured to be generated from each source does not support the concept of local fiscal self-reliance and fiscal autonomy. This is especially the case in regard to the system of intergovernmental grant allocation, which does not include mechanisms to ensure fiscal equality. Grant allocations are often delayed, unpredictable, and (for some grant types) susceptible to political intervention. Many national laws concerning local fiscal issues are outdated and work against the potential for generating new revenue sources. At the local level, glaring problems are in evidence regarding LAO lack of capacity to develop efficient fiscal management systems, inadequate training and/or competence among local administrators, inadequate budget auditing, and lack of public participation in local budgeting processes.

Assessment of local government and decentralisation in Thailand

The decentralisation plan initiated in 1999 has not reached even minimal goals in transferring responsibilities and resources from the central government to local jurisdictions.[50] In recent years there have been calls for more decentralisation from diverse groups, including supporters of the "self-governing Chiang Mai" movement,[51] supporters of the Student and People Network for Thailand's Reform (STR), the People's Democratic Reform Committee (PDRC),[52] and local government politicians and officers who have rallied in the street to ask for a higher share of the national revenue.[53] Detailed proposals by these groups have been diverse, but they shared a belief that decentralised local governance is key for national development and a democratic Thailand.

The 2014 coup put an end to a prolonged political turmoil and to the diverse calls for more decentralisation. This coup differs from the preceding one in many aspects, one of which is about decentralisation. While the 2006 coup and its short-lived interim civilian government did not alter the structure of local governance, the most recent coup and the resultant military junta have had dramatic impacts. MOI has re-asserted its authority through manipulation of the decentralised decision-making structure that was introduced during the 1990s. Paradoxically, citizens at the local level demonstrate acceptance of this re-centralisation of power through the structures of decentralisation. Evidence indicates that since the decentralisation moves of the 1990s, citizens feel that government action at the local level is more transparent and responsive than was previously the case. They also feel that local administration requires the competence and resources of the central government because such does not exist at the local level.[54]

The centuries-old problems related to local governance have remained in place in Thailand today. These include a tug of war between the central and local administrations in terms of authority and capacity to govern, a lack and inequitable distribution of financial resources necessary to govern, a lack of management capacity and competence, and a lack of citizen engagement and trust at the local level.

Notes

1 Thanet Charoenmuang, *A Century of Thai Local Government (1897–1997)* (Bangkok: Khrongkan Chatphim Khopfai, 2007).

2 Chris Baker and Pasuk Phongpaichit, *A History of Thailand*, third edition (Port Melbourne, Australia: Cambridge University Press, 2014).

3 Alex M. Mutebi, "Recentralising While Decentralising: Centre-Local Relations and 'CEO' Governors in Thailand", *Asia Pacific Journal of Public Administration*, Vol. 26, No. 1 (2004), pp. 33–53; Supasaward Chardchawarn, "Local Governance in Thailand: The Politics of Decentralisation and the Roles of Bureaucrats, Politicians, and the People", *VRF* (Chiba: Institute of Developing Economies, Japan External Trade Organisation, 2010).

4 Wathana Wongsekiarttirat, "Central-Local Relations in Thailand: Bureaucratic Centralism and Democratisation", in *Central-Local Relations in Asia-Pacific: Convergence or Divergence?*, edited by Mark Turner (Basingstoke, UK: Macmillan Press, 1999), pp. 71–96; Mutebi, "Recentralising While Decentralising: Centre-Local Relations and 'CEO' Governors in Thailand"; Grichawat Lowatcharin, "Along Came the Junta: The Evolution and Stagnation of Thailand's Local Governance", *Kyoto Review of Southeast Asia* (Kyoto: Center for Southeast Asian Studies, 2014) <http://kyotoreview.org/yav/along-came-the-junta-the-evolution-and-stagnation-of-thailands-local-governance/>.

5 Björn Dressel, "Thailand's Elusive Quest for a Workable Constitution", *Contemporary Southeast Asia*, Vol. 31, No. 2 (2009), pp. 296–325.

6 Bidhya Bowornwathana, "History and Political Context of Public Administration in Thailand", in *Public Administration in Southeast Asia: Thailand, Philippines, Malaysia, Hong Kong, and Macao*, edited by Evan M. Berman (Boca Raton, FL: CRC Press, 2011), pp. 29–52.

7 Ibid.

8 Ibid.

9 Danny Unger and Chandra Mahakanjana, "Decentralisation in Thailand", *Journal of Southeast Asian Economies*, Vol. 33, No. 2 (2016), pp. 172–187.

10 A. Wongpredee and C. Mahakanjana, "Decentralisation and Local Governance in Thailand", in *Public Administration in Southeast Asia: Thailand, Philippines, Malaysia, Hong Kong, and Macao*, edited by Evan M. Berman (Boca Raton, FL: CRC Press, 2011), pp. 53–78.

11 Aaron Schneider, "Decentralisation: Conceptualisation and Measurement", *Studies in Comparative International Development*, Vol. 38, No. 3 (2003), pp. 32–56.

12 Michael H. Nelson, "Thailand: Problems with Decentralisation?" in *Thailand's New Politics: KPI Yearbook 2001*, edited by Michael H. Nelson (Bangkok, Thailand: King Prajadhipok's Institute, 2002), pp. 219–281.

13 Wongpredee and Mahakanjana, "Decentralisation and Local Governance in Thailand".

14 Ibid.

15 Ibid.

16 Ibid.

17 Kovit Puang-ngam, "History and Development of Subdistrict Administrative Organisation", King Prajadhipok's Institute, 2016 <http://wiki.kpi.ac.th/index.php?title=ประวัติและความเป็นมาขององค์การบริหารส่วนตำบล>.

18 Fumio Nagai et al.,"Local Government in Thailand: Analysis of the Local Administrative Organisation Survey", *Joint Research Programme Series* (Chiba: Institute of Developing Economies, Japan External Trade Organisation, 2008); Weerasak Krueathep, "Local Government Initiatives in Thailand: Cases and Lessons Learned", *The Asia Pacific Journal of Public Administration*, Vol. 26, No. 2 (2004), pp. 217–239.
19 Chardchawarn, "Local Governance in Thailand: The Politics of Decentralisation and the Roles of Bureaucrats, Politicians, and the People".
20 Nagai et al.,"Local Government in Thailand: Analysis of the Local Administrative Organisation Survey".
21 Puang-ngam, "History and Development of Subdistrict Administrative Organisation".
22 Nagai et al.,"Local Government in Thailand: Analysis of the Local Administrative Organisation Survey".
23 Ibid.
24 Ibid.
25 Prakorn Siriprakob and N. Joseph Cayer, "The Effects of Decentralisation on Local Governance in Thailand", *School of Public Affairs* (Arizona State University, 2007) <http://search.proquest.com/docview/304896337?accountid=14576>.
26 M. Shamsul Haque, "Decentralising Local Governance in Thailand: Contemporary Trends and Challenges", *International Journal of Public Administration*, Vol. 33, No. 12–13 (2010), pp. 673–688.
27 Unger and Mahakanjana, "Decentralisation in Thailand".
28 Ibid.
29 Ibid.
30 Kovit Puang-ngam, *Thai Local Government: Principles and Future Aspects*, eighth edition (Bangkok: Winyuchon, 2012).
31 Chardchawarn, "Local Governance in Thailand: The Politics of Decentralisation and the Roles of Bureaucrats, Politicians, and the People"; Wongsekiarttirat, "Central-Local Relations in Thailand: Bureaucratic Centralism and Democratisation"; Lowatcharin, "Along Came the Junta: The Evolution and Stagnation of Thailand's Local Governance".
32 Krueathep, "Local Government Initiatives in Thailand: Cases and Lessons Learned"; Haque, "Decentralising Local Governance in Thailand: Contemporary Trends and Challenges"; Christopher J. Rees and Farhad Hossain, "Perspectives on Decentralisation and Local Governance in Developing and Transitional Countries", *International Journal of Public Administration*, Vol. 33, No. 12–13 (2010), pp. 581–587.
33 Nagai et al.,"Local Government in Thailand: Analysis of the Local Administrative Organisation Survey".
34 Haque, "Decentralising Local Governance in Thailand: Contemporary Trends and Challenges"; Chardchawarn, "Local Governance in Thailand: The Politics of Decentralisation and the Roles of Bureaucrats, Politicians, and the People"; Nagai et al., "Local Government in Thailand: Analysis of the Local Administrative Organisation Survey".
35 Chardchawarn, "Local Governance in Thailand: The Politics of Decentralisation and the Roles of Bureaucrats, Politicians, and the People"; Nagai et al., "Local Government in Thailand: Analysis of the Local Administrative Organisation Survey".
36 Peerasit Kamnuansilpa and Supawatanakorn Wongthanavasu, "Management Potential of Tambon Administration Organisations in the Northeast of Thailand", *Humanities and Social Science Journal*, Vol. 20, No. 2 (2003); Peerasit Kamnuansilpa and Supawatanakorn Wongthanavasu, "The Effectiveness of Local Government in Promoting Development: People's View", *Cho Tabaek Journal of Kalasin Rajabhat University*, Vol. 2, No. 1 (2014), pp. 94–106.
37 Sarot Santaphan, "Problem in Decentralising Power to Local Administrative Organisations: A Case Study of School Transferring", *Public Law Net*, 2006 <www.public-law.net/publaw/view.aspx?id=857>; Suraphol Nitikraipot, "Problems and Facts about Transferring Education to Local Administrative Organisations", *Public Law Net*, 2004 <http://public-law.net/publaw/view.aspx?id=127>.
38 Santaphan, "Problem in Decentralising Power to Local Administrative Organisations: A Case Study of School Transferring".
39 Chardchawarn, "Local Governance in Thailand: The Politics of Decentralisation and the Roles of Bureaucrats, Politicians, and the People".
40 Kovit Puang-ngam, "Inter-Local Collaboration in Thailand", *King Prajadhipok's Institute Journal*, Vol. 2, No. 1 (2004), pp. 1–15.
41 Grichawat Lowatcharin, "Local Government Consolidation: Lessons from Selected Countries", *Thai Journal of Public Administration*, Vol. 15, No. 2 (2017), pp. 57–78.
42 Ibid.
43 Grichawat Lowatcharin and Charles David Crumpton, "Intergovernmental Relations in a World of Governance: A Consideration of International Experiences, Challenges and New Directions", 2018.

44 Krueathep, "Local Government Initiatives in Thailand: Cases and Lessons Learned".
45 Natthakorn Withitanon, "A Decade of Decentralisation (2000–2009): A Journey Back to the Beginning", *Prachathai*, 2010 <http://prachatai.com/journal/2010/10/31342>.
46 Sutapa Amornvivat, *Fiscal Decentralisation: The Case of Thailand* (Bangkok, Thailand: Ministry of Finance, 2004).
47 Direk Patmasiriwat, "Fiscal Inequality and Grant Allocation: Provincial Analysis of Thailand's Local Government Finance", in *Local Governments in a Global Context* (Khon Kaen: College of Local Administration, Khon Kaen University, 2012), pp. 91–115.
48 Achakorn Wongpredee and Tatchalerm Sudhipongpracha, "The Politics of Intergovernmental Transfers in Northeast Thailand", *Journal of Developing Studies*, Vol. 30, No. 3 (2014), pp. 343–363.
49 Banjerd Singkaneti and Darunee Pumkaew, "Problems of Local Public Finance in Thailand", *Thai Journal of Public Administration*, Vol. 15, No. 2 (2017), pp. 3–23.
50 Withitanon, "A Decade of Decentralisation (2000–2009): A Journey Back to the Beginning".
51 Chamnan Chanruang, "Disclosed: People's draft of Chiang Mai Metropolitan Bill", *Pub-Law*, 2011 <www.pub-law.net/publaw/view.aspx?id=1543>.
52 "PDRC Calls for Police Reform-No Politicisation", *Dailynews*, 2014 <http://goo.gl/M3ICrK>; "STR's Blueprint for Justice System Reform", *PatNews*, 2014 <http://goo.gl/OyWGFc>.
53 "Local Government Protesters Call for 30% Share", *Isara News Agency*, 2013 <www.isranews.org/community/comm-news/comm-politics/item/21212-newscommu2-140513.html>.
54 Unger and Mahakanjana, "Decentralisation in Thailand"; Thomas Dufhues et al., "The Political Economy of Decentralisation in Thailand: Does Decentralisation Allow for Peasant Participation?" *EAAE 2011 Congress* (Zurich, 2011).

References

Achakorn Wongpredee and Chandra Mahakanjana. 2011. "Decentralisation and Local Governance in Thailand". In *Public Administration in Southeast Asia: Thailand, Philippines, Malaysia, Hong Kong, and Macao*. Edited by Evan M. Berman. Boca Raton, FL: CRC Press, pp. 53–78.

Achakorn Wongpredee and Tatchalerm Sudhipongpracha. 2014. "The Politics of Intergovernmental Transfers in Northeast Thailand". *Journal of Developing Studies*, Vol. 30, No. 3: 343–363.

Baker, Chris and Pasuk Phongpaichit. 2014. *A History of Thailand*, third edition. Port Melbourne: Cambridge University Press.

Banjerd Singkaneti and Darunee Pumkaew. 2017. "Problems of Local Public Finance in Thailand". *Thai Journal of Public Administration*, Vol. 15, No. 2: 3–23.

Bidhya Bowornwathana. 2011. "History and Political Context of Public Administration in Thailand". In *Public Administration in Southeast Asia: Thailand, Philippines, Malaysia, Hong Kong, and Macao*. Edited by Evan M. Berman. Boca Raton, FL: CRC Press, pp. 29–52.

Chamnan Chanruang. 2011. "Disclosed: People's draft of Chiang Mai Metropolitan Bill". *Pub-Law* <www.pub-law.net/publaw/view.aspx?id=1543> (accessed 20 April 2014).

Dailynews. 2014. "PDRC Calls for Police Reform-No Politicisation" <http://goo.gl/M3ICrK> (accessed 20 April 2014).

Direk Patmasiriwat. 2012. "Fiscal Inequality and Grant Allocation: Provincial Analysis of Thailand's Local Government Finance". In *Local Governments in a Global Context*. Edited by Peerasit Kamnuansilpa. Khon Kaen, Thailand: College of Local Administration, Khon Kaen University, pp. 91–115.

Dressel, Björn. 2009. "Thailand's Elusive Quest for a Workable Constitution". *Contemporary Southeast Asia*, Vol. 31, No. 2: 296–325.

Dufhues, Thomas, Insa Theesfeld, Gertrud Buchenrieder and Nuchanata Munkung. 2011. "The Political Economy of Decentralisation in Thailand: Does Decentralisation Allow for Peasant Participation?" In *EAAE 2011 Congress*, Zurich.

Grichawat Lowatcharin and Charles David Crumpton. 2018. "Intergovernmental Relations in a World of Governance: A Consideration of International Experiences, Challenges and New Directions".

———. 2014. "Along Came the Junta: The Evolution and Stagnation of Thailand's Local Governance". *Kyoto Review of Southeast Asia*. Kyoto, Japan: Center for Southeast Asian Studies <http://kyotoreview.org/yav/along-came-the-junta-the-evolution-and-stagnation-of-thailands-local-governance/> (accessed 5 May 2017).

———. 2017. "Local Government Consolidation: Lessons from Selected Countries". *Thai Journal of Public Administration*, Vol. 15, No. 2: 57–78.

Haque, M. Shamsul. 2010. "Decentralising Local Governance in Thailand: Contemporary Trends and Challenges". *International Journal of Public Administration*, Vol. 33, No. 12–13: 673–688.

Isara News Agency. 2013. "Local Government Protesters Call for 30% Share". *Isara News Agency* <www.isranews.org/community/comm-news/comm-politics/item/21212-newscommu2-140513.html> (accessed 10 June 2014).

Kovit Puang-ngam. 2016. "History and Development of Subdistrict Administrative Organisation". King Prajadhipok's Institute <http://wiki.kpi.ac.th/index.php?title=ประวัติและความเป็นมาขององค์การบริหารส่วนตำบล> (accessed 28 July 2017).

———. 2004. "Inter-Local Collaboration in Thailand". *King Prajadhipok's Institute Journal*, Vol. 2, No. 1: 1–15.

———. 2012. *Thai Local Government: Principles and Future Aspects*, eighth edition. Bangkok: Winyuchon.

Mutebi, Alex M. 2004. "Recentralising While Decentralising: Centre-Local Relations and "CEO" Governors in Thailand". *Asia Pacific Journal of Public Administration*, Vol. 26, No. 1: 33–53.

Nagai, Fumio, Nakharin Mektrairat, and Tsuruyo Funatsu. 2008. "Local Government in Thailand: Analysis of the Local Administrative Organisation Survey". *Joint Research Programme Series*. Chiba: Institute of Developing Economies, Japan External Trade Organisation.

Natthakorn Withitanon. 2010. "A Decade of Decentralisation (2000–2009): A Journey Back to the Beginning". *Prachathai* <http://prachatai.com/journal/2010/10/31342> (accessed 10 July 2014).

Nelson, Michael H. 2002. "Thailand: Problems with Decentralisation?" In *Thailand's New Politics: KPI Yearbook 2001*. Edited by Michael H Nelson. Bangkok, Thailand: King Prajadhipok's Institute, pp. 219–281.

PatNews. 2014. "STR's Blueprint for Justice System Reform" <http://goo.gl/OyWGFc> (accessed 19 April 2014).

Peerasit Kamnuansilpa and Supawatanakorn Wongthanavasu. 2003. "Management Potential of Tambon Administration Organisations in the Northeast of Thailand". *Humanities and Social Science Journal*, Vol. 20, No. 2.

———. 2014. "The Effectiveness of Local Government in Promoting Development: People's View". *Cho Tabaek Journal of Kalasin Rajabhat University*, Vol. 2, No. 1: 94–106.

Prakorn Siriprakob and Cayer N. Joseph. 2007. "The Effects of Decentralisation on Local Governance in Thailand". *School of Public Affairs*. Arizona State University <http://search.proquest.com/docview/304896337?accountid=14576> (accessed 28 July 2017).

Rees, Christopher J. and Farhad Hossain. 2010. "Perspectives on Decentralisation and Local Governance in Developing and Transitional Countries". *International Journal of Public Administration*, Vol. 33, No. 12–13: 581–587.

Sarot Santaphan. 2006. "Problem in Decentralising Power to Local Administrative Organisations: A Case Study of School Transferring". *Public Law Net* <www.public-law.net/publaw/view.aspx?id=857> (accessed 14 April 2017).

Schneider, Aaron. 2003. "Decentralisation: Conceptualisation and Measurement". *Studies in Comparative International Development*, Vol. 38, No. 3: 32–56.

Supasaward Chardchawarn. 2010. "Local Governance in Thailand: The Politics of Decentralisation and the Roles of Bureaucrats, Politicians, and the People". *VRF*. Chiba: Institute of Developing Economies, Japan External Trade Organisation.

Suraphol Nitikraipot. 2004. "Problems and Facts about Transferring Education to Local Administrative Organisations". *Public Law Net* <http://public-law.net/publaw/view.aspx?id=127> (accessed 5 May 2017).

Sutapa Amornvivat. 2004. *Fiscal Decentralisation: The Case of Thailand*. Bangkok, Thailand: Ministry of Finance.

Thanet Charoenmuang. 2007. *A Century of Thai Local Government (1897–1997)*. Bangkok, Thailand: Khrongkan Chatphim Khopfai.

Unger, Danny and Chandra Mahakanjana. 2016. "Decentralisation in Thailand". *Journal of Southeast Asian Economies*, Vol. 33, No. 2: 172–187.

Wathana Wongsekiarttirat. 1999. "Central-Local Relations in Thailand: Bureaucratic Centralism and Democratisation". In *Central-Local Relations in Asia-Pacific: Convergence or Divergence?* Edited by Mark Turner. Basingstoke, UK: Macmillan Press, pp. 71–96.

Weerasak Krueathep. 2004. "Local Government Initiatives in Thailand: Cases and Lessons Learned". *The Asia Pacific Journal of Public Administration*, Vol. 26, No. 2: 217–239.

15

THAI STATE FORMATION AND THE POLITICAL ECONOMY OF THE MIDDLE-INCOME TRAP

Apichat Satitniramai

Thailand has long faced two major and unresolvable economic problems. The first is the "middle-income trap" (MIT). Its economy has not been able to raise incomes from middle to higher levels while rapidly entering the state of a hyper-aged society. The second is a very high degree of inequality. This chapter argues that the problems of MIT and economic disparity are profoundly connected with the prominent characteristic of the Thai state, which is "highly centralised but fragmented". This is a result of the history of modern nation-state building. The centralised but fragmented nature of the Thai state has been the major cause of the difficulty for increased industrial productivity to escape the MIT. Meanwhile, the highly centralised state means that the Thai elitist class possesses the power to create enormous economic rents for themselves. This is clearly shown in the concentration of wealth in the hands of a small number of people, or the "1 per cent problem" found in many other rich countries. It is in no way surprising that the centralisation of state power in creating this surplus would be a significant "legacy" that the elite, who have long controlled the state, would not allow to disappear. It is also not at all surprising that they are ready to use almost all forms of political violence, whether internal suppression or military coups, to establish a military dictatorship to control social forces when they feel challenged.

The main argument of this chapter is that the basic characteristic of the Thai state – centralised but fragmented – is a major obstacle in overcoming the MIT problem. In illustrating this argument, the chapter is divided into four sections. The first section discusses MIT and points out gaps in the existing literature. The second deals with process of Thai state formation, which has molded the basic characteristic of the state. The third section argues that this characteristic has essentially been unchanged until today in spite of several bureaucratic reforms and political opportunities. The last section concludes.

Middle-income trap

In general, MIT means the economic condition of a country, which has reached a high rate of economic growth during the initial stage of economic development until it can quickly lift itself out of poverty to become a middle-income country. But the level of economic growth later falls until it cannot shift from a middle-income to a higher-income society. This entraps a country in middle-income status for a long period. Economic growth in Thailand has followed this pattern.[1] However, the criterion or dividing line between middle- and high-income countries is debatable. So is the question of how long a country must be middle income so as to be considered as being in the MIT.[2] But there seems to be agreement among Thai policymakers and economists that Thailand has been caught in the MIT, as shown by the following studies.

The literature on the MIT can be classified into two major categories. First, there are works that focus on technical economic policies to help Thailand escape the MIT, such as Somkiat, Saowaruj and Nuthasid who studied approaches to increasing productivity in the industrial sector by three basic processes including process upgrading, product upgrading and functional upgrading through company case studies.[3] Pokpong and Supanutt attempt to identify several measurements in education reform to improve the quality of teaching and learning in the school system.[4]

Second, previous works that focus on institutional factors and political economy do not reject the policy proposals of the first group of studies.[5] However, they criticise the effectiveness of earlier decade-long technical economic proposals for solving the MIT problem.[6] They instead suggest that one should try to understand the institutional and political factors causing the MIT. Pasuk and Pornthep point out various aspects of high inequality in Thai society and further suggest that if these disparities were solved it would generate a higher rate of growth since there is no trade-off between growth and equity.[7] This work further explains the reasons behind the failure of solving the high level of inequality through an "oligarchy" – rule by the few in some form or another. They argue that the Thai economy has long been dominated by an oligarchy.[8] This was a major cause of inequality and a significant obstacle to dealing with this problem. However, the Thai oligarchy is highly dynamic, evolving and adapting through various mechanisms. It has co-opted leading members of new political forces, and in doing so, it has been successful in deterring and countering demands for resource redistribution. The Thai oligarchy has its roots in the era of the absolute monarchy. It comprises aristocrats, high-level bureaucrats and major Chinese merchants. After 1932, Thai politics shifted from absolutism to a bureaucratic polity. High-level bureaucrats, led by the military, became the leading force. Subsequently, as the Thai economy gradually grew, new national and local business groups proliferated. These new groups were incorporated into the oligarchic circle. After the 2006 coup, the circle of senior judges became a part of the oligarchy.[9]

With regard to institutional factors, Somchai argues that market-enhancing governance (MEG), introduced by the Sarit Thanarat government (1959–1963), is not enough to overcome the MIT because the MEG focuses on letting market forces work rather than pursuing a higher rate of growth by means of public-private collaboration.[10] The MEG mainly calls for market liberalisation, such as reducing the role of the state in producing private goods, and opening the economy for trade, investment and international technology transfers. These strategies were used by many poor countries like Thailand and China. They successfully helped these countries to upgrade to the middle-income level but not to overcome the MIT. Somchai suggests replacing the MEG with growth-enhancing governance (GEG), which is a set of rules to boost economic expansion without adhering solely to market mechanisms. Collaboration between government and the private sector should be the key to accelerating economic growth, such as collaborative projects in producing public or semi-public goods as a foundation for greater economic development in areas like knowledge and technology. The state should provide various forms of support for the private sector, such as financing and protection for the private sector during the technological catch-up process. The government needs to ensure that non-performers are not supported indefinitely.[11] Somchai stresses that one necessary condition of GEG implementation is reduced opportunity for economic rent captured by non-performers and prevention of unproductive economic rents. This means the government needs to have measures to control corruption both within GEG rules and existing law as well as at the policy level in the form of new rules offering benefits to insiders. Somchai also suggests (without elaboration) other broad guidelines including bureaucratic efficiency reform.[12]

This chapter broadly agrees with the ideas of Somchai and Pasuk and Pornthep. However, their works are insufficient. One needs to consider the political economy of the state. Characteristics of the Thai state explain why numerous proposed policies and solutions for the MIT problem suggested by various sectors of Thai society have not been implemented or have been implemented unsuccessfully. This chapter argues that the Thai state is a highly centralised but fragmented state. On the one hand, the Thai state is fragmented, implying that, in general, the Thai state is ineffective and inefficient as a mechanism for development. This characteristic is a major hurdle to solving the MIT problem, especially in the context of industrial upgrading. It is one thing to diversify an economic structure from being predominantly agricultural to manufacturing. But it is altogether another thing to upgrade existing industries to be more productive. The latter task is, in general, far more difficult to implement due mainly to various types of market failures.[13] In this light, the state has to play a leading role in the process of public-private collaboration. The fragmented state, clearly, is not an effective engine for this task. Somchai is aware of this problem, since he proposes bureaucratic reform. However, he pays no attention to analysing the nature and problem of the Thai state. This chapter hence argues that considering the failure of earlier efforts, there is very little chance that bureaucratic reform will be able to turn the Thai bureaucracy into an efficient mechanism to upgrade industrial policy without a major political transition.[14] On the other hand, the political and decision-making powers have been concentrated in the hands of members of a small elite faction in Bangkok who utilised this centralised "public" power as if it were their personal asset. Unproductive rents and the abuse of public power for private gain are thus common in the realm of public policies and programmes. Hence, the chances of successfully implementing the GEG as proposed by Somchai are rare, since one necessary condition is the ability of the state to prevent the creation of unproductive and undue economic rents. There are two major factors that might help create the necessary conditions for implementing GEG. The first is a merit-based professional bureaucracy. The second is a clear boundary between public and private interest. Public policy needs to prioritise public interest over private through an "ideology of publicness".[15] This boundary would help prevent bureaucrats and other state actors from intentionally creating undue and unproductive rents and distributing them to non-performers. However, these two conditions have long been interwoven in the history of state formation. The Thai society has not yet achieved these missions. It is no surprise that the centralised power of the Thai state in creating such rents has been highly protected as if it were the private asset of the Thai oligarchy, whose members are prepared to use any means of violence, such as a coup d'état, to protect their interest if threatened by any other forces. This is one of the reasons why political conflict during the past decade has been very contentious.

The formation and development of the Thai state: the centralised but fragmented patrimonial state

In explaining how the nature of the Thai state has been ineffective in industrial upgrading, this section begins with the formation of the "modern" Thai nation in the second half of the 19th century. The new state continued to preserve the characteristics of a patrimonial state. It had yet to achieve the status of "nation-state", meaning a state with public power separated from both the ruler and the ruled, and constituting the supreme authority within certain defined boundaries. Anderson[16] proposed that the patrimonial nature of the state, a legacy from the early days of state building, was not sufficiently transformed by the People's Party in 1932 to become an efficient mechanism to pursue the state's objectives and policies.[17]

The Thai state formation: a centralised-modernised patrimonial state

The reforms during the reign of King Rama V, Chulalongkorn, were not intended to fundamentally change Siam (the former name of Thailand) into a modern society, as in the Meiji era in Japan. Instead, the major purpose of the king was merely to transform the state so as to centralise political power in two aspects. The first is the territorial dimension. The reforms helped to consolidate power, turning tributary states into provinces of Siam under the administration of bureaucrats from Bangkok. The second is the centralisation of power from formerly powerful nobles to the king. The reforms ended the system of power sharing among kings and nobles and established the first absolutist monarchy in Thai history. The central government in Bangkok managed to have more control over its territory and the population of Siam than earlier traditional states.[18] The Chulalongkorn reforms were confined to the means and mechanisms of public and state administration rather than broader social and economic changes. This limited reform was sufficient for the rulers' demands. In understanding these issues, one needs to be aware that the origins of these reforms were inseparable from the confrontations and connections between Siam and the Western states during the colonisation period in Southeast Asia.

In contrast to earlier explanations by academics of the origin of Siamese modernisation reforms as a reaction against the spread of Western colonisation in order to preserve territorial integrity and political autonomy, Anderson argues that the expansion of Western power in this region reduced the ability of old enemies like Burma and Vietnam to threaten Siam.[19] Without these external threats, Siam enjoyed peaceful conditions it had never before experienced. Even by the 100th anniversary of the modern Thai military (1840–1940) it had not encountered any threats from powerful neighbours as in the early Rattanakosin era. Safe from powerful external threats, the rulers of Siam took the opportunity to implement their "internal colonisation" project, consolidating the power of rural elites and kings of tributary states under the control of Bangkok. The administration of rural areas was connected to central power through new technologies and institutions: railways, the telegraph, a modern bureaucracy, and a modern standing army. King Chulalongkorn learned a lesson about a small but effective standing army from his study tour of Singapore, Batavia and India.

Without any serious challenge from Western colonisers from the beginning, the reforms in Siam were not aimed at resisting the colonisers. Unlike Japan, it was unnecessary for Siam to invest in grand reform. Even efforts to revise the unfair and unequal treaties with other countries were delayed. This revision was completed only after the 1932 revolution.[20] Subsequently, the rulers did not attempt to establish state mechanisms with high capacity and efficiency to extract resources (like direct tax collection) and mobilise human resource for modern economic development. The Siamese rulers did not consider investment in large infrastructure for nation building as urgent. This is reflected in the annual budget of Siam prior to 1932. In the 30 years before 1931, annual state expenditure was focused on the internal colonising project and maintaining the lifestyle of the ruling class, while spending on economic development, especially on education for the citizenry, and agricultural development was minimal.[21] For example, in 1905, 29.07 per cent of total spending went to the Ministry of Defence, 19.43 per cent to the Ministry of Interior, and 14.45 per cent to the royal family.[22] These three agencies consumed 62.95 per cent of the total annual budget. As a result, the government had little budget left for other affairs, especially investment in mass education and other infrastructure for economic development.

In terms of education, official compulsory education was not introduced until the reign of King Rama VI, Vajiravudh. Education for all was implemented very slowly, partly because of the limited budget. Only 4 per cent of the total expenditure went towards education in nearly 30 years from 1900 to 1930. Compulsory education was expanded only in 1921 thanks to a special new individual tax on males in the active labour force. In Japan, by comparison, nearly all children were completing primary education from the early 20th century onwards.[23]

Furthermore, the Siamese state paid little attention to infrastructure investment for economic development. Feeny explains the undeveloped economy of Siam during 1880–1940 as the result of neglect of the agricultural sector by the Siamese government.[24] For example, Siam did not invest in irrigation systems to increase agricultural productivity but prioritised colonising their tributary states. In 1903, Siam rejected the suggestion ofVan der Heide to construct a comprehensive irrigation system. Instead it chose to build a railway line from Bangkok to the north.[25] Limited investment in irrigation systems was located in areas benefiting the ruling class instead of those would benefit from wider economic development. For example, the Pa Sak project, which helped to fill the Rangsit Canal, where land along both sides was owned by members of the royal family and the elite, was chosen instead of the Suphan project, which would have helped to open up new agricultural areas for the people but may have negatively affected ruling groups in the Rangsit area.

Agricultural development projects under Kings Rama V and VI failed to promote economic development. And unlike the experience of neighbouring countries that rapidly developed large-scale commercial farms, government concessions for digging irrigation canals, such as the Rangsit project, had unwanted consequences for the king. Initially, the construction of the canal system was a major factor in expanding the area of paddy fields in response to increasing global demand after the Bowring Treaty, which Siam signed with Britain in 1855. This expansion in international trade later became a crucial economic foundation for Chulalongkorn's consolidation of power. However, the irrigation project created land enclosure and many new large-scale landlords, who developed a new source of wealth through renting out land to farmers. As their influence expanded, the king rejected further concessions for canal digging. This stopped "the trend of the expanding landlord system".[26] After the government had rejected irrigation project proposals by Van der Heide and stopped granting further canal concessions, it ordered the the Royal Irrigation Department to construct smaller irrigation projects. These smaller projects created small-scale farming systems, since new lands were made freely available for slaves and commoners liberated from the feudal system during the 1870s. Rice production in Siam was therefore not based on a "plantation" system with big land owners, as had happened in neighbouring colonised countries, especially Burma and Vietnam, which also exported rice like Siam. Kullada[27] argues that the curtailment of the big landlord system created a different situation in Thailand from other countries: "[Burma and Vietnam] developed a large-scale land clearance and institutionalised credit system which consequently led to complex social stratification of the landowning class, big and small peasant holders and labourers, as well as money lenders".[28] In the same vein, Baker and Pasuk explain that "Siam did not follow the pattern of land development and land-tax of colonial countries, which forced farmers to increase productivity in order to be able to pay more taxes to the government".[29] Thus, unsurprisingly, the rice yield per acre in Siam gradually declined 1 per cent each year from 1920 to 1941.[30]

In terms of state revenue, after the successful centralisation of political power, the Ministry of Finance gained revenue mainly from indirect taxes. Aside from import-export taxes, consumption taxes on opium, gambling, liquor and lotteries from urban Chinese migrant labourers provided 40–50 per cent of annual government revenue throughout the second half of the 19th century.[31] The major reason for Siam's reliance on indirect taxes was the state's low capacity in tax collection, both the capitation tax and farm taxes from rural areas. The work of Seksan offers data on widespread evasion of these two types of tax.[32] Many provinces in northeastern Thailand did not pay farm tax up until 1932. In 1921, around 300,000 people did not pay the capitation tax; this number was 409,852 in the next two years. In 1926–1929, at least 325,000 people did not pay the capitation tax. In comparison to Japan, Seksan points out:

> [Japan] was able to maintain a firm control of the countryside. In the second half of the 19th Century incredible extraction was made on the peasantry by the Japanese state, whereas in the same period the Siamese state was unable to collect even a faction of its revenue from the rural population.[33]

The inability of the state to collect tax from the rural population reflects the low capacity of its mechanisms of control over population and resource mobilisation.

The low capacity of the modern bureaucratic system was caused by various factors. The major one was modernising without meritocracy. The newly constructed bureaucratic system transformed the earlier feudal system of nobles without regular payment into a monthly salary system accountable to rulers. Also, the state invested in a modern education system to transform the heirs of the elite into a new generation of modern bureaucrats via Suan Kulap School and Chulalongkorn University.[34] However, this new bureaucratic system was not merit based. Career paths and benefits still relied on patrimonial relationships and loyalty to the regime. Academic examinations in the recruitment of bureaucrats started only four years before the 1932 revolution – far later than in neighbouring countries.[35] The merit-free system was clear among the military, where blood relationships were the most important factor in the system. In 1910, the military ranks, from field marshal and lieutenant general, were confined to royal family members. Six out of 16 major generals and five out of nine brigade commanders were royal family members. In many cases, royal family members obtained the rank of general in their twenties. Among all the bureaucratic agencies, the Ministry of Defence was mostly occupied by royal family members.[36] Frustration with the privileged patrimonial system of personal connections was the major condition that brought about the 1932 revolution. Bureaucrats from commoner backgrounds called for a merit system in public administration. They demanded the same stipend for the same position (not dependent on the personal favour of the king) and career promotion based on seniority and performance as opposed to personal favouritism or family background. However, King Vajiravudh argued for what he called the "king's principle" or the king's judgment in public administration.[37]

The "king's principle" reflects the foundation and character of the new state constructed by King Chulalongkorn. Although it could be called a "nation-state", as it included the major basic elements of a nation-state (including population, territorial integrity, sovereignty and government), it is fundamentally still considered a "patrimonial state". There was still no separation between private and public interests. Rulers still treated the state as their private asset, particularly with regard to public administration and annual budgets.[38] In other words, since the colonial superpowers did not radically challenge the existence of the Siamese state, Siam consequently had no need to carry out comprehensive reform.

A revolution without mass support and autonomy of state: a fragmented state

Anderson argues that the 1932 revolution was led by commoner bureaucrats but because of the lack of mass support, the fundamental nature of the "centralised" and "patrimonial" state, without separation of the private and public spheres, was not disrupted.[39] He finds that the Thai state after 1932, which Riggs termed a "bureaucratic polity", was actually a mutated type of absolutist state but without the legitimacy such as the monarch used to have.[40] "Riggs's 'bureaucratic polity' was, in fact, the absolutist *moi*-state manqué".[41] It still kept the perspective and traditions of an absolutist monarchy. What was new after 1932 was a "fragmented state", in which the newly constructed bureaucratic system was fragmented in line with the functionalised bureaucracy. This was because in the new state, without the monarch's absolute controlling power as before, other social forces (arising from the 1932 non-mass-based revolution) could not establish control over the different functionalised partitions. Consequently, bureaucrats have a high degree of autonomy from society as a whole. Thus, the individual functional units of this fragmented bureaucratic system fought against one another to seize and to maintain their powers and interests. Riggs explains the post-1932 Thai politics as intra-bureaucratic conflict as follows:

> Cabinet members, for the most part, have been officials . . . [in the] conduct of their roles as members of a ruling circle, cabinet politicians have shown themselves more responsive to the interests and demands of their bureaucratic subordinates than to the concerns of interest groups, political parties, or legislative bodies outside the state apparatus.[42]

This type of polity did not respond to the demands of social forces outside the bureaucratic system. Without legitimacy among the wider public, post-1932 politics was unstable. It also brought about the protracted problem of a "centralised and fragmented state", which persists until today. This bureaucratic system took control over political power but acted as an ineffective and inefficient mechanism in economic development.

Riggs's important study of Thai politics illustrates the ineffectiveness and inefficiency of the Thai bureaucratic system in terms of policy decision-making, policy planning, public finance allocation, human resource management, lines of command and internal controls, the problem of redundancy, and internal and intra-organisation conflicts.[43] Although it looks senseless from a public interest perspective, it makes much sense from the viewpoint of the bureaucrats who gain great benefit from this ineffective system. Because of these incentives, bureaucrats strongly resisted later efforts at reform, especially when outside pressure from the general public was low. In conducting this work, Riggs uses studies from the International Bank for Reconstruction and Development (IBRD), later known as the World Bank, which sent experts to survey Thailand during the period between the second term of Field Marshal Plaek Phibun Songkram's government (1948–1957) and the coup d'état of 1957 to prepare for the drafting of the first economic development plan.[44] The report points out the low capability of the Thai bureaucratic system in economic development.

The IBRD report states that the ineffective and inefficient organisational structure of the Thai bureaucracy was one of the major causes of its low ability to promote economic development. There are four major characteristics of the division of labour among different bureaucratic units. First, there was "duplication of function without collaboration". More than one unit was responsible for the same function. The Ministry of Agriculture and Cooperatives and the Department of Public Welfare both conducted land settlement projects. At the same time, different state enterprises in industry were allocated to different ministries. Several ministries possessed two to three statistics units in the same ministry without coordination among them. An interesting example was the development work in the northeastern region, where more than two ministries worked separately on developing irrigation systems in the same region. The

Royal Irrigation Department was responsible for construction of reservoirs while the Ministry of Agriculture and Cooperatives was responsible for digging canals connecting the reservoirs to farms. Without expertise and capacity in engineering, the Ministry of Agriculture and Cooperatives eventually failed to construct the canal system. Second, the government was unable to abolish a unit which already completed its mission. For example, the Rice Office was established to monopolise international rice trading during the period of global scarcity of rice after the Second World War. However, when this situation had passed and the government allowed private traders to export rice, this office was not abolished until after the coup in 1957. Third, there were several missions without directly responsible units. They became extra work for existing agencies without any incentive to implement those missions. Lastly, there was the distribution of work within the same function among different offices, but there was no incentive for the related offices to collaborate.[45] From the perspective of the public interest, the whole system turned ineffective owing to duplication of staff, needless expense and inefficiency because of senseless rivalry among different offices.[46]

In sum, the major characteristic of the Thai bureaucratic system could be described through fragmentation, duplication, excessive costs, non-collaboration and intra-bureaucratic conflicts which rendered the state purposeless and without direction. This can be explained by the nature of the Thai state after 1932 coup. Fundamentally, it maintained the character of an absolutist state with highly centralised power. The post-1932 civilian governments failed to separate public powers from the elite. On top of that, after the coup the administrative system became fragmented, as bureaucrats were liberated from the control of the absolutist monarchy and wider public pressures. Without controls and a system of checks and balances, different groups within bureaucratic system turned to fight against one another to seize public powers and interests. This turned Thailand into a centralised but fragmented state. This situation continued in spite of subsequent multiple efforts at reform.

Failures to reform the state and the middle-income trap

Following the overthrow of a decades-old authoritarian government by a mass movement on 14 October 1973, extra-bureaucratic forces grew and democratic elections occurred frequently after 1978. However, there was still insufficient social pressure to push successful bureaucratic reform. Since 1990, nearly every government has recognised demands for bureaucratic reform. There have been countless reform policies and at least 23 committees for bureaucratic reform. The continuing problems of weak or no collaboration among different divisions since the 1960s have persisted. Departmental parochialism has triumphed over a comprehensive view of the whole mission. The bureaucratic system faces endemic problems of overlapping, duplication, red tape, inconsistency, inefficiency and ineffectiveness.[47]

The government of Prime Minister Anand Panyarachun (1991–1992) issued the Government Officer Administration Act 1992, which set a limit of zero manpower growth on the entire bureaucracy in order to modernise the bureaucratic system and to reduce the power of elected politicians in the appointment of high-ranking officials. During the following government of Prime Minister Chuan Leekpai (1992–1995), there was a pioneering effort to re-engineer the bureaucratic system in an initial 15 departments through reallocating and reducing the number of government officials and increasing the use of new technology, evaluation processes, training and subcontracting to the private sector. Nevertheless, Chuan did not place them as a priority of the government when these initiatives were resisted by senior officials. The next government under the premiership of Banharn Silpa-archa reduced the research budget for bureaucratic reform initiated by the Chuan government, while the next government led by General Chavalit Yongchaiyudh was barely able to make any further progress in reform in spite of demands from various organisations and the private sector. The 8th National Economic and Social Development Plan (1997–2001) even argued that the weakness of the state sector and bureaucratic system was a major obstacle to achieving the 8th Plan.[48]

The 1997 economic crisis and the 1997 constitution forced the next government to draft major bureaucratic reform plans. In collaboration with and driven by the World Bank (WB) and Asian Development Bank (ADB), the second Chuan government (1997–2001) signed an agreement with the WB to accept US$400 million to promote the reform of the state sector in 1998. Three years later, this collaboration brought about the Public Sector Management Reform Plan (PSMRP), covering various aspects of reform like expenditure management, human resource management and establishment of new autonomous public organisations. However, owing to various problems ranging from low awareness of the demand for reform and resistance from bureaucrats to the short life of governments (Thai coalition governments during the 1990s lasted less than two years), the new efforts begun in 1990 were hardly successful. Borwornsak Uwanno characterised this failure of bureaucratic reform as a "vicious circle".[49]

Despite efforts at reform since 1990, the problems of the Thai bureaucratic system remained unsolved. Fragmentation and non-collaboration among different departments within the same and different ministries continued. Each part functioned as a small empire which competed and struggled for power and resources. The work of individual agencies was dominated by departmental parochialism. Each focused mainly on their limited and narrowly viewed mission, and collaboration was lacking. Their work was fragmented, overlapping and duplicated. The government invented new divisions, departments and committees without abolishing pre-existing agencies with the same work. At the same time, power over the public policy process was still centralised in the hands of senior government officers rather than elected politicians. Senior officers like permanent secretaries still held strong control over policy either in terms of planning, implementation or enforcement. They had the upper hand over elected politicians. The major reason was the short life of unstable coalition governments. Compared to elected governments and politicians, bureaucrats last longer, thus they can delay unwanted policies and even hinder policies proposed by ministers.[50]

The bureaucratic reform of the government of Prime Minister Thaksin Shinawatra was part of the political campaign of the Thai Rak Thai (TRT) party since the 2001 election. Although nearly all previous governments had advocated bureaucratic reform, none of them had successfully pushed forward a policy like the Thaksin government. After its victory in the 2001 general election, it established the Office of the Public Sector Development Commission (OPDC), directly under the command of the Office of the Prime Minister. This reform commission was the first reform body to act independently from the bureaucratic system. Unlike earlier government efforts, which included government officials in the reform process, this committee comprised only elected politicians and non-bureaucrat experts. This shows the superior power the government had over the bureaucratic system.[51]

In 2003, the OPDC launched a five-year master plan, the Public Sector Development Strategy (PSDS). The Thaksin government proposed two new acts: (1) the Reorganisation of Ministries, Bureaux and Departments Act and (2) the Administration of State Affairs Reform Act. These two acts were passed by parliament in October 2003. The Reorganisation of Ministries, Bureaux and Departments Act aimed at reducing duplication among different government units. It restructured all ministries, departments and divisions. For example, 14 ministries with 125 departments increased to 19 ministries with 156 departments. This increase, however, was contrary to the original plan of downsizing public agencies. Suehiro argues that Thaksin was forced to increase the numbers.[52] The Board of Investment (BOI), which had been under the Office of the Prime Minister, was reallocated to the Ministry of Industry. Under the Administration of State Affairs Reform Act, departments with related missions were regrouped into clusters to integrate their work, with deputy permanent secretaries assigned to head the clusters. Furthermore, the cabinet passed an action plan to improve the capacity of government offices. Eleven ministries were forced to formulate action plans to reduce redundant processes and the time frame to complete missions and tasks to 30–50 per cent of the existing time frame.[53] Although the overall result of the Thaksin government's bureaucratic reform was more comprehensive and more effective than previous governments, it was still inadequate for the bureaucracy to conduct the task of industrial upgrading, as will be further discussed below.

Aside from failures of bureaucratic reform, the unsuccessful efforts to promote industrial upgrading are a major cause of why the Thai state has failed to escape the MIT. The study by Lauridsen finds that between 1990 and 2001, prior to the Thaksin government, Thailand did not have a coherent strategic plan for industrial upgrading.[54] On the contrary, the orientation of industrial policy was dominated by neo-liberal ideas and a development strategy implicit in promoting industry through attracting foreign capital. For foreign investors, Thailand had several comparative advantages, particularly cheap labour and abundant natural resources. Thailand still pursued this pattern of development after the 1997 economic crisis. The Chuan government dealt with the crisis by focusing on the financial sector, but it abandoned the real economy. Only between 1998 and 2001 was there a comprehensive set of small and medium enterprise (SME) support and industrial restructuring policy packages. However, this reform initiative was trapped in a vicious circle. Many policy or law initiatives were defeated before they got to the cabinet or were passed to parliament. Any that were passed were canceled and delayed until the next change of government.[55]

In addition to this vicious circle, such industrial policies were generally fragmented, contradictory and inconsistent. Industrial upgrading policy was under the responsibility of relatively weaker government offices, like the Ministry of Industry, the Ministry of Education or the Ministry of Science and Technology, while trade and investment policies were under more powerful agencies like the Ministry of Finance and the BOI. The main problem was that these latter two agencies regularly designed policies only to meet the short-term goals of increasing income and improving the trade balance rather than the long-term goal of industrial upgrading. For example, while the Ministry of Industry launched a domestic automobile supplier development policy, the Ministry of Finance set a tax structure which disadvantaged domestic spare-part supplies with respect to imports.[56]

This fragmented nature was also the result of conflict and competition among agencies in the same ministry and different ministries. The repeated phenomenon is of one mission under responsibility of many offices without either a clear division of labour or collaboration to enhance effectiveness. Moreover, the "real" commanding power in Thai public administration has been based at the level of the department, which is responsible only for a narrow scope of work. Ministries, which are supposed to take care of the overall picture, have been relatively weak owing to the short life of governments and cabinets. This has made the power of ministries weaker than that of departments. The result is that no one has overall power to command all aspects of policy implementation. Hence, any mission that required collaboration among different offices was nearly impossible. In other words, the Thai government did not have strong pilot agencies capable of collaborating policy and work among different government agencies with harmony and without conflict.[57] The situation was not very different from what Riggs found during the 1960s when he observed that the cabinet could not act as a supreme power to exact collaboration among government agencies.[58]

The Thaksin government was the first with a clear and cohesive strategy to promote industrial upgrading. This comprised two major sub-strategies. The first was the "industrial competitiveness upgrading" strategy. In 2003, the government set up the National Competitiveness Committee, to which it allocated 17.6 billion baht in the annual government budget. In promoting industrial competitiveness, the committee chose to support five industrial sectors: processed food; vehicles and parts; fashion (garments, leather and jewelry); tourism; and software (multimedia and animation). The second was a strategy to promote the technology capacity of the industrial sector. In pushing this strategy, several measures were designed. For instance, there were several reforms in the BOI. The BOI was reassigned from the Office of the Prime Minister to the Ministry of Industry in order to harmonise industrial policy. The BOI redesigned its investment promoting measures from a focus on infrastructure investment to supporting investment in "Skill, Technology and Innovation Packages", like tax exemptions for one to two years for entrepreneurs investing in research and development (R&D) or innovation. Furthermore, the BOI put more effort in connecting government R&D agencies with the industrial sector.[59]

In spite of a major new round of reform efforts and initiatives during the Thaksin government, one of the strongest and most powerful elected governments in the Thai contemporary history, Thailand has yet to succeed in industrial upgrading and escaping the MIT. Compared to earlier governments, the Thaksin government was among the most aware of the core problem and designed a clear strategy to deal with it. Nevertheless, it still failed.[60] Doner and Ramsay find that even in an uncomplicated and long-successful industry like sugar, the Thaksin government failed to upgrade productivity.[61] Government agencies related to this sector were fragmented and minimally collaborative. Government agencies' research work on sugarcane and the sugar industry was divided among three different ministries: Agriculture and Cooperatives, Trade, and Industry.[62]

The major obstacle to industrial upgrading during the Thaksin government was still the bureaucratic system. The government did not succeed in bureaucratic reform. It could hardly solve problems of inefficiency, fragmentation, departmental parochialism, overlapping and duplication of work among state agencies. Despite initial efforts, once it faced resistance from within the system, the government turned to repeating old mistakes, like setting up a new agency to undertake the mission without dissolving earlier ineffective and now redundant agencies which the government could not command. For example, in dealing with the fragmentation in labour skills training and development, which had been conducted by separate agencies in nine ministries, the government set up a new Labour Skills Development Promotion Committee of its own rather than push to consolidate the work of existing agencies.[63]

Subsequent governments since the coup against the Thaksin government in 2006, either civilian or military, have not had any strong commitment to bureaucratic reform. Due to the political conflicts, violence and instability during the last decade, no government could pay serious attention to bureaucratic reform, although the current military government has started to pay more attention to the MIT problem and included it as a priority in the mandated National Strategic Plan. Most new laws supporting the strategy will be implemented by future elected governments. The most important is the Eastern Special Development Zone Act passed into law in May 2018. It is far too soon to tell how the projects and measures of this act would help to upgrade Thai industry or to escape the MIT. Judging from previous reform outcomes, there is little hope that the reform under the current military government or the next government would be successful. Although the 2017 constitution requires future governments to reform and to follow the content of National Strategic Plan, the constitution in itself will be the most important obstacle to reform. This is because the constitution is intentionally designed to create weak or impotent elected governments in order to preserve and maximise the power of unelected actors, most of whom will be "bureaucratic-polity actors" led by the same old oligarchs of Thailand. Viewed from this aspect, the prospect of successful state reform is far from reality.

Conclusion

This chapter proposes that the problems of MIT and high levels of inequality now facing Thailand have small chance of a successful solution, at least in the medium term. This is inextricably related to the major characteristic of the Thai state as "centralised but fragmented" with no separation between private and public interest among the ruling elite. This creates a state with a low capacity to upgrade its industrial development and escape the MIT. In the same vein, centralised power without a separation between the public and private spheres among the elite turns the state into a political instrument of the oligarchs to generate undue economic rents which they can share. Unsurprisingly, the degree of inequality in Thailand is ranked among the worst in the world.

Notes

1 Somchai Jitsuchon, *Upgrading Thailand Long-run Economic Expansion: Perspective and Recommendation from Institutional Approach* (Bangkok: Thailand Development Research Institution (TDRI) and Bank of Thailand, 2014), pp. 6–7.

2 Several academics argue that the MIT is only a conjecture rather than a theory. See further detail in B. Eichengreen, D. Park, and K. Shin, "When Fast Growing Economies Slow Down: International Evidence and Implications for the People's Republic of China", *NBER Working Paper 16919*. Cambridge: National Bureau of Economic Research, 2011.

3 Somkiat Tangkitvanich, Saowaruj Rattanakhamfu, and Nuthasid Rukkiatwong, "Toward Innovation Creation and Technological Development of Industrial Sector", Paper presented at the TDRI's Annual Public Conference, "New Development Model: Towards Quality Growth Based on Productivity Improvement", 18 November 2013, Bangkok Convention Centre, Centrara Hotel, Central World.

4 Pokpong Junvith and Supanutt Sasiwuttiwat, "Human Capital Development for Better Productivity", Paper presented at the TDRI's Annual Public Conference, "New Development Model: Towards Quality Growth Based on Productivity Improvement", 18 November 2013, Bangkok Convention Centre, Centrara Hotel, Central World.

5 See Somchai, *Upgrading Thailand Long-run Economic Expansion: Perspective and Recommendation from Institutional Approach*; L.S. Lauridsen, *State, Institutions, and Industrial Development: Industrial Deepening and Upgrading Policies in Taiwan and Thailand Compared*, Vol. 1 (Aachen: Shaker Verlag, 2008); L.S. Lauridsen, *State, Institutions, and Industrial Development: Industrial Deepening and Upgrading Policies in Taiwan and Thailand Compared*, Vol. 2 (Aachen: Shaker Verlag, 2008); L.S. Lauridsen, *The Policies and Politics of Industrial Upgrading in Thailand during the Thaksin Era*, Paper presented at the 10th International Conference on Thai Studies, the Thai Khadi Research Institute, Thammasat University, 9–11 January 2008; Pasuk Phongpaichit and Pornthep Benyaapikul, "Political Economy Dimension of a Middle Income Trap: Challenges and Opportunities for Policy Reform: Thailand", 2013 <www.econ.chula.ac.th/public/publication/books/pdf/Political%20Economy%20Dimension%20of%20aMiddle%20Income%20Trap%20Thailand.pdf>; and R.F. Doner and A. Ramsay, "Growth into Trouble: Institutions and Politics in the Thai Sugar Industry", *Journal of East Asian Studies*, Vol. 4, No. 1 (2004), pp. 97–138.

6 Pasuk and Pornthep, "Political Economy Dimension of a Middle Income Trap", p. 1.

7 Ibid.

8 Ibid., p. 33.

9 Ibid., pp. 34–35.

10 Somchai, *Upgrading Thailand Long-Run Economic Expansion.*

11 Ibid., p. 19.

12 Ibid., pp. 32–33.

13 See further details on the differences between diversification and upgrading in R.F. Doner, *The Politics of Uneven Development: Thailand's Economic Growth in Comparative Perspective* (New York: Cambridge University Press 2009), pp. 74–78.

14 This chapter does not mean by this that there has been no change in the bureaucratic system since it was first created in the second half of the 19th century. Indeed, there was major change in the early 1960s during the military dictatorship (1959–1963) of Field Marshal Sarit Thanarat. Administrative mechanisms and macro-economic policy planning were reformed to support the initial stage of industrial development. It can be argued that this bureaucratic reform was one of the major reasons behind the success in lifting Thailand from a poor to a middle-income country (further details in Apichat, 2013). However, this chapter sees no change in the essential character of the Thai state in qualitative terms.

15 This chapter refers the "ideology of publicness" as a "nationalist ideology" in the sense that the people are truly the owners of the state. This is in line with the modern "nation-state". In this sense, the nation is not private property. According to Anderson (1978), the Thai state that was created between the reign of King Chulalongkorn and 1932 was not a "nation-state" but rather a "patrimonial state". It was still an asset of the rulers. Reform merely introduced a modernised element to maintain territorial integrity and a population.

16 See, B.R.O'G. Anderson, "Studies of the Thai State: The State of Thai Studies", in *The Study of Thailand: Analysis of Knowledge, Approaches, and Prospects in Anthropology, Art History, Economics, History, and Political Science*, edited by Eliezer B. Ayal (Ohio: Ohio University Center for International Studies Southeast Asia Program, 1978).

17 Thanks to Professor Kasian Tejapira, who suggested Anderson (1978) to the author to read and raised awareness of the importance of how the nature of the Thai state originated.

18 An earlier form of states in mainland Southeast Asia was the "mandala". Central state power was limited and was hardly able to penetrate distant localities. The state power was concentrated in the centre rather than the periphery (Kullada Kesboonchoo Mead, *The Rise and Decline of Thai Absolutism* (New York: Routledge Curzon, 2004), pp. 179–180).

19 Anderson, "Studies of the Thai State: The State of Thai Studies", pp. 200–203.

20 Seksan Prasertkul, *The Transformation of the Thai State and Economic Change (1855–1945)*, PhD Thesis, Cornell University, 1989, p. 471, fn. 76.

21 Data on the annual budget of the Siamese state during this period was calculated. See Ibid., pp. 431–440, table 6.1.

22 Annual expenditure on the royal family averaged more than 10 per cent until 1925 and was reduced to below 10 per cent between 1926 and 1930, due to the global economic decline during the Great Depression.

23 Anderson, "Studies of the Thai State: The State of Thai Studies", p. 201, fn. 11.

24 See D. Feeny, "Competing Hypothesis of Underdevelopment: A Thai Case Study", *Journal of Economic History*, Vol. 39, No. 1 (1979).

25 Certainly, the construction of a railway line would indirectly generate economic expansion. However, the major purpose of the railway system was to transport troops and military equipment to fight "rebels" in Lanna state, such as in Phrae province.

26 Christ Baker and Pasuk Phongpaichit, *A History of Thailand* (Bangkok: Matichon, 2014), p. 140.

27 Kullada, *The Rise and Decline of Thai Absolutism*, p. 181.

28 These major differences meant that the revolutionary nationalist movement in 1932 lacked allies and mass supporters, while in Burma and Vietnam, nationalist movements were widely supported by mass movements. The following section will explain how this problem affected the development of the Siamese state.

29 Baker and Pasuk, *A History of Thailand*, p. 145.

30 Feeny, "Competing Hypothesis of Underdevelopment".

31 Anderson, "Studies of the Thai State: The State of Thai Studies", p. 223.

32 Seksan, "The Transformation of the Thai State and Economic Change (1855–1945)", p. 477, fn. 87.

33 Ibid., pp. 477–478.

34 The initial objective of Suan Kulab School was not to provide education for all. It targeted low-ranking royal family members, including Mom Chao and Mom Rajawongse, and discriminated against commoners by increases in tuition fees. However, because of the rapid expansion of the bureaucratic system, it gradually started recruiting more commoners (Kullada, *The Rise and Decline of Thai Absolutism*, pp. 72–77). Suan Kulab School is a good example of how the Siamese rulers had little intention to use education as a social ladder for the masses and to promote a meritocratic system.

35 Anderson, "Studies of the Thai State: The State of Thai Studies", p. 215.

36 Ibid., pp. 206–207, fn. 20.

37 Baker and Pasuk, *A History of Thailand*, pp. 177–178.
38 However, there were several aspects where the Siamese state developed a division between private and public spheres, such as private property rights. There were no reports of the state seizing assets from private or individual between 1855 and 1932; see Seksan, pp. 180–194. There was also a partition between the Privy Purse Bureau taking care of crown property and the State Treasury; see Kullada, *The Rise and Decline of Thai Absolutism*, pp. 20 and 34 for details.
39 Anderson, "Studies of the Thai State: The State of Thai Studies", pp. 225–226.
40 See, F.W. Riggs, *Thailand: The Modernisation of a Bureaucratic Polity* (Honolulu: East-West Center Press, 1966).
41 Anderson, "Studies of the Thai State: The State of Thai Studies", p. 22.
42 Riggs, *Thailand: The Modernisation of a Bureaucratic Polity*, p. 312.
43 Ibid., pp. 311–366.
44 Ibid., pp. 346–358.
45 Ibid., pp. 333–335. Riggs finds that even the cabinet, which should act as the supreme organisation, could hardly coordinate different parts of the bureaucratic system. Instead, cabinet meetings turned into battlefields for different sets of bureaucrats to negotiate for their interests and benefits through the representation of their ministers. Most of the time, meetings ended up with compromises among interest groups rather than collaboration for the public interest.
46 For further details, see Riggs (*Thailand: The Modernisation of a Bureaucratic Polity*, pp. 352–358) in the case of the battle to control rice research between the Royal Irrigation Department and the Rice Department without collaboration, even though both departments were under the same Ministry of Agriculture and Cooperatives. Another piece of evidence is the struggle among different departments under the Ministry of Agriculture and Cooperatives to take control over agriculture promotion programmes. Instead of establishing a core department to conduct extension services comprehensively, each department initiated their own.
47 Apichat Satitniramai, "The Thai State and Economic Reform: From the Genesis of Bankers' Capitalism to 1997 Economic Crisis", *Fahdiewkan* (Bangkok: Fahdiewkan Publishing, 20130, p. 299. Also see Lauridsen, *State, Institutions, and Industrial Development*, Vol. 2, pp. 822–827.
48 Apichat, "The Thai State and Economic Reform", pp. 299–300.
49 Ibid., pp. 300–301. Also Lauridsen, *State, Institutions, and Industrial Development*, Vol. 2, p. 827.
50 Apichat, "The Thai State and Economic Reform", p. 301. Lauridsen, *State, Institutions, and Industrial Development*, Vol. 2, p. 832.
51 Apichat, "The Thai State and Economic Reform", p. 302. Lauridsen, *State, Institutions, and Industrial Development*, Vol. 2, pp. 845–848. Nakharin Mektrairat, Public Sector Reform in the Period of Thaksin Government, 2001–2006: Thai Rak Thai (TRT)'s Initiatives, Technocrat's Support and Bureaucrats' Resistance, Unpublished research report, 2007.
52 A. Suehiro, "Technocracy and Thaksinocracy in Thailand: Reforms of the Public Sector and the Budget System under the Thaksin Government", *Southeast Asian Studies*, Vol. 3, No. 2 (2014).
53 Apichat, "The Thai State and Economic Reform", pp. 302–303. Also see A. Suehiro, "Technocracy and Thaksinocracy in Thailand: Reforms of the Public Sector and the Budget System under the Thaksin Administration", Unpublished paper, 2007, p. 27.
54 Lauridsen, *State, Institutions, and Industrial Development*, Vol. 2.
55 Ibid., pp. 864–866. Also Apichat, pp. 310–311.
56 Ibid., pp. 864–866. Also Apichat, pp. 310–311.
57 Ibid., pp. 864–866. Also Apichat, pp. 310–311.
58 Riggs, *Thailand: The Modernisation of a Bureaucratic Polity*.
59 Apichat, "The Thai State and Economic Reform", pp. 312–313. Also see Lauridsen, *The Policies and Politics of Industrial Upgrading in Thailand during the Thaksin Era,* Paper presented at the 10th International Conference on Thai Studies, The Thai Khadi Research Institute, Thammasat University, 9–11 January 2008.
60 Lauridsen, *The Policies and Politics of Industrial Upgrading in Thailand during the Thaksin Era,* p. 24.
61 R.F. Doner and A. Ramsay, "Growth into Trouble: Institutions and Politics in the Thai Sugar Industry", *Journal of East Asian Studies*, Vol. 4, No. 1 (2004).
62 Ibid., pp. 44–45. Also, Apichat, "The Thai State and Economic Reform", pp. 313–314.
63 Apichat, "The Thai State and Economic Reform", p. 316. Also, Lauridsen, *The Policies and Politics of Industrial Upgrading in Thailand during the Thaksin Era*, pp. 18–20.

References

Anderson, B.R.O'G. 1978. "Studies of the Thai State: The State of Thai Studies". In *The Study of Thailand: Analysis of Knowledge, Approaches, and Prospects in Anthropology, Art History, Economics, History, and Political Science*. Edited by Eliezer B. Ayal. Ohio: Ohio University Center for International Studies Southeast Asia Program.

Apichat Satitniramai. 2013. "The Thai State and Economic Reform: From the Genesis of Bankers' Capitalism to 1997 Economic Crisis". *Fahdiewkan*. Bangkok: Fahdiewkan Publishing.

Baker, Chris and Pasuk Phongpaichit. 2014. *A History of Thailand*. Bangkok: Matichon.

Doner, R.F. and A. Ramsay. 2004. "Growth into Trouble: Institutions and Politics in the Thai Sugar Industry". *Journal of East Asian Studies*, Vol. 4, No. 1: 97–138.

Doner, R.F. 2009. *The Politics of Uneven Development: Thailand's Economic Growth in Comparative Perspective*. New York: Cambridge University Press.

Eichengreen, B., D. Park and K. Shin. 2011. "When Fast Growing Economies Slow Down: International Evidence and Implications for the People's Republic of China". *NBER Working Paper 16919*. Cambridge: National Bureau of Economic Research.

Feeny, D. 1979. "Competing Hypothesis of Underdevelopment: A Thai Case Study". *Journal of Economic History*, Vol. 39, No. 1: 113–127.

Kullada Kesboonchoo Mead. 2004. *The Rise and Decline of Thai Absolutism*. New York: Routledge Curzon.

Lauridsen, L.S. 2008a. *State, Institutions, and Industrial Development: Industrial Deepening and Upgrading Policies in Taiwan and Thailand Compared*. Vol. 1. Aachen: Shaker Verlag.

———. 2008b. *State, Institutions, and Industrial Development: Industrial Deepening and Upgrading Policies in Taiwan and Thailand Compared*. Vol. 2. Aachen: Shaker Verlag.

———. 2008c. "The Policies and Politics of Industrial Upgrading in Thailand during the Thaksin Era". In Paper presented at the 10th International Conference on Thai Studies, The Thai Khadi Research Institute, Thammasat University, 9–11 January.

Nakharin Mektrairat. 2007. Public Sector Reform in the Period of Thaksin Government, 2001–2006: Thai Rak Thai (TRT)'s Initiatives, Technocrat's Support and Bureaucrats' Resistance. Unpublished research report.

Pasuk Phongpaichit and Pornthep Benyaapikul. 2013. "Political Economy Dimension of a Middle Income Trap: Challenges and Opportunities for Policy Reform: Thailand" <www.econ.chula.ac.th/public/publication/books/pdf/Political%20Economy%20Dimension%20of%20aMiddle%20Income%20Trap%20Thailand.pdf> (accessed 15 November 2018).

Pokpong Junvith and Supanutt Sasiwuttiwat. 2013. "Human Capital Development for Better Productivity". Paper Presented in the TDRI's Annual Public Conference, New Development Model: Towards Quality Growth Based on Productivity Improvement, 18 November, Bangkok Convention Centre, Centrara Hotel, Central World.

Riggs, F.W. 1966. *Thailand: The Modernisation of a Bureaucratic Polity*. Honolulu: East-West Center Press.

Seksan Prasertkul. 1989. *The Transformation of the Thai State and Economic Change (1855–1945)*. PhD Thesis, Cornell University.

Somchai Jitsuchon. 2014. *Upgrading Thailand Long-run Economic Expansion: Perspective and Recommendation from Institutional Approach*. Bangkok: Thailand Development Research Institution (TDRI) and Bank of Thailand.

Somkiat Tangkitvanich, Saowaruj Rattanakhamfu and Nuthasid Rukkiatwong. 2013. "Toward Innovation Creation and Technological Development of Industrial Sector". In Paper Presented in the TDRI's Annual Public Conference, New Development Model: Towards Quality Growth Based on Productivity Improvement, 18 November, Bangkok Convention Centre, Centrara Hotel, Central World.

Suehiro, A. 2007. "Technocracy and Thaksinocracy in Thailand: Reforms of the Public Sector and the Budget System under the Thaksin Administration". Unpublished paper.

———. 2014. "Technocracy and Thaksinocracy in Thailand: Reforms of the Public Sector and the Budget System under the Thaksin Government". *Southeast Asian Studies*, Vol. 3, No. 2: 299–344.

16

THAILAND'S CORPORATE SECTOR AND INTERNATIONAL TRADE

Krislert Samphantharak

This chapter provides an overview of firms in Thailand. It first explores salient characteristics of the corporate landscape in Thailand. There is an enormous concentration of corporate wealth where a few large firms occupy a disproportionately large share of the country's total output. There has largely been an absence of indigenous ownership of corporate wealth. Big business corporations have been owned and run by descendants of migrants, mostly with ethnic Chinese origins. Complex business groups are a pervasive form of corporate structure – pyramids and cross shareholdings across firms are used extensively. Government-business symbiotic relationship and corruption are commonly practiced and accepted. State-owned enterprises play a significant role, especially as a revenue source to the government. Multinational enterprises are important in the export sector, especially as they make Thailand a crucial node in the global production networks. And finally, small and medium-sized enterprises – though small in size – are large in number and provide employment to the majority of Thai workers.

The chapter then discusses macroeconomic development episodes of Thailand – namely industrialisation and the Asian financial crisis – and connects them with the micro foundation from a perspective of the corporate sector.

Given that miraculous growth of the Thai economy took place during the export-oriented industrialisation period, the chapter explores firm-level evidence on exports and imports and connects it to aggregate international trade. In doing so, it highlights the following stylised facts: (1) the Thai economy has experienced rapid growth in exports as well as a drastic change in the structure of trade, from traditional commodities to manufacturing products; (2) Thailand has also become an integrated part of global production networks, especially in computer electronics and automobile industries; (3) direct participation in international trade is very limited and there is extreme concentration of trade among those who export or import; and (4) although only a small number of large firms directly engage in international trade, several small and medium-sized enterprises indirectly take part in exporting and importing activities through production networks.

Finally, the chapter ends with a discussion on how the corporate landscape of Thailand handles challenges currently faced by the country: resource misallocation, inefficiency, and low productivity; lack of competitiveness and the middle-income trap; increased inequality; and vulnerability to shocks. These problems raise a concern over the future sustainability of economic performance of the Thai economy.

Corporate sector in Thailand

Thailand's production sector consists of four main types of enterprises: family business groups, state-owned enterprises (SOEs), multinational enterprises (MNEs), and small and medium-sized enterprises (SMEs). Although SMEs account for the largest number of firms, a small number of state-owned enterprises and large family firms in business groups contribute the largest share of the economy's total output.[1] MNEs, on the other hand, dominate the manufacturing export sector. This section highlights some stylised facts about Thailand's corporate sector.

First, there is enormous concentration of the corporate sector in various dimensions. In terms of production, a few large firms occupy a disproportionately large share of the country's total output. In particular, in 2016, the top 5th percentile of Thai firms accounted for approximately 90 per cent of the total revenue of all firms in the country, while the bottom 75th percentile accounted for only about 5 per cent. This extreme concentration is not a new phenomenon, but the corporate sector in Thailand has become even more concentrated over time: the revenue share of the top 5th percentile firms was as high as 85 per cent in 2004. In terms of industry diversification, the top five industries accounted for half of the total revenue of the economy in 2016. These industries are wholesale trade (excluding motor vehicles), retail trade (excluding motor vehicles), motor vehicle manufacturing, petroleum manufacturing, and motor vehicle wholesaling. Finally, in terms of ownership, corporate ownership is limited among certain groups of the Thai population and is also highly concentrated. In 2017, the top 0.1 per cent of shareholders account for 44 per cent of the aggregate equity of all corporate firms in Thailand. Likewise, the top 0.1 per cent earners account for 55 per cent of the aggregate profit generated by all companies.[2]

Second, if we focus on large private business corporations, we find that these firms are mostly family businesses and have been owned and run by descendants of immigrants, mostly with ethnic Chinese origins. There is limited corporate wealth owned by indigenous Thai. In particular, out of the top 150 business groups in Thailand, almost all belong to families whose ancestors migrated from the southeastern coast of China to Thailand during the political unrest of the 19th and 20th centuries.[3] Although a few of these businesses were founded a century ago, their expansion did not take place until after the Second World War. For example, the agro-business empire Charoen Phokphan (CP) Group, despite being founded in the early 1920s, became a dominant actor in the Thai economy only in the 1970s and expanded to other business ventures such as wholesaling, convenience stores, and telecommunication as late as in the 1990s. Many business empires are relatively new, such as those belonging to the Sirivadhanabhakdi family. Of course, there are older families that were once at the epitome of the corporate world in Thailand but later experienced a significant decline, especially during the Asian Financial Crisis of 1997–1998. The Tejapaibul family represents one of them.

In principle, family involvement could provide some benefits to the business, especially in an environment where markets do not perfectly operate. For example, information problems are less severe within the family and trust between family members could help mitigate contracts that would not be legally enforceable otherwise. On the other hand, family involvement could create problems. Sibling rivalry, infighting for inheritance, and conflicts over a firm's strategy and policy among descendants after the founder passes away could be detrimental to the performance of the family firm. In the case of large Thai family businesses, a study shows that family ownership and control in fact creates a governance problem instead of being the second-best solution to market imperfections.[4]

Third, corporate structure of large business empires in Thailand is also different from those in the Anglo-Saxon economies where diversified conglomerates, which are legally a unitary business entity operating in multiple sectors, are common. Like in many other developing economies, Thai businesses are structured as business groups where legally distinct firms are tied together by formal or informal links and coordinate their decision-making. The organisation of groups could be relatively simple or extremely complicated. They could be structured by common ownership (such as firms independently owned by the same person or family) or chain ownership (such as pyramids and cross shareholding). Group size also varies tremendously. While some groups may consist of only two firms, there are some groups with hundreds of companies.[5]

Using business groups as a form of corporate structure has several implications. Since each of the firms in a business group is legally distinct, the compositions of shareholders could be different across member firms even though they belong to the same group. In this respect, this structure allows the controlling family to set up a new firm for each joint venture with different business partners and with different ownership arrangements. It also facilitates fund-raising from outside shareholders while keeping control of each firm within the family. Analysing large business groups in Thailand, a study shows that there are in fact funds transferred between firms in business groups.[6] Another study finds that complex group structures are associated with worse corporate governance, resulting in lower performance of the member firms.[7]

Fourth, there are widespread symbiotic relationships between government and businesses. This phenomenon is commonly known and widely accepted. The relationships are formed in various ways. In many cases, connections to political figures – through kinship or friendship – began long before they entered politics or rose up to power. In other cases, however, political connection is driven by business motives. Family firms sometimes build connections with political elites through friendship, corporate board nomination, or political contribution. At the extreme, the most direct way for a business family to have political connections is to have its family member enter politics and become a member of the parliament or cabinet. This practice is not uncommon in Thailand. The most evident case was Thaksin Shinawatra, a former telecom tycoon who eventually became prime minister in the 2001 general election. Later, his brother-in-law, Somchai Wongawat, and his sister, Yingluck Shinawatra, also entered politics and became prime ministers as well.

Political connection can benefit business in various ways. It can give the business preferential treatments such as contracts, subsidised loans, or a monopoly license from the government. A study of politically connected firm in Thailand shows that this is indeed the case: connected firms outperform their non-connected counterparts. The main channel was through legislation changes such as tax and license fee cuts, new government contracts, and market entry barriers, which allowed the firms to expand their market share at the expense of competitors.[8]

Fifth, another type of large firms in Thailand is state-owned enterprises (SOEs). Currently, there are 58 enterprises in which the state holds more than 50 per cent of voting rights; 33 engage in commercial activities. SOEs in the energy sector are the largest by revenue, with PTT Public Company Limited alone accounting for 49 per cent of total revenue. The others in the top five SOEs by revenue are Electricity Generating Authority of Thailand, Provincial Electricity Authority, Metropolitan Electricity Authority, and Thai Airways International Public Company Limited. With the exception of Thai Airways, four out of the five largest SOEs operate in the energy sector. Most Thai SOEs make positive profit and contribute to the state's revenue. In 2016, the total contribution (including transfers of profit and dividend payment) is 6.8 per cent of total state revenue. The factors that help make SOEs profitable include monopoly power granted by the government, preferential treatment in state procurement, and financial assistance from the government.[9]

Sixth, the final type of large firm in Thailand is multinational enterprises (MNEs). These foreign enterprises operate in Thailand through foreign direct investment (FDI) and enjoy benefits from producing locally, such as lower cost of production and transportation as well as bypassing import tariffs. In practice, these operations could be done through outsourcing or subcontracting rather than direct ownership. However, the reason why these MNCs decide to have ownership over the operation in Thailand (and other host countries) is the advantage from internalisation, which includes more effective protection of intellectual property rights outside their home country, lower within-firm transaction cost, and facilitating investment in firm-specific assets.

At first glance, the role of FDI and MNEs in Thailand seems limited. During the past several decades, the highest annual FDI to private fixed investment ratio had been only 8 per cent (in 1989–1990). Likewise, cumulative FDI had accounted for less than 5 per cent of cumulative private fixed investment during 1965–1996.[10] However, this argument could be misleading. The most significant FDI to Thailand during the 1980s and 1990s had been in the electric machinery and transport machinery manufacturing sectors. The MNCs in these sectors in turn have created upstream industries served by enterprises owned by Thai entrepreneurs.

Finally, although large corporations account for the majority of Thailand's total output, over 99 per cent of the total number of business enterprises in the country is classified as small and medium enterprises (SMEs). Among the SMEs, only a quarter are formally registered as juristic persons while the other three-quarters are informal businesses operated by individuals who are ordinary persons. Since employment by SMEs accounts for 80 per cent of Thailand's total employment, they are an important source of income for Thai households.[11]

Industrialisation and the Asian Financial Crisis

Economic development of Thailand since the end of the Second World War has been marked by rapid industrialisation, especially in the 1980s and early 1990s, and the Asian Financial Crisis that broke out largely unexpectedly in 1997. This section discusses the relationship between the corporate sector and these two phenomena.

Import substitution industrialisation

By the end of the Second World War, the stage had already been set for the growth of businesses that would eventually dominate the post-war corporate sector in Southeast Asia. Ethnic Chinese immigrants were already entrenched in the trading sector. The 1949 communist revolution in Mainland China changed the intention of many of these migrants from sojourners to permanent settlers. However, Thailand's xenophobic nationalist policies following the end of the Second World War made them politically vulnerable, forcing them to respond in several ways. One was assimilation into the Thai society through extensive interracial marriages and name changes, which were well accepted by the government. Most important response, however, was the formation of patronage connections with the military and civilian politicians as well as government bureaucrats. These connections not only provided protection to them but also gave them access to privileges.

Meanwhile, in the late 1950s the Thai government began to pursue industrialisation as the country's development strategy amid the growing concern over communism and the increasing influence from the United States during the Vietnam War. Consistent with nationalist policies, industrialisation was focused on import substitution of manufacturing goods. Connections to the Thai government allowed Chinese entrepreneurs to enter the manufacturing sector. At the same time, foreign companies formed joint ventures with these entrepreneurs to bypass the government's import restrictions. With their background largely exclusive to trading, Chinese entrepreneurs did not possess expertise in manufacturing and had to rely on technology and production process provided by foreign (usually Japanese) companies. Import substitution thus gave rise to joint ventures for domestic assembly with foreign suppliers without genuine domestic production. This pattern was observed in a wide range of industries that included automobiles, electrical appliances, steel products, glass, and chemicals.[12]

Export-oriented industrialisation

The end of import-substitution strategy was caused by the balance of payment crises in the 1970s and 1980s. The appreciation of the Thai baht, which was fixed to the US dollar during the 1970s, reduced the competitiveness of Thai exports. Declining prices of agricultural products throughout the world during this period also put pressure on Thailand's balance of payments and emphasised the need to switch to non-agricultural exports. Finally, the withdrawal of US military operations after the Vietnam War further reduced foreign exchange that Thailand previously received from US foreign aid and military-related spending. To respond to these crises, the Thai government reoriented its industrialisation strategy towards manufacturing exports in the late 1970s and early 1980s.

Coincidently, Thailand's shift to manufacturing export orientation took place at the time when there was a massive relocation of manufacturing firms from Japan to countries with lower labour costs in response to the Plaza Accord and the appreciation of the Japanese yen in 1985. This export-oriented industrialisation led to rapid growth in exports. Compared to the annual growth rate of only 6 per cent per year in the 1960s, export growth increased to 11 per cent in the 1970s and 16 per cent in the 1980s and continued to grow at a high rate into the early 1990s. The Thai economy experienced unprecedented growth, increasing at an average of 7.7 per cent annually during the 25-year period from 1971 to 1995.

Export-oriented industrialisation in Thailand was driven by FDI from MNEs, cheap labour in the region, and domestic SMEs that supplied parts and components to the MNEs. In contrast, the contribution of large family business groups to the economy during this period was questionable. Many of them were operating in less competitive domestic sectors that enjoyed local monopoly power or privileges from the government (such as trade, services, property development, protected manufacturing industries) rather than in the manufacturing export sector which requires efficiency and competitiveness. Domestic firms, both large and small, who partnered with foreign capitalists and operated in manufacturing export industries such as textiles, garments, and electronics had a limited or passive role in technology development and innovation. They continued to depend on cheap domestic labour and foreign technology. It is not surprising that despite decades of rapid economic growth and exposure to foreign markets, there is an absence of globally recognised brands that originated from Thailand or were developed by Thai entrepreneurs.[13]

The Asian Financial Crisis

Not only were large domestic business groups in Thailand disproportionately favoured by industrialisation policies during the second half of the last century, but they also benefited from financial regulation and deregulation in the late 1980s and early 1990s. Many of these groups also owned commercial banks or other financial institutions that allowed them to raise funds from the public to finance the expansion of their business empires. Financial liberalisation, with the development of stock markets, further facilitated fund-raising by large business groups from the public while allowing them to keep control of the firms. Liberalisation of international capital flows further allowed business groups and their banks to borrow money from abroad, with debts denominated in foreign currencies. However, relationship-based capitalism resulted in connected lending and tunneling of funds out of the companies while a complex and non-transparent business group structure helped conceal such behaviors from public awareness and regulators. This fragility made the financial system ripe for crisis. When the public started to doubt the solvency of financial companies, an information cascade rapidly led to banking panics and eventually financial meltdown.

International trade

Given the role of the export sector in Thailand's industrialisation, this section further looks at the corporate sector behind exporting activities and highlights some additional stylised facts.

First, Thailand's export has increased tremendously since the 1980s. While the export to gross domestic product (GDP) ratio was only 23 per cent in 1985, it went up to 39 per cent in 1996 and peaked at 71 per cent in 2008. The structure of trade also changed over time. In 1960, 95 per cent of Thai exports were agricultural products and crude materials while manufacturing exports accounted for only 1 per cent.[14] However, by the early 1990s, 75 per cent of Thai exports were from manufacturing products. In 2017, the top three exports of the country were (1) computers and electronics, (2) motor vehicles, and (3) plastic and rubber.[15] However, the country seems to fail to adjust its manufacturing sector to respond to changes in technologies and global demand. In particular, computer storage devices produced in Thailand are hard disk drives whose industry is in a relative decline, while the country fails to switch to flash and touchscreen technologies used in the rapidly growing tablet and handheld device market.[16]

Second, advances in transportation and communication technologies in recent decades have allowed production to be divided into stages, each performed in different locations. These advances arrived at the time when Thailand shifted its development strategy to export orientation and Japan, and the newly industrialised economies (NIEs) looked for destinations to relocate their production in the 1980s – the two important factors that have made Thailand a crucial node in regional production networks. In particular, computers and electronics and automobiles have become the country's major production exports, rapidly replacing traditional agricultural and mineral products. The role of Thailand in the global production networks was confirmed by the OECD's Global Value Chain Participation Index, which ranked the country third in the world for electrical equipment and third in Asia for transport equipment in 2009.[17]

As part of the global production network, Thailand's gross export value embeds components that are imported from elsewhere. This is reflected in the growth of imports to the country during the period of export expansion. Once taking this issue into consideration, however, Thailand's value added in export accounts for only 62.7 per cent of total gross export value in 2014, implying that a large share of Thailand's export is not from the contribution of the country's factor of production.[18] For machinery, computers and electronics, and motor vehicles, the two major industries in the global value chains, the shares of domestic value added in gross export value are both less than 50 per cent.[19]

Third, direct engagement in international trade is extremely rare and highly concentrated. Participation in international trade is uncommon. Only 5.7 per cent of all registered firms in Thailand export while just over 10 per cent import. Once further taking into account that many firms are both importers and exporters, only 12 per cent directly engage in either import or export. Trading firms are also special. They differ substantially from firms purely producing for the domestic market and tend to be larger, more capital intensive, more productive, and utilise more external finance. Among export and import firms, trade is extremely concentrated. The top 5 per cent of export firms account for 88 per cent of total Thai exports in 2015. Similarly, the top 5 per cent of import firms account for 90 per cent of all imports. Most importers as well as exporters tend to trade relatively few products and engage in trade with a relatively small number of countries. However, the small number of firms with the greatest product and trading-partner intensity account for a large share of both exports and imports.[20]

Finally, although firms directly engaging in international trade tend to be large, SMEs have played a significant role as suppliers of intermediate inputs in the production chain. The automobile industry vividly illustrates this point: while MNEs and large domestic firms dominate the downstream production, SMEs are involved in the upstream. Specifically, there are 18 car assemblers, which are joint ventures between foreign MNEs and large domestic companies. Tier-1 suppliers consist of 710 auto parts companies (58 per cent owned by foreign majority, 3 per cent joint ventures, and 39 per cent Thai majority). Some of these firms are large while others are SMEs. Finally, over 1,700 companies are Tier-2 and Tier-3 suppliers, most of which are SMEs.[21]

Current and future challenges

We end this chapter with a discussion on how the corporate landscape of Thailand handles challenges currently faced by the country: (1) resource misallocation, inefficiency, and low productivity; (2) lack of competitiveness and the middle-income trap; (3) increased inequality; and (4) vulnerability to shocks.

First, the landscape of Thailand's corporate sector discussed in this chapter suggests that there seems to be resource misallocation and inefficiency, as several firms enjoy privileges and do not operate in a competitive environment. Many large private domestic firms benefit from their connection with the government. Given the unstable nature of Thai politics, several businesses diversify and form relationships with various political parties, military personnel, and government officials in order to ensure their preferential treatments regardless of who is in power, exacerbating wasteful rent-seeking behaviours. Many SOEs continue to benefit from government protection and assistance. Since the government receives significant transfers from SOEs, competition policies that would increase the efficiency of resource allocation in the economy but at the cost of lowered profits of SOEs are unlikely to be implemented.

Many export-oriented industries in Southeast Asia are unskilled labour intensive. They attract investment from MNEs thanks to low domestic wages. These industries often have low investment in physical capital and are relatively mobile, making the host countries vulnerable to the relocation of the production to somewhere else when domestic wage rates become higher over time. These industries are known as *footloose industries*, with examples including garment and footwear. Given that the industries are unskilled labour intensive and require minimal transfers of technology or know-how, the benefits to the host countries are not likely to be long-lasting once the MNEs relocate their production facilities elsewhere.

Since the domestic sector is dominated by politically connected business groups who enjoy privileges granted by the governments and have little incentive to innovate or improve efficiency and the export sector is driven by MNEs where technology transfer is minimal, Thailand's industrialisation has been characterised as technology-less. This was one of the possible explanations for Thailand's low productivity.

Second, demographic transition affects the production of the Thai economy. Given that several manufacturing industries are labour intensive, shortage of the working-age population due to the ageing society has driven the wage rate higher. This increased labour cost in turn undermines the comparative advantage of Thailand in those industries, making the country less competitive when compared with lower-income countries. At the same time, the lack of technology improvement prevents Thailand from climbing up the value chain and becoming a high-income country. This situation, known as the middle-income trap, has raised concerns over Thailand's long-term economic development. One possible solution is to allow foreign labour to work legally. However, this idea has been resisted. Meanwhile, illegal workers are in high demand, paving the way for bribes and human rights violation, as they are not protected by the Thai laws.

Third, increasing concentration of the corporate sector suggests that inequality seems worsened. Many SMEs cannot grow and mature to become large firms, while large firms expand more disproportionately. There are several constraints faced by SMEs. Since they are young, they do not have the credit history required for loan applications. Being small often means that they do not have a large amount of assets that could be used as collateral. Thus SMEs are more likely than large firms to have financial constraints. In addition, unlike large firms, SMEs do not have the advantage of economies of scale due to their small size, limiting them to adopt certain technology or access certain markets (including foreign) that require substantial initial investment. Given that SMEs are an important source of employment of Thailand's labour force, this problem consequently has an implication on income inequality of Thai households.

Finally, the high level of concentration of production at the firm level has important implications for risk and shock transmission. It implies that idiosyncratic shocks specific to a particular firm can have large impacts on the aggregate output of the economy. Likewise, extremely high levels of export and import concentration imply that shocks on specific exporters or importers could create significant impacts on the Thai economy. This implication raises a concern on the vulnerability of Thailand as an export-dependent economy from a micro perspective in addition to the traditional macro argument that is based on external dependency on particular trading partners or on the global economy.

In summary, examining the Thai economy from the micro perspective of the corporate sector suggests that, although the Thai economy has experienced extraordinary growth in the past several decades, it is fundamentally inefficient, unequally distributed, and fragile. These problems raise concern over the sustainability of economic performance of Thailand.

Notes

1 Based on Banternghansa, Paweenawat, and Samphantharak (2019), the largest company in Thailand is PTT Public Company Limited, which operates in the energy sector. It was an unincorporated state-owned enterprise before being partially privatised and listed in the Stock Exchange of Thailand in 2001. Thailand's Ministry of Finance remains the largest shareholder, owning about 51 per cent of the company.

2 For concentration of economic activities among few large firms, see Banternghansa, Paweenawat, and Samphantharak (2019). For concentration of corporate ownership among few shareholders, see Banternghansa and Samphantharak (2019).

3 A notable exception is the business empire owned by the Crown Property Bureau. Examples of the very few prominent business families that are not Chinese descendants are the Heinecke (American), the Link (German), and the Narula (Indian) families; see Brooker Group (2003). For a more detailed discussion on family business groups in Thailand, see Samphantharak (2019a).

4 Bertrand, Johnson, Samphantharak, and Schoar. 2008.

5 Based on a detailed study, Suehiro (1989) finds that the Bangkok Bank Group owned 83 domestic and 38 overseas affiliates in the 1980s. Studying the large 93 family business groups in Thailand in 1996, Bertrand, Johnson, Samphantharak, and Schoar (2008) find that the average number of member firms per family business group was 6.6, with one group having at least 58 firms. They also find that the deepest group had eight tiers of chain shareholdings. One firm in their sample owned 23 other firms while another firm was owned by seven other firms. Banternghansa and Samphantharak (2019) present recent findings based on the data of all registered firms in Thailand in 2018.

6 Samphantharak (2006).

7 Bertrand, Johnson, Samphantharak, and Schoar (2008).
8 Bunkanwanicha and Wiwattanakantang (2009).
9 Nikomborirak (2017).
10 Ramstetter (1997).
11 Samphantharak (2019).
12 Studwell (2007), p. 31.
13 For example, recent rankings by Interbrand, BrandZ, and Forbes have no brand from Thailand in its top 100 global brands. Red Bull, a beverage brand originated in Thailand, is 61st in the 2018 Forbes list but through Red Bull GmbH, an Austrian company that is a joint venture between Austrian entrepreneur Dietrich Mateschitz and Thai businessman Chaleo Yoovidhya.
14 Ajanant (1989), p. 474.
15 The Observatory of Economic Complexity, Massachusetts Institute of Technology <https://atlas.media.mit.edu/en/profile/country/tha/#Exports> (accessed 27 August 2018).
16 Cheewatrakoolpong, Potipiti, Satchachai, Bunditwattanawong, and Nopparattayaporn (2015).
17 OECD Global Value Chain Indicators, May 2013 <https://stats.oecd.org/Index.aspx?DataSetCode=GVC_INDICATORS#> (accessed 27 August 2018).
18 OECD Quarterly International Trade Statistics 2018 <https://data.oecd.org/trade/domestic-value-added-in-gross-exports.htm> (accessed 27 August 2018).
19 Figure 3.1 in Sessomboon (2016).
20 For more details about Thailand's international trade at the firm level, see Apaitan, Disyatat, and Samphantharak (2019).
21 Punyasavatsut (2008).

References

Bertrand, Marianne, Simon Johnson, Krislert Samphantharak, and Antoinette Schoar. 2008. "Mixing Family with Business: A Study of Thai Business Groups and the Families Behind Them". *Journal of Financial Economics*, Elsevier, Vol. 88, No. 3 (June): 466–498.

Brooker Group. 2003. *Thai Business Groups: A Unique Guide to Who Owns What*, Fifth Edition. Bangkok, Thailand.

Chaiyuth Punyasavatsut. 2008. "SMEs in the Thai Manufacturing Industry: Linking with MNEs". In *SME in Asia and Globalization*. Edited by H. Lim. ERIA Research Project Report 2007–05, pp. 287–321.

Chanont Bunternghansa and Krislert Samphantharak. 2019. "Understanding Corporate Thailand II: Ownership". Unpublished Manuscript, University of California San Diego.

Chanont Banternghansa, Archawa Paweenawat and Krislert Samphantharak. 2019. "Understanding Corporate Thailand I: Finance". Discussion Paper No. 112. Puey Ungphakorn Institute for Economic Research, Bangkok, Thailand.

Duenden Nikormborirak. 2017. "SOE Reform in Thailand: Preparing for Free Trade Agreements". In *Comprehensive Development Strategy to Meet TPP*. Edited by Daisuke Hiratsuka. Bangkok, Thailand: Bangkok Research Center, IDE-JETRO.

Juanjai Ajanant. 1987. "Trade Patterns and Trends of Thailand". In *Trade and Structural Change in Pacific Asia*. Edited by Colin I. Bradford, Jr. and William H. Branson. Chicago, IL: University of Chicago Press.

Kornkarun Cheewatrakoolpong, Tanapong Potipiti, Panutat Satchachai, Nath Bunditwattanawong and Arpakorn Nopparattayaporn. 2015. "The New Normal of Global Trade: The Impact on Thailand's Export Structure and Structural Change". Presentation at the Bank of Thailand Symposium, 17–18 September.

Kritlert Samphantharak. 2006. Internal Capital Markets in Business Groups. Unpublished Manuscript, University of California San Diego.

Krislert Samphantharak. 2019a. "Family Business Groups and Economic Development in Southeast Asia". Unpublished Manuscript, University of California San Diego.

Krislert Samphantharak. 2019b. "SMEs and Economic Development in Southeast Asia". Unpublished Manuscript, University of California San Diego.

Pramuan Bunhanwanicha and Yupana Wiwattanakantang. 2009. "Big Business Owners in Politics". *Review of Financial Studies*, Vol. 22, No. 6.

Punyawich Sessomboon. 2016. Decomposition Analysis of Global Value Chain's Impact on Thai Economy. Unpublished Manuscript, Thammasat University.

Ramstetter, Eric D. 1997. "Prospects for Foreign Multinationals in Thailand in the Late 1990s: Another Test for Thailand's Economic Policy Makers". December.

Studwell, Joe. 2007. *Asian Godfathers: Money and Power in Hong Kong and Southeast Asia*. New York, NY: Atlantic Monthly Press.

Suehiro, Akira. 1989. *Capital Accumulation in Thailand, 1855–1985*. Tokyo: Centre for East Asian Cultural Studies.

Tosapol Apaitan, Piti Disyatat and Krislert Samphantharak. 2019. "Dissecting Thailand's International Trade: Evidence from 88 Million Export and Import Entries". *Asian Development Review*, Vol. 36, No. 1 (March): 20–53.

17

INDUSTRIALISATION, TECHNOLOGICAL UPGRADING, AND INNOVATION

Patarapong Intarakumnerd

An overview: industrialisation with limited technological upgrading and innovation

Since the late 1950s, Thailand's economy has undergone a substantial structural change from agriculture to export-oriented manufacturing alongside rapid economic growth per annum of approximately 7 per cent. It has done so while integrating key manufacturing production, particularly in automobiles and electronics, into regional and global value chains. Thailand's industrialisation can be divided into three periods: import substitution (late 1950s to 1970s), export promotion (1980s to mid-1990s), and liberalisation (late 1990s onwards). The contribution of the aquaculture sector to gross domestic product (GDP) has significantly decreased, from 44 per cent in 1951 to 8.7 in 2015, while the share of manufacturing increased markedly from 13 per cent to 27.5 in the same period. Nonetheless, in terms of export, while the role of primary products has declined relative to that of manufacturing, agriculture itself has diversified significantly as Thailand has become one of the world's top exporters of a wide range of primary or primary-based products including not only rice, rubber, sugar, and cassava but also prawns and canned pineapple.

At the same time, the growth and diversification of manufactured exports, in sectors ranging from textiles, automobiles and parts to electronic and electrical components, has also been impressive. For example, the shares of exports of electronic and automotive products increased from 0.04 and 0.25 in 1970 to 25.20 and 6.68 in 2006, respectively.[1] Thailand's economic status has changed from a lower-income to an upper-middle-income country since 2003. Behind this success lies rather prudent macro-economic management, early adoption of export and foreign direct investment (FDI) promotion policies, and investment in physical infrastructure and expansion of school and university enrolment.[2]

Nonetheless, some scholars such as Yoshihara Kunio have critically questioned the sustainability of Thailand's economic prosperity.[3] He describes the Thai economy as "ersatz Capitalism". Unlike Western countries, Japan, and first-tier East Asian newly industrialised economies (NIEs), the Thai economy grew by overcoming its bottlenecks with foreign technology and capital without making serious efforts to increase its own savings and upgrade technology. He argues that this type of capitalism could not keep expanding. Kunio's argument came true. The country experienced a major economic crisis in 1997. Since then, the economic growth rates have decreased substantially to 3–4 per cent annually, and even further down to 2–3 per cent on average after 2014 when the military took over the country. This growth rate has become the new normal for Thailand. The country's once rising-star and labour-intensive sectors (such as textiles, garments, toys and shoes) have lost their competitive edge to lower-wage countries. The concern regarding the middle-income trap is widespread among Thai policymakers, scholars, and the public at large. More specifically, the concern is about the limited intensity of technology development in industry which has contributed to that competitive weakness. This reflected in a number of key economic indicators, especially the growth of total factor productivity (TFP). TFP's growth explains other reasons for a country's economic growth beyond the growth of capital, labour, and land. Apart from education and other social capital and institutional factors, it includes the progress of science and technology and innovation. Although Thailand's economic growth rate in the past 50 years is rather impressive, this has been achieved largely by utilisation of factor inputs. Between 1987 and 1995, as the Thai economy grew at the rate almost 10 per cent, the TFP growth rate was only around 1.5 per cent.[4] The failures in technological upgrading and innovation are derived from ineffective government policies and behaviours of firms in Thailand. We will examine these two aspects in the subsequent sections.

Passive technological upgrading and innovation of firms in Thailand

Several studies of firms in Thailand conducted since the 1980s demonstrate that most firms have grown without deepening their technological capabilities in the long run, and their technological learning has been very slow and passive.[5] According to a study commissioned by the World Bank in 2000, only a small minority of large subsidiaries of transnational corporations (TNCs), large domestic firms and small and medium-sized enterprises (SMEs) have capability in research and development (R&D), while the majority are still struggling with increasing their design and engineering capabilities.[6] For a very large number of SMEs, the key issue is hugely concerned with building up more basic operational capabilities together with craft and technician capabilities for efficient acquisition, assimilation and incremental upgrading of fairly standard technology.

The slow technological capability development of Thai firms is quite different from what characterised Japan, Korea, and Taiwan. Firms in these countries moved rather rapidly from mere imitators to innovators. As early as the 1960s, Japanese firms became more innovative, investing heavily in R&D and relying less on importation of foreign technologies.[7] In general, firms in Korea and Taiwan, where industrialisation (beginning with import substitution) started more or less in the same period as in Thailand, were more successful in increasing absorptive capacity (of foreign technology) and deepening indigenous technological capabilities in several industries.[8] In the electronics industry, for instance, Korean and Taiwanese firms were able to climb technological ladders by exploiting institutional mechanisms like providing assembly services as original equipment manufacturers (OEMs) and/or providing design as own brand manufacturers (OBMs) to transnational corporations. As a result, latecomer firms in those countries could acquire advanced technology and access demanding foreign markets.[9]

Nonetheless, after the economic crisis in 1997, there were a few interesting positive changes in industrial sectors in Thailand:

1 Several large conglomerates, such as the CP Group and Siam Cement Group, increased their R&D activities. One large conglomerate alone invested 500 million baht in R&D in 1999. This is because the crisis compelled executives of those companies to consider that

long-term survival depended on deepening their technological and innovative capabilities. They could not simply rely on importing off-the-shelf technologies and knowledge necessary for simple production as before.

2 A number of smaller companies recently increased their technological efforts by collaborating with universities' R&D groups in order to stay ahead in the market or to seize the most profitable market section.

3 Several subcontracting suppliers in the automobile and electronics industries were lately forced by their TNCs' customers/partners to strengthen their efforts to modify product design and improve efficiency and to be able to absorb the design and know-how from foreign experts.

4 There were emerging new start-up firms (with fewer than 50 employees) relying on their own design, engineering, or development activities. These companies were managed by entrepreneurs, having acquired a strong R&D background, while studying or working abroad. Many of them are "fabless" companies.[10] Nonetheless, the pool of potential entrepreneurs is relatively small, as the rate of enterprise creation per capita is relatively low, and scientists, engineers, and managers prefer to work in public agencies or large businesses.[11]

The low level of technological and innovative capabilities and passive learning of Thai firms can be illustrated by R&D and Innovation Community Surveys. The surveys have been carried out by the National Science and Technology Development Agency (NSTDA) and later, the National Science Technology and Innovation Policy Office. They are carried out every year, except the innovation surveys which were conducted in 2003, 2008, 2011, and 2014. The number of R&D performing and innovating firms, both in manufacturing and service sectors, were 27 and 23 per cent in 2014, respectively. R&D performing and innovating firms were around 6 per cent in 2003. This shows moderate improvement in innovation performance of firms in Thailand, which correspond to the positive changes after the Asian Financial Crisis of 1997, as mentioned earlier.

Nonetheless, Thailand's performance is still comparatively poor in relation to more successful Asian countries. Comparison between Thailand and Korea Innovation Surveys conducted during 2011–2012 illustrates the differences in terms of innovative capabilities of these two countries. Companies in Thailand lag far behind those in Korea with respect to innovation. More than 40 per cent of firms in Korea carried out innovations versus just about 6 per cent in Thailand. It is also striking that a much higher share of companies in Korea carries out product innovations. This could be an indication that companies in Thailand are at a stage where they would rather use their resources to improve the production process than the product itself, which in turn could hint at an OEM-oriented economy. Transnational corporations (TNCs) and joint ventures in Thailand operate at the low end of the global value chain. Most of their products (67 per cent) are manufactured according to the design specifications of parent companies or those provided by external buyers and traders. Similarly, most of Thai-owned firms' products (59 per cent) are manufactured at the low end of the global value chain. Meanwhile, very few companies in Thailand do both product and process innovations, which is very common in Korea. This reflects the more advanced innovation behaviour of companies in Korea.

In terms of size, smaller firms tend to engage less in R&D and innovation activities than larger ones. And when they do so, their activities tend to be less sophisticated. Conducting quality control or testing activities is quite common in Thailand; over two-thirds of surveyed firms carried out these activities in 2011. Smaller firms were also less receptive and had fewer capabilities in absorbing external knowledge and technology. In general, a small number of firms conducted sophisticated activity of R&D. Only 10 per cent of SMEs performed in-house R&D, while more than 25 per cent of large firms did.

In terms of R&D intensity, TNCs and Thai-owned firms are similar. They spent only around 0.1 per cent of their total sales on R&D. This is relatively low compared to firms in other Asian NIEs. The propensity of firms doing R&D varied across sectors. The leading sectors were science-based industries such as chemical and electronics as well as resource-based industries like food and rubber. In-house R&D expenditure was largely devoted to the development of new or improved products (65 per cent) rather than processes (22 per cent). Almost one-fifth of manufacturing firms had innovation, compared to only 5 per cent of service firms.

The main barriers to innovation were the lack of qualified personnel, the high cost of innovation, and limited access to information on technology and markets. The cost of innovation was more of an important obstacle to innovation for smaller firms. Interestingly, the main sources of information used for innovation were those entities interacting with firms on a regular basis: customers, parent firms, suppliers, and the Internet. More sophisticated sources of information such as patent disclosures, public research institutions, and universities or business service providers are much less important.

Regarding external collaboration, horizontal relationships between firms in the same or related industries are viewed as unimportant by the surveyed firms. Cooperative consortiums among competing firms, as seen in Japan or Taiwan, to research a particular technology or products are very rare in Thailand. Also, as the intra-firm technological capabilities themselves are weak, as already mentioned, the innovation-centre interaction generated from such links is therefore limited. On the contrary, firms tend to have more vertical collaboration with their customers and suppliers.

University-industry linkages (UIL) in Thailand are weak. Firms do not regard university and public research institutes as important sources of information and knowledge. They do not collaborate actively with local universities and public research institutes. They also perceive that technical supports from local universities and public research institutes are weak. Thus, most UIL projects are limited to consulting and technical services. More advanced projects only occur in some outstanding cases.

Ineffective government policies on technological upgrading and innovation

Many policymakers equate promotion of technology and innovation capability of the country with promotion of R&D investment. The ratio of gross expenditure on R&D (GERD) to GDP is one of the most significant leading indicators used for formulating

Science, Technology and Innovation (STI) policy in the country. For example, according to National Science, Technology and Innovation Policy and Plan 2012–2021, the Thai government has set a target to achieve 1 and 2 per cent of GERD to GDP by the year 2016 and 2021, respectively.

Before the government of Prime Minister Thaksin Shinawatra (January 2001–September 2006), the scope of Science and Technology policy in Thailand was narrow. It covered only four conventional functions: research and development, human resource development, technology transfer, and S&T infrastructure development. This narrow scope of S&T was very much based on the so-called technology-push R&D model, or linear model of innovation. For example, the results of R&D will be readily designed and engineered to become new processes and/or products set to be sold in the market. The model was popular between the First World War and the 1960s, but it has faded away in other countries. Academics and policymakers in more successful places like Korea, Taiwan, and Singapore realise that the innovation process is not automatic and the failure rate can be high. It is also necessary to effectively manage all actors (government, private sector, funding institutes and market, academia) to participate in all relevant functions from R&D to design, engineering, testing, and marketing, and the forward and backward interactions between these functions.

The aforementioned World Bank study on Thailand stipulates that only a small minority of large subsidiaries of transnational corporations (TNCs), large domestic firms, and SMEs have capability in R&D, while the majority are still struggling with increasing their design and engineering capabilities. More important than R&D as inputs for technological progress of a country like Thailand are technology absorption capacity, design activities, engineering developments, experimentation, training and exploration of markets for new products, and so forth. Government policies should be geared towards enhancing firms' capabilities in these areas. As late as 2015, the tax incentives provided by government agencies started to cover non-R&D innovation activities. Tax incentives offered by the Department of Revenue increased from 200 to 300 per cent and expanded to cover firms' expenditure on licensing in foreign technology to advance their product and process innovations. The Board of Investment's (BOI) new "merit-based" investment promotion scheme introduced in the year 2017 also covered non-R&D technological upgrading activities including product design, packaging design, advance technology training, licensing fees of intellectual property rights, collaboration with universities, and development of local suppliers.[12]

At the end, it is the firms that have to compete internationally, not universities or public research institutes. However, due to the influence of the linear model of innovation, the dominant orientation of policy and resource allocation for building technology development capabilities since the 1960s has been on the capabilities and resources of scientific, technological, and training institutions that were intended to undertake technological activities "on behalf of firms". Conversely, policy measures and resource allocations designed to strengthen the technological learning, technological capabilities, and innovative activities "within firms" and knowledge flow among firms and between firms and other actors in innovation processes were minimal and ineffective.[13]

Unlike Japan, Korea, and Taiwan, science and technology elements were not part of broader economic policies, such as industrial policy, investment policy, trade policy, and to a lesser extent, education policies.[14] The Ministry of Science and Technology, which was not considered an economic ministry until as late as 2016, has more roles in promoting technology development than economic agencies such as the Ministry of Industry.[15] This imbalance is very different from NIEs and Japan, where economic organisations the like Ministry of International Trade and Industry (MITI) of Japan, the Economic Development Board (EDB) of Singapore, and the Economic Planning Board (EPB) of Korea have significant roles in policy and institutional support for industrial technology development.[16]

Trade policy, with tariffs being the most important instrument in Thailand, was not implemented strategically to promote technological learning like in NIEs. Instead, trade policy was much influenced by macro-economic policy, for instance to reduce domestic demand for imports at the time of balance of payment deficit. The Ministry of Finance, the dominant agency controlling the policy, had little knowledge or experience of industry and industrial restructuring.[17] The industrial policy of Thailand did not pay enough attention to the development of indigenous technological capability as an integral factor in the process of industrialisation. In 2016, the Thailand 4.0 Plan was introduced. It aims to change the country to become a value-based and innovation-driven economy by emphasising on promoting technology, creativity, and innovation in focused industries. Subsequently, the Law on National Competitive Enhancement for Targeted Industries was enacted. The law is designed to promote investment in line with Thailand 4.0. Incentives are given to promote projects of the targeted industries. Remarkably, apart from tax incentives, the Fund for Enhancement of Competitiveness for Targeted Industries was established with government seed money of US$285 million for investment projects in research and development or human resource development in specific areas.

Nonetheless, with the exception of the automotive industry, there were no reciprocal performance-based criteria (such as export and local value added and technological upgrading targets) set for providing state incentives like in Korea and Taiwan.[18] Investment promotion privileges, for example, were given away once approved.

The National Research and Innovation Policy Council (NRIC), chaired by the prime minister, was created in 2016 to integrate once separated research policy with STI policy; implant science, technology and innovation issues to broader economic policies; and enhance cross-ministry coordination. Members of this council comprise not only the minister of science and technology but also ministers from key economic ministries. The National Research Council of Thailand (NRCT) under the Office of the Prime Minister and the National Science, Technology and Innovation Policy Office (STI) under the Ministry of Science and Technology work together as a joint secretariat. It is too early to evaluate whether this council will achieve its objective. However, previous supra-ministerial committees and councils failed to achieve their goals since the prime minister himself did not really chair the meeting. The meetings were infrequent and there was a lack of mechanisms to execute, monitor, and evaluate resolutions once approved by such committees.

Economic policies were heavily influenced by the World Bank's "market-friendly" approach to industrialisation. Given the neo-classical economic inclination of leading Thai technocrats, they were limited to the so-called functional intervention such as promoting infrastructure building, general education, and export push in general. There were virtually no selective policy measures such as special credit allocation and special tariff protection targeting particular industries or clusters, since they were regarded as a market distortion by mainstream economists. The exception was the automobile industry. Despite a relatively liberal policy on the automotive industry, the Thai government successively raised its local content requirements for automobile manufacturers investing in Thailand.

A major change in policy came under the Thaksin government. It was the first time the Thai government produced serious "selective" policies addressing specific sectors and clusters. The government declared five strategic sectors which Thailand should pursue: automotive, food, tourism, fashion, and software. Clear visions were given to these five sectors: Detroit of Asia (automotive cluster), Kitchen of the World (food cluster), Asia Tourism Capital, Asia Tropical Fashion, and World Graphic Design and Animation Centre (software cluster). The cluster concept was introduced, which went far beyond the linear model of innovation, since it focused on interactive and collective learning among firms and between firms and other actors in close geographical proximity. Thailand was divided into 19 geographical areas. Each area had to plan and implement its own cluster strategy focusing on a few strategic products or services. It was supervised by the so-called CEO governors who were given authorities by the central government to act like provincial chief executive officers (CEOs). At the local level, the cluster concept was applied to increase the capacity of the grassroots economy in the name of "community-based clusters", especially to make "One-Tambon-One Product" (OTOP) a success. Nonetheless, the actual implementation of the concept has had mixed results because of the misinterpretation of the concept of policy practitioners at the implementation level, policy discontinuity, inadequate trust and participation of concerned actors, and lack of champions in the private sector in several cases.[19] Furthermore, the Thaksin government did not pay enough attention to long-term industrial upgrading beyond short-term and politically branded schemes. For example, it scraped the most ambitious upgrading plan, the Industrial Restructuring Project (IRP), which was initiated by the earlier government and went through extensive consultation processes with the private sector. The IRP targeted at upgrading 13 sectors with eight sets of measures ranging from equipment modernisation and labour skills to product design.[20]

In 2015, the BOI's "Super Cluster" incentive scheme was introduced to upgrade the existing five industries and encourage the emergence of five new industries for the future development of Thailand. Two new cluster-like mega projects were implemented: the Eastern Economic Corridor (EEC) and Food Innopolis. The Eastern Economic Corridor consists of the three eastern provinces of Rayong, Chonburi, and Chachoengsao with a combined area of 13,285 km^2. The Eastern Economic Corridor will see investments of US$43 billion during the next five years, mostly through FDIs. The EEC is intended to accommodate investment in ten targeted industries important for Thailand's future: next-generation cars, smart electronics, affluent medical and wellness tourism, agriculture and biotechnology, food, robotics for industry, logistics and aviation, biofuels and biochemical, digital industry, and medical services. Private enterprises investing in the EEC will receive a super incentive promotion package which goes far beyond the current BOI's incentives, including very preferable corporate and personnel income tax privileges, a longer-term land lease for investors, a fast-tracked environmental impact assessment, and using foreign currencies in trade directly without having to exchange them into Thai baht.

Food Innopolis is located at the Thailand Science Park (TSP) under the National Science and Technology Development Agency. This project aims to position Thailand as a global food innovation hub in the global food industry. The expected availability of resources for the Food Innopolis includes 3,000 researchers, 10,000 students in food science and technology, 9,000 food factories, 150 food research laboratories, 20 pilot plants, and 70 universities. Food Innopolis belongs to one of the BOI's super clusters. Tax-based incentives include the exemption of corporate income tax for up to eight years, with an additional 50 per cent reduction for five years, and the exemption of import duty on machinery. Non-tax incentives include the permission to own land and facilitation on visas and work permits. However, whether these two initiatives will be successful or not depends on implementation, which is usually problematic in Thailand due to a lack of long-term commitment and coordination between concerned agencies.

Due to a lack of introduction of selective policies, there are very few institutions set up to support development of indigenous technological and innovative capabilities of firms in specific sectors. Most of research institutes in the country are "jack of all trades but master of none". They have too many missions: assisting industry, building up S&T manpower, educating the general public on S&T, helping disadvantaged groups of society, and so on. They usually cover a broad range of technologies with no specific targets for particular industries, and their linkages with the industry are weak. Moreover, sectoral promotion institutes under the Ministry of Industry, like the Thai Textile Institute, Thailand Automotive Institute, and National Food Institute, are preoccupied with their own financial survival since, due to shortsighted policy design, they have to be financially independent after having been set up as public organisations for five years. As a result, they need to rely on short-term and money-making activities, like training, to generate quick income at the expenses of activities promoting long-term capability development of firms in the sector. The situation in Thailand is quite different from Taiwan and Korea, where there are many government research institutes with clear-cut missions dedicated to strengthening technological capabilities of firms in particular sectors and sub-sectors or even specific products.

There is an institutionalised belief among policymakers that the main target of government policies should be Thai-owned firms, especially SMEs. Beyond providing tax incentives to attract FDI to bring in foreign exchanges and generate employment, TNCs should be left alone.

Like Thailand, Singapore is another country where FDI has been very much encouraged. However, it has specific government measures in generating spillover effects from FDI in terms of development of local technological capabilities. Initiated as early as the 1970s, the Local Industry Upgrading Programme executed by Singapore's Economic Development Board (EDB), for instance, specifically aims at exploiting TNCs' knowledgeable and experienced engineers to train employees of local firms in developing skills considered "critical" for technologically upgrading of high-priority industrial sectors.

Unfortunately, partially due to conventional wisdom on the roles of TNCs, there was no such explicit and proactive link between promoting FDI and upgrading of local technological capability in Thailand. Until as late as 2004, the BOI launched the "Skill, Technology and Innovation" (STI) policy incentive for firms investing in R&D, employing university graduates in S&T, and training their personnel and those of suppliers. Even so, the number of projects approved under the STI scheme has been small and the incentive for training of suppliers' employees – the most deliberate attempt to generate spillover impacts from FDI – was abolished. Hopefully the new BOI's merit-based incentive scheme, introduced in 2017 as mentioned earlier, could somewhat bring closer knowledge-intensive collaboration between TNCs and local firms.

In other Asian NIEs like Taiwan, Singapore, and Malaysia, grants were used effectively to promote "specific" activities. In Thailand, on the contrary, grant schemes to promote specific or targeted activities aiming at enhancing technological learning of firms remained limited. This is because, as mentioned earlier, the dominance of the notion of market distortion and obstructed and rigid government regulations emerged out of the fear of corruption and cronyism. Therefore, Thailand is missing opportunities to employ an effective and more targeted policy tool and has to rely only on tax incentives – a blunter but easier to handle instrument.

Policymakers, especially those who came from scientific disciplines at universities, strongly believe that the most critical issue in S&T human resource development is to significantly increase the numbers of master's and PhD graduates. This might be true for other reasons, such as teaching and basic research at universities and public research institutes, but several studies have confirmed that firms in Thailand, local and foreign, do not considerably need graduates at the postgraduate level.[21] Their main concern, instead, is on the quantity of "qualified" bachelor's degree and vocational certificate holders. Not only production-based firms, but also those conducting R&D basically require only first-degree graduates.

The overemphasis on the university's postgraduate level is at the expense of others. What has been largely neglected by policymakers is the quality of vocational education. As an industrialising latecomer country, Thailand has an opportunity to exploit and upgrade technologies already developed by forerunner countries. Nonetheless, to seize such an opportunity, the qualified engineers and technicians at the shop-floor level are necessary inputs for firms' technological absorption capacity and "incremental" innovation at the time of technological catching up. Even though the Vocational Education Act and relevant laws exist, the lack of focus and the negative societal value towards vocational education deters sufficient accumulation of vocational students and technicians in the manufacturing sector. Vocational students and graduates are perceived as inferior human resources to students and graduates in general studies. This is very different from Japan, Taiwan, and Korea, where the importance of vocational education is highly recognised by their governments and viewed positively by society, especially during their technological catching-up period when innovations were mostly incremental and emerged from factories' shop floors. "Project-execution" capabilities were important for latecomer firms to enter new industries.[22]

Conclusion

Since the 1950s, Thailand has embarked on industrialisation. It started with import substitution and shifted to export promotion during the 1980s. FDIs played very important roles in the process. After the Asian Financial Crisis in 1997, transnational corporation and large domestic firms began to invest considerably in more technologically sophisticated activities like advanced engineering, design, and R&D. Several SMEs that survived the crisis emerged much stronger and paid more attention to technological development. There was also an emergence of new start-up firms with high technological capabilities and product or process innovations. Though there was significant technological upgrading and innovation, the majority of Thai-owned firms have remained technologically weak. The challenge is how to create a critical mass of innovative firms to spearhead the next round of industrialisation amid much fiercer international competition, more demanding customers, and global rules. More proactive government policies focusing on technological upgrading and innovation within firms are also needed. Policies should be more sector specific, risk taking, and encompassing various ministries and several types of firms, from TNCs to startups.

Notes

1 Shahid Yusuf and Kaoru Nabeshima, *Tiger Economies under Threat: A Comparative Analysis of Malaysia's Industrial Prospects and Policy Options*, World Bank Publications No. 2680 (Washington DC: World Bank, 2009), p. 20.
2 World Bank, *The East Asian Miracle: Economic Growth and Public Policy* (New York: Oxford University Press, 1993), pp. 115–149.
3 Yoshihara Kunio, *The Rise of Ersatz Capitalism in Southeast Asia* (New York: Oxford University Press, 1988), p. 12.
4 National Economic and Social Development Board (NESDB), *National Productivity Enhancement Plan*, Unpublished PowerPoint Presentation on 26 January 2007, p. 7
5 See for example, Erik Arnold, Martin Bell, John Bessant, and Peter Brimble, "Enhancing Policy and Institutional Support for Industrial Technology Development in Thailand", in *Volume 1, The Overall Policy Framework and the Development of the Industrial Innovation System* (Washington, DC: World Bank, December 2000); Martin Bell and Don Scott-Kemmis, "Technological Capacity and Technical Change", *Draft Working Paper No. 1, 2, 4 and 6.* Report on Technology Transfer in Manufacturing Industry in Thailand, Science Policy Research Unit, University of Sussex, Brighton, UK. 1985; Nit Chantramonklasri, *Technological Responses to Rising Energy Prices: A Study of Technological Capability and Technological Change Efforts in Energy-Intensive Manufacturing Industries in Thailand.* Unpublished PhD Thesis. Science Policy Research Unit, University of Sussex, Brighton, UK, 1985; Sanjaya Lall, "Thailand's Manufacturing Competitiveness: A Preliminary Overview", Unpublished paper for Conference on Thailand's Dynamic Economic Recovery and Competitiveness, Session 4. Bangkok, 20–21 May 1998; Yada Mukdapitak, *The Technology Strategies of Thai Firms.* Unpublished PhD Thesis. Science Policy Research Unit, University of Sussex, Brighton, UK, 1994; Chatri Sripaipan, Sutham Vanichseni, and Yada Mukdapitak, "Technological Innovation Policy of Thailand" (Thai version) (Bangkok: National Science and Technology Development Agency, 1999).
6 Arnold, Bell, Bessant, and Brimble, "Enhancing Policy and Institutional Support for Industrial Technology Development in Thailand", p. 45.
7 Hiroyuki Odagiri and Akira Goto, "The Japanese System of Innovation: Past, Present and Future", in *National Innovation Systems: A Comparative Analysis,* edited by Richard Nelson (Oxford: Oxford University Press, 1993), p. 77.
8 See for example, Alice Amsden, *Asia's Next Giant: South Korea and Late Industrialisation* (New York: Oxford University Press, 1989); Linsu Kim, "National System of Industrial Innovation: Dynamics of Capability Building in Korea", in *National Innovation System*, edited by Richard Nelsolson (Oxford: Oxford University Press, 1993); Sanjyaya Lall, *Learning from the Asian Tigers: Studies in Technology and Industrial Policy* (London: Macmillan Press, 1996); Michael Hobday, *Innovation in East Asia: the Challenge to Japan* (Aldershot: Edward Elgar, 1995); Lin-Su Kim, "National System of Industrial Innovation:

Dynamics of Capability Building in Korea", in *National Innovation System*, edited by Richard Nelsolson (Oxford: Oxford University Press, 1997).

9 Hobday, *Innovation in East Asia: The Challenge to Japan*, pp. 15–53.

10 Patarapong Intarakumnerd, Pun-arj Chairatana, Tippawan Tangjitpiboon, "National Innovation System in Less Successful Developing Countries: the Case of Thailand", *Research Policy*, Vol. 31, No. 8–9 (2002), p. 1453.

11 Organisation for Economic Cooperation and Development (OECD), *Thailand: Key Issues and Policies*, OECD Studies on SMEs and Entrepreneurship (Paris: OECD, 2011), p. 25.

12 Hirunya Suchinai, "New Chapter of Investment Promotion", a presentation given by Hirunya Suchinai, Secretary General, Board of Investment, Bangkok, Thailand, 15 February 2017 <www.boi.go.th/upload/New%20Chapter%20of%20Investment%20Promotion_EN%20by%20Hirunya%20Suchinai_1234343.pdf> (accessed 5 March 2017), pp. 2–3.

13 Arnold, Bell, Bessant and Brimble, "Enhancing Policy and Institutional Support for Industrial Technology Development in Thailand", p. ix.

14 Patarapong, Pun-arj and Tippawan, "National Innovation System in Less Successful Developing Countries", p. 1452.

15 Arnold, Bell, Bessant and Brimble, "Enhancing Policy and Institutional Support for Industrial Technology Development in Thailand", p. vii.

16 See for example, Hajoon Chang, "Institutional Structure and Economic Performance: Some Theoretical and Policy Lessons from the Experience of the Republic of Korea", *Asia Pacific Development Journal*, Vol. 4, No. 1 (1997), pp. 39–56; C. Johnson, *MITI and the Japanese Miracle: The Growth of Industrial Policy, 1925–1975* (Stanford: Stanford University Press, 1982); Poh-Kam Wong, "National Innovation Systems for Rapid Technological Catch-up: An Analytical Framework and a Comparative Analysis of Korea, Taiwan, and Singapore", a paper presented at DRUID's summer conference, Rebild, Denmark, 1999.

17 Laurid Lauridsen, "Coping with the Triple Challenge of Globalisation, Liberalisation and Crisis: The Role of Industrial Technology Policies and Technology Institutions in Thailand", *The European Journal of Development Research*, Vol. 14, No. 1 (2002), p. 110.

18 See for instance, Alice Amsden, *Asia's Next Giant: South Korea and Late Industrialisation.* New York: Oxford University Press, 1989); Alice Amsden, *The Rise of the Rest: Challenges to the West from Late-Industrialising Economies* (New York: Oxford University Press, 2001); Alice Amsden and Wan-wen Chu, *Beyond Late Development: Taiwan's Upgrading Policies* (Cambridge, MA: MIT Press, 2003).

19 See Patarapong Intarakumnerd, "Thailand's Cluster Initiatives: Successes, Failures and Impacts on National Innovation System", Paper presented at International Workshop's Programme Industrial Clusters in Asia: Old and New Forms, Lyon, France, 29–30 November and 1 December 2006.

20 Richard Doner, "Politics and the Growth of Local Capital in Southeast Asia: Auto Industries in the Philippines and Thailand", in *Southeast Asian Capitalists*, edited by R. McVey. Southeast Asia Program (SEAP) (Ithaca: Cornell University Press, 1992), p. 240.

21 See for example, Thailand Development Research Institute (TDRI), "Human Resource Development for Competitiveness in Industry", a final report submitted to National Economic and Social Development Board, August. Bangkok: Thailand Development Research Institute (in Thai), 2004; Chalamwong, "A Study on application pattern of Human Resource Development Plan for Demand in Industrial Sector", a final report submitted to Department of Industrial Economics, Ministry of Industry, Bangkok: Thailand Development Research Institute (in Thai), April 2007; National Economic and Social Development Board (NESDB). The Growth Potentials of Targeted Industries in the Next Five Years (2007–2011), an unpublished power point presentation, July, Bangkok, Thailand, 2007.

22 A. Amsden and T. Hikino, "Project Execution Capability, Organisational Know-How and Conglomerate Corporate Growth in Late Industrialisation", *Industrial and Corporate Change*, Vol. 3, No. 1 (1994), p. 120.

References

Amsden, Alice. 1989. *Asia's Next Giant: South Korea and Late Industrialisation.* New York: Oxford University Press.

———. 2001. *The Rise of the Rest: Challenges to the West from Late-industrialising Economies.* New York: Oxford University Press.

Amsden, Alice and T. Hikino. 1994. "Project Execution Capability, Organisational Know-how and Conglomerate Corporate Growth in Late Industrialisation". *Industrial and Corporate Change*, Vol. 3, No. 1: 111–147.

Amsden, Alice and Wan-wen Chu. 2003. *Beyond Late Development: Taiwan's Upgrading Policies*. Cambridge, MA: MIT Press.

Anupap Tiralap. 1990. *The Economics of the Process of Technological Change of the Firm: The Case of the Electronics Industry in Thailand*. Unpublished PhD Thesis. Science Policy Research Unit, University of Sussex, Brighton, UK.

Arnold, Erik, Martin Bell, John Bessant, and Peter Brimble. 2000. "Enhancing Policy and Institutional Support for Industrial Technology Development in Thailand". Volume 1, The Overall Policy Framework and the Development of the Industrial Innovation System. December. Washington, DC: World Bank.

Bell, Martin and Scott-Kemmis, Don. 1985. "Technological Capacity and Technical Change". *Draft Working Paper No. 1,2,4 and 6*. Report on Technology Transfer in Manufacturing Industry in Thailand, Science Policy Research Unit, University of Sussex, Brighton, UK.

Chang, Ha-joon. 1994. *The Political Economic of Industrial Policy*. London: Macmillan.

———. 1997. "Institutional Structure and Economic Performance: Some Theoretical and Policy Lessons from the Experience of the Republic of Korea". *Asia Pacific Development Journal*, Vol. 4, No. 1: 39–56.

Chatri Sripaipan, Sutham Vanichseni and Yada Mukdapitak. 1999. "Technological Innovation policy of Thailand" (Thai version). Bangkok: National Science and Technology Development Agency.

Doner, Richard. 1992. "Politics and the Growth of Local Capital in Southeast Asia: Auto Industries in the Philippines and Thailand". In *Southeast Asian Capitalists*. Edited by R. McVey, Southeast Asia Program (SEAP). New York: Cornell University Press, pp. 223–256.

Goto, Akira and Hiroyuki Odagiri. 1996. *Technology and Industrial Development in Japan: Building Capabilities by Learning, Innovation and Public Policy*. Oxford: Oxford University Press.

Hirunya Suchinai. 2017. "New Chapter of Investment Promotion". A presentation given by Hirunya Suchinai, Secretary General, Board of Investment, Bangkok. 15 February <www.boi.go.th/upload/New%20Chapter%20of%20Investment%20Promotion_EN%20by%20Hirunya%20Suchinai_1234343.pdf> (accessed on 5 March 2017).

Hobday, Michael. 1995. *Innovation in East Asia: the Challenge to Japan*. Aldershot: Edward Elgar.

Hobday, Michael and Howard Rush. 2007. "Upgrading the Technological Capabilities of Foreign Transnational Subsidiaries in Developing Countries: The Case of Electronics in Thailand". *Research Policy*, Vol. 36, No. 9: 1335–1356.

Johnson, C. 1982. *MITI and the Japanese Miracle: The Growth of Industrial Policy, 1925–1975*. Stanford: Stanford University Press.

Kim, Lin-Su. 1993. "National System of Industrial Innovation: Dynamics of Capability Building in Korea". In *National Innovation System*. Edited by Richard Nelsolson. Oxford: Oxford University Press, pp. 357–383.

———. 1997. *Imitation to Innovation: The Dynamics of Korea's Technological Learning*. Boston, MA: Harvard Business School Press.

Kunio, Yoshihara. 1988. *The Rise of Ersatz Capitalism in Southeast Asia*. New York: Oxford University Press.

Lall, Sanjyaya. 1996. *Learning from the Asian Tigers: Studies in Technology and Industrial Policy*. London: Macmillan Press.

———. 1998. "Thailand's Manufacturing Competitiveness: A Preliminary Overview". Unpublished paper for Conference on Thailand's Dynamic Economic Recovery and Competitiveness, Session 4. Bangkok, 20–21 May.

Lauridsen, Laurid. 1999. "Policies and Institutions of Industrial Deepening and Upgrading in Taiwan III-Technological Upgrading". International Development Studies Working Paper No. 13, Roskilde University, Roskilde, Denmark.

———. 2002. "Coping with the Triple Challenge of Globalisation, Liberalisation and Crisis: The Role of Industrial Technology Policies and Technology Institutions in Thailand". *The European Journal of Development Research*, Vol. 14, No. 1 (June 2): 101–125.

———. 2008. "Industrial Upgrading: Industrial Technology Policy in Taiwan". In *State, Institutions and Industrial Development: Industrial Deepening and Upgrading Policies in Taiwan and Thailand Compared*. Edited by Laurid Lauridsen. Aachen: Shaker Verlag, pp. 442–513.

National Economic and Social Development Board (NESDB). 2007a. *National Productivity Enhancement Plan*. Unpublished power point presentation on 26 January.

———. 2007b. *The Growth Potentials of Targeted Industries in the Next Five Years* (2007–2011). Unpublished power point presentation, July, Bangkok, Thailand.

Nit Chantramonklasri. 1985. *Technological Responses to Rising Energy Prices: A Study of Technological Capability and Technological Change Efforts in Energy-Intensive Manufacturing Industries in Thailand*. Unpublished PhD Thesis. Science Policy Research Unit, University of Sussex, Brighton, UK.

Organisation for Economic Cooperation and Development (OECD) 2011. *Thailand: Key Issues and Policies*. OECD Studies on SMEs and Entrepreneurship. OECD: Paris.

Odagiri, Hiroyuki and Akira Goto. 1993. "The Japanese System of Innovation: Past, Present and Future". In *National Innovation Systems: A Comparative Analysis*. Edited by Richard Nelson. Oxford: Oxford University Press, pp. 76–114.

Patarapong Intarakumnerd. 2006. "Thailand's Cluster Initiatives: Successes, Failures and Impacts on National Innovation System". Paper presented at International Workshop's Programme Industrial Clusters in Asia: Old and New Forms. Lyon, France, 29–30 November and 1 December.

Patarapong Intarakumnerd, Pun-arj Chairatana and Tippawan Tangjitpiboon. 2002, "National Innovation System in Less Successful Developing Countries: The Case of Thailand". *Research Policy*, Vol. 31, No. 8–9: 1445–1457.

Thailand Development Research Institute (TDRI). 2004. "Human Resource Development for Competitiveness in Industry". A final report submitted to National Economic and Social Development Board, August. Bangkok: Thailand Development Research Institute (in Thai).

Wong, Poh-Kam. 1999. "National Innovation Systems for Rapid Technological Catch-up: An Analytical Framework and a Comparative Analysis of Korea, Taiwan, and Singapore". A paper presented at DRUID's summer conference, Rebild, Denmark.

World Bank. 1993. *The East Asian Miracle: Economic Growth and Public Policy*. New York: Oxford University Press.

Yada Mukdapitak. 1994. *The Technology Strategies of Thai Firms*. Unpublished PhD Thesis. Science Policy Research Unit, University of Sussex, Brighton, UK.

Yongyuth Chalamwong, et. al. 2007. "A Study on Application Pattern of Human Resource Development Plan for Demand in Industrial Sector". A Final Report Submitted to Department of Industrial Economics, Ministry of Industry, Bangkok: Thailand Development Research Institute, April (in Thai).

Yusuf, Shahid and Kaoru Nabeshima. 2009. *Tiger Economies under Threat: A Comparative Analysis of Malaysia's Industrial Prospects and Policy Options*. World Bank Publications No. 2680. Washington, DC: World Bank.

18

TRANSPORT AND THE THAI STATE

Saksith Chalermpong

As an upper-middle-income country, Thailand's transport system and its transport policy are problematic in many aspects. Traffic congestion is legendary in its capital, Bangkok, and the situation in the provinces is rapidly deteriorating. Thai roads are the deadliest in the world, and traffic accidents and fatalities are prevalent among the young male population. Closely related is the negligence on the part the government in providing efficient public transport in much of the country outside of the capital, forcing a large part of the population to rely on private modes of transport – cars for the relatively well-off and motorcycles for the less well-off – aggravating the problems of traffic congestion and traffic accidents, respectively. Unintegrated and imbalanced transport policies often produce stand-alone and inefficient infrastructure projects that benefit only certain population groups, further widening the economic gap between the rich and the poor, urban and rural population, and motorists and non-motorists. While huge public investments have poured into Bangkok in the form of elevated highways, expressways, and more recently urban railways, none of the other major cities has received public transport investment from the government. Recently, politicians from all sides have enthusiastically pursued an extravagant high-speed rail project while intercity rail services in most of the country are still operated by old locomotives on single tracks with primitive signaling systems. Many major transport projects were ill-conceived and politically motivated, and they often created a huge financial burden or operational problems. Examples of poor planning abound, including train stations with no access roads, paralleled highways and expressways competing for traffic, and an undermaintained rail transit system prone to breakdowns. This chapter provides an overview of Thailand's transport systems, with an analysis of the state's contributions to the performance of the Thai transport infrastructure and services.

Overview of Thailand's transport infrastructure

Transport infrastructure is essential for any country's economic competitiveness. In Thailand, roads constitute the largest bulk of transport infrastructure, with the total of length of paved roads and highways of approximately 218,000 km. The rail infrastructure is much less developed, with 4,230 km of urban and intercity railways, most of which are single and unelectrified track, although a substantial amount of investment in rail has been made recently, especially in urban rail, and many new rail lines will be in operation in the next few years. Thailand has six international airports in Bangkok and major regional cities as well as 30 domestic airports. Although water transport was historically important thanks to numerous navigable rivers and canals, the predominance of land-based transport today implies much less development of water transport infrastructure, with an exception of deep-sea ports on which the country's vital export industry depends. In general, most transport infrastructure is owned, operated, and maintained by the government or state-owned enterprises. Some types of infrastructure are privately owned or concessionary, particularly urban railways and expressways. The transport sector plays an important part in the economy, contributing to about 8 per cent of the GDP from 2007 to 2016.[1]

Performance of Thailand's transport infrastructure

The Thai government has consistently allocated between 5 and 10 per cent of the national budget to transport infrastructure development and maintenance, placing it in the top five items of the national budget that include education, health care, defence, and agriculture.[2] However, despite receiving the government's high priority, the country's transport system performs relatively poorly compared to its regional neighbours. Bangkok's traffic congestion has always been notorious, invariably charting at or near the top in the list of congested cities according to various sources.[3] The country as a whole ranked first in the world in terms of average amount of time spent in peak congestion.[4] Thailand's road safety record is appalling. With an estimate of 24,000 fatalities per year or 36.2 fatalities per 100,000, it was ranked as the world's second deadliest country in terms of road traffic accidents in a survey of 180 countries by the World Health Organisation (WHO) – the economic losses of which are estimated between 3 and 5 per cent of GDP.[5] The performance of Thailand's freight transport system is also inefficient. The country's costs of transport and logistics account for over 14 per cent of the GDP in 2015, compared to 8 per cent in the United States and Singapore, about 10 per cent in the European Union, and 11.4 per cent in Japan.[6] In 2016, the World Bank ranked Thailand 43rd out of 160 countries in terms of its Logistics Performance Index (LPI).[7] Overall, the performance of Thailand's transport infrastructure is middling when compared with its Southeast Asian neighbours. According to the World Economic Forum's Global Competitiveness Report 2017–2018, Thailand's road infrastructure is ranked fourth among nine ASEAN countries; its rail infrastructure is ranked fifth among seven ASEAN countries. Airport and port infrastructure both ranked third among the nine countries scored.[8] See Table 18.1 for details.

A question must be asked: why is there such poor performance of the transport infrastructure in Thailand? As transport infrastructure displays some characteristics of public goods, classical microeconomic theory suggests that the free market will not provide it efficiently and the government must take the leading role in transport investment.[9] Also, transport generates various kinds of negative externalities, including pollution and congestion, and these problems require government interventions, such as regulation and taxation, to ensure an optimal level of transport consumption.[10] Thus the quality of transport infrastructure and services depends on the effectiveness of institutions that carry out their investments and regulations, which in turns depend on well-designed institutional arrangements.[11] The lacklustre performance of Thailand's transport system can be attributed significantly to the Thai government's policies and the effectiveness of its agencies, which vary greatly by modes of transport. In the following sections, the involvement of the contemporary Thai state in each mode of transport as well as mode-specific issues and problems will be examined in turn.

Table 18.1 WEF Global Competitiveness ranking of transport infrastructure in ASEAN countries

Road			*Rail*			*Airport*			*Port*		
Rank	*Country*	*Score*	*Rank*	*Country*	*Score*	*Rank*	*Country*	*Score*	*Rank*	*Country*	*Score*
1	Singapore	6.3	1	Singapore	5.9	1	Singapore	6.9	1	Singapore	6.7
2	Malaysia	5.3	2	Malaysia	5	2	Malaysia	5.7	2	Malaysia	5.4
3	Brunei	4.8	3	Indonesia	4.2	3	Thailand	5.2	3	Thailand	4.3
4	Thailand	4.3	4	Vietnam	3	4	Indonesia	4.8	4	Indonesia	4
5	Indonesia	4.1	5	Thailand	2.6	5	Brunei	4.5	5	Brunei	3.9
6	Vietnam	3.4	6	Philippines	1.9	6	Vietnam	3.8	6	Vietnam	3.7
7	Cambodia	3.4	7	Cambodia	1.6	7	Laos	3.8	7	Cambodia	3.7
8	Laos	3.3				8	Cambodia	3.7	8	Philippines	2.9
9	Philippines	3.1				9	Philippines	2.9	9	Laos	2.8

Source: The Global Competitiveness Report 2017–2018, World Economic Forum, Schwab (2017)

Urban transport

The governance of urban transport in Thailand is characterised by highly centralised and fragmented organisational arrangements, which create difficulties in coordination among numerous agencies, particularly in the Bangkok Metropolitan Region (BMR).[12] Historically, two rival ministries, the Ministry of Transport (MOT) and Ministry of Interior (MOI), compete for the responsibility of urban transport in Thailand. The government reorganisation in 2002 consolidated most transport-related responsibilities under several government agencies and state-owned enterprises in the MOT, but the municipal governments retain many important responsibilities in the urban transport systems. In addition, the Traffic Police (TP), a division of the Royal Thai Police (RTP), which reports directly to the Office of the Prime Minister (OPM), is also a key agency that regulates urban transport activities.

Thailand's national transport planning organisation is the Office of Transport Planning and Policy (OTP), a department-level agency under the MOT. Since there is no metropolitan-level transport planning organisation, the OTP also functions as such for the BMR as well as for large municipalities in other provinces, carrying out the tasks of planning and designing urban transport systems, including urban bus and rail master plans and urban road and highway network master plans. Housed within the MOT, the Department of Land Transport (DLT) is responsible for public bus licensing and regulation, the Department of Highways for highways, and the Department of Rural Roads for river-crossing bridges in Bangkok. In addition, several government-owned state enterprises under MOT are involved in urban transport in Bangkok, including the Expressway Authority of Thailand (EXAT), the Mass Rapid Transit Authority (MRTA), the Bangkok Mass Transit Authority (BMTA), and the State Railway of Thailand (SRT), which are in charge of the construction of infrastructure and delivery of expressways, metro rail, urban bus, and commuter rail services, respectively. Transport-related agencies outside of the MOT include municipal governments and the police. For example, the Bangkok Metropolitan Administration (BMA) is responsible for local road construction and maintenance, while the TP is in charge of traffic control and traffic law enforcement. Since each of these agencies reports to different ministers in the government, they are independent in preparing budgets and have different policy goals and priorities which are not necessarily in line with those of the OTP. These arrangements make it extremely difficult to implement complex urban transport projects which require a high degree of coordination among agencies at the physical, budgetary, and operational levels.

The current emphasis of urban transport policy in BMR is on urban rail development. Under the current Rail Transit System in the BMR 20-year Master Plan (2010–2029), 509 km of urban railways, including 12 lines with 310 stations, will be completed by 2029, with a total investment of over 830 billion baht (US$24.4 billion at the exchange rate of 34 baht per US$1, in 2009 constant prices).[13] Again, these new urban railways will be implemented by MRTA and SRT and will likely be stand-alone with limited integration with pedestrian and cyclist facilities and urban bus services, which fall under the responsibilities of other agencies. In the BMR, urban buses remain a neglected mode of transport, operated by the debt-ridden BMTA whose ancient fleet averaged over 20 years old and whose total debt in 2018 amounted to over 100 billion baht (US$3 billion at the exchange rate of 33 baht per US$1 in 2018).[14] In the provincial cities, public transport is left primarily to the private sector without any help from the government. In major provincial cities like Khon Kaen, groups of local businesses chip in to invest in public transport with no financial support from the government.[15] In smaller cities with little demand, public transport is virtually non-existent, forcing citizens to rely on private modes of transport, especially motorcycles.

The lack of coordination among transport agencies manifests itself in a variety of examples of transport problems in Bangkok, especially the very unintegrated nature of transport systems. For example, there are currently three urban rail agencies in Bangkok – the MRTA subway, the BMA's Bangkok Transit System (BTS), and the SRT Airport Rail Link (ARL) – each uses a different fare structure and different electronic fare payment systems which are not compatible with one another. The BMTA buses use a paper-based fare that is not transferable to any of the rail systems. The same problem exists between the two road agencies. EXAT's expressways and DOH's tollways use different electronic fare payment systems. The integration between railways and roads is also problematic, particularly between SRT rail stations and access roads. Since SRT has its own right of way, unlike other urban rail lines which were built along existing roads, the SRT stations tend to have very difficult approaches due to the lack of access roads. Similarly, bus feeder services are not planned and implemented in coordination with metro rail projects, and the debt-ridden BMTA has neither the will nor the resources to reroute its services to feed passengers to metro rail stations. Likewise, park and ride facilities receive low priority for resources from the rail agencies, and neither BMA nor other transport agencies are interested in developing a system that will benefit other agencies.

The traffic congestion problem that results in the severe degradation of mobility of Bangkok residents stems partly from the lack of coordination. As mentioned earlier, BMA is responsible for local road construction, including the installation of traffic signal and control equipment, road marking, and on-street parking regulation. However, it is the Traffic Police Division of the Metropolitan Police Bureau under the RTP that actually controls the traffic and enforces traffic rules, including parking bans. Since these two agencies are independent, the BMA is headed by the elected governor, while the RTP is headed by the police chief who reports directly to the prime minister – they have little incentive to cooperate and often attribute the causes of traffic problems to the other. For example, the BMA's policy to build at-grade crosswalks with automatic traffic signals to facilitate pedestrian movement in the central Bangkok area is met with opposition from the Traffic Police, which feels the crosswalks will slow down the traffic and should be replaced by pedestrian bridges. Similarly, the police's lax enforcement of bus and high-occupancy vehicle (HOV) lanes, as well as the on-street parking ban, means that BMA's policy which favours a more sustainable mode of transport over cars, such as bicycle lanes, cannot be implemented effectively.

Road transport

Several agencies are responsible for road infrastructure in Thailand. Three agencies in the MOT are responsible for construction, maintenance, and operation of the road infrastructure: the Department of Highways (DOH) for the national highway network, the Department of Rural Roads (DORR) for secondary and rural roads, and the Expressway Authority of Thailand (EXAT) for urban and intercity expressways. Road projects are generally financed by the government's budget or loans. But several toll road projects were also financed by the private sector in various forms of public-private partnership (PPP), most prominently the build-operate-transfer (BOT) model. Due to the limited legal expertise of government agencies involved, many disputes occurred in some of these PPP projects, resulting in failure of some projects[16] and monetary penalties imposed by the government.[17] Municipal governments are responsible for local roads in their jurisdictions. Road traffic control and traffic law enforcement, however, are the responsibility of the Royal Thai Police, which operates through the traffic police division in each police jurisdiction as well as a specific division for the national highways, namely the Highway Police Division.

As far as road-based transport operation, the private sector plays the main role as road transport operator with the public sector as the regulator. The public sector plays a limited part as operator through the Transport Company Limited (TCL), which is solely licensed to operate intercity bus service along the routes between Bangkok and other provinces. Private bus operators in these routes must subcontract from the TCL. As for freight transport, the now defunct Express Transportation Organisation (ETO) was a state enterprise that provided road-based service, but it was unable to compete with private operators and closed down in 2006. The Department of Land Transport (DLT) in the MOT regulates various aspects of road transport, including vehicle inspection and licensing, driver training and licensing, and passenger and freight transport operator licensing. The department is also in charge of passenger bus and freight terminals, mainly through issuance of licenses to private operators or municipalities.

Road infrastructure is a critical component of the transport system, linking a variety of transport modes to provide an integrated transport network. However, the institutional arrangements of road agencies in Thailand have not been successful in promoting such integration. While the OTP is supposedly the national planning authority of all types of transport infrastructure, it has little control over the road agencies due to the bureaucratic structure of the MOT. The DOH, DORR, and EXAT are all at the same hierarchical level as OTP, but they are older and more prestigious. The heads of these agencies carry more clout than the OTP director and have their own plans and priorities. With ineffective master planning, the development of road transport network in Thailand is often on a stand-alone project basis, with limited consideration of integration with other components of the network. Some highway projects by rival agencies run in parallel, such as DOH's Motorway 7 (Bangkok-Chonburi Motorway) and EXAT's Burapha Withi Expressway (Bang Na-Bang Pakong), and DOH's Don Muang Tollway and EXAT's Si Rat Expressway (Second Stage Expressway). These projects compete for vehicle traffic, resulting in a lower than expected rate of return on investment. Recently, the redundancy has also extended to competition between highways and railways, such as the DOH's Motorway 6 (Bang Pa In-Nakhon Ratchasima Intercity Motorway) and the Bangkok-Nakhon Ratchasima High-Speed Rail projects.

The most pressing issue in Thailand's road transport is traffic safety, which has been well recognised by the government and the public. Campaigns to reduce road accidents began in earnest in the early 2000s but have thus far been unsuccessful in containing traffic fatalities, the majority of which have been motorcycle drivers and passengers. According to the World Health

Organisation, 83 per cent of traffic fatalities are among motorcyclists, bicyclists, and pedestrians, including 73 per cent among motorcyclists.[18] Despite extensive investment in roads, little effort has been expended to improve public transport services, non-motorised transport, and pedestrian facilities, hence limiting the benefits of roads to the population groups without private motor vehicles. Consequently, a large portion of the population, especially youths and young adults, relies on motorcycles for personal mobility, subjecting these vulnerable road users to disproportionate exposure to traffic accident risk. In addition, lax traffic rule enforcement fails to curb unsafe driving behaviors.[19] Speeding, overloading, riding a motorcycle without wearing a helmet, and carrying passengers in the uncovered back of a pickup truck are all common, especially outside of Bangkok, contributing to the high rate of traffic accidents.

Weak law enforcement and corruption also generate several other problems in road transport in Thailand. Trucks are regularly overloaded to minimise transport cost, but overloading causes substantial damages to roads, putting a huge burden on MOT's annual budget.[20] Despite the government effort to crack down on a widely known system of bribery by truck operators to the police and the authority, the bribery system is becoming more elusive with the use of mobile technology. The regular bribery payment provides protection of penalty not only for overloading of trucks but also for other types of violations, such as failure to adhere to emission standards. Regular bribes are also common among passenger transport operators, particularly informal transport such as passenger vans[21] and motorcycle taxis.[22]

Rail transport

The State Railways of Thailand (SRT) is the state enterprise under the MOT that owns railway infrastructure and operates intercity passenger and freight services as well as limited commuter services. As a state-owned enterprise, it is the government's policy that SRT is expected to be financially independent, generating a sufficient amount of revenue not only to cover its operations but also to pay for infrastructure maintenance and repair. However, the SRT has maintained a staunch system of bureaucracy – a legacy from the previous era when it was still a government agency. Unlike road agencies, the SRT has not enjoyed generous budget allocation from the government until recently. Owing to continued negligence, the rail system is in serious need of upgrade. The majority of the network is in poor condition, unelectrified, and single tracked; crossings are mostly at grade; the diesel-engine locomotives are old and dilapidated; and the signaling systems are primitive. As a result, the performance of SRT is largely very poor, with the maximum speed of the fastest express service being 90 km/h, accidents involving at-grade crossing crashes are frequent and derailments are common.[23] The majority of the SRT services is subsidised third-class passenger services. Freight constitutes a small fraction of SRT's services due to the limitations of the rail system.[24] The SRT suffers a huge of amount of losses each year, with cumulative debt of over 150 billion baht (US$4.5 billion at the exchange rate of 33 baht per US$1) as of early 2018.[25]

In recent years, realising that the high cost of transport and logistics in Thailand stems partly from overdependence on road transport, the government's transport policy has placed more emphasis on railways. Unfortunately, the national railway policy is extremely confusing, the decision-making uninformed, and the policy based on form over substance. For example, the High-Speed Rail (HSR) project was proposed and approved for the first time in 1994 by Chuan Leekpai's government. Thaksin Shinawatra's government in 2004 expanded the scope of the project, and the idea of having the prestigious HSR in Thailand has caught the imagination of the general public ever since – so much so that successive governments from all sides, including the current military junta, have continued to push the HSR project, which has become highly politicised.

One of the most memorable episodes about HSR occurred in 2014, when the Constitutional Court struck down Yingluck Shinawatra's government-proposed bill that would authorise borrowing funds to build the HSR project. In highly controversial footage of the court hearing, a judge opined that Thailand had no need for the HSR yet, not until all dirt roads are paved – an opinion that was later widely mocked. But the whole debate over the HSR missed many critical points and never recognised the fact that the system must be built entirely from the ground up, that the existing meter-gauge tracks cannot be used for the HSR, and that the HSR cannot be used for freight transport so it will not lower the country's logistics costs. It should also be noted that while the Constitutional Court and those opposing the Yingluck government vehemently disapproved the HSR project, few critics have voiced argument against the military junta's version of the HSR project, which would require borrowing even more funds.

That HSR investment is difficult to justify economically is well known. Nash argued that three conditions for successful HSR investment include (1) a dense population corridor that can generate high travel demand; (2) flat, non-mountainous terrain to ensure lower construction costs; and (3) a high-income population that can afford the relatively high fare.[26] In Japan, where the high-speed Shinkansen (also known as the bullet train) has been in operation since the 1960s, only the operations between Tokyo and Osaka, which meet all three conditions, are profitable. In Thailand, none of the three conditions applies. Also, lost in the discussion are the long overdue improvements on the conventional rail network, including laying double and triple tracks and constructing grade-separated crossings, upgrading signaling and train control systems, electrifying tracks, and procuring a sufficient number of locomotives – all of which could vastly improve the average speed, energy efficiency, safety, and overall rail service quality for both passenger and freight transport at a fraction of the cost of the HSR project. The current junta-led government seems more preoccupied with the security issues and international politics than the economic justification of the HSR. Thus, with exceptionally close ties with the government of the People's Republic of China, the huge exporter of steel and HSR technology, pushing HSR forward in Thailand seems like a good idea.

Despite the Thai government's enthusiastic push for the HSR, it is doubtful that the SRT will be capable of taking the leading role in implementing such a large and complex project. Discussion of SRT reform to make it more efficient has been ongoing for many years.[27] A proposal for reform is to reorganise the railway agencies by splitting the SRT into three agencies: one government agency for railway infrastructure investment and maintenance, one train operator that must compete with other private operators, and one independent regulator of railway operators.[28] However, the opposition from the SRT labour union to the proposed reform has been fierce. As of early 2018, the draft railway transport act under consideration by the Council of State will not significantly change the SRT's organisation but will only establish a new policy committee and a regulatory body of railway operation, with the proposed Department of Railway Transport as their secretariat to be established by the end of 2018.[29]

Air transport

Several government agencies serve in different capacities relating to air transport. Until 2017, the department within the Ministry of Transport that regulated all matters concerning commercial aviation and airports was the Department of Commercial Aviation, renamed the Department of Air Transport in 2002 and the Department of Civil Aviation in 2009. These successive departments were also responsible for the operations of small, domestic (often loss-making) commercial airports. Airports of Thailand PCL (AOT), a publicly listed company with 70 per cent of stocks held by the Thai government, operates six highly profitable international airports: two in

Bangkok and four in major provincial cities. Air traffic control services are provided by Aeronautical Radio of Thailand Company Limited, another state-owned enterprise with airlines as minority shareholders.

Thai Airways International PCL is the national flag carrier and also a state-owned enterprise. Thai Airways and other full-service carriers are operated out of Bangkok Suvarnabhumi Airport while a number of low-cost carriers (LCCs), notably Thai Air Asia and Nok Air, are based out of Bangkok Don Muang Airport. With intense competition from LCCs, Thai Airways regularly shows poor financial performance. LCCs provide frequent and cheap services both domestically and internationally and capture a large share of the growing aviation market. In response to competition from LCCs, Thai Airways' subsidiary, Thai Smile, was launched in 2012 to serve domestic and short-haul international routes. Thai Airways' inability to compete may stem partly from the company's highly bureaucratic management and the limited industry expertise on the corporate board, which is often filled with high-ranking government and military officials.[30] Also, Thai Airways' large, strong labour union often makes it difficult to carry out major organisational reforms that are necessary for cutting operational costs.

The organisational arrangements of the MOT's air transport agencies proved problematic, particularly the Department of Civil Aviation (DCA), which oversaw both airlines regulation and airports. As a government agency, the department's remuneration and incentive schemes were unattractive compared with those of other employers in the aviation industry, and it was unable to hire a sufficient number of qualified personnel to carry out the regulatory tasks. In June 2015, the International Civil Aviation Organisation (ICAO), the United Nations regulatory body of the aviation industry, placed Thailand on the list of red-flagged states with substandard airline regulation due to non-conforming arrangements of the regulatory system and the lack of technical and personnel capacity of the DCA.[31] This decision severely limited the ability of airlines licensed in Thailand to modify their service routes or open new ones. The junta-led government hastened to fix the problem by reorganising the aviation agencies, splitting the DCA into the Department of Airports (DOA) and the Civil Aviation Authority of Thailand (CAAT). The DOA inherited responsibilities relating to airport development, maintenance, and operations from the DCA. The CAAT is classified as the government's "independent administrative organisation" which is neither a government agency nor a state-owned enterprise, allowing it to establish its own lucrative remuneration schedule to attract qualified personnel. As an independent organisation, the CAAT took over all aviation regulatory responsibilities with the organisational structure that is in line with international practices. The reorganisation was successful in restoring Thailand's credibility in aviation regulation, and in October 2017 the ICAO removed Thailand from the list of red-flagged states.[32]

Water transport

The MOT's Marine Department (MD) is the government agency in charge of water transport infrastructure, vessels and ports regulation, and licensing of water transport operation. In 2002, as part of the government's reorganisation initiated by the Thaksin administration, the Marine Department was renamed the Department of Water Transport and Merchant Marine, to be in line with the departments responsible for other transport modes, such as the DLT. In 2009, after Thaksin's administration was overthrown in the 2006 coup and the Democrat government led by Abhisit Vejjajiva, the department was renamed the Marine Department again, ostensibly in grateful recognition of King Chulalongkorn (Rama V), who conferred the original name to the agency. In addition to the MD, the Port Authority of Thailand (PAT), a state-owned enterprise under MOT, also plays an important part in the development and operations of port infrastructure that is of critical importance to Thailand's export economy. PAT itself and private concessionaire operators manage Laem Chabang Port and Bangkok Port, the country's two largest ports. The government's participation in water transport operations is limited to the regulations of private operators. Since the Thai Maritime Navigation Company Limited, the state-owned enterprise that provided shipping services, closed down in 2011, water transport operations are mainly under private responsibilities.

Maritime transport and containerised freight are crucial for Thailand's export economy, and the Thai government invested heavily in maritime transport infrastructure, especially deep seaports. Laem Chabang Port in the eastern province of Chonburi has been the key component of the Eastern Seaboard initiative by Chatichai Choonhavan's government, which contributed significantly to Thailand's rapid industrialisation and economic growth in the late 1980s and early 1990s. But PAT's rigid bureaucracy, strong labour union, and poor management make it difficult for Thai ports to compete with those in the neighbouring countries, particularly the Port of Singapore and Port Klang and Tanjung Pelepas of Malaysia. Poor integration with landside transport infrastructure, especially intermodal transport and railway facilities, further undermines the competitiveness of Thai ports.

Another disadvantage of Thai ports is that they are situated out of the way from major ocean shipping routes which pass through the Straits of Malacca. The digging of a canal across Kra Isthmus that would shorten the sea route between East Asia and the Middle East by 1,200 km could put Thailand on the world's map of global shipping. It has been proposed since the 17th century but has never gained the government's support. Recently, as part of China's One Belt One Road (OBOR) initiative, the idea of Chinese-financed construction of Kra Canal was reported by the media.[33] In early 2018, the military government again formally dismissed the feasibility of the project, citing national security concerns in the troubled southern provinces that would be physically separated from the rest of the country by the canal.[34]

Transport by inland waterways was historically important in Thailand. But because of the modern Thai state's policy which placed an emphasis on road transport, investments in inland waterway infrastructure for domestic transport have traditionally received low priority. As a result, despite the relatively low cost of transport, the amount of domestic freight in tonne-kilometer carried by water transport is less than 5 per cent of that carried by road transport.[35] The share of intercity passenger transport in inland waterways is also small compared to other modes of transport. In Bangkok, where canals historically formed a significant part of the city's transport system, river and canal ferry services run by private operators provide reliable alternatives to congested roads. As in the case of road transport, the government regulates waterway passenger services but also controls fares without any financial support, resulting in low quality of service and poor safety performance.

Conclusion

The Thai government has invested heavily in transport infrastructure, but the country's current performance in transport has much room for improvement. While different sector-specific issues vary among modes of transport, the Thai state's involvements in different modes of transport share a few common characteristics. First, the government's transport agencies tend to favour stand-alone projects that can be implemented independently without interfacing with other agencies because the organisational structure of transport governance and the budgetary process do not promote such integrated projects. Second, Thai policymakers often show a penchant for capital-intensive prestige projects including motorways, urban rail, airports, and high-speed rail, and they often nurture ambitions to create a regional transport hub. For these reasons, large investments in transport projects may not necessarily generate sufficient social benefits that can justify the use of public funds. Third, there is a clear policy bias towards private cars and a limited effort to promote sustainable mobility, which requires integrated facilities and services for pedestrians and non-motorised transport users. Due to these characteristics of the Thai state's roles in transport development, it is unsurprising that the efficiency of public investments in transport is the exception rather than the rule. In addition, the unbalanced transport policy has disproportionately benefited certain groups of the population at the expense of others, thus intensifying social inequality. To improve the country's performance in transport, the Thai government needs to address the institutional and organisational shortcomings by implementing major reforms to improve transport governance so that incentives are in place for different agencies to work together, that transport decision-making is well-informed and unpoliticised, and that considerations for the groups of population are carefully weighed and balanced.

Notes

1 National Statistical Office, *Gross National Product, Gross Domestic Product and National Income at Current Market Prices by Economic Activities Year: 2007–16*, 2017 <http://statbbi.nso.go.th/staticreport/page/sector/en/10.aspx> (accessed 24 June 2018).
2 Pichet Kunadhamraks, *Thailand's Transport Infrastructure Development Strategy, Office of Transport Planning and Policy* <www.otp.go.th/uploads/tiny_uploads/PolicyPlan/1-PolicyPlan/M-MAP2/25600629-PDF1-Dr.Pichet.pdf> (accessed 24 June 2018).
3 TomTom International BV, *TomTom Traffic Index: Measuring Congestion Worldwide*, 2016 <www.tomtom.com/en_gb/trafficindex/> (accessed 24 June 2018). See also INRIX, *INRIX Global Traffic Scorecard*, 2017 <http://inrix.com/scorecard/> (accessed 24 June 2018).
4 Yasmin Lee, "Thailand Has the World's Most Congested Roads: Survey", *The Strait Times*, 22 February 2017.
5 World Health Organisation, *Strengthening Road Safety in Thailand*, 2018 <www.searo.who.int/thailand/areas/roadsafety/en> (accessed 24 June 2018).
6 Policy and Strategy Bureau, *Strategic Plan of Ministry of Transport B.E. 2560–2564* (Bangkok: Office of the Permanent Secretary, Ministry of Transport, 2016).
7 Alonggot Limcharoen, Varattaya Jangkrajarng, Warisa Wisittipanich, and Sakgasem Ramingwong, "Thailand Logistics Trend: Logistics Performance Index", *International Journal of Applied Engineering Research*, Vol. 12, No. 15 (January 2017) pp. 4882–4885.
8 Klaus Schwab, *The Global Competitiveness Report 2017–2018, World Economic Forum* (Geneva: World Economic Forum, 2017), p. 286.
9 Joseph Berechman, *The Evaluation of Transportation Investment Projects, Routledge Advances in Management and Business Studies* (New York: Routledge, 2009), p. 18.
10 Kenneth J. Button, "Market and Government Failures in Transportation", in *Handbook of Transport Strategy, Policy, and Institutions*, edited by Kenneth J. Button and David A. Hensher (Amsterdam: Elsevier, 2005).
11 Christophe Feder, "Decentralisation and Spillovers: A New Role for Transportation Infrastructure", *Economics of Transportation*, Vol. 13 (2018), pp. 36–47.
12 World Bank, *Strategic Urban Transport Policy Directions for Bangkok* (Washington, DC: The World Bank, 2007).
13 TEAM Consulting Engineering and Management Company Limited, TEAM Logistics and Transport Company Limited, and Daorerk Communications Company Limited, *Master Plan of Rail Transit System in the Bangkok Metropolitan Region* (Bangkok: Office of Transport Planning and Policy, 2010).
14 Public Debt Management Office, *Non-Financial State Enterprise Debt (Guaranteed Debt)*, 2018 <www.pdmo.go.th/en/popup_money_data2.php?m=money&ts2_id=10&ts3_id=28> (accessed 24 June 2018).
15 Khon Kaen City Development Company Limited, *Khon Kaen Smart City Project Phase 1*, 2018 <www.khonkaenthinktank.com/> (accessed 24 June 2018).
16 Andrew Chetham and William Barnes, "Hopewell to Sue Bangkok over Failed Transit Project", *South China Morning Post*, 4 November 1998.
17 Fernando Carbrera Diaz, "German Investor Awarded 29 Million Euros in Claim against Thailand over Highway Concession", *ITN*, 11 May 2010.

18 World Health Organisation, *Strengthening Road Safety in Thailand*, 2018 <www.searo.who.int/thailand/areas/roadsafety/en> (accessed 24 June 2018).
19 Jon Fernquest, "Thailand's Deadly Roads Stem from Law Enforcement Failure", *Bangkok Post*, 8 November 2015.
20 The Nation, "Bribery and the Burden on our Roads", *The Nation*, 24 November 2016.
21 Saksith Chalermpong, Apiwat Ratanawaraha, and Suradet Sucharitkul, "Market and Institutional Characteristics of Passenger Van Services in Bangkok, Thailand", *Transportation Research Record: Journal of the Transportation Research Board*, No. 2581 (2016), pp. 88–94.
22 Apiwat Ratanawaraha and Saksith Chalermpong, "Monopoly Rents in Motorcycle Taxi Services in Bangkok, Thailand", *Transportation Research Record: Journal of the Transportation Research Board*, No. 2512 (2015), pp. 66–72.
23 Saritdet Marukatat, "SRT Going Off the Rails as Aged Route Rots", *Bangkok Post*, 22 July 2013.
24 Asian Development Bank, *THAILAND: Supporting Railway Sector Reform Final Report* (Manila: Asian Development Bank, 2013).
25 Public Debt Management Office, *Non-Financial State Enterprise Debt (Guaranteed Debt)*, 2018 <www.pdmo.go.th/en/popup_money_data2.php?m=money&ts2_id=10&ts3_id=28> (accessed 24 June 2018).
26 Christopher Nash, "When to Invest in High Speed Rail", in *Roundtable on The Economics of Investment in High-Speed Rail*, New Delhi, India, 18–19 December 2013.
27 Asian Development Bank, *THAILAND: Supporting Railway Sector Reform Final Report* (Manila: Asian Development Bank, 2013).
28 Chadchart Sittipunt, *High-Speed Rail: Government's Duty to Provide Network Service and ASEAN Community*, 1 November 2014 <www.youtube.com/watch?v=QERx3Kprb78> (accessed 24 June 2018).
29 Om Jotikasthira, "Deal Signed to Set Thai Rail Standards", *Bangkok Post*, 10 March, 2018.
30 Duenden Nikomborirak and Saowaluk Cheevasittiyanon, "Corporate Governance among State-Owned Enterprises", in *Corporate Governance in Thailand*, edited by Sakulrat Montreevat (Singapore: Institute of Southeast Asian Studies, 2006), p. 60.
31 Chalunthip Pradubponsa, "From ICAO's Reg Flag to FAA CAT2 followed by Green Light from EASA: Aviation Review", *Trainer* (2016), pp. 44–45.
32 Matthew Tostevin, "Aviation Red Flag Removed against Thailand, Airline Shares Jump", *Reuters*, 7 October 2017.
33 Lam Peng Er, "Thailand's Kra Canal Proposal and China's Maritime Silk Road: Between Fantasy and Reality?" *Asian Affairs: An American Review*, Vol. 45, No. 1 (February 2018), pp. 1–17.
34 Channel News Asia, "Proposed Kra Canal not Current Government Project: Thailand", *Channel News Asia*, 12 February 2018.
35 Pichet Kunadhamraks, *Thailand's Transport Infrastructure Development Strategy, Office of Transport Planning and Policy* <www.otp.go.th/uploads/tiny_uploads/PolicyPlan/1-PolicyPlan/M-MAP2/25600629-PDF1-Dr.Pichet.pdf> (accessed 24 June 2018).

References

Alonggot Limcharoen, Varattaya Jangkrajarng, Warisa Wisittipanich, and Sakgasem Ramingwong. 2017. "Thailand Logistics Trend: Logistics Performance Index". *International Journal of Applied Engineering Research*, Vol. 12, No. 15: 4882–4885.

Apiwat Ratanawaraha and Saksith Chalermpong. 2015. "Monopoly Rents in Motorcycle Taxi Services in Bangkok, Thailand". *Transportation Research Record: Journal of the Transportation Research Board*, No. 2512: 66–72.

Asian Development Bank. 2013. *THAILAND: Supporting Railway Sector Reform Final Report*. Manila: Asian Development Bank.

Berechman, Joseph. 2009. *The Evaluation of Transportation Investment Projects, Routledge Advances in Management and Business Studies*. New York: Routledge.

Button, Kenneth J. 2005. "Market and Government Failures in Transportation". In *Handbook of Transport Strategy, Policy, and Institutions*. Edited by Kenneth J. Button and David A. Hensher. Amsterdam: Elsevier.

Carbrera Diaz, Fernando. 2010. "German Investor Awarded 29 Million Euros in Claim against Thailand over Highway Concession". *ITN*. 11 May <www.iisd.org/itn/2010/05/11/german-investor-awarded-29-million-euros-in-claim-against-thailand-over-highway-concession/> (accessed 24 June 2018).

Chadchart Sittipunt. 2014. "High-Speed Rail: Government's Duty to Provide Network Service and ASEAN Community". <www.youtube.com/watch?v=QERx3Kprb78> (accessed 24 June 2018).

Chalunthip Pradubponsa. 2016. "From ICAO's Reg Flag to FAA CAT2 followed by Green Light from EASA: Aviation Review". *Trainer*, No. 30: 44–45.

Channel News Asia. 2018. "Proposed Kra Canal Not Current Government Project: Thailand". *Channel News Asia*, 12 February <www.channelnewsasia.com/news/asia/proposed-kra-canal-not-current-government-project-thailand-9950434> (accessed 24 June 2018).

Chetham, Andrew and William Barnes. 1998. "Hopewell to Sue Bangkok over Failed Transit Project". *South China Morning Post*. 4 November <www.scmp.com/article/261048/hopewell-sue-bangkok-over-failed-transit-project> (accessed 24 June 2018).

Duenden Nikomborirak and Saowaluk Cheevasittiyanon. 2006. "Corporate Governance among State-Owned Enterprises". In *Corporate Governance in Thailand*. Edited by Sakulrat Montreevat. Singapore: Institute of Southeast Asian Studies.

Feder, Christophe. 2018. "Decentralisation and Spillovers: A New Role for Transportation Infrastructure". *Economics of Transportation*, Vol. 13: 36–47.

Fernquest, Jon. 2015. "Thailand's Deadly Roads Stem from Law Enforcement Failure". *Bangkok Post*. 8 November.

INRIX. 2017. "INRIX Global Traffic Scorecard" <http://inrix.com/scorecard/> (accessed 24 June 2018).

Khon Kaen City Development Company Limited. 2018. "Khon Kaen Smart City Project Phase 1" <www.khonkaenthinktank.com/> (accessed 24 June 2018).

Lam, Peng Er. 2018. "Thailand's Kra Canal Proposal and China's Maritime Silk Road: Between Fantasy and Reality?" *Asian Affairs: An American Review* Vol. 45, No. 1:1–17.

Lee, Yasmin. 2017. "Thailand Has the World's Most Congested Roads: Survey". *The Strait Times*. 22 February <www.straitstimes.com/asia/se-asia/thailand-has-worlds-most-congested-roads-survey>.

Nash, Christopher. 2013. "When to Invest in High Speed Rail". *Roundtable on The Economics of Investment in High-Speed Rail*. New Delhi, India, 18–19 December 2013.

National Statistical Office. 2017. "Gross National Product, Gross Domestic Product and National Income at Current Market Prices by Economic Activities Year: 2007–16" <http://statbbi.nso.go.th/staticreport/page/sector/en/10.aspx> (accessed 24 June 2018).

Om Jotikasthira. 2018. "Deal Signed to Set Thai Rail Standards". *Bangkok Post*, 10 March <www.bangkokpost.com/news/transport/1425527/deal-signed-to-set-thai-rail-standards>.

Pichet Kunadhamraks. 2017. "Thailand's Transport Infrastructure Development Strategy" <www.otp.go.th/uploads/tiny_uploads/PolicyPlan/1-PolicyPlan/M-MAP2/25600629-PDF1-Dr.Pichet.pdf> (accessed 24 June 2018).

Policy and Strategy Bureau. 2016. *Strategic Plan of Ministry of Transport B.E. 2560–2564*. Bangkok: Office of the Permanent Secretary, Ministry of Transport.

Public Debt Management Office. 2018. "Non-Financial State Enterprise Debt (Guaranteed Debt)" <www.pdmo.go.th/en/popup_money_data2.php?m=money&ts2_id=10&ts3_id=28>.

Saksith Chalermpong, Apiwat Ratanawaraha, and Suradet Sucharitkul. 2016. "Market and Institutional Characteristics of Passenger Van Services in Bangkok, Thailand". *Transportation Research Record: Journal of the Transportation Research Board*, No. 2581: 88–94.

Saritdet Marukatat. 2013. "SRT Going off the Rails as Aged Route Rots". *Bangkok Post*. 22 July <www.bangkokpost.com/news/local/360955/srt-going-off-the-rails-as-aged-route-rots> (accessed 24 June 2018).

Schwab, Klaus. 2017. *The Global Competitiveness Report 2017–2018, World Economic Forum*. Geneva: World Economic Forum.

TEAM Consulting Engineering and Management Company Limited, TEAM Logistics and Transport Company Limited, and Daorerk Communications Company Limited. 2010. *Master Plan of Rail Transit System in the Bangkok Metropolitan Region*. Bangkok: Office of Transport Planning and Policy.

The Nation. 2016. "Bribery and the Burden on our Roads". *The Nation*. 24 November <www.nationmultimedia.com/detail/opinion/30300681> (accessed 24 June 2018).

TomTom International BV. 2016. "TomTom Traffic Index: Measuring Congestion Worldwide" <www.tomtom.com/en_gb/trafficindex/>.

Tostevin, Matthew. 2017. "Aviation Red Flag Removed against Thailand, Airline Shares Jump". *Reuters*. 7 October <www.reuters.com/article/thailand-aviation/aviation-red-flag-removed-against-thailand-airline-shares-jump-idUSL4N1MK0OV> (accessed 24 June 2018).

World Bank. 2007. *Strategic Urban Transport Policy Directions for Bangkok*. Washington, DC: The World Bank.

World Health Organisation. 2018. "Strengthening Road Safety in Thailand" <www.searo.who.int/thailand/areas/roadsafety/en> (accessed 24 June 2018).

PART III

The social development

19

THAI IDENTITY AND NATIONALISM

Saichol Sattayanurak

Thai identity and Thai nationalism are two sides of the same coin. The perception of elements of "Thainess" reinforces and makes the "Thai nation" real and sensible in the Thai people's imagination. The meaning of Thainess is the root of national art and culture, which gave "the Land in the Shape of a Golden Axe", where the Thai people live, not only monarchs and Buddhism but also valuable Thai art and culture. That arouses in the Thai people a love of and pride in the Thai nation and induces them to cherish the land of the Thais by placing great importance on territorial integrity. Furthermore, many dimensions of Thainess, such as loyalty to the king, a belief in Buddhism, use of the Thai language, and admiration of Thai beauty or art, also "adorn the temperament" of Thai people to have the "Thai mind", allowing them to accept a hierarchical social structure and centralised political structure to be correct and meritorious by the way of thinking that "Thainess makes Thailand good". This way of thinking compels Thais to think that they should maintain spiritual Thainess (by accepting only material development and technology from the West), which would make "Thailand good" forever.[1]

Thai identity and nationalism arose from the definition of the Thai nation and Thainess at the end of the 19th century. For more than a century now, although many groups have defined the Thai nation and Thainess differently, the meanings created by the political elite still have an immense influence on Thai identity and nationalism. It is because not only can the meanings be adjusted to respond to changes in the political and social context by being able to maintain the basic ideological concept, but other meanings and concepts can also be co-opted and blocked, and there are also many state mechanisms and institutions to reproduce the meaning intensely.[2]

Under the absolute monarchy

When Thailand was ruled by absolute monarchy, intellectuals, who were the political and cultural elite, crafted a strict definition and meaning of the Thai nation and Thainess in order to manage political problems in many dimensions. These dimensions included especially the ordering of society after the abolition of the system of commoners and slaves (but still with the need to preserve a hierarchical social structure); the building of a conscious sense of unity among people from all regions (this was the first time they were directly governed by a central administration in a clearly demarcated and politically centralised state); the fight against nationalism disseminated by the Chinese and the middle class (who were expanding their economic and political roles); and the resistance against new economic and political doctrines (such as republicanism, democracy, socialism, and constitutionalism).

Thainess as it was constructed during the absolute monarchy had forms and contents that covered a variety of meanings, for example, "Thai-style government", "Thai monarchy", "Thai-style Buddhism", "Thai virtues", "Thai ceremonies", "Thai traditions", "Thai language", "Thai literature", "Thai art", and "Thai manners". These meanings have been created and reproduced by various organisations and media, such as schools, books, museums, architecture, monuments, drama, inspirational music, exhibitions, and speeches or royal addresses on numerous occasions. Meanwhile, King Vajiravudh (Rama VI) established a new form of "communities", called "people's groups" (คณชน), such as family names, the Wild Tiger Corps, the Boy Scouts, and various associations. Royal speeches, ceremonies, and songs given to and performed in front of these people's groups all constructed the meaning of "nation, Buddhism, monarch" with an emphasis on their profound inter-relationship. This included a creation of a meaning for Thai virtue based on loyalty and sacrifice. The king showed that Thainess also possessed a genuinely global core that could be compared to civilisations of the European nations. Buddhism, meanwhile, was made superior to other religions because of the rationality of *dharma* and the birth status of the Buddha. He constructed "the other" by focusing on the Chinese, dissecting them into different categories, as well as condemning them as the "true Chinese" who refused to "turn into Thais". He wanted different ethnic groups to speak the Thai language, which would make them "Thai" both "in taste and thought".[3]

Prince Damrong Rajanubhab inculcated historical knowledge to "adorn the temperament" of the Thai people to have national "habits" or "virtues". These were "to be loyal to the independence of the nation, to be free of malice, and to be intelligent in coordinating benefits". There was a sense that the kings of the Chakri dynasty were the leaders in these virtues, successfully leading the Thai nation with sovereignty, peace, and prosperity. To oppose the growing calls for a constitutional monarchy, Vajiravudh pointed out the advantages of the tradition of Thai-style government, reflected in the manner of "a father governing his children". Accordingly, he moulded the concept of *Phra Piya Maharat* to mean the great king beloved by his people, because the king "belonged to the people" and governed "for the people".[4] At the same time, the meaning of the Thai nation and Thainess was juxtaposed with the Buddhist concepts through the production by Supreme Patriarch Vajirananavarorasa of a great number of dharma books for use in the monks' education system. They are still used for the moral education of male and female youth in schools today.[5]

After the 1932 revolution

After the 1932 revolution, the state and state intellectuals adjusted the meanings of the Thai nation and Thainess with a number of objectives, such as creating legitimacy for the new regime, debunking the former meaning which upheld the absolute monarchy, controlling of Thai sentiment with the warning not to "go too fast or go the wrong way", and constituting new cultures to respond to the policy of "elevating the Thai nation into becoming a great power in Laem Thong" (literally the "Golden Peninsula", a name for a greater Thailand). To achieve these objectives, an emphasis was placed on racism, so the Thai nation meant the nation of the Thai race, by defining the Thai race as including the Laotians, Cambodians, Burmese, Mon, and Vietnamese. They constructed linguistic and etymological knowledge to convince the Chinese that their ancestors once belonged to the Thai race but were assimilated by China thousands of years ago. They changed the name of the country to become analogous to the name of the race. They propagated inspirational music and fascist art to construct the meaning of the nation and instilled in the people the love of and pride in the nation. They invented a bushido-like character or morality, called *wiradharm* (chivalry), with slogans such as "Thais love the nation more than their lives", "Thais are first-class warriors", "Thailand is a nation that worships Buddhism more than life", and "Thailand is a nation whose citizens follow their leader". They stressed a new kind of behaviour by focusing on citizens' economic duty through the discourse of "progress loving", whether in trade or construction, while defining it as "real Thai" behaviour. They also promoted a role for Thai women in caring for their family and raising children to be good citizens of the nation.[6] This new type of nationalism enabled Field Marshal Phibun Songkhram to turn the army into his power base under the policy of uniting all territories of the Thai race into one land.[7]

In the wake of the 1932 revolution, although the new meaning of Thainess was meant to illustrate the ideal of the new regime, such as "equality", through the promotion of the Thai language and contemporary art, equality did not really occur because there was no reform in the economic and political structure. The Thai state was still totalitarian, so the former meaning of the Thai nation and Thainess that maintained a hierarchical social structure and a dictatorial state structure was reproduced. Although the role of the commoners in history was underlined, it was aimed at building a collective memory that would make Thai people "good citizens" in a totalitarian system. This included, for example, respect for laws which were orders of the sovereign to do one's duty to the nation with harmony and sacrifice without interfering in the duty of others. The prestige of Phibun was reinforced through poetry such as the praise for his virtues as a leader (similar to the Ten Kingly Virtues), and his "ethical behaviour" shone through his "integrity at heart and honesty", his perseverance, and his mercy.[8] This was the beginning of the creation of the idea of governance by a "good man" who was a commoner.

The early 1950s era

Since the early 1950s, the original meaning of the Thai Nation and Thainess from the period of the absolute monarchy had regained its force as a result of the United States supporting the traditional institutions in informing the Thai people of the danger of communism. Many conservative intellectuals embarked on glorifying the sacredness of the "nation, religion, monarch" to oppose Phibun. They reconstructed the meaning of "monarch" by reiterating the royal duties in conserving the high art and culture of the Thai nation both in the past and the present.[9] This kind of propaganda ingrained into Thais that danger of communism could destroy Buddhism, the monarchy, and the ownership of land and assets passed on from one's ancestors –who had sacrificed their blood to preserve the nation – to later generations.[10] Knowledge was created in Thai studies, including art history, to resurrect the memory of the monarch and Buddhism.

To contain communism and liberalism, Phraya Anuman Rajadhon formed knowledge about the countryside to bring about the concept of "Thai peasant society" so abundant that it "has fish in the river, rice in the fields". Villagers' beliefs in ghosts and Buddhism entrenched a valuable tradition for life and brought peace to society. But the government needed to develop rural areas to improve the backward condition that could turn the people into victims of communism. Phraya Anuman Rajadhon, hence, underlined the importance of class in which culture could be classified as high and low. For example, a belief in Buddhism by "wisdom" or "reason" was superior to that by faith or mood, seen mostly in villagers. He also recommended the integration of various cultures of those in the outlying locations for the sake of national security. Most of the ideas from Phraya Anuman Rajadhon conformed to mainstream intellectuals, such as the emphasis on the role of the elite in taking a lead in cultural progress through preservation of the legacy of the past, so that social change would be gradual or evolutionary. Accordingly, "class" became a main focus. Meanwhile, "patriotism" was inculcated, such as the love of artistic and cultural legacy of the nation and the traditions related to the monarchy and Buddhism (not the love of the people who were equal and an integral part of the Thai nation).

In the Cold War period, the approach to national development that conveniently combined the old and the new and always connected the present with the past was reproduced and assisted by various art and media platforms, through the making of sculptures that reflected Thai architecture, and through music, such as the promotion of the *Suntharaphon* band which mixed Western music with lyrics from Thai poems.[11] The idea of gradual change by preserving the Thai mind has been prominent and influential until today. Moreover, state organisations and intellectuals created and disseminated knowledge accumulated from different localities and combined into the one Thai nation, such as the knowledge of good things in localities including archaeological sites and antiques, heroes and heroines, to foster a sense of significance of localities as a major part of the Thai nation. Besides, the good things in localities were developed as a part of the civilisation of the centre (such as the acceptance of Buddhism and Thai writing from Sukhothai).[12]

In the period of the Cold War, intellectuals who referred to royalist ideology intensively revived the meanings of the Thai nation and Thainess constructed during the period of the absolute monarchy by adapting to some extent the meaning to cope with the changing social context. M.R. Kukrit Pramoj played an important role in reproducing the idea of Thai identity which highlighted the monarch and Buddhism as the heart of Thainess, and Thai-style government, where one man held absolute power, as a source of production of good quality of life and justice for the people.[13] This Thai-style government was accompanied by a Thai-style leader who was a good person replete with knowledge, ability, and morality. If any leader was found not to fall into this type, the monarch who was imbued in the Ten Kingly Virtues (a teaching from Buddhism) would look after the "state's arms and legs" (administration and government service) to prevent the incorrect use of power, such as seeking personal benefit or oppressing the people. Kukrit emphasised that the monarch had a close relationship with the people and was an institution which belonged to the people. In the book *Thai Character*, Kukrit portrayed that the kings of the Chakri dynasty bestowed rights and freedoms on the people, created prosperity in trade, and made Thailand a place where democracy was bestowed. It was clearly seen that Kukrit employed a middle-class value system to define and create loyalty.[14] These ideas influenced the middle class to believe that the stability of the monarchical institution and the choice of a good person to govern were more important than the parliament and constitution based on the principle that sovereign power belonged to the people.

To maintain Buddhism's credibility among modern Thais, Kukrit used dharma to explain the biological evolution of living things while revealing its scientific nature. He talked about the complexity of karma and the results of karma to preserve the belief of "as you sow, so you shall reap" and to convince Thais that Buddhism was not an obstacle to the country's development. For example, he stressed that Thai-style Buddhism was a "worldly dharma". Kukrit spoke about an unworldly dharma when he wanted to confirm the depth of Buddhism. Even the idea of "detachment" he interpreted as "satisfaction with what we have in accordance with what we can earn, looking for more without exploiting others".[15] The book *Japanese Scenes* confirmed that Japan developed because of Buddhism. It asserts, "Everything that is good, beautiful and admirable in Japan is solely because it comes from the Buddha's teachings".[16]

The influence of Thainess on the thoughts and minds of the Thai people, apart from arising as a result of its meaning changes to respond effectively to the shifting environment, also stemmed from the refutation and exclusion of new ideas. The government and various conservative intellectuals rejected the definition of the Thai nation and Thainess of a group of Marxist-Maoist writers, who attacked the Thai nation as oppressive. Information was created about the rural areas to prevent a revolution through the strategy of "the forest surrounds the city", and to fight against progressive writers who defined the Thai nation as a nation of many ethnic groups possessing valuable arts and culture of their own.[17]

Kukrit responded to and marginalised the ideas of those who favoured Marxist-Maoist doctrine and liberals through the *Phai Daeng* (Red Bamboo) stories narrating a young communist sympathiser being turned into a humorous figure and finally surrendering to an abbot. In the 1970s, Kukrit taught Thai civilisation, a subject which allowed him to reproduce the meaning of the monarchy and Buddhism as the essence of Thainess and to inculcate that Buddhism made the Thai state a "compassionate state" and Thai society a "compassionate society".[18] In many of his writings, Kukrit demonstrated that Thai society was divided into classes. The upper class did not oppress and exploit, but supported the lower classes who had fallen into the cycle of "foolishness, poverty, suffering". If there were problems, the Thai people did not have to rise up in revolution since "Buddhism has established a society which is an alternative society inside Thai society" (for example, ordination) so as to prevent Thais from supporting a revolution of the working class.[19] Kukrit explained that Buddhism enabled the Thai people to accept suffering calmly without struggling to change their status. This "gave birth to a kind of tranquillity in society that other societies unfortunately do not have". He further stated, "The labour system in Thailand" was not oppressive. Rather it had a system of "mutual assistance" because "the mind of Thai society had mercy and generosity in helping one another, as Buddhism always teaches".[20]

Luang Wichitwathakan responded to the concept of "liberalism" by disclosing the disadvantages to the nation because individualism "always leads to the formation of egoism".[21] Communist doctrine turned citizens into mere "employees of the state" and "if the communists come in this must lead to the loss of a government with a revered monarch and religion, traditions, ideals, and many ways of life which we wish to preserve".[22] Therefore, we should maintain the doctrine of economic nationalism which is "taking the middle path between liberalism and socialism" in order to avoid "disgracing the nation".[23]

Meanwhile, Phraya Anuman Rajadhon reacted to the concepts of liberalism and socialism by elucidating that social and cultural stratification was an ordinary and natural matter and that equality "exists only in words". He also explained, "Do not misunderstand that monks take advantage from the society, they are the ones who produce food for the mind".[24] Social critic Sulak Sivaraksa opposed capitalism-consumerism and republicanism by raising the importance of the institutions of the monarchy and Buddhism and proposed that democracy should be practiced when it serves as a form of government that exists in the monastic culture.[25] Many other intellectuals have widely accepted the concept that democracy exists in Buddhism and that only the "Buddhist-style mind" facilitates a democratic system.

To reduce the influence of new concepts and to construct a meaning of Thainess that was civilised, mainstream intellectuals have incorporated some of these concepts as a component of Thainess, such as constructing the meaning of "Thai gentleman" by merging it with the Western concept of the gentleman into loyalty to the nation and the monarchy,[26] and creating the concept of Thai-style democracy decorated with a constitution and elections but reducing the authority of parliament.[27] They also stressed,

> There is an absolute necessity of a monarch as head of state because the monarch has the status of a monarch who reigns by invitation of or through selection by the people and rules by the "Ten Kingly Virtues". The characteristics of democracy latent in royal qualities and royal conduct are therefore the reasons for the deep reverence of the people. Thai-style democracy will have to offer a prerogative for the monarch to participate in governing the country and the civilian population as appropriate.[28]

The incorporation of the meanings of the Thai nation and Thainess into democracy is the subjection of the citizenry in the name of democracy. In the 1970s, mass organisations were established under the name "Voluntary Development and Self-Defense Villages" based on the concept of "secure nation, wealthy people". The Ministry of Interior was assigned to provide the rural residents an economic security and to approach democracy on a foundation of love for "nation, Buddhism and monarchy". In 1994, the Department of Local Administration claimed that there were 9,850,736 people in these organisations. Monks implementing the "Dharma Ambassadors Project" combined the meaning of Thainess with new ideas on the quality of life, teaching them, while integrating the northeast residents and hill tribes as a part of the Thai nation. Prime Minister Thanin Kraivichien later proposed the concept of "democratic socialism", which partly limited political rights and liberties in order to generate economic equality in a "democratic system with the monarch as head of state".[29] In the process, the principle of the "three marks of existence" received a new interpretation in order to conform to the state and the capitalist system.[30] For example, "It is not meant that property and power are impermanent, suffering and non-self, and therefore we ignore them and do not manage them. We must know how to use them to create what is good and beneficial".[31] This change in the meaning and role of the traditional institutions, both the monarchy and Buddhism can be considered as a part of the integration process of "the old with the new", which showed a great success in reproducing the meaning of the Thai nation and Thainess to strengthen a royal hegemony, which in turn gave the meaning of Thainess a continuing force.

In opposing new economic and political ideas, mainstream intellectuals continued to exploit the method of explaining that in the Thai nation and Thainess, certain characteristics already existed. For example, a "socialist" form of society already existed in northeastern Thai villages.[32] The Ramkhamhaeng Inscription is a "constitution".[33] The Thai nation had owned the concept of electing administrations since ancient times. This can be seen from the concept of "a monarch who reigns by invitation of or through selection by the people" and the coronation ceremony where representatives of various groups of people offer the kingdom to a new king.[34] Regarding socialism, King Bhumibol asserted that it could be cherished in Thailand. For instance, Hup Kaphong village had a cooperative, had activities through which the villagers carried out together as joint owners, and thus demonstrated the most perfect form of socialism. Villages of the hill peoples in the north are the same.[35]

When the idea of the "New Left" arrived in the 1970s, Kukrit analysed this concept as having "anti-establishment characteristics". One should not be excited about the concept of the "New Left".[36] He also asserted that when the Committee of Democracy Development proposed political reforms to the president of parliament in 1995, it still reiterated that nation, religion, and monarchy were of utmost importance to the integrity and peace of the nation and the good qualities that existed in religion, and the monarchy could give birth to a good democracy.[37]

The method of cooperating parts of new ideas into Thainess and affirming that they already existed in Thainess also broadens the words and modernises the meanings of the Thai nation and Thainess in adjusting to the changing context. An example of success of this adjustment is the implementation of the policy for communists to return "to join in developing the Thai nation" in 1980 under the concept of preservation of Thai identity and the "democratic system of government with the monarch as head of state" by ensuring economic, social and political justice and peace. Even though some adjustments failed, the meaning of the Thai nation and Thainess did not remain intact, but was continuously adjustable to remain as a constant force in defining Thai identity and Thai nationalism.

Thai identity and Thai nationalism from the 1980s until the present

The meaning of the Thai nation and Thainess was modified to respond to the changing context and the diverse discursive practices over the past century. Together with the changing meaning of the Thai nation and Thailand, Bhumibol put in practice the "moral standards of Thai-style leadership", which no leaders must violate; they helped expand the royal projects and bestow the sufficiency economy philosophy. The royal attempt was echoed in the principle of "interdependence between the monarchy and the people", which means "the monarchy and the people rely on each other".[38] The meaning of "monarch" was changed to "father of the nation" which gave rise to the motto of "doing good for father". What is more important is the existence of a consciousness among the middle class in which the monarchy was the greatest protector of the people. So the middle class had to protect the "democratic system with the king as head of state" in order to rely on royal influence in various matters, most especially in protecting the Thai nation from national leaders who lacked the virtue of Thai-style leadership.

At the same time, the idea was repeated that the monarchy and Buddhism were sources of valuable and proud Thai art and culture and Thainess. All this helped maintain the spiritual Thainess of the Thai people. As long as they could maintain Thainess, Thailand would be good forever. But the influence of the West, if too overwhelming, would spark problems as described by a royalist, Sumet Tantivejkul: "Increasing changes following the trends of the western world are alien to Thai culture, which is modest, sufficient, respectful, humble and generous. Virtue declines. There is dishonesty and corruption".[39]

In addition, from the 1980s onwards, when the development of the country à la the West produced a greater effect on the countryside, academics and non-governmental organisation (NGO) development workers joined together to engineer the concept of Thai community culture to oppose capitalism and individualism. They attached a greater importance to the Thai local wisdom of villagers to carry out the development by conserving spiritual Thainess that was part of Thai community culture. They, too, created knowledge and initiated a large number of activities to revive the moral standards that existed in community culture. Monastic intellectuals shaped the meaning of "gift" into "sharing", which fitted the value system of community culture.[40] Prawase Wasi emphasised the sustainable development that must link the economy, spirituality, and the community or society together. When combined with the movement of NGOs, development workers, and academics in pushing for "community rights" and "equal rights" of all ethnicities, the concepts of "community" and "community rights" gained broad acceptance leading to provisions in the 1997 constitution.

For many decades, there have been groups of people who deny hierarchical society and dictatorship. They have continuously reconstructed the meaning of Thai nation and Thainess. But the former meaning still persists. For example, in the humanities, knowledge has been constructed to confirm that Thai art and culture was equal or superior to that of the West because the spirit of democracy had existed since the beginning of the Rattanakosin era.[41] Many films and plays were produced on issues that underlined the importance of the monarchy and the role of commoners who sacrificed their lives to protect the nation. Thai visual arts continue to express Thainess. The new Supreme Court and parliament buildings clearly have characteristics of Thai art.

From 2000 to 2010, opposition to "capitalist democracy" led to a concept of governance by good people which made a confident comeback. A well-respected monk even stated, "You have to be good to be great".[42] However, the concept of governance by good people and the combination between Western democracy with Thainess, in many ways, had both supporters and opponents. The supporters of integrating the old with the new coined a new political word under the ideology of Buddhist nationalism by introducing the word *dharm*, which means the teachings of the Buddha, with a wide application. For example when the word *tulakanphiwat* was coined to mean "judicialisation of politics", Thirayuth Boonmee suggested another word with the same meaning, *nithithamphiwat*.[43] The word *nithitham* was invented to be used in lieu of the "rule of law", since the Thai elite did not favour the use of *nithirat* (legal state) even though Thais have used "civil law" as in other European countries, not the term "common law" or the principle of *nithitham*, as used in the United Kingdom.[44]

While well-respected legal experts interpreted the dharma of the monarch or the characteristics of a *thammaracha* (dharma monarch) in order to define the "democratic system of government with the monarch as head of state",[45] Phra Bhramagunabhorn confined the word *thammathipatai* (rule of dharma) to a meaning of good governance by asserting that rulers who gained power from elections and the people who went to elect them should uphold the criteria of *thammathipatai*. Phra Bhramagunabhorn identified *thammathipatai*-style democracy as a better form of government than any others, but oligarchy and dictatorship could be *thammathipatai*.[46] The word *thammaphiban* was translated as "good governance". The Office of the Civil Service Commission regulated that *thammaphiban* had six features, the most important of which was ethics and the next was rule of law.[47] This emphasis from many sides on dharma was an emphasis on governance by good people itself.

But Thai nationalism has not specifically targeted opposing enemies or the other. For the most part, the other has only been painted for domestic political conflict. Even the opposition to Japan and communism was part of shoring up power by the army and legitimising the military leaders as leaders of the nation. In 1980, the army employed a strategy of inviting communists to join in developing the Thai nation. Even though nationalism engendered a negative attitude towards neighbours and ethnic groups in the country who were not Thais, the importance of revenue from tourism compelled Thais in general to accept ethnic diversity. Therefore, the dependency of the people's status on ethnicity was lower than on class, which was raised increasingly in relation to their economic status.[48] And since the beginning of the ASEAN Community in 2015, as Thai capitalist groups invested hugely in neighbouring countries, the government's promotion of unity and tourism in ASEAN gained solid support among Thais. The negative attitude towards nations who had been others was therefore greatly reduced. Meanwhile, Thai art and culture was turned into a commodity sold to other nations. At present, the political importance of Thainess lies mostly in guaranteeing legitimacy for state power.

In the case of the Chinese in Thailand, before the 1960s, on the one hand, there was an opinion that they were the other, and on the other hand they were people of the Thai race who were colonised and "swallowed" by China.[49] At the beginning of the Cold War, the Chinese increasingly earned the status of the other.

> Conditioned by the Cold War politics, the production of knowledge [on the Chinese] was dominated by Skinner's assimilation paradigm portraying the Chinese as the other that needed to be assimilated into Thai society. However, with the rise of Thai intellectual nationalism in the 1970s, Thai scholars began to challenge the authority of American scholarship on Thailand. Sets of knowledge redefining the Thai nation as a country formed out of the admixture of various ethnicities started to be produced by Thai scholars. This led to the identification of the Chinese as the Chinese of Thailand, underscoring their status as a significant part of the Thai nation.[50]

Since the Chinese had been made the other in the past as a danger to the Thai nation and Thainess, the process of becoming Thai had a role in private capital groups and the middle class with Chinese ancestral heritage benefiting from Thai state policy. But in the last three decades the Chinese have benefited from a dual identity, such as being both Thai and Chinese at the same time. In some contexts, this dual identity was transformed into an identity that is more Chinese, for example in the political movements in 2008 as "Chinese descendants save the nation" or "Chinese descendants love the nation", as they claimed to eliminate "evil capital", which was a danger to the Thai nation and Thainess.

It is noticeable that the meaning of the Thai nation and Thainess was propagated to sustain a state structure where the elite monopolise power. Yet there are also many cases where this meaning has been used by the middle class as an ideological principle to eliminate the power of certain leaders, for example in attacking leaders for not loving the nation, for not being loyal to the monarchy, for lacking Buddhist morality, and for trying to change the system of government into a republic. At the same time, there is an emphasis on "relying on the prestige" of the monarch in political struggles, where middle-class movements under this nationalist ideology successfully removed many national leaders from power.[51]

Conclusion

The importance of the monarchy and Buddhism has shaped Thai nationalism into being both royalist nationalism and Buddhist nationalism. These two dimensions are closely intertwined. The construction of the meaning of the Thai nation and Thainess emerged about at the same time. Thais have the idea and the feeling that they belong to the same nation thanks to the shared sentiment about Thainess. They still believe that Thai-style governance, which is "apolitical", is a good form of governance because Thai-style leaders are good people who adhere to dharma. Influenced by Thainess, which is the basis of national culture, a Thai-style mind is crafted on the belief that social relations that divide people according to a hierarchy of class (which is the basis of a centralised political system) are correct and proper. Thais therefore do not give importance to the sovereign power of the people but rather to territorial integrity so that "future generations will have a home".[52]

The Thai economic and political structure makes "becoming Thai" the source of power and benefits, such as the right to enter government service and the right to own land, and the expression of loyalty to the monarchy and patronage of Buddhism still raise one's social status. Many ethnic minorities, including the hill peoples, must therefore become Thai, at least in certain dimensions. The only exception is the Malay Muslim community. It can be seen that these ethnic minorities began to recognise their own identity within the framework of Thainess partly to gain benefits from having a Thai identity. At the same time, the Thai state has enjoyed both power and the means to reproduce its meaning, while conservative intellectuals in the reformist wing play a role in the redefinition of the Thai nation and Thainess to better respond to the changing social and political context.

In the last three decades, there have been new groups in the Thai social structure, struggling to hegemonise in the cultural field. The Thai identity and Thai nationalism therefore remain important parts of the political struggle in the country. With regard to relations with other countries, foreign investment and revenue from tourism has greatly changed the face of Thai nationalism, which previously had a negative view of the other, while Thainess in many dimensions has been transformed into a sign of commodities.

However, the fact that the state and mainstream intellectuals respond to challenges to the meaning of the Thai nation and Thainess in a variety of ways, including the use of artistic and ceremonial media which affect public sentiment, means that mainstream nationalism is still a dominant force. Even though parts of Thainess are fluid and the former meanings have been rejected by many groups of people, some of the important dimensions of the Thai identity and Thai nationalism are still influential in society and continue to be exploited by the elite.

Notes

1 See details in Saichol Sattayanurak, *10 Siamese Intellectuals*, Vols. 1–2 (Bangkok: Openbooks, 2014) [in Thai].
2 Ibid.
3 Saichol Sattayanurak, *10 Siamese Intellectuals*, Vol. 1 (Bangkok: Open Books, 2014), pp. 138–295 [in Thai].
4 Ibid. pp. 314–394.
5 Ibid. pp. 404–461.
6 Saichol Sattayanurak, *Change in the Construction of the "Thai Nation" and "Thainess" by Luang Wichitwathakan* (Bangkok: Matichon, 2002), pp. 76–78 [in Thai].
7 Saichol Sattayanurak, *10 Siamese Intellectuals*, Vol. 2 (Bangkok: Open Books, 2014), pp. 82–85 [in Thai].
8 Cherry Kasemsuksamran, "Wannakadisan and the Construction of Field Marshal Plaek Phibunsongkhram's Images and Prestige", *Humanities Journal*, Vol. 19, No. 2 (July–December 2012).
9 The journal of the Fine Arts Department published these stories. For example, in 1951, it published an article about Rama IV's donation of money to the Fine Arts Department to support national arts and cultural conservation work.
10 Office of the Prime Minister, "Report of the Office of the Prime Minister", *Fine Arts Journal* (April 1950) [in Thai].
11 Chatree Prakitnontakarn, *Politics and Society in Architecture, Siam Period Adapted Thailand and Nationalism* (Bangkok: Matichon, 2004), pp. 448–449 [in Thai].
12 This information can clearly be seen in the Silpakorn Journal and Bulletin of History and Archaeological Documents.
13 Saichol Sattayanurak, *Kukrit and the Invention of "Thainess"*, Vol. 2 (Bangkok: Matichon, 2007), p. 301 [in Thai].
14 Ibid. pp. 404–413.
15 M.R. Kukrit Pramoj, "Moral Principles to be Taken into Consideration in Providing Education", *Thammakhadi* (Bangkok: Dokya, 2001), pp. 40, 44–45 [in Thai].
16 M.R. Kukrit Pramoj, *Japanese Scenes* (Bangkok: Dokya, 2001), p. 61 [in Thai].
17 See Sopha Chanamool, *"Thai Nation" in the Perspective of Progressive Intellectuals* (Bangkok: Matichon, 2007) [in Thai].
18 Saichol, *Kukrit and the Invention of "Thainess"*, p. 296.
19 Ibid., p. 301.
20 Ibid., pp. 297–300.
21 Luang Wichitwathakan, *Future* (Bangkok: Kasembannakij, 1957), pp. 154–155 [in Thai].
22 Luang Wichitwathakan, *In Memory of Wichit* (Bangkok: Office of the Prime Minister, 1962), p. 251 [in Thai].
23 Ibid., p. 26.
24 Saichol, *10 Siamese Intellectuals*, Vol. 2, pp. 118–127.
25 Ibid., pp. 448–450.
26 Saharot Kittimahacharoen, *"Gentleman" in the Literary Works of King Rama VI and Sriburapha*, PhD Dissertation in Arts, Chulalongkorn University, 2008 [in Thai].
27 *Thai-Style Democracy and Thoughts on the Constitution* (Phranakhon: Chokchai Books, 1965), pp. 212–219 [in Thai].
28 Ibid. pp. 195–196.
29 Michael K. Connors, *Democracy and National Identity in Thailand* (London: Routledge, 2003), ch. 4–5.
30 In Buddhism, the three marks of existence are the three characteristics of all existence and being, namely impermanence, suffering, and non-self.
31 Phra Bhramagunabhorn (Prayudh Payutto), *Political Science of the Rule of Thamma* (Bangkok: Thammasapa, 2011), pp. 49–50 [in Thai].
32 Prince Damrong Rajanubhab, *Archaeological Tales* (Bangkok: Banakhan 1971), pp. 378–380 [in Thai].

33 M.R. Seni Pramoj, *Collected Writings* (Bangkok: Bamrungsan 1966) [in Thai].
34 M.R. Kukrit Pramoj, *Thai Character* (Bangkok: Thai Watana Panich, 1982) [in Thai].
35 Quoted in Saichol, *Kukrit and the Invention of "Thainess"*, Vol. 2, pp. 225–226.
36 Ibid., pp. 303–304.
37 Committee of Democracy Development, *Proposals on Thai Political Reform* (Bangkok: Thailand Research Fund, 1997) [in Thai].
38 Saichol, *Kukrit and the Invention of "Thainess"*, Vol. 2, pp. 271–304.
39 Sumet Tantivejkul, "Thailand's Problems across the Stages of Transformation", *Thai Rath*, 19 February 2018 <www.thairath.co.th/content/1207302> (accessed 24 March 2018).
40 Phra Bhramagunabhorn (Prayudh Payutto), *Political Science of the Rule of Thamma*, 3rd printing (Bangkok: Thammasapa, 2011), pp. 101–109 [in Thai].
41 See examples in the many works of Chetana Nagavajara, a highly respected and influential humanities scholar.
42 Phra, *Political Science of the Rule of Thamma*, p. 42.
43 Saichol Sattayanurak, "The Historical Legacy and the Origin of Judicialisation in the Thai State", *Journal of Law and Social Sciences*, Vol. 9, No. 1 (January–June 2016) [in Thai].
44 Worachet Pakeerut, *Principles of Nithirat and Principles of Nithitham*, 2010 <www.pub-law.net/publaw/view.aspx?id=1431> (accessed 30 September 2012).
45 Borwornsak Uwanno, *"Ten Principles of the Righteous King and the King of Thailand"*, Faculty of Law, Chulalongkorn University, Bangkok, 2006.
46 Phra Bhramagunabhorn (Prayudh Payutto), "The Rule of Thamma Has Not Come, So Democracy Cannot Be Found" (Convergence of Political Science and Juristic Science), *Questions and Answers* (February 2006) [in Thai].
47 Thawinwadi Burikun, "Thammaphiban: A New Principle for the Administration of Public Affairs", *King Prajadhipok's Institute Journal*, Vol. 1, No. 2 (May–August) 2003 [in Thai].
48 Saichol Sattayanurak, "The Struggle on the Field of Memory of People of Diverse Ethnicities in Thailand", in *Thai History and Thai Society: Family, Community, Ordinary Life, Memory and Ethnic Identity*, Collected Research Articles (Chiang Mai: Chiang Mai University Press, 2015), pp. 185–186 [in Thai].
49 Saichol, *10 Siamese Intellectuals*, Vol. 2, pp. 6–85 [in Thai].
50 Sittithep Eaksittipong, *Textualising the "Chinese of 'Thailand": Politics, Knowledge and the Chinese in Thailand during the Cold War*, PhD Dissertation, National University of Singapore, 2017, p. x. <http://scholarbank.nus.edu.sg/handle/10635/139087>.
51 Saichol Sattayanurak, "Nationalist Ideology and Thai Politics", Opening Address at the Annual Meeting of the Japanese Society for Thai Studies at the Graduate School of Asian and African Area Studies, Kyoto University, 5–6 July 2014 [in Thai].
52 *"We Fight" Song*, August 2000 <www.baanjomyut.com/library_4/music_thailand/345.html> (accessed 24 March 2018).

References

Borwornsak Uwanno. 2006. *Ten Principles of the Righteous King and the King of Thailand". Faculty of Law*. Bangkok: Chulalongkorn University.

Chatree Prakitnontakarn. 2004. *Politics and Society in Architecture, Siam Period, Adapted Thailand and Nationalism*. Bangkok: Matichon.

Cherry Kasemsuksamran. 2012. "Wannakadisan and the Construction of Field Marshal Plaek Phibun Songkhram Images and Prestige". *Humanities Journal*, Vol. 9, No. 2 (July–December).

Committee of Democracy Development. 1997. *Proposals on Thai Political Reform*. Bangkok: Thailand Research Fund.

Connors, Michael K. 2003. *Democracy and National Identity in Thailand*. London: Routledge.

Kukrit Pramoj, M.R. 1982. *Thai Character*. Bangkok: Thai Watana Panich.

———. 2001. *Japanese Scenes*. Bangkok: Dokya.

———. 2001. "Moral Principles to be Taken into Consideration in Providing Education". In *Thammakhadi*. Bangkok: Dokya.

Luang Wichitwathakan. 1957. *Future*. Bangkok: Kasembannakij.

———. 1962. *In Memory of Wichit*. Bangkok: Office of the Prime Minister.

———. 2011. *Political Science of the Rule of Thamma*. Bangkok: Thammasapa.

Office of the Prime Minister. 1950. "Report of the Office of the Prime Minister". *Fine Arts Journal* (April).
Phra Bhramagunabhorn (Prayudh Payutto). 2006. "The Rule of Thamma Has Not Come, so Democracy Cannot Be Found" (Convergence of Political Science and Juristic Science). *Questions and Answers* (February).
Prawase Wasi. 1997. *Conceptual Journey: Political Reform*. Bangkok: Moh-chao-Ban Publishing House.
Prince Damrong Rajanubhab. 1971. *Archaeological Tales*. Bangkok: Banakhan.
Saharot Kittimahacharoen. 2008. *"Gentleman" in the Literary Works of King Rama VI and Sriburapha*. PhD Dissertation in Arts, Chulalongkorn University.
Saichol Sattayanurak. 2002. *Change in the Construction of the "Thai Nation" and "Thainess" by Luang Wichitwathakan*. Bangkok: Matichon.
———. 2007. *Kukrit and the Invention of "Thainess"*, Vol. 2. Bangkok: Matichon.
———. 2014. *10 Siamese Intellectuals*, Vol. 1–2. Bangkok: Openbooks.
———. 2014. "Nationalist Ideology and Thai Politics". Opening Address at the Annual Meeting of the Japanese Society for Thai Studies at the Graduate School of Asian and African Area Studies, Kyoto University, 5–6 July, 2014.
———. 2015. "The Struggle on the Field of Memory of People of Diverse Ethnicities in Thailand". In *Thai History and Thai Society: Family, Community, Ordinary Life, Memory and Ethnic Identity*. Collected Research Articles. Chiang Mai: Chiang Mai University Press.
———. 2016. "The Historical Legacy and the Origin of Judicialisation in the Thai State". *Journal of Law and Social Sciences*, Vol. 9, No. 1 (January–June).
Seni Pramoj, M.R. 1966. *Collected Writings*. Bangkok: Bamrungsan.
Sittithep Eaksittipong. 2017. *Textualising the "Chinese of Thailand": Politics, Knowledge, and the Chinese in Thailand during the Cold War*. Department of History, National University of Singapore.
Sopha Chanamool. 2007. *"Thai Nation" in the Perspective of Progressive Intellectuals*. Bangkok: Matichon.
Sumet Tantivejkul, 2018. "Thailand's Problems across the Stages of Transformation". *Thai Rath*, 19 February <www.thairath.co.th/content/1207302> (accessed 24 March 2018).
Thai-style Democracy and Thoughts on the Constitution. 1965. Phranakhon: Chokchai Books.
Thawinwadi Burikun. 2003. "Thammaphiban: A New Principle for the Administration of Public Affairs". *King Prajadhipok's Institute Journal*, Vol. 1, No. 2 (May–August).
"We Fight" Song. 2000. August <www.baanjomyut.com/library_4/music_thailand/345.html> (accessed 24 March 2018).
Worachet Pakeerut. 2010. *Principles of Nithirat and Principles of Nithitham*. <www.pub-law.net/publaw/view.aspx?id=1431> (accessed 30 September 2012).

20

MONEY AND POLITICS IN BUDDHIST SANGHA IN MODERN THAILAND

Sara E. Michaels and Justin Thomas McDaniel

In June 2013, photographs and YouTube clips of a Thai Buddhist monk wearing designer sunglasses, carrying a Louis Vuitton bag, and flying with another monk on a private jet were circulated on the Internet. The news embarrassed Buddhist devotees and was derided by non-Buddhists in chat rooms and Facebook feeds, causing the monk to come under fire on social media. Soon after, a photograph of him lying next to a naked woman began to circulate and he was accused of fathering two children. Thailand's Anti-Money Laundering Office ordered the monk's bank accounts to be frozen as they discovered daily transactions of 200 million baht (US$6.25 million). The monk, Luang Pu Nen Kham (lay name: Wirapol Sukpol), had a humble background from a poor family in rural northeast Thailand; however, he became known as the "Gucci Monk" and upon investigation it was found that he was part of a large embezzlement scheme that netted him and a few close associates almost US$32 million over five years, a private jet, 30 Mercedes-Benz vehicles and lavish vacations. He received donations through his practice of magic and fortune telling. While most of his funds were seized by the Thai government in a raid, he managed to escape to France, where he ran a short-lived meditation course, and then to California where he lived lavishly. After three years, he was arrested on suspicion of sexually assaulting a 15-year-old girl, and he was extradited to Thailand in 2016 in a joint effort between the Department of Special Investigation (DSI) in Thailand and the FBI in the United States.[1] This scandal occurred in an interim period when the Thai Sangha (the umbrella organisation of all nuns and monks) did not have a supreme patriarch, Sangharat, who leads the Sangha Council (Mahatheresamakhom) and when the National Office of Buddhism was itself under suspicions of financial impropriety.[2] However, while this lack of oversight and leadership in the Thai Buddhist Sangha may have played a small part in Luang Pu Nen Kham's excessive corruption, controversies like this, while still relatively rare, have been a part of Thai Buddhism long before and after the Gucci Monk. Most interestingly, the case's chief investigator of the DSI has depicted this scandal, saying, "Over the years, there have been several cases of men who abused the robe, but never has a monk been implicated in so many crimes. We have never seen a case this widespread, where a monk has caused so much damage to so many people and to Thai society", thereby admitting that Luang Pu Nen Kham's graft was not an isolated incident. Critics claim that the monk is an example of the Buddhism crisis. Specifically, Buddhism had become "marginalised by a shortage of monks and an increasingly secular society".[3] Similar to

Nen Kham, in the last three years, over 300 monks in Thailand were reprimanded by Sangha leadership and some were disrobed for violating their vows.

This short overview of the modern Thai Sangha examines the relationship between Buddhism and politics in modern Thailand.[4] Foreign and domestic scholars and social critics of the Buddhist Sangha in Thailand often depict it as a tool of the state and/or royal family. This outcry has become particularly acute in recent years with the public commentaries and books on the lèse-majesté law in Thailand, the royal family's role in the coups of 1973, 1977, 1992, and 2006, the recent Sangha administrative act, the intense propaganda campaign by the Thai Privy Council for the 80th birthday of King Bhumibol Adulyadej (Rama IX), the recent heated debates over abuses of the lèse-majesté law, and the financial corruption of several large monasteries. However, the commentary often treats Thai people as a herd of uninformed, blind worshippers and monks as either political opportunists or pawns. In reality, it is well-known that every day Thai people trade in rumors about the royal family, openly share risqué photographs, protest government policies, and question both the royal family's and the government's role in development, politics, and social welfare. Perhaps the self-congratulatory attitude scholars have that they are the only people who question the wisdom of kings and politicians needs to be reconsidered.

While there are certainly serious challenges facing the Sangha, it is not in a constant state of crisis, and reports of its demise have been overstated. Many Thai nuns, monks, and lay Buddhists are deeply invested in trying to maintain and improve the health of the institution. They continue to support their local monastery and look to nuns and monks as teachers, mentors, and protectors. However, salacious and popular stories of sexual and financial scandal overshadow news of the Buddhist publishing industry, the launch of new monasteries, the large number of monastic fairs and festivals, monks and nuns working to help environmental causes, the expansion of Buddhist charitable work at home and abroad, the support of orphanages, hospices, meditation and rehabilitation centres, the active presence of many Buddhist teachers online and on television, and the expansion of new monastic schools.[5]

The "demise" and "crisis" industry in the study of Thai Buddhism

For as long as there has been something known as Thai Buddhist studies, there has been regular reports on the crisis of Thai Buddhism ranging from historians and sociologists to royal and governmental reformers to social critics and public scholars. Saying that Buddhism is declining in Thailand has been a constant refrain in local history. This makes perfect sense: religious leaders usually state that religion is in crisis in order to justify more donations, building projects, new schools, and encourage people to be "better" followers. This is common rhetoric in all religions.

In the 19th century, the Siamese kings, Mongkut (Rama IV) and Chulalongkorn (Rama V), made great efforts to formalise the Buddhist ecclesiastical system and educational practices in Siam and in their spheres of influence (or vassal states) in the north, northeast, and south. This was part of the nation-building and social control process to suppress regionalism, strengthen the country against foreign missionary influence, formalise the curriculum, and modernise the entire education system. Siamese ecclesiastical ranks, textbooks printed in Siamese script, monastic examinations, the Pali Buddhist canon, and teachers approved from Bangkok and Central Siam were disseminated to the rural and urban areas in Siam and its holdings. Monks from the recently pacified north, northeast and south were brought to Bangkok to study in two new monastic universities (Mahachulalongkorn and Mahamakut). Localised forms of expression, language, curricula, and script were considered irrelevant to this formalisation and centralisation. One of the most significant features of Buddhism in modern Thailand is its well-organised and centralised institutional structure. Since Siam (later Thailand) is the only country in Southeast Asia that was never officially colonised, the nation-building project, in which religious reform played a major part, could be considered a success. Although the Buddhist ecclesia has grown in wealth and numbers since 1902, there are deep fissures in Thai Buddhism that existed before 1902 and still persist today.

Royal reform of Buddhism is not particularly modern. Consistently from the earliest 13th-century records to 1902, Siamese kings and high ranking monks had seen it as their duty to collect and edit Buddhist texts, rewrite Buddhist history, purge the community of monks (Sangha) of corrupt persons, and rein in renegade, independent-minded practitioners. By 1902, these techniques had become more efficient and widespread. In 1902, King Chulalongkorn and Prince Wachirayan, who was an ordained monk and who had become the supreme patriarch of the entire Thai Buddhist Sangha, appropriated the role of the Sangha to educate the Thai people and regulated the organisation of monastic education. Along with a half-brother, Prince Damrong (the minister of the interior), King Chulalongkorn and Prince Wachirayan released the Act on the Administration of the Sangha ("Acts of the Administration of the Buddhist Order of Sangha of Thailand: 2445 [1902], 2484 [1941], 2505 [1962]").[6] Before this Sangha Act, monastic education and administration in Thailand was neither formal nor centralised.[7] It depended largely on the aims of the monks of each monastery. The Sangha Act was designed to make those residing in a monastery a "service to the nation" and to deflect criticism from European missionaries who denounced the poor and idiosyncratic state of Thai Buddhist education and organisation.

Still in effect today, the details of the Sangha Act represented largely administrative rules dividing the Buddhist ecclesia into formal ranks, assigning national, provincial and district heads of the Sangha.[8] Each of the regions (north, south, central, and northeast) has a formal hierarchy of monks and they all report to the *Mahathera Samakom* (Council of Elder Teachers) headed by the supreme patriarch. Individual monasteries are still run by abbots (*chao awat*) and deputy abbots (*rong chao awat*), but after 1902 they had to report regularly to their district and regional heads. All monks had to be registered with a particular monastery and were issued identification numbers and cards. Prince Wachirayan commented on the act, saying, "Although monks are already subject to the ancient law contained in the Vinaya [Buddhist Book of Precepts], they must also subject themselves to the authority which derives from the specific and general law of the State".[9]

In 1902, around 80,000 monks became subject to the law of the royal government of Siam. They controlled the admission to monkhood, the right to ordain, the size and status of monastic ground, and the ranking of monks. There was certainly sporadic resistance in the form of renegade monks in the north, like Krupa Siwichai, and rebellions of holy men in the northeast until 1924. The state-centred and sponsored reform movements of King Mongkut, Prince Wachirayan, King Chulalongkorn, and Prince Damrong among others portrayed Thai Buddhism as overly corrupted by those claiming magical and fortune-telling powers and, thus, in need of renewal. Prince Wachirayan in particular believed that there was an ideal past when Buddhist practice did not involve protective magic, when all monks studied Pali for many years, and when the Buddhist ecclesia and benevolent Buddhist kings worked together to care for the common people. His reform efforts had mixed results.

It is not simply foreign critics of Thai Buddhism who present Thai Buddhists as victims, but modern Thai liberal social, internationally connected commentators (often writing in English books and newspapers) do as well. Describing Thai Buddhism and culture as being in crisis is also an effective political tool promoted by the Royal Privy Council and politicians from multiple Thai parties on the right and left. Blaming global forces, the World Bank, scantily clad tourists, and Muslims for the demise of Thai Buddhism has been de rigueur in Thai public discourse for decades. Thailand, according to most statistics, is between 92 and 96 per cent Buddhist. This does not mean that everyone is following the same basic tradition with a few minor idiosyncrasies. Of course, the Sangha as a body of monks is often presents itself as possessing an ideology of sameness and employs the technologies of sameness (to the casual observer they have relatively standard sets of robes, bowls, shaven heads, rules, and the like). However, it is a misconception to assume more broadly that local diversity is being homogenised in the modern period with the rise of mass media, language standardisation, and organisational efficiency. In fact, the great rise in the followings of Mae Nak, Ganesha, Somdet To, King Chulalongkorn, Maechi Sansanee, Jatukham Ramathep, Kuan Im, Luang Phu Man, Khruba Siwichai, Luang Pho Doem, Buddhadasa Bhikkhu, and many other saints and deities over the last century shows that modern communication, transportation technology, and the mechanical reproduction of visual images has led to more diversity in Thai religiosity. From manuscript, archaeological, and epigraphic evidence, it is clear that there has always been diversity. Models of religious decline and the corruption of orthodoxy do not match the facts. However, whereas in the 19th century, the rituals and images connected to a community saint like Luang Pho Si or Shiva in a single village or valley would have remained local, now millions throughout the country can hear the miracle stories of a particular saint and see the photographs of his or her amulets and relics. They can order these objects through eBay or over the phone with numbers provided in mass-produced amulet magazines. They can board a relatively inexpensive bus to another village, provincial capital, or Bangkok to visit a saint's home monastery or a ghost's main shrine. The corruption of a monk like Luang Pu Nen Kham is not necessary more common nowadays, but the Internet allows it to be more widely known today.

The political managing of crisis in Thai Buddhism

A visit to the historical past is imperative. Siam changed its name to Thailand in 1939, after changing from an absolute monarchy to a constitutional monarchy in 1932. Since 1932, the state has sought to gain more direct control over the Sangha than even the absolute monarchy. In 1934 an independent government committee was established to examine Sangha finances and, in 1941, the new Sangha Act effectively gave the government control over the internal Sangha organisation and executive offices. The various military and elected governments over the past 60 years have generally asserted that the Sangha should play a more "practical" role in modern Thai society and become more "socially engaged". This social engagement advocated by many Thammayut monks, the government, Christian missionaries and Western critics of the Sangha led the two major monastic universities to institute programmes in which urban educated monks would provide social services for the poor, especially in the north, northeast, and south. The role of the monastery in health and education was seen as part of the original vision of the Buddha, even though there were no monasteries at the time of the Buddha and little textual evidence showing that ancient nuns and monks were encouraged to be socially engaged.

In 1966, Mahachulalongkorn Monastic University stated that their students should provide "voluntary social services, both material and spiritual, for the welfare of the community, such as giving advice and help in the event of family problems and misfortunes".[10] Monks from these universities were assigned to be teachers in rural areas, volunteers in hospitals, and assistants in community economic development. They also established numerous new monastic schools designed to prepare rural students for future training at the monastic universities in the city, and they disseminated textbooks printed for urban monastic students. The universities provided training courses to prepare their "volunteer" students. The monks who were sent out also became part of the nation-building process to help incorporate the rural areas in the north, northeast, and south. Most monks were sent to the northeast, the place of most communist activity and ethnic Lao residents, causing the movement to become anti-communist. A report of their progress in 1969 confirms their role in combating communist activity that they believed "caused the people great suffering".[11] These monks were working side by side with anti-communist government policies and the US military, which had thousands of troops in the region. Monks were also involved in helping the government pacify and incorporate hill tribes and Muslim populations in the north and south of the country.

One well-known monk who led the anti-red brigade was Phra Kittivuddho. He resided at the famous Wat Paknam in Thonburi across the Chao Phraya River from Bangkok and later moved to teach at the Monastic University on the grounds of Wat Mahathat. He argued for greater social engagement of monks and novices. He also called for an aggressive conversion campaign to increase the moral coffers of the country. He believed superstitious practices of protective magic by village *mo wiset* or *mo du* (magicians), and the millenarian movements that predicted the coming of the future Buddha were being influenced by communist propaganda. He was also suspicious of the allegiance of the ethnic Lao associated with these practices and their possible connections to the communists in Laos proper.

Due to increasing student activism in the early 1970s, partly in response to the war in Vietnam and the military dictatorship in Thailand, Phra Kittivuddho believed that communism would destroy the country and ban true Buddhist practice. Therefore, he believed that the Sangha had to defend Thailand from foreign invaders and the threat of communism. In one often-quoted sermon, he claimed that killing communists was not demeritorious because they were "enemies" of Buddhism and the Thai state.[12] Furthermore, communists were "bestial" and not fully human. He even called for monastic participation in protests against left-leaning student groups and to disrobe monks suspected of harboring communist sympathies. Many monks, including the supreme patriarch, believed that Kittivuddho was becoming too involved in politics and social reform, but no disciplinary action was taken against him.[13]

Phra Kittivudho, although radical in his views, was not the only monk who viewed Thailand, capitalism, and Buddhism as related entities. Luang Ta Mahabua, a well-known student of Phra Ajarn Man believed that Buddhism could be used to support the national market economy. While not anti-communist (which of course involves much more than rejecting a particular economic system), Mahabua asserted that after the Asian Financial Crisis of 1997, the subsequent devaluing of the Thai baht, and the massive recession, Buddhists should donate money to the government. He set up his monastery in Udon Thani province in the rural northeast as a broker for the donations The campaign, whose moniker translates as "Thais Help Thais", collected over US$4.3 million and 1,457 kg of gold in donations that were given to the Ministry of Finance. Mahabua believed that the money wasted on protective amulets (*phra kreuang*) and fortune-tellers could be put to better use helping the nation.

After the money was collected and handed over to the ministry, Mahabua inquired how it was serving the country and was not satisfied by the answers he received. He started referring to the ministry as "ever-hungry ghouls seeking to eat people's guts" and "idiots who think of nothing but cheating on the country". The minister of finance, Tarrin Nimmanahaeminda, was seen as the thief of the money. Mahabua's followers collected over 50,000 signatures in a drive to impeach Tarrin. Many believe that enemies of the Democrat Party, of which Tarrin is a well-known member, fanned the flames of controversy in order to bolster their own party – the "Thai Rak Thai" (Thais Love Thais) Party. Despite the possible trickery and alleged corruption of the Ministry of Finance, Mahabua was widely criticised for being becoming over-involved in the world of political corruption. The problem with the money is seen as a problem of his own creation. Monks, many believe, should "be above" these worldly (*lokiya*) matters. In general, the criticism of Kittivuddho and Mahabua reflects a common, modern reaction of Buddhist laypeople and monks to "social engagement". This social engagement is often seen as "Western", Bangkok-centric, elitist, and potentially destructive to the supramundane status of monkhood. Mahabua, being a part of a lineage of forest monks, was seen as sullying the purity of the Sangha.

Money, corruption, and excess have dominated international news about Thai Buddhism in the 25 years since the Mahabua controversy. However, this is not a new phenomenon and the Gucci Monk will not be the last. Besides the Gucci Monk, there have been a series of other scandals. For example, Phra Buddha Issara is an activist monk who until recently was the abbot of Wat Or Noi in Nakhon Pathom (a suburb of Bangkok). He controversially supported the military coup by now Prime Minister Prayuth Chan-ocha and has led protests against financial corruption at Wat Dhammakaya and what he sees as the decline of Buddhism in modern society. However, despite his frequent television appearances and his accusations against corrupt monks, he was recently arrested and disrobed for his own financial crimes and offenses to the royal family. The former Phra Buddha Issara was taken into custody despite his connection to Prayuth. The former monk was charged with using royal initials stamped onto amulets he made without permission from the royal family. He was also accused of robbery and running an illegal secret society for political and financial purposes. The charge relates to an incident in 2014 when Phra Buddha Issara led an operation that saw the shutdown of Chaeng Wattana Road as part of the protests by the People's Democratic Reform Committee (PDRC). During the protests, he allegedly mobilised protesters to attack special-branch policemen, who were seriously injured following the assault. Following news of the allegations, heavily armed policemen arrested the former monk. Shortly after, a video clip of his arrest circulated around social media, probing citizens to question why there was excessive force against a "man of religion". The Thai prime minister, who is accused of having ties to the Issara, made a rare apology for the rough handling of the monk. However, when old pictures, which suggest the pair were close, began to be widely spread on social media, the prime minister denied he had ties to the former monk.

The crackdown on a monk supposedly protected by the prime minister can be seen as part of a larger wave of recent investigations into corrupt, largely urban monasteries. The police force's Counter Corruption Division (CCD) has increased its investigations into temples accused of wrongdoing, shining light on the problem of widespread corruption in large monasteries in Thailand. News of the CCD's investigation comes after five senior monks were accused of temple fund embezzlement and money laundering. At Wat Sa Ket, three assistant abbots were arrested while abbot Phra Phrom Sitthi could not be found; at Wat Sam Phraya, abbot Phra Phrom Dilok and his secretary were taken into custody. Meanwhile, similar raids were occurring at Wat Samphanthawong and Wat Or Noi, where Phra Buddha Issara was arrested.

Following news of the scandals, activists worry the raids will damage people's belief in Buddhism and are demanding the "Sangha council [to] promptly issue regulations". However, in reality, the Buddhist faith has not been swayed by the allegations, and followers continue making merit and associating with other monks. In one example, a visitor of Wat Sa Ket explained, "I put my faith in the religion, not the temple or any individual monk". Interestingly, others are professing their loyalty to former monks accused of wrongdoing. One woman defended abbot Phra Phrom Sitthi, saying, "I feel connected to him. He is very compassionate. I have faith in him". The widespread admiration has caused police to threaten legal action against anyone who provides shelter for rogue monks.[14]

The Dhammakaya movement

The largest political scandal that has faced Thai Buddhism in the past 25 years involves one of the fastest growing religious movements in the world: the Dhammakaya. Probably the most politically, economically, educationally, and socially "engaged" of monks in the modern period has been Phra Dhammajayo. Dhammajayo is the head of the well-known Dhammakaya movement in Thailand. Born as Chaiboon Sutipol in 1944, he was heavily influenced by the teachings of the nun, Khun Yai, who herself was the primary student of the founder of the Dhammakaya movement – Luang Po Sod (1906–1959). Luang Po Sod (or Mongkolthepamuni/Sod Mikaeonoi) was the abbot of Wat Pak Nam, a large and well-stocked Bangkok monastery. He started instructing his students, both lay and ordained, in a new type of meditation which he called "Dhammakaya". He believed that if meditators concentrated with the proper intention and repeated the mantra "*samma arahan*" (right attainment), they would see in their mind's eye a luminescent Buddha figure growing in their stomach. This figure could grow up to 20 cm tall and could be seen by anyone in only 20–30 minutes of practice.

Although criticised by some who see Dhammakaya meditation as psychological manipulation or implantation, the movement under Luang Po Sod and Khun Yai grew quickly, especially after Dhammajayo started encouraging college students in Bangkok to attend classes with Khun Yai in 1969. The next year a wealthy laywoman donated a large tract of land in the suburbs and funds to build a meditation centre. Dhammajayo glorified Khun Yai with a large photo above the front door and started to arrange mass ordinations for college graduates. He appealed to college students and young professionals by stating that there was no need for elaborate Buddhist ritual, protective magic, fortune telling, and the tools, images, and money that went towards them. He also started university Dhammakaya study groups for well-educated nuns, monks, and laypeople all over the country and obtained support from Princess Sirinthorn and the National Sangha Council to build the Dhammakaya monastery.[15] By the early 1990s the monastery had grown significantly. During those years, one encountered colossal billboards for the Dhammakaya movement on the northern tollway coming out of Bangkok advertising its "convenient" (*saduak*) weekend retreats for working adults and students and the talks sponsored by the movement held at Chulalongkorn University.

Up until today, these events have grown even larger despite charges that the movement oversimplifies spiritual attainment in Buddhism. At Dhammakaya retreats, which were attended by up to 10,000 people a week until controversies started piling up in 2015–2016, participants are asked to meditate for two 30-minute sessions and develop self-confidence to help them succeed in family and business matters. They are offered a meditation "kit" for sale, including a set of white robes, a plastic umbrella, a bottle of water, and a small pamphlet. Dhammakaya adherents emphasise that magic, amulets, and even ordination are arbitrary for the path to personal well-being and financial and spiritual security. Thais are told of their need to return to the original body of the Buddha and cast off the accouterments that have polluted his or her body and message over the centuries. The iridescent Dhammakaya body growing in a person's belly was the true body of the Buddha. This message has attracted the growing well-educated and entrepreneurial Thai middle class who lack the time to ordain, study large tomes of Buddhist scripture, or mediate for several hours a day in the forest. Dhammakaya also enjoined the sponsorship of influential politicians like General Chavalit Yongchaiyudh and major financial institutions like the Siam Commercial Bank and the Bangkok Bank. By some estimates Dhammakaya coffers have grown to over US$50 million. By supporting the state and capitalist ideals, the Dhammakaya movement has been very successful thus far.

While still very popular and powerful, Dhammajayo's fortunes started to change in 2015 when news came to light that between 2009–2011 over 11 billion baht (US$30 million) was stolen slowly from the Klongchan Credit Union and donated in a series of cashier's checks to Wat Dhammakaya and many personally to Dhammajayo.[16] While his followers acknowledge the donations and have raised money to replace the funds and reimburse the credit union, they deny that Dhammjayo or any monastic official knew of the illegal source of the donations. This suspicion of extreme corruption combined with the fact that the two deposed and disgraced former prime ministers, Thaksin and his sister Yingluck Shinawatra, and were open supporters of Dhammajayo, led the new military government under General Prayuth to invoke Article 44 of the interim constitution to declare Wat Dhammakaya a "special control zone". Under this new power, between 16 and 18 February 2017, they raided the temple with hundreds of armed police, drones, and armored vehicles searching for Dhammajayo, who refused to report to the police to face charges. In a bizarre incident that could be a scene from a film, the police pulled back the covers of his bed to discover pillows arranged to look like sleeping person; they never found him after long searches of the compound and massive protests by thousands of Dhammajayo's followers, who vehemently defended him on social media and in person, claiming that they would defend him "in this life and the next" (*chat nii lae chat naa*). Despite not being able to arrest the 74-year-old monk, the power of the Dhammakaya movement was seriously hurt. One of the leading voices calling for Dhammajayo's arrest was ironically Phra Issara, who was later arrested and disrobed for financial corruption himself despite being a vocal supporter of military dictator Prayuth.[17]

The case of the missing supreme patriarch

In July 2017, after a four-year hiatus, the new king, Maha Vajiralongkorn, chose Somdet Phra Maha Muniwong, abbot of Wat Ratchabophit, to be the 20th supreme patriarch (Sangharat) of Thailand. This was no simple decision and it was no accident that the post was left empty for four years. Somdet Muniwong, an 89-year-old member of the Thammayut order (the order traditionally supported by the Chakri dynasty despite its small size of less than 5 per cent of all Thai monks), was the safe choice after controversies swirling around the most senior monk (and interim patriarch) in the country: Somdet Chuang (Somdet Phra Maha Ratchamangalacharn). Thailand's last Buddhist patriarch, Somdet Phra Nyanasamvara, died in 2013 when he was 100 years old. He had been largely bedridden and without the power of speech for the last several years of his reign, so there was already a vacuum of leadership. The new king was permitted to choose the patriarch directly after a law was changed that allowed him to bypass the Mahatherasamakhom (the executive monk's council, comprising the most senior monks in the country). The Mahatherasamakhom had chosen Somdet Chuang, but General (and now Prime Minister) Prayuth denied him the power to take the position because he had been under investigation since 2013 for a tax scam involving an antique 1953 Mercedes-Benz (worth US$250,000 and closely resembling the one the previous king, Bhumibol, drove when he was first crowned). However, his denial was most likely related to his previous support for deposed Prime Minister Thaksin and his association with Dhammajayo and Wat Dhammakaya. It seems that mere association with the disgraced Dhammajayo proved to be dangerous, even to an elderly monk. Again, Phra Issara was involved in this. Working with Prayuth and Paiboon Nititawan (a former senator and coup leader), he sparked protests against Somdet Chuang and brought him up on the tax evasion charges.[18]

The Bhikkhuni controversy and the politics of being a Thai Buddhist woman

One of the longest political and social debates in modern Thai Buddhism is the case of the role of the Bhikkhuni. Unlike many other Buddhist countries, there are no official "fully ordained" (although that is a controversial term and there is no exact Thai language equivalence for it) nuns in Thailand that could be seen as equal in ritual and social status to the role of Bhikkhu (monk) for men.[19] One major Thai voice that has protested the industry is Chatsumarn Kabilsingh, a foreign-educated and former associate professor of philosophy at Thailand's elite Thammasat University, who became the first fully ordained *bhikkhuni* (nun) in Thailand. As of 2018, there are 270 *bhikkhunis*, with the numbers still rising, thanks in many ways to the work of Kabilsingh.[20] Kabilsingh believes that there is a general lack of female voices of authority in the Thai Sangha. Her protest is part of a larger project to open up the opportunity for women in Thailand to be fully ordained nuns (*bhikkhuni*) and to bring a Franco-American type of women's rights movement to Thailand.[21] She founded a branch of the Sakyadhita Foundation for women in Thailand. She also helps promote the ordination of women as nuns, the training of *mae chee* (women who lead a religious life, take the first 8–10 precepts, shave their head, wear white robes, and live monastic lives) at the Dhammajarinee Wittaya (the first Buddhist school especially for women in Thailand), and calls for an end to the prostitution industry in Thailand. Kabilsingh asserts that if poor women had greater opportunities for education and a legitimate place in the Sangha, then prostitution would not be their only option to escape poverty. Furthermore, a greater number of female voices in the Sangha may inspire women to turn from prostitution to religion. Poverty and socio-cultural factors, such as Thai notions of "the ideal male as conqueror" and the great value placed on beauty, are very difficult to change, but if women had a religious voice and space then these factors may be less oppressive.[22] She is also against the production of "love potions" by Buddhist magicians and the perpetuation of poverty by offering women false hopes of financial success through the lottery, promoting a belief in spirits that give magical gifts of money, for example. Since they cannot become nuns, women are reduced to being slaves to male magicians and false beliefs. The official reason the Thai Sangha has not re-instituted the full ordination of female renunciates from the time of the Buddha is because the lineage of Theravada nuns literally died out in Sri Lanka around the 13th century. Since the Buddha himself ordained the first nun, Mahapajapati, any new lineage of nuns would thus not be initiated by the Buddha and, therefore, of questionable validity.

Kabilsingh instead believes that the reason the Thai all-male Sangha has resisted the full ordination of women is because it would threaten their own financial support. It is clear from a visit to any Thai monastery that women are the primary lay supporters. Renewing the lineage of nuns could provide counseling for prostitutes, offer an alternative career path, and institutionalise free women's education at monasteries. Currently as it stands in Thailand, women are barred from becoming monks and are instead allowed to become white-cloaked nuns. The role of women in Buddhism is heavily debated. The Sangha has argued to ban Sri Lankan clergy from "coming into the country following an ordination". *Bhikkhunis* explain that instead of crushing women's aspirations, the clergy should focus on cleaning up Buddhism in Thailand to restore declining public faith.[23] Over recent years, *bhikkhunis* have been growing in numbers and gaining public support even though the Sangha Council openly rejected their plea for equality. A recent article in *Asia Review* highlights how the women are "becoming agents for reform of a religious establishment plagued by corruption and lurid scandals".[24] Women have been able to gain support from people such as local villagers, who explain that the women monks behave more properly than men. Though there has been support, *bhikkhunis* are sometimes targeted. For example, *bhikkhunis* in Rayong were forced out of their temple when they were the target of an arson attack. They were previously taunted by death threats.[25] Dennis Grey of *Asia Review* shed more light on the scandal by focusing on the Sangha's reaction to the rise of *bhikkhunis*. Grey highlights how in 1962, a Sangha act excluded "female monastics from healthcare coverage, public funding for monasteries and other benefits enjoyed by monks". In more recent times, the Sangha barred nearly 100 female clergy from paying respects to the body of late King Bhumibol, stating that it was illegal for women to wear ochre robes.[26]

Kabilsingh and many others argue that new lineage of nuns would also allow Thai Buddhism to return to the past where, she believes, women were treated soteriologically and institutionally as equals. She justifies this ideal past by drawing on canonical Pali texts in her research. She asserts that the concept of prostitution and the status of women "in the Buddha's time differs radically" from today.[27] While she does acknowledge that there was still some androcentrism and misogyny in the times of the Buddha, women had more choices (although from archaeological and epigraphical evidence in India, the opposite seems to be true). Citing the work of Sulak Sivaraksa, she sees modernity as removing those choices. Sulak himself claims that *bhikkhunis* have not yet been accepted in Thailand because of the monks' "self interest". He depicts that the monks fear *bhikkhunis* would challenge the monkhood because they follow the rules more closely than the current monkhood, which has been plagued by sex scandals.[28]

Kabilsingh has been successful in helping establish a school for *mae chee* in Thailand and publishing feminist Buddhist texts in Thai and English. She was also ordained as a *bhikkhuni* herself on 28 February 2003 in Sri Lanka, but her ordination is not officially recognised (or particularly protested) in Thailand. Kabilsingh's ordained name is Bhikkhuni Dhammananda, and she resides at Wat Songdhamma Kalayani, which was founded in 1971 when her mother Voramai became the first Thai woman to be ordained as a nun in the Mahayana tradition of Taiwan. Nevertheless, the argument in favour of female ordination in Thailand and the condemnation of the prostitution industry have largely fallen on deaf ears, as the Thai Sangha and scholar-activists (besides Suwanna Satha-anand) have generally avoided these subjects. Moreover, the ban has been largely avoided in newspapers and academic studies in Thailand and in 2014, the Sangha Supreme Council's decision reiterated the ban on the ordination of female monks. Campaigner, Sutada Mekrungruengkul challenged the ban and he was supported by a Constitution Drafting Committee member, who said the order was "contrary to rights and civil liberties recognised by the chapter, [and] also against Lord Buddha's permission for females to be ordained".[29]

Kabilsingh, while perhaps the leading and most vocal political voice for the Bhikkhuni movement, she is not the only voice. Maechee Sansanee Sathirasut, for example, is a wealthy Bangkok woman who recently began her own monastery and meditation centre in northern Bangkok. This centre hosts its own television programme and discusses ways in which meditation, physical exercise, a healthy cuisine, and active study of Buddhist texts can help reduce stress and lead to a productive and altruistic life (personal communication). She does not claim that this is the way ancient Buddhists lived, that she is the practitioner of the true Thai Buddhism, or that she feels victimised by the male Thai Sangha. Women and men practice together at the centre, and she maintains the medium between social activist and ascetic/scholarly recluse, as most monks and nuns in Thailand do. In fact, nuns and monks rarely choose to be completely isolated or completely socially active.

Modern scholarship on women and Thai Buddhism often ignores the important role of *mae chee* in Thai society and instead focuses on the issue of opportunity and equality which reveals an assumption that Franco-American ideals of human rights, social equality, and the arbitrariness of gender differences are somehow universal. *Mae chee* like Sansanee Sathirasut, Pairor Thipayathasana, and Jutipa Tapasut are recognised scholars and teachers working in major Thai monasteries. Thai universities, especially departments of philosophy, literature and religious studies, generally have a majority of female teachers and students. Meditation centres and lay Pali and textual study classes at urban and rural monasteries are also mostly populated by women. Women play an often dominant role in modern Thai lay Buddhist scholarship and practice. We also saw the central role women like Khun Yai have played in changes in modern Thai Buddhist practice. Although an important movement, concentrating on the fact that Thai women have never been fully ordained *bhikkhunis* ignores women's contributions and vilifies the role of the *mae chee*.

Conclusion

So is the Thai Buddhist Sangha a corrupt institution that has to be completely reformed and reimagined? No. The Thai Sangha is undoubtedly facing challenges and the news of financial corruption, sexual misconduct (if space were unlimited, this chapter could have discussed several modern sexual scandals involving monks like Phra Yanta Amaro and others), and political involvement. However, this is not particularly new and moreover, the "Sangha" is not equal to Thai Buddhism and there are many ways to define the Sangha. Sangha in Pali simply means community and historically was often seen as "fourfold" (*bhikkhuni*, or nuns; *bhikkhu*, or monks; *upasika*, or dedicated laywomen; and *upasaka*, or dedicated laymen). There were no provisions in early Buddhism or in the Vinaya/Patimokkha monastic code of conduct for nationally, politically, or royally sanctioned institutions like the Thai Sangha and the Department of Religion in the Thai Ministry of Culture, or the Office of National Buddhism in Thailand. Thai Buddhist practice has thrived without the need for national organisations such as the Sangha and most education, ritual, meditation, religious artistic expression, and practice takes place without the need of the national Sangha and without much oversight.[30] Moreover, most studies of Buddhism and politics are of large, wealthy urban monasteries in Bangkok and the thousands of largely non-corrupt and non-political Thai monasteries and monks are largely left out of these studies. Thai Buddhists are often depicted as "impacted" by globalisation or "rampant" modernisation. It sees the "traditional Buddhist worldview" challenged by technology and the global market with "dramatic consequences". This chapter warns that this type of approach, which has been taken by most Buddhist Studies scholars over the past century, establishes a dichotomy of victim/victimiser among the Thai Buddhist community. It suggests that Thai Buddhism was a static entity that existed in a pristine state before modernisation, commercialisation, and politics assaulted it. This chapter instead sees Thai Buddhists as dynamic arbiters and sponsors of ideology and innovation on the global stage. Thai Buddhists are not simply the supine receivers of modernisation who choose to profit from it or be overrun. There are many who have developed unique and new practices outside of negotiations with non-Thai Buddhist groups and non-Buddhists. Many who are both supportive and resistant to the state's role in religion negotiate with, rather than blindly embrace or passively ignore, the modern age. These figures dynamically respond and adjust according to the times rather than becoming either the victimisers or victims of modernity.

Notes

1 There are dozens of stories in English and Thai on this controversy. Here are a few that summarise the case: <www.bangkokpost.com/learning/advanced/1041794/fugitive-monk-nen-kham-arrested-in-us>; <www.sandiegouniontribune.com/sdut-scandal-of-the-jet-setting-monk-rivets-thailand-2013jul18-story.html>; <www.nydailynews.com/news/world/thailand-hunts-fugitive-jet-setting-monk-article-1.1402055>; <www.bangkokpost.com/learning/learning-news/712928/jet-setting-monk-gets-new-life-in-us>.

2 <www.bbc.com/news/world-asia-40678511>.

3 <www.telegraph.co.uk/news/worldnews/asia/thailand/10124885/Thai-Buddhist-monks-criticised-over-private-jet.html>. See also a criticism of foreign coverage in Thai <www.posttoday.com/world/235339>.

4 This chapter largely looks at politics and the Sangha over the past twenty years. For studies before that, see especially Peter Jackson, "Withering Centre, Flourishing Margins: Buddhism's Changing Political Roles", in *Political Change in Thailand: Democracy and Participation*, edited by Kevin Hewison (London: Routledge. 1997), pp. 75–93. While this is a good summary for the Thai case, the numbers of studies about "Buddhism and the State" published in Southeast Asia is truly staggering. See a list of these studies and commentary in Justin Thomas McDaniel, *The Lovelorn Ghost and the Magical Monk* (New York: Columbia University Press, 2011). It seemed that scholars have said all that could be said on the topic.

However, Harris's (2007) edited collection of essays, *Buddhism, Power, and Political Order* (London: Routledge, 2007) proves there is considerably more research to be done on the topic. Most of the articles in this volume do not reify the dichotomy between the elite and the non-elite in Southeast Asia that has often dominated approaches to the subject.

5 I have discussed "development" monks and others working for social causes in "Buddhism in Modern Thailand", in an edited volume titled *Buddhism in World Cultures*, edited by Steven Berkwitz (Santa Barbara: ABC-CLIO, 2006), pp. 101–128. See also James Taylor, "'Thamma-chaat': Activist Monks and Competing Discourses of Nature and Nation in Northeastern Thailand", in *Seeing Forests for Trees: Environment and Environmentalism in Thailand*, edited by Philip Hirsch (Chiang Mai: Silkworm Books, 1997), pp. 37–50, and *Forest Monks and the Nation-State: An Anthropological and Historical Study in Northeastern Thailand* (Singapore: Institute of Southeast Asian Studies, 1999). See also the recent book by Sue Darlington, *The Ordination of a Tree: The Thai Buddhist Environmental Movement* (Albany: SUNY Press, 2012).

6 See specifically the act *Phraratchāban-yat Laksana Pokkrōng Khana Ratanakōsinsok*, No. 121, cited in Yoneo Ishii, *Sangha, State, and Society: Thai Buddhism in History*, translated by Peter Hawkes (Honolulu: University of Hawaii Press, 1986), p. 68.

7 See Wutichai Mulasilpa, *Kan Patirup Kanseuksa Nai Sami Phrapatsomdet Phracjulachomklaochao Yu Hua* (Bangkok: Thai Wattana Panich, 1996); and Craig Reynolds, *Buddhist Monkhood in Nineteenth Century Thailand*, PhD Dissertation, Cornell University, 1973.

8 Ishii, *Sangha, State, and Society*, p. 69.

9 Ibid., p. 70.

10 Ibid., p. 138.

11 Ibid., p. 139. See also Reynolds, *Buddhist Monkhood in Nineteenth Century Thailand.*

12 Charles Keyes, "Moral Authority of the Sangha and Modernity in Thailand: Sexual Scandals, Sectarian Dissent, and Political Resistance", in *Socially Engaged Buddhism for the New Millennium: Essays in Honour of the Venerable Phra Dhammapitaka (Bhikkhu P.A. Payutto) on his 60th Birthday Anniversary*, edited by Pipob Udomittipong (Bangkok: Sathirakoses-Nagapradipa Foundation; Berkeley: Parallax Press, 1999), p. 153.

13 Ibid., p. 152.

14 These stories of raids of monasteries and Phra Buddha Issara and Phra Dhammajayo's cases have been heavily covered in Thai language press; see particularly <www.matichon.co.th/prachachuen/news_470906>; <www.thairath.co.th/content/1299239>; <www.ejan.co/news/5b07ba4ca45af>; <www.khaosod.co.th/special-stories/news_221318>; <https://news.mthai.com/tag/พระธัมมชโย>; <www.matichon.co.th/prachachuen/news_470906>; English readers can see the many stories in the Bangkok Post and other news sources this past year. These were all accessed in May and June 2018. See particularly <www.reuters.com/article/us-thailand-buddhism/activist-monk-seeks-buddhism-overhaul-in-thailand-over-corruption-fears-idUSKBN0LX13Q20150301>; <www.bangkokpost.com/opinion/opinion/1477885/#cxrecs_s>; <www.nationmultimedia.com/detail/national/30346206>; <www.khaosodenglish.com/news/crimecourtscalamity/2017/02/24/tiger-temple-reopen-9-months-raid/>; <www.bangkokpost.com/news/general/440290/11-arrested-in-temple-casino-raid>; <www.bangkokpost.com/news/general/1467478/maid-admits-role-in-temple-fund-fraud>; <www.bangkokpost.com/news/general/1483081/#cxrecs_s>; <www.bangkokpost.com/news/crime/1473429/cops-urge-no-shelter-for-rogue-monks#cxrecs_s>; <www.bangkokpost.com/news/general/1482169/#cxrecs_s>; <www.bangkokpost.com/news/general/1482189#cxrecs_s>; <www.bangkokpost.com/news/crime/1471861/#cxrecs_s>; <www.bangkokpost.com/news/general/1480749/warrants-out-for-ex-monks-accomplices#cxrecs_s>; <www.bangkokpost.com/news/crime/1479277/asst-abbot-of-wat-sa-ket-faces-human-trafficking-charge#cxrecs_s>; <www.bangkokpost.com/news/crime/1478065/wat-samphanthawong-ex-assistant-abbot-arrested-in-germany#cxrecs_s>; <www.bangkokpost.com/news/crime/1479993/police-chief-returns-from-frankfurt-alone-after-ex-monk-claims-asylum#cxrecs_s>; <www.bangkokpost.com/news/crime/1480577/arrest-warrants-out-for-5-who-helped-fugitive-ex-monk-escape#cxrecs_s>; <www.bangkokpost.com/news/crime/1473929/police-to-probe-more-temples#cxrecs_s>; <www.bangkokpost.com/news/general/1473365/pm-apologises-for-rough-handling-of-phra-buddha-isara-arrest#cxrecs_s>; <www.bangkokpost.com/news/politics/1474461/#cxrecs_s; <www.bangkokpost.com/news/general/1472801/#cxrecs_s>.

15 Marja-Leena Heikkila-Horn, "Two Paths to Revivalism in Thai Buddhism: The Dhammakaya and Santi Asoke Movements" *Temenos*, Vol. 32 (1996), pp. 93–96.

16 <www.huffingtonpost.com/nicholas-liusuwan/prosecution-of-buddhist-m_b_10122666.html>. Dhammjayo's followers have argued that the temple received donations in the open public, and it would not be appropriate for temples to ask donors about the source of their donations. Therefore, since the donations were transparent, his followers argue there is no proof the temple knew the donations were embezzled.

17 <www.bangkokpost.com/learning/learning-news/749184/phra-dhammachayo-faces-theft-money-laundering-charges>. This short article by the *Bangkok Post* highlights the main facts of the embezzlement scandal and therefore avoids any bias. This article was written in 2015, when the government began to investigate Wat Dhammakaya, specifically Phra Dhammachayo. See <www.bangkokpost.com/learning/learning-news/749184/phra-dhammachayo-faces-theft-money-laundering-charges>; <www.straitstimes.com/asia/se-asia/thai-police-order-thousands-of-worshippers-out-of-scandal-hit-dhammakaya-temple-to>. See also <www.latimes.com/world/la-fg-thailand-monk-20170310-story.html> and <www.huffingtonpost.com/nicholas-liusuwan/prosecution-of-buddhist-m_b_10122666.html>. Also <www.reuters.com/article/us-myanmar-buddhism/scandal-hit-thai-temple-helps-to-stage-mass-buddhist-event-in-myanmar-idUSKBN1FA0LS>. In light of recent controversies in the Burmese Sangha related to the Rohingya refugee crisis, it is striking that the abbot received overwhelming support from Myanmar. The newspaper, Reuters, recounts how thousands of Buddhist monks gathered for an event hosted by the Dhammakaya temple. The event was hosted to help organise the mass almsgiving for 20,000 monks. The temple was able to use this event to their advantage after a year of terrible press; the Dhammakaya Foundation press release explained that the event, "aimed to tighten the relationship between both Myanmar and Thailand (and) unite the Theravada monkhood" <www.dhammakayauncovered.com/facts/2016/7/5/everything-you-need-to-know-about-the-dhammakaya-scandal-a-short-fact-sheet>. This piece was a fact sheet in which it tried to answer basic and more complex questions about the Wat Dhammakaya embezzlement scandal. What struck me about this article, and why I chose to include it here is its bias; and therefore, I was able to draw a connection between this article and the previous ones that discuss his devotees. The entire website is devoted to clearing the abbot's name; it talks about him almost as a god and says he is wrongly accused. Most notably, the article declares: "the Abbot is clearly innocent, yet DSI continues to pursue criminal charges against him in a manner that has been suspicious, unjust, and illegal".

18 <www.theguardian.com/world/2016/mar/29/top-thai-buddhist-monk-somdet-chuang-investi gated-over-vintage-mercedes-benz>; <www.channelnewsasia.com/news/asiapacific/the-politics-of-thailand-s-buddhist-supreme-patriarch-7598378>; <www.bangkokpost.com/learning/advanced/1194432/somdet-phra-maha-muniwong-new-supreme-patriarch>; <www.reuters.com/article/uk-thailand-buddhism/thailands-king-appoints-new-buddhist-patriarch-idUKKBN15M15A>. This was granted power to appoint the patriarch, which reverts back to a system used in 1992. The article mentions that the Sangha Supreme Council proposed their favoured candidate, Somdet Chuang, however the junta vetoed him, and instead the parliament granted the king the power to nominate his own candidate.

19 Steven Collins and Justin Thomas McDaniel wrote a long summary of the scholarly, social, and political approaches to this issue in "'Buddhist nuns' (mae chee) and the Teaching of Pali in Contemporary Thailand", *Modern Asian Studies*, Vol. 44, No. 6 (2010), pp. 1373–1408. This subject has been scrutinised by many scholars over the past 30 years. Some of the best studies are by Sanitsuda Ekachai, "Crusading for Nun's Rights", *Newsletter on International Buddhist Women's Activities*, Vol. 13, No. 2 (1997), pp. 16–19 (Originally published in the *Bangkok Post*, 4 September 1996); Monica Lindberg Falk, *Making Fields of Merit: Buddhist Female Ascetics and Gendered Orders in Thailand* (Copenhagen: NIAS Press, and Seattle: University of Washington Press, 2007); Tomomi Ito, "Buddhist Women in *Dhamma* Practice in Contemporary Thailand: Movements Regarding Their Status and World Renunciates", *Journal of Sophia Asian Studies*, Vol. 17 (1999), pp. 147–181; Marjorie Muecke, "Female Sexuality in Thai Discourses about *Mae Chee* [Lay Nuns]", *Culture, Health & Sexuality*, Vol. 6, No. 3 (2004), pp. 221–238. Martin Seeger has published extensively on this topic. See his most recent book: *Gender and the Path to Awakening: Hidden Histories of Nuns in Modern Thai Buddhism* (Chiang Mai: Silkworm Books/Copenhagen: NIAS Press, 2018).

20 <www.npr.org/2018/01/06/576197738/the-female-monks-of-thailand>.

21 <http://theweek.com/captured/721705/thailands-female-rebel-monks>.

22 Chatsumarn Kabilsingh, *Thai Women in Buddhism* (Berkeley: Parallax Press, 1991), p. 80.

23 <www.scmp.com/news/asia/southeast-asia/article/1857608/rise-bhikkhunis-how-thai lands-rebel-female-buddhist-monks>.

24 <https://asia.nikkei.com/Life-Arts/Life/Women-in-Ocher-Thailand-s-rebel-nuns-gain-ground>.
25 <www.lionsroar.com/thai-bhikkhuni-monastery-attacked/>.
26 <https://asia.nikkei.com/magazine/20170824/Life-Arts/Thailand-s-Buddhist-nuns-fight-for-equality>.
27 Kabilsingh, *Thai Women in Buddhism*, p. 71.
28 <www.npr.org/2018/01/06/576197738/the-female-monks-of-thailand>.
29 <www.pressreader.com/thailand/bangkok-post/20150720/281517929808350/TextView>.
30 For the decentralisation and lack of national oversight of Thai Buddhist education, see Justin McDaniel, *Gathering Leaves and Lifting Words: Histories of Buddhist Monastic Education in Laos and Thailand* (Seattle: University of Washington Press, 2008). For one of the best studies of the complex relationship between local and translocal politics in Thai Buddhism and politics, see Thomas Borchert's recent *Educating Monks: Minority Buddhism on China's Southwest Border* (Honolulu: University of Hawaii Press, 2018).

References

Borchert, Thomas. 2018. *Educating Monks: Minority Buddhism on China's Southwest Border*. Honolulu: University of Hawaii Press.

Darlington, Sue. 2012. *The Ordination of a Tree: The Thai Buddhist Environmental Movement*. Albany: SUNY Press.

Ekachai Sanitsuda. 1997. "Crusading for Nun's Rights". *Newsletter on International Buddhist Women's Activities*, Vol. 13, No. 2: 16–19.

Falk, Monica Lindberg. 2007. *Making Fields of Merit: Buddhist Female Ascetics and Gendered Orders in Thailand*. Copenhagen: NIAS Press and Washington, DC: University of Washington Press.

Heikkila-Horn, Marja-Leena. 1996. "Two Paths to Revivalism in Thai Buddhism: The Dhammakaya and Santi Asoke Movements". *Temenos*, Vol. 32: 93–111.

Ishii, Yoneo. 1986. *Sangha, State, and Society: Thai Buddhism in History*. Translated by Peter Hawkes. Honolulu: University of Hawaii Press.

Ito, Tomomi. 1999. "Buddhist Women in *Dhamma* Practice in Contemporary Thailand: Movements Regarding Their Status and World Renunciates". *Journal of Sophia Asian Studies*, Vol. 17: 147–181.

Jackson, Peter. 1997. "Withering Centre, Flourishing Margins: Buddhism's Changing Political Roles". In *Political Change in Thailand: Democracy and Participation*. Edited by Kevin Hewison. London: Routledge, pp. 75–93.

Kabilsingh, Chatsumarn. 1991. *Thai Women in Buddhism*. Berkeley: Parallax Press.

Kaeodaeng, Rung. 2542 [1999]. *Rai Ngan Sathiti Dan Sasana Khong Brathet Thai 2542 Samnakngan Gana Kammakan Seuksa Haeng Chat Samnak Naiyok Ratamontri*. Bangkok.

Keyes, Charles. 1999. "Moral Authority of the Sangha and Modernity in Thailand: Sexual Scandals, Sectarian Dissent, and Political Resistance". In *Socially Engaged Buddhism for the New Millennium: Essays in honor of the Venerable Phra Dhammapitaka (Bhikkhu P.A. Payutto) on his 60th Birthday Anniversary*. Edited by Pipob Udomittipong. Bangkok: Sathirakoses-Nagapradipa Foundation; Berkeley: Parallax Press, pp. 121–147.

McDaniel, Justin. 2002. "The Curricular Canon in Northern Thailand and Laos". *Manusya: Journal of Thai Language and Literature*, Special Issue: 20–59.

———. 2008. *Gathering Leaves and Lifting Words: Histories of Buddhist Monastic Education in Laos and Thailand*. Seattle: University of Washington Press.

———. 2011. *The Lovelorn Ghost and the Magical Monk*. New York: Columbia University Press.

Muecke, Marjorie A. 2004. "Female Sexuality in Thai Discourses about *Mae Chii* [Lay Nuns]". *Culture, Health & Sexuality*, Vol. 6, No. 3: 221–238.

Mulasilpa, Wutichai. 2539 [1996]. *Kan Patirup Kanseuksa Nai Sami Phrapatsomdet Phracjulachomklaochao Yu Hua*. Bangkok: Thai Wattana Panich.

Reynolds, Craig. 1973. *Buddhist Monkhood in Nineteenth Century Thailand*. PhD Dissertation, Cornell University.

Santikaro Bhikkhu. 1996. "Buddhadasa Bhikkhu: Life and Society through the Natural Eyes of Voidness". In *Buddhist Liberation Movements*. Edited by Sallie King and Christopher Queen. Albany: State University Press of New York, pp. 147–189.

Seeger, Martin. 2018. *Gender and the Path to Awakening: Hidden Histories of Nuns in Modern Thai Buddhism*. Chiang Mai: Silkworm Books/Copenhagen: NIAS Press.

Suwanna Satha-Anand. 1999. "Looking to Buddhism to Turn Back Prostitution in Thailand". In *The East Asian Challenge for Human Rights*. Edited by Joanne Bauer and Daniel Bell. New York: Cambridge University Press, pp. 193–211.

Swearer, Donald. 1996. "Sulak Sivaraksa's Buddhist Vision for Renewing Society". In *Buddhist Liberation Movements*. Edited by Sallie King and Christopher Queen. Albany: State University of New York Press, pp. 195–230.

Taylor, James. 1997. "'Thamma-chaat:' Activist Monks and Competing Discourses of Nature and Nation in Northeastern Thailand". In *Seeing Forests for Trees: Environment and Environmentalism in Thailand*. Edited by Philip Hirsch. Chiang Mai: Silkworm Books, pp. 37–50.

———. 1999. *Forest Monks and the Nation-State: An Anthropological and Historical Study in Northeastern Thailand*. Singapore: Institute of Southeast Asian Studies.

Yukio, Hayashi. 2003. *Practical Buddhism among the Thai-Lao: Religion in the Making of a Region*. Kyoto: Kyoto University's Center for Southeast Asian Studies.

21
BUDDHISM AND POWER

Edoardo Siani

Buddhist kings rule by the Dharma. Saffron-robed monks cheer massacre. Business tycoons preach asceticism. Spirit mediums distribute lottery numbers. Astrologers advise soldiers on staging a coup. With roughly 95 per cent of its population identifying as Buddhist, Thailand is sometimes thought to be the country with the highest percentage of Buddhists in the world.[1] Observed on the ground, however, Buddhism as predominantly practiced in Thailand appears to be more preoccupied with securing political influence and material wealth than peace and detachment. As such, it is hardly reminiscent of the doctrine of self-restraint that many celebrate worldwide.

This chapter frames Thai Buddhism so as to account for its intimacy with power. Drawing from current scholarly debates, it demonstrates the obvious, yet obscure(d): Thai Buddhism cannot be conceived of as disfranchised from the accumulation of political influence and material wealth, as power is a constitutive part of its history, makeup, continued appeal and relevance in the contemporary world.

Thai Buddhism and the question of power

Buddhism is extremely diverse worldwide. So varied are its beliefs and practices that some scholars warn against generalisations and advocate speaking about *Buddhisms* in the plural.[2] "Thai Buddhism" is an umbrella category that designates the form or forms of Buddhism prevalently practiced in the kingdom. It belongs, officially if not unproblematically, to the Theravada tradition. It is also influenced by Hinduism, Brahmanism and Mahayana Buddhism, and includes spirit cults shared by other peoples in South and Southeast Asia. Since at least the 19th century, Thai Buddhism has additionally incorporated discourses of proper religiosity, rationality and modernity that draw from Christianity and modern epistemologies. Scholars have long attempted to make sense of how these various elements – call them creeds, worldviews or cosmologies – interrelate or how they do not in Thai religion.[3] Power has played a significant role in shaping the discussion.

Earlier scholars commonly differentiated between Buddhism on one side and Hinduism, Brahmanism and local spirit cults on the other. They made the distinction precisely on the basis of what they deemed to be conflicting attitudes to power. Melford Spiro proposed to differentiate between Buddhism and local spirit cults.[4] Although he acknowledged that the existence of spirits is upheld by the canonical texts, he maintained that the scriptures show that Buddhism is interested in salvation in the form of Nirvana (*nipphan* in Thai). As such, it is about attaining detachment from this world and indeed from the cycle of rebirths (Samsara in Sanskrit, and *watta songsan* in Thai), rather than doing well in it. Spirit cults, to the contrary, offer strategies to prosper in this world by means of appeasing invisible beings in exchange for their favour. Other scholars argued for a similar power-based distinction between Buddhism and Brahmanism or Hinduism.[5]

Scholars began challenging the tendency to equate Buddhism in its entirety to the sacred scriptures in the early 1970s. Stanley Tambiah, who worked ethnographically in the Thai north-eastern region of Isan and wrote extensively on Buddhist kingship, critiqued it as reductionism. Such an approach, he argued, took the scriptural tradition at face value, overlooking the infinite facets that Buddhism acquires in the lived experience of its practitioners.[6] Tambiah thus advanced that a fixation with the scriptures is of limited usefulness for the study of Thai Buddhism as an ever-changing social phenomenon. This paved the way for an analysis of power.

Social scientists today tend to embrace the dynamic character of Thai Buddhism.[7] Among other things, they study how its practices transform to respond to the needs of a changing society.[8] These needs, of course, intrinsically relate to power. Contemporary ethnographies portray Buddhist monks, spirit mediums and astrologers as coexisting as they engage in practices, ranging from the sale (*chao*) of amulets (*phra khrueang*) to the divination of lottery numbers and lucky dates, which address the most pressing concerns of laypeople in the era of neoliberalism.[9] While prosperity-oriented practices may not seem orthodox, they are, after all, realistically more central to scores of Thai Buddhists than the reading of the sacred texts.[10]

In fact, power makes the scriptures themselves subject to change. Thai monarchs gain legitimacy by regarding themselves as the protectors of Buddhism. They have the duty to adapt the sacred texts so as to ensure that religious practice in the kingdom is always aligned with the ultimate cosmic Truth.[11] Fearing a decline in the observance of the *Tripitaka*, the canon of the Theravada tradition, 14th-century King Lithai compiled the *Traibhumi*, a cosmological treatise.[12] Under colonial pressures, King Mongkut (r. 1851–1868) overwrote the *Traibhumi* by commissioning textbooks that advanced a more modern cosmology. He and King Chulalongkorn (r. 1868–1910) further undermined the authority of the *Vessantara Jataka*, a key text from the *Tripitaka* itself.[13] The late King Bhumibol Adulyadej (r. 1946–2016) then resurrected the *Vessantara Jataka* and modified it to be in line with his contemporary model of kingship.

But the assumption that Buddhism should be unrelated to power is problematic on yet another account. In Thailand, Buddhism – including those forms that dictate its orthodoxy – has never really made a secret of its intimacy with power. In fact, it has openly conflated the ability to amass political influence and wealth with notions of virtue.

Cosmological notions of power

Thai Buddhist cosmology conceives power as a force, which flows from the celestial realms into this world. Ambitious individuals must accumulate (*sa som*) this kind of power and preserve (*raksa*) it. In the Thai language, several terms designate cosmological notions of power. The most central one is *barami*. The term appears in the aforementioned text of the *Vessantara Jataka*, which tells the story of the last incarnation of the Buddha as a royal before being reborn as Gautama. In its original scriptural context, *barami* refers to the ten virtues that are mastered by the Buddha-to-be and, by extension, by any legitimate Buddhist king.[14] In contemporary Thai society, however, the term has come to designate a kind of power that anyone may accumulate through moral behaviour.

Human beings in Thai Buddhist cosmology operate within a moral universe, in which any action (*kan kratham*) has a moral dimension and, as a result, karmic consequences. This universe is governed by the Dharma (*thamma*), an ultimate cosmic law that is simultaneously moral and natural. When humans behave virtuously – that is, in accordance with the Dharma (*yang mi sin tham*), and thus morality and nature – they automatically accumulate positive karma, or merit (*bun*). When they behave unvirtuously (*tham bap*), they are penalised with negative karma, or demerit (*kam*). *Barami* designates an individual's total reservoir of merit, accumulated through past good deeds, also in previous lives (*chat*).

Barami has the key faculty of attracting anything that is desirable. Primarily, it enables whoever accumulates it in extraordinary quantities to attract a proportionally extraordinary number of followers (*phu tam*).[15] People are said to "flock" spontaneously to meritorious individuals.[16] Because of this, *barami* is the quintessential quality of leaders (*phu nam*) – be it kings, politicians, corporate managers, vendors, teachers, or parents.[17] Extraordinary quantities of *barami* also attract proportionally extraordinary quantities of wealth. A Thai religious specialist explains that the latter faculty of *barami* derives from the former.[18] In his reasoning, when an individual behaves morally, he or she attracts followers, who in turn demonstrate their gratitude (*khwam kathan yu*) through material gifts.

The exemplary leader is generous, as the upmost virtue of a Buddhist is selfless giving (*dana*).[19] The more a Thai Buddhist acts for the sake of others, the more influence and wealth he or she accumulates, becoming a rich leader. In a society where desire (*khwam yak*) is frowned upon, following the basic doctrinal tenets of Buddhism, it is those who master the virtue of selflessness who receive all the goods. The ascetic becomes powerful because he or she is disinterested in power while the power-crazed remain empty-handed. As political influence and wealth flow naturally to those moral people who do not really want them, *barami* is then a very ambiguous form of power. It attracts anything that is desirable – and yet not desired.

From this perspective, the attainment of Nirvana easily takes the back seat in favour of worldly power. The latter narrative gained popularity in recent times. Bhumibol's model of kingship rested on the idea that a moral leader should renounce Nirvana for the sake of his or her followers rather than renounce worldly power for his personal salvation. The monarch's revised version of the *Vessantara Jataka* echoed this precept. Following the modification, the Buddha-to-be chose to forgo Nirvana to remain on the throne and enlighten the masses.[20] To be sure, Bhumibol's revision of the text was not a substantial departure from the original scriptural tradition, according to which Gautama could have become either a Buddha or a king, his decision to pursue Nirvana being "essentially a career choice".[21]

The relationship between *barami* and the accumulation of wealth also underwent some transformations during King Bhumibol's 70-year reign. As Thailand allied with the United States in the war against communism in Southeast Asia, state forces promoted capitalism with reference to Buddhist cosmology. Christine Gray shows that they actively encouraged investments into the business ventures of the crown as virtuous – monetary returns being associated with returns in the realm of merit.[22] The notion that cosmic power attracts wealth likely already existed in Thailand as it did elsewhere in the region.[23] Regardless, the public relations campaign reinforced the idea that financial success is an expression of individuals' moral stature.

Because *barami* is acquired through meritorious conduct, all good people will necessarily become powerful at some stage, or at least in some future rebirth. By the same token, all the existing powerful must necessarily be good, or at least they must have been good in the past, even if in a past life. If moral behaviour brings about political influence and wealth, and if political influence and wealth are in turn signs of moral behaviour, however, we arrive at a chicken-or-egg situation, in which individuals in a position of privilege are automatically legitimised.

This turns notions of power like *barami* into formidable instruments of legitimacy, which can be invoked to seal social inequality on celestial grounds.

Social stratification in the Buddhist polity

A Buddhist king is regarded as the most meritorious individual in the kingdom, with unmatched quantities of *barami*, accumulated in previous lives. A king's heightened cosmological status legitimises his political role of sovereign as divinely sanctioned. It also legitimises his economic stature. Like him, if on a comparatively smaller scale, any powerful person (*phu yai*) may support his or her right to prestige with claims of exceptional past moral conduct.

The Thai Buddhist polity resembles a fractal, in which the king-led social pyramid is replicated ad infinitum, from big to small, with different individuals performing the role of the ultimate moral leader over their affiliates.[24] The relationship between the meritorious monarch and his less meritorious subjects is mirrored, at village level, by the relationship between the village head and the villagers; at the level of a school, by the relationship between the teacher and the pupils; at the level of a corporation, by that between the president and the subordinates; at the level of a household, by that between parents and children; and so on.

While the workings of the moral universe grant a leader – a king rather than a parent – exceptional political and economic power, their affiliates expect them to always act for the benefit of the community rather than their own.[25] The role of a leader is to selflessly share assets, including wealth, in order to allow their followers to benefit from their virtue. The followers, in return, demonstrate their gratitude by giving a portion of their earnings or labour to their leader. Giving to a leader amounts to an act of merit-making for the affiliates, as it enables them to contribute to the leader's inherently moral activities, partaking in his or her virtuous conduct.[26]

In theory, followers who behave morally will be able to climb the social pyramid themselves one day. Unlike the caste system of contemporary Hinduism, class (*chon chan*) in Thailand is not prescriptive. This means that while bad deeds performed in previous lives determine one's social conditions at birth, mobility is still possible through moral action. The ability to improve one's condition is in fact widely regarded as a sign of exceptional virtue.[27] In practice, however, an unwelcoming political and business environment makes upward mobility extremely difficult for those who are not in a position of power to start with, resulting in social exclusion.

The cosmology-based model also affects political participation. This is evident in the crisis that has characterised the Thai new millennium. Supporters of the military juntas that installed themselves in power with coups d'état in 2006 and 2014 cheered the suspension of the electoral system by invoking international notions of "good governance". They nevertheless subsumed them to a Thai Buddhist framework. They therefore suggested that letting individuals with insufficient quantities of *barami* partake in political decision-making inevitably leads to "corruption" (*korapchan*).[28]

The military governments themselves advocated the idea of an administration of handpicked "good people" (*khon di*), as opposed to delegating political decisions to the masses. Their rhetoric built on existing discourses, which advance the necessity of replacing "Western" democracy with a "Thai democracy" or "dharmacracy": a nebulous mode of rule by an enlightened few people. Ironically, even Thaksin Shinawatra, the prime minister whose repeated electoral victories prompted the coups, once juxtaposed his ideal of "Buddhist politics" to his "Western" counterpart. "Politics which has the *thamma* is politics of men or moral integrity", he said. "We ourselves must understand that that politics in our country has been influenced by British politics where they argue against each other like lawyers. This may conflict with [. . .] the parliament as a gathering of men of moral integrity".[29]

Discourses of exclusive sovereignty that draw from Buddhist cosmology have long existed in Thailand. In the Cold War, propaganda compared communist sympathisers to the beasts that inhabit the lower realms of the Buddhist cosmos. As such, the insurgents were deprived of the most basic rights and protections that should come with citizenship. A monk, Kittiwuttho Bhikkhu, even preached that killing Thai communists was a virtuous action, amounting to an act of merit-making.[30] During the massacre of pro-election protesters in 2010, celebrity monk Wo Wachiramethi similarly commented that "killing time is more sinful than killing people".[31]

The uncertainties of power

Despite its multiple advantages, *barami* is finite. As an individual stays in a position of power, he or she consumes his or her existing reservoir (*chai barami*), incurring the risk of running out. Because one's remaining levels of *barami* are unknown,[32] the only way to avoid falling from grace is to continuously engage in moral behaviour, producing new *barami* (*sang barami*). The constant cultivation of morality is also crucial, as there is a delay between the performance of virtuous action and the accumulation of *barami*. While an individual may engage in virtuous behaviour today, the acquired merit (*bun*) may not be spendable in terms of *barami* for days, months, years, decades, or even several rebirths.

The original intimacy between morality and power does not preclude the possibility of immoral uses of power either. While power is moral in the sense that it is accumulated by means of virtuous action, its exercise is a different matter. Anybody who has authority (*amnat*) necessarily owes their position to *barami*, or past merit. Even immoral leaders like gangsters (*nak leng*) or corrupt politicians must have behaved morally in the past in order to be in a position of power now.[33] Having *barami* is nevertheless no guarantee of moral conduct in the present or the future, as the authority that stems from it may be exercised either way.[34] When a leader exercises his or her authority (*chai amnat*) morally, he or she produces new merit (*bun*), resulting in an increase in *barami* and, as a result, popularity and political longevity. When he or she exercises it immorally, however, he or she spoils *barami* (*sia barami*). As a result, his or her followers will be estranged, and his or her period of power will be shortened.

Leaders engage in conspicuous acts of virtue to reassure followers of their continued morality. This serves as a relatively good guarantee that they are trustworthy and cautious not to abuse their authority. It also suggests that their levels of *barami* are not likely to subside any time soon, enabling leaders to offer protection to existing or prospective followers well into the future. The publication of photographs and footage showing political leaders engaging in lavish rituals of merit-making (*kan tham bun*), like freeing fish or financing the construction of monasteries, serves this dual purpose.

Depending on the circumstances, even the use of violence may be regarded as an expression of a leader's virtue, and thus compared to an act of merit-making. In times when the nation or the religion is under threat, for instance, Buddhist kings must wage war against the enemies.[35] The sacred scriptures themselves prescribed that legitimate kings turn into warriors for the protection of the common good.[36] Applauded by some and abhorred by others, the belligerent words of the monks quoted above adhere to these notions of virtuous violence.

In Thai Buddhism, power is therefore something that may be both gained and lost, in accordance with the moral and natural law of the Dharma. Power may also be used for either good (*khwam di*) or bad (*khwam chua*), and the difference between the two is as open to debate as it is in other societies. These elements complicate the picture with uncertainty and ambiguity. They create the ideal terrain for contestation and resistance.

Contesting power

This chapter opened with a word of caution regarding the tendency to understand Buddhism in the singular. It proposed that there are multiple *Buddhisms*, both worldwide and within Thailand itself, as individuals continuously innovate for its beliefs and practices to be relevant and in line with their most pressing concerns. Are there not then cosmologies "from below"? The cosmological model presented so far, after all, does not respond effectively to the necessities of the masses. On the contrary, it supports a hierarchy that penalises them. The model also appears somewhat static and rigid rather than dynamic and negotiable.

Contemporary Thai Buddhists engage to a great degree in cosmological speculation. In an era when people from all walks of life have access to an unprecedented wealth of information – also from other societies and worldviews – individuals venture in very personal interpretations of the workings of the cosmos. Insights from books read at university, rather than from documentaries watched on TV or conversations with foreign friends, merge creatively with Buddhist notions of morality, producing new ideas of virtue.[37]

Given that cosmology has to do with power, speculation about cosmic matters is an inherently political practice with an enormous potential for resistance. People's freedom to engage in cosmology-making, however, has certain limits. These coincide with the karma-shaped hierarchy, which is normally unquestioned, along with notions of power like *barami*.[38] The hesitance to discuss critically these issues inhibits the emergence of solid counter-narratives. This does not mean that resistance is absent. Instead of challenging the cosmological model, some try to use it to their advantage. They therefore question the morality of leaders they are in conflict with, arguing that their authority rests merely on *barami* accumulated in previous lives. On top of this, they depict *themselves* as the ultimate moral people, attempting to eclipse those in power by advancing their own relatively greater right to leadership.[39]

Historically, such narratives of resistance have allowed individuals from the margins to rise to authority. Especially before centralisation, charismatic leaders able to attract followers by making credible claims of exceptional *barami* could set up competing kingdoms.[40] Regarded as "men of merit" (*phu mi bun*), they were usually religious masters like monks, spirit mediums and diviners. Some of them became so powerful that they posed a threat to the Bangkok government.[41] Their emergence was suppressed with the use of force.

Acting in their capacity of the protectors of Buddhism, kings often inaugurate their reigns by curbing charismatic individuals who may compete for power (*khaeng barami*). In the past century, kings did so by drawing a line between religion and "superstition" (*ngom ngai*). In some instances, they outlawed religious leaders like spirit mediums and diviners. At times, they even punished them by death.[42] The practices of mediums and diviners were not illegitimate per se; divination, for instance, was regularly practiced at the royal court.[43] Rather, the issue was determining who could engage in them and who could not – who had true *barami* and who did not.[44] In the 19th century, the monastic community was additionally brought to a greater degree under the control of the state,[45] The move stressed the primacy of standardised liturgy over the authority of individual charismatic monks.[46]

Like elsewhere in the world, politics and economy in today's Thailand take on a secular façade. As a result, popular religious masters are less likely competitors than before in the greater contest for power. Buddhist idioms of power nevertheless continue to form an underlying (if fundamental) leitmotif, as politicians and businesspeople seek legitimacy by suggesting exceptional quantities of *barami*. Among those who attempted to install themselves to power as modern "men of merit" was Field Marshal Phibun Songkhram, one of the rebels who overthrew absolutism in 1932.[47] Some understand that business tycoon Thaksin was also a "man of merit".[48]

In the contemporary era, state authorities have invoked discourses of good governance to wage their crusades against "meritorious individuals" like Phibun and Thaksin, both of whom ended up in exile. These discourses establish the old line between those who can legitimately make claims of exceptional *barami* and those who cannot, while appealing to idioms of power and statecraft that have an international resonance and modern appeal. Their usage conceals the role played by Buddhist notions of power in Thai politics, tucking it under a veneer of secularism.

Conclusion

The intimacy between Buddhism and power goes often unacknowledged, obscured by a celebration of the doctrine's emphasis on self-restraint. In Thailand, the Buddhist notion of asceticism is actually interlocked with the pursuit of worldly power, as people hold that political influence and material wealth flow naturally to those who behave selflessly, transforming them into rich leaders. This understanding of the workings of the cosmos enables individuals who are in a position of privilege to legitimise their existing status. Historically, people from the margins, like the "men of merit", have nevertheless also managed to rise to positions of authority by turning the rules that govern the Buddhist cosmos to their advantage.

Today, Buddhism continues to be a key arena for the contestation of power in Thailand, the mastery of its practices and idioms remaining critical for those who are on a quest for influence and wealth. As talks of a "dharmacracy" still loom over domestic politics and exorbitant amounts of money are passed around among the kingdom's most ascetic politicians, business tycoons and military men, Buddhism will persist in being a fundamental ally of power as well as an instrument for its maintenance and assertion.

Notes

1 See Peter Friedlander, "Buddhism and Politics", in *Routledge Handbook of Religion and Politics*, edited by Jeffrey Haynes (London and New York: Routledge, 2009), p. 16.

2 Gustaaf Houtman, "How a Foreigner Invented Buddhendom in Burmese: From Tha-tha-na to Bok-da Ba-tha", *Journal of the Anthropological Society of Oxford*, Vol. 21, No. 2 (1990), pp. 113–128.

3 Talal Asad (2002) problematised religion as a category that is produced by secular modernity. Daniel (2000) demonstrated that the concept of "religion" is of little use for the study of societies like India, whereby what we refer to as Hinduism actually encompasses all spheres of life. In the context of Theravada Southeast Asia, Houtman (1990, pp. 113–117) argued that a term for "Buddhism" did not exist until the colonial period. Previously, the Burmese phrase *bokda thathana* (*sasana phut* in Thai) designated the teachings of the Buddha, while *bokda dama* (*thamma* in Thai) designated the Dharma as the ultimate Truth. See Talal Asad, "The Construction of Religion as an Anthropological Category", in *A Reader in the Anthropology of Religion*, edited by M. Lambek (Oxford: Blackwell Publishing, 2002), pp. 114–132; E. Valentin Daniel, "The Arrogation of Being: Revisiting the Anthropology of Religion", in *Macalester International*, Vol. 8, No. 17 (2000), pp. 171–191; Houtman, "How a Foreigner Invented Buddhendom in Burmese: From Tha-tha-na to Bok-da Ba-tha", pp. 113–128.

4 Melford Spiro, *Burmese Supernaturalism* (Philadelphia, PA: ISHI, 1967), pp. 247–280. Spiro's work on religion in Burma had a strong influence on the scholarship on Thai Buddhism.

5 See, for example, Phraya Hanuman Rachathon, *Essays on Thai Folklore* (Bangkok: Duang Kamol, 1988).

6 Stanley Tambiah, *Buddhism and the Spirit Cults in North-East Thailand* (Cambridge: Cambridge University Press, 1970).

7 A fixation with scriptural Buddhism is additionally problematic as it may overlook different religions' approaches to soteriology. In Judaism, Christianity and Islam, the Truth is "revealed" from God to the faithful through the sacred scriptures. As a result, the texts are the ultimate authoritative vehicle on this world for salvation. The Buddha, however, was a (very special, to be sure) human being who worked his way up towards Enlightenment through the practice of virtuous conduct. As a result, some

contemporary Buddhists place greater emphasis on virtuous practice (*kan patibat tham*) as opposed to the scriptures. In this context, the practice of correct behaviour – whether by means of upholding precepts, worship, meditation and so on – may trump the scriptures as the ultimate vehicle of salvation. Reading the scriptures, if anything, may be one kind of correct practice.

8 Pattana Kitiarsa, "Beyond Syncretism: Hybridisation of Popular Religion in Contemporary Thailand", *Journal of Southeast Asian Studies,* Vol. 36, No. 3 (2005), pp. 461–487.

9 Ibid. See also Justin McDaniel, *The Lovelorn Ghost and the Magical Monk: Practicing Buddhism in Modern Thailand* (New York: Columbia University Press, 2011).

10 See Peter Jackson, "The Enchanting Spirit of Thai Capitalism: The Cult of Luang Phor Khoon and the Post-Modernisation of Thai Buddhism", *South East Asia Research,* Vol. 7, No. 1 (1999), pp. 5–60.

11 See Christine Gray, "Buddhism as a Language of Images, Transtextuality as a Language of Power", *Word & Image: A Journal of Verbal/Visual Enquiry,* Vol. 11, No. 3 (1995), pp. 225–236.

12 See Frank Reynolds and Mani Reynolds, *Three Worlds According to King Ruang: A Thai Buddhist Cosmology* (Berkeley: University of California, 1982).

13 Patrick Jory, *Thailand's Theory of Monarchy: The Vessantara Jataka and the Idea of the Perfect Man* (Albany: SUNY Press, 2016), pp. 107–126.

14 Known in contemporary Thai as *thotsabarami*, the Ten Perfections are giving, moral conduct, renunciation, wisdom, energy, patience, truthfulness, resolution, loving-kindness and equanimity. See Jory, *Thailand's Theory of Monarchy: The Vessantara Jataka and the Idea of the Perfect Man*, p. 18.

15 The key feature of *barami* is likely imputable to the pre-modern period, when Southeast Asian kings used to rule over a borderless territory, their success being measured in their ability to exercise control over numerous people rather than a clearly delineated land. See Stanley Tambiah, *World Conqueror and World Renouncer: A Study of Buddhism and Polity in Thailand against a Historical Background* (Cambridge: Cambridge University Press, 1976).

16 This recurrent use of the word "flocking" in Thai Buddhist narratives is highlighted in Christine Gray's *Thailand – The Soteriological State in the 1970s,* PhD Dissertation at University of Chicago, 1986.

17 In theory, both women and men can have *barami* in Thai Buddhism. The relationship between gender and power, however, is more problematic than this. While both males and females can accumulate *barami*, only men have access to the routes of monkhood and kingship, which grant the highest cosmological and social statuses a human being can aim for. Because monks and kings are in a unique position to maintain a virtuous conduct, women are also banned from the most effective ways of accumulating *barami*.

18 The author's own fieldwork data, collected with diviners, their clients and their detractors in Bangkok between 2014 and 2016. See Edoardo Siani, *The Eclipse of the Diviners: Sovereign Power and the Buddhist Cosmos at the End of Thailand's Ninth Reign*, PhD Dissertation at SOAS, University of London, 2017, p. 117.

19 In the *Vessantara Jataka, dana,* or selfless giving, is the utmost virtue to be perfectioned by a Buddhist king. See Jory, *Thailand's Theory of Monarchy: The Vessantara Jataka and the Idea of the Perfect Man*, p. 18.

20 Ibid., p. 3.

21 Craig Reynolds, "Power", in *Critical Terms for the Study of Buddhism*, edited by Donald Lopez Jr. (Chicago: University of Chicago Press, 2005), pp. 211–228.

22 Gray, *Thailand – The Soteriological State in the 1970s*.

23 See, for instance, Benedict Anderson, "The Idea of Power in Javanese Culture", in *Culture and Politics in Indonesia*, edited by Claire Holt, Benedict Anderson and James Siegel (Ithaca: Cornell University Press, 2007), pp. 1–70.

24 See Lucien Hanks, "The Thai Social Order as Entourage and Circle", *Change and Persistence in Thai Society*, edited by G. William Skinner and A. Thomas Kirsch (Ithaca: Cornell University Press, 1975), pp. 197–218.

25 See Siani, *The Eclipse of the Diviners: Sovereign Power and the Buddhist Cosmos at the End of Thailand's Ninth Reign*, p. 123.

26 See Gray, "Buddhism as a Language of Images, Transtextuality as a Language of Power", pp. 225–236.

27 Lucien Hanks, "Merit and Power in the Thai Social Order", *American Anthropologist, New Series,* Vol. 64, No. 6 (1962), pp. 1247–1261; and Charles Keyes, "Millennialism, Theravada Buddhism, and Thai Society", *The Journal of Asian Studies,* Vol. 36, No. 2 (1977), p. 286.

28 See Daena Funahashi, "Rule by Good People: health Governance and the Violence of Moral Authority in Thailand", *Cultural Anthropology,* Vol. 31, No. 1 (2016), pp. 107–130.

29 Pasuk Phongpaichit and Chris Baker, *Thaksin: The Business of Politics in Thailand* (Chiang Mai: Silkworm Books, 2004), p. 137.

30 Donald Swearer, *The Buddhist World of Southeast Asia,* second edition (Albany: SUNY Press, 2010), pp. 137, 273.
31 Suluck Lamubol, "Understanding Thai-style Buddhism", *Prachathai*, 28 February 2014 <https://prachatai.com/english/node/3883> (accessed 30 July 2018).
32 Keyes, "Millennialism, Theravada Buddhism, and Thai Society", p. 286.
33 Charles Keyes, "Economic Action and Buddhist Morality in a Thai Village", *Journal of Asian Studies,* Vol. 42, No. 4 (1983), pp. 851–868.
34 The author's use of the term *amnat*, which is based on the author's own fieldwork data, collected with diviners, their clients and detractors in Bangkok between 2014 and 2016, differs from Peter Jackson's. Jackson writes that *amnat* "denotes raw, amoral power that may be used for either good or evil and which is accumulated and maintained by sheer force", while "in contrast, *barami* denotes the charismatic power possessed by morally upright, righteous people". See Peter Jackson, "Markets, Media and Magic: Thailand's Monarch as a 'Virtual Deity'", *Inter-Asian Cultural Studies*, Vol. 10, No. 3 (2009), p. 363.
35 Edoardo Siani, "Purifying Violence: Buddhist Kingship, Legitimacy and Crisis in Thailand's Ninth and Tenth Reigns", in *King, Coup, Crisis: Time of a Critical Interregnum in Thailand*, edited by Pavin Chachavalpongpun (NUS Press, forthcoming 2019).
36 Suwanna Satha-Anand, "The Question of Violence in Thai Buddhism", in *Buddhism and Violence: Buddhism and Militarism in Modern Asia*, edited by V. Tikhonov and T. Brekke (London and New York: Routledge, 2013), pp. 178–185.
37 Edoardo Siani, "Stranger Diviners and their Stanger Clients: Popular Cosmology-Making and its Kingly Power in Buddhist Thailand", *South East Asia Research*, Vol. 26, No. 4 (2018), pp. 416–431.
38 Ibid.
39 Ibid.
40 Tambiah, *World Conqueror and World Renouncer: A Study of Buddhism and Polity in Thailand against a Historical Background.*
41 See Patrice Ladwig, "Millennialism, Charisma and Utopia: Revolutionary Potentialities in Pre-modern Lao and Theravada Buddhism", *Politics, Religion & Ideology,* Vol. 15, No. 2 (2014), pp. 308–329.
42 Pattana Kitiarsa, *You May Not Believe, But Never Offend the Spirits: Spirit Medium Discourses and the Postmodernisation of Thai Religion,* PhD Dissertation at the University of Washington, 1999, pp. 67–68.
43 Peter Jackson, *Buddhism, Legitimation, and Conflict: The Political Functions of Urban Thai Buddhism* (Singapore: ISEAS, 1989), pp. 58–59.
44 See Christine Gray, "Hegemonic Images: Language and Silence in the Royal Thai Polity", *Man, New Series*, Vol. 26, No. 1 (1991), p. 45.
45 The effectiveness of King Rama IV's reforms in standardising Buddhist practice in the kingdom is questioned by scholars like Justin McDaniel in *The Lovelorn Ghost and the Magical Monk: Practicing Buddhism in Modern Thailand.*
46 See Tiyavanich Tiyavanich, *Forest Recollections: Wandering Monks in Twentieth-Century Thailand* (Honolulu: University of Hawaii Press, 1997).
47 See Gray, *Thailand – The Soteriological State in the 1970s*, pp. 366–370.
48 See Serhat Unaldi, *Working Towards the Monarchy: The Politics of Space in Downtown Bangkok* (Honolulu: University of Hawaii Press, 2016), pp. 53–85.

References

Anderson, B. 2007. "The Idea of Power in Javanese Culture". In *Culture and Politics in Indonesia*. Edited by C. Holt, B. Anderson, and J. Siegel. Ithaca: Cornell University Press, pp. 1–70.

Asad, T. 2002. "The Construction of Religion as an Anthropological Category". In *A Reader in the Anthropology of Religion*. Edited by M. Lambek. Oxford: Blackwell Publishing, pp. 114–132.

Daniel, E.V. 2000. "The Arrogation of Being: Revisiting the Anthropology of Religion". *Macalester International*, Vol. 8, No. 17: 171–191.

Friedlander, P. 2009. "Buddhism and Politics". In *Routledge Handbook of Religion and Politics*. Edited by J. Haynes. London and New York: Routledge, pp. 11–25.

Funahashi, D. 2016. "Rule by Good People: Health Governance and the Violence of Moral Authority in Thailand". *Cultural Anthropology*, Vol. 31, No. 1: 107–130.

Gray, C. 1986. *Thailand – The Soteriological State in the 1970s*. PhD Dissertation, University of Chicago.

———. 1991. "Hegemonic Images: Language and Silence in the Royal Thai Polity". *Man, New Series*, Vol. 26, No. 1: 43–65.

———. 1995. "Buddhism as a Language of Images, Transtextuality as a Language of Power". *Word & Image: A Journal of Verbal/Visual Enquiry*, Vol. 11, No. 3: 225–236.

Hanks, L. 1962. "Merit and Power in the Thai Social Order". *American Anthropologist, New Series*, Vol. 64, No. 6: 1247–1261.

———. 1975. "The Thai Social Order as Entourage and Circle". In *Change and Persistence in Thai Society*. Edited by G.W. Skinner and A.T. Kirsch. Ithaca: Cornell University Press, 1975, pp. 197–218.

Houtman, G. 1990. "How a Foreigner Invented Buddhendom in Burmese: From Tha-tha-na to Bok-da Ba-tha". *Journal of the Anthropological Society of Oxford*, Vol. 21, No. 2: 113–128.

Jackson, P. 1989. *Buddhism, Legitimation and Conflict: The Political Functions of Urban Thai Buddhism*. Singapore: ISEAS.

———. 1999. "The Enchanting Spirit of Thai Capitalism: The Cult of Luang Phor Khoon and the Post-Modernisation of Thai Buddhism". *South East Asia Research*, Vol. 7, No. 1: 5–60.

———. 2009. "Markets, Media and Magic: Thailand's Monarch as a 'Virtual Deity'". *Inter-Asian Cultural Studies*, Vol. 10, No. 3: 361–380.

Jory, P. 2016. *Thailand's Theory of Monarchy: The Vessantara Jataka and the Idea of the Perfect Man*. Albany: SUNY Press.

Keyes, C. 1977. "Millennialism, Theravada Buddhism, and Thai Society". *The Journal of Asian Studies*, Vol. 36, No. 2: 283–302.

———. 1983. "Economic Action and Buddhist Morality in a Thai Village". *Journal of Asian Studies*, Vol. 42, No. 4: 851–868.

Ladwig, P. 2014. "Millennialism, Charisma and Utopia: Revolutionary Potentialities in Pre-modern Lao and Theravada Buddhism". *Politics, Religion & Ideology*, Vol. 15, No. 2: 308–329.

McDaniel, J. 2011. *The Lovelorn Ghost and the Magical Monk: Practicing Buddhism in Modern Thailand*. New York: Columbia University Press.

Pasuk Phongpaichit and C. Baker. 2004. *Thaksin: The Business of Politics in Thailand*. Chiang Mai: Silkworm Books.

Pattana Kitiarsa. 1999. *You May Not Believe, But Never Offend the Spirits: Spirit Medium Discourses and the Post-modernisation of Thai Religion*. PhD Dissertation, University of Washington.

———. 2005. "Beyond Syncretism: Hybridisation of Popular Religion in Contemporary Thailand". *Journal of Southeast Asian Studies*, Vol. 36, No. 3: 461–487.

Phraya Hanuman Rachathon. 1988. *Essays on Thai folklore*. Bangkok: Duang Kamol.

Reynolds, C. 2005. "Power". In *Critical Terms for the Study of Buddhism*. Edited by D. Lopez, Jr. Chicago: University of Chicago Press, pp. 211–228.

Reynolds, F. and M. Reynolds. 1982. *Three Worlds According to King Ruang: A Thai Buddhist Cosmology*. Berkeley: University of California.

Siani, E. 2017. *The Eclipse of the Diviners: Sovereign Power and the Buddhist Cosmos at the End of Thailand's Ninth Reign*. PhD Dissertation, SOAS, University of London.

———. 2018. "Stranger Diviners and their Stranger Clients: Popular Cosmology-Making and its Kingly Power in Buddhist Thailand". *South East Asia Research*, Vol. 26, No. 4: 416–431.

———. 2019 (forthcoming). "Purifying Violence: Buddhist Kingship, Legitimacy and Crisis in Thailand's Ninth and Tenth Reign". In *Coup, King, Crisis: Time of a Critical Interregnum in Thailand*. Edited by Pavin Chachavalpongpun. Singapore: NUS Press.

Spiro, M. 1967. *Burmese Supernaturalism*. Philadelphia: ISHI, pp. 251–252.

Suluck Lamubol. 2014. "Understanding Thai-style Buddhism". *Prachathai*, 28 February <https://prachatai.com/english/node/3883> (accessed 30 July 2018).

Suwanna Satha-Anand. 2013. "The Question of Violence in Thai Buddhism". In *Buddhism and Violence: Buddhism and Militarism in Modern Asia*. Edited by V. Tikhonov and T. Brekke. London and New York: Routledge, pp. 175–193.

Swearer, D. 2010. *The Buddhist World of Southeast Asia*, second edition. Albany: SUNY Press.

Tambiah, S. 1970. *Buddhism and the Spirit Cults in North-East Thailand*. Cambridge: Cambridge University Press.

———. 1976. *World Conqueror and World Renouncer: A Study of Buddhism and Polity in Thailand against a Historical Background*. Cambridge: Cambridge University Press.

Tiyavanich Tiyavanich. 1997. *Forest Recollections: Wandering Monks in Twentieth-Century Thailand*. Honolulu: University of Hawaii Press.

Unaldi, S. 2016. *Working Towards the Monarchy: The Politics of Space in Downtown Bangkok*. Honolulu: University of Hawaii Press.

22

SECULARISATION, SECULARISM, AND THE THAI STATE

Tomas Larsson

Debates about secularisation (a historical process) and secularism (an ideology) have played a prominent role in social science debates. In the scholarship of the 1960s and 1970s, it was assumed that secularisation was a precondition for political modernity as well as a historical inevitability. Scholars today have abandoned both these preconceptions in favour of the view that there are multiple viable modernities and that religion is not going away. While much of the scholarship on secularisation and secularism remains centred on European and North American (and therefore Christian) experiences, more recent contributions have explored the politics of secularity "beyond the West", including in Southeast Asia.[1] It is in this spirit that this chapter uses the concepts of secularisation and secularism to illuminate religion-state relations in Thailand.

Secularisation can refer to several distinct processes. Here attention will be focused on secularisation as it relates to the state. A central aspect of this is what Smith referred to as polity secularisation: "The process by which a traditional system undergoes radical differentiation, resulting in separation of the polity from religious structures, substitution of religious modes of legitimation, and extension of the polity's jurisdiction into areas formerly regulated by religion".[2] Secularism, in turn, is an ideology that seeks to justify "the separation of religious and political authority, the expulsion of religious law from the legal system, and sometimes even the exclusion of religion from the public sphere".[3]

It must be recognised, however, that secularism as state practice rarely results in a clean separation of religion and politics or church and state. On the contrary, some have argued that the modern secular state is characterised less by its separation from than its expanded role in managing "religion".[4] According to this line of argument, secularism as political practice should be understood as the modern state's assertion of "sovereign power to reorganise substantive features of religious life, stipulating what religion is or ought to be, assigning its proper content, and disseminating concomitant subjectivities, ethical frameworks, and quotidian practices".[5] While scholars may disagree on how secularisation and secularism should be understood in relation to the state, they recognise that the politics of secularisation and secularism pivots around struggles that seek to define, to blur, or to deny the very existence of boundaries separating a "religious" domain from that which stands outside it ("the secular"). Gorski helpfully highlights how such boundary struggles typically involve efforts to change either the location of activities from one domain to the other or the degree of permeability between the religious and secular domains.[6]

In light of these conceptual clarifications, Thailand is an ambiguous and contradictory case. It is possible to find evidence of secularisation processes leading to a significant degree of differentiation and separation of religion and state. However, the overarching ideological framework within which a "secular" Thai state has emerged in the course of the 19th and 20th centuries reflects a fundamentally "Buddhist" worldview. Thailand, in short, has adopted a predominant mode of secularism that translates into extensive state control of the religious field.

Before proceeding, it should of course be noted that secularisation may be understood in less state-centric ways: as a decline in popular religiosity, as observed in many parts of Europe, or as the privatisation of religion, causing religion to cease being a public concern.[7] I set aside the question of religious decline because of a lack of detailed sociological data on religious beliefs and behaviour in Thailand. What we do know is that most Thais are "religious", and that the vast majority – close to 94 per cent, according to the 2010 national census – identify as Buddhist. Only a minuscule proportion of the country's population (0.07 per cent) claims they have "no religion".[8] In one recent survey, 56.6 per cent responded that religion plays a "very important" role in their lives, and 31.1 per cent responded that it plays a "rather important" role.[9] As this evidence suggests, economic and social modernisation has not triggered a decline of religious belief, practice, and identity in Thailand. For similar reasons I also set aside the question of secularisation as "privatisation" of religion. The growing popularity of meditation among laypersons otherwise provides a striking example of how a religious practice traditionally reserved for Buddhist monks has been privatised and disconnected from its roots in the Theravada Buddhist institutional and cultural context.[10]

Thai conceptions of religion and secularity

So far I have treated the categories of "religion" and "secular" as rather unproblematic. As recent scholarship has emphasised, they are anything but.[11] Defining what is and is not "religion" is a fraught scholarly endeavour and a matter subject to intense political and legal contestation. This is true even in the most secular of states, where constitutional guarantees of religious freedom compel legislatures and courts to specify what counts as "religion" and as such deserves protection.

Analytically, religion can be defined as "a complex of prescribed practices that are based on premises about the existence and nature of superhuman powers".[12] Thai Buddhism, broadly conceived, is characterised by an impressive ability to incorporate within its scope an abundance of such superhuman powers – including but not limited to the impersonal powers of *thamma* and *kam* and a diverse and ever-expanding collection of deities, spirits, and ghosts.[13] However, the country's religious and political authorities do not necessarily recognise practices relating to the latter as "religious", often viewing them instead as deplorable "superstitions" that distract from the pursuit of enlightenment.[14] The Thai term for "religion" is *satsana*. While originally referring to the teachings of Gautama Buddha, its use as a generic term that applies also to, for example, Hinduism, Christianity, and Islam, has deep historical roots.[15]

As far as I am aware, a Thai conceptual history of the secular has yet to be written. Today, the most common way of expressing the idea of secularity is by reference to *kharawat*, denoting "householders", such as laypersons, as distinct from ordained monks (*bhikkhu*). The dichotomy between monastic and non-monastic is therefore central to abstractions such as the idea of a secular state (*rat kharawat*, or lay state), secularism (*kharawatniyom*, lay-ism), and secularisation (*kanplian pen kharawat*, to make lay). In Thai academic debates another neologism, *lokwisay* (world-character), is sometimes also used in reference to secularity, but it is more rarely found in popular discourse. The term *kharawat* remains closely associated with its traditional referent, the devout lay supporter of the Buddhist *sangha*; it does not connote someone or something separated from "religion". Suraphot has noted that Thai terms such as *kharawat* and *lokwisai* fail to capture the meaning of "secular" in Western philosophical thought.[16] That may be correct, but it does not adequately recognise the ambiguity and complexity of the concept of the secular also in the West.

Rather than relying on these neologisms, Thai discourse about the relationship between religion and the state still tends to rely on a traditional Buddhist idiom. Theravada Buddhist polities have long emphasised the distinction between a "religious" sphere (*satsanachak*) and a "worldly" sphere (*anachak*). These represent two aspects of power – the two "wheels" of *thamma*. In sociological and legal terms, *satsanachak* refers to the Buddhist monastic order that is governed by ecclesiastic laws laid down by the Buddha (*vinaya*). *Anachak*, in turn, refers to Buddhist kings and the secular (e.g. non-monastic) and universal law of the *thammasat* as well as king-made law. While the spheres are distinct, they also overlap, with the worldly sphere ultimately subsumed within the religious sphere. This can be seen not least in the fact that worldly rulers are expected to govern in accordance with Buddhist morality. The Jataka stories, which have greatly influenced the Thai political imagination, place emphasis on the importance of the ten royal virtues (*thotsaphit ratchatham*), which are moral ideals constraining "proper" Buddhist rulers.[17] It would therefore be wrong to think that *anachak* corresponds to the secular, if by that we mean something that stands "outside" religion. In fact, the *thammaracha* (righteous Buddhist ruler) is expected to govern in accordance with *thamma*, such as to administer "worldly" affairs in appropriate ways and to act as guarantor for the good governance of the Buddhist monastic order and the preservation of Buddha's teachings. Adopting a strategy that can be categorised as neo-traditionalist,[18] the modern Thai state has continued to shoulder that ancient responsibility of Buddhist kingship.[19]

Arguably, the Thai conception of *anachak* is somewhat analogous with pre-modern European conceptualisations of the secular and secularisation. These were theological concepts, grounded in a distinction between two kinds of clergy, religious (withdrawn from "the world") and secular (living in "the world"). "Secularisation" thus referred first to clergy and later to assets, functions, and so forth, leaving monasteries and entering "the world". However, while "the secular" may have denoted that which is outside the monastery walls, it did not, in medieval Christendom, in any way stand in contrast to or opposed to "religion", given that the surrounding society and the state were equally Christian. In the hitherto dominant Thai religiopolitical imaginary, likewise, *satsanachak* and *anachak* are two separate but equally Buddhist domains.

Religion and state formation

In the earliest states in the territory that corresponds to contemporary Thailand, Buddhism and Brahmanism served important legitimating roles, exalting the head of state as (some combination of) a *chakraphattirat* (a Buddhist universal monarch), *thewaracha* (god-king), and *thammaracha* (righteous ruler), and by associating king-made law with the revealed universal legal principles of the *thammasat*.[20] The cosmological treatise *Traiphum phraruang* ("Three worlds according to King Ruang"), which dates back to the 1340s, served as a central ideological pillar of Thai kingship into the 19th century, when its worldview was increasingly challenged by the arrival of a rival epistemology, Western in origin and scientific in character.[21]

The 19th-century encounter with the West, not least in the shape of Christian missionaries, served as catalyst for both religious and political reforms, with implications for the role of religion within the kingdom. Thai Buddhism and the Thai state were both subjected to "rationalising" reforms led by Kings Mongkut, Chulalongkorn, and Vajiravudh, and in the process Buddhism was turned into a pillar of both the Thai state and of official Thai nationalism.[22] As a consequence of these developments, Buddhism shifted from being a matter of belief and ritual to being a component of an ethno-religious nationalist ideology in which an abstracted Buddhist identity was an essential marker of "Thainess" (*khwam pen thai*).

The modernisation of the Thai state in the late 19th and early 20th centuries required a radical transformation of the Thai legal system, separating law from its religious moorings.[23] In contemporary Thailand, therefore, religion does not serve as a source of secular law. The main exception relates to Islam. A 1901 law decreed that Muslim minorities in the four southern provinces should (in accordance with the then prevailing custom) be governed by Islamic family law and Islamic courts.[24] Although the government of Phibun Songkhram made an attempt to abolish Islamic law and courts, this caused such discontent that Bangkok in 1946 reinstated Islamic family law and Islamic courts for the southern provinces.[25]

Rather than Buddhism serving as a source of secular law, secular lawmakers have occasionally aspired to fill perceived gaps in the *vinaya*. Most notably, canonical Buddhism provides no basis for structures that facilitate centralised governance and administration of the *sangha*. In Thailand this supposed lacuna has been remedied by a series of Sangha Acts, the first of which was enacted in 1902. These have established a nationally integrated hierarchical *sangha* administration and regulated its relationship with the secular (e.g. non-monastic) bureaucracy. The Sangha Acts have sought to shape the *sangha*'s governance structures in ways that closely mirror those of the secular bureaucracy. The 2018 Sangha Act significantly increased the Thai king's discretionary power over ecclesiastic affairs and appointments "in accordance with ancient royal traditions".

While it was absolute monarchs who fused *sangha*, state, and nation, subsequent political changes have not threatened the trinity. In 1932 a small group of bureaucrats, mostly French-trained, toppled the absolute monarchy and replaced it with a constitutional monarchy. Sovereignty was now transferred, at least in theory, to the people. However, the revolutionaries faced no opposition from the *sangha*, and they accordingly could see no compelling reason to overturn the well-established, symbiotic pattern of *sangha*-state relations. All post-1932 governments have continued to act as patrons and protectors of Buddhism, and they have exploited opportunities to instrumentalise religion for the purpose of state building. In the early 1930s, for instance, the government asked the monkhood to participate in elaborate rituals and propaganda campaigns intended to legitimise and, indeed, sacralise the new political order.[26] In the early 1940s, a hyper-nationalist government sought to impose a greater degree of religious uniformity on the population and, especially, on the state bureaucracy. Because religious identity was regarded as a test of loyalty to the state, Christian and Muslim civil servants were forced to "return" to Buddhism in order to keep their jobs.[27] During the Cold War, military governments mobilised the *sangha* in the struggle against communism, with clandestine US support. Clergy suspected of secretly harbouring radical sympathies were purged from the *sangha*, while other monks, and especially those belonging to the elite Thammayut sect, joined state-led Buddhist missionary and local development efforts intended to inoculate the impoverished peasantry in the country's north and northeast against communism.[28]

Since the mid-1970s, Thai state elites have intensified the public cultivation of a form of hyper-royalism with distinctly "religious" overtones. The royalist cult has been centred on king Bhumibol Adulyadej (r. 1946–2016) and articulated in terms of *barami* (Buddhist royal virtue). As a consequence, the Thai royal family has effectively been turned into "sacred beings and royalism [into] a religion".[29] Thai police and courts protect the image of the *thammaracha* by enforcing astonishingly harsh laws that treat signs of royal disrespect as blasphemous and heretical crimes that undermine the cultural foundations of the Buddhist-monarchical state.[30] This illustrates how, in the dominant political imagination, *satsanachak* and *anachak* are understood as separate domains that are equally infused with Buddhist sacrality.

The introduction of the principle of modern constitutionalism should of course be understood as a move away from religious modes of political legitimation towards a political order where legitimacy rests on principles of citizenship and public participation. However, the modernisation of the Thai polity has been led by political elites who have, for the most part, been keen to ensure that the legal and political order remains an extensively Buddhicised one. This can most readily be discerned by noting the frequency with which *tham/thamma* serves as the etymological root for the neologisms of modern Thai statecraft: constitution (*ratthathammanun*), rule of law (*nittitham*), and good governance (*thammaphiban*).[31] It is therefore virtually impossible to discuss (in Thai) the governance of the modern Thai state without invoking pre-modern conceptions of Buddhist moral order. In a similar manner, Thai state elites have gone to great lengths to infuse nominally secular public-policy frameworks with religious and specifically *thammic* meaning.[32]

The Buddhicisation of the Thai constitutional order is also reflected in the legal position of the monarch. While there have been very many constitutions since the first one was introduced in 1932, they have all emphasised that the monarch is a Buddhist and acts as the "supreme defender" of religion in the realm.[33] While Thailand therefore is obliged to have a Buddhist head of state, Thai constitutions have never explicitly declared Buddhism the official state religion, despite increasingly strident demands that it be recognised as such. While still falling short in that regard, the 2017 constitution nevertheless places an unprecedentedly heavy burden of responsibility on the Thai state with regards to the management of religion in general and Buddhism in particular. It decrees as follows:

> The state shall patronise and protect Buddhism and other religions. With a view to patronising and protecting the Buddhism that has long been professed by the majority of the Thai people, the state shall promote and support education in and Spropagation of the principles of Theravada Buddhism for the purpose of mental and intellectual development, and must establish measures and mechanisms to prevent the desecration of Buddhism in any form. The state shall also encourage the participation of all Buddhists in the application of such measures and mechanisms.[34]

This section of the constitution empowers well-established mechanisms for extensive, and increasingly intrusive, bureaucratic control of Thai religiosity. Apart from the state-backed Buddhist ecclesiastical hierarchy, the National Office of Buddhism, the Ministry of Culture's Religious Affairs Department, and two state-funded Buddhist universities today play important roles in controlling and promoting religious activities.[35] Over the past few decades, furthermore, Thai governments have devoted a rapidly increasing share of their budgets to this religious bureaucracy.[36] In this context it is important to acknowledge that Thai state agencies seek to manage not only Buddhism, but also – but generally with a much lighter touch – the officially recognised minority religions (Christianity, Islam, Hinduism, and Sikhism). In addition, a separate arm of the state, the Ministry of Interior, administers religious cults and *latthi* (doctrines) that are not recognised as belonging to the category of *satsana*. Chinese temples and religious associations provide a prominent example of such non-*satsana* religiosity.[37]

Thus, the secular Thai state is for a host of constitutional, legal, and administrative reasons obliged to stipulate what counts as "religion" and to define what is and is not "Theravada Buddhism". It is furthermore obligated to ensure that the population embraces religious ethics, holds "correct" religious beliefs, and engages in "proper" religious practices. There are, perhaps unsurprisingly, serious doubts about the capacity of the Thai state to actually deliver on many of these fronts.[38]

Boundary struggles

Boundary struggles are, as mentioned earlier, central features of the politics of secularisation and secularism. Here two issue areas – education and political citizenship – will serve as brief illustrations of how the boundary between the religious and the secular has been constructed and negotiated. Education provides a paradigmatic example of how two institutionally differentiated spheres – the religious and the secular – emerged in Thailand. The denial of full political citizenship to religious persons, in contrast, is a classic case of the state regulating the degree of permeability between religious and political domains. Adopting the Thai conceptual scheme, these are struggles over the location of activities in the realm of *satsanachak* or *anachak*, and over the ease with which activities and persons are allowed to move between these realms.

Education

In pre-modern Siamese society, educational functions were located in temples, with monks serving as teachers in religious as well as more worldly realms of knowledge (such as medicine). Once ruling kings embraced policies seeking to modernise the Thai education system around the turn of the 20th century, this gradually changed. Because of limited state capacities, the blow of secularisation was, however, softened. The new public schools were often located on temple grounds, and monks continued to play a prominent role in the education of children. However, as educational ambitions grew, political dissatisfaction with the limited ability of monks to provide instruction in modern scientific subjects caused the state to marginalise monks and to rely, increasingly, on public school teachers who had earned qualifications from the state's secular (non-monastic) teacher-training programmes.

In parallel with its takeover of public education, the Thai state simultaneously seized firmer control over much of what counts as religious education. King Chulalongkorn, notably, established two Buddhist universities in Bangkok, one each for the two most prominent Buddhist sects (Thammayut and Mahanikai), introduced a new national curriculum and monastic examination system, and developed a system of royal honours and ecclesiastic promotions designed to incentivise monks to take Buddhist scholarship seriously. While this monastic education system remains in place today, its impact should perhaps not be exaggerated. McDaniel has emphasised that the early 20th-century efforts to nationalise, centralise, and formalise monastic education have had limited success in imposing a standardised education on the Thai monkhood.[39]

The boundary between secular and religious education is not impermeable. Indeed, in recent years Buddhist tenets and Buddhist monks have increasingly penetrated the secular education system, as governments have turned to traditional sources of moral authority and instruction in an effort to halt what has been perceived as the degeneration of the country's youth. Since the early 2000s, the Thai state has therefore encouraged public schools to apply Buddhist principles (*rongrian withiphut*) and to use monks rather than laypersons as teachers of civic morality.[40] The school curriculum has also been reformed so as to give religiously based and especially Buddhist ethics a more prominent position, with the option to study secular ethics removed in 2002.[41] Conversely, the Thai state has over the past century sought to reform monastic education in order to make it more "useful". As a consequence, secular subjects have come to play a more prominent role within the monastic curriculum, so as to make graduates more able to exercise "secular" leadership in local communities[42] and to make them attractive in the job market once they return to "the world".[43] Also with an eye towards enhancing the secular utility of monastic education, a military-appointed legislature has recently put forward the controversial proposal that monastic examinations in Pali language and Buddhist doctrinal studies should be regarded as equivalent to secular academic qualifications. While the secular and religious education systems remain institutionally separate, the flow of ideas, persons, and resources between them increasingly blurs the boundary.

Political citizenship

Section 96 of the 2017 constitution of the Kingdom of Thailand prohibits four distinct categories of person from exercising the right to vote. These are the mentally infirm, convicts, those whose political rights have been revoked, and finally, Buddhist monks, novices, ascetics, and clergy. Similar prohibitions can be found in other laws relating to the exercise of political citizenship, such as election laws and political party laws. Buddhist monks are therefore legally banned from voting in elections, standing as candidates for public office, and establishing or joining political parties.[44] The exclusion of Buddhist monks from formal political participation is not anti-clerical in intent, but rather motivated by a desire to protect the *sangha* from being tainted by any association with "dirty" partisan struggles over worldly power.[45]

Religious disenfranchisement provides perhaps the most explicit illustration of the normative position that those who renounce "the world" and enter *satsanachak* should abandon politics. Further seeking to reinforce this norm are regular reminders from senior *sangha* ecclesiastics that monks also should not exercise political citizenship in more informal ways, such as by joining "political" protests or in other ways expressing political opinions.[46] The supreme patriarch has further decreed that Buddhist temple grounds and facilities must not be used for any "political" purposes, such as electioneering.[47] However, such ecclesiastic injunctions do not have the force of law, and it is noteworthy that there are no *secular* legal provisions that prevent temples and monks from participating in electoral campaigns, publicly indicating their support for or opposition to particular parties and candidates, for example.[48] The enthusiasm with which some monks participate in informal politics serves as a reminder that the ideal of the apolitical *bhikkhu* is not universally shared within the Thai monkhood.

The impermeability of the boundary that excludes the monkhood from formal political participation is, it must be acknowledged, premised on the boundary between *satsanachak* and *anachak* being an extremely permeable one in other ways. Thai men can both enter and exit the monkhood at great ease, with no stigma attached to those who abandon the yellow robe and return to "the world" (at which point political citizenship is restored).

Secularisation without secularism

The main strategy adopted by Thailand's religious and lay elites has been to posit Buddhism as eminently compatible with all relevant demands that modernity has placed upon their state. The approach was pioneered by King Mongkut, who founded and led a Buddhist reform movement that sought to "rationalise" Thai Buddhism by discarding its cosmological and supernatural tenets so as to make Thai Buddhism compatible with the epistemology of modern empirical science.[49] Following in Mongkut's footsteps, Thai elites – monarchs or commoners, clergy and laypersons – have tended to assert the compatibility between Thai Buddhism and the modern nation-state, capitalist development, socialism, and democratic politics. Likewise, Buddhism has been presented as a viable basis of religious toleration. In all these respects, the religious and philosophical basis of the Buddhist state has been posited as equal or indeed superior to "Western" political secularism, whether as practiced in Europe and North America or as attempted in neighbouring Asian countries.

This strategy has been remarkably successful. At the very least, Thai history has yet to arrive at an impasse where institutional Buddhism and its fusion with the state are perceived by the country's elites as a major obstacle to progress or to public order. Secularism, as an ideology, has therefore appeared superfluous to many Thai intellectuals. "Proper" Thai Buddhism, it is argued, is modern, universal, and rational, and as such is able to deliver what secularism promises and more thereto.

If anything, critics have charged that Thai Buddhism only too readily has accommodated itself to modernity, especially to the modern state and to the market economy, with claiming that Thai Buddhism has been turned into a prosperity religion.[50] Some religious conservatives, on the other hand, bemoan what they perceive as the ecclesiastic and political establishment's excessive tolerance towards religious pluralism. Partly in response to proposed "secularising" reforms of the national education system that were initiated in the late 1990s, they have mobilised to pressure political elites to more forcefully assert the hegemonic position of Buddhism within the polity.[51] In recent years, therefore, religious pluralism has increasingly come to be viewed less as a point of Buddhist pride than as a flaw that threatens to undermine the security of the *satsana*.[52] Both of these lines of critique have generated political demands for state-led initiatives designed to "strengthen" Thai Buddhism. Even those who advocate for the state to stop meddling in religious affairs recognise that this, ironically, will only be possible through state-led reform of existing laws and institutions.

In Thai intellectual history, proposals for a more radical separation of *sangha* and state were pioneered by Marxist intellectuals who regarded Thai Buddhism as an opiate of the people.[53] In more recent years, a number of Thai intellectuals have begun advocating for secularism from a more liberal ideological position, as they view the state's control over the religious sphere as a serious obstacle for the consolidation of liberal-democratic political institutions.[54] The continuing violence in the Malay-Muslim provinces in southern Thailand also casts considerable doubt on the capacity of the Thai-Buddhist state as currently fashioned to foster peaceful relations with religious minorities.[55]

Conclusion

The Thai state presents us with a perplexing case. On the one hand, there is clear evidence for the emergence of a distinct "religious" domain and a separate (secular) public sphere of science, economy, and politics, with the boundaries between the two policed by ecclesiastic and lay elites, and occasionally subject to heated political contestation. On the other hand, dominant state ideology rejects any such separation, insisting that that which to the uninitiated might appear as "secular" is subordinated, ultimately, to *thamma*.

Notes

1 See Clemens Six, *Secularism, Decolonisation, and the Cold War in South and Southeast Asia* (Abingdon: Routledge, 2018).

2 Donald E. Smith, "Religion and Political Modernisation: Comparative Perspectives", in *Religion and Political Modernisation*, edited by Donald Eugene Smith (New Haven: Yale University Press, 1974), p. 4.

3 Mirjam Künkler and Shylashri Shankar, "Introduction", in *A Secular Age beyond the West: Religion, Law and the State in Asia, the Middle East and North Africa*, edited by Mirjam Künkler, John Madeley and Shylashri Shankar (Cambridge: Cambridge University Press, 2018), p. 3.

4 See Talal Asad, *Formations of the Secular: Christianity, Islam, Modernity* (Stanford: Stanford University Press, 2003).

5 Saba Mahmood, *Religious Difference in a Secular Age: A Minority Report* (Princeton: Princeton University Press, 2015), p. 3.
6 Philip S. Gorski, "Secularity I: Varieties and Dilemmas", in *A Secular Age beyond the West: Religion, Law and the State in Asia, the Middle East and North Africa*, edited by Mirjam Künkler, John Madeley, and Shylashri Shankar (Cambridge: Cambridge University Press, 2018), p. 49.
7 José Casanova, *Public Religions in the Modern World* (Chicago: University of Chicago Press, 1994), pp. 19–20.
8 National Statistical Office, *Statistical Tables*, Table 4: Population by Religion, Sex, and Area <www.nso.go.th/sites/2014en/Pages/Census/Population%20and%20Housing%20Census/The%202010%20Population%20and%20Housing%20Census/Year%202010/Statistical%20Tables.pdf> (accessed 3 July 2018).
9 See World Values Survey, *Survey Wave 6: 2010–2014 – Thailand* <www.worldvaluessurvey.org/wvs.jsp> (accessed 3 July 2018).
10 Brooke Schedneck, *Thailand's International Meditation Centres: Tourism and the Global Commodification of Religious Practices* (London: Routledge, 2015).
11 See Asad, *Formations of the Secular*.
12 Christian Smith, *Religion: What It Is, How It Works, and Why It Matters* (Princeton: Princeton University Press, 2017), p. 3.
13 *Thamma* (Thai) refers to universal law and to Buddha's teachings; it may also be spelled *dharma* (Pali) or *dhamma* (Sanskrit). *Kam* (Thai) refers to intentional action which affects the cycle of rebirth; it may also be spelled *kamma* (Pali) or *karma* (Sanskrit).
14 Erick White, "The Cultural Politics of the Supernatural in Theravada Buddhist Thailand", *Anthropological Forum*, Vol. 13, No. 2 (2003), pp. 205–212.
15 Thongchai Winichakul, "Buddhist Apologetics and a Genealogy of Comparative Religion in Siam", *Numen*, Vol. 62, No. 1 (2015), p. 88.
16 Suraphot Thaweesak, *Chak Phraphutthasatsana Haeng Rat Su Anakhot Phutthasatsana Thi Pen Isara Chak Rat* [From State Buddhism to a Future Buddhism Liberated from the State] (Bangkok: Siam Parithat, 2017), p. 151.
17 Patrick Jory, *Thailand's Theory of Monarchy: The Vessantara Jātaka and the Idea of the Perfect Man* (Albany: SUNY Press, 2016).
18 See André Laliberté, "Something Got Lost in Translation: From 'Secularism' to 'Separation Between Politics and Religion' in Taiwan", in *Secular States and Religious Diversity*, edited by Bruce J. Berman, Rajeev Bhargava and André Laliberté (Vancouver: UBC Press, 2013).
19 P.A. Payutto (Phrathammapidok), *Rat Kap Phraphutthasatsana: Thueng Wela Chamra Lang Rue Yang* [State and Buddhism: Is It Time for a Complete Cleansing?], third printing (Bangkok: Munithi Phutthatham, 1994).
20 Chris Baker and Pasuk Phongpaichit, *The Palace Law of Ayutthaya and the Thammasat: Law and Kingship in Siam* (Ithaca: SEAP Publications, Cornell University, 2016).
21 Thongchai, "Buddhist Apologetics"; Frank E. Reynolds and Mani B. Reynolds. *Three Worlds According to King Ruang: A Thai Buddhist Cosmology* (Berkeley: Asian Humanities Press/Motilal Banarsidas, 1982); Charles F. Keyes, "Buddhist Politics and Their Revolutionary Origins in Thailand", *International Political Science Review*, Vol. 10, No. 2 (1989), pp. 121–142.
22 Fred R. von der Mehden, "Secularisation of Buddhist Polities: Burma and Thailand", in *Religion and Political Modernisation*, edited by Donald Eugene Smith (New Haven: Yale University Press, 1974).
23 David M. Engel, *Law and Kingship in Thailand During the Reign of King Chulalongkorn* (Ann Arbor: Center for South and Southeast Asian Studies, University of Michigan, 1975). Andrew J. Harding, "The Eclipse of the Astrologers: King Mongkut, His Successors, and the Reformation of Law in Thailand", in *Examining Practice, Interrogating Theory: Comparative Legal Studies in Asia*, edited by Penelope Nicholson and Sarah Biddulph (Leiden: Martin Nijhuis Publishers, 2008).
24 Tamara Loos, *Subject Siam: Family, Law, and Colonial Modernity in Thailand* (Ithaca: Cornell University Press, 2006), p. 92.
25 Ibid., p. 94.
26 Puli Fuwongcharoen. "'Long Live Ratthathammanūn!': Constitution Worship in Revolutionary Siam", *Modern Asian Studies*, Vol. 52, No. 2 (2018), pp. 609–644.
27 Shane Strate, "An Uncivil State of Affairs: Fascism and Anti-Catholicism in Thailand, 1940–1944", *Journal of Southeast Asian* Studies, Vol. 42, No. 1 (2011), pp. 75–76.

28 Eugene Ford, *Cold War Monks: Buddhism and America's Secret Strategy in Southeast Asia* (New Haven: Yale University Press, 2017).
29 Thongchai, "Buddhist Apologetics", p. 9; Christine Gray, *Thailand: the Soteriological State in the 1970s*, PhD Dissertation, University of Chicago, 1986; Peter A. Jackson, "Virtual Divinity: A 21st-Century Discourse of the Royal Influence", in *Saying the Unsayable: Monarchy and Democracy in Thailand*, edited by Søren Ivarsson and Lotte Isager (Copenhagen: NIAS Press, 2010); Christine Stengs, *Worshipping the Great Moderniser: King Chulalongkorn, Patron Saint of the Thai Middle Class* (Singapore: NUS Press, 2009).
30 David Streckfuss, *Truth on Trial in Thailand: Defamation, Treason, and Lèse-Majesté* (London: Routledge, 2010).
31 Eugénie Mérieau, "Buddhist Constitutionalism: When Rajadhamma Supersedes the Constitution", *Asian Journal of Comparative Law*, National University of Singapore, 2018.
32 Daena Aki Funahashi, "Rule by Good People: Health Governance and the Violence of Moral Authority in Thailand", *Cultural Anthropology*, Vol. 31, No. 1 (2016), pp. 107–130.
33 Yoneo Ishii, *Sangha, State, and Society: Thai Buddhism in History* (Honolulu: University of Hawaii Press, 1986), p. 38.
34 Section 67 of the 2017 Constitution of the Kingdom of Thailand. Author's translation from Thai.
35 Tomas Larsson, "Buddhist Bureaucracy and Religious Freedom in Thailand", *Journal of Law and Religion, OnlineFirst*, 13 September 2018 <https://doi.org/10.1017/jlr.2018.27>.
36 Tomas Larsson, "The Political Economy of State Patronage of Religion: Evidence from Thailand", *International Political Science Review, OnlineFirst*, 20 June 2018 <https://doi.org/10.1177/0192512118770178>.
37 Tatsuki Kataoka, "Religion as Non-religion: The Place of Chinese Temples in Phuket, Southern Thailand", *Southeast Asian Studies*, Vol. 1, No. 3 (2012), pp. 461–485.
38 Larsson, "Buddhist Bureaucracy and Religious Freedom in Thailand". Also Tomas Larsson, "Keeping Monks in Their Place?" *Asian Journal of Law and Society*, Vol. 3, No. 1 (2016), pp. 17–28.
39 Justin Thomas McDaniel, *Gathering Leaves and Lifting Words: Histories of Buddhist Monastic Education in Laos and Thailand* (Seattle: University of Washington Press, 2010).
40 Hidetake Yano, "Religious Activities of Administrative Agencies and the Relation Between Religion and the State in Modern Thailand", Paper presented at APISA, Naresuan University, 20–21 October 2017.
41 Hidetake Yano, *Buddhism in Public Sphere and Concept of Religion: State-Religion Relationship in Contemporary Thailand*, 2011.
42 Ford, *Cold War Monks: Buddhism and America's Secret Strategy in Southeast Asia*, pp. 115–116.
43 Khammai Dhammasami, "Idealism and Pragmatism: A Dilemma in the Current Monastic Education Systems of Burma and Thailand", in *Buddhism, Power and Political Order*, edited by Ian Harris (London: Routledge. 2007).
44 These legal prohibitions also apply to female renunciants (*mae chee*).
45 Tomas Larsson, "Monkish Politics in Southeast Asia: Religious Disenfranchisement in Comparative and Theoretical Perspective", *Modern Asian Studies*, Vol. 49, No. 1 (2015), pp. 40–82; Larsson, "Buddhist Bureaucracy and Religious Freedom in Thailand"; Larsson, "Keeping Monks in Their Place?"
46 Thomas Borchert, "On Being a Monk and a Citizen in Thailand and China", in *Buddhism and the Political Process*, edited by Hiroko Kawanami (New York: Palgrave Macmillan, 2016).
47 *Daily News*. 2018. "'Phrasangkharat' Ok Kot Ham Chat Kitchakan Kanmeuang Nai Wat" ['Supreme Patriarch' Issues Decree Banning Political Activities in Temples], *Daily News*, 31 August 2018 <www.dailynews.co.th/politics/663605> (accessed 21 September 2018).
48 The Thai case can be contrasted with the United States, which does not ban clergy from voting or running for election, but where the tax authorities regulate the informal political activities of churches (Gorski, "Secularity I: Varieties and Dilemmas", p. 50).
49 Thongchai, "Buddhist Apologetics"; Thongchai Winichakul, "Coming to Terms with the West: Intellectual Strategies of Bifurcation and Post-Westernism in Siam", in *The Ambiguous Allure of the West: Traces of the Colonial in Thailand*, edited by Rachel Harrison and Peter Jackson (Ithaca: Southeast Asia Program Publications, Cornell University. 2010).
50 Pattana Kitiarsa, "Buddha Phanit: Thailand's Prosperity Religion and its Commodifying Tactics", in *Religious Commodifications in Asia: Marketing Gods*, edited by Pattana Kitiarsa (Abingon: Routledge. 2008); Peter A. Jackson, "The Supernaturalisation of Thai Political Culture: Thailand's Magical Stamps of Approval at the Nexus of Media, Market and State", *SOJOURN: Journal of Social Issues in Southeast Asia*, Vol. 31, No. 3 (2017), pp. 1–55.

51 Katewadee Kulabkaew, *In Defence of Buddhism: Thai Sangha's Social Movement in the Twenty-first Century*, PhD Dissertation, Waseda University, 2013.
52 Suwanna Satha-anand. "Buddhist Pluralism and Religious Tolerance in Democratising Thailand", in *Philosophy, Democracy, and Education*, edited by Philip Cam (Seoul: Korean National Commission for UNESCO, 2003). Also Michael K. Jerryson, *Buddhist Fury: Religion and Violence in Southern Thailand* (New York: Oxford University Press, 2012).
53 Suraphot, *Chak Phraphutthasatsana Haeng Rat Su Anakhot Phutthasatsana Thi Pen Isara Chak Rat*, p. 104.
54 Ibid. Phiphat Phasuthanchat, *Rat Kap Satsana: Botkhwam Wa Duai Anachak Satsanachak Lae Seriphap* [State and Religion: Essays on the Wordly Realm, the Religious Realm, and Freedom] (Bangkok: Siam, 2010). Also Vichak Panich, *Rat-tham-nua* [State-dharma-confusion] (Bangkok: Matichon, 2015).
55 Manuel Litalien, "The Changing State Monopoly on Religion and Secular Views in Thailand", in *Secular States and Religious Diversity*, edited by Bruce J. Berman, Rajeev Bhargava and André Laliberté (Vancouver: UBC Press, 2013). See also Duncan McCargo, "The Politics of Buddhist Identity in Thailand's Deep South: the Demise of Civil Religion?" *Journal of Southeast Asian Studies*, Vol. 40, No. 1 (2009), pp. 11–32.

References

Asad, Talal. 2003. *Formations of the Secular: Christianity, Islam, Modernity*. Stanford: Stanford University Press.

Baker, Chris and Pasuk Phongpaichit. 2016. *The Palace Law of Ayutthaya and the Thammasat: Law and Kingship in Siam*. Ithaca: SEAP Publications, Cornell University.

Borchert, Thomas. 2016. "On Being a Monk and a Citizen in Thailand and China". In *Buddhism and the Political Process*. Edited by Hiroko Kawanami. New York: Palgrave Macmillan.

Casanova, José. 1994. *Public Religions in the Modern World*. Chicago: University of Chicago Press.

Daily News. 2018. "'Phrasangkharat' Ok Kot Ham Chat Kitchakan Kanmeuang Nai Wat" ['Supreme Patriarch' Issues Decree Banning Political Activities in Temples]. 31 August <www.dailynews.co.th/politics/663605> (accessed 21 September 2018).

Engel, David M. 1975. *Law and Kingship in Thailand During the Reign of King Chulalongkorn*. Ann Arbor: Center for South and Southeast Asian Studies, University of Michigan.

Ford, Eugene. 2017. *Cold War Monks: Buddhism and America's Secret Strategy in Southeast Asia*. New Haven: Yale University Press.

Funahashi, Daena Aki. 2016. "Rule by Good People: Health Governance and the Violence of Moral Authority in Thailand". *Cultural Anthropology*, Vol. 31, No. 1: 107–130.

Gorski, Philip S. 2018. "Secularity I: Varieties and Dilemmas". In *A Secular Age beyond the West: Religion, Law and the State in Asia, the Middle East and North Africa*. Edited by Mirjam Künkler, John Madeley, and Shylashri Shankar. Cambridge: Cambridge University Press.

Gray, Christine. 1986. *Thailand: the Soteriological State in the 1970s*. PhD dissertation, University of Chicago.

Harding, Andrew J. 2008. "The Eclipse of the Astrologers: King Mongkut, His Successors, and the Reformation of Law in Thailand". In *Examining Practice, Interrogating Theory: Comparative Legal Studies in Asia*. Edited by Penelope Nicholson and Sarah Biddulph. Leiden: Martin Nijhuis Publishers.

Ishii, Yoneo. 1986. *Sangha, State, and Society: Thai Buddhism in History*. Honolulu: University of Hawaii Press.

Jackson, Peter A. 2010. "Virtual Divinity: A 21st-Century Discourse of the Royal Influence". In *Saying the Unsayable: Monarchy and Democracy in Thailand*. Edited by Søren Ivarsson and Lotte Isager. Copenhagen: NIAS Press.

———. 2017. "The Supernaturalisation of Thai Political Culture: Thailand's Magical Stamps of Approval at the Nexus of Media, Market and State". *SOJOURN: Journal of Social Issues in Southeast Asia*, Vol. 31, No. 3: 1–55.

Jerryson, Michael K. 2012. *Buddhist Fury: Religion and Violence in Southern Thailand*. New York: Oxford University Press.

Jory, Patrick. 2016. *Thailand's Theory of Monarchy: The Vessantara Jātaka and the Idea of the Perfect Man*. Albany: SUNY Press.

Khammai Dhammasami. 2007. "Idealism and Pragmatism: A Dilemma in the Current Monastic Education Systems of Burma and Thailand". In *Buddhism, Power and Political Order*. Edited by Ian Harris. London: Routledge.

Kataoka, Tatsuki. 2012. "Religion as Non-religion: The Place of Chinese Temples in Phuket, Southern Thailand". *Southeast Asian Studies*, Vol. 1, No. 3: 461–485.

Katewadee Kulabkaew. 2013. *In Defence of Buddhism: Thai Sangha's Social Movement in the Twenty-first Century*. PhD Dissertation, Waseda University.

Keyes, Charles F. 1989. "Buddhist Politics and Their Revolutionary Origins in Thailand". *International Political Science Review*, Vol. 10, No. 2: 121–142.

Künkler, Mirjam and Shylashri Shankar. 2018. "Introduction". In *A Secular Age Beyond the West: Religion, Law and the State in Asia, the Middle East and North Africa*. Edited by Mirjam Künkler, John Madeley and Shylashri Shankar. Cambridge: Cambridge University Press.

Larsson, Tomas. 2015. "Monkish Politics in Southeast Asia: Religious Disenfranchisement in Comparative and Theoretical Perspective". *Modern Asian Studies*, Vol. 49, No. 1: 40–82.

———. 2016a. "Buddha or the Ballot: The Buddhist Exception to Universal Suffrage". In *Buddhism and the Political Process*. Edited by Hiroko Kawanami. New York: Palgrave Macmillan.

———. 2016b. "Keeping Monks in Their Place?" *Asian Journal of Law and Society*, Vol. 3, No. 1: 17–28.

———. 2018a. "Buddhist Bureaucracy and Religious Freedom in Thailand". *Journal of Law and Religion*, OnlineFirst, 13 September <https://doi.org/10.1017/jlr.2018.27>.

———. 2018b. "The Political Economy of State Patronage of Religion: Evidence from Thailand". *International Political Science Review*, OnlineFirst, 20 June <https://doi.org/10.1177/0192512118770178>.

Laliberté, André. 2013. "Something Got Lost in Translation: From 'Secularism' to 'Separation Between Politics and Religion' in Taiwan". In *Secular States and Religious Diversity*. Edited by Bruce J Berman, Rajeev Bhargava and André Laliberté. Vancouver: UBC Press.

Litalien, Manuel. 2013. "The Changing State Monopoly on Religion and Secular Views in Thailand". In *Secular States and Religious Diversity*. Edited by Bruce J. Berman, Rajeev Bhargava and André Laliberté. Vancouver: UBC Press.

Loos, Tamara. 2006. *Subject Siam: Family, Law, and Colonial Modernity in Thailand*. Ithaca: Cornell University Press.

Mahmood, Saba. 2015. *Religious Difference in a Secular Age: A Minority Report*. Princeton: Princeton University Press.

McCargo, Duncan. 2009. "The Politics of Buddhist Identity in Thailand's Deep South: the Demise of Civil Religion?" *Journal of Southeast Asian Studies*, Vol. 40, No. 1: 11–32.

McDaniel, Justin Thomas. 2010. *Gathering Leaves and Lifting Words: Histories of Buddhist Monastic Education in Laos and Thailand*. Seattle: University of Washington Press.

Mérieau, Eugénie. 2018. "Buddhist Constitutionalism: When Rajadhamma Supersedes the Constitution". *Asian Journal of Comparative Law*, National University of Singapore, Vol. 13, No. 2 (December): 283–305.

National Statistical Office. *Statistical Tables*. Table 4: Population by Religion, Sex, and Area. <www.nso.go.th/sites/2014en/Pages/Census/Population%20and%20Housing%20Census/The%202010%20Population%20and%20Housing%20Census/Year%202010/Statistical%20Tables.pdf> (accessed 3 July 2018).

P.A. Payutto (Phrathammapidok). 1994. *Rat Kap Phraphutthasatsana: Thueng Wela Chamra Lang Rue Yang* [State and Buddhism: Is It Time for a Complete Cleansing?]. Bangkok: Munithi Phutthatham. Third printing.

Pattana Kitiarsa. 2008. "Buddha Phanit: Thailand's Prosperity Religion and its Commodifying Tactics". In *Religious Commodifications in Asia: Marketing Gods*. Edited by Pattana Kitiarsa. Abingon: Routledge.

Phiphat Phasuthanchat. 2010. *Rat Kap Satsana: Botkhwam Wa Duai Anachak Satsanachak Lae Seriphap* [State and Religion: Essays on the Wordly Realm, the Religious Realm, and freedom]. Bangkok: Siam.

Puli Fuwongcharoen. 2018. "'Long Live Ratthathammanūn!': Constitution Worship in Revolutionary Siam". *Modern Asian Studies*, Vol. 52, No. 2: 609–644.

Reynolds, Frank E. and Mani B. Reynolds. 1982. *Three Worlds According to King Ruang: A Thai Buddhist Cosmology*. Berkeley, CA: Asian Humanities Press/Motilal Banarsidas.

Schedneck, Brooke. 2015. *Thailand's International Meditation Centres: Tourism and the Global Commodification of Religious Practices*. London: Routledge.

Six, Clemens. 2018. *Secularism, Decolonisation, and the Cold War in South and Southeast Asia*. Abingdon: Routledge.

Stengs, Christine. 2009. *Worshipping the Great Moderniser: King Chulalongkorn, Patron Saint of the Thai Middle Class*. Singapore: NUS Press.

Smith, Christian. 2017. *Religion: What It Is, How It Works, and Why It Matters*. Princeton: Princeton University Press.

Smith, Donald E. 1974. "Religion and Political Modernisation: Comparative Perspectives". In *Religion and Political Modernisation*. Edited by Donald Eugene Smith. New Haven: Yale University Press.

Strate, Shane. 2011. "An Uncivil State of Affairs: Fascism and Anti-Catholicism in Thailand, 1940–1944". *Journal of Southeast Asian Studies*, Vol. 42, No. 1: 59–87.

Streckfuss, David. 2010. *Truth on Trial in Thailand: Defamation, Treason, and Lèse-majesté*. London: Routledge.

Suraphot Thaweesak. 2017. *Chak Phraphutthasatsana Haeng Rat Su Anakhot Phutthasatsana Thi Pen Isara Chak Rat* [From State Buddhism to a Future Buddhism Liberated from the State]. Bangkok: Siam Parithat.

Suwanna Satha-anand. 2003. "Buddhist Pluralism and Religious Tolerance in Democratising Thailand". In *Philosophy, Democracy and Education*. Edited by Philip Cam. Seoul: Korean National Commission for UNESCO.

Thongchai Winichakul. 2010. "Coming to Terms with the West: Intellectual Strategies of Bifurcation and Post-Westernism in Siam". In *The Ambiguous Allure of the West: Traces of the Colonial in Thailand*. Edited by Rachel Harrison and Peter Jackson. Ithaca: Southeast Asia Program Publications, Cornell University.

———. 2015. "Buddhist Apologetics and a Genealogy of Comparative Religion in Siam". *Numen*, Vol. 62, No. 1: 76–99.

———. 2016. "Thailand's Hyper-royalism: Its Past Successes and Present Predicament". *Trends in Southeast Asia*, No. 7. Singapore: ISEAS-Yusof Ishak Institute.

Vichak Panich. *Rat-tham-nua* [State-dharma-confusion]. Bangkok: Matichon, 2015.

von der Mehden, Fred R. 1974. "Secularisation of Buddhist Polities: Burma and Thailand". In *Religion and Political Modernisation*. Edited by Donald Eugene Smith. New Haven: Yale University Press.

White, Erick. 2003. "The Cultural Politics of the Supernatural in Theravada Buddhist Thailand". *Anthropological Forum*, Vol. 13, No. 2: 205–212.

World Values Survey. Survey Wave 6:2010–2014 – Thailand. <www.worldvaluessurvey.org/wvs.jsp> (accessed 3 July 2018).

Yano, Hidetake. 2011. *Buddhism in Public Sphere and Concept of Religion: State-Religion Relationship in Contemporary Thailand*. < http://repo.komazawa-u.ac.jp/opac/repository/all/30441/rbk029-05.pdf> (accessed 14 August 2019).

———. 2017. "Religious Activities of Administrative Agencies and the Relation Between Religion and the State in Modern Thailand". Paper presented at APISA, Naresuan University, 20–21 October.

23
THE SOUTHERN CONFLICT

Anders Engvall and Magnus Andersson

This chapter introduces the southern border provinces affected by a long-running violent conflict by analysing the unique character of the area, an overview of the violent conflict, its impact and the ongoing political process towards conflict resolution. The violent conflict in southern Thailand has taken more than 6,000 lives, and yet the underlying causes of the conflict in southern Thailand remain disputed.

The southern context

Thailand is marked by substantial ethnic and religious variation. While all 13 provinces in southern Thailand have substantial Thai-speaking Muslim populations, the three provinces of Yala, Pattani and Narathiwat and four districts of neighbouring Songkhla province are dominated by a Malay-Muslim population. The southern border region is predominantly Malay and Muslim, giving it a unique character within a Thai-speaking and largely Buddhist country. The local language, religion and culture are akin to those of the Malay Muslims in neighbouring Malaysia. While Malay Muslims form a majority in the southern border area, making up about 80 per cent of the population in the region, they are a small minority in Thailand as a whole. The southern border area is a transition zone at a boundary between religions, languages and cultures. The transition is not clear-cut and does not conform to the borders of modern-day nation-states.

It is commonly observed that the southern border provinces form a majority Muslim area within a largely Buddhist state. The picture that emerges is of large local variation in religion. While a core area in the central parts of the region is almost universally Muslim, there are many majority Buddhist sub-districts, particularly on the periphery of the three provinces.

There is no official religion in Thailand and religious freedom is maintained, with a long tradition of inter-religious coexistence and a state that has been accommodating towards Muslims.[1] But despite a history of harmony between Buddhists and Muslims, there is a clear religious dimension to the conflict in southern Thailand. During the conflict, religious leaders from both sides have become targets of violence from insurgents and the security agencies. Islamic leaders who have been taken into custody by security agencies have disappeared or been extrajudicially executed.[2] Insurgent attacks include decapitations of unarmed Buddhist monks.[3] There is evidence that the systematic militarisation of Buddhist temples, many of which have been turned into military posts, and the practice of allowing soldiers to ordain as military monks while remaining armed have increased religious tensions in the region.[4] This gives an indication that religious tolerance has declined in southern Thailand – something that might affect trust in government among adherents to Islam.

Ethnic relations have shaped relations between the state and the local population. The main part of the population in the southern border region is ethnically Malay, which is manifested in some unique cultural characteristics.[6] This may lead to resistance against a state that represents a single cultural norm. This introduces frictions in their relations with a state that, to a great extent, is built on Thai ethnic identity.

The role of language use goes beyond its link to ethnic identity. More than half of the population in the region exclusively speaks Pattani Malay at home, while just above 20 per cent exclusively rely on Thai. Since Pattani Malay is distinctively different from Thai, this creates a significant barrier for interaction with the state and its representatives, where Thai is the only accepted language. Conservative Thai language policy fails to create opportunities for mutual understanding. The failure of the Thai state to accommodate local language use clearly hampers the ability of the government to respond to the needs of the population in the Malay-speaking parts of the southern border region.[7]

The linguistic pattern is similar to the religious, which is unsurprising given the close connection between speaking Malay and being of Muslim faith. Areas with high levels of violence tend to have high rates of Malay speakers, while the reverse holds for areas with low violence. This confirms the commonly held view that one source of friction in the region is the conservative Thai language policy that is perceived as discriminatory towards local minority languages.

The statistics give an indication of the division between the large share of the population (more than half) that only speaks Pattani Malay at home, and about a fourth that only speak Thai. Malay use is highest in a core part of the region made up of Pattani and northern parts of Yala and Narathiwat. The use of Thai is higher in peripheral areas along the southern border to Malaysia and in sub-districts close to Songkhla province to the northwest (see Map 23.1). Almost 80 per cent of the population adheres to Islam. Islamic faith is high throughout the region, at more than 90 per cent in most sub-districts (see Map 23.1). The share is lower in peripheral areas along the border to Malaysia.

The Malay-Muslim region is also marked by substantial variation in religion and language use. The northern part of the region and urban centres such as Yala's Muang district has large Thai-speaking Buddhist populations, many of them descendants of Chinese immigrants. Rural areas, particularly in the interior part of the region, are more homogenously Malay Muslim. Interesting to note is that the four maps show an overlap between the number of population sub-districts using Thai language with Buddhist faith on one hand and population speaking Pattani Malay and belonging to the Islamic faith. The number of sub-districts with an equal mix of the population are low, indicating segregated spatial patterns. However, there are a number of sub-districts with mixed population; these are located in the centre of the region, mainly in Yala and Narathiwat provinces.

Education has for long been regarded as a key instrument both for economic development and for integration of minorities throughout the country.[8] But the response from the population in the southern border region to government education has been mixed. Many opt out of secular education and enrol their children into *pondok* (traditional Islamic boarding schools) or private Islamic schools.[9] Pattani Malay is the main language of instruction in *pondok* and students rarely develop proficiency in Thai. The religious curriculum does not prepare the students for formal employment outside the local villages. Private Islamic schools are run on a dual-curriculum basis, with both religious teaching and secular education similar to that offered in Thai government run schools. Receiving education at *pondok* or private Islamic schools outside the

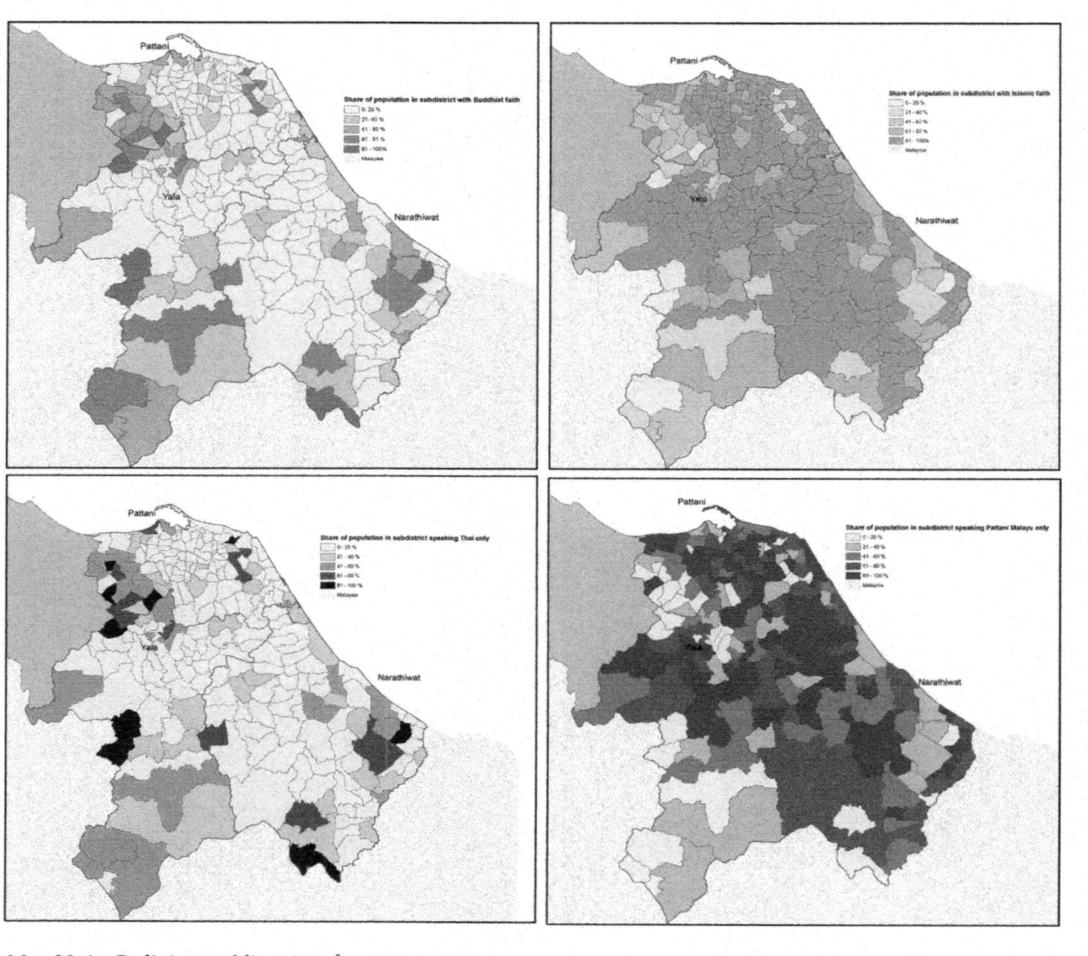

Map 23.1 Religion and linguistics[5]

government run system may reinforce a perception of the Thai government as unable to fulfil the individual's educational needs. It can also increase the perceived distance towards the Thai state and its institutions. The underdevelopment is confirmed by the fact that more than a third of the population lacks formal education and about 15 per cent was unemployed or outside the labour force at the time of the outbreak of the violent conflict.[10]

The region has large variation in natural conditions, with implications for livelihoods and economic opportunities. Coastal regions have diversity in both natural conditions and income opportunities, with fisheries, agriculture and rubber farming, and trading opportunities. The interior areas are dominated by rubber monoculture, which is highly dependent on world market prices. The natural geography of the region has also affected the character of the violent conflict, through the impact on both insurgent and state strategy – lowland areas with larger population density and easier access by a relatively dense road and railway network. The most southern parts of the region neighbouring Malaysia are elevated and covered by forest. Particularly, Yala province contains elevated areas without road access. Malaysia can be reached at any of the border crossings along the border, with two larger border crossings connecting the two countries railway systems.

The southern border provinces are among the poorer regions of Thailand and are substantially less economically developed than other parts of the south. Poverty incidence is on par with poor and isolated provinces in the northeast of the country and household income is well below the average of rural Thailand. In addition to the low aggregate level of economic development, there are also persistent economic cleavages within the area, as the Sino-Thai merchants that dominate the urban economy and the Thai Buddhists that make up a substantial share of government officials enjoy higher standards of living than Malay Muslim villagers. This division between economic activities among the ethnic groups also provides a spatial structure of the ethnic settlements where the urban locations also provide higher income levels.

The poverty rate in the Malay-Muslim south is larger than comparable areas in the upper south as well as the neighbouring areas in Malaysia. All of the three provinces are among the poorest provinces of Thailand. This is both driven by lower income among the economically active and higher numbers of dependents.

A history of rebellion

The region has been claimed as a vassal state by Thai kingdoms since the 15th century.[11] However, the region still retained a great deal of autonomy under Thai suzerainty. Thai provincial administration was heavily decentralised prior to reforms at the end of the 19th century.[12] With the reforms, the Bangkok government made efforts to bring about assimilation and increased central control of the southern provinces.[13] As a result, Pattani was incorporated into the Thai administrative system around 1901–1902, resulting in the abolishment of the local sultans. The centralisation brought about the first revolts against Siamese rule in 1903 due to the resentment of the local aristocracy.[14] In 1906, Bangkok made a policy reversal and gave traditional ruling families a greater role in governing the area. Expanding colonial powers created formal Thai hegemony over the region, and the Anglo-Siamese Treaty of 1909 fixed the current Thailand-Malaysia border.[15] With the treaty, some Malay Muslims were placed under Siamese sovereignty while the majority came under British jurisdiction, later forming an independent Malaysia. The system of indirect rule was retained until 1933, after Thailand's transition from absolute monarchy to constitutional rule.

For a long time the Bangkok government was content with maintaining authority and central control over the southern border provinces without assimilating its population, and the

Malay Muslims kept their separate religious and ethnolinguistic identity.[16] Yet the local elite gradually lost its position in the provincial administration to Thai Buddhists from the centre.[17] The policy of cautious integration changed when a military-led nationalistic regime came to power in the late 1930s. The administration attempted to forcibly assimilate the Malay Muslim population.[18] Prime Minister Field Marshal Phibun Songkhram implemented nationalist policies under which Malay Muslims were forced to assimilate with the predominantly Buddhist Thailand. The Malays were forbidden to wear Malay clothing, use Malay names, and speak and learn the Malay language. Phibun also abolished the Islamic courts, which had been in charge of cases relating to family and inheritance of Muslims in the four southernmost provinces since 1901. Broad public resentment grew as the government removed local laws and discriminated against the use of the Pattani Malay language.[19] During the first wave of resistance there were two parallel movements: the overseas movement led by descendants of former sultans, and a domestic movement led by the religious leader Haji Sulong Abdul Kadir demanding autonomy. The resistance met with strong suppression. Haji Sulong was imprisoned for treason and later mysteriously disappeared in 1954. His disappearance put an end to the largely peaceful resistance.

During the ensuing decade Malay Muslim resistance continued, but at somewhat lower intensity.[20] The late 1960s saw another increase in separatism as a succession of separatist groups carried out a range of bombings, arson attacks, and shootings, targeting representatives of the Thai government.[21] The insurgent activities continued throughout the 1970s. Many of the armed movements that have fought for independence over the years have emerged as reactions against recurring efforts by Bangkok to exert increased authority over the region. The 1970s and 1980s saw an extended separatist campaign by the Pattani United Liberation Organisation (PULO), which relied on traditional guerrilla warfare conducted from jungle bases. This was effectively suppressed by a combination of conventional military campaigns and amnesty programmes. Following the decline of PULO, Barisan Revolusi Nasional (BRN) or the National Revolutionary Front emerged as the main insurgent group, and the movement made a number of strategic shifts away from its predecessors' failures. BRN also focused on initially conducting a systematic mass indoctrination of the local southern population in order to build a solid political base before eventually launching its violent struggle.

The current wave of violence erupted in 2004 with a bold raid on the Chulabhorn military camp, where the separatists made away with a large weapons cache. More violence followed in the early hours of 28 April in the same year, when simultaneous attacks were launched on a dozen checkpoints throughout the region, including a symbolic storming of the Kru Se Mosque. Many of the militants were only armed with sticks or knives and 105 were killed by the security agencies, which only suffered five casualties. On 25 October 2004, a demonstration outside Tak Bai police station in Narathiwat got out of hand and left seven demonstrators killed at the site, with another 78 casualties claimed from suffocation during transport to an army camp.[22]

Most casualties have been claimed in a continuous stream of attacks using light weapons with a small number of victims in each attack. There have also been a few spectacular and coordinated acts of violence including bombings. The history of relations between the south and the central government in Bangkok show that violent opposition against the state has escalated at times when central control over the area has increased and when systems for local resolution of grievances has been absent. In particular, the latest outbreak of violence is associated with the dismantling of a system of governance that had guaranteed relative stability in the region since the 1980s.

BRN has developed a refined organisational strategy that directs the group's activities. In the past, Pattani insurgent groups claimed responsibility for violent attacks, a practice which proved fatal in the end as it allows security agencies to target them effectively. The centralised administrative structure is led by the Party Leadership Council under which there are military and political wings. The two wings serve as a link between leaders and the general population at the village level. The political strategy of BRN is primarily centred on building mass support among the general population in the region. Having village-level support is a precondition for insurgent activity. A primary aim of the group is to ensure that the political wing gains control over the population and destroys the state's legitimacy among the Malay Muslims in the region through continued subversion. Ideally, members of the communities recruit additional supporters. Therefore BRN tries to win support from local religious leaders, who are well placed to take on this role.

The militants are organised into Runda Kumpulan Kecil, or RKK squads, each with six fighters. These squads function as small-group assault units and are organised as largely independent cells. Larger operations may be carried out through cooperation between two or more RKK squads. Such coordinated military action is planned and executed by commanders. Being based in the villages, members of the BRN military wing are amateurs and may switch between their roles as combatants and civilians. This makes counter-insurgency very difficult for the Thai authorities.

The insurgency displays some clear strategic patterns, such as the targeting of representatives of the Thai state, notably military, police, and civil servants. The targets extend to locals collaborating with or working for the government, including village headmen and teachers. Moreover, the strategy includes attempts to provoke violent reactions from the security forces to generate sympathy for the insurgents and legitimise their use of force. The selection of the highly symbolic Krue Se Mosque for a hostage siege was an example of this strategy. The Thai state has responded with violent suppression of the insurgency and with increased presence of police and military. Security agencies have also resorted to extrajudicial killings and abductions.[23] The government has also promoted paramilitary groups, such as village defence volunteers and rangers.[24]

The impact of violent conflict

The dramatic resurgence of armed resistance in 2004 took the Thai state by surprise. While Thailand has experienced armed rebellion in the southernmost region for several decades, the scale and intensity of violence in the post-2004 period are unprecedented. Government agents or those perceived to be the symbol of Buddhist-dominated Thai state – security forces, civil servants, government-sponsored militias, Buddhist monks, and schoolteachers – have been targeted in insurgent attacks. Civilians have suffered the heaviest casualties.

Instances of violence tend to follow linguistic and religious patterns, reinforcing the view that southern insurgents rely on ethnic and religious identities for mobilisation. The central Thai government has been largely ineffective at handling the violence in the south. Efforts to mediate in the conflict are hampered by the hyper-secrecy maintained by BRN leaders and the state's unwillingness to make any concessions.

There has been a marked reduction in the number of incidents and dead since 2012. There was also a fall in the level of violence in mid-2014 (the time of the military coup). Since then, there have not been any marked changes in the level of violence; incidents, dead, and injured remained at constant levels for the last two and a half years covered in the data.

Map 23.2 provides an overview of dead persons due to insurgency-related violence during the period January 2004 to June 2017 divided into sub-districts. There is a low number of sub-districts without any casualties, mainly in very remote located areas on the border to Malaysia.

When disaggregating the violence into different categories, we find that the spatial pattern from Map 23.2 is almost identical. The number of females and civilians as shown in Map 23.3

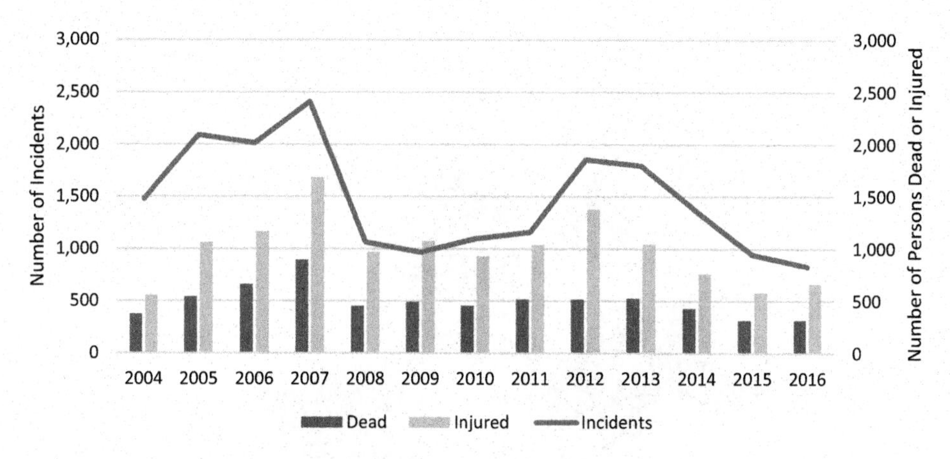

Figure 23.1 Annual violence levels, 2004–2016[25]

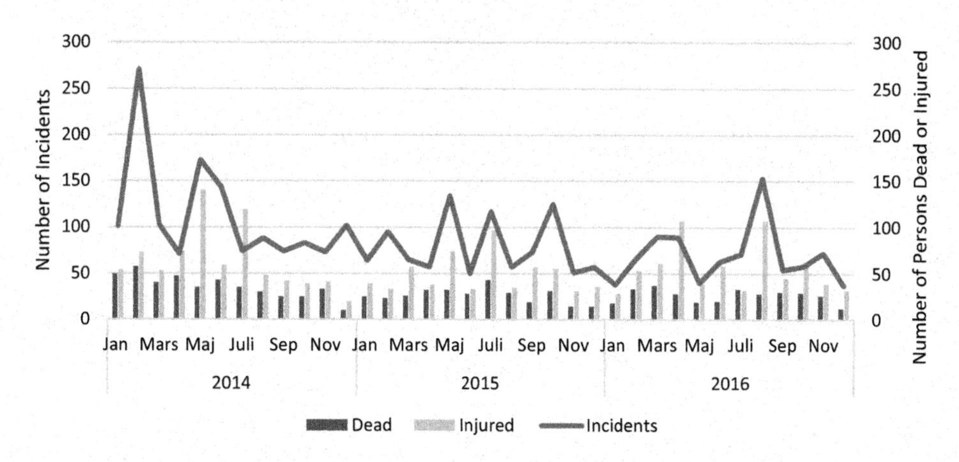

Figure 23.2 Monthly violence levels, 2014–2016

provide similar patterns, with casualties being most severe in the central parts of the region. Also the type of violence divided into shootings or bombings as shown in Map 23.3 indicates similar spatial patterns.

The widespread use of bombs, with their indiscriminate impact, has led to large civilian casualties. Civilians also form the majority of dead throughout the conflict. The most violent years, such as 2007 and 2011, also saw the highest shares of civilians among those killed. While there has been a decline in violence since 2014, this has not been marked by any change in the composition of civilian versus non-civilian deaths.

The victims of are predominately male. The share of killed males was more than 90 per cent during the first three years of conflict, with subsequent years seeing a larger share of women affected by violence. The share of female victims had a first peak in 2007 and a second in 2014. While Buddhists formed the majority of dead in the first year of the conflict, the following years saw a shift to a higher share of Muslim victims. Muslims form the majority of dead at a stable level of approximately 60 per cent during the past six years.

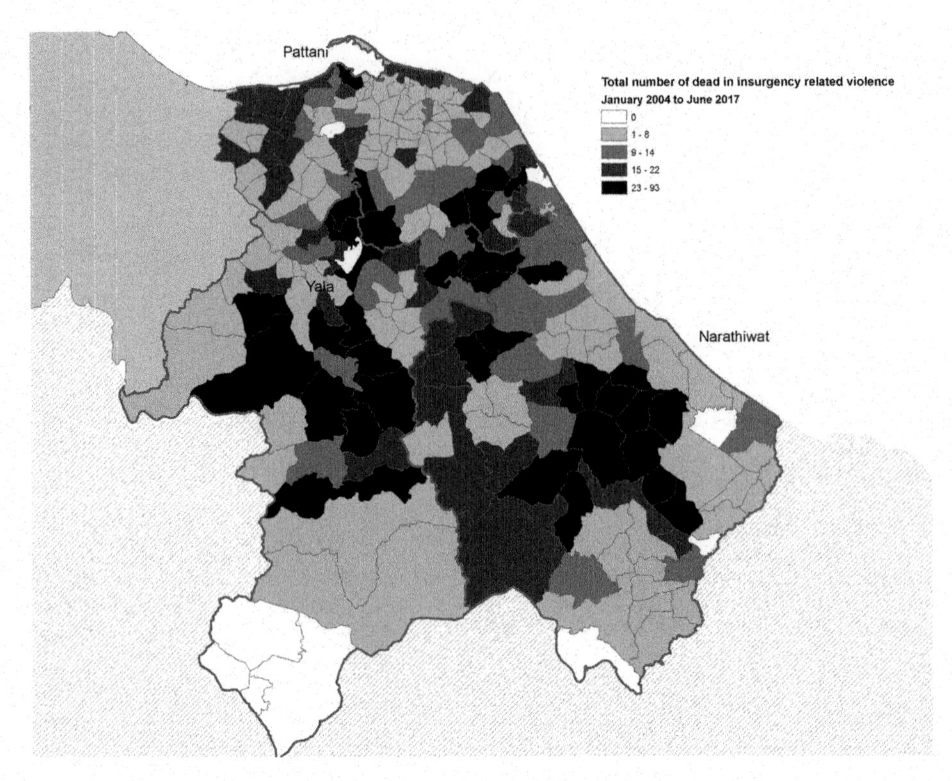

Map 23.2 Conflict-related deaths to illustrate the general spatial pattern of violence

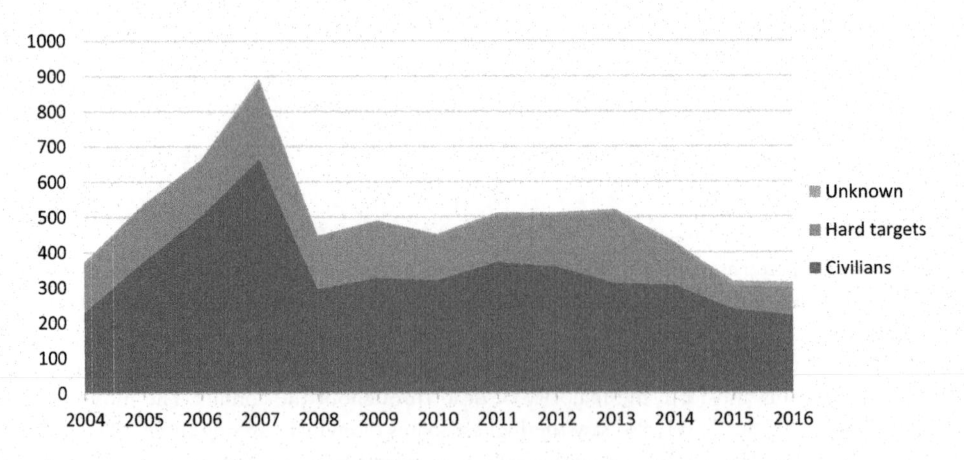

Figure 23.3 Number of dead, civilians and hard targets, 2004–2016

Towards a resolution

From 2004 to 2012, conflict management in southern Thailand was primarily led by classical counter-insurgency operations, which focused on military suppression and development assistance. The military's response to the sharp rise of violence in 2004 suggests its inefficiency,

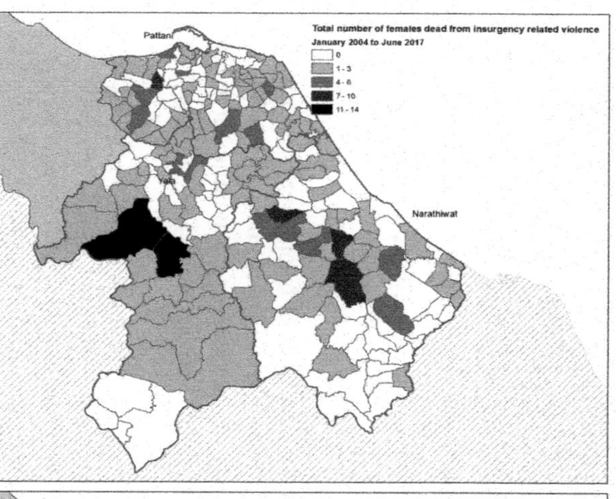

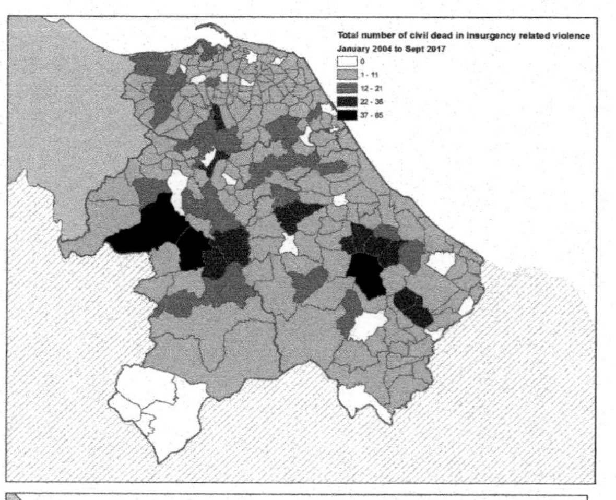

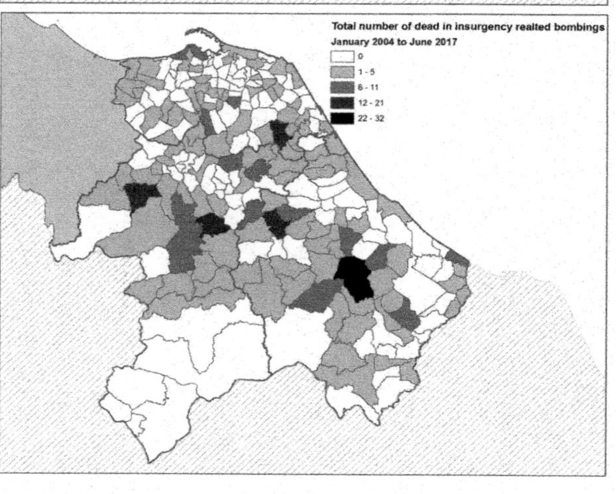

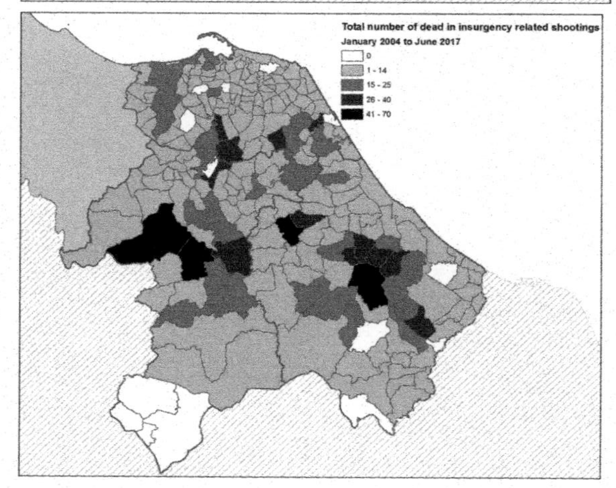

Map 23.3 In-depth spatial pattern of violence

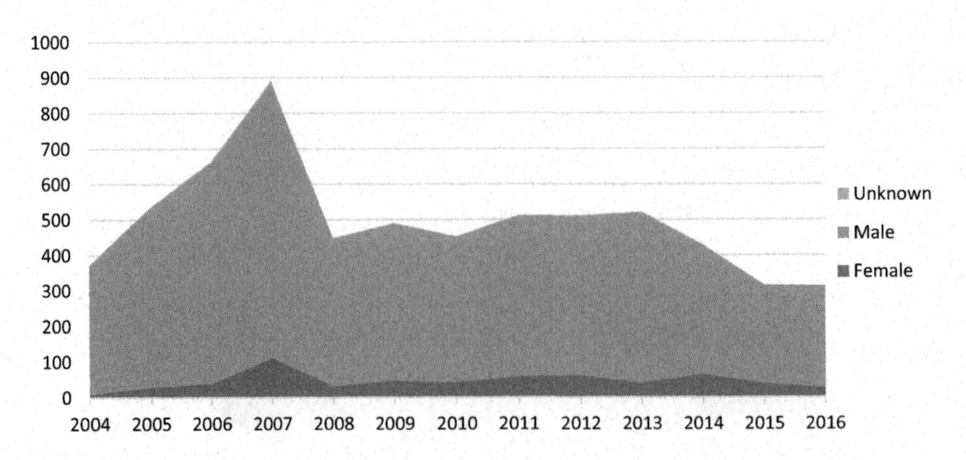

Figure 23.4 Number of dead, female and male, 2004–2016

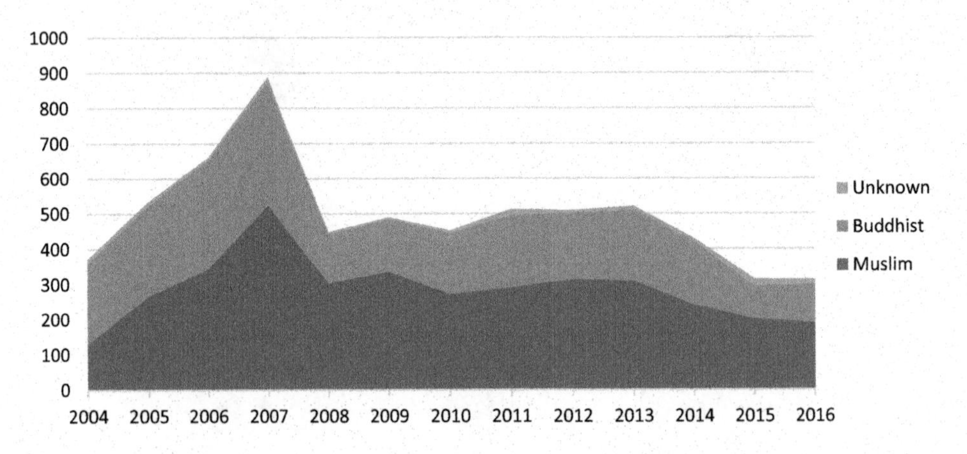

Figure 23.5 Number of dead by religion, 2004–2016

poor intelligence, and lack of proper training. Ten of thousands of troops were dispatched to the Deep South following the 4 January 2004 raid. The Thaksin government responded with heavy-handed suppression, which exacerbated the situations. Since no group came out to claim responsibility for the violence, the prime minister ordered security forces to hunt down the perpetrators.

The counter-insurgency doctrine in Thailand has largely been influenced by the strategies developed during the fight against the Communist Party of Thailand in the 1960s–1980s. It focuses on a two-pronged approach of security and development. The doctrine suggests that the suppression of insurgents need to go hand in hand with development assistance, which enables the military to conduct propaganda operations in order to win the people's hearts and minds. Since 2004, the government has poured in billions of baht to fight against the Malay-Muslim insurgency through military suppression and development assistance, but its effectiveness in neutralising the insurgency remains in doubt.

Between 2004 and 2012, peace dialogue was at the periphery of conflict resolution in southern Thailand. For many years, the highly secretive nature of the insurgency gave the Thai state an excuse to refuse to hold formal peace talks with the liberation movements, citing the elusive identities of violence perpetrators. The southern Thai conflict has been described as Thai security agencies "fighting with ghosts" due to the elusive nature of the militants operating in the southern border provinces.[26] The leadership of these organisations are equally difficult to understand for both outsiders and those seeking a negotiated end to hostilities. The leadership of the main separatist group, BRN, is largely collective and fractionalised. Smaller separatist groups are led by single charismatic individuals who are often engaged in conflicts with other groups claiming to fight for self-determination for the Malay Muslims. This creates challenges which have led to multiple failures in the efforts to end armed hostilities in the region. The role of international actors in efforts to resolve the conflict has been limited, as the Thai state tried at all costs to keep the southern conflict an internal affair. It feared that the involvement of international organisations would help elevate the insurgents' political standing and heighten the risk of secession.

There have been several attempts to find negotiated solutions to the armed conflicts in southern Thailand. Successive Thai governments in the first nine years of the conflict initiated secret talks, only to terminate them without any concrete results, until an official peace process was officially launched in 2013. There were at least four major initiatives prior to 2013, but all failed to develop into a formalised peace process. Malaysian Prime Minister Mahathir Mohamad took the first initiative, the Langkawi Talks. It brought together exiled leaders of the separatist movements with senior Thai security officials in November 2005 and February 2006. The meeting did not have representatives from BRN, seriously limiting the prospects for any major impact on the conflict. The initiative produced recommendations submitted to the Thaksin government, but no serious consideration was given to the proposals.

A second initiative was the Bogor Talks led by Indonesian Vice President Yusuf Kalla. He organised a two-day meeting between Thai officials and separatist leaders at the presidential palace in Bogor, West Java, in September 2008. The talk was put to an end after media revealed the supposedly secret meetings. The Organisation of Islamic Cooperation (OIC) was behind a third attempted to explore the possibility of mediating the southern conflict by holding two parallel meetings with exiled leaders in Malaysia and Saudi Arabia in September 2010. This was met with fierce opposition from the Thai government. While continuing to support the establishment of peace process, the OIC did not make any further attempt to become a mediator.

Another initiative was led by the HD Centre, called the Geneva Process. From the early phase of the violent conflict, HD Centre attempted to establish channels of communication between the Thai government and insurgent leaders. Its effort came to focus on the National Security Council on the Thai government side and the PULO faction led by Sweden-based Kasturi Makota. The Geneva Process appeared to make progress during the government of Abhisit Vejjajiva when the prime minister secretly instructed the NSC to formalise the peace dialogue. However, this process was abandoned after Yingluck came to power in 2011.

The peace process was held back by several factors. A prime reason was the fact that the southern violence has coincided with more than a decade of national-level political turmoil. This has disrupted initiatives to start a sustained peace process. In addition, the Thai military, which plays a dominant role both in conflict management in southern Thailand and national politics, has opposed any sincere peace talks by civilian governments and only supported secret negotiations aiming to co-opt or seed divisions among the insurgent groups. Finally, influential Thai technocrats have opposed any external support to ensure that the southern conflict remains a solely internal affair, making it difficult for international organisations to support peace initiatives.

An important change towards a negotiated resolution came during the Yingluck government, when a formal peace dialogue process with BRN was initiated in February 2013. With

Malaysia acting as facilitator, Thai representatives met with BRN counterparts in Kuala Lumpur in a series of talks. This was the very first time that the BRN expressed their demands to the public. The Yingluck government faced military opposition and General Prayuth Chan-ocha, then the army commander, publicly stated his disagreement with the BRN's demands. A major breakthrough was made as the Thai government and BRN agreed on a 40-day ceasefire during Ramadan in 2013, but this quickly turned into a failure as the Thai military decided to continue their combat operations in the south and in effect spoil both the agreement and the prospects for further talks.

The BRN's military wing was also unwilling to pursue the talks unless its five demands were endorsed in principle. The process was brought to a final end in November as the Yingluck government was put under pressure from growing anti-government protests in Bangkok and BRN announced their withdrawal from the talks. As the political turmoil spiralled out of hand in Bangkok, Yingluck was ousted from office by the Constitutional Court before General Prayuth took power in a military coup against the caretaker government on 22 May 2014. This change of government was bound to have a significant impact on the direction of the peace dialogue.

Somewhat surprisingly, given the military's earlier stance, the junta-led government announced its willingness to resume a Malaysia-facilitated peace dialogue. Nevertheless, substantial changes were made, particularly the representatives of both parties. The new structure overseeing policies on the southern insurgency was dominated by the military. All Thaksin-allied civilians who previously dominated the Thai delegation to the talks were removed and General Aksara Kerdpol was appointed to lead the Thai delegation to the talks. The resumption of peace dialogue officially began when Prayuth met Malaysian Prime Minister Najib in Kuala Lumpur in December 2014.

Decades of mistrust of the Thai military made the separatist leaders wary of taking part in a military-dominated peace process. Under substantial Malaysia pressure, BRN joined other five other organisations in forming a new umbrella group, Mara Pattani, to represent a united front in the talks that resumed in March 2015.

The discussion between the government and Mara Pattani came to an end with the October 2018 appointment of General Udomchai Thammasarorat as the head of the Thai delegation to the peace process. Under Udomchai, the government has opened up to talks with all separatist groups willing to engage with the Thai state. It remains unclear if this effort to broaden the scope of the talks will be more successful than earlier rounds of talks. It seems unlikely that the peace dialogue will bring any settlement in the near term.

Notes

1. Michael Jerryson, "Appropriating a Space for Violence: State Buddhism in Southern Thailand", *Journal of Southeast Asian Studies*, Vol. 40, No. 1 (2009), pp. 33–57.
2. Human Rights Watch, "*It Was Like Suddenly My Son No Longer Existed*": *Enforced Disappearances in Thailand's Southern Border Provinces* (New York: Human Rights Watch, 2007).
3. Human Rights Watch. *No One is Safe: Insurgent Attacks on Civilians in Thailand's Southern Border Provinces* (New York: Human Rights Watch, 2007).
4. Jerryson, "Appropriating a Space for Violence: State Buddhism in Southern Thailand", pp. 33–57.
5. The maps have been developed using the Thai administrative unit sub-district. The choice of level of geographical analysis relate to the relatively small size of sub-districts. Sub-districts are subdivided into administrative villages as the lowest administrative subdivision. Maps showing the three provinces of Yala, Pattani and Narathiwat consist of 244 sub-districts.
6. Thomas M. Fraser, *Fishermen of South Thailand: The Malay Villagers* (Holt, Rinehart and Winston, 1966).
7. William A. Smalley, *Linguistic Diversity and National Unity: Language Ecology in Thailand* (University of Chicago Press, 1994).

8 Astri Shurke. "The Thai Muslims: Some Aspects of Minority Integration", *Pacific Affairs*, Vol. 43, No. 4 (1970), pp. 531–547.
9 Joseph Chinyong Liow and Don Pathan, *Confronting Ghosts: Thailand's Shapeless Southern Insurgency* (Lowy Institute, 2010).
10 National Statistics Office, *2003 Census Project in the Southern Region* (Bangkok: National Statistics Office, 2004).
11 David Wyatt, *Thailand: A Short History* (New Haven: Yale University Press, 2003).
12 Michael Vickery, "Thai Regional Elites and the Reform of King Chulalongkorn", *The Journal of Asian Studies*, Vol. 29, No. 4 (1970), pp. 863–881.
13 Tej Bunnag, *The Provincial Administration of Siam 1892–1915: The Ministry of the Interior under Prince Damrong Rajanubhab* (London: Oxford University Press, 1977).
14 Surin Pitsuwan, *Islam and Malay Nationalism: A Case Study of the Malay-Muslims of Southern Thailand* (Bangkok: Thai Khadi Research Institute, Thammasat University, 1985).
15 Ira Klein, "Britain, Siam and the Malay Peninsula, 1906–1909", *The Historical Journal*, Vol. 12, No. 1 (1969), pp. 119–136.
16 Andrew Forbes, "Thailand's Muslim Minorities: Assimilation, Secession, or Coexistence?" *Asian Survey*, Vol. 22, No. 11 (1982), pp. 1056–1073.
17 Shurke, "The Thai Muslims: Some Aspects of Minority Integration", pp. 531–547.
18 Forbes, "Thailand's Muslim Minorities: Assimilation, Secession, or Coexistence?" pp. 1056–1073.
19 Virginia Thompson and Richard Adloff, *Minority Problems in Southeast Asia* (Stanford: Stanford University Press, 1951).
20 Kadir Che Man, *Muslim Separatism: The Moros of Southern Philippines and the Malay of Southern Thailand* (Singapore: Oxford University Press, 1990).
21 Forbes, "Thailand's Muslim Minorities: Assimilation, Secession, or Coexistence?" pp. 1056–1073.
22 Senate, "Violent Incident at Tak Bai and Problems of Human Security in the Area of Three Southern Border Provinces", *Report of Senate Committee on Social Development and Human Security* (Bangkok, 2005).
23 Amnesty International, *Thailand: Torture in the Southern Counter-Insurgency* (London: Amnesty International, 2009).
24 International Crisis Group, *Southern Thailand: The Problem with Paramilitaries* (Brussels: International Crisis Group, 2007).
25 Authors' calculations based on Deep South Watch (2018).
26 Chinyong Liow and Pathan, *Confronting Ghosts: Thailand's Shapeless Southern Insurgency*.

References

Amnesty International. 2009. *Thailand: Torture in the Southern Counter-Insurgency*. London: Amnesty International.

Deep South Watch. 2018. *Deep South Incident Database*. Prince of Songkhla University, Pattani Campus.

Forbes, Andrew D.W. 1982. "Thailand's Muslim Minorities: Assimilation, Secession, or Coexistence?" *Asian Survey*, Vol. 22, No. 11 (November): 1056–1073.

Fraser, Thomas M. 1966. *Fishermen of South Thailand: The Malay Villagers*. New York: Holt, Rinehart and Winston.

Human Rights Watch. 2007a. "It Was Like Suddenly My Son No Longer Existed". In *Enforced Disappearances in Thailand's Southern Border Provinces*. New York: Human Rights Watch.

———. 2007b. *No One Is Safe: Insurgent Attacks on Civilians in Thailand's Southern Border Provinces*. New York: Human Rights Watch.

International Crisis Group. 2007. *Southern Thailand: The Problem with Paramilitaries*. Brussels: International Crisis Group.

Jerryson, Michael. 2009. "Appropriating a Space for Violence: State Buddhism in Southern Thailand". *Journal of Southeast Asian Studies*, Vol. 40, No. 1 (2009): 33–57.

Klein, Ira. 1969. "Britain, Siam and the Malay Peninsula, 1906–1909". *The Historical Journal*, Vol. 12, No. 1: 119–136.

Liow, Joseph Chinyong and Don Pathan. 2010. *Confronting Ghosts: Thailand's Shapeless Southern Insurgency*. Sydney: Lowy Institute.

Man, Kadir Che. 1990. *Muslim Separatism: The Moros of Southern Philippines and the Malays of Southern Thailand*. Singapore: Oxford University Press.

National Statistics Office. 2004. *2003 Census Project in the Southern Region*. Bangkok: National Statistics Office.

Senate. 2005. "Violent Incident at Tak Bai and Problems of Human Security in the Area of Three Southern Border Provinces". *Report of Senate Committee on Social Development and Human Security*, Bangkok.

Smalley, William A. 1994. *Linguistic Diversity and National Unity: Language Ecology in Thailand*. Chicago: University of Chicago Press.

Suhrke, Astri. 1970. "The Thai Muslims: Some Aspects of Minority Integration". *Pacific Affairs*, Vol. 43, No. 4: 531–547.

Surin Pitsuwan. 1985. *Islam and Malay Nationalism: A Case Study of the Malay-Muslims of Southern Thailand*. Bangkok: Thai Khadi Research Institute, Thammasat University.

Tej Bunnag. 1977. *The Provincial Administration of Siam 1892–1915: The Ministry of the Interior under Prince Damrong Rajanubhab*. London: Oxford University Press.

Thompson, Virginia McLean and Richard Adloff. 1970. *Minority Problems in Southeast Asia*. New York: Russell & Russell.

Vickery, Michael. 1970. "Thai regional Elites and the Reforms of King Chulalongkorn". *The Journal of Asian Studies*, Vol. 29, No. 4: 863–881.

24

CLASS, RACE, AND UNEVEN DEVELOPMENT IN THAILAND

Jim Glassman

Thailand's social, economic, political, and geographical disparities have long been a topic of conversation among academics and development planners. The first reports on national income distribution, from the 1960s, already indicated relatively high levels of inequality.[1] Over the next half century, as the Thai economy boomed, income disparity increased overall, and recent studies have documented not only stubbornly high levels of income inequality but even higher and more intransigent disparities in the distribution of wealth.[2] In addition, as a variety of studies have shown, Thailand's inequalities have a very strong spatial gradient, with economic growth being centred in the Bangkok region – especially among privileged groups – to such an extent that the country's levels of urban economic and demographic primacy are among the highest in the world.[3]

More important than the simple empirical facts of socio-economic and socio-spatial disparity, studies of Thailand's uneven development show it to be a contingent outcome, reflecting the success of Thai elites, privileged social groups, and their international backers in producing development agendas that promote – rather than curtail – unevenness.[4] Moreover, these same groups have also worked – conspicuously in recent years – to prevent the disadvantaged from effectively using political-electoral means to alleviate socio-economic inequality. In short, Thailand's uneven development reflects the success of dominant classes in skewing Thai development agendas and priorities in their own interest.

The fact that the uneven development story has a strong geographical dimension has led some to attempt to naturalise the phenomenon – and thus to obscure the class dimension – by rendering it a story of abstracted spatial dynamics, based largely on forces such as pre-existing agglomeration economies in the Bangkok area.[5] But this story of inequality in abstract space misses the deep socio-spatial realities of uneven development, since Thailand's class structures themselves have a strong socio-spatial dimension, one that helps explain regionally uneven development as a class phenomenon and also provides leverage for class-based analysis of recent political conflicts.[6]

In the discussion of uneven regional development, including accounts of these recent political conflicts, the issue of ethnolinguistic difference is often raised, though it is also frequently sequestered from class analysis. By contrast, I suggest that the story of socio-spatial inequality in Thailand is, integrally, a story of both regionally uneven class development and geographically/historically constructed ethnolinguistic difference. Indeed, to connect this discussion with a broader literature on racial formation and the racialisation of class structures, I will refer to Thailand's uneven development as a function of the racialisation of Thailand's interconnected class transformation and territorialisation processes. Ethnogenesis and racialisation can, of course, be distinguished from one another, but there are fundamental overlaps and connections between the two in the social formation processes I am describing here. As such, I will use the terms somewhat interchangeably, differentiating between them as needed.[7]

I present these issues in four parts. In the first section of the chapter, I outline major tenets of approaches that foreground racial formation and the racialisation of class. In the second section, I trace some of the 19th- and early 20th-century processes of territorialisation, racialisation, and class formation that shaped and mobilised the forms of uneven development already evident in Thailand by the early post–World War II period. In the third section, I note some continuities and discontinuities in the regionalisation and racialisation of class process over the last half century. In the final section, I briefly summarise what this sort of perspective on inequality suggests about attempts to make Thailand more egalitarian.

Territory, race, and class: conceptualising interconnections

Theories of racial formation have often focused on American and Atlantic experiences.[8] Yet, the notion that race, like ethnolinguistic identity, is not "natural" but rather the outcome of complex processes of social construction is today a commonplace, shaping studies of racial and ethnolinguistic difference in all corners of the world, across vast swaths of history.[9] To be sure, a certain focus on the Atlantic world is inevitable and warranted, given that modern conceptions of race are very much a product of the Atlantic slave trade and European colonising ventures, including in the Americas.[10] Yet the same broad dynamics of racial formation can be witnessed elsewhere – not least as a reflection of the ways racial ideas propagated in Europe and North America took hold and were deployed by leaders elsewhere, including, for example, in Japan and Thailand.[11]

Given the contexts of colonial geographic expansion in which so much racial and ethnolinguistic formation has taken shape, it is also commonplace to analyse the connections between processes of territorialisation – for example, the production of national boundaries and related forms of sovereignty – and racial/ethnolinguistic formation. This has regularly been done in the case of Thailand, as I note below. But it is equally important to draw out connections between territorialisation, racial formation, and *class formation* – as, for example, Anievas and Nişancıoğlu show in their discussion of the transatlantic foundations of European capitalism.[12]

To set up the discussion that follows, I clarify here what I mean by constructing territorialisation and racialisation in class terms. Classes exist stretched across space in different patterns and forms, which is part of why class can by subjected to differing forms of social analysis.[13] Territorialisation refers not only to the construction of political borders but to all processes that are productive of exclusive claims over space (such as drawing municipal and other juridical boundaries and making exclusive claims to commodity distribution rights in particular areas.). Territorialisation, in this sense, has been an ongoing, if uneven, phenomenon over the capitalist longue durée. Classes and class fractions within capitalism have always formed through complex interconnections with territorialisation processes, not simply within pre-given, hermetically sealed spaces. Thus, the common practice of referring to classes as if they were inherently national in scale, formed and existing most fundamentally within national territorial units, is misleading insofar as it neglects the geographical-historical dynamics through which territories and class structures are formed. Moreover, class analyses that are "territorially trapped" frequently neglect the transnational and sub-national dimensions of actually existing class structures, substituting for these spatial complexities an ideal type model of classes that at best partially approximates reality. Class formation is better seen as a fungible, potentially shifting process, one that leads to a variety of class and class-fractional coalitions rather than to the simple, clean class divisions that are sometimes imposed by mechanical analysts on the vagaries of geographical-historical reality.

In contrast to the relationship between territorialisation and class, the relationship between territorialisation and racial formation has frequently been foregrounded, not least because both ethnogenesis and racial formation have been rightly seen as dramatic manifestations of the geographical expansion and dispersion of different groups of people over long periods of history.[14] Race and ethnolinguistic identity have thus typically been understood to be inherently spatial phenomena. Yet in a parallel to "territorially trapped" conceptions of class, some analysts of ethnicity and race – and especially nationalist political leaders – have attempted to impose upon the socio-spatial complexity of racial formation a simplifying narrative. In this nationalist narrative, no matter the vagaries of a distant past, the present is marked by the consolidation of a single "people", ideally a people with a common ethnolinguistic and/or biological origin. As Anderson notes, these kinds of constructions of the nation are frustrating to scholars because of the wide chasm separating reality from nationalist depictions of it; yet for all of the scholarly and empirical weaknesses of these simplifying nationalist narratives, they have had remarkable political power, not least in the last century.[15] Importantly for the case discussed here, highly racialised – indeed racist – conceptions of "the nation", usually with origins in European and North American racial thought, were frequently adopted by Asian leaders attempting to spur processes of "modernisation" like those of Western industrialising (and imperialist) societies. In this process, territorialisation, racialisation, and nationalism came to play important roles in 19th- and 20th-century social transformations within Asia.

Since both class formation and racialisation/ethnogenesis have been strongly territorialised over recent centuries, it follows that the class formation process being discussed here is racialised and vice versa. That this is the case does not imply that there is any *specific form* that the racialisation of class necessarily takes. Class and race (and ethnicity) are best seen as articulated, joined together in a fashion that is contingent.[16] To say this, however, is *not* to say that there might potentially exist pure class or race structures; class is always necessarily racialised and race is always necessarily classed, but the specific forms of these articulations are contingent. For this reason, the core of any analysis of the articulations between race and class is conjunctural and empirical, not simply logical or macro-theoretical. Thus, the analysis of uneven development in Thailand that follows focuses on the geographical-historical dynamics of racialised Thai class formation in two periods, from the mid-19th to the mid-20th centuries, and from the mid-20th century to the present.

Thai nationalism and the formation of a racialised class structure under the absolute monarchy

The starting point for understanding the territorialisation process that formed modern Thailand between the mid-19th century and the 20th century is to acknowledge, as Ferrara notes, that as of the beginning of the Chakri dynasty in the late 18th century the Siamese Kingdom, based in Bangkok, controlled less than half of the territory that today constitutes Thailand.[17] Moreover, it is something of a misnomer to say that Siam controlled the "territories" over which the Chakri dynasty exercised sovereignty since absolutist, monarchical sovereignty was not at that point territorialised and marked by internationally recognised boundaries, as Thongchai's study has famously emphasised.[18]

The territorialisation process that got underway in the 19th century was a direct response to the threats posed by European colonialism in Asia. Recent Thai historical scholarship has rightly pointed out the ways in which the Chakri elite – particularly under the fourth and fifth reigns – prevented themselves from being simple victims of European colonisation (though having to give up a significant amount of economic sovereignty in the 1855 Bowring Treaty), by exercising what has been called "semi-colonialism", "crypto-colonialism", or simply "colonialism" in their relations with tributary states.[19] The incorporation of what was to become the upper north, the northeast (Isan), and the Deep South (predominantly Muslim), were among the crucial outcomes of the semi-colonial territorialisation process, a process that was met frequently with rebellion and resistance by a variety of actors in those regions.

Both racialisation and class formation were integral aspects of semi-colonial territorialisation. Racialisation was itself a direct response to European colonial ventures, with Chakri elites adopting and adapting European racial categories and deploying them as part of the project of building a modern Thai nation. The contours of this project have been detailed by Thongchai, Streckfuss, and others. One crucial aspect of the project has been that, on the one hand, it has attempted to erase pre-existing conceptions of ethnolinguistic difference – such as between Isan (modern Thailand's northeast region) and Siam – substituting for these a constructed national sense of "Thainess" while at the same time producing a very clear hierarchy of sub-national ethnic and territorial groupings, with Bangkok (and the central Thai language) at the top and regional identities/languages such as Isan, the north (and *kam meuang*, the northern dialect), and the "Muslim" provinces (as four provinces in the south are known) further down the hierarchy.[20]

This project has been problematic in a number of ways, one of which has to do with how it deals with "Chineseness". On the one hand, as an immigrant group from a place clearly distinguishable from Siam, Chinese were constructed, early on, as a separate race.[21] But the deep interpenetration of Chinese and Siamese economic elites throughout the 19th and 20th centuries, along with the increasing assimilation of working-class immigrant Chinese, generated a Sino-Thai identity that has neither been easy to fully sublate within Thainess nor to exclude as foreign. The complexities of this situation were exacerbated in the early 20th century by Rama VI's claim that Chinese in Thailand functioned in the manner of the Jews in Europe – a full-throated adaptation of anti-Semitic European thought that was, however, rarely made into consistent and coherent practice in Siam.[22]

The racial position of the Sino-Thai within late 19th- and early 20th-century Siam was made additionally complex by the fact that recent Chinese immigrants filled out a preponderant share of both the industrial capitalist and merchant classes, generally in cooperation with Thai royals and various other local actors, and the nascent industrial working class, performing much of the labour (often bonded) in Bangkok ports, southern tin mines, and other industrial ventures.[23] As such, Sino-Thais in Siam constituted the leading edge of capitalist class transformation, comprising a likely majority of both the exploiting and the exploited classes within what existed of the capitalist mode of production.

The processes of class transformation in the formerly "Lao" areas of what was to become Thailand (the upper north and Isan) can be noted here for what they illustrate about the class transformations occurring within the process of racialised territorialisation. As these populations were being incorporated into Thailand and slotted into lower positions within the hierarchical national ethnolinguistic order, their class structures gradually changed. Early in the process, rebellions were often launched by actors such as downwardly mobile lower nobility – as with the Phraya Phap rebellion of 1889–1890 near Chiang Mai – allied with peasants disgruntled by the monetisation of the economy through the new taxes collected by Sino-Thai tax farmers[24] and/or by communities of petty producers and local leaders (including religious figures) rebelling against these new taxes and other extra-local impositions, as with the 1900–1902 *phu mi bun* uprising in Isan.[25] The rebellions had a class basis in the resistance of nominally "free" agricultural producers (*phrai*) to new impositions by the territorialising state and its capitalist (and proto-capitalist) backers. The rebellions were also met with the class forces of the Thai state, which mobilised military repression (including via British weapons) that was beyond the means of peasants armed with machetes.[26]

In this sense, the territorialisation process of the late 19th and early 20th centuries produced not only a new national state, with an internal ethnolinguistic hierarchy and an administratively semi-colonial structure, but an evolving class structure in which Bangkok elites and Sino-Thai merchants (including tax farmers) prospered by imposing on new state subjects an array of taxes that forced increasing numbers of agrarian producers into monetised production for the market, the opening wedge of a process of "primitive accumulation".[27] In much of the north and northeast as well as the south, primitive accumulation moved very slowly throughout the first half of the 20th century. As of the end of the Second World War, the majority populations in the upper north and northeast still counted agriculture as their primary household occupation. But the agriculture in which they engaged was increasingly mediated by cash, involving among other activities increased production of rice for the market – a process that had already accelerated dramatically in the regions near Bangkok by the early 20th century.[28]

Though the kinds of national income data used today to gauge economic inequality were not available in this period, there are good reasons to believe that many of the large socio-economic and socio-spatial disparities quantified in post–World War II studies were already in formation through these processes of semi-colonial uneven development. Most important, what the pre–World War II historical geography of Thai state formation shows is that these economic inequalities were not inevitable or naturally occurring but were the result of class-based and racialised projects of state formation and repression. Such projects would evolve further throughout the next half century.

Racial formation and the racialisation of the Thai class structure under developmental and post-developmental authoritarianism

After the end of the absolute monarchy, and especially with the entry of Thai leaders into the US Cold War alliance after the Second World War, Thai governance structures came to be pervasively marked by militarised authoritarianism, often taking the form of military dictatorship. This authoritarianism, which deeply conditioned the forms taken by development projects, was leavened by sporadic and sometimes intense outbursts of popular antagonism to military rule.[29] Like the late 19th- and early 20th-century rebellions against semi-colonial incorporation into the Thai state, many of the post–World War II rebellions against authoritarian state practice have had a strongly regionalised and racialised (or "ethnicised") class basis.[30] I note several examples here while tracing some of the developments in inequality to which they relate.

Between 1944 and 1947, the Thai government in which Pridi Banomyong played a leading role attempted a number of democratic reforms intended to undermine the strength of the military leadership that had collaborated with Japan during the Second World War. Pridi's regime was receptive to the concerns of labour unions, and it had particularly strong support among representatives from Isan.[31] In addition, the Pridi government favoured an approach to post-war reconstruction in East and Southeast Asia that was consistent with some of the non-aligned projects favoured by leaders in Burma, Indonesia, and Indochina.[32] These features of the 1944–1947 government placed it at odds with both Thai elites, including royalists and military leaders, and the US Cold War state. When Pridi was ousted in a coup d'état in 1947, the United States quickly threw support behind the military dictatorship led by Phibun Songkhram, helping to cement an authoritarian alliance that was particularly strong until the end of the US war in Vietnam 25 years later. The dictatorship systematically silenced political voices like those of the Isan representatives who had supported Pridi,[33] and it also presided over a capital-friendly and labour-repressive regime powerful enough to produce substantial growth in the Thai economy with virtually no attendant growth in wages over the entire 25 years.[34] Meanwhile, through policies such as a tax on rice exporters (in effect from 1955 to 1985), who passed the costs down to rural producers, it extracted surplus from the countryside and hastened the exodus to the city of young people looking for more remunerative job opportunities.[35]

All of this helped produce a widening income and wealth gap across Thailand, and the inequalities were marked, in addition, by a shifting racialisation of the Thai working class. While industrial labourers in the early 20th century had been disproportionately Sino-Thai, after the Second World War an increasing share of the urban labour force comprised recent migrants from the countryside, most of whom were ethnically Thai or (increasingly by the end of the 1970s and 1980s) from Isan and northern Thailand.[36] Sino-Thais remained important members of the urban labour force, but they were increasingly likely to be petty capitalists (small business owners), while the richest and most powerful Sino-Thais took on increasingly influential roles within the Thai ruling class.[37] The urban-industrial working class thus became decreasingly "Sinicised" (in relative terms) and increasingly "Thai-ified", but with much of the "Thai" working class actually comprising recent rural in-migrants from ethnolinguistic groups lower in the national hierarchy than Central Thais.

In the period 1973–1976, when there was another important outburst of popular activism that temporarily sidelined the military dictatorship, labour activists demanding better wages and working conditions, peasant and village activists demanding land rent control, and students demanding democratisation combined into a powerful force promoting a more egalitarian society. This movement was met by Thai elites with massacre, repression, and a new round of military dictatorship, and though the forms and intensity of dictatorship varied over the next decade, the period between 1976 and 1988 was again a period in which elites could rule without popular forces derailing their agendas.[38]

This marginalisation of non-elite groups conditioned the growth boom that began in 1986–1987 and lasted until the economic crisis of 1996–1997. The boom disproportionately benefited the richest members of Thai society and exacerbated existing income and wealth gaps, including along regional lines.[39] Moreover, polarisation was once again a contingent matter of policy and not inevitability. For example, the lightly regulated lending policies of private banks, owned by the richest Sino-Thais,[40] allowed them to extract savings from areas such as Isan and the north without having to re-lend comparable funds to those regions, the funds being instead loaned out to support business expansion in the Bangkok region.[41]

The result of the uneven economic boom was not only increasing disparity but an increasingly racialised influx of workers to the Bangkok region, with workers from Isan forming the backbone of the evolving and highly exploited labour force.[42] Supplementing this relocated upcountry labour force was an increasingly large immigrant labour force from neighbouring countries, with these foreign workers typically occupying an even lower rung on the racialised social ladder than workers defined as Thai.[43] Moreover, the Thai labour force was increasingly gendered, and while men and women comprised comparable shares of the workforce by the 1990s, women were particularly heavily represented in the fastest-growing export-oriented industries, where they generated high profits for owners while suffering sometimes debilitating working conditions.[44]

The political upheavals of the early 1990s temporarily put the military on the back foot, but the complex events that have ensued since the 1990s economic crisis have ultimately provided ruling class and military forces with a new lease on life.[45] The crisis response and "populist" policies of the Thaksin Shinawatra government had some salutary effects on outcomes such as the health of poorer communities,[46] along with the overall distribution of income in the country.[47] They also represented some effort at regionalised redistribution, though they were also clearly designed to shield basic structures of power from attempts at radical transformation.[48] Nonetheless, they antagonised elite groups in Thailand, the military leadership, the monarchy, and ultimately Thaksin's erstwhile supporters among the Sino-Thai business elites. Thus, when Thaksin was ousted in the 2006 military coup, it was difficult to miss the very obvious class basis of elite discontent. Moreover, as the Red Shirt movement formed during 2009 and began mass protests in 2010, in an attempt to counter the military's reassertion of power, the strong regional and ethnic basis of the Red Shirts in the communities of Isan and northern Thailand made it clear that the racialisation of class formation and uneven development in Thailand has led to a political cul-de-sac.[49] Falling prey to the contradictions of their own successful, polarising development agenda and the aspirations it has stoked in exploited and repressed social groups, Thai elites and the military could only respond to the Red Shirt protests of April–May 2010 with yet another massacre of their political opponents.[50]

Implications of Thailand's racialised class formation and uneven development

As Keyes notes, Yellow Shirt supporters of the military regime dominating Thailand's political landscape since 2006, including Sino-Thai business elites who have helped organise Yellow Shirt efforts, haven proven remarkably forthright in their expression of classist-racist contempt for Red Shirts – sometimes employing measures such as portraying Red Shirt protestors as water buffalo.[51] Underlying these bigoted portrayals of rural Thais as slow, dim-witted beasts of burden appears to be a deep inability on the part of Thai elites to accept the changes that have occurred in Thai rural society, including the changing consciousness of those who fill out the ranks of the Red Shirt movement.[52] Thus, in spite of the rich documentation that now exists of the ways in which villagers from places such as Isan and the north have exploded old urban-elite myths about their character and aspirations, politically dominant groups in Thailand continue to portray themselves as the bearers of civilisation and their class subordinates from the provinces as social inferiors. The transformed position of Sino-Thai elites – who have increasingly been seen as models to emulate in recent decades – contributes to this phenomenon and complicates the racialised class story.[53] Sino-Thai elites today stand shoulder to shoulder with other Thai elites, bearing no taint of being "the Jews of Asia", and feeling empowered to direct prejudicial epithets at the former peasants from places such as Isan who labour in their enterprises and yield up the surpluses that make elites wealthy.

These complexities do not obscure a fairly straightforward political conclusion. If race and class are neither separate nor reducible to one another but articulated, and if their historical articulation has been a powerful force in the enrichment of Thailand's elites through its effects on uneven development, then we should not expect these elites to abandon racialised class projects of their own accord. Moreover, if Thailand's unevenness is a matter of the articulation of race and class, then neither relying simply on compensatory economic projects (regional or otherwise) nor simply preaching racial tolerance is likely to have much effect, since inequality is both a class-based and a racialised phenomenon. Race-class articulations that disadvantage groups like villagers from Isan can, very likely, only be contested through the resistance of those groups themselves. This is precisely what has animated the Red Shirt movement, but Red Shirt efforts have been met by the usual elite repression – and the usual impunity for Thai figures that violently repress opposition groups.[54] If this pattern is not challenged – including by an "international community" that typically shows little interest in these kinds of realities in Thailand – then racialised class inequality in Thailand may have a long and evolving future, just as it has had a long and evolving past.

Notes

1 Suganya Hutaserani and Somchai Jitsuchon, "Thailand's Income Distribution and Poverty Profile and their Current Situations", 1988 TDRI Year-End Conference on Income Distribution and Long-term Development (Bangkok: Thailand Development Research Institute, 1988); Jim Glassman, *Thailand at the Margins: Internationalisation of the State and the Transformation of Labour* (Oxford: Oxford University Press 2004), p. 162.
2 Kevin Hewison, "Considerations on Inequality and Politics in Thailand", *Democratisation*, Vol. 21, No. 5 (2014), pp. 846–866; Pasuk Phongpaichit, "Inequality, Wealth and Thailand's Politics", *Journal of Contemporary Asia*, Vol. 46, No. 3 (2016), pp. 407–413; see also Alain Mounier and Voravidh Charoenloet, "Thailand: Labour and Growth after the Crisis: New Challenges", *Journal of Contemporary Asia*, Vol. 40, No. 1 (2010), pp. 123–143.
3 Glassman, *Thailand at the Margins*, p. 112.
4 Kevin Hewison, *Bankers and Bureaucrats: Capital and the Role of the State in Thailand* (New Haven, CT: Yale University Southeast Asia Studies, Yale Center for International and Area Studies, 1989); Akira Suehiro, *Capital Accumulation in Thailand, 1855–1985* (Tokyo: Centre for East Asian Cultural Studies, 1989); Glassman, *Thailand at the Margins*; Jim Glassman, *Bounding the Mekong: the Asian Development Bank, China, and Thailand* (Honolulu: University of Hawaii Press, 2010a).
5 Eliezer Ayal, "Thailand's Development: The Role of Bangkok", *Pacific Affairs*, Vol. 65, No. 3 (1992), pp. 353–367.
6 Jim Glassman "'The Provinces Elect Governments, Bangkok Overthrows Them': Urbanity, Class and Post-Democracy in Thailand", *Urban Studies*, Vol. 47, No. 6 (2010b), pp. 1301–1323; Jim Glassman, "Cracking Hegemony in Thailand: Gramsci, Bourdieu, and the Dialectics of Rebellion", *Journal of Contemporary Asia*, Vol. 41, No. 1 (2011), pp. 25–46.
7 See also David Streckfuss, "The Mixed Colonial Legacy in Siam: Origins of Thai Racialist Thought, 1890–1910", in *Autonomous Histories, Particular Truths: Essays in Honor of John R.W. Smail*, edited by Laura J. Sears, Monograph No. 11 (Madison: University of Wisconsin Center for Southeast Asian Studies, 1993), pp. 123–154; David Streckfuss, "An 'Ethnic' Reading of 'Thai' History in the Twilight of the Century-Old Official 'Thai' National Model", *South East Asia Research*, Vol. 20, No. 3 (2012), pp. 305–327.
8 For example, Michael Omi and Howard Winant, *Racial Formation in the United States: From the 1960s to the 1980s* (New York: Routledge and Kegan Paul, 1986).
9 For example, Kees van der Pijl, *Nomads, Empires, States* (London: Pluto Press, 2007).
10 Ivan Hannaford, *Race: The History of an Idea in the West* (Washington, DC and Baltimore, MD: Woodrow Wilson Center Press and Johns Hopkins University Press, 1990); Paul Gilroy, *The Black Atlantic: Modernity and Double Consciousness* (London: Verso, 1991); Alexander Anievas and Kerem Nişancıoğlu, *How the West Came to Rule: The Geopolitical Origins of Capitalism* (London: Pluto Press, 2015), pp. 123–134.
11 John Dower, *War without Mercy: Race & Power in the Pacific War* (New York: Pantheon 1986), pp. 262–290; Streckfuss, "The Mixed Colonial Legacy in Siam: Origins of Thai Racialist Thought, 1890–1910"; "An 'Ethnic' Reading of 'Thai' History in the Twilight of the Century-Old Official 'Thai' National Model"; Thongchai Winichakul, "The Quest for 'Siwilai': A Geographical Discourse of Civilisational Thinking in the Late Nineteenth and Early Twentieth-Century Siam", *Journal of Asian Studies*, Vol. 59, No. 3 (2000a), pp. 528–549; Thongchai Winichakul, "The Others Within: Travel and Ethno-Spatial Differentiation of Siamese Subjects 1885–1910", in *Civility and Savagery: Social Identity in Tai States*, edited by Andrew Turton (Richmond, Surrey: Curzon, 2000b), pp. 38–62.
12 Anievas and Nişancıoğlu, *How the West Came to Rule: The Geopolitical Origins of Capitalism*, fn. 134.
13 Eric Sheppard and Jim Glassman, "Social Class", in *Handbook of Geographical Knowledge*, edited by John Agnew and David Livingstone (Thousand Oaks: Sage Publications, 2011), pp. 405–417.
14 Hannaford, *Race: The History of an Idea in the West*; Pijl, *Nomads, Empires, States*; Anievas and Nişancıoğlu, *How the West Came to Rule: The Geopolitical Origins of Capitalism*.
15 Benedict Anderson, *Imagined Communities: Reflections on the Origin and Spread of Nationalism*, second edition (London and New York: Verso, 1991).

16 Stuart Hall, "Race, Articulation and Societies Structured in Dominance", in *Sociological Theories: Race and Colonialism*, edited by UNESCO (Paris: UNESCO), pp. 305–345; Nancy Fraser, "Roepke Lecture in Economic Geography – From Exploitation to Expropriation: Historic Geographies of Racialised Capitalism", *Economic Geography*, Vol. 94, No. 1 (2018), pp. 1–17.
17 Federico Ferrara, *The Political Development of Modern Thailand* (Cambridge: Cambridge University Press, 2015), p. 42.
18 Thongchai Winichakul, *Siam Mapped: A History of the Geo-Body of a Nation* (Honolulu: University of Hawaii Press, 1994).
19 Chaiyan Rajchagool, *The Rise and Fall of the Thai Absolute Monarchy: Foundations of the Modern Thai State from Feudalism to Peripheral Capitalism* (Bangkok: White Lotus, 1994); Thongchai, *Siam Mapped: A History of the Geo-Body of a Nation*; Tamara Loos, *Subject Siam: Family, Law, and Colonial Modernity in Thailand* (Chiang Mai: Silkworm Books, 2002); Tamara Loos, "Competitive Colonialisms: Siam and the Malay Muslim South", in *The Ambiguous Allure of the West: Traces of the Colonial in Thailand*, edited by Rachel Harrison and Peter Jackson (Hong Kong and Chiang Mai: Hong Kong University Press and Silkworm Books, 2011), pp. 75–91; Michael Herzfeld, "The Conceptual Allure of the West: Dilemmas and Ambiguities of Crypto-Colonialism in Thailand", in *The Ambiguous Allure of the West: Traces of the Colonial in Thailand*, edited by Rachel Harrison and Peter Jackson (Hong Kong and Chiang Mai: Hong Kong University Press and Silkworm Books, 2011), pp. 173–186.
20 Tanet Charoenmuang, "When the Young Cannot Speak their Own Mother Tongue: Explaining a Legacy of Cultural Domination in Lan Na", in *Regions and National Integration in Thailand, 1892–1992*, edited by Volker Grabowsky (Wiesbaden: Harrassowitz Verlag, 1995), pp. 82–93; Streckfuss, "An 'Ethnic' Reading of 'Thai' History in the Twilight of the Century-Old Official 'Thai' National Model", pp. 312–313.
21 Maurizio Peleggi, *Thailand: The Worldly Kingdom* (London: Reaktion Books, 2007), pp. 44–47, 198–202.
22 Ibid., p. 200.
23 Chaiyan, *The Rise and Fall of the Thai Absolute Monarchy*, pp. 44, 50, 57–58; Chatthip Nartsupha, *The Thai Village Economy in the Past* (Chiang Mai: Silkworm Books, 1999), p. 64; Suleemarn N. Wongsuphap, "The Social Network Construction of the Baba Chinese Business in Phuket", in *Dynamic Diversity in Southern Thailand*, edited by Wattana Sugannasil (Pattani and Chiang Mai: Prince of Songkla University and Silkworm Books, 2005), pp. 275–298; Peleggi, *Thailand: The Worldly Kingdom*, pp. 69–70, 198–199.
24 Chaiyan, *The Rise and Fall of the Thai Absolute Monarchy*, pp. 94–95.
25 Charles Keyes, *Finding their Voice: Northeastern Villagers and the Thai State* (Chiang Mai: Silkworm Books, 2014), pp. 37–47.
26 Chaiyan, *The Rise and Fall of the Thai Absolute Monarchy*, p. 92; Keyes, *Finding their Voice*, pp. 45–46.
27 Jim Glassman, "Primitive Accumulation, Accumulation by Dispossession, Accumulation by Extra-Economic Means", *Progress in Human Geography*, Vol. 30, No. 5 (2006), pp. 608–625.
28 Chaiyan, *The Rise and Fall of the Thai Absolute Monarchy*, pp. 48, 57–58, 71.
29 Peter F. Bell, "'Cycles' of Class Struggle in Thailand", *Journal of Contemporary Asia*, Vol. 8, No. 1 (1978), pp. 51–79; Thak Chaloemtiarana, *Thailand: The Politics of Despotic Paternalism*, Bangkok: Social Science Association of Thailand and Thai Khadi Institute, Thammasat University, 1979); Chairat Charoensin-olarn, *Understanding Postwar Reformism in Thailand: A Reinterpretation of Rural Development* (Bangkok: Editions Duang Kamol, 1988); Philip Hirsch, *Development Dilemmas in Rural Thailand* (Singapore: Oxford University Press, 1990); Glassman, *Thailand at the Margins*.
30 See Streckfuss, "An 'Ethnic' Reading of 'Thai' History in the Twilight of the Century-Old Official 'Thai' National Model", pp. 316–323.
31 Glassman, *Thailand at the Margins*, p. 80; Keyes, *Finding their Voice*, p. 68.
32 Keyes, *Finding their Voice*, p. 69.
33 Ibid., pp. 72–75.
34 Glassman, *Thailand at the Margins*, pp. 105–106.
35 Ibid., pp. 60–61.
36 Ibid., pp. 117–119; Peleggi, *Thailand: The Worldly Kingdom*, p. 78; Keyes, *Finding their Voice*, pp. 149–154.
37 Hewison, *Bankers and Bureaucrats*; Pellegi, *Thailand: The Worldly Kingdom*, p. 203; Glassman, *Bounding the Mekong*.
38 Benedict Anderson, "Withdrawal Symptoms: Social and Cultural Aspects of the October 6 Coup", *Bulletin of Concerned Asian Scholars*, Vol. 9, No. 3 (1977), pp. 13–30; Andrew Turton, "The Current Situation in the Thai Countryside", *Journal of Contemporary Asia*, Vol. 8, No. 1 (1978), pp. 104–142; Andrew Turton, "Rural Social Movements: The Peasants' Federation of Thailand", in *Production, Power, and Participation*

in Rural Thailand: Experiences of Poor Farmers' Groups, edited by Andrew Turton (Geneva: UN Research Institute for Social Development, 1987), pp. 35–43; Anan Ganjanapan, "Conflicts over the Deployment and Control of Labour in a Northern Thai Village", in *Agrarian Transformations: Local Processes and the State in Southeast Asia*, edited by Gillian Hart, Andrew Turton, and Benjamin White (Berkeley: University of California Press, 1989), pp. 98–122; Katherine Bowie, *Rituals of National Loyalty: An Anthropology of the State and the Village Scout Movement in Thailand* (New York: Columbia University Press, 1997); Tyrell Haberkorn, *Revolution Interrupted: Farmers, Students, Law, and Violence in Northern Thailand* (Madison: University of Wisconsin Press, 2011).

39 Glassman, *Thailand at the Margins*, pp. 162–165.

40 Hewison, *Bankers and Bureaucrats*.

41 Somrudee Nicrowattanayingyong, *Development Planning, Politics, and Paradox: A Study of Khon Kaen, A Regional City in Northeast Thailand*, PhD Dissertation, Syracuse University, 1991; Glassman, *Thailand at the Margins*, p. 121.

42 Keyes, *Finding their Voice*, pp. 149–157.

43 Dennis Arnold and Kevin Hewison, "Exploitation in Global Supply Chains: Burmese Migrant Workers in Mae Sot, Thailand", *Journal of Contemporary Asia*, Vol. 35, No. 3 (2005), pp. 319–340; Piya Pongsapa and Michael Smith, "Political Economy of Southeast Asian Borderlands: Migration, Environment, and Developing Country Firms", *Journal of Contemporary Asia*, Vol. 38, No. 4 (2008), pp. 485–514; Glassman, *Thailand at the Margins*, pp. 150–157.

44 Peter F. Bell, "Thailand's Economic Miracle: Built on the Backs of Women", in *Women, Gender Relations and Development in Thai Society*, edited by Virada Somswasdi and Sally Theobald (Chiang Mai: Women's Study Center, Chiang Mai University, 1997), pp. 55–82; Jim Glassman, "Women Workers and the Regulation of Health and Safety on the Industrial Periphery: The Case of Northern Thailand", in *Geographies of Women's Health*, edited by Isabel Dyck, Nancy Lewis, and Sarah McLafferty (London and New York: Routledge, 2001), pp. 61–87.

45 Duncan McCargo, "Thaksin and the Resurgence of Violence in the Thai South: Network Monarchy Strikes Back?" *Critical Asian Studies*, Vol. 38, No. 1 (2006), pp. 39–71; Pavin Chachavalpongpun (ed.), *"Good Coup" Gone Bad: Thailand's Political Developments Since Thaksin's Downfall* (Singapore: ISEAS/Yusof Ishak Institute, 2014); Paul Chambers and Napisa Waitoolkiat, "The Resilience of Monarchised Military in Thailand", *Journal of Contemporary Asia*, Vol. 46, No. 3 (2016), pp. 425–444; Eugenia Mérieau, "Thailand's Deep State, Royal Power and the Constitutional Court (1997–2015)", *Journal of Contemporary Asia*, Vol. 46, No. 3 (2016), pp. 445–466; Prajak Kongkirati, "Thailand's Failed 2014 Election: The Anti-Election Movement, Violence, and Democratic Breakdown", *Journal of Contemporary Asia*, Vol. 46, No. 3 (2016), pp. 467–485.

46 Jonathan Gruber, Nathaniel Hendren and Robert N. Townsend, "The Great Equaliser: Health Care Access and Infant Mortality in Thailand", *American Economic Journal of Applied Economics*, Vol. 6, No. 1 (2014), pp. 91–107.

47 Pasuk Phongpaichit, "Inequality, Wealth and Thailand's Politics", *Journal of Contemporary Asia*, Vol. 46, No. 3 (2016), pp. 405–424.

48 Jim Glassman, "Recovering from Crisis: The Case of Thailand's Spatial Fix", *Economic Geography*, Vol. 83, No. 4 (2007), pp. 349–370.

49 Chairat Charoensin-olarn, "A New Politics of Desire and Disintegration in Thailand", in *Bangkok, May 2010: Perspectives on a Divided Thailand*, edited by Michael J. Montesano, Pavin Chachavalpongpun, and Aekepol Chongvilaivan (Singapore: Institute for Southeast Asian Studies, 2012), pp. 87–96.

50 Glassman, "Cracking Hegemony in Thailand: Gramsci, Bourdieu, and the Dialectics of Rebellion", 2011 pp. 25–46; Claudio Sopranzetti, "Burning Red Desires: Isan Migrants and the Politics of Desire in Contemporary Thailand", *South East Asia Research*, Vol. 20, No. 3 (2012), pp. 361–379; Tyrell Haberkorn, *In Plain Sight: Impunity and Human Rights in Thailand* (Madison: University of Wisconsin Press, 2018), pp. 190–191.

51 Keyes, *Finding their Voice*, pp. 188–189.

52 Elliott Elinoff, "Smouldering Aspirations: Burning Buildings and the Politics of Belonging in Contemporary Isan", *South East Asia Research*, Vol. 20, No. 3 (2012), pp. 381–397; Sopranzetti, "Burning Red Desires: Isan Migrants and the Politics of Desire in Contemporary Thailand"; Andrew Walker, *Thailand's Political Peasants: Power in the Modern Rural Economy* (Madison: University of Wisconsin Press, 2012); Keyes, *Finding their Voice*; Somchai Phatharathananunth, "Rural Transformations and Democracy in Northeast Thailand", *Journal of Contemporary Asia*, Vol. 46, No. 3 (2016), pp. 504–519.

53 Peleggi, *Thailand: The Worldly Kingdom*, pp. 203–205.

54 Haberkorn, *In Plain Sight: Impunity and Human Rights in Thailand.*

References

Anan Ganjanapan. 1989. "Conflicts over the Deployment and Control of Labour in a Northern Thai Village". In *Agrarian Transformations: Local Processes and the State in Southeast Asia*. Edited by Gillian Hart, Andrew Turton, and Benjamin White. Berkeley: University of California Press, pp. 98–122.

Anderson, Benedict. 1977. "Withdrawal Symptoms: Social and Cultural Aspects of the October 6 Coup". *Bulletin of Concerned Asian Scholars*. Vol. 9, No. 3: 13–30.

———. 1991. *Imagined Communities: Reflections on the Origin and Spread of Nationalism*, second edition. London and New York: Verso.

Anievas, Alexander and Kerem Nişancıoğlu. 2015. *How the West Came to Rule: The Geopolitical Origins of Capitalism*. London: Pluto Press.

Arnold, Dennis and Kevin Hewison. 2005. "Exploitation in Global Supply Chains: Burmese Migrant Workers in Mae Sot, Thailand". *Journal of Contemporary Asia*, Vol. 35, No. 3: 319–340.

Ayal, Eliezer B. 1992. "Thailand's Development: The Role of Bangkok". *Pacific Affairs*, Vol. 65, No. 3: 353–367.

Bell, Peter F. 1978. "'Cycles' of Class Struggle in Thailand". *Journal of Contemporary Asia*. Vol. 8, No. 1: 51–79.

———. 1997. "Thailand's Economic Miracle: Built on the Backs of Women". In *Women, Gender Relations and Development in Thai Society*. Edited by Virada Somswasdi and Sally Theobald. Chiang Mai: Women's Study Center, Chiang Mai University, pp. 55–82.

Bowie, Katherine A. 1997. *Rituals of National Loyalty: An Anthropology of the State and the Village Scout Movement in Thailand*. New York: Columbia University Press.

Chairat Charoensin-olarn. 1988. *Understanding Postwar Reformism in Thailand: A Reinterpretation of Rural Development*. Bangkok: Editions Duang Kamol.

———. 2012. "A New Politics of Desire and Disintegration in Thailand". In *Bangkok, May 2010: Perspectives on a Divided Thailand*. Edited by Michael J. Montesano, Pavin Chachavalpongpun, and Aekepol Chongvilaivan. Singapore: Institute for Southeast Asian Studies, pp. 87–96.

Chaiyan Rajchagool. 1994. *The Rise and Fall of the Thai Absolute Monarchy: Foundations of the Modern Thai State from Feudalism to Peripheral Capitalism*. Bangkok: White Lotus.

Chambers, Paul and Napisa Waitoolkiat. 2016. "The Resilience of Monarchised Military in Thailand". *Journal of Contemporary Asia*, Vol. 46, No. 3: 425–444.

Chatthip Nartsupha. 1999. *The Thai Village Economy in the Past*. Chiang Mai: Silkworm Books.

Dower, John. 1986. *War without Mercy: Race & Power in the Pacific War*. New York: Pantheon.

Elinoff, Elliott. 2012. "Smouldering Aspirations: Burning Buildings and the Politics of Belonging in Contemporary Isan". *South East Asia Research*, Vol. 20, No. 3: 381–397.

Ferrara, Federico. 2015. *The Political Development of Modern Thailand*. Cambridge: Cambridge University Press.

Fraser, Nancy. 2018. "Roepke Lecture in Economic Geography – From Exploitation to Expropriation: Historic Geographies of Racialised Capitalism". *Economic Geography*, Vol. 94, No. 1: 1–17.

Gilroy, Paul. 1991. *The Black Atlantic: Modernity and Double Consciousness*. London: Verso.

Glassman, Jim. 2001. "Women Workers and the Regulation of Health and Safety on the Industrial Periphery: The Case of Northern Thailand". In *Geographies of Women's Health*. Edited by Isabel Dyck, Nancy Lewis, and Sarah McLafferty. London and New York: Routledge, pp. 61–87.

———. 2004. *Thailand at the Margins: Internationalisation of the State and the Transformation of Labour*. Oxford: Oxford University Press.

———. 2006. "Primitive Accumulation, Accumulation by Dispossession, Accumulation by Extra-Economic Means". *Progress in Human Geography*, Vol. 30, No. 5: 608–625.

———. 2007. "Recovering from Crisis: The Case of Thailand's Spatial Fix". *Economic Geography*, Vol. 83, No. 4: 349–370.

———. 2010a. *Bounding the Mekong: the Asian Development Bank, China, and Thailand*. Honolulu: University of Hawaii Press.

———. 2010b. "'The Provinces Elect Governments, Bangkok Overthrows Them': Urbanity, Class and Post-Democracy in Thailand". *Urban Studies*, Vol. 47, No. 6: 1301–1323.

———. 2011. "Cracking Hegemony in Thailand: Gramsci, Bourdieu, and the Dialectics of Rebellion". *Journal of Contemporary Asia*, Vol. 41, No. 1: 25–46.

Gruber, Jonathan, Nathaniel Hendren, and Robert N. Townsend. 2014. "The Great Equaliser: Health Care Access and Infant Mortality in Thailand". *American Economic Journal of Applied Economics*, Vol. 6, No. 1: 91–107.

Haberkorn, Tyrell. 2011. *Revolution Interrupted: Farmers, Students, Law, and Violence in Northern Thailand*. Madison: University of Wisconsin Press.

———. 2018. *In Plain Sight: Impunity and Human Rights in Thailand*. Madison: University of Wisconsin Press.

Hall, Stuart. 1980. "Race, Articulation and Societies Structured in Dominance". In *Sociological Theories: Race and Colonialism*. Edited by UNESCO. Paris: UNESCO, pp. 305–345.

Hannaford, Ivan. 1996. *Race: The History of an Idea in the West*. Washington, DC and Baltimore, MD: Woodrow Wilson Center Press and Johns Hopkins University Press.

Herzfeld, Michael. 2011. "The Conceptual Allure of the West: Dilemmas and Ambiguities of Crypto-Colonialism in Thailand". In *The Ambiguous Allure of the West: Traces of the Colonial in Thailand*. Edited by Rachel Harrison and Peter Jackson. Hong Kong and Chiang Mai: Hong Kong University Press and Silkworm Books, pp. 173–186.

Hewison, Kevin. 1989. *Bankers and Bureaucrats: Capital and the Role of the State in Thailand*. New Haven, CT: Yale University Southeast Asia Studies, Yale Center for International and Area Studies.

———. 2014. "Considerations on Inequality and Politics in Thailand". *Democratisation*, Vol. 21, No. 5: 846–866.

Hirsch, Philip. 1990. *Development Dilemmas in Rural Thailand*. Singapore: Oxford University Press.

Keyes, Charles. 2014. *Finding Their Voice: Northeastern Villagers and the Thai State*. Chiang Mai: Silkworm Books.

Loos, Tamara. 2002. *Subject Siam: Family, Law, and Colonial Modernity in Thailand*. Chiang Mai: Silkworm Books.

———. 2011. "Competitive Colonialisms: Siam and the Malay Muslim South". In *The Ambiguous Allure of the West: Traces of the Colonial in Thailand*. Edited by Rachel Harrison and Peter Jackson. Hong Kong and Chiang Mai: Hong Kong University Press and Silkworm Books, pp. 75–91.

McCargo, Duncan. 2006. "Thaksin and the Resurgence of Violence in the Thai South: Network Monarchy Strikes Back?" *Critical Asian Studies*, Vol. 38, No. 1: 39–71.

Mérieau, Eugenia. 2016. "Thailand's Deep State, Royal Power and the Constitutional Court (1997–2015)". *Journal of Contemporary Asia*, Vol. 46, No. 3: 445–466.

Mounier, Alain and Voravidh Charoenloet. 2010. "Thailand: Labour and Growth after the Crisis: New Challenges". *Journal of Contemporary Asia*, Vol. 40, No. 1: 123–143.

Omi, Michael and Howard Winant. 1986. *Racial formation in the United States: From the 1960s to the 1980s*. New York: Routledge and Kegan Paul.

Pasuk Phongpaichit. 2016. "Inequality, Wealth and Thailand's Politics". *Journal of Contemporary Asia*, Vol. 46, No. 3: 405–424.

Pavin Chachavalpongpun, ed. 2014. *"Good Coup" Gone Bad: Thailand's Political Developments Since Thaksin's Downfall*. Singapore: ISEAS/Yusof Ishak Institute.

Peleggi, Maurizio. 2007. *Thailand: The Worldly Kingdom*. London: Reaktion Books.

Pijl, Kees van der. 2007. *Nomads, Empires, States*, Vol. I of Modes of Foreign Relations and Political Economy. London: Pluto Press.

Piya Pongsapa and Michael Smith. 2008. "Political Economy of Southeast Asian Borderlands: Migration, Environment, and Developing Country Firms". *Journal of Contemporary Asia*, Vol. 38, No. 4: 485–514.

Prajak Kongkirati. 2016. "Thailand's Failed 2014 Election: The Anti-Election Movement, Violence, and Democratic Breakdown". *Journal of Contemporary Asia*, Vol. 46, No. 3: 467–485.

Sheppard, Eric and Jim Glassman. 2011. "Social Class". In *Handbook of Geographical Knowledge*. Edited by John Agnew and David Livingstone. Thousand Oaks, CA: Sage Publications, pp. 405–417.

Somchai Phatharathananunth. 2016. "Rural Transformations and Democracy in Northeast Thailand". *Journal of Contemporary Asia*, Vol. 46, No. 3: 504–519.

Somrudee Nicrowattanayingyong. 1991. "Development Planning, Politics, and Paradox: A Study of Khon Kaen, A Regional City in Northeast Thailand". PhD Dissertation, Syracuse University.

Sopranzetti, Claudio. 2012. "Burning Red Desires: Isan Migrants and the Politics of Desire in Contemporary Thailand". *South East Asia Research*, Vol. 20, No. 3: 361–379.

Streckfuss, David. 1993. "The Mixed Colonial Legacy in Siam: Origins of Thai Racialist Thought, 1890–1910". In *Autonomous Histories, Particular Truths: Essays in Honor of John R.W. Smail*. Edited by Laura J. Sears. Madison: University of Wisconsin Center for Southeast Asian Studies Monograph No. 11, pp. 123–154.

———. 2012. "An 'Ethnic' Reading of 'Thai' History in the Twilight of the Century-Old Official 'Thai' National Model". *South East Asia Research*, Vol. 20, No. 3: 305–327.

Suehiro, Akira. 1989. *Capital Accumulation in Thailand, 1855–1985*. Tokyo: Centre for East Asian Cultural Studies.

Suganya Hutaserani and Somchai Jitsuchon. 1988. "Thailand's Income Distribution and Poverty Profile and their Current Situations". In *1988 TDRI Year-End Conference on Income Distribution and Long-term Development*. Bangkok: Thailand Development Research Institute.

Suleemarn N. Wongsuphap. 2005. "The Social Network Construction of the Baba Chinese Business in Phuket". In *Dynamic Diversity in Southern Thailand*. Edited by Wattana Sugannasil. Pattani and Chiang Mai: Prince of Songkla University and Silkworm Books, pp. 275–298.

Tanet Charoenmuang. 1995. "When the Young Cannot Speak Their Own Mother Tongue: Explaining a Legacy of Cultural Domination in Lan Na". In *Regions and National Integration in Thailand, 1892–1992*. Edited by Volker Grabowsky. Wiesbaden: Harrassowitz Verlag, pp. 82–93.

Thak Chaloemtiarana. 1979. *Thailand: The Politics of Despotic Paternalism*. Bangkok: Social Science Association of Thailand and Thai Khadi Institute, Thammasat University.

Thongchai Winichakul. 1994. *Siam Mapped: A History of the Geo-Body of a Nation*. Honolulu: University of Hawaii Press.

———. 2000a. "The Quest for 'Siwilai': A Geographical Discourse of Civilisational Thinking in the Late Nineteenth and Early Twentieth-Century Siam". *Journal of Asian Studies*, Vol. 59, No. 3: 528–549.

———. 2000b. "The Others Within: Travel and Ethno-Spatial Differentiation of Siamese Subjects 1885–1910". In *Civility and Savagery: Social Identity in Tai States*. Edited by Andrew Turton. Richmond, Surrey: Curzon, pp. 38–62.

Turton, Andrew. 1978. "The Current Situation in the Thai Countryside". *Journal of Contemporary Asia*, Vol. 8, No. 1: 104–142.

———. 1987. "Rural Social Movements: The Peasants' Federation of Thailand". In *Production, Power, and Participation in Rural Thailand: Experiences of Poor Farmers' Groups*. Edited by Andrew Turton. Geneva: UN Research Institute for Social Development, pp. 35–43.

Walker, Andrew. 2012. *Thailand's Political Peasants: Power in the Modern Rural Economy*. Madison: University of Wisconsin Press.

25

THE POLITICS, ECONOMICS AND CULTURAL BORROWING OF THAI HIGHER EDUCATION REFORMS

Rattana Lao

This chapter analyses the logic of the Thai state for selectively borrowing educational models from abroad and adapting them to the local situation. Although Thailand has taken pride as the only Southeast Asia country not to have officially been colonised by the West, Western models and values have had enormous influences on the formation of the Thai state and the making of Thai universities.[1] It is an attempt to bridge together two strands of thought. On the one hand, this chapter relies on the literature on Thai studies and the studies of the Thai state as the basis of analysis. On the other hand, it looks into different facets of comparative education to analyse the Thai case. While a critical analysis on the colonial legacy on the trajectory of Thailand development has been well documented in the studies of the Thai state, it deserves further attention in the field of education studies.[2] As things stand, the existing literature on the history of Thailand's education system has viewed this dual process through a normative lens and portrayed it as a successful strategy of Thai elites to weather external imposition, in this case colonisation.[3] This article critically analyses the process of selective borrowing as a historical manoeuvre to refer to external forces that have been used as a political, economic and cultural strategy for the survival of the Thai state and the development of Thai higher education. The threat of colonisation, reliance on American economic assistance and scandalisation of globalisation forces showcase this. The chapter begins with the basic tenets of the logic of the Thai state. Then it traces the historical development of the Thai higher education during the Europeanisation and Americanisation period. Finally the chapter discusses the rise of globalisation and internationalisation to exemplify this selective borrowing process.

Understanding the paradoxes of the Thai state: westernisation and localisation

The logic of the Thai state rests upon three main premises: externalisation strategy, aspiration to modernity and selective borrowing. From the historical period until contemporary development, the Thai state has often used an "externalisation strategy", which is a reference to education reforms elsewhere, to legitimise contested reforms at home to serve the political, economic and cultural agenda. Historically, the Thai state has been an active borrower and importer of European ideas in various realms ranging from the restructuring of royal households to public administration and higher education policymaking. Second, "modernity" has served as the most important motivation driving the emulation, adoption and borrowing of foreign-based examples for Thai elites.[4] There existed a strong obsession to achieve modernity through active emulation of Western values and models.[5] Since the late 19th century, Thai policy elites have strived and aspired to become "Siwilai" – a term transliterated from the word "civilised".

Despite the embracement of Western models and images, Thai policymakers have taken pride in being able to selectively and strategically borrow foreign policy. The well-known historian of the 19th century, Prince Damrong Rajanubhab, argued:

> The Tai [Thai] knows how to pick and choose. When they saw some good feature in the culture of other peoples, if it was not in conflict with their own interests, they did not hesitate to borrow it and adapt it to their own requirements.[6]

The quotation from Prince Damrong reflects how selective borrowing of foreign concepts, cultures and policy has been an integral part of the formation, evolution and development of the Thai state. Thailand's aspiration to become modern is not a straightforward account of copying others or shared international norms and values. Thai ruling elites overtly highlighted their ability to selectively borrow and promote Western culture.[7] Thai policymakers have highlighted the flexibility of the policy agents to "pick and choose" Western elements to fit their own agenda.[8] Selective borrowing and local adaptation have been extensively used in public policies to exemplify the uniqueness of Thai policy elites to become modern while maintaining Thai and Buddhist traditions. Although such discourse has been used to herald Thailand's independence and her uniqueness, it is important to read beyond the normative and nationalistic discourse. Nuances are necessary to understand the westernisation/localisation of Thailand.

Despite the absence of an official "coloniser", the political, economic and cultural influences of the Western empires were paramount in the formation of the Thai state and construction of "Thainess". Jackson points out the growing researches to indicate that the development of Thailand's "economy, polity, culture, and social structure were all deeply impacted by western imperialism in ways very similar to the situation in direct colony".[9] Thailand might have escaped direct colonisation, but the remnants of Western political, economic and cultural influences are paramount in all realms. Nevertheless, by being officially independent, the Thai state has had relative autonomy to select and borrow what they deem necessary. The integral relationship between being "non-colonised" by the West but heavily influenced by its models and values has resulted in the creation of what Herzfeld called the "living paradoxes" of dependency and independence. Herzfeld argues, "They are nominally independent, but their independence comes at the price of a sometimes humiliating form of effective dependence".[10] Thailand's struggle to emulate Western models and its obsession to emphasise local idiosyncrasy reflects the inherent logic of the Thai state to maintain "relative superiority".[11] On the one hand, the acquisition and appropriation of Western elements enabled the Thai elites to paint itself as being "civilised"/"siwilai" – an equal partner to the West. On the other hand, the selective westernisation and localisation discourse served to elevate Bangkok elites as more advanced and more modern than the rest of the country as well as better than other colonised countries in the region.

The Europeanisation of Thai higher education: from 1889 to the 1940s

The foundation of the Thai higher education system must be understood in relation to the larger national attempt to "modernise" and "westernise" the Thai nation. The Thai king undoubtedly used the threat of colonisation as a premise to push for multiple reform initiatives at the local level. During the early days, European models – particularly of French and British values – dominated the perspective of Thailand's political elites on how to create and structure higher education. Peleggi persuasively argued that the early period of modernisation was the equivalent of "European periodisation".[12]

The first reform that laid groundwork for higher education system in Thailand was founded during the period 1889–1909.[13] In order to equip the princes and ruling elites for modern bureaucracy, Rama V established eight different professional schools such as civil servants, medical, educational and legal schools. These professional schools imitated the British systems of tutoring. Not only were foreign tutors hired to conduct training, but the Thai trainers must also have had Western training.[14]

However, it was not until the reign of King Vajiravudh (Rama VI), in which the first university was established. Influenced by the French concept of *la grande école* ("the great school") and the German idea of the "school of science", the first university, Chulalongkorn, was founded on 28 March 1917.[15] The university had the sole purpose of training Thailand elites to serve in the modern bureaucracy and nation building. There were four faculties including medical, political science, engineering, and arts and science. The French concept of the great school went through local interpretation and adaptation by Thai ruling elites. The lack of resources, national politics and hierarchical structure had a significant impact on translating the concept of *la grande école* in Thailand. First, Chulalongkorn University was developed by integrating four existing educational structures into one. These four institutions included the Medical School, the Teachers' Training School, the Law School and the Royal Pages' School founded in 1889, 1892, 1897 and 1902, respectively (Ministry of University Affairs, 2003). Although the Thai elite envisioned the first university to be modelled after the British, French and German concepts of higher education, the translation of the ideas was limited by the lack of resources and the urgent needs of the nation to have trained civil servants.[16] The following excerpt captures Phraya Thammasak Montri's assessment of Chulalongkorn:

> If we use Oxford of Cambridge as the standards, we are not yet ready to establish any university. We are lack of tremendous resources: financially and academically. However, if we lowered our standards to be just like those newly established universities that are mushroomed in the West and the in East, we are capable to do something. . . . the old style British universities provide stamp of approval that these individuals are capable to do anything. They are full gentlemen. Whereas the newly established universities can only stamp that these individuals can have professions in engineering, medicine, lawyer, science and arts.[17]

Thai elites were pragmatic in using university to serve the needy demands of the rapid national development. Accordingly to the Ministry of University Affairs, "The need for trained human resources was the most pressing issue. Thai universities strive to become the 'university' in the western concept".[18]

The establishment of Thammasat University also illustrated the politics of borrowing from a foreign model to serve the political purposes in Thailand. The purposes were driven purely by local political needs. The prime objectives in creating this second university were to create greater access to higher education for the masses and to train a new type of bureaucracy that would favour service to the democratic state rather than to the monarchy.[19] Unlike Chulalongkorn, the rationale for creating Thammasat University was to "train a new kind of bureaucrat for the post-absolutist age".[20] The principles underlined the establishment of Thammasat University were coined between modernisation, democratisation and equal educational opportunities.

The origin of Thammasat has had a strong link between this university with the Ministry of Public Justice and the Department of Public Administration. Similar pattern of university–civil servant linkages can be seen in the creation of three other specialised universities in this period. After Thammasat was established, specialised universities were created in the Bangkok metropolitan area: Silapakorn University was opened in 1934 and affiliated with the Department of Fine Arts, Mahidol was created in 1942 and linked to the Ministry of Public Health and Kasertsart University was created in 1943 and linked to the Ministry of Agriculture. Later on in 1964, the Ministry of Education established a college of Education at the Prasarnmitr in order to train teachers and educators.

The bureaucratic norms and attitudes were heavily embedded in Thai higher education institutions. On the one hand, the number of seats available was determined by the demands of each newly established ministry.[21] Manpower planning was the key factor determining the number of students being accepted to each faculty. On the other hand, the bureaucracy continued to be the most powerful employer of graduates. Public sector accounted for 60 per cent employment for university graduates in the mid-1970s. Higher education was the engine behind the creation of this "civil service mystique".[22] The domination of the bureaucracy reinforced the limited purpose of higher education in Thailand, which was to serve the bureaucratic apparatus rather than pursue research and development.

There is a political dimension to the policy borrowing of a European-style higher education system in Thailand. An "externalisation" strategy to the external threat to "maintain its own sovereignty" from colonisation was used to justify the creation and establishment of the first university in Thailand, Chulalongkorn. Meanwhile, university, as a European model, was perceived as a significant mean for Thailand to become a "modern state". A similar concern has been echoed by professor Charas Suwanwela, a respectable medical doctor and former president of Chulalongkorn University:

> The primary aim of higher education was to develop manpower to serve national management purposes, and as a means of importing existing knowledge from the West. Unlike institution of higher learning in classical Europe, universities were not aimed at the pursuit of truth or the refinement of the mind. The approaches in the university were more or less copy of the western style.[23]

US assistance and national development: from the 1940s to 1980s

The rationale behind the increasing involvement of the United States in Thailand was based on the proliferation of US power after the Second World War, increasing attention to preventing the spread of communism and the growing conflict in Vietnam. With the threat of communism in Southeast Asia, Thailand became a strategic partner of the United States. Between 1966 and 1971, US military assistance and World Bank development loans accounted for one-third of Thailand's total public expenditure.[24] During the two decades of US involvement in Thailand, it was estimated that at least US$2 billion were poured into the country.[25] Not only did loans help to improve military hardware, but US cash and technology also helped expand the Thai bureaucracy. During this period, the total size of Thailand's bureaucracy expanded from 75,000 individuals in 1944 to 250,000 individuals by 1965. Subsequently, Thailand signed three important bilateral agreements with the United States to cooperate in terms of economics, the military and culture. Under this cooperation, several agreements were directly related to education. The relationship between the United States and the Thai higher education system occurred at multiple levels. From a bilateral economic assistance, the US Agency for International Development (USAID) had invested enormously to expand higher education accesses in the regional area of Thailand. Other philanthropic organisations such as the Ford, the Rockefeller and the Fulbright Foundations have also played significant roles. Under the University Development Programme (UDP) founded in 1961, the Rockefeller laid a fundamental foundation for three main universities: Thammasat, Mahidol and Kasetsart.[26] Harrison argued that there was "a radical reorientation and reconfiguration of Thai political, economic and cultural relations from Europe to North America".[27]

The infiltration of American influences into Thailand's political economy resulted in a significant change in Thai public policy. Higher education policy was no exception. Sinlarat describes higher education reform during this period as "responded to the demands of society for economic development and industrialisation according to the US Guidelines".[28] There are four important aspects that reflected the US legacy in Thailand's higher education reform. First, rather than sending students to Europe, the state and private households began to send more young people to receive education in the United States. Between 1954 and 1962, 150 students were supported by USAID to obtain higher education at Indiana University.[29] Between 1951 and 1985, at least 1,500 Thai students went to the United States under the Fulbright and Ford Foundations and other scholarships. The numbers proliferated substantially. By the 1980s, there were at least 7,000 Thai students in the United States for higher education.[30] Given that the Thai elites had been trained for the most part in Britain and France, sending a new cadre to the United States was seen as a significant shift in both Thai international relations and changing perceptions of the upcoming elites. According to Thanet, not only have the majority of social scientists in Thailand been trained in the United States, but its educational system and social science model have permeated and dominated Thai academics.[31]

Second, there was the expansion of higher education in the regional areas of Thailand. Up until this period, all five universities were located in Bangkok; there were no universities in the provinces. Three more universities expanded into three different provinces in three regions to accommodate the growing demands for higher education. According to the First National Economic and Social Development Plan (1961–1966), Field Marshall Sarit Thanarat augmented the policy to expand higher education into the provinces in the regional areas. Chiang Mai, Khon Kean and Prince of Songkhla Universities were founded in 1964, 1965 and 1968, respectively.[32]

Unlike the specialised universities to fill bureaucratic and ministerial needs in the first period of higher education, new universities were founded on the American concept of a comprehensive university. They were not linked to particular ministries or ministerial mandates and they offer multiple and wide range of faculties in order to directly contribute to the national and economic development of the country.

The rationale to create universities in the regional areas of Thailand came from the US concept. The centralised state of Thailand would help other regions through education. Higher education has a high impact. Otherwise, students who graduated from high school would go to study in other provinces and would not return to their homeland.[33] The expansion of higher education into the regional areas and the embracement of US doctrine can also be explained through the concept of the politics and economics of borrowing. Politically speaking, the expansion of educational opportunities is in fact intended to mitigate the spread of communism in the regions. The increasing economic and political power of the United States in Thailand made it no coincidence that Thailand began to follow US models instead of the European influences. Furthermore, the excerpts highlighted an intricate relationship between external influences and existing power relations in the country. Not only was the external assistance welcomed, but it was also used to serve the socio-political needs of the country.

The third influence of US thinking in Thai higher education policymaking is the inclusion of the private sector. This is one aspect of the US model of higher education as "the encouragement of both private and state universities; open access to all who have finished secondary level successfully".[34] Prior to 1969, higher education was limited to state control and there were only public universities. The Private College Act was promulgated in 1969, enabling private institutions to provide educational services. Nevertheless, the government imposed rigid requirements and enforced strict control over all aspects of private higher education.

The US influence in Thai higher education is also evident in the expansion of graduate studies. To respond to the rapid national development, the government began to pay more attention to the need for graduate studies and students. Founded in 1962, the National Institute of Development Administration (NIDA) received substantial economic assistance and technical assistance from various US-related organisations such as the Fulbright, Rockefeller and Ford Foundations and the United States Overseas Mission.[35] In *University Development in the Third World: The Rockefeller Foundation Experience*, Coleman and Court also outline the impacts of the Rockefeller Foundation University Development Programme (UDP), which later was renamed to the Education for Development (EFD) Programme at three universities in Thailand: Mahidol, Thammasat and Kasertsart.[36] Inevitably, the UDP/EFD programme has a significant impact on the development of research and graduate studies in Thailand. The development of Mahidol University's Faculty of Science is also the by-product of the two-decade programme. Thirty high-calibre scholars received sponsorship to pursue their doctorate degree in the United States, Meanwhile, some Rockefeller Foundation members assisted in setting up the structure for the Faculty of Science at Mahidol University. Eventually, Mahidol University received approximately US$11 million – the highest recipient of all UPD/EFD global programmes.

Additional to the expansion of access, another significant change in higher education policy during this period was the move away from bureaucratic structure to the rising needs of the markets. Pasuk and Baker aptly painted the political economy of Thailand during this time: "The ethos of the civil servant has been overwhelmed by the spirit of the marketplace". The expansion of the bureaucracy, which had absorbed the graduates of higher education, ended by the 1980s. It was reported that the annual growth of the bureaucracy plummeted from 10 per cent to only 2 per cent. In contrast, the demand for white-collar workers, engineers and businessmen grew exponentially. The changing political economy in Thailand towards a more open economy by the end of the 1970s to the 1980s required a reconfiguration of the higher education to serve the new economy. From being dominated by a military dictator, Thailand from the mid-1970s onward opened doors for other interest groups to enter politics, such as businessmen, intellectuals and social activists.[37] The changing economic landscape of the country contributed significantly to the changing mission of the Thai university. It is argued that the objectives of the university increasingly expanded from solely serving the bureaucracy.

Pad argued that this period witnessed a significant shift from state-directed higher education policy to market-directed reform.[38] The role of market demand on Thai higher education can be understood from two strands. On the one hand, the market represents the external demands resulting from the socio-economic change of the country. The change was felt in terms of quantitative expansion and the establishment of new universities and new programmes to accommodate the rising needs of the business sector. On the other hand, the market force is understood through increasing internal demands to reform the administrative structure of Thai universities. Since the government promulgated the Private Education Act of 1969, this new legal framework allowed the establishment of privately funded higher education institutions, triggering the state to begin re-thinking its own internal administrative strategies, such as encouraging public universities to seek external funding, increasing internal efficiency and flexibility, and creating self-supporting programmes within public universities. Given that most of the universities during that time were founded by state-directed policy, Pasuk and Baker argued that the transition of mission and purpose from the demand of the state to that of the market was "ill-prepared".[39] Higher education to train engineers, mechanics and scientists was not the same as training social scientists. It required well-equipped laboratories and expensive equipment which were not available to the country overnight. It was imperative to change the mission of Thai higher education to fit economic demands. The domination of the economic principles of Thai educational reform is the underlying principle of Thai education.[40]

Globalisation and internationalisation

Similarly to how "globalisation" has replaced the European/American as the point of reference, so has the term "internationalisation". In fact, Yang persuasively argued that for the non-Western universities, especially those of Asia, the internationalisation of higher education should be understood in relations to the long-standing Western domination and conception of higher education in the region: "Within the contemporary context of western dominance, internationalisation of higher education in non-western societies necessarily touches on longstanding knotty issues and tensions between westernisation and indigenisation".[41] According to an interview with Thailand's expert on internationalisation of higher education, conducted on 9 April 2014: "Westernisation, modernisation and internationalisation are all the same things. It is the process of internal adjustment to respond to external demands".

Undoubtedly, at the national level, the Thai policy elites became aware of the increasing importance of globalisation and used it as the main rationale to justify the policy need for internationalisation. Nilphan argued:

> The discourse of "global challenges" has become prevalent in the Thai bureaucracy; the MUA's long-term plan is peppered with such terms such as global awareness, economic competitiveness, international level competence and specific skills.[42]

Accordingly, the excerpt below was a keynote speech given by Wichit Srisa-an, the former secretary-general of the Ministry of University Affairs and the leading policy figure in Thailand higher education. It addresses the integral relationship between globalisation and the need for Thai higher education to become more "internationalised":

> The scientific and technological advancement and advent of information technology has intensified the need for Thailand to work and collaborate with other countries in terms of its economic, social and political dimension. Such "international" characteristic requires higher education and academic community to realise and understand its changing roles and responsibility in order to respond to the changing global context and challenges of the ever-growing interconnectedness. Given these reasons, one of the main goals of Fifteen Years Long Range Plan 1990–2005) and the 7th Higher Education Development Plan (1992–96) is to stress the importance of Thai higher education to become "internationalised".[43]

The internationalisation of higher education in Thailand has been justified by both external and internal factors. On the one hand, external factors, which refer to the international agencies and bilateral and multilateral agreements, play pivotal roles to promote internationalisation in Thailand. On the other hand, the changing socio-economic conditions of Thailand also required students to have English language proficiency and international understanding skills necessary to respond to the economy. Supaporn supports this point.[44] The spread of English as the international language contributes significantly to the growth of international programmes in Thailand. The following paragraphs outline the key policy efforts by Thai policymakers and its collaboration with regional and international agencies to promote internationalisation in Thailand.

The first university to establish an international programme in Thailand was Assumption University. In 1972, it was the first programme to offer the Bachelor of Business Administration. Another early development of internationalisation of Thai higher education is SASIN, which is a Master of Business Administration, and created a joint programme between Chulalongkorn University, the J.L. Kellogg Graduate School of Management of Northeastern University and the Wharton School of the University of Pennsylvania. These are considered the pioneers of the internationalisation process in Thai higher education.[45]

However, there has been a mushrooming of international programmes in Thailand which use English as medium of instruction. According to the First Long Range Plan (1990) and the 8th Higher Education National Development Plan (1996), the role of international programmes is mentioned as a pertinent factor to the development of internationalisation policy in Thailand. By having more international programmes, it is believed that Thai students will have greater chances to obtain English proficiency skills and international outlook. Although private universities, such as Assumption University, were the pioneers of the international programme in Thailand, in recent years it has been the public universities that have led the internationalisation process. Many international programmes have been created inside public universities. The Office of Higher Education Commission has collected systematic data on the international programmes in Thailand, which reflects the importance of the topic viewed by the Thai state.

The Ministry of University Affairs reported that there were approximately a hundred international programmes in 1992: 58 bachelor's degree programmes, 55 master's degree programmes and 19 doctoral programmes for public and private universities.[46] Within the course of six years, the number of international programmes in Thailand has doubled at all levels. While there were only 465 programmes in 2004, there are total of 1,017 international programmes as of 2012. The numbers have dropped because some programmes were unsuccessful. There are currently 769 international programmes in Thailand: 249 are bachelor's degrees, 290 are master's degrees, 224 are doctoral degrees, and 6 are other degrees. As of the most recent data, there are at least 18,814 international students in Thailand.

Although Thailand has been receptive to the influence of globalisation and the expansion of international programmes, research by Lavankura and Lao exemplified the Thai struggle to adopt and adapt "international-ness" into "Thainess".[47] There are two examples that exemplify the influence ofThai local characteristics in the development of these international programmes. First, these international programmes, although taught in English, continued to maintain the Thai-style teaching of rote learning and memorisation. There has been quantitative expansion of international programmes without ensuring the efficiency and effectiveness of the programmes being taught. The curriculum, the pedagogy and classroom management does not reflect the international dimension of education. Rather, it is a Thai curriculum that translates into English. Second, the focus of these international programmes is on having westerners as the instructors rather than having an "international cadre of academics" who can be Thai or other nationalities. There is a mentality in Thailand that "West is best", as illustrated in the beginning of this chapter. Such a mentality has carried on in the expansion of international programmes. Given that there is a dearth of foreign and Western academics in Thailand, this demand has compelled universities to opt for part-time academics who come to Thailand for a short period of time. This lacks continuity and commitment in the learning process. The superficiality of these international programmes is telling and dangerous if Thailand truly aspires to become an international hub for education. Diversity in content knowledge is more essential than the accent and nationality of the academics.

Conclusion

This chapter has analysed a hundred years of Thai higher education through the view of selective borrowing. Although it has been heralded as a "successful development strategy", it is important to take a critical view to analyse and understand this approach in relation to the politics, economics and cultural dimension of the country. Since its inception, Thai elites have integrated their strategy with European influences of higher education. The lack of resources inhibited Thai policymakers to fully adopt European-style higher education. Instead, they produced a civil-servant training school modelled after European schools. Through the influence of Americanisation, Thai policymakers welcomed economic assistance of US aid and implemented comprehensive universities in the regional area. In the era of globalisation, Thai elites embraced the internationalisation of higher education. The expansion of international programmes in the last 20 years is a case in point. Despite quantitative expansion, the method, curriculum and pedagogy remained very Thai. The only thing international about Thai higher education is the use of English and the excitement of Western academics. By using selective borrowing as a lens to understand Thai higher education, one is able to see that "picking and choosing" different elements of the Western modality is limited and conditioned by local characteristics. It is not an independent and free will strategy as the nationalist discourse would endorse. The amalgam of Thai studies and comparative higher education enabled one to see the dynamics and contours of Thai higher education.

Notes

1 Keith Watson, *Educational Development in Thailand* (Hong Kong: Heinemann Educational Books, 1980).
2 Benedict Anderson, "Studies of the Thai State: The State of Thai Studies", in *The Study of Thailand: Analyses of Knowledge, Approaches, and Prospects in Anthropology, Art History, Economics, History and Political Science* (Athens: Ohio University, 1978). Also, Peter A. Jackson, "Autonomy and Subordination in Thai History: The Case of Semicolonial Analysis", *Inter-Asia Cultural Studies*, Vol. 8, No. 3 (2017), pp. 329–348.
3 See Watson, *Educational Development in* Thailand; Gerald Fry, "The Evolution of Educational Reform in Thailand", *Second International Forum on Education Reform* (Bangkok: Office of the National Education Commission, 2002) <www.worldedreform.com/intercon2/fly.pdf>; Susan Jungck and Boonreang Kajornsin, "Thai Wisdom and Glocalisation: Negotiating the Global and the Local in Thailand's National Education Reform", in *Local Meanings, Global Schooling: Anthropology and World Culture Theory*, edited by Kathryn Anderson-Levitt (New York: Palgrave Macmillan, 2003); David K. Wyatt, *The Politics of Reform in Thailand: Education in the Reign of King Chulalongkorn* (New Haven: Yale University Press, 1969).
4 Maurizio Peleggi, *Lords of Things: The Fashioning of the Siamese Monarchy's Modern Images* (Honolulu, HI: University of Hawaii Press, 2002).
5 Anderson, "Studies of the Thai State: The State of Thai Studies".
6 Maurizio Peleggi, *Lords of Things: The Fashioning of the Siamese Monarchy's Modern Images* (Honolulu, HI: University of Hawaii Press, 2002), p. 12.
7 Ibid., p. 16.
8 Rachel V. Harrison, "The Allure of Ambiguity: The West and the Making of Thai Identities", in *The Ambiguous Allure of the West: Traces of the Colonial in Thailand*, edited by Rachel V. Harrison and Peter A. Jackson (Hong Kong: Hong Kong University Press, 2010), p. 15.

9 Peter A. Jackson, "Autonomy and Subordination in Thai History: The Case of Semicolonial Analysis", *Inter-Asia Cultural Studies*, Vol. 8, No. 3 (2007), p. 331.
10 Michael Herzfeld, "The Absence Presence: Discourses of Crypto-Colonialism", *The South Atlantic Quarterly*, Vol. 101, No. 4 (2002), p. 901.
11 Ibid., p. 537.
12 Peleggi, *Lords of Things*.
13 Ministry of University Affairs, *Three Decades of the Ministry of University Affairs* (Bangkok: Ministry of University Affairs, 2003).
14 Wyatt, *The Politics of Reform in Thailand*, pp. 376–385.
15 Watson, *Educational Development in Thailand*; Varuni Osatharom, *Analytical Study of the Historical Policy Development of Higher Education in Thailand: Its Implications for Current Development and Future Trends* (Bangkok: Ministry of University Affairs, 1990).
16 Varuni, *Analytical Study of the Historical Policy Development of Higher Education in Thailand.*
17 Ministry of University Affairs, *Two Decades of the Ministry of University Affairs* (Bangkok: Ministry of University Affairs, 1992), p. 27.
18 Ibid., p. 30.
19 Chris Baker and Pasuk Phongpaichit, *A History of Thailand* (Singapore: Cambridge University Press, 2009), p. 164.
20 Ibid., p. 123.
21 Varuni, *Analytical Study of the Historical Policy Development of Higher Education in Thailand.*
22 Keith Watson, *Educational Development in Thailand*, p. 136.
23 Charas Suwanwela, *Southeast Asian Universities and the Challenges of the Twenty-First Century*, Presented at the ASAIHL Lecture, the Association of Southeast Asian Institutions of Higher Learning General Conference and Seminar on the role of ASAIHL Universities in the Transfer of Technology, Jakarta, 1988, p. 2.
24 Alasdair Bowie and Daniel Unger, *The Politics of Open Economies: Indonesia, Malaysia, the Philippines and Thailand* (Cambridge: Cambridge University Press, 1997), p. 139.
25 Baker and Pasuk, *A History of Thailand.*
26 James S. Coleman and David Court, *University Development in the Third World: The Rockefeller Foundation Experience* (Oxford: Pergamon Press, 1993).
27 Rachel V. Harrison, "The Allure of Ambiguity: The West and the Making of Thai Identities", in *The Ambiguous Allure of the West: Traces of the Colonial in Thailand*, edited by Rachel V. Harrison and Peter A. Jackson (Hong Kong: Hong Kong University Press, 2010), p. 10.
28 Paitoon Sinlarat, "Thai Universities: Past, Present and Future", in *Asian Universities: Historical Perspectives and Contemporary Challenges*, edited by Philip G. Altbach and Tōru Umakoshi (Baltimore: Johns Hopkins University Press, 2004), p. 209.
29 Ministry of University Affairs, *Three Decades of the Ministry of University Affairs.*
30 Baker and Pasuk, *A History of Thailand.*
31 Thanes Wongyannava, "Wathakam: The Thai Appropriation of Foucault's Discourse", in *The Ambiguous Allure of the West: Traces of the Colonial in Thailand*, edited by Rachel V. Harrison and Peter A. Jackson (Chiang Mai: Silkworm Books, 2010), p. 156.
32 Watson, *Educational Development in Thailand.*
33 Paitoon Sinlarat, *Thai Higher Education System after World War II: Changes for Modern Society* (Bangkok: Chulalongkorn University Press, 2005), p. 62.
34 Keith Watson, "The Higher Education Dilemma in Developing Countries: Thailand's Two Decades of Reform", *Higher Education*, Vol. 10, No. 3 (1991), p. 300.
35 Ministry of University Affairs, *Three Decades of the Ministry of University Affairs.*
36 Coleman and Court, *University Development in the Third World: The Rockefeller Foundation Experience.*
37 Dhiwakorn Kaewmanee, *The Evolution of the Thai State: The Political Economy of Formative and Transformative External Influences* (Berlin: Dissertation.de, 2007).
38 Pad Nilphan, *Internationalising Thai Higher Education: Examining Policy Implementation,* Doctoral dissertation, University of Leeds, England, 2005.
39 Baker and Pasuk, *A History of Thailand*, p. 146.
40 Watson, *Educational Development in Thailand.*
41 Rui Yang, "China's Strategy for the Internationalisation of Higher Education: An Overview", *Frontiers of Education in China*, Vol. 9, No. 2 (2014), pp. 151–162.
42 Pad Nilphan, *Internationalising Thai Higher Education: Examining Policy Implementation*, p. 107.

43 Ministry of University Affairs, *National Seminar Report on Internationalisation of Thai Higher Education* (Bangkok: Ministry of University Affairs, 1991), p. 36.
44 Supaporn Chalapati, *The Internationalisation of Higher Education in Thailand: Case Studies of Two English-Medium Business Graduate Programmes*, Doctoral Dissertation, RMIT University, Australia, 2007.
45 Ministry of University Affairs, *National Seminar Report on Internationalisation of Thai Higher Education*, pp. 8–11.
46 Ministry of University Affairs, *Three Decades of the Ministry of University Affairs*.
47 Pad Lavankura and Rattana Lao, "Second Order Change without First Order Change: A Case of Thai Internationalisation of Higher Education", in *Education and Globalisation in Southeast Asia: Issues and Challenges,* edited by Lee Hock Guan (Singapore: ISEAS, 2017).

References

Anderson, Benedict. 1978. "Studies of the Thai State: The State of Thai Studies". In *The Study of Thailand: Analyses of Knowledge, Approaches, and Prospects in Anthropology, Art History, Economics, History and Political Science*. Athens: Ohio University.

Baker, Chris, and Pasuk Phongpaichit. 2009. *A History of Thailand*, 2nd edition. Singapore: Cambridge University Press.

Bowie, Alasdair, and Daniel Unger. 1997. *The Politics of Open Economies: Indonesia, Malaysia, the Philippines and Thailand*. Cambridge: Cambridge University Press.

Charas Suwanwela. 1988. "Southeast Asian Universities and the Challenges of the Twenty-First Century". Presented at the ASAIHL Lecture, the Association of Southeast Asian Institutions of Higher Learning General Conference and Seminar on the role of ASAIHL Universities in the Transfer of Technology, Jakarta, Indonesia.

Coleman, James S., and David Court. 1993. *University Development in the Third World: The Rockefeller Foundation Experience*. Oxford: Pergamon Press.

Dhiwakorn Kaewmanee. 2007. *The Evolution of the Thai State: The Political Economy of Formative and Transformative External Influences*. Berlin: Dissertation.de.

Fry, Gerald. 2002a. "The Evolution of Educational Reform in Thailand". *Second International Forum on Education Reform*. Bangkok: Office of the National Education Commission <www.worldedreform.com/intercon2/fly.pdf>.

Harrison, Rachel V. 2010. "The Allure of Ambiguity: The West and the Making of Thai Identities". In *The Ambiguous Allure of the West: Traces of the Colonial in Thailand*. Edited by Rachel V. Harrison and Peter A. Jackson. Hong Kong: Hong Kong University Press.

Herzfeld, Michael. 2002. "The Absence Presence: Discourses of Crypto-Colonialism". *The South Atlantic Quarterly*, Vol. 101, No. 4: 899–926.

Jackson, Peter A. 2007. "Autonomy and Subordination in Thai History: The Case of Semicolonial Analysis". *Inter-Asia Cultural Studies*, Vol. 8, No. 3: 329–348.

Jungck, Susan, and Boonreang Kajornsin. 2003. "*Thai Wisdom* and Glocalisation: Negotiating the Global and the Local in Thailand's National Education Reform". In *Local Meanings, Global Schooling: Anthropology and World Culture Theory*. Edited by Kathryn Anderson-Levitt. New York: Palgrave Macmillan.

Ministry of University Affairs. 1991. *National Seminar Report on Internationalisation of Thai Higher Education*. Bangkok: Ministry of University Affairs.

———. 1992. *Two Decades of the Ministry of University Affairs*. Bangkok: Ministry of University Affairs.

———. 2003. *Three Decades of the Ministry of University Affairs*. Bangkok: Ministry of University Affairs.

Pad Lavankura and Rattana Lao. 2017. "Second Order Change without First Order Change: A Case of Thai Internationalisation of Higher Education". In *Education and Globalisation in Southeast Asia: Issues and Challenges*. Edited by Lee Hock Guan. Singapore: ISEAS.

Pad Nilphan. 2005. "Internationalising Thai Higher Education: Examining Policy Implementation". Doctoral dissertation, University of Leeds, England.

Paitoon Sinlarat. 2004. "Thai Universities: Past, Present and Future". In *Asian Universities: Historical Perspectives and Contemporary Challenges*. Edited by Philip G. Altbach and Tōru Umakoshi. Baltimore: Johns Hopkins University Press.

———. 2005. *Thai Higher Education System after World War II: Changes for Modern Society*. Bangkok: Chulalongkorn University Press.

Pasuk Phongpaichit and Chris Baker. 1998. *Thailand's Boom and Bust*. Chiang Mai: Silkworm Books.

Peleggi, Maurizio. 2002. *Lords of Things: The Fashioning of the Siamese Monarchy's Modern Images*. Honolulu, HI: University of Hawai'i Press.
Supaporn Chalapati. 2007. "The Internationalisation of Higher Education in Thailand: Case Studies of Two English-Medium Business Graduate Programmes". Doctoral dissertation, RMIT University, Australia.
Thanes Wongyannava. 2010. "Wathakam: The Thai Appropriation of Foucault's Discourse". In *The Ambiguous Allure of the West: Traces of the Colonial in Thailand*. Edited by Rachel V. Harrison and Peter A. Jackson. Chiang Mai: Silkworm Books.
Varuni Osatharom. 1990. *Analytical Study of the Historical Policy Development of Higher Education in Thailand: Its Implications for Current Development and Future Trends*. Bangkok: Ministry of University Affairs.
Watson, Keith. 1980. *Educational Development in Thailand*. Hong Kong: Heinemann Educational Books.
———. 1989. "Looking West and East: Thailand's Academic Development". In *From Dependency to Autonomy*. Edited by Philip G. Altbach and Wiswanathan Selvaratnam. Dordrecht, The Netherlands: Kluwer Academic.
———. 1991. "The Higher Education Dilemma in Developing Countries: Thailand's Two Decades of Reform". *Higher Education*, Vol. 10, No. 3: 297–314.
Wyatt, David K. 1969. *The Politics of Reform in Thailand: Education in the Reign of King Chulalongkorn*. New Haven: Yale University Press.
Yang, Rui. 2014, "China's Strategy for the Internationalisation of Higher Education: An Overview". *Frontiers of Education in China*, Vol. 9, No. 2: 151–162.

26

THE STATE OF HUMAN RIGHTS IN THE AFTERMATH OF THE 2014 MILITARY COUP D'ÉTAT

Titipol Phakdeewanich

Over the course of the Thai democratic development, since the Siamese revolution on 24 June 1932, democracy and human rights have been straggling to become fully fledged in Thailand, fighting against multiple military interventions. The Thai elites and the conservative group forcefully argued that democracy and human rights must conform to the existing Thai cultural heritage while highlighting the necessity of compromising on the fundamental principles of democracy, especially liberty and freedom. In the wake of the 2014 coup d'état, led by the National Council for Peace and Order (NCPO), the future prospect for a full-fledged democracy has been significantly undermined, while millions of Thais believe that returning to democracy is neither in the best interests of the country nor the Thai people as a whole.

In over a decade now, with political divide and the disruptions of multiple street protests in addition to endemic problems of corruption, some Thais have lost their faith in representative democracy and now look up to the authoritarian style of government under the military. This provides a space for the military to reinforce its role and institutionalise itself as a correcting mechanism when democracy fails. The role of the military has been well accepted by millions of Thais, especially since Thaksin Shinawatra took office in a landslide victory in 2001. The 2006 coup removed Thaksin from office, who now lives in exile. In 2014, his sister, Yingluck, the then prime minister, was also removed from power by a coup. This demonstrates the rejection of voting rights of the rural population.

A widespread suppression of freedoms has been observed and documented since the latest coup in 2014 – an arbitrary measure that was imposed in the name of "national security" in order to protect the sovereignty of Thailand. Furthermore, the NCPO has highlighted that the concept of "Thainess" is important for the construction of Thai democracy, while the concept itself remains vague and rather unclear for both Thais and foreign observers. This chapter attempts to define so-called Thainess with respect to its utilisation in recent Thai history and its implication on the promotion and protection of human rights within the country. It also examines whether Thainess can support and reinforce the process of democratisation within Thailand.

The notion of Thainess has raised questions among Thai political observers in terms of whether there is such thing as so-called Thainess and whether the term undermines democratic progress and the promotion and protection of human rights in Thailand. In the recent Thai political discourse, the term has been generally employed to justify the state of democracy and human rights within Thailand, specifically in relation to Thailand's failure to synchronise with the universal concepts of democracy and human rights. This was made particularly clear since the mass street protects and the "Shut Down Bangkok" campaign of the People's Democratic Reform Committee (PDRC) during 2013–2014, which culminated in the 2014 coup.

The utilisation of the notion of Thainess can be explained by the concept of cultural relativism. Thainess is employed by the Thai political leadership and conservative Thais when the issues of democracy and human rights are raised, at the expense of intellectual honesty, and ultimately serve to maintain their control of the population and ensure their obedience. Therefore, this chapter primarily discusses the state of human rights and freedoms under the NCPO, which was restricted by the promotion of Thainess.

Thailand's international obligations on human rights: discussing the Thai sovereignty and universality of human rights

In 1948, Thailand was among the first 48 countries to endorse the United Nations Universal Declaration of Human Rights (UDHR). There are seven core international human rights instruments, including the International Covenant on Civil and Political Rights (ICCPR), the International Covenant on Economic, Social and Cultural Rights (ICESCR) and the Convention on the Elimination of All Forms of Discrimination Against Women (CEDAW) and its Optional Protocol, to which Thailand has been a state party.[1]

In May 2016, in the Universal Periodic Review (UPR) Working Group report on the review of Thailand (Second Cycle) during the 25th Session of the UPR Working Group in Geneva, Switzerland, Thailand demonstrated a significant progress with respect to its willingness to abide by its international commitments. The Thai delegation immediately supported 181 of the 249 recommendations (72.69 per cent) received during the second cycle. It is a significant step forwards compared to 58.24 per cent received in the first cycle.[2] Nevertheless, these challenges remain under question if Thailand can make progress on its human rights under the leadership of the military government, and the NCPO, specifically with respect to political rights and freedom of expression, despite being categorised by Freedom House as a "Not Free" country since 2015 as a result of the 2014 military coup, after being in the category of "Partly Free" between 2007–2014.[3]

Since their successful seizure of power, the NCPO claims the necessity for the suppression of freedom expression and the violations of political rights of Thai people to maintain peace and order, which claimed to be part of the NCPO's reconciliation process to reform Thai politics. The NCPO continued to utilise the protection of sovereignty as a pretext against criticism from international communities and the UN, with respect to the state of human rights within the country. Nevertheless, Thailand is not the only country to make such claim in order to overrule the universality of human rights while neglecting its duty to protect human rights, abiding by its international commitments and obligations.

In the contemporary political discourse, the supremacy of power of the state and state sovereignty has been challenged and rejected. Pollis argues, "It is frequently argued that the functions of the state have been, at a minimum, transformed, often facilitate globalisation. State sovereignty is no longer absolute". At the same time, Pollis highlights that the promotion and protection of human rights is being interpreted differently, despite the proclamation of the universal human rights by the Vienna Declaration of 1993 and the legal instruments adopted over decades by the United Nations.[4]

According to Forsythe, the notion of state sovereignty is a social construct and can evolve over time and circumstances. To a certain extent, the adjustment of the international norms is a result of technological advancement, globalised markets and many other causes. As a result state sovereignty is not constant, as he argues that "state sovereignty was no longer what it once was".[5] Therefore the notion of state sovereignty cannot be utilised as a de jure mechanism to deny the universality of human rights and undermine its applicability to the citizen, because human rights are birthrights and it is a duty of the state to promote and protect those rights.

However, the Thai elites and political establishments have been in denial to accept the declining power and influence of the state in changing global circumstances. They continue to control the structure of power and reiterate the superiority of being Thai in order to justify the construction of Thai-style democracy in the post-1932 Siamese Revolution. In addition, the notion of Thainess has been highlighted and included as a core element of Thai identity and sovereignty in order to deny the universal principle of democracy and the universality of human rights in the construction of Thai-style democracy.

Under the leadership of the NCPO, the notion of Thainess has been included in the NCPO's "Twelve Core Values for a Strong Thailand", claiming to be important for the NCPO's reform process for a better Thailand. This emphasis on the conformity to the Thai cultural heritages and norms enables the military and the NCPO to reinforce the significance of Thai-style democracy, which primarily denies embracing the universality of human rights as an important element.

Indeed, Thailand is similar to other member states of ASEAN, which consider the supremacy of culture, norms and state sovereignty and cannot be overruled or undermined by the promotion of democracy and human rights. ASEAN has been willing to neglect the universality of human rights in order to enshrine a general notion of "non-intervention of internal matters of other states", which is highlighted as a core element of ASEAN. Chayes and Chayes argue that the legitimacy of norms is primarily based on the non-discrimination, minimum substantive standards of fairness and equality.[6] Thus Thailand and ASEAN cannot disregard the universality of human rights. After all, Thailand and its ASEAN friends have declared to create an equal society for their citizens.

Human rights in Thailand: universalism versus cultural relativism

One of the ongoing challenges against the promotion and protection of human rights is the notion of cultural relativism, which argues that every society is unique and functions under different circumstances with different cultural heritages. Therefore, it is important to acknowledge and respect the cultural differences, including a different set of social values, norms and beliefs, when discussing the applicability of universal values. Thus the notion of cultural relativism gains support from human rights activists, largely because they argue that respecting cultural differences is not a mere translation of cultural supremacy.[7] Despite a focus on respecting cultural differences, the implementation of this concept is somewhat different. As observed in many cases, including Thailand, the arguments regarding the uniqueness of the Thai cultural heritages have been deployed to overrule the universality of human rights. This delays the promotion and protection of human rights in the country.

For example, at the Second Cycle of the Universal Periodic Review (UPR) in 2016, Thailand noted but did not support the recommendations from Belgium, Norway, the United States and other nations with respect to the amendment of Article 112 of Thailand's Criminal Code.[8] The Thai delegation and the Thai government explained the reasons why maintaining this article is important to the 25th Session, highlighting the important of the institution of the monarchy. Thailand's consideration to comply with international obligations is normally based on interests, norms and cultural heritages. Nevertheless, it is important to note that the use of

Article 112 has significantly declined after King Vajiralongkorn ascended to the throne following the death of his father, the late King Bhumibol Adulyadej. It is been observed that similar cases have been charged by the Thai Computer Crime Act instead.

Despite being a member of the UDHR of the United Nations since its inception, Thailand has continued to deny the full embracement of human rights as a universal concept while asserting the significance of its cultural heritages as a precondition for the implementation of the universality of human rights. Throughout its stop-start democratic progress with military interventions, Thailand has often denied to fully enshrine the universal concept of human rights despite including it in the country's constitutions.

In theory, the constitution should be able to control actions of the state, because it is the highest law. Therefore, the value of law must be accepted and constrain the interpretation of the constitution, because the constitution maintains it is predominant as law, "which means that courts must construe it through a process of reasoning that is replicable, that remains fairly stable and that is constantly applied", as Post argues.[9]

According to Article 6 of the 1997,[10] 2007,[11] and 2017[12] constitutions, "The Constitution is the supreme law of the State. The provisions of any law, rule or regulation, which are contrary to or inconsistent with this Constitution, shall be unenforceable". The drafters of the 1997, 2007 and 2017 constitutions claimed that "Section 3: The Rights and Liberties of the Thai People" was primarily written with reference to the foundation of human rights of the UDHR. As a result, the interpretation of Section 3 in the Thai constitutions must be consternated by the universality of human rights, as accepted as international standards by the UDHR, instead of implementing it within limitations of the Thai cultural heritages.

Prior the military coup d'état on 22 May 2014, the protests of the People's Democratic Reform Committee (PDRC) between 2013 and 2014 demonstrated the rejection of liberty and equality of human beings. The PDRC campaigned to undermine voting rights of rural voters, arguing that they have a low level of educational attainment, therefore their voting rights should be reconsidered and perhaps taken away. Indeed, the rejection of universal suffrage in order to eradicate the problem of vote-buying in Thai politics is an ongoing fallacy promoted by the Thai elites and political establishment in order to maintain their grip on power.

However, the 2014 coup demonstrated that the Thai cultural heritages would continue to overrule Thailand's commitments to the universal concept of human rights, as observed in the implementation of the military's "Twelve Core Values". Thus it is unlikely that that the interpretation of the 2017 Thai Constitution will be constrained by the value of law. This will continue to undermine the promotion and protection of human rights within Thailand.

Thainess imposes restrictions to human rights

Over a decade of the Thai political divide, the term "Thainess" lacks clear definition and the concept itself remains vague and rather unclear for both Thais and foreign observers, although it has been utilised more in the daily life of Thais, as a means to enshrine the superiority of being Thai as a nation and also to reject the universal principles of democracy and human rights. The NCPO has highlighted that the concept of "Thainess" is important for the construction of Thai-style democracy. Nevertheless, a question arises whether Thainess can support and reinforce the process of democratisation and promote human rights within Thailand.

This chapter specifically discusses the notion of Thainess or "Khwampenthai" with respect to the hierarchical structure listed in "No. 8: the Maintaining discipline, respectful of laws and the elderly and seniority" of the NCPO's Twelve Core Values for a Strong Thailand.[13] From a cultural perspective, showing respect to the elderly and seniority is common in most Asian societies and accepted as good manners. However, in the Thai societal and political context, the connotation of respecting seniority is beyond simply showing respect, but it conveys a message of being "obedient" for those who are younger or inferior. As a result, Thai society has been characterised as a hierarchical society.

The connotation of being obedient enables the Thai political establishment and elite to dominate the power structure, preventing Thailand from making progress on the promotion and protection of human rights. For example, through the formal educational system, Thai students are taught to understand human rights within the Thai context instead of accepting them as a universal principle. In addition, a discussion with respect to the hierarchical structure within Thailand cannot disregard the influence of Buddhism, because it is the main religious belief in the country. Respect of the elderly and seniority is heavily referenced by the concept of "Bunk-hun" in Buddhist teaching, which means children are obliged to pay back their parents. Indeed, from a moral perspective, this form of social expectation is common and humane, but in Thai society this obligation is almost absolute and children are told not to question if they should do it, regardless of whether they were treated badly by their parents.

With reference to this, the promotion of Thainess demonstrates a lack of understanding of individual rights and duty, as highlighted in the universal concept of human rights – "All human beings are born free and equal in dignity and rights". As a result, the interpretation of rights in the Thai context must be revisited under the influence of Buddhism. General Prayuth Chan-ocha, prime minister and leader of the NCPO, announced "human rights" as Thailand's national priority in 2018, while his government and the military continued to deny the universal principle of human rights and employ the notion of Thainess as a means to alienate Thailand from internationally accepted human rights practices.

How voting rights are undermined by the Thai elite in the lead-up to the 2014 coup

Both the Pheu Thai Party and its predecessor have consistently proven themselves to be rather more accommodating and responsive than the Democrat Party with regard to the call for a greater degree of autonomy at the local level. This has been the case in relation to candidate selection and in tailoring a message to their grassroots. With Thailand's anticipation for the 2019 general elections, the NCPO is now following the footsteps of the Pheu Thai Party in many ways. Significantly, because of the emergence of Thaksin in the early 2000s and his appeal to the historically disregarded and disenfranchised rural poor demographic, through his populist policies, the Thai social-political dynamic has continued to take on an increasingly competitive aspect.

Just as tens of millions of rural poor from across Thailand discovered their right to become increasingly assertive in taking their opportunity to utilise political mobilisation and empowerment, in order to bring into effect the kinds of changes which might redress their legitimate grievances, now it is the middle class – albeit very much in the lower millions – who are expecting that the Thai political system must now respond to their arguably rather less pressing demands.

This demonstrates that rural populations understand the notion of representative democracy and how "representative democracy" can function in their best interests. However, the Thai political establishment and elite remains in denial to accept that rural people are political aware and active in Thailand's representative democracy, and they continue to undermine the rights to vote of rural people because of the low level of educational attainment. According to statements made by Chitpas Krisdakorn, a core member of the People's Democratic Reform Committee

(PDRC), in December 2013, in the early days of the Bangkok protests, which were considered noteworthy enough that they were reported in prominent international publications, including the *Japan Times*:

> "We're not taking away democracy. We just need some time to reform the country before we can move on to democracy", she said, explaining that problems such as corruption and vote-buying must be tackled before free and fair elections can be held. The problem, she added, is that many Thais lack a "true understanding of democracy . . . especially in the rural areas".

This "we" could be a reference to (1) the PDRC; (2) all who seek the removal of Shinawatra-led governments, (3) all and anyone who agrees with her; (4) her and her family; (5) those who share her social-political-economic status; (6) the dominant establishment or elite faction who are in a position to arbitrarily disregard the supposed primacy of the democratic process; or perhaps (7) all "right-thinking" Thais. It follows, therefore, that "we" cannot mean "those" (presumably wrong-thinking) Thais, who only very recently overwhelmingly voted for the very same government, which "we" sought the removal of.

At the last general election, which was held in February 2014, Yingluck and the Pheu Thai Party received a clear majority of the votes cast, with 59.35 per cent of the vote (according to state estimates), even though only 47.72 per cent of potential voters (according to state estimates) were recognised as having voted. (It is noteworthy that voter turnout in the previous several elections had been consistently higher.) This result was then annulled in March 2014 by the Thai Constitutional Court, several weeks after the official result had been declared, after the Bangkok protests and after significant intimidating picketing of polling stations in many provinces (in violation of the then Thai Constitution) had evidently failed to prevent an "undesired" electoral outcome.

The result of 2014, in favour of Pheu Thai, was one of the most compelling democratic mandates in Thailand's democratic history. Any such perception would likely be, at least in part, because of the lengths that the anti-Thaksin coalition went to in order to attempt to either prevent or subvert an unwelcome electoral mandate and undermine the voting rights of rural people. This evidence, in itself, would be a clear enough indication that an even more compelling result would likely have been expected.

After five years under the NCPO, a large number of Thais remain convinced that voting rights are not for all, instead these rights should only granted to those with a high level of education, while rural people are now waiting to have their voices heard once again at the polls, once Thailand returns to democracy. In considering the trajectory of current Thai politics, we may once again be reminded of the famously instructive words attributed to the 18th-century philosopher and writer, François-Marie Arouet, otherwise known as Voltaire: "Those who can make you believe absurdities, can make you commit atrocities".

Five years of human rights violations and the future prospect of human rights: discussing the state of academic freedom and freedom of expression

Between 2014–2019, Thailand has undergone a so-called process of reform by the military, through the NCPO, as a consequence of the 2014 coup. Nevertheless, there is evidently a general lack of understanding in relation to the importance of promoting academic freedom as something which is intimately and intrinsically linked to freedom of expression; and, moreover, how this can significantly contribute towards a more favourable process of democratic reform within Thailand. Despite the NCPO's claim, a vast majority of Thais in the country question if Thailand wasted its five years under the military jurisdiction without making any democratic progress.

Accordingly, it is incumbent that all stakeholders become more fully aware of the important correlation between democracy and human rights, in addition to supporting the promotion of human rights per se. According to the rules and regulations of the NCPO and the government, any academic event concerning the issue of democracy and freedom of expression and human rights is subject to permission of the NCPO and the military. Therefore, the raising of questions relating to the state of human rights within Thailand, given the current climate, must of course be necessarily approached with the requisite degree of sensitivity and caution.

Ever since the 2014 coup, academics and activists who have been working within the area of democracy and human rights have often been closely followed and/or monitored by the NCPO. Such actions by the NCPO will be regarded – at least implicitly, and by many – as restrictions that can act as obstacles to any positive process of reform within Thailand. In addition, such actions may well help to further raise the question as to whether non-democratic reforms can ultimately assist Thailand as it undertakes to restore unto itself a semblance of a "democracy" or any more evidently "democratic" values.

For instance, the Faculty of Political Science at Ubon Ratchathani was invited to the Royal Thai Military base in Ubon Ratchathani in order to discuss, with senior military officers, the position and role of Ubon Ratchathani University in relation to the possible student and academic activities which are planned to take place within Isan (the northeastern region of Thailand). Furthermore, the NCPO raised concern about the role of the political scientists who are based at Thailand's universities in the context of the social-political climate after the coup. The discussion also explored the important question of how much academic freedom there is, under the NCPO, in the undertaking to both promote and improve the study of political science, as this is a discipline which retains its importance especially at this critical juncture in the evolution and reform of Thailand's political system.

Further questions arise in relation to freedom of expression, and, in addition, the individual safety of those who undertake efforts to work within the area of democracy and human rights within Thailand. For as long as the state and its institutions stay under the effective control of the NCPO and the military government, this awareness will remain heightened; but nevertheless, such concerns will more than likely retain their validity – even with any possible return to a civilian, democratic form of government.

Despite the rhetoric of "inclusiveness" and "participation", which was repeatedly introduced by the NCPO and the current government in the aftermath of the coup d'état, and although these issues were discussed at certain select events, including a well-financed and publicised seminar conducted in English and intended for a foreign audience titled "On the Path to Reform",[14] for the most part the reality outside the capital of Bangkok runs contrary to the rhetoric of the authorities and what it intends to purport. This self-evident contradiction is clearly indicated by the military suppression surrounding many human rights activities and events held in the provinces outside Bangkok.

In general, the NCPO grants permissions to organise human rights activities and events with the "guidelines" for the event from the Thai military, and more specifically, the demand for "no political discussion", by very strong implication, the speakers and participants were not allowed to in any way criticise the NCPO or the current Thai government. It was ensured by the event organisers that everything that took place during the event was recorded in order that it would be uploaded onto the Internet, including through online social media.

This can assist in helping to mitigate the very real problem of "interpretation" by government representatives or operatives who included an officer from the "Special Branch Division" of the Royal Thai Police, as well as a number of unidentified individuals who were evidently in attendance at the event. It follows that by making the event freely available on the Internet, including through websites such as YouTube, Twitter, and Facebook, this can provide the wider public with access to the event, including what was discussed; this can give others the opportunity to make more independent and well informed assessments and judgements.

It has been observed that supports from and collaborations with the international community have been a significant contributing factor for the consideration of the NCPO to grant permissions to human rights activities. The NCPO imposed a constraint on liberty and freedom and claimed its action as necessity to maintain peace and security because Thailand has a different set of cultural heritages and norms when compared with the West. Thus criticisms on the state of human rights within Thailand from the West were often rejected and discredited by the NCPO.

As a matter of principle, all voices should be equally heard by the government and the authorities of Thailand. With regard to the current political circumstances, there are millions who support the present role of the military, and millions who are in opposition to it. Notwithstanding the very evident and extreme differences of opinion within Thailand, it is very clear that those who are not in favour of the role that is being played by the Thai military have been strongly suppressed – and this will only act to further deepen the schism of an already heavily divided society.

Therefore, the continued effort to promote democracy and human rights – including freedom of expression – will indeed act to show the Thai population that the seemingly intractable problem of societal disagreement can yet be resolved by a constructive discussion rather than resorting to violent acts of suppression. This will assist Thailand in the historical and profound endeavour to become a country where individual rights are increasingly emphasised and respected. Despite the oft-repeated claim to be promoting a so-called Thai-style democracy and emphasising the significant of Thainess, while announcing "human rights" on the national agenda in 2018, the military must nevertheless acknowledge that universal democratic and human rights principles cannot be arbitrarily disregarded and overruled by the notion of Thainess. For example, "freedom of expression" is commensurate with the freedom to be able to act to empower the citizenry in order that they can hold government accountable. In the context of either an academic public forum or the wider public debate, this freedom should not be seen to provide for the sense of entitlement to freely insult others, at least not without further poisoning and damaging the public discourse as a consequence.

However, the NCPO proved that the announcement of human rights as a national priority was only intended to mitigate pressure from the international community. In 2018, the NCPO continued to suppress voices of the opposition and young people, as observed during the visit of General Prayuth to Ubon Ratchathani University for the cabinet meeting on 24 July 2018.

The military is deeply concerned about the possibility of protests and demonstrations against the prime minister and his cabinet. Therefore, the military has listed names of university lecturers perceived to be closely monitored by the military. They seek collaborations from the university in order to ensure that there would be no protest or demonstrations against the prime minister and the cabinet during their visits to Ubon Ratchathani University. The 22nd Army Military Circle at Ubon Ratchathani was primarily assigned to ensure a "smooth trip" for the prime minister and was ordered to take full responsibility if anything should happen. As a result, the military have attempted to "meet", "inform" and "seek collaboration" with those deemed necessary of close monitoring.

This demonstrated that the state human rights were significantly jeopardised by the NCPO. Indeed, it is important for the NCPO and the military to embrace differences rather than suppress voices dissenting against the military and NCPO in order to exhibit their support for the promotion and protection of human rights. Unfortunately, the NCPO and the Thai military are willing to sacrifice and ignore the universality of human rights and allow the notion of Thainess to overrule Thailand's international commitments on human rights.

Notes

1 Permanent Mission of Thailand to the United Nations, "Human Rights and Social Issues", April 2015 <www.thaiembassy.org/unmissionnewyork/en/relation/80917-Human-Rights-and-Social-Issues.html> (accessed 26 June 2018).
2 Ministry of Foreign Affairs, "Press Releases: The Adoption of the Universal Periodic Review (UPR) Working Group Report on the Review of Thailand (Second Cycle) during the 25th Session of the UPR Working Group in Geneva, Switzerland", May 2016 <www.mfa.go.th/main/en/media-center/14/66810-The-Adoption-of-the-Universal-Periodic-Review-(UPR.html> (accessed 26 June 2018>.
3 <https://freedomhouse.org/report/freedom-world/2018/thailand>.
4 Adamantia Pollis, "A New Universalism", in *Human Rights: News Perspectives, New Realities,* edited by Adamantia Pollis, Perter Schwab (London: Lynne Rienner Publishers, 2000).
5 David P. Forsythe, *Human Rights in International Relations*, Fourth Edition (Cambridge: Cambridge University Press, 2018), p. 31.
6 Abram Chayes and Antonia Handler Chayes, *The New Sovereignty: Compliance with International Regulatory Agreements* (Cambridge, MA: Harvard University Press, 1995), p. 127.
7 Michael Freeman, *Universalism of Human Rights and Cultural Relativism* (London: Routledge, 2013), p. 51.
8 <www.upr-info.org/database/index.php?limit=0&f_SUR=173&f_SMR=All&order=&orderDir=ASC&orderP=true&f_Issue=17&issueFilterType=OR&searchReco=&resultMax=300&response=&action_type=&session=&SuRRgrp=&SuROrg=&SMRRgrp=&SMROrg=&pledges=RecoOnly>.
9 Robert C. Post, *Constitutional Domains: Democracy, Community, Management* (Cambridge, MA: Harvard University Press, 1995), p. 30.
10 Constitution of the Kingdom of Thailand, B.E. 2540.
11 Constitution of the Kingdom of Thailand, B.E. 2550.
12 Constitution of the Kingdom of Thailand, B.E. 2560.
13 <http://thainews.prd.go.th/banner/en/Core_Values/>.
14 Seminar titled "On the Path to Reform", held on 3 December 2014, organised by the Ministry of Foreign Affairs, the Secretariat of the House of the Senate, and the Secretariat of the House of Representatives. The participants and audience in attendance were at the event by private invitation, and the majority of these were representatives from the foreign diplomatic community.

References

Chayes, Abram and Antonia Handler Chayes. 1995. *The New Sovereignty: Compliance with International Regulatory Agreements*. Cambridge, MA: Harvard University Press.

Constitution of the Kingdom of Thailand, B.E. 2540 (1997).

Constitution of the Kingdom of Thailand, B.E. 2550 (2007).

Constitution of the Kingdom of Thailand, B.E. 2560 (2017).

Forsythe, David P. 2018. *Human Rights in International Relations*, fourth edition. Cambridge: Cambridge University Press.

Freeman, Michael. 2013. "Universalism of Human Rights and Cultural Relativism". In *Routledge Handbook of International Human Rights Law*. Edited by Scott Sheeran and Sir Nigel Rodley. Oxon: Routledge.

Ilias, Bantekas, and Lutz Oette. 2013. *International Human Rights Law and Practice*. Cambridge: Cambridge University Press.

Ministry of Foreign Affairs. 2016. "Press Releases: The Adoption of the Universal Periodic Review (UPR) Working Group Report on the Review of Thailand (Second Cycle) during the 25th Session of the UPR Working Group in Geneva, Switzerland" <www.mfa.go.th/main/en/media-center/14/66810-The-Adoption-of-the-Universal-Periodic-Review-(UPR.html> (accessed 26 June 2018).

Nanchanok Wongsamuth. 2015. "The Army Camp Creeps on to the Uni Campus". *Bangkok Post: Spectrum*, Vol. 8, No. 47. 22 November: 3–5.

Neher, Clark. D. 1994. "Asian Style Democracy". *Asian Survey*, Vol. XXXIV, No. 11 (November): 949–961.

Parichart Siwaraksa, Chaowana Traimas, and Ratha Vayagool. 1997. *Thai Constitutions in Brief.* Bangkok: Institute of Public Policy Studies.

Pasuk Phongpaichit. 1999. *Cultural Factors That Shape Governance in South-East Asia.* Essay written for UNESCO.

———. 1999. *Civilising the State: State, Civil Society, and Politics in Thailand.* Amsterdam: Centre for Asian Studies Amsterdam.

Pasuk Phongpaichit and Chris Baker. 1999. *Thailand: Economy and Politics.* Kuala Lumpur: Oxford University Press.

Permanent Mission of Thailand to the United Nations. 2015. "Human Rights and Social Issues" <www.thaiembassy.org/unmissionnewyork/en/relation/80917-Human-Rights-and-Social-Issues.html> (accessed 26 June 2018).

Pollis, Adamantia. 2000. "A New Universalism". In *Human Rights: News Perspectives, New Realities.* Edited by Adamantia Pollis and Peter Schwab. London: Lynne Rienner Publishers.

Post. Robert C. 1995. *Constitutional Domains: Democracy, Community, Management.* Cambridge, MA: Harvard University Press.

Universal Periodic Review (UPR), Database of Recommendations <www.upr-info.org/database/index.php?limit=0&f_SUR=173&f_SMR=All&order=&orderDir=ASC&orderP=true&f_Issue=17&issueFilterType=OR&searchReco=&resultMax=300&response=&action_type=&session=&SuRRgrp=&SuROrg=&SMRRgrp=&SMROrg=&pledges=RecoOnly> (accessed 20 June 2018).

Freedom House. 2015. Thailand <https://freedomhouse.org/report/freedom-world/2015/thailand> (accessed 2 February 2019).

———. 2016. Thailand < https://freedomhouse.org/report/freedom-world/2016/thailand> (accessed 2 February 2019).

———. 2017. Thailand < https://freedomhouse.org/report/freedom-world/2017/thailand> (accessed 2 February 2019).

———. 2018. Thailand < https://freedomhouse.org/report/freedom-world/2018/thailand> (accessed 2 February 2019).

———. 2019. Thailand < https://freedomhouse.org/report/freedom-world/2019/thailand> (accessed 2 February 2019).

Seminar titled "On the Path to Reform", held on 3 December 2014, and organised by the Ministry of Foreign Affairs, the Secretariat of the House of the Senate, and the Secretariat of the House of Representatives. The participants and audience in attendance were at the event by private invitation, and the majority of these were representatives from the foreign diplomatic community.

27
SEX AND GENDER DIVERSITY

Douglas Sanders

There is a tolerance for sex/gender diversity in Thailand that is unique in Asia. The country has no criminal laws targeting same-sex acts. Most of Asia now has no such laws, but nowhere else in Asia is there such an open array of commercial venues focused on gay men and transwomen. Sometimes there are one or two events or venues for lesbian women. There are gay host "go go" bars, in parallel to the bars with female "service workers" that cater to heterosexual men. There are gay massage parlours, though none as large as some of their heterosexual equivalents. *Babylon* is famous as the largest gay male sauna in Asia. There are large transgender cabaret theatres, aimed at tourists, in Bangkok, Chiang Mai, Pattaya and Phuket. There are local, national and international transgender beauty pageants. *Mr. Gay Thailand* is now an annual showy event.[1] Bangkok is famous for internationally regarded trans medical services, including genital reconstruction surgery. Local drag performers feature on *Drag Race Thailand*[2] on the Line TV digital station and serve as the hosts of regularly scheduled large weekly and monthly parties held in particular upmarket hotel premises in Bangkok. When Singapore banned the very successful *Nation* circuit party, it moved to Phuket in 2005.[3] Now major gay oriented "circuit" parties are held in Thailand over three days at both Songkran (Thai New Year) and the international New Year holidays. They draw visitors from the region and beyond. Thai gay/trans movies regularly gain commercial release and show in local theatres, along with foreign LGBT offerings. Lesbian characters occasionally appear in Thai films. The Tourist Authority of Thailand has an advertising campaign expressly aimed at attracting gay and lesbian tourists –a first in Asia. A government promotional video from 2013 calls out "Go Thai. Be Free", which is also now an appealing website.[4] The difficulty in holding LGBTI[5] events in other Asian countries has drawn business to Thailand. Two Asian regional conferences of the International Lesbian, Gay, Bisexual, Trans and Intersex Association (ILGA) have been held in Thailand as well as the largest of its world conferences in November 2016 (unproblematic even during the official mourning period after the death of King Bhumibol Adulyadej that October). Various foreign embassies sponsor or co-sponsor public events related to LGBTI issues. Thai LGBTI people do not report conflict with Buddhism, the most widely practiced religion. No conservative, evangelical or fundamentalist religious groupings publicly campaign against LGBTI rights. The government's Museum of Siam in 2018 mounted a show on the history of LGBT issues and activism in Thailand. The Bangkok Art and Culture Centre has hosted events marking the International Day Against Homophobia and Transphobia. It will host a large show of

LGBTQ-related contemporary Asian art in November 2019. The police do not raid or harass gay venues.

These patterns of tolerance have persisted through periods of both elected and military governments. It was the military government that came to power after the coup in May 2014 that enacted the Gender Equality Act, aimed at protecting individuals from discrimination (a legislative proposal it inherited from previous governments).

Yet, most Thai parents resist recognising or accepting sex and gender variation on the part of their children. In turn, most lesbian and gay individuals are not "out" to their birth parents or close relatives. Maybe they will confide in a trusted brother or sister. They are typically not "out" on the job, either to co-workers or employers.[6] "Social stigma and family discrimination remain as running themes in many local movies with homosexual characters".[7] Discussions of sexuality in society are still taboo and there is limited sex education in schools.[8] Bullying in schools seems common (as it is elsewhere).[9] The most active discrimination centres on trans individuals, whose gender expression makes their difference more obvious. There have been no "out" politicians, established academics or major entertainment figures. No political parties have a history of advocacy of LGBTI rights. Thai universities do not provide "support or encouragement" for their academics to do research on LGBTI issues, even when individuals have done such work during graduate studies.[10] Older Thais often see sex/gender variation as the results of bad karma from sexual indiscretions in previous lives.[11] Trans individuals cannot get recognition of their gender identity on official documents. Their national identity cards, driver's licenses and passports indicate their sex at birth. Repeatedly there have been issues of trans students being told to wear the school uniforms that accord with their birth sex, particularly for examinations or graduation ceremonies. At some institutions, individuals can petition for an exemption. Activist LGBTI rights organisations are small and poorly funded. Visible spokespeople are few in number. The only Thai organisations with offices and paid staff are, at their core, health organisations concerned with HIV/AIDS. "Pride" events are held indoors. There has been no "pride parade" in Bangkok for over a decade. Phuket, a tourist destination, with a continuous history of parades since 1999, took a break in 2018. In early 2019, Chiang Mai held its first pride parade in a decade. Visitors see the openness of Thai society. Locals know the constrictions.

An open letter to Thai authorities in 2012 identified 15 targeted killings of "lesbians and tomboys" in Thailand over the previous six years. The letter said the police downplayed the incidents as "crimes of passion, love gone wrong or the fault of the victims".[12]

How to cast an overview of seemingly contradictory patterns? Two scholars comment:

> Thailand is perceived to be among the more tolerant nations in South-East Asia with respect to same-sex attraction and cross-national comparisons in the region support this notion. Nevertheless, in the World Values Survey of 2007, 33% of surveyed Thais mentioned "homosexuals" as a group they would not want as neighbours, and the respondents on average ranked the justifiability of homosexuality at 3.1 on a scale from 1 (lowest) to 10. This combination of low acceptance and relatively low rejection of gay or lesbian neighbours supports Jackson's (1990) description of Thailand as tolerant but not accepting of homosexuality.[13]

Discrimination

There is no policy of exclusion of homosexuals from the civil service or the military. It is widely known that gay males have served in high positions, including as prime minister. Some historians have identified King Vajiravudh, or Rama VI (r. 1910–1925), as homosexual.[14] His homosexuality is an open secret, well-known in Thailand but only explicitly referred to these days in academic publications. Gay males are not excluded from the Buddhist Sangha (monkhood). All monks are required to be celibate. There has been unease over photographs of effeminate young monks cuddled together, and recently of bodybuilders, but any sexual activity is forbidden.

In 2007, during a previous military-installed government, there was an attempt to expand the equality provision in the constitution to have it specifically refer to "diverse sexualities" or "sexual identity". There was debate and three separate votes. New language was not added, but an "intentions" document was issued explaining the Thai term *phet*, used in the particular constitutional provision (and variously translated as "sex" or "gender"). The document said the word covered "the differences between individuals in sexual identity or gender or sexual diversity, which may be different from the *phet* in which the person was born". In other words, in the view of the constitutional drafters, LGBT were already covered by the constitutional provision. Additional language was not necessary.[15]

There have been certain specific well-publicised cases of discrimination involving transgender women. In 2007, Sutthirat Simsiriwong was barred from entering a nightclub in a Novotel Hotel. Management publicly apologised.[16] Transwomen were excluded from parades organised by the Chiang Mai provincial government. In 2010 the Administrative Court ruled in their favour.[17]

Kathawut "Kath" Khangpiboon, a transwoman and one of the founders of the Thai Transgender Alliance, had been a guest lecturer at Thammasat University for five years. In June 2014, she was offered a full-time position in the university's Faculty of Social Administration. The university then rescinded the job offer on the basis that Kath had used inappropriate language and postings on social media. The rector said a posted photo of a penis-shaped lipstick, an amusing gift, was a sign of Kath's improper conduct online, indicating that she would not be a good teacher.[18] In 2018 the Administrative Court ruled that the social media postings had no bearing on her qualifications. Thammasat was ordered to hire her and the university complied. There was no ruling that Thammasat had discriminated on grounds of gender identity. Yet, clearly, gender identity was being treated as irrelevant in the hiring process.[19] The scandalous tube of lipstick was a featured item in the Museum of Siam's LGBT history show that opened in May 2018.

The Gender Equality Act (GEA) was enacted in March 2015 and officially launched at a public event in September 2015, with representatives of the government, the United Nations Development Programme (UNDP), Rainbow Sky Association, the Lawyer's Council and the Association of Women Lawyers. Individuals associated with the Thai Transgender Alliance and Rainbow Sky are members of the Committee on the Consideration of Unfair Gender Discrimination established under the act. When Prime Minister General Prayuth Chan-ocha spoke at the UN General Assembly as head of the Thai military government, he specifically mentioned the GEA as an accomplishment of his government.[20] The GEA seems to be the legislative implementation of the *phet* equality provision in the constitution, including the elaboration in the "intentions" statement. A UNDP/Thai government report gives a different origin.

> The Act . . . originated from obligations under the ratification of the Convention on the Elimination of All Forms of Discrimination Against Women (CEDAW) which not only protects women but also those persons whose gender expression does not match their sex assigned at birth. A representative from the Ministry of Social Development and Human Security gave unofficial commentary that sexual orientation is not covered under this ground. This interpretation excluding sexual orientation runs contrary to several key recommendations issued by the Committee on the Elimination of All

> Forms of Discrimination against Women recognising discrimination faced by women as a result of sexual orientation and gender identity. Other wider work by UN Women also promotes the rights of lesbian, bisexual, transgender and intersex people, within the context of gender.[21]

On a finding of unfair gender discrimination, an order can require government agencies, private organisations or individuals to take remedial action. Failure to comply can result in criminal charges.[22]

The process has proven slow and bureaucratic. The first case to be resolved, after three years of operation, told the University of Payao to allow a trans/intersex student, whose birth certificate indicated "male", to receive her law diploma while presenting in her photograph as a woman. Tawanchai Wittiya boarding school in Korat had a policy of not accepting LGBT students. The activist TEA Group sought to have the Gender Equality Act apply to the school. The relevant GEA committee said it was necessary to have an affected individual as complainant. It declined to intervene.[23]

Associate Professor Matalak Orungrote, Faculty of Law, Thammasat University, said the act's legal processes needed to be improved (one of many critics saying that the act was not working well).[24] The UNDP has commissioned a study of the working of the act. National Human Rights Commissioner Angkhana Neelapaijit said the act should be amended.[25]

The World Bank study, *Economic Inclusion of LGBT Groups in Thailand*, was released in March 2018. Of responses to an online survey of 2,302 LGBTIQ individuals, 77 per cent of transgenders, 49 per cent of gay males and 62.5 per cent of lesbians believed job applications had been refused because they were LGBT.[26] Such figures are suggestive but statistically quite unreliable. It is not possible to get a representative sample of the populations involved. LGBT would not be "out" at work and LGBT individuals in stable jobs are unlikely to respond to such surveys. As with so many things about LGBTI groupings, our information is quite limited.

Anti-discrimination laws, in general, are largely symbolic in their prohibition of discrimination on LGBTI grounds. Most employers are unlikely to expressly reject applicants on grounds of sexual orientation. Many individuals will not challenge a rejection, for that would require them to be open about their sexual orientation.

Media

In Thailand, small-format gay magazines begin with *Mituna* in 1983. In Asia, only Japan has an earlier history of gay print magazines. Over a thousand of the early Thai magazines have been digitalised in a project funded by the British Library's Endangered Archives Programme.

There have been a couple of sequences of regular magazine-size Thai language publications, one or two of which looked as if they could mature into stable commercial publications. They, and others with partial nudity, did not achieve regular distribution. None survives. Later, two promising Thai language magazines emerged. The first, *@Tom Act*, appealed to a lesbian audience and survived for perhaps five years, with interruptions, new beginnings and slightly variant titles. In 2011 the leading British gay magazine, *Attitude*, began a Thai edition –its first overseas venture. Half of the content was locally generated. It had mainstream distribution, but, like *@ Tom Act*, never achieved an adequate advertising base. It ceased publication in February 2018. Free, small-format English-language magazines for tourists began with *Thai Guys* over a decade ago. There have been five or so iterations of this model, supported by bar ads. One, *Thai Puan*, survives, with gay tourist maps for Bangkok, Chiang Mai, Pattaya, Phuket and Samui (and some English-language commentary by expats).

A study of content in newspapers was published in 2017, based on a survey of six publications over a one-year period, including the high-circulation *Thai Rath*. In general it found quite hostile and discriminatory reporting. But two English-language newspapers, *The Nation* and the *Bangkok Post*, are decidedly supportive. *Bangkok Post* has a monthly LGBTI page plus frequent stories by staff reporter Melalin Mahavongtrakul, who for four or so years has been the main author of Thai LGBTI news stories and commentary. Kong Rithdee, senior columnist, often writes long reviews of particular Thai language LGBTI films. Both papers regularly reprint wire service stories of developments abroad. The 2017 study noted: "Most high[er] quality news was found in the English-language press or were derived from international news outlets".[27]

News and advertising (and dating and cruising) have moved online, and this has negatively affected both print publications and some of the commercial social venues. The only places in Asia with surviving print publications seem to be Taiwan and Japan.

Gay films are flourishing. In recent years Thailand has seen the commercial release of more locally made gay and trans films than any other jurisdiction in Asia. Many are low-budget comedies or stories of young peoples' loves and foibles. A pattern of having ridiculous stereotypes of effeminate males as comedy figures has not gone away, but offerings increasingly include interesting productions. In 2007 two films were released that seemed to mark a kind of mainstreaming transition: *Bangkok Love Story* and *Love of Siam*. The first focused on a hit man and a young police detective. The second was of a tentative romance between two teenage students, set in Bangkok middle-class life. In 2008 *Love of Siam* was Thailand's submission to the US Academy Awards for best foreign-language film (the decision made by the Federation of Thai Film Producers). It was included among a listing of the 70 best films released during the long reign of King Bhumibol.[28] Another very successfully executed Thai-language gay film was *P'Chai My Hero*, or *How to Win at Checkers (Every Time)*, the official Thai submission to the Academy Awards for 2015. A third gay-themed film, *Malila, The Farewell Flower*, was the submission for 2018.

A few of the films are interesting "art theatre" productions, notably *Tropical Malady* (2004), directed by Apichatpong Weerasethakul.[29] The film featured a charming rural gay romantic story. The second part has a mysterious jungle sequence in which a male character transforms into a tiger – hence the Thai title of *Sat Pralaad* (A Queer Animal). It won a special jury award at the Cannes Film Festival. It did poorly at the Thai box office. *Malila, The Farewell Flower* (2018) told of the return of a tragic figure, now dying of cancer, to his home village and former lover. He was skilled in making *bai sri*, which are elaborate multi-tiered floral ornaments using jasmine flowers and banana leaves. Baisri symbolise the fleetingness of life, and they soon fade. After the man's death, the former lover becomes a monk, travelling with a mentor monk in a dangerous jungle area near a border. They stop to meditate over a rotting corpse –an insurgency victim. The slow-paced film was directed by Anucha Boonyawatana who had earlier made the mysterious gay film, *The Blue Hour*. Kong Rithdee, writing in the Bangkok Post, called *Malila* "truly remarkable" and "one of the best films of the year".[30]

A very surprising 2018 release was *Toot Tee Kee Chart* (The Last Heroes), a big-budget historical drama of Thai villagers fighting against a Burmese military invasion (a familiar Thai nationalist narrative). A small group of villagers infiltrate the Burmese Kingdom to learn of battle plans. The group includes two handsome straight muscular males, whose wives remain behind – plus four or five effeminate gay/*kathoey* comedy figures, including regular character actors from previous Poj Arnon films. So a slapstick trans comedy is inserted between the opening idyllic scenes of traditional village life and the final battle (in which the villagers are killed and their homes torched). Three recent small LGBTI film festivals have been held in Bangkok in 2015, 2016 and 2018. Sponsorship and location have varied each time.

Important now are television and online series, which often reach large audiences. LGBTI focus or content is notable in productions such as *Love Sick* (second season of 36 episodes), *Hormones* (three seasons), *Diary of Tootsies* and *Gay OK Bangkok*.[31] The Thai Boy Love *SOTUS, the Series* of 2016 became highly popular in other parts of Asia, with a new series planned, *My Skyy*.

The show *Gender Illumination* opened at the Museum of Siam on 17 May 2018, the International Day against Homophobia and Transphobia.[32] The Thai-language title *Ying-Chai Sing Samut* translates as "female-male are imagined categories". The show had three distinct parts. In the first part, panels urged young people to reject being "just a copy of someone else. Be you. Be original. Let's find your own way. Gender is diversity".

> Roles aren't definite. Get to know your own mind, body, organs, desire and drive. Don't just follow and repeat. Be yourself. Go with your own flow. When it comes to gender identity, only you know what gender you are. When it comes to gender expression, how you act, what you wear, and how you style your hair is up to you. Your biological sex can only be male, female and intersex, but your sexual orientation can be whatever. You can love anyone, whatever their gender and expression. You can be as fluid and eclectic as you want.

This section went on to define basic terms, including the newer "non-binary" and "genderqueer". There were three streamed videos. The material made no references to Thai identity terms or local understandings. It presented a contemporary international activist discourse on sex and gender diversity. The highly individualistic approach was restated in the final room of the show: "Be you, like you want to be. Be fluid. Be Diverse. Don't label yourself". Pan Pan Narkprasert, a Thai hairdresser, declared in a video that "gender is limitless".

The second part of the show featured a ten-minute video of the Buddhist Monk Phra Shine Waradhammo, known as Phra Chai. For a number of years, Phra Chai has attended various public events concerned with LGBTI issues. For a period he held film showings and discussions. Phra Chai uses the standard terms gay, lesbian and *kathoey* (for transwomen). In the video, he expressed support for all-gender bathrooms and said the monkhood is (should be) open to individuals of all genders and orientations. For Phra Chai, sex and gender diversity constitute aspects of personal identity. Buddhism, he said, teaches us to overcome identity. That means treating gender as a "nothingness". "The physical appearances of sex or gender are not real because real sex/gender does not exist". There was no attempt to relate his views to the messages of the other parts of the show.

The third part of the show was titled "Siam Gender Records". It presented a detailed overview of Thai gay history, identifying magazines, films, events and campaigns – and even a bit of royal history. There were many "coming out" stories from named individuals. Short texts had personal accounts and perhaps identified some artifact from their personal history – a costume, a song, a rejected application to register a marriage, a tube of lipstick. The message was the acceptance of the sex and gender characteristics and practices of the individuals. It supported what is often called "identity politics". Sex/gender variation was real and identities were to be embraced. At the end of the show, visitors were invited to experiment. There were wigs, makeup, underwear, padded genitals and a flashy shirt. Visitors could view their altered image in a full-length three-part mirror. A sign on one mirror asked: "Who is that man?" It was a selfie moment.

Trans*[33]

Thai people are aware of many trans, or gender-crossing individuals, living in the society. The older term, *kathoey*, coexists with the newer "transwoman". "Lady boy" is a common term for young effeminate males. The term for masculine women, "tom", continues in use. Transmen are now more visible and have an organisation. Drag queens have emerged as an entertainment category separate from cabaret performers and beauty pageant contestants. The large cabaret shows, pulling in on tour buses, are famous. Two are in Pattaya: Tiffany (from 1974) and Alcazar (from 1981). In Bangkok we have Calypso (from 1988) now at the tourist destination Asiatique, and Playhouse at Asia Hotel. Miracle in Chiang Mai is recent. Tiffany, the matriarch, holds lavish annual national and international trans beauty pageants. A transgender woman was named one of the prime ministerial candidates of Mahachon Party in the March 2019 election (one of 46 parties with PM nominees).[34]

Two recent stories have been important. At age 20, Thai males must report for the military draft, run on a lottery basis. Individuals with genital surgery were exempted from the lottery as suffering from a permanent mental disorder. Since military conscription documents are regularly presented in employment situations, having such a dismissal was a serious problem. A lawsuit was launched against the military, backed by activists and the National Human Rights Commission. The initiative led to prolonged negotiations. Eventually the military agreed to use the then standard international medical classification of "gender identity disorder". This was confirmed in a consent judgment by the Administrative Court in 2011. While still stigmatising, it was much more satisfactory.[35] International terminology has since shifted to "gender dysphoria", still unsatisfactory language. International advocacy seeking non-pathological terminology continues.[36] Thai military may now be using the formula that the individual's "gender does not match their sex at birth" in exemption documents.[37] Reliance on genital surgery for the exemption also goes against newer international practice.

In Thailand, an individual's name is required to be one that indicates whether the person is male or female. Individuals are unable to have the sex/gender marker on official documents changed, even after genital surgery and sustained presentation in the desired sex. Changed documents are possible in the West and Latin America (often, now, without the requirement of a medical diagnosis or genital surgery). In Asia, changed documents are possible in China, Indonesia, Japan, Singapore, South Korea and Taiwan. Taiwan is the first in Asia to drop the requirements of a medical diagnosis and surgery.

A public forum was held in Bangkok on 14 May 2017 on a possible law to allow changed sex/gender markers. Government drafting was at an early stage for a law establishing a committee to certify the changed "gender" of individuals. Applicants had to be at least 20. Criticism was voiced about the binary thinking of only two "genders". There was criticism of having a government committee make the decision, rather than the state accepting individual self-determination, as in newer reform laws abroad. A UNDP/Thai government report gives background.

> In 2016, the Department of Women's Affairs and Family Development . . . partnered with the Faculty of Law at Thammasat University to present a study they had completed on legal gender recognition. This study analysed the gender recognition laws of different countries around the world in order to provide a basis for a draft Thai law. A proposed draft Gender Recognition Act was unofficially released to the public early in 2017. On 14 March 2017, For-SOGI, a network of 28 LGBTI CSOs organised a public forum to comment on the leaked draft Gender Recognition Act. Two days after the public forum, the network submitted a joint statement to the Department of Women's Affairs and Family Development outlining concerns about restrictive eligibility criteria (such as the need to have surgery) contained in the draft Gender Recognition Act and requested the public's and especially the transgender community's involvement in the development of the Act. The public forum received much attention

> from the media, which reported the concerns outlined by civil society at the forum. As a result of the concerns expressed by CSOs and the media about the restrictive criteria contained in the law, as well as the need for a more consultative process in drafting the law, the Director General of the Department of Women's Affairs and Family Development. . . . decided to halt the study and the development of the draft Act for the time being.[38]

The UNDP and the Asia Pacific Transgender Network in 2017 published *Legal Gender Recognition: A Multi-country Legal and Policy Review in Asia.* The more technical legal follow-up for Thailand was a joint study by the Thai government's Department of Women's Affairs and Family Development and the UNDP, *Legal Gender Recognition in Thailand: A Legal and Policy Review.* It was released at a multi-stakeholder national roundtable discussion on 2–3 May 2018. The two-day roundtable meeting in Bangkok was charged with developing a draft gender recognition act. A UNDP press release described the event:

> One of the key goals of the two-day meeting (2–3 May) is to contribute to the ongoing discussions by the government and civil society groups in Thailand to develop a draft gender recognition act. The roundtable will also facilitate south-south learning, with government and civil society representatives in attendance from Cambodia and Vietnam, who are in various stages of developing their own gender recognition laws. Discussions will be held on good practices and common difficulties in developing and implementing related legislation and policy, with inputs from international experts from Argentina and Malta, as well as from the World Professional Association for Transgender Health and APTN.[39]

We await developments.

Tangerine Community Health Centre opened in 2015, located on the second floor of the Thai Red Cross AIDS Research Centre in central Bangkok. It is the first facility to provide trans-specific health care and counseling services.[40]

Recognising relationships

In 2012, long-time activist Natee Theerarojanapong and his partner applied to register a same-sex marriage. Registration was refused. Natee had successfully sued Chiang Mai province twice in the Administrative Court over discrimination against transgender women in the staging of local festivals. He now asked the National Human Rights Commission to send the issue of equal marriage to the Constitutional Court. Dr. Tairjing Siripanich, one of the commissioners, accepted the complaint. He indicated support for the rights claim, and (in what must have been a pre-planned strategy), forwarded the issue to the House of Representatives Committee on Legal Affairs, Justice and Human Rights. The Committee named three representatives of LGBTI organisations as advisors. Four official Committee hearings were held in different regions, with a final one back in Bangkok at the Thai parliament on 19 April 2013. The Committee was chaired by a member of parliament for the then governing Pheu Thai party. Committee member Wirat Kalayasiri, the head of the opposition Democrat party's working group on legal affairs, was responsible for drafting, working with the Rights and Liberties Protection Department (RLPD), a unit within the Ministry of Justice. No express endorsement of the project came from either the Pheu Thai or Democrat Party leadership, though they allowed the hearings and drafting to proceed.

A draft bill emerged to establish a system of "civil unions". On registration, the rights and obligations of marriage would apply to same-sex couples, mutatis mutandis. That Latin legal phrase, familiar to lawyers, troubled the activists involved for its lack of specificity.[41] The age of eligibility to register was 20 (the voting age) rather than 17, the age for legal marriage (perhaps the most obvious inequality in the proposal). There were no express provisions relating to adoption or child custody.

The final committee hearing, held in the legislative building in Bangkok on 19 April 2013, was dramatic. In a panel presentation, Kerdchoke Kasamwongjit, deputy director-general of the Rights and Liberties Protection Department, identified himself as gay (a very un-Thai public self-outing by a government official) and spoke of his commitment to move the issue forward. He has continued to be the key activist within the government pushing reform.

At the hearing the draft bill was supported by Tairjing of the National Human Rights Commission and Nicholas Booth of the UNDP (which was actively backing LGBTI rights in Asia). But there was strong dissent. Weaknesses in the draft bill were noted by Anjana Suvarndananda, who like Natee had been a pioneer activist as a founder of the lesbian Anjaree group. The bill's silence on adoption and custody issues were a particular concern of lesbians. Over time, Anjana has generally promoted amendment to the marriage provisions in the Civil and Commercial Code. Because there had been no endorsement from the government or opposition parties, it was not clear whether the draft legislation would get legislative consideration. Getting a "citizens' initiative" draft bill into the legislative process required the backing of 20 individual members of parliament or 10,000 signatures of voters. The Rights and Liberties Protection Department referred their draft bill to various government agencies for comments –a required step.

Certain LGBTI organisations, with leadership from the Foundation for SOGI Rights and Justice (FOR-SOGI), worked with the Law Reform Commission of Thailand, a separate government agency, on an alternative "civil society" draft. They sought a registration system open to any couple, whether same-sex or opposite-sex (as in France). It would specifically include rights of adoption, custody and access.

A joint LRC–FOR-SOGI drafting committee started functioning in early 2014. After the military coup on 24 May 2014, FOR-SOGI stated it would not lobby the military government, seeing it as illegitimate. The Law Reform Commission continued drafting until its work, in general, was halted by the new government. It had not completed a final draft by that date (still consulting with various government departments).

With little visibility, the Rights and Liberties Protection Department (RLPD) in the Ministry of Justice continued drafting. In late April 2018, Kerdchoke told media that the drafting committee would meet on 4 May to finalise the text before presenting it to the cabinet through the Minister of Justice. It seems that did not happen. Same-sex couples who registered would be termed "life partners". The focus had been narrowed to "asset management of same-sex life partners to avoid large-scale impacts and the need to amend multiple laws".[42] This narrow focus was new and sure to provoke LGBTI objections. Enactment could take place within the current legislative session (it was said). This news release included two personal stories – one of a 20-year relationship, another of a denial of access to a partner in a coma in intensive care.

On 4 May, Kerdchoke stated that the proposed law followed a "French model", in that it provided for registered civil partnerships, and could be followed later by opening marriage.[43] The French law had offered the least rights of any of the European registration systems. A news report commented:

> The bill is modeled after a similar French law that allows couples – same-sex and otherwise – tax and pension rights. A draft has been unofficially released online, with

> strong emphasis on financial aspects of the relationship. It does not mention anything about adoption or the rights to establish one's family. "To put it bluntly, this current draft looks more like a contract for a financial transaction. It has nothing to do with rights, or establishing a family, or for equality and protection of LGBTI people whatsoever", said gender activist Chumaporn Taengkliang.[44]

In May 2018, speakers from three political parties (Chartthaipattana, Bhumjaithai and Pheu Thai) indicated support for some form of legal recognition of same-sex relationships, with the first supporting equal marriage.[45] Also in May, the Democrat Party home page displayed the rainbow flag and quoted party leader Abhisit Vejjajiva as saying he was ready to push for relationship recognition.[46]

On 29 June 2018, FOR-SOGI announced a change in strategy. It saw its advocacy of the "civil society" draft as having come to a standstill. Gathering signatures for a citizens' initiative had not been possible under the military government. Additionally, the research study by Chawinroj Terapachalaphon had shown how to introduce gender-neutral language in the marriage sections of the Civil and Commercial Code. References to "a man and a woman" would become "two persons"; "husband and wife" would become "spouses". This would give a same-sex partner the well-established status of being a "spouse" rather than the novel identity of being a "life partner". The detailed studies which had bogged down the completion of the RLPD and Law Reform Commission drafts could be avoided. FOR-SOGI now supported amending the marriage and inheritance parts of the Civil and Commercial Code.[47]

A draft "life partnership" bill was finally released by the Ministry of Justice on 5 November 2018. The *Bangkok Post*, the leading English language newspaper, called the bill "a slap in the face to the LGBT community" for having much more limited rights than earlier drafts.[48] Public consultations were held in five parts of the country, all completed within one week. The Bangkok hearing on 12 November was a half-day event, and most of that time was devoted to a panel explaining the "step by step" approach that justified a modest "first step" registration law. The panelists acknowledged that the bill was quite limited and that it focused on financial issues. The panelists described it as a beginning that could be expanded in the future. One hour was allocated for statements from the audience, most of which were critical. There was praise in a *Bangkok Post* opinion piece in the name of the Swedish and Canadian ambassadors and a representative of the UNDP.[49] There were calls for expansion of the rights involved and opposition from Muslims in the far south of Thailand.[50]

On 25 December, a revised draft gained preliminary approval from the cabinet. The draft was forwarded to the Council of State for the routine vetting. The Council posted the draft on its website and declared a short period during which it would receive comments, suggesting a substantive review of the initiative (not just an appraisal of its relationship to the Constitution and existing laws). The Thai Council of State had been established by King Chulalongkorn in 1874, copying the French *Conseil d'État* organised by Napoleon in 1799. In France, as a general rule, the advice of the *Conseil d'État* is followed. This seems true in Thailand as well. By mid-February, the Council had not named the chair of the committee to review the bill. Legislative activity ended in March, ahead of the national election. The bill, with whatever changes will have been suggested by the Council, may be taken up by the new post-election government.

On issues of custody and adoption of children we seem to have limited information. Joint adoption, including second partner adoption, is not available to same-sex couples. Lesbian women and lesbian couples are often raising children conceived in a previous heterosexual relationship or sometimes placed with them by relatives through extended family patterns. One of the partners may have legal custody.[51] Surrogacy services in all of Asia seem now limited to non-commercial arrangements made by married heterosexual couples.

Observations

As we have seen, reform efforts in Thailand have been limited or long stalled. So far, the Gender Equality Act has been a disappointment. The military still uses pathologising language when exempting transwomen from conscription (but more moderate language than before). Efforts to write a gender identity law seem suspended. A registration bill for same-sex couples officially emerged in November 2018, and had rushed public hearings in five places in one week. It might get to the legislature in 2019.

A Thai norm continues: hands off on gay venues, a warm welcome to gay tourists, limited legal reforms.

Notes

1 See https:// mrgaythai.com.
2 See https:// tv.line.me/dragracethailand.
3 See Nation Parties, "The Singapore LGBT Encyclopaedia Wiki" <Wikia.org> (accessed 15 August 2019).
4 See <www.gothaibefree.com>, showing eight different same-sex couples enjoying themselves at different sites in the kingdom. The government's Tourist Authority of Thailand hosted a two day LGBT+ Travel Symposium at the So Sofitel hotel, a gay-friendly venue, in the central business district, 29–30 June 2018, and earlier in the year had been the presenting LGBTI partner at ITB Berlin 2018: Xingi Liang-Pholsena, "Finally, Thailand comes out of closet", Travel Trade Group, 2 July 2018.
5 The acronym is for lesbian, gay, bisexual, trans, intersex. There are many versions of the acronym. LGBTI is used in the UN documents. That usage is followed here.
6 UNDP, USAID, 2014, p. 33.
7 Kong Rithdee, "The Darkest Hours", *Bangkok Post*, 7 August 2015, p. 5.
8 UNDP, USAID, 2014, p. 6.
9 Ibid., p. 36; Taylor McAvoy, "Bullying of LGBT Youth Still Pervasive in Thai Schools", *Khaosod English*, 22 June 2018.
10 Peter A. Jackson and Narupon Duangwises, *Review of Studies of Gender and Sexual Diversity*, International Conference of Thai Studies, Chiang Mai, July 2017.
11 UNDP, USAID, 2014, pp. 7, 30.
12 The letter and press release are on the website <www.outrightinternational.org>. See, Melalin Mahavongtrakul, "Stop the Attacks, Both Physical and Verbal", *Bangkok Post*, 30 July 2018, p. 8, which recounts police taking an abusive "tom" partner into custody. On a fatal assault, see also "Slain Thai 'Beaten by Lesbian Lover'", *Bangkok Post*, 9 September 2018, p. 3.
13 Jan de Lind van Wijngaarden and Timo Ojanen, "Identity Management and Sense of Belonging to Gay Community among Young Rural Thai Same-Sex Attracted Men", *Culture, Health & Sexuality*, (2015), pp. 1–14. Sex and gender diversity, like prostitution, is very visible to foreign visitors. Prostitution is decried by Thais and Thai governments while clearly tolerated and facilitated. Raids on heterosexual massage parlors offering sexual services focus on underage girls and illegal foreign workers and incidentally on police corruption. Prostitution is another example of a sexual pattern that is tolerated but not accepted.
14 Chris Baker and Pasuk Phongpaichit, *A History of Thailand*, 3rd edition, (Cambridge: Cambridge University Press, 2014), p. 105; Tamara Loos, *Subject Siam: Family, Law, and Colonial Modernity in Thailand*, (Ithaca: Cornell University Press, 2002), pp. 116, 171.
15 See Douglas Sanders, "The Rainbow Lobby", in *Queer Bangkok*, edited by Peter A. Jackson, (Hong Kong: Hong Kong University Press, 2011), pp. 229, 240–244.
16 Ibid., p. 239.
17 See, *International Commission of Jurists, Sexual Orientation, Gender Identity and Justice: A Comparative Law Casebook*, no date, p. 165.
18 Sasiwan Mokkhasen, "Thammasat Cites Risqué Lipstick for Rejecting Transgender Professor", *Khaosod English*, 14 October 2015; Melalin Mahavongtrakul, "It's That Time of Year", *Bangkok Post*, 30 April 2015.
19 Chularat Saengpassa, "Thammasat Ordered to Rehire Transgender Lecturer", *The Nation*, 9 March 2018, p. 1A.

20 Prayuth Chan-ocha, "PM Lauds Partnership to Tackle Inequality", *Bangkok Post*, 29 September 2015, p. 11.
21 Department of Women's Affairs and Family Development, UNDP, *Legal Gender Recognitio in Thailand: A Legal and Policy Review*, May 2018, pp. 28–29.
22 Ibid., pp. 29–30.
23 Pratch Rumvanarom, "Equality Group Files Complaint after Korat School Bans LGBT Pupils", *The Nation*, 4 June 2017, p. 2.
24 Om Jotikasthira, "Law to Let Thais Choose Their Sexual Identity Gains Traction", *BangkokPost*, 26 July 2017, p. 2.
25 Wichit Chaitrong, "Drive for Gender Inclusion", *The Nation*, 27 March 2018, p. 1B.
26 Patpon Sabpaitoon, "LGBTI Discrimination 'Still Prevalent'", *Bangkok Post*, 27 March 2018, p. 4.
27 Burapa, UNDP, A Tool for Change, 19July 19, 2017, p. 22.
28 Monthly Snippets, *Bangkok Post*, 30April 30, 2018, p. 8.
29 See Arnika Fuhrmann, *Ghostly Desires*, (Duke, 2016), ch. 3.
30 Kong Rithdee, "The Inevitability of Farewell", *BangkokPost*, 16 February 2018, p. 11. The film won seven prizes in the annual national Subhanahongsa film awards in March 2019; Parinyaporn Pajee, "Pure Love Conquers All", *TheNation*, 5 March 2019, p. 8B.
31 Jamie James, "Boys in Love", *New Yorker*, 8 January 2016; Melalin Mahavongtrakul, "Re-characterising HIV and AIDS", *Bangkok Post*, 28 March 2016, p. 10.
32 Melalin Mahavongtrakul, "Paradise or Paradox?" *Bangkok Post*, 15 May 2018, p. 1.
33 Trans* (with the asterisk or a plus sign) is a relatively new usage designed to indicate a range of patterns being signified, displacing older more specific terms such as transvestite and transsexual.
34 Nattaya Chetchotiros, "Unlocking the Potential Within", *Bangkok Post*, 25 February 2019, p. 2. The nominee is a former sports promoter, widely recognised in Thailand. The party fielded 199 candidates for the 500-seat House. It held only one seat after the 2011 election. One party policy is to legalise prostitution.
35 UNDP, UNAID, 2014, p. 7.
36 The WHO now proposes a move of the category out of the lists of pathologies, but will include "gender incongruence" in a separate section of factors relevant to mental health. Again, the change is an improvement, but it still holds up a norm of congruence.
37 UNDP, Ministry of Social Development, *Legal Gender Recognition in Thailand*, May 2018; Victor Madrigal-Borloz, *UN Independent Expert, Protection against Violence and Discrimination Based on Sexual Orientation and Gender Identity*, 12 July 2018, p.A/73/152, para.63.
38 Department of Women's Affairs and Family Development (Thailand), UNDP, *Legal Gender Recognition in Thailand: A Legal and Policy Review*, 2018, p. 32. And see Melalin Mahavongtrakul, "Newly Released Research Provides a Framework for Gender Recognition Law", *Bangkok Post*, 26 September 2016; "Our Genders Need no Approval from the State", *Prachatai*, 13 March 2016; Om Jotikasthira, "Law to Let Thais Choose Their Sexual Identity Gains Traction", *BangkokPost*, 26 July 2017, p. 2.
39 UNDP Press Release, *New Study Highlights Need for Legal Protections to Reduce Marginalisation of Transgender People in Thailand*, 2 May 2018.
40 Melalin Mahavongtrakul, "Transitioning Healthcare", *Bangkok Post*, 5 January 2016, p. 12.
41 In the context, it would have meant that the marriage provisions in the Civil and Commercial Code would be applied to the civil registration, while making the necessary adjustments, say in gendered language.
42 "New Law to Pave Way for Same-Sex Partnerships", *TheNation*, 25 April 2018, p. 1A.
43 King-Oua Laohong, "Same-Sex Civil Partnerships Law Hoped for End of Year", *Bangkok Post*, 5 May 2018, p. 1.
44 Melalin Mahavongtrakul, "Paradise or Paradox?" *Bangkok Post*, 15 May 2018, p. 1.
45 Aekarach Sattaburuth, "Parties Broaden Policy Offerings to Public", *Bangkok Post*, 20 May 2018, p. 6. Political party activity was still prohibited, so the statements were not official party views.
46 ILGA Asia, *Recap of the International Day Against Homophobia, Transphobia, and Biphobia, 2018 around Asia Media*.
47 FOR-SOGI, "Change in Strategy in Campaigning for the Civil Society Draft of the Civil Partnership Bill", 29 June 2018, posted on the Facebook page of FOR-SOGI, translation supplied to the author by Timo Ojanen.
48 "LGBT Bill Still Gets it Wrong", *Bangkok Post*, 10 November 2018, p. 8.
49 Deirdre Boyd, Staffan Herrstrom, and Donica Pottie, "Thailand at Forefront of Push for LGBTI Inclusion", *Bangkok Post*, 16 November 2018, p. 9.

50 King-Oua Laohong and Reuters, "Calls Mount to Amend LGBT Bill", *Bangkok Post*, 23 November 2018, p. 2; Piyanuch Thannukasetchai, "Muslim Leaders Oppose LGBT Partner Bill", *The Nation*, 23 November 2018, p. 1A.
51 Aree Iam-Aoran, "The Parent Trap", *Bangkok Post*, 10 November 2016, p. 1.

References

Burapha University and UNDP. 2017. *A Tool for Change: Working with the Media on Issues Relating to Sexual Orientation Gender Identity, Expression and Sex Characteristics in Thailand.*

Department of Women's Affairs and Family Development (Thailand), UNDP. 2018. *Legal Gender Recognition in Thailand: A Legal and Policy Review*, May.

Jackson, Peter. 2011. *Queer Bangkok: 21st Century Markets, Media, and Rights.* Hong Kong: Hong Kong University Press.

Ojanen, Timo T., Rattanakorn Ratanashevorn, and Sumonthip Boonkerd. 2016. "Gaps in Responses to LGBT Issues in Thailand: Mental Health Research, Services and Policies". *Psychology of Sexualities Review*, Vol. 7, No.1: 41.

UNDP, USAID. 2014. *Being LGBT in Asia: Thailand Country Report*, UNDP, Bangkok.

World Bank Group. 2018. *Economic Inclusion of LGBTI Groups in Thailand*, March, World Bank, Bangkok.

28
THE SOCIAL MEDIA

Wolfram Schaffar

In absolute figures, Bangkok is the city with the largest number of Facebook users worldwide – 22 million, according to the recent global statistics of Digital 2018. Since the city has about nine million inhabitants, this proves that many Bangkokians defy Facebook's real-name policy and have more than one account – usually two or three.[1] This in itself indicates that social media in Thailand plays a more important role than in most other countries in the world. Thais spent an average of 9 hours 38 minutes per day on the Internet.[2] With 3 hours 10 minutes per day on social media (on any device), Thailand ranks fourth worldwide – after the Philippines, Brazil and Indonesia – and is part of the global trend in the increasing use of social media, a trend which Southeast Asia is leading.[3]

Thailand ranks eighth worldwide concerning social media penetration (with 67 per cent penetration rate of active accounts per capita). Facebook is the leading social media platform, not only globally but also in Thailand, which ranks eighth worldwide in terms of active Facebook users.[4] Facebook started in 2004 and quickly spread throughout the United States in the following years. Since about 2010, it has enjoyed continuous double-digit year over year growth rates in Asia, despite the fact that it is banned in China. What is behind these figures is the fact that, starting in 2010, entire countries were penetrated by Facebook within only a few years. In January 2010, Facebook counted about two million users in Thailand and ranked only fourth, after Google+, the local Hi5 (four million users), and Life.com. By early 2011 it had reached ten million users. And in 2014, it reached 24 million, leaving all competitors far behind.[5] Since then, Bangkok has held the status of the city with the highest density of Facebook accounts worldwide, a success which the company supported by opening a regional office there in 2015.

While there is still a huge gap between cities and the countryside – with almost half of Thai Facebook accounts registered in Bangkok[6] – the rising penetration rate, the growing number of people from different generations starting to use Facebook and the continuously increasing average time spent on the Internet and on social media indicate that digital gaps within the country are growing less and that access to Facebook is becoming more generalised. Since the spread of smartphones and mobile Internet devices, and after Facebook's successful launch of its Free Basics programme via Internet.org[7] – providing free Internet access to selected Internet services through Facebook – it is safe to say that, for many Thais, Facebook has become a synonym for the Internet as such. It is used not only as a social network. Indeed, different functions are converging: socialising, dating, exchanging messages, information sharing, news consumption and shopping.[8]

Initially, social media such as Facebook were welcomed by utopians as the final fulfilment of the promises of the Internet and were expected to help develop a realm of democratic deliberation and free social and political exchange.[9] The reality of Facebook in Thailand, however, looks quite different.

Witch-hunt and surveillance on Thai social media

On 20 October 2016, Sarun Chuichai, also known as Aum Neko, a Thai student and political activist in France, barely escaped the physical attack of a violent mob of fellow Thais in a restaurant in Paris.[10] The attack was organised and coordinated through Facebook by a network of right-wing hyper-royalist Facebook groups, such as the Rubbish Collector Organisation, with the self-declared aim of cleansing Thailand of social rubbish. According to the group's own definition, this means individuals who are Red Shirts and not loyal to the monarchy.[11] Membership of this group, which owes its prominence to the public relations activities of its leader, Rienthong Nanna,[12] reached its first climax in summer 2015 with 250,000 followers.[13] Apart from this, there are minor groups, more or less institutionalised, run by monarchist activists in Thailand as well as abroad, which sometimes directly collaborate with Rienthong and at other times only in a loose connection with him and his project.

Aum Neko is a political activist addressing gender and LGBTIQ* issues as well as those of democracy and human rights.[14] After the coup d'état in 2014, she fled to France along with many other pro-democracy activists who were summoned to report at tribunals for what the junta called "attitude adjustment". When King Bhumibol Adulyadej passed away on 13 October 2016, Aum Neko used her Facebook account to post several videos through which she commented on the king's death in language deemed insulting by Thai standards. Immediately, she became the target of severe hateful verbal attacks and the outrage of several people who demanded her extradition to Thailand or proposed hiring a contract killer to kill her in Paris.

Her case was taken up by Rienthong Nanna himself and his Facebook followers as an "emergency case" (Postings of Rienthong Nanna, 14 and 15 October 2014). Her address in Paris was published on Facebook and people were encouraged to go there to "deal with her" (Postings of Rienthong Nanna, 14 and 15 October 2014). Under the pressure of severe threats, Aum Neko left Paris for a couple of days, hoping that the witch-hunt would fade away. However, when she was spotted by fellow Thais in Paris, her picture and the address of the restaurant were posted on Facebook, and the homepage of Rienthong was used to mobilise a crowd of other Thais, who happened to be in Paris, to come to this place. It was only by chance that Aum Neko became aware early enough of the activities behind her back that she managed to escape.

Through groups such as the Rubbish Collector Organisation and affiliated groups, Facebook became an arena where political violence is planned, performed and shared/consumed. For Thais, whether living in Thailand or abroad, apart from its other functions, it also became the realm of tight surveillance and fear.[15]

Approaches to social media in Thailand

The broad penetration of social media and Facebook in Thailand has attracted a number of studies. Nevertheless, because of the recent phenomenon and the unprecedented pace and comprehensiveness of Facebook penetration, social media and Facebook is still a fragmented and under-researched field, especially when it comes to phenomena such as the witch-hunts outlined earlier.

Recent analyses dealing with Facebook include normative assessments, which diagnose Facebook "overuse" and warn against a broad range of distortions, including "Facebook addictedness"[16] and even premature pregnancy.[17] Another strand of literature, coming from media studies, draws on network analysis or other mixed approaches to describe and explain patterns of Facebook use, often with the aim of conducting basic research for applied online marketing. Drawing on theories of intercultural communication, Götzenbrucker et al. study how different cultural concepts of friendship can explain the different use patterns of Thai and Austrian young Facebook users.[18] McCargo traces back some characteristics of Thai Facebook use to older structures in print and other media and argues that these patterns are now augmented by social media.[19] The metaphor of augmentation is also behind Grömping's study of Facebook functioning as echo chambers for partisan groups.[20] These studies follow, broadly speaking, a paradigm of technology in use, a constructivist approach which assumes that the actual use of technical innovations and appliances is the outcome of a process of social appropriation.

Political science approaches which address political effects of Facebook, for mobilisation of social movements for instance, often depart from an implicit technical decisionist/liberal utopian viewpoint. They assume that social media opens a new sphere which, due to the technical features of the application, has the potential to be a realm of non-hierarchical democratic deliberation and an unspoiled sphere of (a liberal imaginary of) civil society.[21] The fact that this social media (cyber)space appears to be the very opposite of this utopian sphere is explained by state encroachment and state-organised censorship. Janjira, for example, speaks of the "manipulation of the civil society space".[22]

Such a perspective, however, is problematic for two reasons. First, it draws on a conceptual dichotomy between a primary, material offline world and a hybrid or "cyber" online world. Second, it draws on the dichotomy between the state and civil society, with the liberal assumption that (middle class) civil society is intrinsically democratic and progressive. What the case of social media in Thailand shows, however, is that privately organised Facebook groups, such as civil society groups, play a pivotal role in political violence on Facebook.

This chapter will depart from a relational approach following Poulantzas's notion of the state as a social relation which is substantially brought about by social conflict.[23] Moreover, it will draw on Gramsci's notion of an integral state, which treats civil society as an organic part of the state, parallel to state institutions and apparatuses. From this perspective, neither the dichotomy between the offline and online world nor the view that Facebook constitutes a new space/cyberspace which is appropriated by society or encroached by the state makes much sense. Indeed, rather than being appropriated by society, Facebook is viewed as constitutive of Thai society. It transforms society and the state or even re-configures society and the state, as will be shown, with the tendency towards a fascistisation of society.

Political polarisation

The emergence of social media as an arena of political violence is closely intertwined with the political polarisation between two political camps – the Red Shirts and the Yellow Shirts – which has been deepening since 2005–2006. In all elections since 2001, the bloc of Thaksin (and the parties affiliated to him) has been able to achieve a clear electoral majority, especially among the poor and lower-middle-class population in the north and the northeast. Whenever a Red Shirt party was elected to power, however, the Yellow Shirt bloc, consisting of the Bangkok upper-middle class and royalist elites, mobilised demonstrations in Bangkok and used their exclusive access to the judiciary and military in order to oust the Red Shirt–supported government and install a royal-conservative administration. This confrontation formed the ideological and strategic identity of the two blocs and has led to a stalemate whereby no bloc is able to achieve hegemony.[24]

Whereas the actual actors in Thailand seem to be unique, the reality of an increasing political polarisation matches a current global trend.[25] It was also typical of the situation in Europe in the 1920s and 1930s, at the advent of fascist regimes.[26] What is important as regards the character of social media in Thailand is the fact that the fiercest periods of confrontation coincided with a steep rise in the penetration rate of social media, and in particular the rise of Facebook accounts. In 2006–2007, with the coup against Thaksin, YouTube became popular. Facebook had a first peak in new accounts in 2010 coinciding with the street protests of the Red Shirts and their violent crackdown by the military. A second acceleration of its rise was in 2013–2014, when the Yellow Shirt camp staged their campaign to oust Yingluck, which led to the next coup. As will be shown, the political conflict was not only augmented by but to a large extent happened in and through cyberspace.[27] Moreover, the actors of the confrontation did not simply use Facebook in order to organise better, but they were formed through conflicts on and about Facebook, and Facebook thus also evolved as a sphere to perform political identities.

Singaporean-style "calibrated coercion"

What is crucial for the emergence of social media and later, Facebook, as central arenas for political struggle and as stages for performing political identities are the censorship measures of the Thai authorities which, in a first step, were modelled after Singaporean policies. Singapore was in many respects a laboratory for Internet censorship and served as a role model for many countries in the world. Being the most technically advanced country in Southeast Asia and at the same time a city-state, the economy of which is crucially dependent on its position as a global city, Singapore, starting in 1996, set out to prove that it is possible to promote 100 per cent broadband penetration while keeping the authoritarian People's Action Party (PAP) in power. What the Singaporean government described as a "light touch"[28] policy was built on "calibrated coercion",[29] strategies to trigger self-censorship and foster what can be called governmentality (Foucault) rather than direct blocking of Internet sites. Instrumental to this policy were draconian defamation laws which would be applied to the provider of a platform, which would ruin the person socially and financially, and so-called red lines, taboos which remained largely unspecified but which would trigger the application of defamation laws.

Legal background and political practice of Internet censorship in Thailand

After the coup in 2006, the Thai military government also implemented harsh censorship measures tailored after the Singaporean model.[30] Posting comments on newspaper blogs, as well as phone-in radio shows and TV programmes, were very popular in Thailand and served as major fora for political debate up until 2006.[31] The Computer Crime Act, actually drafted by the Thaksin administration, was implemented in 2006 and used as a tool to suppress any political organising against the military. Following the Singaporean model, it built on the provision that the owner of a blog or webpage with a commentary function is liable for the content. In 2016, the Computer Crime Act was amended and the government acquired even more competence to check private accounts.

Central to the suppression of political debates in Thailand and an equivalent to the Singaporean "red lines" is Article 112 in the Criminal Code, which penalises lèse-majesté, alleged disrespectful behaviour towards the monarchy, with up to 15 years in prison. Charges of lèse-majesté were increasingly used as a weapon against political enemies. The number of lèse-majesté charges skyrocketed from a couple of cases in the years up to 2006 to more than 500 in 2006 and 2007, and again several hundred after the coup in 2014. This political abuse and instrumentalisation is made possible through the opaque formulation of Article 112. It does not give any clear definition of what counts as lèse-majesté, and court trials are held behind closed doors. What is evident from recent court rulings, however, is the fact that the scope of Article 112 was considerably broadened to cover not only the king and the heir to the throne, but also the entire royal family, the pet dogs of the king, former kings and even kings of various ancient dynasties.

The imposition of the Computer Crime Act in 2006–2007 quickly displayed its desired effect but also came with a quite unwanted side effect. After the detention of prominent people, such as the director of a popular blog, *Prachatai*, most political blogs fell silent and newspapers closed down their commentary functions altogether or severely censored them. In this situation, political activists first shifted to dating platforms to escape the radar of the military.[32] Later, social media, initially YouTube, evolved as an arena for symbolic battles over freedom of expression and was used to publish openly insulting comments about the monarchy. Legal standards of freedom of speech rested on the US law, where freedom of expression ranks high and is limited only by US "red lines", such as notions of hate speech or related issues. The anti-monarchy clips, which were uploaded onto YouTube, however, used expressions and pictures which were highly culturally encoded and which, from a Western or American viewpoint, did not constitute any serious offence.

The increasing use of lèse-majesté allegations against political opponents not only led to an atmosphere of fear. The unwanted side effect was that it shifted lèse-majesté into the centre of the political struggle over freedom of expression, and when lèse-majesté was declared an issue of national security by a ruling of the Constitutional Court in 2012, it was drawn into the centre of politics as such. Since the 2014 coup, and especially since the new king, Vajiralongkorn, ascended the throne in December 2016, Article 112 has been used even more excessively against political opponents who speak out against the military, and almost all cases relate to postings, likes or links on social media. In April 2017, the junta issued a decree that all online interactions with two scholars in exile (Somsak Jeamtheerasakul, Pavin Chachavalpongpun) and one journalist (Andrew MacGregor Marshall) will be considered as offences under the Computer Crime Act, a step directed against the large number of Facebook friends and followers of the three. The following month, in May 2017, saw a first victim of this new provision. Prawet Prapanukul, a human rights lawyer, who was arrested and faced up to 171 years in prison for ten times sharing an Internet link of Somsak Jeamteerasakul – an act which is considered an instance of lèse-majesté.[33] After his case was taken up in the international media, the major charges against him were dropped in August 2018.

Vigilante groups and cyber-paramilitary

The Thai administration tried hard to get a grip on platforms like YouTube to force them to obey Thai legal standards of lèse-majesté and to rank it as similar to the hate speech concept. In this process, YouTube was made inaccessible in Thailand in early 2007. It was only after YouTube decided to bow to Thai pressure, partly apply Thai standards and delete video clips with anti-monarchy content, that it was able to return to Thailand.[34] Yet the victory of the Thai government was only temporary, since a sheer mass of defamatory clips soon flooded YouTube, and any attempt to delete these would have strained the capacity of the YouTube administration. At the same time, in the wake of the deepening political polarisation and spurred by the Red Shirt demonstrations in 2009 and 2010, Facebook grew into the most prominent social network in Thailand.

When it became increasingly clear that Singaporean-style censorship was almost impossible, the Thai administration changed its strategy and implemented a programme, modelled after more recent and more sophisticated know-how from China, which relied on Internet vigilantism.[35] In 2010, the Ministry of Interior took the initiative of setting up vigilante groups to help scan information flows on the Internet. The Cyber Scout programme was aimed at teenage schoolchildren and had as its major component a nationwide course of training in schools on Internet technologies and on strategies of how to detect defamatory content. Initially, the Royal Thai National Police offered 500 baht (US$15) to anyone providing information on anti-coup or other postings or instances qualifying as lèse-majesté that could result in an arrest.[36] Under the administration of Yingluck Shinawatra, the programme was reduced, but after the 2014 coup it was re-introduced by the Ministry of Information and Communication Technology (ICT) in cooperation with 200 schools. An estimated 120,000 cyber scouts spanning 88 schools were active by 2015.[37] Today's Cyber Scouts, however, do not seem to receive any money. Instead they earn points for successfully incriminating neighbours, in the hope of being featured on the Cyber Scouts website.[38]

The Cyber Scout programme paved the way for another development. From 2011 onwards, privately organised groups on Facebook started copying the work of the Cyber Scouts. The first groups were run anonymously in a guerrilla style and carried names such as Witch Hunt or Social Sanction.[39] The abbreviation used by the group was "SS", and it was not chosen by accident but indicates the process of fascistisation which took place.

In 2013, the Rubbish Collector Organisation (RCO) was founded by Rienthong Nanna, a medical doctor in his fifties, and the group soon became the central campaigning agent on Facebook.[40] The activities of Rienthong actually fused two different strands. He brought together state-sponsored vigilante groups (Cyber Scouts) with the idea of private initiatives such as the Social Sanction (SS) group. While the Cyber Scouts and the Social Sanction group operated anonymously, Rienthong invited journalists to the opening ceremony of the RCO, gave interviews to mainstream media and newspapers, and presented his personality as part of the project.[41] A former major general now running the family-owned hospital in central Bangkok, he portrayed himself as apolitical but with a deep sense of injustice, embodying a combination of the features of a *Wutbürger* (enraged, angry citizen) with the determination and ruthlessness of a soldier. With this strategy, he managed to gain the open support of many well-known and influential people from the urban middle class and royalist elites in Bangkok, including professors from large traditional universities, celebrities of the pop culture industry, businesspeople and politicians. At its height, the RCO counted more than 250,000 followers on Facebook.

Rienthong promoted the idea that lèse-majesté was not only an issue of public security but also of public mobilisation, because the authorities under the Yingluck administration allegedly failed to prosecute offenders and were actually turning a blind eye to the lèse-majesté incidents. Against this background, he vowed that within two years, the Rubbish Collector Organisation would cleanse Thailand of "social rubbish" and announced that he was working on the establishment of a "people's army to protect the monarchy". This self-perception as a vigilante group taking the law into their hands, and Rienthong's support from the elites, is reminiscent of Fascist groups in Europe in the 1920s and 1930s. Back then, various European countries' political landscapes, too, were characterised by a stalemate between two antagonistic blocs: socialist parties and organised workers on one side and conservative capitalist elites on the other. In this stalemate, and under the specter of a global economic crisis, the middle classes/bourgeoisie employed vigilante groups to bring about change by intimidating organised labour and creating chaos, thereby legitimising the dissolution of the parliamentarian system and the establishment of an authoritarian regime. This is exactly what was later achieved by the coup in May 2014. Vigilante groups in Thailand working in the aforementioned way, however, are not a new phenomenon. Village scouts, part of the anti-insurgency policy of the Royal Thai Army's Internal Security Operations Command (ISOC), were active in the coup in 1976, and it is not by chance that the Cyber Scouts were named after these paramilitary forces.[42]

The creeping fascistisation of Thai society through these kinds of Facebook groups became obvious in the in the days and weeks after the death of King Bhumibol in October 2016. On 17 October, a women in Nonthaburi was spotted and exposed on Facebook wearing a red dress in the town hall, inside the room where a book of condolence was opened. According to the etiquette, only black-coloured dresses were allowed to be worn in the first days after the death. Her picture was shared widely and commented on with vulgar language and threats, such as "Can someone please rid the country of this rubbish?" Speculations about her identity as an activist of the anti-military New Democracy Movement were exchanged between members of different Facebook groups, and in the course of this witch-hunt, several women who might have matched her identity were attacked.

On 18 October, a 19-year-old man was dragged out of his bedroom and onto the street by a furious mob and forced to kneel down in front of the picture of the king while he was beaten and kicked at his head several times. He was accused of having posted disrespectful comments about the king in a conversation with a friend. The humiliation was posted on YouTube and went viral through various Facebook groups. The campaign against Aum Neko, which was described in the introduction to this chapter, happened at the same time.

What is remarkable in these campaigns is the level of violence which was unleashed against the victims, as well as the often informal nature of the mob which performed the witch-hunts and broadcasted and published them through Facebook. In the case of Aum Neko, what is striking is the transnational and yet ethnicised character of the network. The campaigns reached beyond the national boundaries of Thailand through a transnational Facebook-mediated coordination between the actors. The targets of this bullying, however, are restricted to Thai nationals. This shows that the moral basis of this bullying relies on an essentialist concept of Thainess which, on Facebook, is bound to language, insofar as only individuals outside Thailand are being attacked when their name is Thai and they are running a Thai-speaking Facebook account.

Intimidation and mass mobilisation

The ritualised performance of indignation, followed by hate speech and the documentation of actions, under the guidance of a fatherly but uncompromising and rigorous leader like Rienthong was increasingly combined with calls for, and the documentation of, the mass mobilisation of members demonstrating their loyalty to the monarchy. On the way to the proclaimed objective of cleansing Thailand of social rubbish, Rienthong launched campaigns to declare certain sectors of society as "clean" by asking specific sub-groups of his followers to declare themselves, on their Facebook accounts, loyal to the monarchy.

In this respect too, the RCO page constitutes a new development. Whereas several other Facebook pages were used as fora for the documentation of private royalist initiatives – such as public singing performances of the royal anthem – the RCO, with its prominent individual members and its mass membership, triggered a new effect. State-organised mass events were advertised on RCO with an almost coercive effect on members to, at the very least, click the "like" button or post greetings such as the expected "Long live the King!" One such example is the campaigns "Bike for Mom" and "Bike for Dad" in 2015, which aimed to promote the then heir to the throne, Vajiralongkorn, as dutiful son and legitimate successor. For this reason, Vajiralongkorn invited Thai citizens to join a public cycling event in Bangkok and other provincial capitals on the occasion of his parents' birthdays in August and December 2015. Apart from public TV channels and the websites of the mainstream media, participants were also mobilised through Facebook groups such as the RCO.[43] Group members posting selfies showing them on bicycles wearing T-shirts and caps which were sold in masses.

These events strikingly resembled fascist mobilisations in Europe in the 1920s and 1930s. Back then, after taking power, the regimes were consolidated not only by mutual mass surveillance but also by a policy of activation of the people through coerced mobilisations – emotive mass events like parades and torch marches, which were featured in newsreels. In Thailand today, Facebook is playing a crucial role in organising and publicising such kinds of events. In the days and weeks after the death of Bhumibol, this comprehensive and coercive effect of Facebook reached its climax. Active signs of public mourning, such as using black banners or switching to black and white images were expected and, in the fearful atmosphere, amounted to an ubiquitous group pressure. Facebook became the forum for what Pitch Pongsawat called "competitive grief", where people, wearing black, posted tearful messages of condolence and shared links with hagiographic content about the former king.[44] At the same time, they closely watched whether the other's performance conformed to the expectations and attempted to surpass these expectations.

Big data analytics and future developments

The development of social media in Thailand showed a characteristic merger of civil society and the state apparatus. Another important aspect, however, is the increasingly blurred boundary between the economic interests of companies and the interests of political actors and the state apparatus, a development which has the potential to result in an intensified totalitarian grip on the population.

In its endeavour to censor the activities of YouTube users, Thailand came into conflict with providers like YouTube and Google. Initially, YouTube played the role of a defender of liberal rights but eventually bowed to the pressure of the Thai authorities in order not to lose the Thai market. When the Thai authorities targeted Facebook, this company, too, officially announced that they would not cooperate with the Thai authorities. When in October 2017 the founder and CEO of Facebook, Mark Zuckerberg, visited the company's regional office in Bangkok, Prime Minister Prayuth publicly announced that he would meet him and discuss pending issues of lèse-majesté. Zuckerberg – very much to the embarrassment of Prayuth – denied that such a meeting was scheduled.[45] This incident was very telling, since Zuckerberg's visit was covered by the mainstream media in much the same terms as the visit of the head of a sovereign state, so expressing the political importance ascribed to him. Zuckerberg's snub to Prayuth was described as a diplomatic affront against the Thai junta chief. Zuckerberg, for his part, played the card of the young and rebellious start-up pioneer who does not bow to state authorities.

Reality, however, is different. As early as the occasion of the death of Bhumibol in October 2016, Facebook cooperated with the Thai authorities and, without the users' consent, inserted geo-blocking into the privacy settings of individual accounts, so that comments like the ones of Aum Neko became invisible in Thailand. In May 2017, a video showing the new king in a shopping mall in Germany, wearing a crop-top revealing his distinctive tattoos and accompanied by one of his mistresses, went viral on Facebook. After the Thai authorities publicly threatened to take legal measures against Facebook, an agreement was reached and this video, too, was geo-blocked in Thailand.[46]

Despite these publicly performed standoffs, the interests between Facebook and the Thai authorities might not be as contradictory as they seem. The vast amount of data collected by Facebook and other social media is an invaluable asset. Economically, this is already used in business operations of private rating companies like CredoLab and Lenddo, operating from Singapore, which offer their credit scoring services to insurance firms and banks. They are active in Thailand and use Bangkok, the city with the highest numbers of Facebook accounts, in order to develop their business models.[47]

The political value and use of big data from Facebook, however, became known to the public in early 2018 with the scandal surrounding Cambridge Analytica, a company which was involved in tailoring election campaigns, including the one for Donald Trump in the United States, on the basis of data harvested from Facebook.[48] This provided a first glimpse into the far-reaching possibilities and practice of political manipulation through the use of social media. The Chinese government combined both the commercial and political aspects in its social credit system, which has been developed and implemented in a fast-tracked process since 2014. The aim is to assign social credit ratings to every citizen based on their social and economic status and political conduct. Failure to achieve a sufficient score results in denial of access to transportation services like trains or airplanes or exclusion from other public services.[49] The social reputation scores are based on a variety of data, including data collected through social media. The companies involved in this project include Sesame Credit, a credit rating company for banks, which is run by the eight biggest Chinese social media platforms (Alibaba).

The implementation of this vast merging of data into a social reputation rating system has very aptly been described as the final step in the implementation of a totalitarian surveillance system. As it seems, the Thai authorities are flirting with such a kind of system, too. The chief of the Crime Suppression Department announced that he will employ new technology to be able to scan big data.[50] The military government is continuously pushing to realise its project of a single-node policy, which would channel all traffic through one node where all content can then be collected. If this policy is successful, it would represent the logical conclusion of the knowledge transfer of surveillance technology from Singapore and China to Thailand.

Outlook

The process of reconfiguration of the society and the state through social media like Facebook is not uncontested. Different from China, where the implementation of the social credit system has not triggered any opposition, netizens in Thailand are more alert and critical. Social media was used as a tool for mobilisation against the government's plan to set up a single node and channel all Internet traffic through a single gate in 2015. The same kind of movement, however, failed when the government pushed for the amendment of the Cyber Crime Act.[51] Despite of these movements on social media addressing specific single issues, the deep split within society and the transformation, militarisation and maybe even fascistisation of society through social media will be difficult to overcome. Whatever the future of the public sphere in Thailand and social movements will look like, social media will be a constitutive element of it. Politics will not look like offline activism as we know it, with an add-on of social media-based networking, but it will have to be genuinely social media-based way(s) of doing politics.

Notes

1 Duncan McCargo, "New Media, New Partisanship: Divided Virtual Politics in and Beyond Thailand", *International Journal of Communication,* Vol. 11 (2017), pp. 4138–4157.
2 We Are Social/Hootsuite, *Digital 2018 Global Overview*, 30 January 2018 <https://wearesocial.com/blog/2018/01/global-digital-report-2018> (accessed 9 July 2018).
3 Ibid.
4 While the number of Facebook users is "only" 2.2 billion monthly active users (1.45 billion daily active users) among 4.1 billion Internet users worldwide (Internet World Stats, 2018), it should be noted that there is a clear pattern. If a country shows considerable rate of Internet penetration, and if Facebook is accessible, then it usually holds a quasi-monopoly. The high number of Internet users who are not using Facebook can be explained by the Internet policy of China, where Facebook is illegal and a clone version, Renren.com or Weibo.com, is used instead.
5 The main competitors in 2014 were Instagram with 1.5 million users and Twitter with 4.5 users. In early 2018, Facebook reached 50 million users, Instagram counted 13 and Twitter 12 million. In the United States, Facebook users seem to be decreasing and losing users to Instagram and Twitter. However, this trend does not seem to have reached Thailand. And since Instagram belongs to the Facebook group, it can still count as a quasi monopoly.
6 Zephoria Digital Marketing, "The Top 20 Valuable Facebook Statistics – Updated June 2018", July 2018 <https://zephoria.com/top-15-valuable-facebook-statistics/> (accessed 9 July 2018); Aim Sinpeng, "Participatory Inequality in Online and Offline Political Engagement in Thailand", *Pacific Affairs,* Vol. 90, No. 2 (2017), pp. 253–274.
7 "Where We Have Launched", *Internet.org* <https://info.internet.org/en/story/where-weve-launched/> (accessed 9 July 2018).
8 Noppamash Suvachart, "An Exploratory Study of Behaviour-Based Segmentation Typology of Facebook Users in Thailand", *Asian Social Science,* Vol. 12, No. 3 (2016), pp. 140–151; Vikanda Pornsakulvanich and Nuchada Dumrongsiri, "Internal and External Influences on Social Networking Site Usage in Thailand", *Computers in Human Behaviour,* Vol. 29 (2013), pp. 2788–2795. This trend of convergence is neither specific to Facebook nor to Thailand. It is also visible in China and seems to be a general feature of social media, independent of the concrete application.
9 Manuel Castells, "The New Public Sphere: Global Civil Society, Communication Networks, and Global Governance", *The ANNALS of the American Academy of Political and Social Science,* Vol. 616, No. 1 (2008), pp. 78–93.
10 "Ultra-royalist Calls for Lèse-majesté Purge in Paris", *Prachatai*, 18 October 2016 <https://prachatai.com/english/node/6658> (accessed 9 July 2018); Nicola Smith, "Exiled Thai Activist in Paris Threatened over 'Insulting the Royal Family'". *The Telegraph*, 18 October 2016.
11 Chaiyot Yongcharoenchai, "Doctor Sick of All the 'Trash'". *Bangkok Post*, 3 August 2014.
12 Kultida Samabuddhi and Patsara Jikkham, "Monarchists Vow to Fight 'Armed Threat'". *Bangkok Post,* 20 April 2014.
13 Wolfram Schaffar, "New Social Media and Politics in Thailand: The Emergence of Fascist Vigilante Groups on Facebook", *Austrian Journal of South-East Asian Studies,* Vol. 9, No. 2 (2016), pp. 215–234.
14 Political Prisoners Thailand, *Aum Neko*, 2016 <https://thaipoliticalprisoners.wordpress.com/pending-cases/aum-neko/> (accessed 9 July 2018).
15 Schaffar, "New Social Media and Politics in Thailand"; Pinkaew Laungaramsri, "Mass Surveillance and the Militarisation of Cyberspace in Post-Coup Thailand", *Austrian Journal of South-East Asian Studies*, Vol. 9, No. 2 (2016), pp. 95–214; Pirongrong Ramasoota, "Online Social Surveillance and Cyber-Witch Hunting in Post-2014 Coup Thailand", in *Globalisation and Democracy in Southeast Asia*, edited by C.B. Wungaeo et al., 2016, pp. 269–288.
16 Nitt Hanprathet et al., "Facebook Addiction and Its Relationship with Mental Health among Thai High School Students", *Journal of the Medical Association of Thailand,* Vol. 98, No. 3 (April 2015), pp. 81–90; Jiraporn Khumsri et al., "Prevalence of Facebook Addiction and Related Factors Among Thai High School Students", *Journal of the Medical Association of Thailand,* Vol. 98, No. 3 (April 2015), pp. 51–60; Sirada Chittiwan and Pramote Sukanich, "Relationship between Facebook Addiction and Abnormal Eating Attitudes and Behaviors among Female Adolescents in Patumthani Province, Thailand", *Journal of the Psychiatric Association of Thailand,* Vol. 62, No. 2 (2017), pp. 117–128.
17 National Economic and Social Development Board (NESDB), "ภาวะสังคมไทยไตรมาสสีและภาพรวมปี 2555" ["Social Situation and Outlook"], 2012 <http://social.nesdb.go.th/social/Portals/0/Documents/

Social%20outlook%20Q4-55%20และภาพรวมปี%202555_165.pdf> (accessed 9 July 2018); Saksith Saiyasombut, "Did Thai Report Really Say Facebook 'Causes Teen Pregnancy'?" *Asian Correspondant*, 5 March 2012 <https://asiancorrespondent.com/2012/03/did-a-thai-report-really-say-facebook-causes-teen-pregnancy/#AkPS6SItYoUHLsFy.99> (accessed 9 July 2018).

18 Gerit Götzenbrucker and Margarita Köhl, "Online Relationship Management, Friendship Cultures, and Ego-Networks ofYoung People in Thailand and Austria", *Asia Europe Journal*, Vol. 12, No. 3 (September 2013), pp. 265–283; Margarita Köhl and Gerit Götzenbrucker, "Networked Technologies as Emotional Resources? Exploring Emerging Emotional Cultures on Social Network Sites such as Facebook and Hi5: A Trans-cultural Study", *Media, Culture & Society*, Vol. 36, No. 4 (2014), pp. 508–525.

19 McCargo, "New Media, New Partisanship: Divided Virtual Politics in and Beyond Thailand".

20 Max Grömping, "'Echo Chambers' Partisan Facebook Groups during the 2014 Thai Election", *Asia Pacific Media Educator*, Vol. 24, No. 1 (2014), pp. 39–59.

21 Castells, 2008, fn. 9.

22 Janjira Sombatpoonsiri, "Manipulation des Zivilgesellschaftlichen Raums: Cybertrolle in Thailand und den Philippinen", *GIGA Focus Asien*, No. 3 (June 2018) <www.giga-hamburg.de/de/publikation/cybertrolle-in-thailand-und-den-philippinen> (accessed 9 July 2018).

23 Bob Jessop, *State Power: A Strategic-relational Approach* (Cambridge: Polity Press, 2008); Alex Demirovic, Stephan Adolphs and Serhat Karakayali (eds.), *Das Staatsverständnis von Nicos Poulantzas: Der Staat als Gesellschaftliches Verhältnis* (Baden-Baden: Nomos, 2010).

24 Michael John Montesano, Pavin Chachavalpongpun and Aekapol Chongvilaivan (eds.), *Bangkok May 2010: Perspectives on a Divided Thailand* (Singapore: Institute of Southeast Asian Studies, 2012); Pavin Chachavalpongpun (ed.) *"Good Coup" Gone Bad. Thailand's Political Developments since Thaksin's Downfall* (Singapore: ISEAS, 2014).

25 The deepening polarisation in Turkey between secularists and AKP conservatives; in Egypt between the secular military and the Muslim brotherhood; and the polarisation of society in the United States are only the most iconic examples of a worldwide trend.

26 Richard Saage, *Faschismus. Konzeptionen und Historische Kontexte. Eine Einführung* (Wiesbaden: VS-Verlag, 2007); R.J.B. Bosworth (ed.), *The Oxford Handbook of Fascism* (Oxford: Oxford University Press, 2009).

27 For a similar argument concerning the relation between the 2005 and 2006 protests and conventional media, see McCargo Duncan, "Thai Politics as Reality TV", *Journal of Asian Studies*, Vol. 68 (2009), pp. 7–19.

28 Cherian George, *The Limits of Singapore's Light Touch Web Regulation*, Manuscript 2015, Hong Kong Baptist University <www.researchgate.net/profile/Cherian_George3> (accessed 9 July 2018).

29 Cherian George, "Consolidating Authoritarian Rule: Calibrated Coercion in Singapore", *The Pacific Review*, Vol. 20, No. 2 (2007), pp. 127–145.

30 Thai Netizen Network, "*Thailand Internet Freedom and Online Culture Report 2011*", Bangkok: Thai Netizen Network, 2012 <https://thainetizen.org/docs/netizen-report-2011-en/> (accessed 9 July 2018); Thai Netizen Network, "*Thai Netizen Report 2013*", Bangkok: Thai Netizen Network, 2014 <https://thainetizen.org/wp-content/uploads/netizen-report-2013.pdf> (accessed 9 July 2018).

31 McCargo "New Media, New Partisanship: Divided Virtual Politics in and Beyond Thailand".

32 Pitch Pongsawat. "Internet Politics in Thailand since the 19 August 2006 Coup: Self-Censorship, the Empire Strikes Back, and a New Hope?" Paper presented at the International Convention of Asia Scholars (ICAS) 5, Institute of Occidental Studies (IKON), 2–5 August 2007, Kuala Lumpur, Malaysia.

33 Metta Wongwat, "Prawet Prapanukul: Lèse majesté Suspect who Fights Like an 'Underdog'", *Prachatai*, 10 May 2018 <https://prachatai.com/english/node/7740> (accessed 9 July 2018).

34 Seth Mydans, "Agreeing to Block Some Videos, YouTube Returns to Thailand", *The New York Times*, 1 September 2007.

35 Lennon Y.C. Chang, Lena Y. Zhong, and Peter N. Grabosky, "Citizen Co-production of Cyber Security: Self-Help, Vigilantes, and Cybercrime", *Regulation and Governance*, Vol. 12, Issue 1 (August 2016), pp. 101–114.

36 Kay Yen Wong et al., "The State of Internet Censorship in Thailand", *Open Observatory of Network Interference*, 20 March 2017 <https://ooni.torproject.org/post/thailand-internet-censorship/#censorship-and-surveillance> (accessed 9 July 2018); Freedomhouse, "Freedom of the Net, Thailand, Country Profile" <https://freedomhouse.org/report/freedom-net/2016/thailand> (accessed 9 July 2018).

37 Saksith Sayasombuth, "Thailand Junta Reactivates 'Cyber Scout' Programme to Curb Online Dissent", *Asian Correspondent*, 7 August 2014.

38 David Gilbert, "Thailand's Government is Using Child 'Cyber Scouts' to Monitor Dissent", *VICE News*, 20 September 2016.
39 Thai Netizen Network, "*Thailand Internet Freedom and Online Culture Report 2011*".
40 Schaffar, "New Social Media and Politics in Thailand: The Emergence of Fascist Vigilante Groups on Facebook".
41 Chaiyot Yongcharoenchai, "Doctor Sick of All the 'Trash'"; Kultida Samabuddhi and Patsara Jikkham, "Monarchists Vow to Fight 'Armed Threat'".
42 John Draper, "Steady Rise of Fascism Here Is Terrifying", *Bangkok Post*, 28 April 2014. Nicolas Farrelly, "From Village Scouts to Cyber Scouts", *New Mandala*, 2 July 2010.
43 "Crown Prince Leads 'Bike for Mom' Cycling Event", *Bangkok Post*, 16 August 2015; Wassana Nanuam, "Prince to Lead Thais in 'Bike for Dad' Event", *Bangkok Post*, 28 September 2015.
44 Nicola Smith, "Boost for the 'Black Market' as Thais Don Mourning Clothes", *The Telegraph*, 15 October 2016.
45 "No Face-time for Prayuth", *Bangkok Post*, 18 Oct 2017.
46 "Facebook Is Censoring Posts in Thailand that the Government Has Deemed Unsuitable", *TechCrunch*, 12 January 2017 <https://techcrunch.com/2017/01/11/facebook-censorship-thailand/> (accessed 9 July 2018).
47 "Social Credit: How Fintech Uses Facebook to Give You a Loan", *Bangkok Post*, 5 August 2016.
48 "The Cambridge Analytica Files", *The Guardian* <www.theguardian.com/news/series/cambridge-analytica-files> (accessed 9 July 2018).
49 Alexandra Ma, "China Ranks Citizens with a Social Credit System – Here's What You Can Do Wrong and How You Can Be Punished", *The Independent*, 10 April 2018.
50 Wassayos Ngamkham, "Maitri Wrestles with Mafia Networks", *Bangkok Post*, 30 April 2018.
51 Janjira Sombatpoonsiri, "Growing Cyber Activism in Thailand", <http://carnegieendowment.org/2017/08/14/growing-cyber-activism-in-thailand-pub-72804> (accessed 30 April 2018).

References

Aim Sinpeng. 2017. "Participatory Inequality in Online and Offline Political Engagement in Thailand". *Pacific Affairs*, Vol. 90, No. 2: 253–274.

Chang, Lennon Y.C., Lena Y. Zhong, and Peter N Grabosky. 2016. "Citizen Co-production of Cyber Security: Self-help, Vigilantes, and Cybercrime". *Regulation and Governance*, Vol. 12, No. 1 (August 2016): 101–114.

Freedom House. 2017. "Freedom of the Net, Thailand, Country Profile" <https://freedomhouse.org/report/freedom-net/2016/thailand> (accessed 9 July 2018).

George, Cherian. 2007. "Consolidating Authoritarian Rule: Calibrated Coercion in Singapore". *The Pacific Review*, Vol. 20, No. 2 (2007): 127–145.

Götzenbrucker, Gerit, and Margarita Köhl. 2013 "Online Relationship Management, Friendship Cultures, and Ego-networks of Young People in Thailand and Austria". *Asia Europe Journal*, Vol. 12, No. 3: 265–283.

Grömping, Max. 2014. "'Echo Chambers' Partisan Facebook Groups during the 2014 Thai Election". *Asia Pacific Media Educator*, Vol. 24, No. 1 (2014): 39–59.

Janjira Sombatpoonsiri. 2018. "Manipulation des Zivilgesellschaftlichen Raums: Cybertrolle in Thailand und den Philippinen". *GIGA Focus Asien*, No. 3 (June) <www.giga-hamburg.de/de/publikation/cybertrolle-in-thailand-und-den-philippinen> (accessed 9 July 2018).

Köhl, Margarita and Gerit Götzenbrucker. 2014. "Networked Technologies as Emotional Resources? Exploring Emerging Emotional Cultures on Social Network Sites such as Facebook and Hi5: A Trans-cultural Study". *Media, Culture & Society*, Vol. 36, No. 4, pp. 508–525.

McCargo, Duncan. 2009. "Thai Politics as Reality TV". *Journal of Asian Studies*. Vol. 68: 7–19.

———. 2017. "New Media, New Partisanship: Divided Virtual Politics in and Beyond Thailand". *International Journal of Communication*, Vol. 11: 4138–4157.

Montesano, Michael John., Pavin Chachavalpongpun, and Aekapol Chongvilaivan. (eds.). 2012. *Bangkok May 2010: Perspectives on a Divided Thailand*. Singapore: Institute of Southeast Asian Studies.

Noppamash Suvachart. 2016. "An Exploratory Study of Behaviour-Based Segmentation Typology of Facebook Users in Thailand". *Asian Social Science*, Vol. 12, No. 3: 140–151.

Pavin Chachavalpongpun. (ed.). 2014. *"Good Coup" Gone Bad: Thailand's Political Developments since Thaksin's Downfall*. Singapore: ISEAS.

Pinkaew Laungaramsri. 2016. "Mass Surveillance and the Militarisation of Cyberspace in Post-Coup Thailand". *Austrian Journal of South-East Asian Studies*. Vol. 9, No. 2: 95–214.

Pirongrong Ramasoota. 2016. "Online Social Surveillance and Cyber-Witch Hunting in Post-2014 Coup Thailand". In *Globalisation and Democracy in Southeast Asia*. Edited by Chantana Banpasirichote Wungaeo, Surichai Wun'gaeo, and Boike Rehbein. London: Palgrave Macmillan, pp. 269–288.

Schaffar, Wolfram. 2016. "New Social Media and Politics in Thailand: The Emergence of Fascist Vigilante Groups on Facebook". *Austrian Journal of South-East Asian Studies*, Vol. 9, No. 2: 215–234.

Thai Netizen Network. 2012. "*Thailand Internet Freedom and Online Culture Report 2011*". Bangkok: Thai Netizen Network.

———. 2014. *Thai Netizen Report 2013*. Bangkok: Thai Netizen Network.

Vikanda Pornsakulvanich and Nuchada Dumrongsiri. 2013 "Internal and External Influences on Social Networking Site Usage in Thailand". *Computers in Human Behaviour*, Vol. 29: 2788–2795.

Wong, Kay Yen et al. 2017. "The State of Internet Censorship in Thailand". *Open Observatory of Network Interference*, 20 March <https://ooni.torproject.org/post/thailand-internet-censorship/#censorship-and-surveillance> (accessed 9 July 2018).

29
NGOS AND CIVIL SOCIETY IN THAILAND

Kanokwan Manorom

NGOs (non-governmental organisations) and CSOs (civil society organisations) have repeatedly played important roles in the development of Thai society. In part, they represent key actors who are able to influence individual perspectives and institutions involved in many sustainable development initiatives and policy formulations. Such initiatives cover health promotion and prevention, natural resource management, human rights, alternative agriculture, political and economic issues, local wisdom preservation, community empowerment and social welfare. They also offer some alternative solutions to the government through their advocacy, wide dissemination of information, networks, action groups, monitoring, knowledge and evidence contribution.

Over the period 1973–2018, both domestic and international socio-economic and political contexts have influenced and stimulated the growth and works of Thai NGOs and CSOs.[1] They include centralised development policy, democratic transition, authoritarian governments, the global human rights movement, neoliberalist policy, international support and globalisation. These factors, in one way or another, have influenced NGOs' and CSOs' ideologies, perspectives and practices.

The freedom and independence of NGOs and CSOs highly depends on the political situation. During a democratic transition, NGOs and CSOs have a better opportunity to run their activities. As stated earlier, specific contexts of socio-economic and political situations in the country inexorably influence the work, expression and demand of NGOs and CSOs. During a democratic period, NGOs and CSOs enjoy freedom of expression. People's centres, human rights and democratic movements are notably pursued. For example, CSOs such as the Small-Scale Farmers' Assembly of Isan (SSFAI) in the northeast region (locally called Isan), have successfully organised a democratic movement to tackle particular problems including land access and environmental destruction caused by state-initiated development projects, failed agricultural products and market prices. This indicates that CSOs confront the state and strengthen their rights based on a democratic perspective.[2]

On the other hand, they have limited freedom of expression during a dictatorship period. Their space for action is restricted by law and various state apparatus, especially the space for those who work on human rights, people's empowerment, democratic campaigns and sustainable natural resources management issues. In 2012, Thailand was ranked by the World Bank as a middle-income country just two years before the coup of 2014.[3] International support for

Thai NGOs and CSOs has since steadily declined. It has become difficult for them to undertake their activities and hire staffs because of the lack of international support, particularly financial.[4]

The chapter is organised by four core areas: the emergence of NGOs and CSOs; legal entities; approaches of NGOs/CSOs on development; and their challenges. Most examples illustrated in the chapter are related to the environmental issue on which the author has worked during the past two decades. Data presented here was from both literature and personal interviews with representatives of the NGO and CSO representatives conducted in early 2018.

Birth of NGOs and CSOs

Non-profit organisations were first established long before CSOs. NGOs were originally religious based. They earlier acted on behalf of a philanthropic organisation and focused more on social welfare. Monks, temples and churches played great roles in supporting the poor on health and education.[5] After the first three decades of the national plans (1961–1986), first focusing on an imported-substitution industrialisation and later shifting to export-oriented industrialisation in the later plans, Bangkok residents and the neighbouring provinces had gained much more income and enjoyed more prosperity than the rest whose residents had lived in the rural areas, especially those in the northeast region.[6] Many local NGOs emerged to provide alternative development to local, poor, marginalised and vulnerable people who were excluded from the benefits of government's development programmes. They mainly supported local communities to achieve self-reliance and to maintain community culture. From the 1980s to the late 1990s, the poverty rate in Thailand dropped and the economy grew steadily. The country enjoyed an average economic growth rate of 6.2 per cent per annum. During this period, economic achievements were obtained. Most evidently, the country saw a notable decline of the poverty rate: 40 per cent of the Thai population emerged out of poverty.[7]

However, the lack of trickle-down of income under high economic growth was witnessed, and large regional disparities persisted.[8] Natural resources including forest, land, water and rivers were grabbed by powerful elite groups with the support of government policy. Thus, in the 1990s, a large number of local NGOs emerged. By the year 1989, 12,000 local NGOs were set up, of which 44 per cent were working in the field of welfare and development.[9] They supported local people to realise their rights to protect natural resources for local livelihood and mobilised them to protest and negotiate with the government for sustainable resources management and making environmental preservation a top political agenda.[10]

Unlike NGOs, CSOs have become known only after the democratic transition in the 1990s. With high economic growth based on an export-oriented economy driven by neo-liberalism, CSOs, with the support from local NGOs, mainly addressed negative impacts and unintended consequences of economic growth and resisted many development projects that harm their local livelihood and environment. CSOs launched campaigns and organised protests against inequality, which led to an unequal share of development benefits between urban and rural people. Overall, CSOs worked on similar themes as NGOs, including poverty reduction, grassroots awareness, participatory development, preserving community wisdom, folkways and basic rights of the poor and the disadvantaged groups in both rural and urban areas.[11] Table 29.1 demonstrates that Thai NGOs and CSOs are varied, dynamic and diverse and in parallel with the economic and political contexts that have changed over time. They also vigorously work in the fields of public participation, freedom of expression, requirements for support from the government and institutions, and maintenance of rights of communities.

Table 29.1 Chronological emergence of NGOs and CSOs in Thailand

Period	*Socio-economic and political contexts and key development issues/approaches*	*Key NGOs/social groups/CSOs*
1890–1930	– Absolute monarchy – Centralisation of development – Initiating political reform by intellectuals – The CPT revolutionary party – Socio-economic development groups; and student groups. – Social welfare for the poor	*Private voluntary organisations (PVOs) by the Thai elite and foreign mission organisations*[12] – 1890: Orphanage (*Sathan songkhroh dek kamphra*) founded by the Princess Suddhasininat in 1890 – 1893: Thai Red Cross (*Sapha kachat thai*), established under the patronage of Queen Sripatcharinthra in 1893 – 1908: The Makane Institute was founded in Chiang Mai by the Presbyterian Mission to care for lepers – 1931: Missions Étrangères de Paris (MEP), the Sacred Heart and Regina and Montford on health care and education
1932–1972	– Change from Absolute Monarchy to Constitutional Monarchy, under Pridi Banomyong – Country's first Social and Economic Development Plan in 1961by adopting capitalist framework – Post–World War II – Threat of communism especially since the Chinese revolution of 1949 – 1946 abolished the Anti-Communist Act – Unsettling political situation ruled by the system of bureaucratic polity as a form of military authoritarianism. – Prevailing poverty in rural areas – Promoting the citizen's welfare supported by urban elites – Poverty reduction and a discussion about socio-economic problems of the poor based on Marxist approach by intellectuals, especially left-wing journalists, writers and university students – Community development approach – Government promoted and conserve Thai cultural heritage and prevent of communist doctrine – The National Cultural Act of 1942 was launched to provide legislation for the establishment of Foundations and Associations[13]	*Rising of several secular organisations*[14] – The Protestant Church of Christ Foundation (CCF) – Philanthropic social groups – 1958: The Catholic Council of Thailand for Development (CCTD) – Setting up Central Labour Union (CLU), Bangkok Labour Union (BLU), the Student Group of Thailand (SGT) and the Thai Youth Organisation (TYO) – 1966: the Graduate Volunteer Project (GVP) (*Khrongkan bandit asasamak*) founded by Puey Ungpakorn[15] – The Thai Rural Reconstruction Movement (TRRM) – Foundations and associations were founded by the Thai government to preserve Thai culture
1973 uprising	– Establishment of the people-centred NGOs – Alternative development strategies for marginalised people in urban and rural areas – Authoritarian government	*Student uprising* 1974: the Mae Klong Rural Development Project with the cooperation of three leading universities in Bangkok, Kasetsat, Thammasat and Chulalongkorn, initiated by Puey Ungpakorn

Period	Socio-economic and political contexts and key development issues/approaches	Key NGOs/social groups/CSOs
1973–1980	– Semi-democracy – Influence of the Mae Klong Rural Development Project Negative impacts of growth model that Thai government had adopted on local people Strategies of the Thai NGOs[16] – People participation in development – Alternative for social development – Decentralisation of power to community – Creation of international alliance	*Roles of universities and influences of Puey Ungpakorn*[17] – Thai Rural Reconstruction Movement (TRRM), by Puey Ungpakorn – Komol Khimthong Foundation (KKF) – More active role of the Catholic Council of Thailand for Development (CCTD)
1980–1990	– Democratic contexts and flourishing of NGOs – Cold War – Financially supported by foreign donors – Rural development – People's centre – Alternative development – Community culture approach – End of end of semi-democracy regime	*Formal NGOs' establishment* – 1984: Local Development Assistance Programme (LDAP) – 1985: NGO Coordinating Committee on Rural Development (NGO-CORD) – 1984: The Foundation for Women (FFW) – 1985: Yadfon (Raindrops) Association
1990–1999	– Wave of economic boom[18] – Development of firmed parliamentary politics – Short interruption by the coup by National Peace Keeping Council (NPKC) – Globalisation and Governance – The 1997 East Asian economic, known as *Tom Yum Kung* crisis – Strength of civil society organisations and NGOs – Human rights and environmental movement – Mekong integration – Globalisation – Energy conservation – Rights of consumers	*The Rise of CSOs*[19] – CSOs formation – Founding of the Assembly of the Poor (AoP) – Thailand Environmental Fund – 1994: The Foundation for Consumers (FFC) – Andaman Organisation for Participatory Restoration of Natural Resource – MAP Foundation (Migration) – Women, Friends of Women Foundation – Foundation for Slum Child Care – The Mirror Foundation – World Vision Thailand – Raks Thai Foundation – Small Scale Farmers' Assembly of Isan (SSFAI)
2000–2009	– Democratic governments – Human rights – Environmental movement – Health issues – Freedom of expression	*Establishment of public organisations* – 2003: Rainbow Sky Association of Thailand – Establishment of key public organisations 1 The Community Organisations Development Institute (CODI) for urban poor development 2 Thai Health Promotion Foundation 3 The Thai National Human Rights Commission – Mekong Energy and Ecology Network (MEE Net)

(*Continued*)

Table 29.1 (Continued)

Period	*Socio-economic and political contexts and key development issues/approaches*	*Key NGOs/social groups/CSOs*
2010–2018	– Middle-income country announced by the World Bank – Globalisation, Mekong integration and ASEAN community – Law related human rights – River and livelihood protection – System of semi-democracy – Authoritarian government (Censorship and restrictions on free expression policy under the military government (2015-present)	*Rise and fall of democratic system*[20] – EnLAW foundation (formerly Environmental Litigation and Advocacy for the Wants) – People Movement of Just Society (P-Move), formerly the Assembly of the Poor – 2012: Living River Siam Association – Thai Mekong People's Network from Eight Provinces – Mekong and Lanna Natural conservation and culture network – Community Resource Centre (CRC) (Lawyer working with CSOs on Mekong dam) – Se Bai River Conservation Network – Khon Taam Association (Wetlands and People association)

Legal entities

To obtain legal entities, NGOs and CSOs have to register either as foundation/charity (*mulaniti*) or association (*samakom*). For foundation, they need to register with the Ministry of Interior and the National Cultural Commission, under Civil and Commercial Code Section 81–97, National Cultural Act 1942. If they choose to register as an association, they have to do it with the director-general of the Department of Police and the National Cultural 1942 Commission, under the Civil and Commercial Code Section 1247–1297, National Cultural Act 1942.[21] NGOs and CSOs are not always required to register for a specific purpose if they do not wish to retain legal status. They can be named as project or working groups, units and forums. They are small and temporal. They can work under particular umbrella projects of colleges and universities.[22]

Approaches of NGOs/CSOs in development

There are four main approaches guiding NGOs and CSOs in working with local people, government and other stakeholders. They are (1) rights based, (2) knowledge based, (3) livelihood based and concepts of participation and (4) alternative development and community culture.

Rights based

The rights-based approach is a mainstream paradigm that NGOs have promoted worldwide since the late 1990s. It emerged as a new development discourse that combines human rights with poverty reduction and development. It underlines policy formulation and implementation to ensure that rights of the poor are met when development projects are implemented. The poor fully participate in the projects from the start. Accountability, governance, reduction of discrimination and ownership of the development must be realised.[23] In Thailand, the rights-based approach has been extensively employed by both NGOs and CSOs, especially on natural resources management. Natural resources including forest, land, marine, wild plants and animals and coastal resources have been heavily exploited over the past 50 years as economic inputs to pursue economic development and prosperity as well as human needs. Now Thailand is facing adverse impacts of unsustainable resource usage on local livelihood, ecological services and well-being.[24] The approach is put forward to ensure that the "intrinsic" benefits of basic human rights and the "instrumental" benefits are provided for the poor and marginalised people. There are some challenges found at its implementing process. For example, in the case of the Forest Community Bill, the powerful private sector is favoured over the rights of the community living in the forest reserve areas. The CSOs have negotiated and defended communities' rights of forest and land access in the public domain and political agenda.[25]

The rights-based paradigm is currently applied with other themes such as legal aspect and lawsuit. For example, the CSOs, NGOs and villagers working on energy and hydropower located in the Mekong River filed a lawsuit on the Xayaburi dam project against five Thai government bodies, including the Electricity Generating Authority of Thailand (EGAT), which agreed to buy 95 per cent of the hydropower from the dam. In February 2013, the case was rejected by the Administrative Court of Thailand.[26] The CSOs insisted on resubmitting their case to another court. Accordingly, they filed an appeal on 21 March 2013. On 24 June 2014, the Thai Supreme Court accepted the case.[27] The CSOs rely more on courts because they have realised that a lawsuit opens a greater opportunity to legally address environmental issues and problems than they had ever observed before. They are more confident to take the court verdict to mobilise, campaign and educate people in a wider society on environmental awareness.[28]

Knowledge based

The knowledge-based approach is another key paradigm that NGOs and CSOs have utilised to propose sustainable resources management. The CSOs have produced their own knowledge called Thai Baan Research, also known as Citizen Science.[29] It is a new approach to conventional research carried out by scholars in the academic institutes. Through academic advocacy and that of NGOs, the research is supported by local people who have suffered from developmental impacts. This research was originally initiated by academics at Chiang Mai University, the Southeast Asia River Network and villagers on the controversial Pak Mun dam. Villagers who have formed themselves into groups such as the Assembly of the Poor (AoP) with support of the NGOs and advocating academics have documented their livelihood related to river resources. It was quite a successful research tool to unite local people and to inform political debates and decision-making.[30] It was also treated as a political agenda and campaign to ensure that the voices of affected people are heard in the debates and political arenas. Thai Baan Research has been accepted nationwide and in the Mekong region.[31]

Also, knowledge produced in the form of systematic research could be helpful to engage local people in protecting the environment and to educate the middle class and other social groups about environmental problems.[32] Nowadays, knowledge, information and activities inculcated by the CSOs and NGOs are often presented through social media. They have created their own varieties of social media platforms including websites, YouTube, LINE and Facebook to communicate with society at large. Also, some CSOs have combined local knowledge with scientific data supported by academics to inform decision makers.[33]

Livelihood based

The livelihood-based approach is influenced by political, economic, cultural, biophysical and psychological components to construct and constitute the life of people.[34] The Thai CSOs and NGOs have increasingly been concerned about the worsening livelihood of the local people. They have worked against many state-led projects to protect their livelihood and resources. For example, coal power plants are the second main energy source after natural gas. EGAT attempted to build a plant in the southern region,[35] known as the Thepha coal-fire power plant. It was planned for construction in Thepha district of Songkhla province. The project was strongly resisted by local residents and the CSOs, as it was expected to destroy marine resources on which the local people have depended to sustain their livelihood. In January 2018, the prime minister ordered the Ministry of Energy to suspend the project.[36] Finally, the energy minister decided to end the protest by concluding a memorandum of understanding (MOU) with the protesters. It means that the Environmental Health Impact Assessment (EHIA) would be carried out in the next two years.[37]

Given the past experiences of protests against the first proposed hydropower dam, EGAT hesitated to build hydropower dams in Thailand. The CSOs, environmental groups and activists collectively opposed the first proposed Nam Joan dam some 30 years ago, to be built within the vicinity of the core protected Thung Yai Naresuan Wildlife Sanctuary. Years later, there were two key dams sparking controversies. First was the Keng Sua Ten dam on the Mae Yom River in Prae province; the impact of the construction would have inundated the last important remaining teak forest in Thailand. Second was the Mae Wong dam, which was expected to seriously flood Huay Kha Khaeng Wildlife Sanctuary, an important tiger habitat. The protest of Mae Wong dam was led by an NGO, the Seub Nakasathien Foundation.[38] The most controversial constructed dam in Thailand, the Pak Mun dam on the Mun River, near the confluence with the Mekong River, has persistently been protested by local NGOs and the Assembly of the Poor (AoP) due to possible severe impacts on local livelihoods and degradation of fishery resources.[39]

Another case concerns marine resources. The Coastal Resources Management Network through Community-Based Learning Centres was recently set up to conserve mangroves based on community participation. On 20–21 January 2018, the network, the government agencies and other networks jointly held the first meeting on "participation in marine and coastal resources management". The goal of the network is to empower existing community-based coastal resources management networks, in terms of knowledge transmission and improved management practices, to ensure the sustainable management of coastal ecosystem and the livelihood of local communities in the coastal areas are protected and improved. This project is designed to strengthen the newly established community-based coastal resources management network in six sub-districts of Trat province using best practices developed in several communities, especially the Community-Based Learning Centre (CBLC) Pred Nai as a pilot project to enhance sustainable coastal resources management processes within the network and selected communities.[40]

People's participation, alternative development and community culture approaches

People's participation, alternative development and community cultures or popular wisdom were dominant and popular concepts practiced by the NGOs and CSOs to counterbalance the penetration of development in the 1980s.[41] At that time, many large-scale infrastructure development projects were initiated and built by the Thai government, mostly in the rural areas, such as hydropower, irrigation schemes, telecommunication and highways in order to attract foreign investors. The NGOs used the above concepts to mobilise local people to realise the impacts of large-scale projects that would damage resources on which the people depended.[42] However, cultural-oriented approaches, especially community culture, have gradually weakened because the concepts separated practice from theory and culture from politics. They fail to empower local people at the political level.[43]

Current challenges of NGOs/CSOs

There are several significant constraints facing the NGOs and CSOs. These restrictions are affected by economic and political contexts, as explained below.

Censorship and restriction under the military government

The NGOs and CSOs have had limited social and political platforms to maintain free speech under authoritarian state since 2014. In 2015, the NCPO (National Council for Peace and Order) enacted the Martial Law Act of 1914 and implemented Section 44 of the 2014 constitution to apply censorship and restrictions on freedom of expression to the public. Also, SLAPP (Strategy Legality Against Public Participation) is another powerful legal instrument exploited by the NCPO to prevent freedom of expression by the NGOs and CSOs. The constitution gives the NCPO absolute power to disrupt or suppress.[44] The CSOs and NGOs have been restricted in mobilisation and criticism as well as in participation in environmental management. Mass protests against the government have been monitored and censored.[45]

Another case is about an exemption of EIA process. The 1992 National Environmental Quality Promotion and Protection Act enforces private and government projects to carry out proper environmental impact assessments (EIAs) for 35 types of projects to minimise their adverse environmental impacts. However, in 2016, the NCPO issued a decree on exemption of the construction of buildings in new Special Economic Zones and other key economic activities such as mining in the northeast region. This exemption is applied to a variety of economic activities including coal and nuclear power plants, water treatment plants, garbage disposal and collection plants, mining, recycling plants and gas processing plants. Private firms can be chosen to initiate projects even before an EIA is approved. Over 40 CSOs requested the NCPO to halt the March 2016 decree, but the government has not yet revised the decree because the military government has wanted to accelerate the economic growth as planned. From the government's point of view, shortening the process of EIA would expedite all mega development projects such as dual rail tracks, high-speed trains, seaports, natural gas and large-scale irrigation schemes and hydropower.[46]

Forest reclamation is another example showing the failure of the CSOs and NGOs in negotiating with the government for the poor people who have lived in the areas of the implemented policy. In 2014, the NCPO's forest reclamation policy was well intentioned to achieve more forest reservation areas. It is a good policy for the involved government agencies especially the Department of National Parks, Wildlife and Plant Conservation as they are strongly backed up by the absolute power and order of the NCPO to expand national parks, to protect against the greenhouse effect and to reduce the impact of climate change.[47] However, the effect of the forest reclamation is far-reaching once the two orders have been put into action. The practice of orders 64/2557 (2014) and 66/2557 (2014) were irrelevant to the reality on the ground. The CSOs and NGOs supported affected villagers to submit their complaints to the National Human Rights

Commission of Thailand. Now there are nearly 100 complaints by people who have been affected by the NCPO's crackdown on forest encroachment since 2014.[48]

Limited budget support

Thailand has transitioned to a middle-income country. Available international donors are scant, except for some NGOs and CSOs working on human rights–related environmental issues. They can secure funding from foreign donors such as the Open Society Foundation and the National Endowment for Democracy (NED).[49] For many CSOs working on development, well-being and related environmental issues, they have sought funding from key public organisations established by the government including the Thai Health Promotion Foundation (known as Sor Sor Sor), funded by the state revenues. The sin taxes from the alcohol and tobacco trade are one key domestic donor.[50] The Thailand Research Fund (TRF) is another public organisation, established in response to the 1992 Research Endowment Act. TRF is another autonomous organisation even if it is overseen by the government body. Unlike Sor Sor Sor, TRF (in the Thai language as Sor Kor Wor) is not granted by the sin taxes. With flexible administrative bureaucracy, TRF is able to be efficiently in research support. The main mission of TRF is to support the people at both the local and national levels to create a knowledge-based society to tackle societal problems. Many of TRF's research grants are given to works on environmental issues.[51] On environmental management, civil society groups, registered as either a foundation or association, can seek funding from the Thailand Environmental Fund (TEF), founded in 1992. The CSOs are calling the TEF to adjust its financial regulations and procedures to ease the process of application and selection.[52] Thailand has environmental funds to provide for sustainable environmental management. Hence, there are key public organisations that have budget available to support the CSOs in running environmental management, including the Thailand Environmental Fund, Thai Health Promotion Foundation and Thailand Research Fund. A few CSOs working on human rights–related environmental issues have received funding from the Open Society Foundation and the National Endowment for Democracy (NED).[53]

Lack of legal mechanism at regional level

The CSOs and NGOs working at the Mekong regional level are facing obstacles in sustainable resource management. They are eager to have a legal mechanism to support their movement and demand for the inter-governments in the Mekong to consider transboundary impacts of hydropower dams on local livelihood and river resources. Thailand is an energy-hungry country. The estimation of energy demand is based on the average growth of projected long-term Thai gross domestic product (GDP), estimated by the National Economic and Social Development Board (NESDB) at 3.94 per cent. Thus, the saving energy would be 89,672 GWh in 2036. According to the plan, energy sources are mainly from natural gas, followed by coal and imported hydropower from neighbouring countries.[54] However, it is unclear whether this excessive energy reservation is necessary.[55] Law and regulations pertaining to hydropower that inform decision-making in the Mekong are highly centralised. It is complicated for the CSOs and NGOs to engage both within and outside the courts when they contest hydropower development at the regional scale, as they have to challenge national sovereignty.[56] Hence, states do not focus on trans-boundary impact and do not recognise regional impacts and solutions because they tend to pursue national interests on top of the lack of regional regulations such as Transboundary Environmental Impact Assessments (TbEIA). Inter-governments in the Mekong also lack a regional legal mechanism on monitoring and challenge transboundary impacts.[57]

Chasing government-led development projects

"We are chasing development and it will be very hard for us to catch up", said one representative from the CSOs working on environmental and development issues.[58] In 2015, the Thai military government has commenced several new projects which would last over the next few years. They initiated the establishment Special Economic Zones (SEZs) in ten provinces located along the borders around the country. The SEZs has two phases. The first phase comprises five provinces including Tak, Mukdahan, Sa Kaeo, Trat and Songkhla. The second phase includes five other provinces: Nong Khai, Narathiwat, Chiang Rai, Nakhon Phanom and Kanchanaburi. The total area is about 6,220 km^2. The ultimate goals of the establishment of SEZs are to promote concrete and balance development for emerging border economic areas as well as to improve quality of life and solve security problems. The specific purposes of the established SEZs are to build production bases connecting with those of ASEAN countries and to promote the economic development of the provinces of Thailand along its border areas. The cabinet appointed the National Committee on Special Economic Zone Development (NC-SEZ) to implement the policy.[59] Furthermore, to accelerate growth and development in the already developed areas, in 2017 the government commenced the Eastern Economic Corridor (EEC) to implement the Thailand 4.0 policy in the Eastern Aerotropolis in U-Tapao; a high-speed train from Bangkok to U-Tapao; sea ports of Laem Chabung, Map-Taput and Sattahip; industries including electric vehicles, robotics, medical and aircraft; and Future Cities and the Digital Park.[60] With limited resources and staff, the CSOs and NGOs are unable to monitor these new development projects. Strengthening the existing CSOs network, moving quickly with accurate key solutions supported by substantial information and budget as well as increasing staff would be highly anticipated by the Thai CSOs and NGOs.[61]

Summary

The NGOs and SCOs in Thailand are highly varied, dynamic and complex by nature. The NGOs originally emerged as philanthropic organisations to support the poor and disadvantaged groups in the country on social welfare. The CSOs were established in the 1990s. Both of them have worked in wide varieties of areas and activities at the community-based or grassroots levels. Their development practices and inspiration are often guided by key approaches including rights-based, knowledge-based, livelihood-based and cultural-oriented and participation approaches. They aim to achieve policy advocacy, rights of communities or disadvantaged groups and sustainable livelihood of local people.

Notes

1 The October uprising took place in 1973.

2 Somchai Phatharathananunth, *Civil Society and Democratisation: Social Movements in Northeast Thailand* (Copenhagen: NIAS Press, 2006).

3 "The Twelfth National Economic and Social Development Plan (2017–2021)", Office of the National Economic and Social Development Board, 2017 <www.nesdb.go.th/nesdb_en/ewt_w3c/ewt_dl_link.php?nid=4345>.

4 NGOs, pers. comm., May 2018.

5 Amara Pongsapich, "Defining the Non-profit Sector: Thailand", Working papers, The Johns Hopkins Comparative Nonprofit Sector Project, No. 11, edited by L.M. Salamon and H.K. Anheier (Baltimore, MD: The Johns Hopkins Institute for Policy Studies, 1993).

6 Q. McQuistan, Saowalak Markphaengthong and Jitkasem Permpatr, *Non-government Organisations in Thailand: A General Overview with Emphasis on Northeastern Thailand, Current Status and*

Potential for Development Cooperation in the Greater Mekong Subregion <www.mekonginfo.org/assets/midocs/0002330-society-non-government-organisations-in-thailand-a-general-overview-with-emphasis-on-north-eastern-thailand-current-status-and-potential-for-development-coo.pdf>.

7 Somchai Jitsuchon, *Thailand in a Middle-income Trap*, 2012 <https://tdri.or.th/wp-content/uploads/2012/12/t5j2012-somchai.pdf>.

8 Nanak Kakwani, "Income Inequality and Poverty: Methods of Estimation and Policy Applications", in *A World Bank Research Publication* (Oxford: Oxford University Press, 1980).

9 Organisation for Economic Co-operation and Development, "Towards Asia's Sustainable Development: The Role of Social Protection: Emerging and Transition Economies Series", Paris, 2002, p. 139.

10 Bruce D. Missingham, *The Assembly of the Poor: From Local Struggle to National Protest Movement* (Chiang Mai: Silkworm Books, 2003).

11 S. Pednekar, *NGOs and Natural Resource Management in Mainland Southeast Asia*, Bangkok, 1995 <www.gdrc.org/ngo/thai-ngo.html)>.

12 R. Quinn, *NGOs, Peasants and the State: Transformation and Intervention in Rural Thailand 1970–1990*, PhD Dissertation, Australian National University, 1997.

13 Ibid.

14 Ibid., 1997.

15 Dr. Puey Ungphakorn was as a Thai economist who served as Governor of the Bank of Thailand and Thammasat University's rector.

16 McQuistan, Saowalak and Jitkasem, *Non-government Organisations in Thailand.*

17 Quinn, *NGOs, Peasants and the State: Transformation and Intervention in Rural Thailand 1970–1990.*

18 Pasuk Phongpaichit and Chris Baker, "Power in Transition: Thailand in the 1990s", in *Political Change in Thailand: Democracy and Participation*, edited by Kevin Hewison (London: Routledge, 1997).

19 *Civil Society Brief: Thailand*, ADB <www.adb.org/sites/default/files/publication/29149/csb-tha.pdf>.

20 NGOs, pers. comm., 22 May 2018.

21 Vitit Muntarbhorn, "Occidental Philosophy, Oriental Philology: Law and Organised Private Philanthropy in Thailand", in *Philanthropy and the Dynamics of Change in East and Southeast Asia*, edited by Barnett F. Baron, Occasional Papers of the East Asian Institute, Columbia University, 1991. The National Cultural Act of 1942 established the National Cultural Commission, which is responsible for both establishment and oversight of foundations and associations.

22 Amara, "Defining the Nonprofit Sector: Thailand".

23 Shannon Kindornay, James Ron, and Charli Carpenter, "Rights-based Approaches to Development: Implications for NGOs", *Human Rights Quarterly*, Vol. 34, No. 2 (2012), pp. 472–506.

24 Pasuk Phongpaichit and Pornthep Benyaapikul, *Locked in the Middle-Income Trap: Thailand's Economy between Resilience and Future Challenges*, Friedrich Ebert Stiftung, 2012 <http://library.fes.de/pdf-files/bueros/thailand/09208.pdf>.

25 J. Craig and T. Forsyth, "In the Eyes of the State: Negotiating a 'Rights-based Approach' to Forest Conservation in Thailand", *World Development*, Vol. 30, No. 9 (2002), pp. 1591–1605 <DOI: 10.1016/S0305–750X(02)00057–8>.

26 The main reasons that the court dismissed the case were that (1) the plaintiffs do not consider injured persons as conditions and compliances set by the cabinet before concluding the power purchase agreement is considered a part of the internal administrative process; (2) the power purchase agreement is binding for contractual parties, such as EGAT and the Xayaburi Power Company, therefore third parties like the plaintiffs are not considered injured persons; and (3) although the defendants did not comply with PNPCA, such process is not considered an administrative act and therefore the court is not able to hear the case. See Pianporn Deetes, *Thai Villagers File Lawsuit on Xayaburi Dam,* 2012 <www.internationalrivers.org/blogs/254/thai-villagers-file-lawsuit-on-xayaburi-dam>.

27 Ibid.

28 NGOs, pers. comm., May 2018.

29 Thai Baan means villages.

30 J. Kirchherr, "Strategies of Successful Anti-Dam Movements: Evidence from Myanmar and Thailand", *Society & Natural Resources*, Vol. 31, No. 2 (2018), pp. 166–182 <DOI: 10.1080/08941920.2017.1364455>.

31 K. Herbertson, *Citizen Science Supports a Healthy Mekong*, 2012 <www.internationalrivers.org/resources/citizen-science-supports-a-healthy-mekong-7759>.

32 CSOs, pers. comm., 2018.

33 NGOs, pers. comm., 2018.

34 J. Walker, Bruce Mitchell and Susan Wismer, "Livelihood Strategy Approach to Community-based Planning and Assessment: A Case Study of Molas, Indonesia", *Impact Assessment and Project Appraisal*, Vol. 1, No. 4 (2001), pp. 297–309 <DOI: 10.3152/147154601781766925>.
35 EGAT (Electricity Generating Authority of Thailand) is the only state enterprise that is mandated to provide electricity in Thailand.
36 "PM Suspends Thepha Coal-fired Power Plant Project", *Thai PBS*, 2018 <http://englishnews.thaipbs.or.th/pm-suspends-thepha-coal-fired-power-plant-project/>.
37 "Coal-fire Electricity Projects Deal Struck", *The Nation*, 2018 <www.nationmultimedia.com/detail/national/30339339>.
38 R. Mather, D. Constable, N.D. Tu, V. Lou, J. Brunner, A. Starr, S. Kampongsun, A.C. Joehl, L. Sylavong, G. Martin and Y. Zhang, *Critical Ecosystem Partnership Fund: Long-term Vision for the Indo-Burma Hotspot*, IUCN Asia Regional Office, Bangkok, 2017.
39 T. Foran and K. Manorom, "Pak Mun Dam: Perpetually Contested?" in *Contested Waterscapes in the Mekong Region: Hydropower, Livelihoods and Governance*, edited by F. Molle, T. Foran and M. Ka Ko Nen (London: Earthscan, 2009), pp. 55–80.
40 See <www.mangrovesforthefuture.org/grants/large-grant-facilities/thailand-large-projects/strengthening-the-community-based-coastal-resources-management-network-through-community-based-learning-centers-in-six-sub-districts-of-trat-province-thailand/>.
41 Quinn, *NGOs, Peasants and the State: Transformation and Intervention in Rural Thailand 1970–1990.*
42 Seri Phongphit (ed.) *Back to the Roots: Village and Self-Reliance in a Thai Context, Culture and Development,* Series 1 (Bangkok: RUDOC, 1986); Seri Phongphit and K.J. Hewison, *Thai Village Life: Culture and Transition in the Northeast* (Bangkok: Mooban Press, 1990).
43 Quinn, *NGOs, Peasants and the State: Transformation and Intervention in Rural Thailand 1970–1990*, p. 106.
44 *Thailand: Interim Constitution Provides Sweeping Powers*, Human Right Watch, 24 July 2014 <www.hrw.org/news/2014/07/24/thailand-interim-constitution-provides-sweeping-powers>.
45 CSOs, pers. comm., 24 May 2018.
46 CSOs, pers. comm., 18 May 2018.
47 E. Gershkovich, Thailand's Deforestation Solution, 2014 <www.worldpolicy.org/blog/2014/10/16/thailands-deforestation-solution>.
48 "Forest Clampdown Hurts Poor", *Bangkok Post*, 2016 <www.bangkokpost.com/print/1083356/>.
49 CSOs, pers. comm., 24 May 2018.
50 Mather, Constable, Tu, Lou, Brunner, Starr, Kampongsun, Joehl, Sylavong, Martin, and Zhang, *Critical Ecosystem Partnership Fund.*
51 <www.trf.or.th/eng/>.
52 CSOs, pers. comm., 24 May 2018.
53 NGOs, pers. comm., 24 May 2018.
54 *Thailand Power Development Plan (2015–2036): PDP2015*, Ministry of Energy, 2015 <www.egat.co.th/en/images/about-egat/PDP2015_Eng.pdf>.
55 C. Weatherby and B. Eyler, *Letters from the Mekong Power Shift: Emerging Trends in the MGS Power Sector*, Stimson Center, 2017.
56 B. Boer, P. Hirsch, J. Fleur, B. Sual and N. Scurrah, *Socio-legal Approach to River Basin Development* (Abingdon and Oxon: Routledge, 2016).
57 CSOs, pers. comm., 11 March 2018.
58 CSOs, pers. comm., 11 March 2018.
59 *Thailand Special Economic Zones,* Office of the National Economic and Social Development Board, 2016 <www.nesdb.go.th/ewt_dl_link.php?nid=5194>.
60 *Thailand Economic Monitor,* Digital transformation, WB group, 2017. <http://pubdocs.worldbank.org/en/823661503543356520/Thailand-Economic-Monitor-August-2017> (accessed 20 February 2018).
61 CSOs, pers. comm., 16 March 2018.

References

Amara Pongsapich. 1993. "Defining the Non-profit Sector: Thailand". Working Papers of the Johns Hopkins Comparative Non-profit Sector Project. Edited by L.M. Salamon and H.K. Anheier. Baltimore: The Johns Hopkins Institute for Policy Studies, No. 11.

Foran, T. and K. Manorom. 2009. "Pak Mun Dam: Perpetually Contested?" In *Contested Waterscapes in the Mekong Region: Hydropower, Livelihoods and Governance*. Edited by F. Molle, T. Foran and M. Ka Ko Nen. London: Earthscan, pp. 55–80.

Kakwani, Nanak. 1980. "Income Inequality and Poverty: Methods of Estimation and Policy Applications". In *A World Bank Research Publication*. Oxford: Oxford University Press.

Missingham, Bruce D. 2003. The Assembly of the Poor: From Local Struggle to National Protest Movement. Chiang Mai, Thailand: Silkworm Books.

Somchai Phatharathananunth. 2006. *Civil Society and Democratisation. Social Movements in Northeast Thailand*. Copenhagen: NIAS Press.

30

ENVIRONMENTAL POLITICS IN THAILAND

Pasts, presents, and futures

Eli Elinoff and Vanessa Lamb

In this chapter, we offer a critical review of Thailand's environmental politics, beginning with the 1970s and tracing these struggles through to the present.[1] Instead of a complete review, we outline the roots and trajectories of Thai environmental politics.[2] To start, we explore the crosscutting political-economic currents that frame contemporary environmental struggles. We follow this by showing how such struggles played out in development policy debates related to the dispossession of farmers and forest-dwellers as well as a separate "politics of water". The final sections cover both emerging pathways of scholarship for investigating environmental struggles raising questions related to urban environments, disaster, climate change, and gender.

In Thailand, struggles over the environment have been as fundamental to understanding the national landscape as its more traditional scenes of volatility over the nature and organisations of state institutions over the past 50 years. Variously understood as forging cross-class ties, mobilising political sensibilities in remote places, extending (or deferring) democracy, dividing and uniting urbanites with rural people, and exposing persistent sites of inequality, environmental politics have been critical in the struggles over power even when formal democratic politics were suspended. At the same time, the internal composition of overtly environmental movements and their implicit imaginaries of the nature-culture division reinforced some pervasive differences within the Thai polity along ethnic, spatial (rural-urban), regional, and class lines.

We argue that environmental struggles are best understood as unfolding within Thailand's changing socio-economic, political, and spatial landscape. That is to say, struggles over rights to forests, to fish, to land, or to clean water and city space have always been struggles over more than resources. These struggles have also raised questions about the political systems that govern these resources that organise and extract capital and the associated uneven distribution of environmental harms. Moreover, they have also implied broader questions about the reach and limitations of political voices – those often poor, ethnically non-Thai, and more remote subjects making those political claims. Within this context, environmental movements and the efforts to extract, govern, manage, and improve the environment have often raised the spectre of democracy but not necessarily advanced it in straightforward ways. In the following review, we situate the emergence of environmental politics within its larger contexts, describing the origins and future trajectories of both environmental struggle and efforts to understand these struggles.

Developing the nation, making the environment

To start, we consider the origins of a specifically environmental politics in two ways, epistemologically and political-economically. The former approach emphasises the emergence of "nature" (*thammachat*) and "the environment" (*singwaetlawm*) as emergent objects that reorganised human relationships with space, with their non-human, bio-physical surroundings, and with one another. The second point of departure – political economy – emphasises the ways the emergence of a distinctly *environmental* politics came within the context of rapid development and the attendant forms of political, economic, and socio-spatial disruption associated with Thailand's breakneck economic transformation. This latter approach also highlights the ways environmental issues mobilised a broad range of people, including the urban and rural poor, the middle-class elites, and other actors into complex new coalitions. Rather than partition these approaches, we see the two as interrelated.

Many scholars locate the origins of "environmental politics" in Thailand within the 1970s. Quigley, for example, argues that there were no environmental struggles prior to this period.[3] Although this may be accurate, strictly speaking, the point ignores how what Vandergeest and Peluso call "state territorialisation" enfolded environmental politics into the heart of Thai politics – especially struggles over inequality, development, and democracy emerging from Thailand's provinces and its poorest communities.[4]

In their now classic argument, Vandergeest and Peluso[5] demonstrate how across the 19th and 20th centuries successive Siamese/Thai governments exerted state control over space in three ways: first, through the expansion of civil administration in Siam/Thailand; second, through land titling; and, third, through the demarcation of large swaths of land as what they later called "political forests", which tied together militarised efforts to dismantle counter-insurgencies while also distributing access rights to natural resources within these forest lands.[6] On one hand, these processes of state territorialisation extended Thai hegemony, enclosing the nation's borders by doing the violent work of making good on the bounded vision of the "geo-body".[7] On the other, territorialisation also enabled land dispossession and commoditisation, which continue to structure uneven development.[8] The effect of these processes is that prior to the 1970s, while struggles on the ground over resources may not appear as specifically environmental, these early struggles were fundamental to struggles later taken up by environmental movements since the 1950s.

The process of spatial transformation was political but also had important epistemological effects. Stott argues the traditional Tai city-state or *muang* was organised by a tripartite division of space.[9] Within this model the *muang* was surrounded by cultivated rice fields (*na*) and associated irrigation works (*muangfai*) and served as a moral, social, and political centre point.[10] Beyond the reach of the *muang* are the "wild forests" (*pathuen*), which house malevolent spirits and wild animals. Davis pointed out that within this schema, forests were considered spaces of immorality, home to both wild animals and wild people.[11]

It bears noting, however, that these schematic descriptions of the ordering force of the *muang* and the disordering forces of *pathuen* reflect ideal types rather than actually existing socio-ecological renderings of life within pre-modern Siam.[12] Nevertheless, these pre-modern forms – ideal and actual – reveal a complex localised set of epistemological understandings of the division between natural and human worlds. Later reflections on the everyday forms of Buddhism demonstrate a rather different sensibility towards nature reflecting the cross-cutting, entangled and embeddedness of humans within non-human worlds.[13] In both approaches, monks emerge as kinds of environmental mediators, emblematic of ideal relations between humans and non-humans, about what the environment was, and how it should be organised prior to territorialisation. With the onset of territorialisation and commodification, these ontological relations were reorganised but also laid the groundwork for the coming environmental movements.

Beginning in the 1950s and intensifying rapidly into the 1980s and 1990s, Thailand saw a radical transformation of its economy and its social life. Cold War geopolitics, rapid economic growth, and the shift from agricultural to industrial production had massive social and spatial implications for the country that led to the emergence of a distinctly environmental politics.[14] There was a massive increase in deforestation,[15] industrialisation, and associated urban population growth, particularly in Bangkok.[16] Urbanisation, the expansion of industrial agriculture and fishing, and a transformation of the remote hinterlands into spaces of extraction and energy infrastructure were all indicative of the intensity of environmental change during this period.[17]

Hirsch and Lohman note that critical environmental struggles took place prior to the economic growth of the 1980s, particularly how poaching within the Thung Yai Narasuan Wildlife Sanctuary galvanised the public and motivator for the student-led democratic revolt of 1973.[18] The event not only highlighted corruption within the regime but also demonstrated an endemic, uneven application of conservation regulations which target the poor actors while enabling wealthier and more powerful actors to continue to exploit resources. This continues to be a major theme in contemporary environmental politics. Further struggles over forests and the peasantry waged beyond the capital highlight the ways in which the Communist Party of Thailand (CPT) took up environmental issues by linking them to broader demands for land.[19] These struggles had important legacies, including the founding of the Peasant Federation of Thailand, whose organisational structure and membership became the backbone of a number of environmental groups in the 1980–1990s that linked environmental struggles to issues of livelihood.[20]

With the violent ouster of the Seni Pramoj government in 1976 by the military, students were banned from rural areas. Many activists fled to the jungle; some joined the CPT.[21] This shifted the terms upon which struggles over politics and therefore struggles over the environment played out. It was not until the activists' return in the 1980s that much of this activism reclaimed the environment as a means of struggling against the imbrications of the market and the state and as a means of pushing ahead political claims made by the poor.[22]

These early examples of environmental struggle not only demonstrate the ideological roots of later environmental movements within crucible of Thailand's mid-1970s politics but also the yoking together of different perspectives on these politics from the countryside and the city. Middle-class students cast themselves as educators and mobilisers of the poor. Within rural areas, environmental politics linked land reform to questions related to political participation, environmental degradation, and material demands.[23] Forsyth notes that internal class politics often shaped both the visibility of different struggles and the ways in which these struggles have been understood.[24] It was within this period that environmentalism was resignified as fundamental to disrupting industrial development as well as a mechanism of democratisation.[25] Of course, this claim was tested and critiqued, but nevertheless it reflects a kind of aspirational hope that environmental politics and democratic politics might converge.[26]

The end of the 20th century saw the collision between the intensification of the spatial transformation as the economy boomed and imploded. The volatile shifts in the economy transformed social life in the countryside from centring on subsistence agriculture and smallholder farming towards new economies that linked existing forms of rural production to seasonal and permanent labour migration from the countryside to the city.[27] This period also witnessed the emergence of a new discursive rendering of rural life as authentically Thai, if always, already lost.[28] By the end of the 1990s, we also see the rise of the Assembly of the Poor and that environmental groups became prominent actors within Thailand's national political struggles.

Linked to this "rural ideal" and returning to the question that started this section, it is also within this period that *thammachat* as a category emerged as a specifically elite resignification that transformed the wild forest into a more manageable form: nature.[29] These epistemological shifts extended cross-class differences around livelihoods and also re-scripted the environment as something to be not feared but enjoyed. While later environmental movements particularly in the 1990s were fomented around cross-class collaborations in the name of protecting nature, the internal class-based disjunctures within these movements and the epistemological shift from wild nature to be controlled to domesticated nature to be enjoyed seems fundamental to the tensions that animate contemporary environmental struggles and their discontents in the early 20th century. Indeed, the next section highlights how questions of conservation and livelihoods exist in environmental movements in contrasting ways. The various genealogies of the environment and understanding its changing condition during a transformative 50 years remain unsettled and contested, a point returned to in the conclusion.

Forests and fields: landscapes of exclusion and dispossession

The centrality of state territorialisation for resource control and governance in the 19th and 20th centuries in Thailand underlines how, going forward, the most foundational environmental struggles played out over the governance, ownership, and access to natural resources. Further, divisions between upland and lowland peoples especially related to the practice of forest governance, swidden agriculture, and the broader effects of deforestation have structured key debates surrounding the politics of forests and who benefits (and suffers) under emerging regimes of dispossession. These processes and discourses have also been fundamental to the ways in which environmental movements have been mobilised and towards what kinds of ends.

The 1950s and 1960s established new codes of governance surrounding private land ownership and forestry in Thailand. The 1954 Land Code, for example, created new sorts of land titles and began the slow yet intensive effort to issue private title across the country. Slowed only by "the capacity of the Department of Land to carry out land surveys and issue titles", by 1998, 19 million of the country's 26 million parcels of non-state-owned land were titled.[30] In northern Thailand, the effect of this was to pit local tenure systems against national systems, with complex outcomes, transforming kinship relations and upsetting previous practices of inheritance.[31] At the same time, laws surrounding forestry expanded the domain of the state's broad control over land. The 1964 Forest Reserve Act transferred 45 per cent of the national territory under forest authority. The results of these dual processes reflect a variety of processes of exclusion and dispossessions.

The closure of the frontier fundamentally impacted Thailand's highland communities. While deforestation emerged as an issue in the 1970s and 1980s, prior to the 1980s the dispossession of highlanders by lowlanders enabled access to teak and other non-timber products.[32] Vandergeest notes that forest space and forest policy is thus highly racialised.[33] By this he means that upland people are partitioned and "fixed" in time and space as non-Thai and tribals, with tribals often understood as either implicitly or explicitly subordinate to lowlander "Thai" wet-rice agriculturalists. Swidden agriculture has been a key indicator of these internal political differences. Further to this point, Pinkaew described earlier how the designation of particular areas as protected "watersheds" enabled lowlanders to take control of upland resources, both redistributing access to forest resources and propelling highland dispossession.[34] These processes of highland dispossession were also driven by forest policy geared towards industrialists, and lowland migration occurred alongside an increase in the securitisation of forests that occurred during the Cold War.

These forces drove deforestation and dispossession among highland communities throughout the 20th century.[35] These processes, Vandergeest notes, are highly racialised in the sense that they only make sense when considering the way that "mapping procedures that produced highland reserves forests and protected areas as uplands, and cadastral space in the lowlands".[36] These conflicts between lowlanders and highlanders account for both the omnipresent "spectre" of state-led eviction and also the ongoing threat of actual eviction, as lowland villages have frequently been involved in the forced removal of highland settlements.[37]

The terms of logging changed in 1989 when, following a massive mudslide that killed 251 people the previous year, the government revoked all logging licenses and banned further logging while also declaring the closure of the land frontier. Delang and others emphasise that the 1989 ban and the closing of the frontier did not constitute an environmental turn in government policy, nor were they reflective of the population reaching a "natural" limit.[38] Instead, these policy shifts served to benefit entrenched interests, propelling landlessness and driving processes of proletarianisation. Usher notes that the ideas of "uninhabited wilderness" which shaped Thai forestry policy were rooted in the United States, where landscapes were not empty of humans but had to be made so by population relocation.[39] Such shifts were critical moments of spatial transformation.[40] In short, 20th-century land policy in Thailand is, like elsewhere in the region, tied to the processes of primitive accumulation and accumulation through dispossession.[41]

For highlanders and ethnic minorities in particular, this process sparked new modes of governance and struggle. As Forsyth and Walker note, highlanders found themselves positioned as either "forest guardians" or "forest destroyers", which narrowed their room to maneuver economically and politically.[42] That these discourses overlap tightly with the pervasive conflation between highlanders, hill tribes, and community forestry, discourses about indigenous guardianship or degradation of forest areas are deeply implicated in the forms of environmental knowledge produced about highlands ecologies, what threatens them, how to conserve them, and the people who live there.[43] Moreover, as Pinkaew demonstrates, emerging attachments to *thammachat* have not resulted in a more integrated view of humans and nature but rather the opposite, as the politics of resource extraction and "upper class elitist nature inspiration" have converged in the dispossession of highland communities.[44] As a result, it is clear that highlanders are caught between environmentalisms.

Because of the close associations between titling, resource extraction, conservation, and land dispossession, forest-based movements were fundamental to other kinds of environmental politics. The most notable example beyond the historical land reform movements described above are the movement against the Khor Jor Kor resettlement scheme, which later influenced the Assembly of the Poor's movements against eviction.

Khor Jor Kor is an acronym for a policy passed in 1990 called the "Land Distribution Programme for the Poor living in Degraded National Forest reserves in the Northeast of Thailand". Organised by the military government, the policy aimed to reorganise every forest reserve in northeastern Thailand, which led to the eventual resettlement of approximately one million people.[45] In a classic example of environmental eviction, the policy was sold as a means of conserving the region's forests and extending land to the poor. Pye argues, however, that the policy was organised to shore up support for the military government through development initiatives, including the Pak Mun dam and broader Kong-Chi-Mun scheme, which is described in the next section.[46] Khor Jor Kor resulted in the eviction of about 2,000 families but also sparked a massive backlash.[47] In 1991–1992, affected communities organised alongside NGOs to develop networks and strategies. The groups held demonstrations in Khon Kaen in March of 1992 which brought together affected communities with other grassroots groups facing similar problems from development projects. Khor Jor Kor was dealt a major blow by pro-democracy movements that led to the fall of the military government in May 1992. Nevertheless, the group continued confrontational tactics including initiating a long march, several road blocks, and not acceding to the project's temporary suspension.[48]

The Khor Jor Kor movement demonstrated that rural politics could in fact influence state policy *and* the ways that such influence could be exerted through on-the-ground organising via the creation of new political relationships. As such, it laid the groundwork for the Assembly of the Poor and its occupation of the Government House for 99 days in 1997. The confluence of democratic uprising alongside these focused environmental livelihood projects became a significant political factor in the 1990s, explored next.

Rivers: environmental struggle and knowledge within and beyond borders

Few environmental issues in Thailand have garnered as much national and international attention as the social movements that have emerged from efforts to dam rivers across Thailand. Within the broader economic, political, and environmental transformations described above, the practice of the use, distribution, and management of water was also transformed. Prior to the 1980s, dam building in Thailand was bound up with the extension of development and Cold War geopolitics. The Bhumibol dam, named after King Rama IX, was commissioned in 1964 and shortly followed by 21 other dams throughout Thailand, including Queen Sirikit's namesake dam. There is little written about early opposition to such dams due to the political context within which they were built. As part of larger efforts to "develop" northeastern Thailand in the shadow of the Cold War, dam building proceeded throughout that region with three notable structures being finished in the 1970s.[49]

These dams combined the Cold War development politics of securitisation with efforts of rural electrification in the name of development. Efforts by villages to remain in place seem to have been signs of resistance in a time when formal resistance would have been conflated with communist insurgency. Indeed, as the anti-dam movement gained momentum in the 1990s, people displaced by the Sirindhorn dam in the late 1960s and early 1970s also joined as prominent voices within these movements, demanding compensation for their losses decades earlier.

The Nam Choan dam was first proposed in 1966 as part of a larger scheme for development of the upper Kwae Yai River in western Thailand. The proposed dam was sited in the middle of a CPT-controlled area, and so this kind of project was not only envisioned as a means of expanding the grid but also as a mode of counter-insurgency and state territorialisation.[50] Although the other projects in the scheme were completed, Nam Choan was successfully blocked by activists, garnering global attention for their work and shaping anti-dam activism moving forward, especially as activists shifted their approach towards making arguments through the mobilisation of environmental knowledge.

Following three rounds of environmental impact assessment (EIA), the project was originally slated to begin in the first half of the 1980s with funding from the World Bank and the Japanese Overseas Economic Cooperation Fund.[51] Opponents however, successfully mobilised an argument that continues to be made in good effect in contemporary Thailand, that the dam was being pushed ahead by a small number of actors for their own benefit.[52] Many scholars point to this moment as a major milestone in the nation's environmental politics, as it reflected shifting middle-class priorities towards the protection of nature, characterising this as a major epistemological turning point in which urban life and the forest were no longer opposed but increasingly seen as interrelated.[53] At the same time, in addition to this reconfiguration of the urban middle class as environmental actors, these experiences revealed a way to do "environmental politics" both safely and effectively: through a focus on knowledge (such as assessments), which also appealed to and aligned with a broader public than earlier pro-poor struggles over land had accomplished. It also cannot be understated how essential it was to broad-based mobilising that this project did not bear a royal moniker.

Following Nam Choan, the struggle over the Pak Mun dam in Ubon Ratchathani relied on similar strategies, and while it was less successful in terms of halting the project, it was perhaps more transformative politically. Proposed in 1982 and commissioned in 1994, the Pak Mun dam, located near the confluence of the Mun and Mekong Rivers, carried with it similar multipurpose promises as Nam Choan. The dam was sold as a vehicle for rural development, electricity generation and, importantly, as a means of irrigation, a cornerstone of the larger Kong-Chi-Mun project that promised to "turn the northeast green" via a second rice crop and radically reconfigure the northeast.[54] It dramatically failed those selling points both in the ways it impacted livelihoods negatively and in its limited electricity generation.

Pak Mun encountered strong resistance from the beginning. Early on, this resistance was highly localised, but building on the success at Nam Choan, villagers began working with activists who mobilised similar sorts of arguments regarding the poor planning, effects on biodiversity, and insufficient impact assessment and knowledge. For instance, villagers dissatisfied with the environmental assessments – and that they occurred *after* project approval – developed their own research methodology called "Villager Research". Villagers and activists collaborated with academics to study and document the dam's effects on fish variety and bio-diversity.[55] A focus on environmental impact assessments, human impact assessments, and "people's knowledge" (Thai *baan* research) was fundamental to these unfolding political terrains.[56] A critical difference at Pak Mun was the way in which such work recast the movement as one not only concerned over fish and rivers for their conservation value but also as one about the livelihoods lost by residents. Though the dam was completed in 1994, subsequent iterations of the movement drew larger and larger swaths of people from the region to the struggle. By 1996, over 12,000 people gathered to demand that then Prime Minister Banharn Silpa-archa set up a participatory committee to solve the issue. These meetings also galvanised over 2,500 villagers affected by the construction of nearby projects, including the Sirindhorn and Rasi Salai dams.

This "red-green" framing, so named because it drew from the legacies of the CPT, combined resistance to the incursive dam project with questions about livelihoods, inequality, and uneven development and positioned the movement as a critical component of the political movements of the 1990s.[57] Indeed, Pak Mun was fundamental to the Assembly of the Poor social movement, which gained prominence after occupying the lawn of the Thai Government House for 99 days in 1997.[58] This struggle not only tied together the question of livelihood with that of the broader environment but also raised the stakes by highlighting the larger framework of transnational institutions behind the dam's construction and the emerging global anti-globalisation movement as a means of pressing their case.[59]

By the early 2000, the villagers and non-governmental organisations (NGOs) were gaining momentum from the World Commission on Dams (WCD) report. This report, which took Pak Mun as a case, highlighted the project's technical and social ramifications as part of a larger argument against large dams globally.[60] Villagers and activists constructed a local protest camp at the dam site, asserting their demands to open the dam's gates to allow for fish migrations. In late 2000, the Electricity Generating Authority of Thailand sent locals to raid the encampment for the first time, burning down shelters and injuring people there; 30 people sent to the hospital.[61] In the face of both intimidation, arrests, and physical violence, the movement pressured the government into a four-year trial of opening the dam's gates during the rainy season to allow fish to spawn. In late 2002, the protest village suffered its second attack, which resulted in its permanent dispersal. Eventually, the movement received some compensation, which was used to construct a local learning centre that featured the research conducted by villagers. The protest struggled to remain cohesive. Nevertheless, villagers from the region often maintain their efforts to keep it open permanently, and several studies have been carried out by villagers on the Mun River system.

Domestically, anti-dam politics continue to mobilise people.[62] However, in considering the legacy of the struggles against damming inside Thailand, there is also an accompanied geopolitical shift which simultaneously saw (1) construction of major dams outside Thailand to generate power for Thai consumption and (2) anti-dam activism and environmental politics mirror the increasingly transboundary and regulatory impacts and context for such developments.[63] While large dams remain tantalising both for what they seem to promise in terms of national development and as a continuing tool for securitisation of tenuous national borders, the transboundary effects of river development, especially on the Mekong[64] and on the Salween River[65] have become a new focus of scholarship and activism that reveal the continued effects of industrial development on the region's rivers and the emerging difficulties of environmental politics at a moment in which impacts and electricity are not confined to national borders.

Future trajectories: urban environments, gendering nature, environmental disasters, and climate change

Tracing the history of environmental political struggles from their political economic origins to debates over land politics through situated struggles over rivers in the previous three sections, we have described the how environmentalism was mobilised in the past 30 years in Thailand. Early struggles revolved around access to resources, even if they were not seen as "environmental". Later, many environmental movements extended this history, as poor communities used the language of the environment to push for autonomy, access, rights, or redress against development projects. Sometimes these movements included the poor in schemes of conservation; other times conservation was used to exclude the poor dependent on resources. Throughout, the ambiguous "good" of the environment made it a contested concept, making environmental politics more complex even as environmental values seem to have become more pervasive. In thinking about the future, we curated the following section to highlight emerging themes within the field that are promising both in terms of opening up new avenues of empirical understanding of Thai politics and reframing what constitutes environmental politics in the first place.

First in terms of gender, Pinkaew points out that the gendered implications of the historical narrative of Thai environmental politics largely ignores the role of women and, more potently, is framed within a gendered binary that is as fundamental to understanding the relationship between nature and culture in Thailand introduced at the start of this chapter. She argues that the forest conservation movement is "exclusively a male story" that positions male conservationists as heroically protecting the forces of wild nature.[66] This framing extends middle-class biases related to the protection of pristine nature while partitioning other issues, like damming, as contaminated with the influences of people.[67] She argues that this gendering of environmental politics accounts for the ways in which certain struggles get prioritised within the hearts and minds of the Bangkok-based middle classes.[68] Others, like Lamb, argue that across Southeast Asia, questioning dispossession and access from a gendered perspective opens up new insights into the implications of the region's changing land politics.[69] Here the material consequences of ongoing land grabs extend and entail different kinds of politics.

Second, urbanisation has been a constant part of Thailand's environmental story, but urban environmental politics are often partitioned from larger environmental struggles,[70] while there have been a number of precursor studies that have engaged the urban environment.[71] As Forsyth notes, "brown environmental" movements organised around industrial pollution occupy an "uncertain place in Thai environmental politics".[72] Other examinations of urban struggles over housing, urban sustainability, and land have revealed both the inherent tensions in framings of sustainability and the ways in which projects that blend urban housing and environmental issues are implicated in ongoing struggles over rights to the city and dispossession.[73] Herzfeld points out the way in which different visions of beauty, space, and heritage are ensnared in struggles over housing.[74]

The city is also ripe for exploring the ongoing effects of environmental disasters. Marks's critical evaluation of the post-2011 flood politics in Bangkok demonstrates the uneven distributions of the effects of the disaster and the uneven recovery after the floodwaters receded.[75] The 2011 Thai floods serve as a "model event" for thinking about post-climate-change futures. In order to understand emerging future-oriented imaginaries, it is necessary to engage with these present machinations on the ground. Indeed, climate change itself is already structuring new kinds of environmental politics.[76] These questions, alongside questions related to damming and the scale of the region's environmental shifts, ask scholars to not only consider environmental politics within Thai borders but the ways in which struggles over the environment now play out across borders and via multinational institutions.[77]

Conclusion

On 4 February 2018, the CEO of the Ital-Thai construction firm was found in the Thungyai Naresuan Wildlife Sanctuary surrounded by firearms and flayed carcasses of deer, game birds, and one rare Indochinese black leopard. The cat was perhaps one of the last 99 in the sanctuary. The public response condemning the mogul was swift and brutal. Images condemning the illegal poaching went viral. Students at Chulalongkorn University gathered, clad in panther masks, to demand justice for the animal. Images of panthers suddenly appeared across social media, on walls, street signs, and even the construction barricades blocking off the ubiquitous Ital-Thai construction sites. The event carries with it the echoes of the poaching event in 1973 that helped usher in that brief but critical moment of democratic governance in Thai history as well as other key environmental mobilisations around conservation and just who retains access to resources in protected areas.

In condensed fashion, the event seemed to encapsulate both how much and how little has changed in Thailand across the last 30 years. As we have reviewed, environmental politics have had important effects across time. Yet, the poaching demonstrates the continued pernicious effects of political inequality and impunity on the environment. It also reveals the ways environmental movements invoke both the possibilities of mass democratic mobilisation while also highlighting the tensions in these movements. Since the 2014 coup, environmental movements have often sparked mobilisations that appear to make the Thai military government's vulnerabilities visible, and some have taken aim at the law and in the courts. In doing so, they extend the both the promise of democratisation that seems to have been embedded in the last half decade of environmental politics (and the scholarly work that attempts to understand it) and the ambiguities of such politics. Environmentalism, as described here, is a moving object and new discourses and practices reframe its politics in different ways while, at times, simultaneously invoking the past. Whatever the future holds for Thailand's national politics, these recent events make clear that environment is both a crucial object of contestation and a source of inspiration and complexity that demands further engagement as we attempt to understand these changing political trajectories.

Notes

1 The 1970s offer a rather conventional starting point for histories of environmental movements in Thailand, but as Peter Vandergeest and Nancy Peluso ("Territorialisaiton and State Power in Thailand", *Theory and Society*, Vol. 24, No. 3 [1995]; "Political Ecologies of War and Forests: Counterinsurgencies and the Making of National Natures", *Annals of the Association of American Geographers*, Vol. 101, No. 3 [2011]) demonstrate, environmental struggles have been fundamental to the unfolding of the national order – often appearing in the form of struggles over territorial control – since the turn of the 20th century. Larry Lohmann ("Visitors to the Commons: Approaching Thailand's "Environmental Struggles from a Western Starting Point", in *Ecological Resistance Movements: The Global Emergence of Radical and Popular Environmentalism*, edited by B.R. Taylor [Albany: State University of New York Press, 1994]) argues that environmental politics in Thailand has been deeply shaped by Western discourses about the relationship between nature and culture, partitioning what counts as an environmental struggle and what does not. Pinkaew Luangaramsri ("Thailand: Whither gender in the environmental movement?" In *Routledge Handbook of the Environment in Southeast Asia.* Edited by P. Hirsch [London: Routledge, 2016, pp. 470–482].) emphasises the ways this conventional telling extends a rather conventional gendered history of environmental politics.

2 There have been several similar reviews over the last three decade (for example, Philip Hirsch, *Political Economy of the Environment in Asia* [Manila: Journal of Contemporary Asia Publishers, 1993]; Philip Hirsch "The politics of environment: Opposition and legitimacy." In *Political Change in Thailand: Democracy and Participation.* Edited by K. Hewison [London: Routledge, 1997, pp. 179–194]); Tim Forsyth, "Environmental Social Movements in Thailand: A Critical Assessment", in *Asian Review 2002: Popular Movements* [Bangkok: Institute of Asian Studies: Chulalongkorn University, 2002]; Tim Forsyth "Are Environmental Movements Socially Exclusive: An Historical Study from Thailand." *World Development.* Vol. 35. No. 12 (2007), pp. 2110–2130; Tim Forsyth. "Environmentalism." In *Handbook of the Environment in Southeast Asia.* [London: Routledge, 2016, pp. 69–81]; Tim Forsyth and Andrew Walker, *Forest Guardians, Forest Destroyers: The Politics of Environmental Knowledge in Northern Thailand* [Chiang Mai: Silkworm Books, 2008]).

3 Kevin F.F. Quigley, "Towards Consolidating Democracy: The Paradoxical Role of Democracy Groups in Thailand", *Democratization,* Vol. 3, No. 3 (1996), pp. 264–286.

4 Vandergeest and Nancy, "Territorialisation and State Power in Thailand", pp. 385–426.

5 Ibid., pp. 391 and 413.

6 Vandergeest and Peluso, "Political Ecologies of War and Forests: Counterinsurgencies and the Making of National Natures", pp. 587–608.

7 Thongchai Winichakul, *Siam Mapped: The History of the Geo-Body of a Nation* (Honolulu: University of Hawaii Press, 1994).

8 Derek Hall, Philip Hirsch and Tania Murray Li, *Powers of Exclusion: Land Dilemmas in Southeast Asia* (Singapore: NUS Press, 2011).

9 Philip Stott, "Mu'ang and Pa: Elite Views of Nature in Changing Thailand", in *Thai Constructions of Knowledge,* edited by M. Chitakasem and A. Turton (Singapore: ISEAS Press, 1991), pp. 145–146.

10 Stott is clear that this description is uses a Lanna-based model which, though reflected in other spatial patterns in Ayutthaya and Khorat, for example, is not the same as spatial patterns in those areas either.

11 See Richard Davis, "Tolerance and Intolerance of Ambiguity in Northern Thai Myth and Ritual", *Ethnology,* Vol. 13, No. 4 (1974), pp. 341–356.

12 Kamala Tiyavanich, *The Buddha in the Jungle* (Seattle: University of Washington Press, 2003).

13 Ibid.

14 Hirsch, *Political Economy of the Environment in Asia.*

15 Jonathan Rigg and Philip Stott, "Forest Tales: Politics, Policy Making, and the Environment in Thailand", in *Ecological Policy and Politics in Developing Countries: Economic Growth, Democracy, and Environment,* edited by U. Desai (Albany, NY: State University of New York Press, 1998).

16 See Pasuk Phongpaichai and Chris Baker, *A History of Thailand* (Cambridge: Cambridge University Press, 2005), p. 201. Current official population projections suggest the population of the city may top 20 million by 2020 <www.100resilientcities.org/wp-content/uploads/2017/07/Bangkok_-_Resilience_Strategy.pdf>.

17 Hirsch, *Political Economy of the Environment in Asia.* Also, Dhira Phantumvanit and Winai Liengcharernsit, "Coming to Terms with Bangkok's Environmental Problems", *Environment and Urbanisation,* Vol. 1, No. 1 (1989), pp. 31–39.

18 Philip Hirsch and Larry Lohmann, "Contemporary Politics of the Environment in Thailand", *Asian Survey*, Vol. 29, No. 4 (1989), pp. 439–451.
19 Oliver Pye, *Khor Jor Kor: Forest Politics in Thailand* (Bangkok: White Lotus Press, 2005), p. 62.
20 Ibid. Also Tyrell Haberkorn, *Revolution Interrupted; Farmers, Students, Law, and Violence in Northern Thailand* (Madison: University of Wisconsin Press, 2011).
21 Hirsch and Lohmann, "Contemporary Politics of the Environment in Thailand", pp. 442–443.
22 Tim Forsyth, "Are Environmental Social Movements Socially Exclusive: An Historical Study from Thailand", *World Development*, Vol. 35, No. 12 (2007), pp. 2110–2130.
23 Haberkorn, *Revolution Interrupted.*
24 Forsyth, "Are Environmental Social Movements Socially Exclusive".
25 Hirsch, *Political Economy of the Environment in Asia.*
26 Forsyth, "Are Environmental Social Movements Socially Exclusive". Also, Tim Forsyth, "Environmental Social Movements in Thailand: A Critical Assessment", *Asian Review 2002: Popular Movements* (Bangkok: Institute of Asian Studies: Chulalongkorn University, 2002).
27 Jonathan Rigg and Albert Salamanca, "Connecting Lives, Living, and Location: Mobility and Spatial Signatures in Northeast Thailand, 1982–2009", *Critical Asian Studies*. Vol. 43, No. 11 (2011).
28 Peter Vandergeest, "Real Villages: National Narratives of Rural Development", in *Creating the Countryside: The Politics of Rural and Environmental Discourse,* edited by M. Dupuis and P. Vandergeest (Philadelphia: Temple University Press, 1999).
29 This reading reflects Raymond Williams's, "Ideas of Nature", in *Problems in Materialism and Culture* (London: Verso, 1980, pp. 67–85).) now classic reading of "the idea of nature" as only understandable with the broader expansion of the classificatory sciences and the growth of industrial capitalism. Williams's work, which mainly focuses on 18th- and 19th-century Europe, offers a useful point of reflection for scholars attempting to consider the ways in which a genealogy.
30 Hall, Hirsch and Li, *Powers of Exclusion*, p. 27.
31 Anan Ganjanapan, "The Northern Thai Land Tenure System: Local Custom versus National Laws", *Law and Society Review,* Vol. 28, No. 3 (1994), p. 614.
32 Claudio Delang, "The Political Ecology of Deforestation in Thailand", *Geography*, Vol. 90, No. 3 (2006), p. 230.
33 Peter Vandergeest, "Racialisation and Citizenship in Thai Forest Politics", *Society & Natural Resources,* Vol. 16, No. 1 (2003), p. 24.
34 Pinkaew Luangaramsri, "The Ambiguity of "Watershed": The Politics of People and Conservation in Northern Thailand", *Sojourn*, Vol. 15, No. 1 (2000).
35 Delang, "The Political Ecology of Deforestation in Thailand", pp. 232–233.
36 Vandergeest, "Racialisation and Citizenship in Thai Forest Politics", p. 26.
37 Andrew Walker and Nicholas Farrelly, "Northern Thailand's Spectre of Eviction", *Critical Asian Studies,* Vol. 40, No. 3 (2008), p. 393.
38 Delang, "The Political Ecology of Deforestation in Thailand". Also Pasuk and Baker, *A History of Thailand,* p. 81.
39 Ann Danaiya Usher, *Thai Forestry: A Critical History* (Silkworm Books, 2009).
40 Robin Roth, "'Fixing' the Forest: The Spatiality of Conservation Conflict in Thailand", *Annals of the Association of American Geographers,* Vol. 98, No. 2 (2008).
41 Hall, Hirsch and Li, *Powers of Exclusion.*
42 Tim Forsyth and Anderw Walker. *Forest Guardians, Forest Destroyers: The Politics of Environmental Knowledge in Northern Thailand* (Chiang Mai: Silkworm Books, 2008).
43 Ibid., pp. 114–116.
44 Pinkaew, "The Ambiguity of 'Watershed'", p. 54.
45 Pye, *Khor Jor Kor: Forest Politics in Thailand,* pp. 109–115.
46 Ibid., p. 119.
47 Ibid., p. 130.
48 Ibid., pp. 201–205.
49 Philip Hirsch, "The Changing Political Dynamics of Dam Building on the Mekong", *Water Alternatives,* Vol. 3, No. 2 (2010), p. 314.
50 Vandergeest and Peluso, "Political Ecologies of War and Forests".
51 Rigg argues that opposition to the project coalesced around three different arguments: The first argument leveled by environmentalists was that the dam would divide the biodiverse zone that bound together Thung Yai Narasuan and Hua Kha Khaeng sanctuaries cutting off migration routes and

providing new access points for illegal loggers and poachers. The second argument was that the dam's overall benefits were misestimated. Third, opponents suggested that the project was unnecessarily detrimental to both local health and cultural heritage. See Jonathan Rigg, "Thailand's Nam Choan Dam Project: A Case Study in the 'Greening' of Southeast Asia", *Global Ecology and Biogeography Letters,* Vol. 1, No. 2 (1991), p. 46.

52 Ibid., pp. 46–51.

53 Ibid. Also see Hirsch, *Political Economy of the Environment in Asia.* And Stott, "Mu'ang and Pa: Elite Views of Nature in Changing Thailand".

54 Francois Molle et al., "The 'Greening of Isaan': Politics, Ideology, and Irrigation Development in Northeast of Thailand", in *Contested Waterscapes in the Mekong Region: Hydropower, livelihoods, and Governance,* edited by F. Molle, T. Foran and M. Käkönen (London: Earthscan, 2009), p. 260.

55 See SEARIN, *The Mun River, the Return of Fishermen: The Findings of Thai Baan Research at Pak Mun Dam* (in Thai), 2002 <www.livingriversiam.org/5pub/print.html>.

56 The Southeast Asia Rivers Network (SEARIN) has extensive documentation (in Thai) of the Thai Baan village research projects on its website <www.livingriversiam.org/5pub/print.html>.

57 Forsyth, "Are Environmental Social Movements Socially Exclusive".

58 Bruce Missingham, "The Village of the Poor Confronts the State: A Geography of Protest in the Assembly of the Poor", *Urban Studies,* Vol. 39, No. 9 (2002); Bruce Missingham, *The Assembly of the Poor: From Local Struggles to National Protest Movement* (Chiang Mai: Silkworm Books, 2003).

59 James Glassman, "From Seattle (and Ubon) to Bangkok: The Scales of Resistance to Corporate Globalisation", *Environment and Planning D: Society and Space,* Vol. 20, No. 5 (2001).

60 Chris Sneddon, "Struggle over Dams as Struggles for Justice: The World Commission on Dams (WCD) and Anti Dam Campaigns in Thailand and Mozambique", *Society and Natural Resources,* Vol. 21, No. 7 (2008).

61 Glassman, "From Seattle (and Ubon) to Bangkok", p. 524.

62 Philip Hisrch, "Thailand's Colliding Mountains and Conflicting Values", *East Asia Forum,* 2013 <www.eastasiaforum.org/2013/10/15/thailands-colliding-mountains-and-conflicting-values/#more-38293> (accessed 31 May 2018).

63 Ian Baird and Noah Quastel, "Re-scaling and Reordering Nature-society Relations: The Nam Thuen 2 Hydropower Dam and Laos-Thailand Electricity Networks", *Annals of the Association of American Geographers,* Vol. 105, No. 6 (2015).

64 Jeremy Allouche, Carl Middleton and Dipak Gyawali, "Technical Veil, Hidden Politics: Interrogating the Power Linkages behind the Nexus", *Water Alternatives,* Vol. 8, No. 1 (2015). And, Jakkrit Sangkhamanee, "From Pak Mun to Xayaburi: The Backwater and Spillover of Thailand's Hydropower Politics", in *Hydropower Development in the Mekong Region: Political, Socio-economic, and Environmental Perspectives,* edited by N. Matthes and K. Geheb (Abingdon: Routledge, 2015).

65 Vanessa Lamb, "Where Is the Border? Villagers, Environmental Consultants, and the 'Work' of the Thai-Burma Border", *Political Geography,* Vol. 40 (2014); Vanessa Lamb "Whose Border? Border Talk and Discursive Governance of the Salween River-Border", in *Placing the Border in Everyday Life,* edited by C. Johnson and R. Jones (London: Routledge, 2016).

66 Pinkaew Luangaramsri, "Thailand: Whither Gender in the Environmental Movement?" in *Routledge Handbook of the Environment in Southeast Asia,* edited by P. Hirsch (London: Routledge, 2016), p. 477.

67 Vanessa Lamb, "Who Knows the River? Gender, Expertise, and the Politics of Local Ecological Knowledge Production of the Salween River", *Gender, Place & Culture,* 2018 <DOI: 10.1080/0966369X.2018.1481018>.

68 Kyoko Kusakabe, "Gender Mainstreaming in Government Offices in Thailand, Cambodia, and Laos: Perspectives from Below", *Gender and Development,* Vol. 13, No. 2 (2005).

69 Vanessa Lamb et al., *Report on Gender and Land Access in Lower Mekong Basin Countries* (Phnom Penh: Oxfam 2015).

70 Tim Forsyth, "Industrial Pollution and Social Movements in Thailand", in *Liberation Ecologies: Environment, Development, Social Movements,* edited by R. Peet and M. Watts (London: Routledge, 2004).

71 Tim Forsyth, "Industrial Pollution and Government Policy in Thailand: Rhetoric Versus Reality", in *Seeing the Forest for the Trees: Environment and Environmentalism in Thailand,* edited by P. Hirsch (Chiang Mai: Silkworm, 1997); Dhira and Winai, "Coming to Terms with Bangkok's Environmental Problems"; Helen Ross, "Bangkok's Environmental Problems: Stakeholders and Avenues for Change", in *Seeing the Forest for the Trees: Environment and Environmentalism in Thailand,* edited by P. Hirsch (Chiang Mai: Silkworm, 1997); Charles Greenberg, "The Varied Responses to an Environmental Crisis in the Extended

Bangkok Metropolitan Region", in *Seeing the Forest for the Trees: Environment and Environmentalism in Thailand,* edited by P. Hirsch (Chiang Mai: Silkworm, 1997); Amrita Daniere and Lois Takahashi, "Poverty and Access: Differences and Commonalities Across Slum Communities in Bangkok", *Habitat International,* Vol. 23, No. 2 (1999).

72 Forsyth, "Industrial Pollution and Social Movements in Thailand", p. 423.

73 There is a rather extensive body of literature on housing, slum politics, and slum improvement in Thailand, but it is rarely framed around the question of the environment per se. The bifurcation here between studies of housing insecurity and environmental politics is revealing of the deeper epistemological schism between nature and culture that shapes inquiries into environmental change in Thailand as well. See Eli Elinoff, "Sufficient Citizens: Moderation and the Politics of Sustainable Development in Thailand", *PoLAR: Political and Legal Anthropology Review,* Vol. 37, No. 1 (2014); Eli Elinoff, "No Future Here: Urbanisation and Hope in Thailand", in *Urban Asia: Essays on Futurity Past and Present,* edited T. Bunnell and D. Goh (Berlin: Verlag/Jovis, 2018).

74 Michael Herzfeld, *Siege of the Spirits: Community and Polity in Bangkok* (Chicago: University of Chicago Press, 2016).

75 D. Marks, "The Urban Political Ecology of the 2011 Floods in Bangkok: The Creation of Uneven Vulnerabilities", *Pacific Affairs,* Vol. 88, No. 3 (2015).

76 Ibid.

77 Jakkrit, "From Pak Mun to Xayaburi". Also, John Dore and Kate Lazarus, "De-marginalising the Mekong River Commission", in *Contested Waterscapes in the Mekong Region: Hydropower, Livelihoods, and Governance,* edited by F. Molle, T. Foran, and M. Käkönen (London: Earthscan, 2009).

References

Allouche, Jeremy, Carl Middleton, and Dipak Gyawali. 2015. "Technical Veil, Hidden Politics: Interrogating the Power Linkages behind the Nexus". *Water Alternatives,* Vol. 8, No. 1: 610–626.

Anan Ganjanapan. 1994. "The Northern Thai Land Tenure System: Local Custom versus National Laws". *Law and Society Review,* Vol. 28, No. 3: 609–622.

Baird, Ian and Noah Quastel. 2015. "Re-scaling and Reordering Nature-society Relations: The Nam Thuen 2 Hydropower Dam and Laos-Thailand Electricity Networks". *Annals of the Association of American Geographers,* Vol. 105, No. 6: 1212–1239.

Baker, Chris. 2000. "Thailand's Assembly of the Poor: Background, Drama, Reaction". *South East Asia Research,* Vol. 8, No. 1: 5–29.

Daniere, Amrita and Lois Takahashi. 1999. "Poverty and Access: Differences and Commonalities across Slum Communities in Bangkok". *Habitat International,* Vol. 23, No. 2: 271–228.

Davis, Richard. 1974. "Tolerance and Intolerance of Ambiguity in Northern Thai Myth and Ritual". *Ethnology,* Vol. XIII, No. 4: 341–356.

Delang, Claudio. 2006. "The Political Ecology of Deforestation in Thailand". *Geography,* Vol. 90, No. 3: 225–237.

Dhira Phantumvanit and Winai Liengcharernsit. 1989. "Coming to Terms with Bangkok's Environmental Problems". *Environment and Urbanization,* Vol. 1, No. 1: 31–39.

Dore, John and Kate Lazarus. 2009. "De-marginalising the Mekong River Commission". In *Contested Waterscapes in the Mekong Region: Hydropower, Livelihoods, and Governance.* Edited by F. Molle, T. Foran, and M. Käkönen. London: Earthscan, pp. 357–382.

Elinoff, Eli. 2014. "Sufficient Citizens: Moderation and the Politics of Sustainable Development in Thailand". *PoLAR: Political and Legal Anthropology Review,* Vol. 37, No. 1: 89–108.

———. 2018. "No Future Here: Urbanisation and Hope in Thailand". In *Urban Asia: Essays on Futurity Past and Present.* Edited by T. Bunnell and D. Goh. Berlin: Verlag/Jovis, pp. 121–134.

Forsyth, Tim. 1997. "Industrial Pollution and Government Policy in Thailand: Rhetoric versus Reality". In *Seeing the Forest for the Trees: Environment and Environmentalism in Thailand.* Edited by P. Hirsch. Chiang Mai: Silkworm, pp. 182–201.

———. 2002. "Environmental Social Movements in Thailand: A Critical Assessment". *Asian Review 2002: Popular Movements.* Bangkok: Institute of Asian Studies: Chulalongkorn University, pp. 104–125.

———. 2004 "Industrial Pollution and Social Movements in Thailand". In *Liberation Ecologies: Environment, Development, Social Movements.* Edited by R. Peet and M. Watts. London: Routledge, pp. 422–438.

———. 2007. "Are Environmental Social Movements Socially Exclusive: An Historical Study from Thailand". *World Development.* Vol. 35, No. 12: 2110–2130.

———. 2016. "Environmentalism". In *Routledge Handbook of the Environment in Southeast Asia*. Edited by P. Hirsch. London: Routledge, pp. 69–81.

Forsyth, Tim and Walker, Anderw. 2008 *Forest Guardians, Forest Destroyers: The Politics of Environmental Knowledge in Northern Thailand*. Chiang Mai: Silkworm Books.

Glassman, James. 2001. "From Seattle (and Ubon) to Bangkok: The Scales of Resistance to Corporate Globalisation". *Environment and Planning D: Society and Space*. Vol. 20, No. 5: 513–533.

Greenberg, Charles. 1997. "The Varied Responses to an Environmental Crisis in the Extended Bangkok Metropolitan Region". In *Seeing the Forest for the Trees: Environment and Environmentalism in Thailand*. Edited by P. Hirsch. Chiang Mai: Silkworm, pp. 167–181.

Haberkorn, Tyrell. 2011. *Revolution Interrupted; Farmers, Students, Law, and Violence in Northern Thailand*. Madison: University of Wisconsin Press.

Hall, Derek., Philip Hirsch, and Tania Murray Li. 2011 *Powers of Exclusion: Land Dilemmas in Southeast Asia*. Singapore: NUS Press.

Herzfeld, Michael. 2016. *Siege of the Spirits: Community and Polity in Bangkok*. Chicago: University of Chicago Press.

Hirsch, Philip. 1993. *Political Economy of the Environment in Asia*. Manila: Journal of Contemporary Asia Publishers.

———. 2010. "The Changing Political Dynamics of Dam Building on the Mekong". *Water Alternatives*. Vol. 3, No. 2: 312–323.

———. 2013. "Thailand's Colliding Mountains and Conflicting Values". *East Asia Forum* <www.eastasiaforum.org/2013/10/15/thailands-colliding-mountains-and-conflicting-values/#more-38293> (accessed 31 May 2018).

Hirsch, Philip and Larry Lohmann. 1989. "Contemporary Politics of the Environment in Thailand". *Asian Survey*. Vol. 29, No. 4: 439–451.

Isager, Lotte and Soren Ivarsson. 2002 "Contesting Landscapes in Thailand: Tree Ordination as Counter-territorialisation". *Critical Asian Studies*, Vol. 34, No. 3: 395–417.

Jakkrit Sangkhamanee. 2015. "From Pak Mun to Xayaburi: The Backwater and Spillover of Thailand's Hydropower Politics". In *Hydropower Development in the Mekong Region: Political, Socio-economic, and Environmental Perspectives*. Edited by N. Matthes and K. Geheb. Abingdon: Routledge, pp. 83–100.

Kamala Tiyavanich. 2003. *The Buddha in the Jungle*. Seattle: University of Washington Press.

Kusakabe, Kyoko. 2005 "Gender Mainstreaming in Government Offices in Thailand, Cambodia, and Laos: Perspectives from Below". *Gender and Development*. Vol. 13, No. 2: 46–56.

Lamb, Vanessa. 2014. "Where is the Border? Villagers, Environmental Consultants, and the 'Work' of the Thai-Burma Border". *Political Geography*. Vol. 40: 1–12.

———. 2016. "Whose Border? Border Talk and Discursive Governance of the Salween River-Border". In *Placing the Border in Everyday Life*. Edited by C. Johnson and R. Jones. London: Routledge, pp. 117–136.

———. (Ed.). 2015. *Report on Gender and Land Access in Lower Mekong Basin Countries*. Oxfam America.

———. 2018 Who Knows the River? Gender, Expertise, and the Politics of Local Ecological Knowledge Production of the Salween River. *Gender, Place & Culture* <DOI: 10.1080/0966369X.2018.1481018>.

Lee, Lily Xiao Hong, Alvin Y. So, and Lee F. Yok-Shiu. 1999. "Environmental Movements in Thailand". In *Asia's Environmental Movements: Comparative Perspectives*. Edited Y.F. Lee and A.Y. So. Armonk: M.E. Sharpe.

Lohmann, Larry. 1994. "Visitors to the Commons: Approaching Thailand's "Environmental Struggles from a Western Starting Point". In *Ecological Resistance Movements: The Global Emergence of Radical and Popular Environmentalism*. Edited by B.R. Taylor. Albany: State University of New York Press, pp. 109–126.

Marks, D. 2015. "The Urban Political Ecology of the 2011 Floods in Bangkok: The Creation of Uneven Vulnerabilities". *Pacific Affairs*, Vol. 88, No. 3.

Middleton, Carl and John Dore. 2015. "Transboundary Water and Electricity Governance in Mainland Southeast Asia: Linkages, Disjunctures, and Implications". *International Journal of Water Governance*. Vol. 3, No. 1: 93–120.

Missingham, Bruce. 2002. "The Village of the Poor Confronts the State: A Geography of Protest in the Assembly of the Poor". *Urban Studies*. Vol. 39, No. 9: 1647–1663.

———. 2003. *The Assembly of the Poor: From Local Struggles to National Protest Movement*. Chiang Mai: Silkworm Books.

Molle, Francois et al. 2009. "The 'Greening of Isaan": Politics, Ideology, and Irrigation Development in Northeast of Thailand". In *Contested Waterscapes in the Mekong Region: Hydropower, Livelihoods, and Governance*. Edited by F. Molle, T. Foran and M. Käkönen. London: Earthscan, pp. 253–282.

Pasuk Phongpaichai and Chris Baker. 2005. *A History of Thailand*. Cambridge: Cambridge University Press.

Pianporn Deets. 2018. *Trans-boundary Rivers and Justice in Relation to the Salween Dams and Thailand's Water Diversion Plans*. Presentation at the Salween Studies Research Workshop Yangon University.

Pinkaew Luangaramsri. 2000. "The Ambiguity of "Watershed": The politics of People and Conservation in Northern Thailand". *Sojourn*. Vol. 15, No. 1: 52–75.

———. 2016. "Thailand: Whither Gender in the Environmental Movement?" *Routledge Handbook of the Environment in Southeast Asia*. Edited by P. Hirsch. London: Routledge, pp. 470–482.

Pye, Oliver. 2005. *Khor Jor Kor: Forest Politics in Thailand*. Bangkok: White Lotus Press.

Quigley, Kevin F.F. 1996. "Towards Consolidating Democracy: The Paradoxical Role of Democracy Groups in Thailand". *Democratization*, Vol. 3, No. 3: 264–286.

Rigg, Jonathan. 1991. "Thailand's Nam Choan Dam Project: A Case Study in the 'Greening' of South-East Asia". *Global Ecology and Biogeography Letters*, Vol. 1, No. 2: 42–54.

———. 1995. "'In the Field There is Dust': Thailand's Water Crisis". *Geography*, Vol. 80, No. 1: 23–32.

Rigg, Jonathan and Philip Stott. 1998. "Forest Tales: Politics, Policy Making, and the Environment in Thailand". In *Ecological Policy and Politics in Developing Countries: Economic Growth, Democracy, and Environment*. Edited by U. Desai. Albany, NY: State University of New York Press.

Rigg, Jonathan and Albert Salamanca. 2011. "Connecting Lives, Living, and Location: Mobility and Spatial Signatures in Northeast Thailand, 1982–2009". *Critical Asian Studies*. Vol. 43, No. 11: 551–575.

Ross, Helen. 1997 "Bangkok's Environmental Problems: Stakeholders and Avenues for Change". In *Seeing the Forest for the Trees: Environment and Environmentalism in Thailand*. Edited by P. Hirsch. Chiang Mai: Silkworm, pp. 147–166.

Roth, Robin. 2008. "'Fixing' the Forest: The Spatiality of Conservation Conflict in Thailand" *Annals of the Association of American Geographers*". Vol. 98, No. 2: 373–391.

SEARIN. 2002. *The Mun River "The return of fishermen": The Findings of Thai Baan Research at Pak Mun dam* (in Thai) <www.livingriversiam.org/5pub/print.html>.

Sneddon, Chris. 2008. "Struggle over Dams as Struggles for Justice: The World Commission on Dams (WCD) and Anti Dam Campaigns in Thailand and Mozambique". *Society and Natural Resources*. Vol. 21, No. 7: 635–640.

Stott, Philip. 1991. "Mu'ang and Pa: Elite Views of Nature in Changing Thailand". In *Thai Constructions of Knowledge*. Edited by M. Chitakasem and A. Turton. Singapore: ISEAS Press.

Thongchai Winichakul. 1994. *Siam Mapped: The History of the Geo-Body of a Nation*. Honolulu: University of Hawaii Press.

Turton, Andrew and Manas Chitakasem (Eds.). 1991. *Thai Construction of Knowledge*. London: School of Oriental and African Studies.

Usher, Ann Danaiya. 2009. *Thai Forestry: A Critical History*. Chiang Mai: Silkworm books.

Vandergeest, Peter. 1999. "Real Villages: National Narratives of Rural Development". In *Creating the Countryside: The Politics of Rural and Environmental Discourse*. Edited by M. Dupuis and P. Vandergeest. Philadelphia: Temple University Press.

———. 2003. "Racialisation and Citizenship in Thai Forest Politics". *Society & Natural Resources*. Vol. 16, No. 1: 19–37.

Vandergeest, Peter and Nancy Peluso. 1995. "Territorialisaiton and State Power in Thailand". *Theory and Society*. Vol. 24, No. 3: 385–426.

Vandergeest, Peter and Nancy Peluso. 2011 "Political Ecologies of War and Forests: Counterinsurgencies and the Making of National Natures". *Annals of the Association of American Geographers*. Vol. 101, No. 3: 587–608.

Walker, Andrew and Nicholas Farrelly. 2008 "Northern Thailand's Spectre of Eviction". *Critical Asian Studies*. Vol. 40, No. 3: 373–397.

PART IV

The international relations

31
THAILAND'S FOREIGN POLICY

Arne Kislenko

A Siamese proverb likens foreign policy to the "bamboo in the wind": solidly rooted but able to bend in any direction to survive.[1] Embedded in Thai culture and Buddhism, the notion pivots on Thailand's success guarding its independence from greater powers. With the "bamboo" precept in mind, a particular historical narrative and Thai identity emerge. The Kingdom of Siam is celebrated for avoiding European colonisation – a fate unique in Southeast Asia. Thailand's alliance with Japan during the Second World War is explained as a "necessary evil". And Bangkok's close Cold War partnership with the United States is seen as a better alternative to communist domination.

But this paradigm has its limitations. While Thailand avoided colonisation, it was not unaffected by it. Nor was foreign policy always so principled. Rather, a complex blend of prudence, pragmatism, and opportunism make up Thailand's international relations history. And although it avoided the struggles that plagued Southeast Asia in the 20th century, it did not emerge unscathed by regional events. In fact, Thailand's foreign policy was, and is, indelibly shaped by global actors and external conflict. In this light, some scholars argue that the "bamboo" notion is somewhat of a national myth: the foundation of a Thai identity carefully scripted over time by political elites. Thongchai Winichakul contends that the idea Thailand was never colonised is a persistent but inaccurate dogma.[2] The perception that it was always successful adapting to changes, and that interaction with Western countries was welcomed and positive, is misleading. There was violent conflict. Western influence on Thailand was far greater than often acknowledged: shaping the country through borders, ideals, and culture. In this analysis, Thailand was at least *semi*-colonised. Some also argue that the "bamboo" notion belies a darker narrative of nationalism integral to Thai identity: evidenced by contentions about "lost territories" in Indochina and Malaya. Instead, great stewardship, artful diplomacy, and emulation of the West are the celebrated hallmarks of Thai history and collectively serve as a source of political legitimacy for the monarchy and its allies.[3] Pavin Chachavalpongpun argues that Thai foreign policy has also conformed to patterns in international relations more than the "bamboo" idea assumes.[4] Its history is better characterised by inconsistencies and an "overarching pragmatism" rather than any absolute diplomatic philosophy.[5] Moreover, particularly in the contemporary, it is essential to examine domestic politics as the primary influence in external relations.[6]

Still, "bamboo" flexibility has been a cornerstone of Thai foreign policy, and Thailand did avoid formal colonisation. That success depended on a unity – preceding the nation-state – premised on Buddhism, the monarchy, and enmity with neighbours.[7] Diplomacy helped early Thai kingdoms overcome regional rivals and keep Europeans at bay.[8] But by the late 19th century Siam was seriously threatened by British and French colonial interests. French ambitions in Laos – for centuries within the Siamese orbit – brought war in 1893, forcing the kingdom to renounce its claims. British indifference revealed that diplomacy alone could not protect the country. Thus the war stimulated reforms aimed at modernisation.[9] King Chulalongkorn (Rama V, r. 1868–1910) emerged at the centre of an enduring national myth that without concessions the whole kingdom would have been lost. The notion that he "sacrificed a finger in order to save the hand" became a powerful trope legitimising the monarchy, subsequently used by royalists and nationalists to maintain authority.[10] Chulalongkorn's embrace of Western ideas helped Siam maintain independence, but it did not avoid more demands. By 1909 Siam was forced to cede additional territories in Laos, Cambodia, and the Malay states: between 1893 and 1910, a total of some 456,000 km^2, or half of the Siamese empire.[11] Serious economic problems followed. Resentment towards both Europeans and the king seethed for decades.[12]

Chulalongkorn and his successor, Vajiravudh (Rama VI, r. 1910–1925), wooed other powers, most notably Germany, to balance British and French designs. Factions within the ruling elites developed over diplomacy towards European states. When the First World War broke out, Siam was neutral, but in hoping to end up on the "right side of history" some leaders pushed for it to join. In this regard the entry of the United States in April 1917 catalysed Siam's declaration of war on the Central Powers.[13] While only 1,300 Siamese soldiers served overseas, a strong identity emerged during the war. Reforms, particularly in the military, gave rise to bureaucratic and economic elites still existent today. Orientation towards the West changed Siamese political culture, illustrated by the adoption of the *Thong Trairong* – the five-striped national flag – bearing the Allies' colours.[14] Vajiravudh knew that joining the Allies could ensure Siam's independence. Focusing on the peace conference in 1918, he told envoys to get rid of the "burdensome provisions of the extremely antiquated treaties" Siam previously accepted.[15] The Siamese delegation lobbied for an end to extraterritoriality and treaty constraints, and ultimately succeeded with crucial American support.[16]

However, the war unleashed tremendous intellectual change in Siam with ideas about democracy, anti-colonialism, and communism. Vajiravudh's rejection of constitutionalism and political reform aggravated many.[17] Some saw the new treaties as failures and harboured ambitions to regain "lost territories". Economic problems plagued the country. His successor, Prajadhipok (Rama VII, r. 1925–1932), was unable to control the tumult, and in 1932 the absolute monarchy ended with a revolution inspired by Western-educated Siamese. The overthrow ushered in a period of instability, and extremist ideologies gained currency. The militant nationalism of Japan was particularly influential. Anti-Western rhetoric struck a deep chord with many Siamese who associated the decline of their empire with European colonialism. By 1938 the country's brief experiment with constitutional democracy ended in a military coup, and Siam – renamed Thailand – headed down a perilous path.

That direction was shaped by its controversial prime minister, Field Marshal Phibun Songkhram (1938–1944 and 1948–1957), who emulated Japanese and German militarism. He pursued a dangerous foreign policy premised on Japan's rise in Asia, intensifying Thai irredentism. As the Second World War broke out, Thailand proclaimed neutrality and sought non-aggression agreements with the British and French, but following the fall of France to German forces, Phibun changed tactics. He approached the Japanese for support retaking the "lost territories". It was in many respects a Faustian pact. As Nicholas Tarling notes, Japan offered "dishonest mediation, succeeding Thailand's dishonest neutrality".[18] Through the German-controlled Vichy government, by September 1940 Japanese forces effectively controlled most of Indochina. Phibun used the context to attack French positions. Thai forces prevailed, but the Franco-Thai War of 1940–1941 was really decided by Japan in brokering negotiations. The results favoured the Thais, but they got only one-third of their territorial demands, mostly because Japan was reluctant to push the Vichy too hard. Phibun was disappointed with Japanese mediation. He knew Tokyo's support would come at a significant price in Japanese access through Thailand to attack British Malaya, Singapore, and Burma. Wary of Japanese invasion if he did not comply, Phibun secretly approached the British and Americans for support.[19] In late 1941 Japanese pressure intensified. The situation was complicated by divisions in government over the direction of Thai foreign policy. Key politicians like former Foreign Minister Pridi Banomyong opposed complicity with Japan and maintained strong personal connections with the West.

Confusion surrounds events that brought Thailand into the war in December 1941. With Phibun away from Bangkok, subordinates refused the Japanese request for transit. When the Japanese attacked, Thai troops resisted. British forces in Malaya crossing into Thailand to stop the Japanese also met Thai resistance.[20] When Phibun returned, he acquiesced to Tokyo's demands, and on 25 January 1942, Thailand declared war on the United States and Britain. Subsequent Japanese victories in Southeast Asia convinced Phibun that the alliance was in Thailand's best interests. However, scholars debate whether he actually favoured Japan or believed that supporting it was the only guarantee of Thai independence. Regardless, foreign relations split the polity. Dissenting officials set up a government in exile in Washington, led by Thai Ambassador Seni Pramoj, and they supported the underground "Free Thai" movement.[21] Fortunately for Thailand, the United States ignored the declaration of war and recognised Seni's government. Many Thais supported the resistance, particularly as Japan's war efforts soured. Phibun's gamble allowed Thailand to retake "lost territories" in Burma and Malaya, but it was a pyrrhic victory. By 1944 the Allies were on the offensive. The war brought severe shortages, high taxes, and inflation. Allied air raids clearly demonstrated that fortunes had changed. That July, following the collapse of Japan's military government, the Thai National Assembly forced Phibun from office to appease inevitable British and French demands.

American support again proved invaluable for Thailand. Washington noted that Thai forces did not fight against the Allies and that some in government never supported the war. When Japan surrendered in September 1945, Pridi, acting as regent, made Thailand's war declarations "null and void" and returned territories to the French and British. Both pushed Thailand for indemnities, but concerned about nationalist movements in Asia and opposed to the restoration of European empires, the United States pressured them to reduce demands.[22]

Thailand's luck stemmed from dramatic changes in the international order. The United States emerged in 1945 as an economic and military superpower. In the context of the rapidly developing Cold War, the United States sought regional allies in Asia as buffers against the spread of communism. Unquestionably, the long civil war in China, won in 1949 by the communists, had the greatest impact shaping American views. The strength of nationalists-cum-communists in Indochina and Malaya also alarmed Washington, leading successive administrations towards economic and military support for allies in Southeast Asia. As Matthew Phillips notes, Americans quickly came to see Thailand as an "oasis on a troubled continent": friendly, independent, and largely "democratic" in spirit.[23]

In this light, the extremism of Thailand's military government was conveniently re-interpreted in Washington. The country was seen as an anti-communist bastion and an increasingly important "forward base" in containment strategy. Political turmoil there accentuated the need for American mentorship. A series of ineffective civilian governments worried US officials, particularly as leftists gained support. Economic difficulties continued. The mysterious death of King Ananda Mahidol (Rama VIII, r. 1935–1946) added to the political drama. Amid rumours of anti-royalist plots and communist influence, the Thai military returned with tacit approval from Washington.[24] In April 1948, a coup restored an exonerated Phibun as prime minister. In September 1949, Washington provided US$75 million for allies in Asia, including Thailand, and freed up nearly US$44 million in Japanese assets owed to Bangkok. A few months later, Phibun's government recognised the US-sponsored regime in South Vietnam and was rewarded with another US$15 million. When the Korean War broke out in June 1950, Washington floated Bangkok US$10 million along with US$25 million in loans from the World Bank.[25] Military assistance followed. By 1953 US aid was more than twice the Thai budget for its military. The Central Intelligence Agency deployed some 200 advisors in the country, fully immersed in the Thai polity. Phibun's harsh restrictions on ethnic Chinese, leftists, and the media further endeared him to Washington.[26] In short, the United States had become Thailand's chief patron, and the Thais positioned themselves as trusted allies on the front lines against communist expansion.

As the Cold War intensified, military authoritarianism in Thailand was entrenched. Despite occasional concerns about its distinctly un-democratic character, successive US presidencies considered Thailand an invaluable ally and even a model for development in the "Third World". In 1954 Bangkok became headquarters of the new Southeast Asian Treaty Organisation (SEATO), a US-inspired alliance designed to protect the region from communism. Indeed, SEATO was premised on the defence of Thailand as a "first priority", demonstrating further its importance in Washington's Cold War calculations.[27] The Thai economy boomed with the influx of US monies. Large-scale infrastructure projects transformed the country with ports, airfields, highways, rail lines, and communications. US technical assistance changed many facets of everyday life. Between 1951 and 1957, Thailand received US$149 million in economic aid and US$222 million in military assistance: both fundamentally shaped the country.[28] But not all was well. Political factions threatened Phibun's rule. Ambitious generals coveted power and siphoned off US aid for themselves. Several had their own backers in Washington. Beginning in 1955, Phibun reversed some earlier policies and relaxed controls at home. Through secret missions to Beijing, he even contemplated changing Thailand's foreign policy orientation. With US blessing, a military coup toppled his government in September 1957.

The main player behind the coup was Field Marshal Sarit Thanarat, who took over as prime minister (1957–1963). He reversed Phibun's liberalisation, annulling parliament, suspending the constitution, and reinstating harsh censorship.[29] Sarit's rule cemented the political and economic interests of the military and developed an even closer relationship with Washington. Thailand became the fulcrum of both overt and clandestine American military efforts in the region, particularly in Laos, where a protracted insurgency and civil war raged. Following the tenets of the "domino theory", President Eisenhower tried to prevent its fall to the communists. That resonated with Sarit, who saw Laos as a "dagger" pointed at Thailand – a conduit for communists in Indochina to strike at the economically disadvantaged Thai northeast. In many ways a "wild frontier", it was a difficult place to foster Thai identity, and fears of an insurgency there were very real. A large ethnically Vietnamese population, many with loyalties to Hanoi, made the region more vulnerable. US military aid was directed there, developing both infrastructure and Thai counter-insurgency capabilities.[30] Part Lao and well-connected to factions there, Sarit was an important player in Laos – something he used as leverage in Washington.

However, Eisenhower's successor, John F. Kennedy, did not share the conviction that Laos was so crucial. Landlocked and small, Laos was, as a Kennedy aide put it, "not all that god-damned important".[31] In addition, Kennedy believed that even with a dramatic military commitment, countering the communist *Pathet Lao* might be very difficult. But abandoning Laos was not possible. US prestige was on the line.[32] Kennedy instead opted for international negotiations. Sarit considered that a sellout, arguing that even if a settlement could be reached, it was not tenable. Although he ultimately agreed to endorse Kennedy's position, Sarit pressed for security guarantees. From Bangkok's perspective, SEATO was a paper tiger. US covert operations in the country provided some protection but were inadequate to defend against any communist invasion. The Kennedy administration considered a formal military alliance an over-extension of US commitments in the region, particularly given priorities in South Vietnam. But a firm demonstration was required to keep Thailand a partner, especially with respect to Laos. In March 1962, during a visit to Washington by Thai Foreign Minister Thanat Khoman, the Kennedy administration issued a joint communiqué reaffirming "the independence and integrity of Thailand as vital to the national interest of the United States and to world peace".[33] Dubbed the Rusk-Thanat Agreement, the statement made it clear that US commitments to Thailand went beyond SEATO. What exactly that entailed remained unclear.

Two months later, *Pathet Lao* forces overran the town of Nam Tha in northwestern Laos. Fearing a communist attack on Thailand, Sarit mobilised the army and threatened to invade Laos. To prevent escalation, Kennedy sent US troops into Thailand to warn the communists, reassure the Thais, and influence negotiations on Laos taking place in Geneva.[34] It was a significant demonstration of American resolve. On 18 May 1962, 6,500 US Marines landed in Thailand – the first overt deployment of US combat soldiers in Southeast Asia since the Second World War, and the first time foreign troops entered Thailand without an invasion.[35] The deployment laid the groundwork for an expanded US presence in Thailand and military engagement in Vietnam, but Kennedy had no intention of keeping troops there permanently. After the final declaration on Laos in July 1962, he pulled the marines out, assuring Sarit that congressional pressure, not a lack of commitment, was the main reason. Sarit was upset and refused to sign the agreement. He eventually acquiesced with reassurances of more assistance from Washington, much of it focused on the northeast. While Thai officials frequently over-estimated its strength to exact more US aid, the insurgency there did represent a threat. By the 1970s, there were approximately 10,000 insurgents in the northeast supported by the *Pathet Lao* and North Vietnam. Sarit and his successor, Field Marshal Thanom Kittikachorn (1963–1973), devised strategies to deal with communist influence by improving economic conditions and fostering a greater sense of Thai identity.[36]

Based on a "despotic paternalism", that identity reinforced the image of the military as "defenders" of a uniquely Thai culture. By carefully promoting the monarchy, particularly the Western-educated King Bhumibol Adulyadej (Rama IX, r. 1946–2016), Sarit cultivated the idea that Thailand was "civilised" and had common ground with "great nations" of the world. That dovetailed with US propaganda efforts to portray Thailand as a kind of Shangri-La: a relative paradise in a troubled region. As Phillips notes, the convergence of US and Thai efforts produced a unique relationship that was more than just political and military, "built upon a complex cultural mode that sat at the apex of America's imperial engagement in the world".[37] The legacies of that relationship ultimately transformed Thailand and remain today at the centre of Thai identity.[38] For example, the military in effect became a separate socio-economic caste, deeply rooted in the mainstream economy and more illicit enterprises. In 1963, Sarit's assets alone were worth nearly 30 per cent of the country's capital budget.[39]

In many respects, the script for these developments was written in Vietnam. As the US war there escalated, Washington and Bangkok expanded their partnership. By the mid-1960s, US officials considered Thailand an "unsinkable aircraft carrier" next to South Vietnam, America's most important regional ally. Nearly 80 per cent of US bombing campaigns against North Vietnam were orchestrated out of Thailand. At the peak of bombing there were 500 US aircraft stationed on Thai soil, flying an average 400 sorties a day, making Thai air space one the most congested on the planet.[40] By 1970, some 140,000 American combat and support troops rotated through the country, making Thailand a major "rest and recreation" destination and an important logistical base for communications and transportation. Moreover, nearly 11,000 Thais, comprising 15 per cent of the country's armed forces, saw combat in Vietnam. Some 23,000 Thai Special Forces fought in Laos, making up the bulk of "secret armies" there.[41] Far from being mercenaries in their support of US military efforts in the region, the Thais risked a great deal. The United States brought massive social, economic, and cultural changes – not all of them good. The military government depended on the United States for its legitimacy. The withdrawal from Vietnam therefore threatened Thailand's stability, particularly following communist victories in Indochina by 1975. A large, affluent Chinese community in Thailand added to fears that Beijing might dominate the region. As Seni warned, "We have let the US forces use our country to bomb Hanoi. When the Americans go away, they won't take that little bit of history with them".[42]

Defeat in Vietnam and the reorientation of US policy towards China ushered in a new era for Thailand. Public discontent with military rule swelled, forcing Thanom's regime to suddenly collapse in 1973. Successor civilian governments tried to distance themselves from the United States and seek a rapprochement with Beijing.[43] When Thanom returned from exile in October 1976, violent protests prompted the military to once again seize power and crack down on dissent.[44] Foreign policy remained surprisingly consistent. Diplomatic overtures to China continued but relations with the United States remained strong. Thailand's economic development continued while independence was preserved, despite communist victories in Indochina.

Relations within the region were, however, decisively strained. Led by the mercurial Norodom Sihanouk, for decades Cambodia was perpetually at odds with Thailand. Bangkok routinely denounced US attempts to woo Sihanouk, convinced he was a communist dupe. The horrors of the communist Khmer Rouge government after 1975 produced a flood of refugees in Thailand, while the Vietnamese invasion of Cambodia in 1978 put one of the world's largest militaries directly on Thai borders. In a bizarre diplomatic arrangement, Thailand found itself in collusion with both China and the United States supporting the Khmer Rouge against Vietnam. Cambodia's agony afforded Thailand a better relationship with Beijing and some protection against potential Vietnamese threats. It also gave Thai elites an opportunity to exploit Cambodia's natural resources.[45] Another coup in Thailand in 1991 shifted the ideological direction of Thai foreign policy towards engagement with Vietnam, again following the lead of both the United States and China. Thailand helped reintegrate its neighbour into the regional economy and sponsored Cambodia for admission to the Association of Southeast Asian Nations (ASEAN) in 1999.

The same engagement applied to Laos. Sino-American détente and Sino-Vietnamese conflict dissipated the threat of a communist invasion of Thailand from there. Isolated and impoverished, Laos remained, as one US diplomat said, the "end of nowhere".[46] But beginning in 1989, Thailand's prime minister, General Chatichai Choonhavan (1988–1991), "welcomed" Indochina back into the Southeast Asian community through his *Suvarnabhumi*, or "Golden Land" concept. Chatichai envisioned Thailand as the regional development hub, determined to "turn Indochina from battlefield to marketplace".[47] Critics pointed out that Chatichai and his associates profited personally from such plans, especially through illicit businesses. They also noted that *Suvarnabhumi* was very similar to the "vassal state" system of ancient Siam, thus representing a new form of Thai colonialism. Several major projects were initiated in the early 1990s as Laos was brought into various development programmes.[48] However, the 1997 financial crisis undermined Thai investment in Laos, as did competition from China and Vietnam. Lao leaders accused Bangkok of supporting rebels opposed to their regime and resented Thai cultural influences as decadent and corrupt.[49]

Similar sentiments were evident in Burma, where Thailand has always been viewed with suspicion. Since independence in 1948, Burma has witnessed near constant revolt and insurgency, some of it supported from Thailand. Thai officials have also long been implicated in the opium trade of the "Golden Triangle" border region. But with *Suvarnabhumi* policies, economic ties with Burma became paramount. In exchange for resource concessions, Thailand withdrew support for insurgents and pushed for constructive engagement. Initially condemned for doing business with such a notorious regime, Thailand can claim to have "led the way" given the recent rush of Western nations to court Burma. Over the past two decades Thailand has promoted Burma's stability and integration with the global community, cognizant that continuing conflict there could dramatically affect its security.

But unquestionably the most significant changes in Thailand's foreign relations since the Cold War are with respect to China. The convergence of geopolitical interests between Beijing and Bangkok after 1975 precipitated closer ties, but China's dramatic modernisation and economic expansion has underpinned the relationship since. In this respect, courting China became the focal point of Thai foreign policy. With a large, affluent ethnic Chinese population, Thailand has helped broker China's economic and political integration with Southeast Asia through both bilateral approaches and multilateral agencies like ASEAN and the Asia Pacific Economic Cooperation. This strategic pivot is consistent with the "bamboo" precept given China's growing international influence. It is also shaped by historical and cultural considerations "since ancient times", as Chinese Premier Wen Jiabao noted when he visited Thailand in 2012.[50]

Conversely, relations between Thailand and the United States have waned. Indeed, some imagine them to be in "terminal decline", considering that as of 2016 nearly 60 per cent of Thai military spending was in China.[51] Thailand's decreased dependency on the United States can be explained in part by the rehabilitation of ASEAN as a regional security and economic forum.[52] But neglect is also behind the decline. Thailand was almost entirely ignored during the presidency of George W. Bush. Perceptions of the United States soured in Thailand over wars in Afghanistan and Iraq. Barack Obama tried to reverse that decline, in part to check Chinese influence, but Thai leaders are wary of Washington. Business with Beijing is better, and perceptions of China are generally positive.[53] Moreover, in contrast to the United States, Chinese leaders avoid commenting on Thailand's domestic affairs.

Contemporary Thai foreign policy is extremely sensitive to the external environment, so much so that its bent "bamboo" may be better described as "swirling in the wind".[54] But in many respects, the greatest challenge for Thai foreign policy is at home. Domestic politics have always influenced foreign relations, but of late the relationship has been more acute, shaping not just policy but the inner workings of the Foreign Ministry. There have been 20 governments since 1991 featuring 18 different prime ministers, and since 1932, Thailand has witnessed 20 constitutions. But over the past decade there has been unprecedented volatility in Thai politics. The protracted crisis surrounding former Prime Minister Thaksin Shinawatra (2001–2006) has left Thai foreign policy largely without coherent direction – the result of both infighting among politicians and the Foreign Ministry's inability to develop long-term strategic policy. Moreover, the crisis has fundamentally reshaped the nature of the Thai polity, having "opened up space for a dramatic increase in Thai political actors who cross the domestic boundary into the unfamiliar domain of foreign affairs and use external issues as a political weapon to eliminate their rivals".[55]

Fashioning himself as a "CEO prime minister", making the country more business-oriented and democratic, Thaksin attacked what Duncan McCargo labels the "network monarchy": elites who intervene in politics as proxies of the king and his top advisors.[56] More than a "populist", Thaksin was a proficient electioneer, connecting to "grassroots" issues, particularly in the northeast. He saw himself as a "bridge" between the processes of globalisation and Thais often ignored or adversely affected by them.[57] But Thaksin aggravated traditional elites, challenging royal hegemony and the military.[58] In foreign affairs he injected a personal, nationalistic agenda with grandiose initiatives designed to make Thailand the dominant regional power. Operationally, he reorganised the Thai Foreign Ministry through appointments and processes, convinced it was antiquated. With a blend of idealistic and pragmatic notions, he ran a "people-centric" diplomacy that undermined the nature and content of traditional foreign policymaking.[59]

His "Forward Engagement Policy" with the region and economic ties with China illustrated a commerce-driven diplomacy. Foreign Ministry staff in effect became "salesmen" charged with showcasing Thailand. However, other core interests of Thai foreign policy were subverted. Democratic principles were secondary. Thaksin avoided questions about human rights when it came to China, Myanmar (the new name of Burma), and his own "war on drugs", which by 2003 resulted in 2,500 deaths.[60] Some of his ideas were ill-formed and poorly executed, and his means were often as opaque as the "old ways" he chided. Most importantly, his "populist foreign policy" undermined the professional diplomatic corps and many of the norms it represents.[61]

Thakin's ouster in 2006 gave rise to a complex political crisis, culminating with the May 2014 military coup. Despite condemnation from many quarters, the military government had curtailed civil liberties, delayed elections, and subverted democratic practices. Adding to the drama was the death of King Bhumibol in October 2016 and the succession of his son, Vajiralongkorn (Rama X, r. 2016–). Questions about the new king vis-à-vis Thai politics have further polarised the nation.[62] Foreign policy has become an extension of these crises. Given factional divide and rapidly changing political spaces, the concept of "national interest" is increasingly hard to define.[63] Thus, mired in domestic crisis, Thailand's regional influence has declined. Deep in this political paroxysm, Thailand's external relations remain in flux. But in many respects the "bamboo" is rotting from within. The craft of Thai foreign policy has been compromised as professional diplomats remain captive to the avaricious nature of domestic politics.

Notes

1 William J. Klausner, *Reflections on Thai Culture* (Bangkok: The Siam Society, 1981), pp. 79–80.
2 Thongchai Winichakul, *Siam Mapped: A History of the Geo-Body of a Nation* (Honolulu: University of Hawaii Press, 1994).
3 Shane Strate, *The Lost Territories: Thailand's History of National Humiliation* (Honolulu: University of Hawaii Press, 2015), pp. 2–15.
4 Pavin Chachavalpongpun, *Reinventing Thailand: Thaksin and His Foreign Policy* (Singapore: Institute of Southeast Asian Studies, 2010), pp. 5–6.
5 Duncan McCargo, preface in Chachavalpongpun (2010), p. vii.
6 Andrew MacGregor Marshall, *A Kingdom in Crisis: Thailand's Struggle for Democracy in the Twenty-First Century* (London: ZED Books, 2014), p. 1.
7 Nicholas Tarling, *Nations and States in Southeast Asia* (Cambridge: Cambridge University Press, 1998), p. 39.
8 David K. Wyatt, *Thailand: A Short History* (New Haven: Yale University Press, 1982), chs. 2–4.
9 B.J. Terwiel, *Thailand's Political History* (Bangkok: River Books, 2005), pp. 204–213.
10 Strate, *The Lost Territories: Thailand's History of National Humiliation*, pp. 9–10.
11 Wyatt, *Thailand: A Short History*, pp. 181–234.
12 Terwiel, *Thailand's Political History*, pp. 223–235.
13 Stefan Hell, *Siam and World War I: An International History* (Bangkok: River Books, 2017), pp. 27–52.
14 Ibid., pp. 11–20.
15 Nicholas Tarling, *Neutrality in Southeast Asia: Concepts and Contexts* (London: Routledge, 2017), pp. 163–164.
16 Hell, *Siam and World War I: An International History*, pp. 46–52.
17 Wyatt, *Thailand: A Short History*, pp. 219–220.
18 Tarling, *Neutrality in Southeast Asia*, pp. 164–165.
19 E. Bruce Reynolds, *Thailand and Japan's Southern Advance* (New York: St. Martin's Press, 1994), pp. 118–120.
20 Wyatt, *Thailand: A Short History*, p. 246.

21 Reynolds, *Thailand and Japan's Southern Advance*, pp. 125–129.
22 Wyatt, *Thailand: A Short History*, pp. 248–253.
23 Matthew Phillips, *Thailand in the Cold War* (Abingdon, UK: Routledge, 2016), p. 1.
24 Daniel Fineman, *A Special Relationship: The United States and Military Government in Thailand, 1947–1958* (Honolulu: University of Hawaii Press, 1997), pp. 58–62.
25 Chris Baker and Pasuk Phongpaichit, *A History of Thailand* (Cambridge: Cambridge University Press, 2005), p. 144.
26 Ibid., pp. 145–146.
27 Damian Fenton, *To Cage the Red Dragon: SEATO and the Defence of Southeast Asia, 1955–1965* (Singapore: National University of Singapore, 2012), p. 115.
28 Wyatt, *Thailand: A Short History*, p. 262.
29 Baker and Pasuk, *A History of Thailand*, pp. 147–150.
30 Arne Kislenko, "A Not so Silent Partner: Thailand's Role in Covert Operations, Counter-Insurgency, and the Wars in Indochina", *Journal of Conflict Studies*, Vol. 24, No. 1 (Summer 2004): 68.
31 As quoted in ibid., p. 70.
32 William J. Rust, *Before the Quagmire: American Intervention in Laos* (Lexington: University of Kentucky Press, 2012), pp. 174–178. See also Joshua Kurlantzick, *A Great Place to Have a War: America in Laos and the Birth of a Military CIA* (New York: Simon and Shuster, 2016), pp. 9–12.
33 Kislenko, "A Not So Silent Partner: Thailand's Role in Covert Operations, Counter-Insurgency, and the Wars in Indochina", p. 71.
34 Nicholas Tarling, *Southeast Asia and the Great Powers* (Abingdon, UK: Routledge, 2010), pp. 152–154.
35 William J. Rust, *So Much to Lose: John F. Kennedy and American Policy in Laos* (Lexington: University Press of Kentucky, 2014), pp. 93–121.
36 Michael Kelly Connors, *Democracy and National Identity in Thailand* (Copenhagen: NIAS, 2007), pp. 60–64.
37 Phillips, *Thailand in the Cold War*, pp. 12–13.
38 Ibid., pp. 9–13. See also Federico Ferrara, *The Political Development of Modern Thailand* (Cambridge: Cambridge University Press, 2015), pp. 183–184.
39 Baker and Pasuk, *A History of Thailand*, pp. 170–171.
40 Sutayut Osornprasop, "Thailand and the Secret War in Laos, 1960–1974", in *Southeast Asia and the Cold War,* edited by Albert Lau (Abingdon: Routledge, 2012), pp. 186–189.
41 Ibid., 201–208.
42 Quoted in Arne Kislenko, "The Vietnam War, Thailand and the United States", in *Trans-Pacific Modernity: An America Asia in the Pacific Century*, edited by Sugita Yone Sugita (Westport: Greenwood Press, 2003), p. 114.
43 Benjamin Zawacki, *Thailand Shifting Ground Between the United States and a Rising China* (London: ZED Books, 2017), pp. 53–54.
44 Baker and Pasuk, *A History of Thailand*, pp. 193–198.
45 Kusuma Snitwongse, "Thai Foreign Policy in the Global Age: Principle or Profit?" *Contemporary Southeast Asia,* Vol. 23 (August 2001): 189–192.
46 Quoted in Kislenko, "The Vietnam War, Thailand and the United States", p. 118.
47 Pongphisoot Busbarat, "A Review of Thailand's Foreign Policy in Mainland Southeast Asia: Exploring an Ideational Approach", *European Journal of East Asian Studies*, Vol. 2 (2012): 127–154.
48 Kusuma, "Thai Foreign Policy in the Global Age: Principle or Profit?" pp. 193–195.
49 Grant Evans, *The Politics of Ritual and Remembrance: Laos since 1975* (Honolulu: University of Hawaii Press, 1998), pp. 88–113.
50 Zawacki, *Thailand Shifting Ground Between the United States and a Rising China*, pp. 10–12, 68–107.
51 Ibid., p. 3.
52 John Blaxland and Greg Raymond, "Tenets of Thailand's ASEAN Engagement", *East Asia Forum*, 28 March 2018 <www.eastasiaforum.org/2018/03/28/tenets-of-thailands-asean-engagement/> (accessed 31 March 2018).
53 Kitti Prasirtsuk, "The Implications of US Strategic Rebalancing: A Perspective from Thailand", *Asia Policy,* Vol. 15 (January 2013): 31–37.
54 Pongphisoot Busbarat, "'Bamboo Swirling in the Wind': Thailand's Foreign Policy Imbalance between China and the United States", *Contemporary Southeast Asia*, Vol. 8, No. 2 (2016): 233–257.
55 Pavin Chachavalpongpun, "Diplomacy under Siege: Thailand's Political Crisis and the Impact on Foreign Policy", *Contemporary Southeast Asia*, Vol. 31, No. 3 (2009): 448–450.

56 Duncan McCargo, "Network Monarchy and Legitimacy Crises in Thailand", *The Pacific Review,* Vol. 18, No. 4 (December 2005): 499–519.
57 Kevin Hewison, "Reluctant Populists: Learning Populism in Thailand", *International Political Science Review,* Vol. 38, No. 4 (2017): 426–440.
58 Kasian Tejapira, "The Irony of Democratisation and the Decline of Royal Hegemony in Thailand", *Southeast Asian Studies,* Vol. 5, No. 2 (2016): 222–223.
59 Pavin, *Reinventing Thailand,* pp. xii, 5–6, 10–11.
60 Ibid., pp. 16–18, 26–29.
61 Donald E. Weatherbee, *International Relations in Southeast Asia: The Struggle for Autonomy* (New York: Rowman and Littlefield, 2009), pp. 40–41.
62 Duncan McCargo, "Thailand in 2017: Politics on Hold", *Asian Survey,* Vol. 58, No. 1 (2018): 181–187.
63 Pavin, "Diplomacy under Siege", pp. 463–464.

References

Baker, Chris and Pasuk Phongpaichit. 2005. *A History of Thailand.* Cambridge: Cambridge University Press.

Blaxland, John and Raymond, Greg. 2018. "Tenets of Thailand's ASEAN Engagement". *East Asia Forum.* 28 March <www.eastasiaforum.org/2018/03/28/tenets-of-thailands-asean-engagement/> (accessed 31 March 2018).

Connors, Michael K. 2007. *Democracy and National Identity in Thailand.* Copenhagen: NIAS.

Evans, Grant. 1998. *The Politics of Ritual and Remembrance: Laos since 1975.* Honolulu: University of Hawaii Press.

Fenton, Damian. 2012. *To Cage the Red Dragon: SEATO and the Defence of Southeast Asia, 1955–1965.* Singapore: National University of Singapore.

Ferrara, Federico. 2015. *The Political Development of Modern Thailand.* Cambridge: Cambridge University Press.

Fineman, Daniel. 1997. *A Special Relationship: The United States and Military Government in Thailand, 1947–1958.* Honolulu: University of Hawaii Press.

Hell, Stefan. 2017. *Siam and World War I: An International History.* Bangkok: River Books.

Hewison, Kevin. 2017. "Reluctant Populists: Learning Populism in Thailand". *International Political Science Review,* Vol. 38, No. 4: 426–440.

Kasian Tejapira. 2016. "The Irony of Democratisation and the Decline of Royal Hegemony in Thailand". *Southeast Asian Studies,* Vol. 5, No. 2: 219–237.

Kislenko, Arne. 2004. "A Not so Silent Partner: Thailand's Role in Covert Operations, Counter-Insurgency, and the Wars in Indochina". *Journal of Conflict Studies,* Vol. XXIV, No. 1 (Summer): 65–96.

Kitti Prasirtsuk. 2013. "The Implications of US Strategic Rebalancing: A Perspective from Thailand". *Asia Policy,* Vol. 15 (January): 31–37.

Klausner, William J. 1981. *Reflections on Thai Culture.* Bangkok: The Siam Society.

Kurlantzick, Joshua. 2016. *A Great Place to Have a War: America in Laos and the Birth of a Military CIA.* New York: Simon and Shuster.

Kusuma Snitwongse. 2001. "Thai Foreign Policy in the Global Age: Principle or Profit?" *Contemporary Southeast Asia,* Vol. 23 (August): 189–192.

Lau, Albert (ed.). 2012. *Southeast Asia and the Cold War.* Abingdon: Routledge.

Marshall, Andrew MacGregor. 2014. *A Kingdom in Crisis: Thailand's Struggle for Democracy in the Twenty-First Century.* London: ZED Books.

McCargo, Duncan. 2005. "Network Monarchy and Legitimacy Crises in Thailand". *The Pacific Review,* Vol. 18, No. 4 (December): 499–519.

———. 2018. "Thailand in 2017: Politics on Hold". *Asian Survey,* Vol. 58, No. 1: 181–187.

Pavin Chachavalpongpun. 2009. "Diplomacy under Siege: Thailand's Political Crisis and the Impact on Foreign Policy". *Contemporary Southeast Asia,* Vol. 31, No. 3: 447–467.

———. 2010. *Reinventing Thailand: Thaksin and His Foreign Policy.* Singapore: Institute of Southeast Asian Studies.

Phillips, Matthew. 2016. *Thailand in the Cold War.* Abingdon, UK: Routledge.

Pongphisoot Busbarat. 2012. "A Review of Thailand's Foreign Policy in Mainland Southeast Asia: Exploring an Ideational Approach". *European Journal of East Asian Studies,* Vol. 2: 127–154.

———. 2016. "'Bamboo Swirling in the Wind': Thailand's Foreign Policy Imbalance between China and the United States". *Contemporary Southeast Asia,* Vol. 38, No. 2 (2016): 233–257.

Reynolds, E. Bruce. 1994. *Thailand and Japan's Southern Advance*. New York: St. Martin's Press.
Rust, William J. 2012. *Before the Quagmire: American Intervention in Laos*. Lexington: University of Kentucky Press.
———. 2014. *So Much to Lose: John F. Kennedy and American Policy in Laos*. Lexington: University Press of Kentucky.
Strate, Shane. 2015. *The Lost Territories: Thailand's History of National Humiliation*. Honolulu: University of Hawaii Press.
Tarling, Nicholas. 1998. *Nations and States in Southeast Asia*. Cambridge: Cambridge University Press.
———. 2010. *Southeast Asia and the Great Powers*. Abingdon, UK: Routledge.
———. 2017. *Neutrality in Southeast Asia: Concepts and Contexts*. London: Routledge.
Terwiel, B.J. 2005. *Thailand's Political History*. Bangkok: River Books.
Thongchai Winichakul. 1994. *Siam Mapped: A History of the Geo-Body of a Nation*. Honolulu: University of Hawaii Press.
Weatherbee, Donald E. 2009. *International Relations in Southeast Asia: The Struggle for Autonomy*. New York: Rowman & Littlefield.
Wyatt, David K. 1982. *Thailand: A Short History*. New Haven: Yale University Press.
Yone, Sugita (ed.). 2003. *Trans-Pacific Modernity: An America Asia in the Pacific Century*. Westport: Greenwood Press.
Zawacki, Benjamin. 2017. *Thailand Shifting Ground between the United States and a Rising China*. London: ZED Books.

32

BEYOND BAMBOO DIPLOMACY

The factor of status anxiety and Thai foreign policy behaviours

Peera Charoenvattananukul

Thailand under the military government led by General Prayuth Chan-ocha has been greatly obsessed with the infamous "20-Year National Strategy Plan". The long-term strategy institutionalised by the junta triggered subsequent trends among ministries to formulate their own 20-year vision. Thailand's Ministry of Foreign Affairs (MFA), a powerhouse and gatekeeper of foreign policy, is no exception to the 20-year revelation. Instead of articulating perennial statements on being friends with all and enemies with none, the MFA's 20-Year Masterplan outlines the "5S Strategy", which enumerates the five concepts to success: Security, Sustainability, Standard, Status, and Synergy.[1] Undoubtedly, each strategy is meaningful to the MFA. However, the acknowledgement of "status" as part of the supreme importance to foreign policy is intriguingly crucial to making sense of Thailand's foreign relations. In other words, it implies that Thailand's long-term strategy will be based on status-seeking foreign policy for at least 20 years. According to the MFA's public declaration, the objective of Thailand's status-based foreign policy is to "enhance prestige and honour in the eyes of the world".[2]

From the outset, the status-driven policy proclaimed by the Thai officials could be a public façade. On closer inspection, however, its implications to Thai foreign policy studies could be significant in two aspects. First, a near consensus and conventional approach to interpret the characters of Thai foreign policy suggests that Thai foreign policy tends to be flexible, adaptable, and adjustable to the dynamic of balance of power and the transformation of the international environment.[3] Such a law-like conviction of Thai foreign policy studies maintains the primacy of national security and survival in the crude realm of international politics over any other factors in policymaking process and descriptive analysis. This so-called "bending with the wind" foreign policy or "bamboo diplomacy" fits nicely within the neo-realist theory of international relations (IR) and remains dominant among Thai academics.[4]

In this regard, there is little room to factor in other elements to understand Thai foreign policy behaviours. But, what if scholars of Thai foreign affairs turn to the assumption that, apart from national security issues, the status factor does matter? Would the change of fundamental assumption reshape how we see Thai foreign policy behaviours? Were there any historical instances in which the status-concern variable influenced Thai foreign policy behaviours?

Second, if the status-concern factor does matter in the Thai case, it could fill the lacuna of the status-related literature in IR. Currently, the IR pieces, which focus on the relationship between status concerns and foreign policy outcomes, often utilise the cases of the Great Powers to prove their theory.[5] The Great Powers fixating on their positional status can be surprising for the IR community. However, a small or middle-sized nation concerning itself with a status-seeking quest would be even more astonishing. Thailand was once small in terms of size in the 19th century and could be considered as a middle-sized nation in the contemporary era.[6] Should there be exemplary cases, they could show that Thailand also cares about its status. This would fill the unanswered gap in the IR literature.[7]

The remaining question is how to find out if the status-concern factor is influential to Thai foreign policy behaviours. This chapter seeks to answer the question by using historical instances to demonstrate the potency and relevance of the status concern and Thai foreign policy. As the literary character Sherlock Holmes famously told his flatmate, "You see, but you do not observe". Likewise, the signs of status concern could have existed in the past. The scholars might see such signs and take them lightly. This chapter, on the contrary, takes the substance of status concern seriously.

The chapter begins with a brief discussion of the conceptual and theoretical tenets of the status-related theory. It seeks to illustrate how status concerns can be influential to Thai foreign policy behaviours in two cases. The first case probes how Siam sought to gain recognition and acceptance from its Western counterparts in the 19th century. Rather than seeing it as a template of diplomatic success, this case will display the underlying logic of status-driven foreign policy. The second case deals with nationalist aspirations during the Second World War. Thailand under the administration of Field Marshal Phibun Songkhram pursued socio-cultural programmes to re-engineer Thai society and culture.[8] The Phibun government (1938–1944) also engaged in the territorial dispute with France. This section aims to elaborate on how a status concern was an implicit source of foreign policy conduct. Last, the chapter concludes by examining an implication to contemporary Thailand. The ending section briefly demonstrates how status concerns are lurking underneath Thailand's armed purchases such as the aircraft carrier and submarines.

A brief version of status theory in IR

The prominence of status concern in IR is nothing new in social sciences. As a matter of fact, the status factor as a source of aggression and interstate conflict was first explored in the 1960s and was placed outside the IR community until the late 20th century.[9] After the turn of the century, there was a resurgent interest of status factor in IR, as one IR theorist notes that there is "the strong belief among scholars that status . . . affects outcomes of importance across international relations".[10] Richard Ned Lebow, for example, finds out that the "standing" motive is responsible for 58 per cent of the outbreak of wars between 1648 and 2003.[11] But what exactly is the concept of status in IR?

In IR, the notion of status is often used interchangeably with standing, prestige, dignity, esteem, pride, honour, and reputation. They are currently treated as inseparable.[12] Some scholars summarise that status "implicates such concepts as prestige, esteem, honour, standing, rank and face".[13] As such, this chapter tends to use these terms interchangeably.

The definitions of status in IR can be classified into three meanings. First, status is *positional*. In the most basic sense, status refers to "standing" or "rank" in the international community.[14] In order for "State A" be cognizant of its own status, it needs a reference group to determine its placement in the international society. For example, in order for Thailand to be able to evaluate its own status, it will need to have reference countries, such as the United States or Myanmar, which are placed positionally higher or lower than itself so that it can identify its positional status.

Second, status entails *identity*. In a general sense, one's specific status confers certain identity and expectation upon him/her. For example, a Great Power status entails a great power identity, which is attached by a set of rights and responsibilities.[15] This trait is connected to the third definition, which views status as *relational*. If status entails identity, it implies that status is contingent on recognition from others. Possessing great military capabilities does not necessarily translate into Great Power status, otherwise North Korea would have joined the club of Great Powers. Similarly, in the 19th century, Prussia under Frederick the Great was recognised as one of the Great Powers even though its geographical size was not as large as France or Russia. Strictly speaking, material conditions alone would not constitute one's status. Recognition from others is a necessary requirement to confirm a nation's status and identity in international politics. As Richard K. Ashley, a post-structural scholar in IR, argues, "[Status is not] in any sense attributable to inherent qualities or possessions. . . . Rather, the power and status of an actor depends on and is limited by the conditions of its recognition within the community as a whole".[16]

If we assume that, just like human beings, states care about their status, it implies that the policymakers could either be contented or dissatisfied with their nation's status.[17] The states that are satisfied with their status are not as puzzling in IR as the states that are dissatisfied with their status. "Status dissatisfaction" is a psychological condition that occurs to the states that possess potentialities and status expectations to achieve a certain status and identity but are not recognised as such.[18] A great indicator of status dissatisfaction is when the states invoke the language of rights to justify their foreign policy. They might, for example, pursue a course of action by claiming their rightful place among their peers, as in the case of Wilhelmine Germany's declaration of the historic *Weltpolitik*, which sought to claim the Germans' "place in the sun".

If the states are dissatisfied with their status, they are more likely to engage in status-seeking behaviours. In the quest for status and recognition, there are several policy choices for the states with status anxiety to select. They can start by attempting to acquire "status symbols" or "status markers" in order to garner recognition from the others. The symbols or markers of status can vary from time to time. For instance, in the colonial periods, having colonies was considered as a marker for a great power status. Or, in the cases of non-Western entities, the "standard of civilisation" was regarded as a status symbol that could earn recognition as an equal and sovereign country which was on par with the West.

By and large, the states' failed bid for status and recognition could fuel more status-seeking behaviours that could be intense and aggressive. The misrecognition from the others could render the status bidders disrespected and humiliated to the extent that the states in question would involve more in "status competitive acts", which could sometimes trump rational, strategic, and material calculus.[19] If the concerns for status matter to the states, and certain behavioural outcomes are determined by such concerns, Thailand's cases are not distinctive in this aspect. The two sections are devoted to explicating how the status factor could play a part in approaching Thailand from a different angle.

Thailand's search of civilisation

The 19th century is conventionally projected by Thailand's mainstream historiography as the period when the country masterfully avoided any attempted colonisation by the Western powers. The survival of the nation, a central element of neorealism, was at stake, and thus bamboo diplomacy has been auspiciously celebrated ever since. Where, then, is the place for the status-concern analysis?

It is no secret that Siam in the 19th century transformed their modus vivendi after the advent of the West in Asia. During the reign of King Mongkut (r. 1851–1868), especially after the adoption of the 1855 Bowring Treaty, which formally integrated Siam into the global economy and legally introduced the principle of extraterritoriality to the kingdom's judicial process, the Siamese leaders strived to gain recognition from the West by undertaking cultural alterations, such as the requirements for courtiers to dress properly when granted an audience to the king,[20] the approval for foreigners to be presented to the monarch without prostrating and crawling,[21] and the hint of slavery abolition.[22] Siam's non-Western traditions were considered backward and barbaric from the standards of civilisation, and its rulers were fully aware of the impetuses to conform to the then preeminent mode of living.

Was the incentive to change driven by the interest to avoid colonisation only as the typical account of Thai diplomatic history has been portraying? Or was the status concern among the Siamese elites also accounted for the total transformations of the society? Thongchai Winichakul, a renowned Thai historian, contends that

> Siam could no longer confirm . . . its meaningful existence by claiming the lineages to the traditional cosmic origins. In order to survive, not from colonialism but from indignity and inferior existence, and to remain majestic, Siam needed a confirmation according to the new ethos of civilisation that it measured up to other leading countries. The desire and anxiety to keep up with the world, not an escape from being colonised was, significant in itself.[23]

In other words, the status and identity of Siam should be judged according to a new standard following the expansion of the Europeans into Asia. The Siamese ruling cliques, devouring a fruit of knowledge from the West, encountered the status anxiety when they were seen as backward by the Western powers. Mongkut, for example, assessed the existence of Siam in comparison to the West and realised that his kingdom was "half-civilised and half barbarian".[24] If the adaptation were prompted by a pretence to avoid colonisation only, how would it explain the subsequent and incessant reforms to gain recognition from the Western counterparts after the end of Mongkut's era?

The overhaul of Siam was systematically and thoroughly executed further during the reign of King Chulalongkorn (r. 1868–1910). The notable examples of such reforms were the creation of modern bureaucracy and ministry, the formation of educational programmes, the modernisation of the Siamese military, and the formal abolition of slavery. The concerns for status and recognition played a vital part in determining such restructuring schemes, as Prince Prisdang commented, "Siam must be accepted and respected by the western powers as a civilised nation. Hence, there is no choice but to bring about a new government modelled after the western pattern".[25]

The disgrace from the "Paknam crisis" in 1893 was similarly motivated by a concern for status. The origin of the dispute began from how the Siamese took the temerity to challenge France in a game of influence over Laos and Cambodia. Why did Siam break away from the "bending with the wind" policy? It was not rational to compete against an imperialist power. The desire to be recognised as an equal country pushed Siam forward, as Noel Alfred Battye articulated:

> Until 1893, it seems, it had been reasonable for Siamese to believe that the West would respect the integrity of earnestly progressive Asian countries. Chulalongkorn believed that Siam had much progress on its own that other nations had nothing to gain by imposing themselves over it politically.[26]

It could be interpreted that since Siam had implemented a number of key reforms, the country sought to prove its status by acquiring a sphere of influence, which was one of the status markers of the 19th century. Thus the Siamese were willing to defy the dictate of France's dominance in the region.

Nonetheless, the humiliation from the Paknam crisis did not deter Siam from seeking more status. On the contrary, the Siamese elite became more addicted to the politics of status seeking. For example, the government dispatched the Siamese representatives to participate in the World's Fairs in 1893 (Chicago), 1889 (Paris), 1900 (Paris), 1904 (St. Louis), and 1911 (Turin). The objective was to enhance the country's profile and status. Thongchai seemingly concurs with the status theory and suggests that the Siamese presence on the global stage was to gain "recognition and elevate their status in the eyes of the world".[27]

Perhaps, there was no other example that was as vivid as the event of Siam's entry into the First World War. When the Great War erupted in 1914, Siam, having no substantial interest in the war, observed strict neutrality. The entire story about Siam finally joining with the Allies was no mystery. But why did King Vajiravudh (r. 1910–1925), son of Chulalongkorn, lead Thailand to side with the Allies before the end of the war?

First, it was not because the tide of the war was turning in favour of the Allies. Prince Devavongse, the Siamese foreign minister, informed Vajiravudh, who personally leaned towards England due to his educational background from Oxford, that the Central Powers were not aggressive towards Siam. As such, the declaration of war would not be justified. Moreover, Devavongse believed, "from the start the Allies would win, but see no good reason for Siam to join in".[28] This statement implicitly hinted that the tide against the Central Powers was not influential to the Siamese leadership.

The concerns for status continued to determine Siam's war stance.[29] One of the worst possibilities of upholding strict neutrality until the end of the war was the exclusion of Siam from the community of nations in the post-war settlement. As a result, Siam's international standing would remain inferior in the eyes of the West. Hence, participating in the Great War would be an enticing option for Siam to improve the country's status and negotiate the unequal treaties with the Western nations.[30] Prince Charoon, a Siamese representative to France, perceived participation in the Great War as the "real opportunity of raising the status of our beloved country".[31]

In this regard, Siam's participation in the war would, according to some foreign opinion, treasure "the reputation of Siam as an advanced country in civilisation".[32] When Siam decided to dispatch the expeditionary force to side with the Allies, the king's message to his men, which was centred on the rationale of status, could not be more apparent: "This was a highly important chance . . . to show other nations which we once feared. But now we are walking uphill. . . . In this time and the next, Siam shall rise to be on par with other nations".[33] This statement indicates how the dynamism of status concerns governed the foreign policy thinking of the Siamese before and after the Great War.

Phibun's nationalism and status anxiety

Phibun's wartime government was notorious for its nationalist policies in several dimensions. One of Phibun's historical legacies was his grand scheme called "State Conventions" (*Ratthaniyom*), which involved a number of socio-cultural engineering projects to improve the lives, cultures, and minds of the Thai people.[34] Broadly speaking, the State Conventions decreed the proper dress that the Thais should follow, the appropriate manners for the Thais in public and private space, and the righteous spirit the Thais should have for their nation.

There might be a set of questions arising from the case of Phibun's nationalism. For example, how is nationalism related to status anxiety? Nationalism, rather cliché as a notion, has been used to explain several events in international politics. What is actually unique about the nationalist factor especially in the case of Phibun's government? It is undeniable that the nationalist fervour seems to be prevalent in political analyses. Yet, there are only few individuals who would identify the mechanism of nationalism and how it could affect foreign policy outcomes. Nonetheless, the nationalist factor is intricately interrelated to the notion of status. Nationalism is inseparable from a certain identity, and the existence of identity is contingent on recognition and acceptance from others. If status entails identity and vice versa, nationalism, which seeks to cherish a specific self-identity to gain recognition from others, is closely associated with the status concerns. Misrecognising one's national identity could be interpreted as an affront to one's status.

The case of Phibun could best exemplify the logic of status anxiety. When Phibun became prime minister in 1938, the first concern that he informed his cabinet was the objective to rid the country of the unequal treaties once and for all.[35] The successful revision of those treaties would be, according to a foreign observer, "a major matter of prestige".[36] Therefore, when the government successfully terminated all the unequal treaties with foreign powers, Phibun held the national festival called "the Celebration Day of the Nation and Treaty Revision" on 24 June 1939. From the outset, such a festivity might seem trivial. Nevertheless, it was significant to the analysis of the status-concern framework because it was the first time that the nation organised an event to ritualise the National Day.[37]

The festival on the National Day marked the turning point of Thailand's status because the treaty revision represented "the recognition of its right as a member of the Family of Nations to negotiate with the various Treaty Powers on a basis of entire equality, full autonomy and complete reciprocity".[38] Thailand, following the revision of the unequal treaties, redefined its position and identity as a country which was equal to others in terms of status. However, the status concern was also evident in Phibun's speech on the National Day: "If we consider it seriously, it can be seen that our civilisation and culture should be improved and enhanced so that we can be on par with other western civilised nations".[39]

Due to the government's status anxiety and fear of misrecognition, Phibun launched the State Conventions, one of his nationalist projects, to enhance status and build prestige as he told the Thais that the "State Conventions are similar to the manners of how the civilised should do.... Same as we have heard stories from the civilised countries".[40] Phibun's nationalist project was inseparable from the status concerns. This could be seen through the government's rationale to transform the dress of the Thai people:

> If we dress like a savage, when foreigners see us, they will insult and try to help us dress.... If foreigners look down on us, it will be a cause for us to be difficult in establishing relations with other nations. As an easy example, suppose we put on silk pants to dine with British diplomats, it is believed that we would not be invited again, and cannot befriend other British.[41]

From this perspective, the status anxieties as seen through the nationalist apprehensions were closely linked to diplomacy. In the case of the Franco-Thai border dispute, the Thai leaders prodded France into returning some parts of Laos and Cambodia to Thailand. Because the Thais understood that Laos and Cambodia were their "lost territories", coupled with the political reality in which Thailand entered a new phase as an equal and sovereign country, the Thai leaders seized the opportunity from the French downfall to negotiate the territorial retrocession.

Thailand's policy gambit stemmed from the aspiration to be recognised as an equal. Phibun admitted his genuine intention as follows: "[The demands] we requested from France, if we are to be frank, we might not receive any land. In fact, our demands are not truly a territorial claim. *It is to symbolise that Thailand has lifted our status already* [my emphasis]".[42] In other words, after June 1939, the Thais took pride in their new status. France, however, was defeated in the European theatre. From the Thai views, if the French empire truly recognised Thailand's new status, they should accede to Thailand's territorial demands. On the contrary, France remained unyielding, which sparked a series of nationalist movements in Thailand.

The emotion of the nationalist movements, which called for the return of the lost territories, was primarily based on the status dissatisfaction. For instance, one leaflet from the demonstration expressed how France's refusal to the territorial demands implied Thailand's inferior status:

> We invite the Thai brethren to support the government policy in territorial retrocession. The Thais should glorify Thailand's prestige to be recognised throughout the world so that they know we are a passionately robust and capable nation. We should all unite to build our nation to the extent that others would worship and respect how civilised we are.[43]

The message from the demonstration could not be any clearer: the concerns for status were central even among the demonstrators. France's denial of Thailand's rights to demand the territories was interpreted as disrespecting the nation's identity and status as a sovereign country which could stand on an equal footing with the others.

In this light, Thailand's military defiance of France could bring about the desired status as a historian argues, "A military victory over France would prove to the country and to the world how far their country had advanced in the past decades".[44] Ultimately, the Thais could achieve some territories from France. Such an epoch-making victory greatly enhanced Thailand's prestige, honour, and status. Such a perception also became widespread after the successful territorial campaign as a newspaper described:

> This success has not only drawn the map of Thailand, it has redrawn the map of our hearts and minds. That is to say, it has made us realise that our beloved nation of Thailand has increased in honour and caused the world to recognise that we are not the same country as 40 years ago. Quite the opposite, other nations are now praising us.[45]

Conclusion: luxurious arms as a status symbol?

In summary, the two historical cases in this chapter described how the concerns for status were no alien to Thai foreign policy. Such concerns for status and recognition have been part of Thailand's foreign considerations since the expansion of the West into Asia, as in the case of Siam in the age of civilisation. The case of Phibun's nationalism also reflects how nationalist policies are inevitably tied to the notion of status anxiety. Although the status theory might be useful to approach Thailand's diplomatic history from a different outlook, what is a wider implication for contemporary Thailand? This concluding section suggests that the status theory can be utilised to examine the case of Thailand's purchases of both the aircraft carrier and submarines.

As mentioned earlier, status symbols could be some international practices such as the standards of civilisation. Likewise, the acquisition of some particular armed weapons could be regarded as an endeavour to attain a certain status. Lilach Gilady, an IR theorist, empirically proposes that the procurement of aircraft carriers could be considered as an effort to lavish financial resources in exchange for prestige.[46] Generally, it is believed that the states are rational and strategic when it comes to key decisions. Nevertheless, if we take the status factor seriously, some key decisions may not be as rationally and strategically optimal as they are supposed to be. Relying on an idea of Thorstein Veblen, Gilady contends that sometimes the states resort to waste money for conspicuous and status-seeking purposes. The purchase for prestige could be identified when the costs of acquiring a particular item far exceed the costs of acquiring other similar items, which could have identical functional utilities. For example, purchasing a Hermès bag would be an expenditure to signify status because, despite the same functional utility with other bags, there is a vast gap in terms of price between this luxurious bag and other non-brand name bags.

The procurement of aircraft carriers is similar to obtaining luxurious items to symbolise status. Thailand, despite its limited role in global affairs, is one of a few countries that possesses an aircraft carrier. Large ships have always been a symbol of greatness from past to present. Before the First World War, Britain and Germany were embroiled in a contest over acquiring the great *Dreadnought*, which offered little strategic values during the war in comparison to smaller ships.[47] The costs of maintaining the aircraft carriers are massive, for instance, governments have to invest in anti-submarine systems and anti-aircraft technologies, otherwise the carriers become vulnerable to attacks.

In the case of the *Chakri Naruebet*, Thailand's aircraft carrier which began operation in 1997, the Thai government was responsible for 50,000 baht for daily maintenance and 2 million baht for a proper care of the fighter helicopters. After the retirement of the carrier in 2006, the repair would approximately cost around 5 million baht.[48] Throughout its service, the *Chakri Naruebet* had no blue-water operation or any other military fighting activities. Thailand after the Cold War has had no serious maritime threat from other nations. Hence, the aircraft carrier project was nothing but another quest for status. An expert comments that without a clear regional threat, "it is hard to see the reasoning behind a navy's build-up except for reasons purely of national prestige".[49]

In 2017, Thailand's military government ventured to purchase three submarines from China. The Sino-Thai naval deal has cost Thailand approximately 36 billion baht.[50] Similar to the inactive aircraft carrier, the cost of maintenance of the submarines could be uneconomical. In terms of the necessity, Thailand is no conflicting party in the South China Sea dispute and receives a rare threat from sea pirates. In this regard, the submarine deal could be another project for prestige.

In making sense of Thai foreign policy, it is not always a fundamental requirement to start from the "bending with the wind" approach or the neo-realist theory of IR. The horizon of IR theory has been extending beyond such two traditional approaches, which are predicated on the survival assumption. This chapter has endeavoured to interpret Thai foreign policy behaviours in a new light. As all the cases have shown, the small or medium-sized state such as Thailand is neither necessarily passive nor docile. The Thai nation could also engage in politics of status-seeking. This chapter is written with the hope that it could inspire and introduce a variety of IR concepts and theories to apply to the case of Thai foreign policy.

Notes

1 Ministry of Foreign Affairs of Kingdom of Thailand, *The 20-Year '5S' Foreign Affairs Masterplan*, 20 August 2018 <www.mfa.go.th/testweb/contents/fles/customize-20180724-142633-552037.pdf> (accessed 1 May 2019).

2 Ministry of Foreign Affairs of Kingdom of Thailand, *Phaen Maebot Dan Kantangpathet Raya 20 Pi (PhoSo 2561–2580)* [Masterplan of Foreign Affairs for Twenty Years (BE 2561–2580)], 11 January 2019 <www.mfa.go.th/main/contents/files/policy-20190111-114336-037494.pdf> (accessed 1 May 2019). It is

important to note here that the content of the Thai version of the policy pamphlet is quite different from the details specified in the English version.

3 See examples from Pavin Chachavalpongpun, "Thailand: The Enigma of Bamboo Diplomacy", in *Routledge Handbook of Diplomacy and Statecraft*, edited by B.J.C. McKerscher (New York and Abingdon: Routledge, 2012), pp. 204–214; Donald E. Neuchterlein, *Thailand and the Struggle for Southeast Asia* (Ithaca: Cornell University Press, 1965); Arne Kislenko, "Bending with the Wind: The Continuity and Flexibility of Thai Foreign Policy", *International Journal*, Vol. 57, No. 4 (December 2002), pp. 537–561.

4 Jittipat Poonkham, "Why Is There No Thai (Critical) International Relations Theory? Great Debates Revisited, Critical Theory and Dissensus of IR in Thailand", in *International Relations as a Discipline in Thailand: Theory and Subfields*, edited by Chanintira na Thalang, Soravis Jayanama, and Jittipat Poonkham (New York and Abingdon, 2019), p. 21.

5 See some examples that emphasises the great power cases from Deborah W. Larson and Alexei Shevchenko, *Quest for Status: Chinese and Russian Foreign Policy* (New Haven: Yale University Press, 2019); Yong Deng, *China's Struggle for Status: The Realignment of International Relations* (Cambridge: Cambridge University Press, 2018); Tadashi Anno, *National Identity and Great-Power Status in Russia and Japan* (New York and Abingdon: Routledge, 2018). There are more theory-oriented pieces which draw on the cases of great powers such as Nazi Germany and the United States. See Steven Ward, *Status and the Challenge of Rising Powers* (Cambridge: Cambridge University Press, 2017).

6 There is no consensus among IR scholars regarding the definitions of the small state and middle-sized power. See examples from Matthias Maass, "The Elusive Definition of the Small State", *International Politics*, Vol. 46, No. 1 (January 2009): 65–83. Regarding the nuancing definitions of the middle power, see David A. Cooper, "Challenging Contemporary Notions of Middle Power Influence: Implications of the Proliferation Security Initiative for Middle Power Theory", *Foreign Policy Analysis*, Vol. 7, No. 3 (July 2011), pp. 317–336.

7 There is an aspiration to answer such a question. See Benjamin de Carvalho and Iver B. Neumann, "Introduction", in *Small State Seeking Status: Norway's Quest for International Standing*, edited by Benjamin de Carvalho and Iver B. Neumann (New York: Routledge, 2015), pp. 1–21.

8 There are various ways to write his name in English. In this chapter, he is referred to as "Phibun".

9 The consideration of the status factor was first pioneered by a peace researcher. See, Johan Galtung, "A Structural Theory of Aggression", *Journal of Peace Research*, Vol. 1, No. 2 (1964), pp. 95–119.

10 Jonathan Renshon, *Fighting for Status: Hierarchy and Conflict in World Politics* (Princeton: Princeton University Press, 2017), p. 15.

11 According to Lebow's findings, which are methodologically questionable, standing (58%), security (18%), revenge (10%), interest (7%), and other (7%) are responsible for the outbreak of war from 1648 to 2003. See Richard Ned Lebow, *Why Nations Fight? Past and Future Motives for War* (Cambridge: Cambridge University Press, 2010), p. 114.

12 Some IR scholars try to distinguish these terms. See Allan Dafoe, Jonathan Renshon and Paul Huth, "Reputation and Status as Motives for War", *Annual Review of Political Science*, Vol. 17 (May 2014), pp. 371–393.

13 William C. Wohlforth and David C. Kang, *Hypotheses on Status Competition*, Paper presented at the Annual Meeting of the American Political Science Association, Toronto, Canada <https://ssrn.com/abstract=1450467> (accessed 2 May 2019).

14 Renshon, *Fighting for Status*, p. 33.

15 Some scholars employ the "social identity theory" to explain the relationship between status and identity. See Deborah Welch Larson and Alexei Shevchenco, "Status Seekers: Chinese and Russian Responses to U.S. Primacy", *International Security*, Vol. 34, No. 4 (Spring 2010), pp. 63–95.

16 Richard K. Ashley, "The Poverty of Neorealism", in *Neo-Realism and Its Critics*, edited by Robert O. Keohane (New York: Columbia University Press, 1986), p. 291.

17 It is inevitable to treat the state as person. See Alexander Wendt, "The State as Person in International Theory", *Review of International Studies*, Vol. 30, No. 2 (April 2004), pp. 289–316.

18 Andrew Q. Greve and Jack S. Levy, "Power Transitions, Status Dissatisfaction, and War: The Sino-Japanese War of 1894–1895", *Security* Studies, Vol. 27, No. 1 (2018), pp. 148–178.

19 Joslyn Barnhart, "Status Competition and Territorial Aggression: Evidence from the Scramble for Africa", *Security Studies*, Vol. 25, No. 3 (2016), pp. 385–419.

20 Chula Chakrabongse, *Lords of Life: The Paternal Monarchy of Bangkok, 1782–1932* (London: Alvin Redman, 1960), p. 183.

21 Ibid., p. 192.

22 Chris Baker and Pasuk Phongpaichit, *A History of Thailand* (Cambridge: Cambridge University Press, 2014), p. 49.
23 Thongchai Winichakul, "The Quest for 'Siwilai': A Geographical Discourse of Civilisational Thinking in the Late Nineteenth and Early Twentieth-Century Siam", *The Journal of Asian Studies*, Vol. 59, No. 3 (August 2000), p. 534.
24 Charnvit Kasetsiri, "Siam/Civilisation – Thailand/Globalisation: Things to Come", *Thammasat Review*, Vol. 5, No. 1 (2000), p. 120.
25 Cited in Eiji Murashima, "The Origin of Modern Official State Ideology in Thailand", *Journal of Southeast Asian Studies*, Vol. 19, No. 1 (March 1988), p. 84.
26 Noel Alfred Battye, *The Military, Government, and Society in Siam, 1868–1910*, PhD Dissertation, Cornell University, 1974, p. 401.
27 Thongchai, "The Quest for 'Siwilai'", p. 540.
28 Walter F. Vella, *Chaiyo! King Vajiravudh and the Development of Thai Nationalism* (Honolulu: University of Hawaii Press, 1978), p. 107.
29 For an argument along this line, see Gregory V. Raymond, "War as Membership: International Society and Thailand's Participation in World War I", *Asian Studies Review*, Vol. 43, No. 1 (2019), pp. 132–147. Although Raymond analyses the case through the English School approach, it is inevitable to set the status concerns from membership in international society.
30 Likhit Dhiravegin, *Siam and Colonialism, 1855–1909: An Analysis of Diplomatic Relations* (Bangkok: Thai Watana Panich, 1964), pp. 59–60.
31 Cited in Stefan Hell, *Siam and the League of Nations: Modernisation, Sovereignty, and Multilateral Diplomacy, 1920–1940*, PhD Dissertation, Leiden University, 2007, p. 25.
32 Cited in Raymond, "War as Membership", p. 138.
33 Cited in Attachak Sattayanurak, *Kanplianplaeng Lokkathat Khong Chonchan Phunam Thai Tangtae Ratchakan Thi 4-Po.So. 2475* [Transformation of the Thai Elite's World View from the Fourth Reign to 1932] (Bangkok: Chulalongkorn University, 1995), p. 204.
34 It was the First State Convention that decreed the change of name from Siam to Thailand.
35 Thailand's Minutes of Cabinet Meeting (TMCM), 1/1938 on 20 December 1938.
36 Judith A. Stowe, *Siam Becomes Thailand: A Story of Intrigue* (Honolulu: University of Hawaii Press, 1991), p. 92.
37 After the Revolution in 1932, the People's Party only held the events to celebrate the Constitution Day on 10 December of each year.
38 Comments by Josiah Crosby. See The National Archives, UK (TNA), FO371/24751, F3326/19/40.
39 Publicity Bureau, *Pramuan Kum Prasai Lae Suntharapojna Khong Nayokratthamontri* [Collections of Words and Speeches of Prime Minister] (Phranakorn: Publicity Bureau, 1940), p. 34.
40 Ibid., pp. 39–40.
41 The National Archives of Thailand (NAT), (3) SoRo 0201.55/7.
42 NAT, (2) SoRo 0201.92.1/7, p. 10.
43 NAT, SoBo. 9 Eak Weesakul.
44 Shane Strate, *The Lost Territories: The Role of Trauma and Humiliation in the Formation of National Consciousness in Thailand*, PhD Dissertation, University of Wisconsin-Madison, 2009, p. 56.
45 Nikorn, cited in Ibid., p. 65.
46 Lilach Gilady, *The Price of Prestige: Conspicuous Consumption in International Relations* (Chicago: University of Chicago Press, 2018).
47 Ibid., p. 59.
48 Ibid., p. 82.
49 Cited in Ibid., p. 82.
50 Wassana Nanuam, "Chinese Submarine Deal Signed", *Bangkok Post*, 5 May 2017 <www.bangkokpost.com/news/security/1244259/chinese-submarine-deal-signed> (accessed 9 May 2019).

References

In English

Anno, Tadashi. 2018. *National Identity and Great-Power Status in Russia and Japan*. New York and Abingdon: Routledge.

Ashley, Richard K. 1986. "The Poverty of Neorealism". In *Neorealism and its Critics*. Edited by Robert O. Keohane. New York: Columbia University Press.

Baker, Chris and Pasuk Phongpaichit. 2014. *A History of Thailand*. Cambridge: Cambridge University Press.

Barnhart, Joslyn. 2016. "Status Competition and Territorial Aggression: Evidence from the Scramble for Africa". *Security Studies*, Vol. 25, No. 3: 385–419.

Battye, Noel Alfred. 1974. *The Military, Government, and Society in Siam, 1868–1910*. PhD Dissertation, Cornell University.

Deng, Yong. 2018. *China's Struggle for Status: The Realignment of International Relations*. Cambridge: Cambridge University Press.

Charnvit Kasetsiri. 2000. "Siam/Civilisation – Thailand/Globalisation: Things to Come". *Thammasat Review*, Vol. 5, No. 1: 114–133.

Chula Chakrabongse. 1960. *Lords of Life: The Paternal Monarchy of Bangkok, 1782–1932*. London: Alvin Redman.

Cooper, David A. 2011. "Challenging Contemporary Notions of Middle Power Influence: Implications of the Proliferation Security Initiative for Middle Power Theory". *Foreign Policy Analysis*, Vol. 7, No. 3: 317–336.

de Carvalho, Benjamin and Neumann, Iver B. 2015. "Introduction". In *Small State Seeking Status: Norway's Quest for International Standing*. Edited by Benjamin de Carvalho and Iver B. Neumann. New York: Routledge.

Galtung, Johan. 1964. "A Structural Theory of Aggression". *Journal of Peace Research*, Vol. 1, No. 2: 95–119.

Gilady, Lilach. 2018. *The Price of Prestige: Conspicuous Consumption in International Relations*. Chicago: University of Chicago Press.

Greve, Andrew Q. and Levy, Jack S. 2018. "Power Transitions, Status Dissatisfaction, and War: The Sino-Japanese War of 1894–1895". *Security Studies*, Vol. 27, No. 1: 148–178.

Hell, Stefan. 2007. *Siam and the League of Nations: Modernisation, Sovereignty, and Multilateral Diplomacy, 1920–1940*. PhD Dissertation, Leiden University.

Jittipat Poonkham. 2019. "Why Is There no Thai (Critical) International Relations Theory? Great Debates Revisited, Critical Theory and Dissensus of IR in Thailand". In *International Relations as a Discipline in Thailand: Theory and Subfields*. Edited by Chanintira na Thalang, Soravis Jayanama, and Jittipat Poonkham. New York and Abingdon: Routledge.

Kislenko, Arne. 2002. "Bending with the Wind: The Continuity and Flexibility of Thai Foreign Policy". *International Journal*. Vol. 57, No. 4: 537–561.

Larson, Deborah W. and Shevchenco, Alexei. 2010. "Status Seekers: Chinese and Russian Responses to U.S. Primacy". *International Security*, Vol. 34, No. 4: 63–95.

———. 2019. *Quest for Status: Chinese and Russian Foreign Policy*. New Haven: Yale University Press.

Lebow, Richard Ned. 2010. *Why Nations Fight? Past and Future Motives for War*. Cambridge: Cambridge University Press.

Likhit Dhiravegin. 1964. *Siam and Colonialism, 1855–1909: An Analysis of Diplomatic Relations*. Bangkok: Thai Watana Panich.

Maass, Matthias. 2009. "The Elusive Definition of the Small State". *International Politics*, Vol. 46, No. 1: 65–83.

Ministry of Foreign Affairs of Kingdom of Thailand. 2018. *The 20-Year '5S' Foreign Affairs Masterplan*. Published online 20 August <www.mfa.go.th/testweb/contents/files/customize-20180724-142633-552037.pdf> (accessed 1 May 2019).

Murashima, Eiji. 1988. "The Origin of Modern Official State Ideology in Thailand". *Journal of Southeast Asian Studies*. Vol. 19, No. 1: 80–96.

Neuchterlein, Donald E. 1965. *Thailand and the Struggle for Southeast Asia*. Ithaca: Cornell University Press.

Pavin Chachavalpongpun. 2012. "Thailand: The Enigma of Bamboo Diplomacy". In *Routledge Handbook of Diplomacy and Statecraft*. Edited by B.J.C. McKerscher. New York and Abingdon: Routledge.

Raymond, Gregory V. "War as Membership: International Society and Thailand's Participation in World War I". *Asian Studies Review*. Vol. 43, No. 1: 132–147.

Renshon, Jonathan. 2017. *Fighting for Status: Hierarchy and Conflict in World Politics*. Princeton: Princeton University Press.

Renshon, Jonathan and Huth, Paul. 2014. "Reputation and Status as Motives for War". *Annual Review of Political Science*, Vol. 17: 371–393.

Strate, Shane. 2009. *The Lost Territories: The Role of Trauma and Humiliation in the Formation of National Consciousness in Thailand*. PhD Dissertation, University of Wisconsin-Madison.

Stowe, Judith A. 1991. *Siam Becomes Thailand: A Story of Intrigue*. Honolulu: University of Hawaii Press.

Thongchai Winichakul. 2000. "The Quest for 'Siwilai': A Geographical Discourse of Civilisational Thinking in the Late Nineteenth and Early Twentieth-Century Siam". *The Journal of Asian Studies*, Vol. 59, No. 3: 528–549.

Vella, Walter F. 1978. *Chaiyo!: King Vajiravudh and the Development of Thai Nationalism*. Honolulu: University of Hawaii Press.

Ward, Steven. 2017. *Status and the Challenge of Rising Powers*. Cambridge: Cambridge University Press.

Wassana Nanuam. 2017. "Chinese Submarine Deal Signed". *Bangkok Post*. Published online 6 May <www.bangkokpost.com/news/security/1244259/chinese-submarine-deal-signed> (accessed 9 May 2019).

Wendt, Alexander. 2004. "The State as Person in International Theory". *Review of International Studies*, Vol. 30, No. 2: 289–316.

Wohlforth, William and Kang, David C. 2019. *Hypotheses on Status Competition*. Paper presented at the Annual Meeting of the American Political Science Association, Toronto, Canada <https://ssrn.com/abstract=1450467> (accessed 2 May 2019).

In Thai

Attachak Sattayanurak. 1995. *Kanplianplaeng Lokkathat Khong Chonchan Phunam Thai Tangtae Ratchakan Thi 4-Po. So. 2475 [Transformation of the Thai Elite's World View from the Fourth Reign to 1932]*. Bangkok: Chulalongkorn University.

Ministry of Foreign Affairs of Kingdom of Thailand. 2019. *Phaen Maebot Dan Kantangpathet Raya 20 Pi (PhoSo 2561–2580)* [Masterplan of Foreign Affairs for Twenty Years (BE 2561–2580)]. 11 January <www.mfa.go.th/main/contents/files/policy-20190111-114336-037494.pdf> (accessed 1 May 2019).

Publicity Bureau. 1940. *Pramuan Kum Prasai Lae Suntharapojna Khong Nayokratthamontri [Collections of Words and Speeches of Prime Minister]*. Phranakorn: Publicity Bureau.

33
THAILAND AND THE GREAT POWERS

Matthew Phillips

Between 1765 and 1767, the Siamese Kingdom of Ayutthaya was defeated by the Konbaung dynasty of Burma. The desecration of the capital was a profound humiliation to the Siamese elite and the memory remains raw today.[1] Yet in the decades that followed, the region entered a new era in which Western rather than Asian powers came to dominate. By the end of the 19th century, the Burmese monarch had been dethroned by British officials, who ran Burma as a province of India. Siam was the only regional polity free from direct imperial rule, governed by a series of Buddhist monarchs who attempted to rule with the absolute power they regarded as their birthright.

While Siam had been successful in avoiding the fate of the Burmese monarchy, the near constant threat of colonisation and overthrow from within had a significant impact on the psyche of the kingdom's elite. The acquisition of large swathes of territory by the West was driven by an imperial worldview that regarded pre-existing forms of government as despotic, inefficient, moribund, or simply tradition bound. As a consequence, Thai leaders became committed to contesting the despotic nature of their own power. Anxious to protect their status as righteous and moral leaders and to inoculate their rule from the humiliation suffered by neighbours, they fostered strong and lasting relations with the Great Powers. In so doing, they entangled themselves in relentless acts of mimesis that risked undermining existing cultural norms as well as the relationship with the people over whom they sought to rule. This chapter argues that sensitivity about Thailand's cultural presence among the Great Powers, and the struggle to balance such concerns with internal political developments, had a major, if definitive, impact on the Thailand's emergence as a modern nation-state.

A world re-ordered

At the beginning of the 19th century, Siam's foreign relations were dominated by an intra-Asian diplomacy run out of the Grand Palace. If there were a relationship to a "great power" it was with China, but the primary focus was on the extraction of tribute from smaller regional kingdoms. When the British Governor of Bengal sent a mission to Siam in 1821, the Siamese appeared indifferent – proof to one British official that they were classic despots. Equally astonishing, the Siamese appeared to interpret the British approach as a sign of their own innate superiority. The lack of interest in free trade and an apparent refusal to show proper deference to British interlopers was proof that Siam was a country "under the most debasing tyranny" and "too low in the scale of nations to be able to form a just estimate of the advantages of friendly intercourse".[2]

In truth, the Siamese were shrewd diplomats and acutely aware of shifts in the balance of power, both regionally and globally. Britain's encroachment on Siamese territory to the south made trade with the newly established port of Singapore an instant priority.[3] The realities were underscored by the first Anglo-Burmese War (1824–1826). Fought with modern weapons including metal gunboats, the war marked a dramatic departure from conflicts of the past. The Burmese king was subject to a crushing defeat and King Nang Klao (r. 1821–1851) was compelled to sign Siam's first treaty with Britain. On his deathbed, he predicted that wars with Vietnam and Burma were things of the past: the threat now came from the West.[4]

In April 1855, King Nang Klao's successor and younger half-brother, King Mongkut (r. 1851–1868), signed the Treaty of Friendship and Commerce between Great Britain and Siam. The Bowring Treaty, as it became known, placed British subjects under the legal jurisdiction of a British consul, making them exempt from local law. They were free to trade and to practice Christianity, and British war ships were allowed to approach the mouth of the Chao Phraya River. Opium, meanwhile, could be imported duty free on condition it was sold to the royally designated "opium farmer" and his agents.

The treaty replicated many of the semi-colonial conditions placed on other Asian polities and fundamentally disrupted existing foreign relations. At the same time, the promise of increased trade, managed from within the palace walls, signaled opportunity for a favoured elite, the most prominent of whom had actively facilitated the agreement.[5]

King Chulalongkorn: an English gentleman?

King Chulalongkorn (r. 1868–1910) succeeded to the throne as a determined youth of 15, trained by his father in ancient and modern statecraft. In early 1872 he visited India, stopping first at Rangoon before travelling on to Calcutta. It was a bold move, a demonstration early in his reign that the British imperial state would serve as a model for his own evolving administration. The visit was also an exquisite piece of political theater that placed the Burmese king, Mindon Min (r. 1853–1878), in negative juxtaposition to his Siamese counterpart. Arriving in Rangoon on a steam yacht, Chulalongkorn entered what was now British sovereign territory. While King Mindon Min had attempted to modernise, his kingdom was much diminished. "British Burma", having been newly integrated into the world economy, had attracted large numbers from the old royal centre in the north, undermining the prestige of the Burmese throne.[6] The British, meanwhile, continued to view the Burmese monarchy as despotic – overly committed to traditional forms of celestial power and, ultimately, unfit to govern. In contrast, British officials who received King Chulalongkorn marvelled at his ability to behave and dress like an English gentleman. It was noted how his entourage shunned "native Court prejudices" and showed little interest in conserving the "manners and customs prevalent amongst neighboring Asiatic cognate races".[7]

At his second coronation in 1873, King Chulalongkorn officially ended the custom of prostration, proclaiming his wish to remove that which was "oppressive and burdensome".[8] Another dramatic and effective act of political theater, the move was intended to impress foreigners and stun local audiences. It was an approach that would become standard practice in his court and in those of his successors.

In embracing "Western" reforms like centralised taxation, and by "opening the country" to foreign trade, King Chulalongkorn courted the approval of British officials while consolidating his power as a *Dhammaraja* (also Thammaracha) or Buddhist "righteous ruler". Later, having secured political power and built up substantial personal wealth, he expanded access to a Western-style education, providing a route for select commoners to enter the royal bureaucracy. At the same time he established yet new forms of hierarchy, offering privileged access to European aristocratic culture – specifically British. Many of his sons were schooled at Eton College. His son and successor, King Vajiravudh (r. 1910–1925), was trained at Sandhurst Royal Military College and studied law at Oxford.

In the manner of princes everywhere, the above served as much as a finishing school as a genuine education. In a world dominated by Europe, notions of civilisational hierarchy determined by relative levels of culture took precedence in the international order. Thongchai Winichakul notes the essential ambiguity in what constituted things *siwilai* in Siam, a term that included almost anything from "new roads, electricity, new bureaucracy, courts and judicial system, law codes, dress codes, and white teeth".[9] Less ambiguous was the fact that civilised lifestyles were intended as the exclusive province of the elite. Vast palaces imitating European environs were constructed in Bangkok, while the etiquette of dress became highly fetishised in the manner of the British aristocracy.[10] The desire to be "up with the times", therefore, did as much to establish new layers of distinction between royalty and commoners (*phrai*) as it did to undermine existing social structures.[11]

A tributary state?

Prior to the onset of European imperialism, the kingdoms of mainland Southeast Asia were organised as celestial, centre-orientated spaces that were in many ways the antithesis of the modern nation-state, defined by clear boundaries.[12] By the late 19th century, a commitment to clear borders had, however, become as much a concern for Siam's royal elite as for the imperial powers. The mini-kingdoms and principalities from which kings had claimed tribute were now integral to the emergent nation-state as well as the maintenance of royal prestige. Initially, the unequal treaties helped extend the power of Bangkok over tributary kingdoms such as Chiang Mai to the north – a territory that had previously fluctuated in and out of control of rival Burmese and Thai rulers. For the British particularly, it was simply easier to recognise Chakri claims of sovereignty over such areas.[13]

Yet in 1893, a crisis erupted when Siamese claims over present-day Laos resulted in French gunboats breaching the mouth of the Chao Phraya River in what became known as the Pak Nam incident. In the short-lived battle that ensued, the French emerged victorious. With French gunboats hovering in sight of his palace, Chulalongkorn was forced to concede territory, agreeing to even more onerous treaty arrangements that expanded French influence. According to Shane Strate, the crisis humiliated Siam's leaders "by exposing the futility of their efforts to achieve parity with the west".[14] Claims that the Siamese king was a universal monarch as dictated by Thai Buddhist tradition were reduced to rubble. At best, he was but a king among many. At worst, he risked appearing the leader of a tributary state under indirect colonial rule.

Royal power weakened even further following King Chulalongkorn's death in 1910. His British education aside, Vajiravudh struggled to present himself as a moderniser. Across the world, absolute monarchies were falling to popular movements and the discourse of progress was becoming ever more linked to democratic government. The crown once again became vulnerable to charges of despotism. Luxury palaces were now exemplars of a moribund and extravagant regime. A revolution in 1932 replaced the so-called absolute monarchy with a constitutional government led by a group of civilian and military officials, their attention fixed on returning the country to their version of a civilised trajectory.

Japan and the greater Thai empire

In the 1930s, Thai foreign policy shifted once again towards Asia. The Great Depression caused the collapse of rice prices, placing the population under considerable financial strain. Accusations of economic mismanagement proliferated, as did the demand for cheap alternatives to British imports. Japanese products began to dominate local markets.[15] In December 1935, Pridi Banomyong, a key architect of the revolution, travelled to Tokyo to discuss the growing trade deficit. While there, the Japanese suggested ways they might help develop the Thai economy. The following year the Japanese sent a trade mission to Siam and investment duly increased.[16] With the outbreak of war in Europe in September 1939, European imports collapsed and Japan once again filled the gap. The war in China also increased Japanese demand for Thai rice, rubber and scrap iron among other commodities, cementing Japan-Thai cooperation.[17] Throughout the 1930s, relations with Japan provided the Thai military and civilian leadership with opportunities to outmanoeuvre existing trading networks, reinforcing the power of the revolutionary regime.

In many respects, the Japanese were natural allies. Under "absolute monarchy", Siam was oriented towards Europe. At the League of Nations, Siamese elites failed to join Japan in demanding a racial equality clause, a carefully considered move that was part political in nature. Following the First World War, Siam's primary focus was the renegotiation of the unequal treaties and Siamese officials were cautious about upsetting the West. In addition, they regarded the meetings as a stage upon which to demonstrate, and hone, their grasp of diplomatic practice.[18] By 1925, Siamese officials had successfully revised the reviled treaties. Yet the memory of "lost" territories and a continued dependence on European knowledge and culture constrained the emergence of a royalist Thai nationalism. These difficulties were fully exploited by the post-1932 government that wove the narrative of national humiliation deep into the fabric of state ideology. Japan, as one of the few non-Western countries to have escaped direct colonial rule and yet have established a degree of parity with the West, made an obvious partner.

In 1939, the nationalist regime of Field Marshal Phibun Songkhram officially changed the name of the country from Siam to Thailand – an insult to royalty. Phibun and his cohort also began to demand that all Thai citizens, regardless of their background, dress in a "civilised" (*siwilai*) manner and eat a more varied diet. The purpose of such policies was threefold: first, they undermined cultural markers like Western dress as the exclusive province of the elite; second, they transformed Thailand into a country with civilised people on the world stage; third, they claimed to liberate the population into a mass nationalism within which all citizens had a stake, opposite to the paternalistic nationalism expounded by both European imperialism and the monarchy.

Following a brief military campaign in Indochina, the Japanese helped the Phibun regime negotiate the ceding of territory in modern-day Cambodia from the Vichy French regime in 1941. In December, Japan invaded Southeast Asia and entered the Pacific war. A brief invasion of Thailand resulted in an alliance of friendship, and on 25 January 1942 Phibun officially declared war on Britain and the United States. Once ensconced in Bangkok, however, the Japanese behaved more like occupiers than allies. Amid all talk of a Greater East-Asia Co-Prosperity Sphere, however, Thai leaders struggled to present the nation on a cultural par with their Japanese brothers.[19]

King Bhumibol and the American century

Serious US interest in Thailand began during the Second World War, when Thailand's declaration of war was somehow never formally handed over to American counterparts. Instead, the Thai ambassador to the United States, Seni Pramoj – himself of royal descent – helped set up the "Free Thai": a group of US-based Thai elite who conspired to overthrow Phibun and reoccupy the country. Operating under the auspices of the Office of Strategic Services (OSS), precursor of the CIA, the Free Thai never formally invaded Thailand. The movement was successful, however, in bringing forces opposed to the militaristic Phibun into temporary coalition. Included were the key royalists, maligned since 1932. In the decades to follow, US officials played a central role in reinforcing the position of royalists within the Thai political system.

Despite having been abruptly removed in 1944, Phibun became prime minister for a second time in 1948. His artful presentation as a staunch anti-communist, for the time being, allowed him to ward off US concerns about his return to power. By the outbreak of the Korean War, US priorities were fixed on preventing the spread of communism in Asia. In this enterprise, Phibun's Thailand became an important ally. US aid to Thailand rose dramatically, bolstering the power of both the Thai armed forces and the Royal Thai Police: bitter rivals and economic competitors.

The United States also supported a raft of enormous development projects that increased central control over the periphery and opened up the economy. The new Friendship Highway, for instance, eventually connected Bangkok to the Laotian border, expanding opportunities for Bangkok investors to cash in on the development-led boom that followed.[20] Another major winner was Police Chief Phao Sriyanond, who worked with the CIA to set up the Border Patrol Police (BPP), and who, as a consequence, gained control over vast opium exports out of the Golden Triangle.[21]

However, Phibun found himself increasingly in competition with the determined King Bhumibol Adulyadej (r. 1946–2016). Like King Chulalongkorn before him, Bhumibol was young when he came to the throne. Born in the United States, having grown up in Switzerland, Bhumibol, like his grandfather, was able to demonstrate knowledge of current world civilisation far superior to that of the military and civilian leaders who headed the post-revolutionary government. A lover of jazz and aficionado of the clarinet, Bhumibol naturally performed an easy cosmopolitanism, appearing at ease when conversing with US officials and wearing an American-style suit. But he also sprung from a royal bloodline. Maddening to Phibun, who worked tirelessly to assert his legitimacy to the local population, the young king was much better placed to present as an authentic Thai-Buddhist ruler in an American age.[22]

Through the 1950s, key architects of US policy in Thailand recognised the role a pro-American monarch might play in entrenching US power for decades. More important, however, was the US public's intoxication with its own imperial ideology. Like the British before them, this ideology presented the world as a hierarchy, the difference being it was now the United States that served as guide to the universal route to modernity.

It was a new worldview exemplified by Rodgers and Hammerstein's popular Broadway musical, *The King and I*. Released on Broadway in 1951 and later, in 1956, as an equally popular motion picture, the production played a vital role in fetishising Thai royal power to a generation of Americans caught in the grip of Cold War ideology. Rather than identifying a classic unreformable despot, *The King and I* showed the historic figure of King Mongkut as a sympathetic yet troubled character forced to manage Siam's transition from tradition-bound society to modern nation on the world stage. Closely reflecting US imperial ideology, the play portrayed modernisation as an inevitable rationalising process, at the same time allowing for select social norms to be recast as cultural characteristics that could be maintained, or even protected, alongside change. As a result, pre-colonial markers of Thai kingship, including full-face prostration in the presence of royals, could now be resurrected as apparently benign expressions of Thai tradition.[23]

In 1957, Phibun was removed from power in a coup. One year later, Field Marshal Sarit Thanarat installed himself as prime minister. Sarit's pro-royalist regime was at once brutal and authoritarian, anti-communist and pro-American. Falling in line with US imperial strategy, Sarit promoted development as a set of processes and knowledges imported from the outside. At the same time, he enabled a royalist vision of Thai culture that actively conservatised nationalist rhetoric. The Tourism Authority of Thailand, established to promote tourism, doubled as a vehicle for royal ceremony and ritual to be reinstated at the centre of national life.[24] King Bhumibol executed his role exquisitely: earnestly performing the ritual duties of a tradition-bound monarch while demonstrating a unique ability to exploit knowledge from the "free world" so as to benefit the Thai people. The effect was to ceaselessly entwine the Cold War ideology of modernisation with his own public persona. In so doing, he bound Thai state ideology to US objectives in Southeast Asia for generations.

Beyond American hegemony

Consummate performances aside, a series of events from the late 1960s altered the nature of Thai-US relations. US rapprochement with China and retrenchment from the conflict in Vietnam reduced US interest in Thailand. By the early 1970s, student protests in Bangkok routinely rallied against the US military presence. In response, the democratically elected government of Kukrit Pramoj moved towards creating a more flexible foreign policy. In July 1975, Kukrit travelled to China to formally normalise relations and in March 1976 he secured the withdrawal of US troops.[25]

Resistance to the new diplomatic environment erupted in late 1976.[26] By then, royalist nationalism was so dependent upon US hegemony that détente was interpreted as a threat to the revival of royal fortunes. Moreover, after decades of propaganda, those who voiced opposition to such hegemony risked becoming a target of anti-communist violence. On 6 October, in a frenzy of anti-communist-fuelled violence, state-sponsored paramilitary groups massacred students at Thammasat University. Many young activists fled to the countryside where they joined the communist insurgency – a move that helped legitimise the maintenance of the Cold War security state. A coup in the aftermath of the massacre saw the installation of a right-wing royalist regime that rejected détente.

However, another coup in 1977 saw the installation of Kriangsak Chamanan, an army general who embraced the changed global environment and recognised the need for closer relations with both China and the Soviet Union. He also moved to de-escalate the Cold War at home. While Kriangsak was replaced as prime minister in 1980, many of his foreign policy priorities survived the decade to come. Between 1980 and 1988, General Prem Tinsulanonda oversaw a "quasi-democracy" in which he gained personal legitimacy by claiming to be a clean, professional commander in chief of the army, a staunch defender of the monarchy, and a champion of a consensus approach to politics.[27] A so-called omnidirectional foreign policy balanced cooperation with China, a regional focus on ASEAN, and strong relations with the United States, particularly in regard to security. The December 1978 Vietnamese invasion of Cambodia brought renewed US interest to the region. The first annual Cobra Gold, a military training exercise intended to demonstrate the US commitment to the Thai armed forces, took place in 1982.

As the 1980s drew to a close, General Chatichai Choonhaven's government (1988–1991) moved to reduce military control over the state and focus instead on expansion into regional markets, including those of former communist enemies. In 1991, a military coup stymied such efforts, but the return of a cold war security state was now rejected. In May 1992, the military gunned down civilian protesters opposed to the military-backed regime. Many felt the United States failed to act quickly enough to restrain their army friends. Returning to civilian rule, Thailand joined the non-aligned movement and renewed efforts to consolidate trading relationships in Asia. By then the country was widely regarded as a primary "tiger" economy, and with the economy booming, the security relationship with the United States looked increasingly redundant.

In the United States, meanwhile, institutions such as the International Monetary Fund (IMF) and the World Bank had become heavily populated with technocrats who favoured the unleashing of markets from "crony capitalism". Cautious about the rapid rise of Asian economies, they warned against coalitions of businessmen and politicians who secured financial returns through "rent seeking" and who were often protected by a strong authoritarian state.[28] When the Asian Financial Crisis hit in 1997, key individuals in Washington interpreted it as proof that this "Asian model of capitalism" had failed.[29] In Thailand, as elsewhere, the IMF tied financial support to structural reforms that removed obstructions to free markets and facilitated foreign investment. By forcing such policies after the depreciation of the baht, they allowed for US businesses to purchase Thai businesses at rock-bottom prices. Critics claimed the crisis was being used by the United States to secure its own interests.[30] In Thailand, anti-American sentiment swelled.

Chinese-Thai relations in a changing world

The Thai elite have always maintained strong links with China. King Taksin (r. 1769–1782), for example, was of Chinese heritage and drew significantly on links with China to rebuild the country in the wake of its defeat by the Burmese. Throughout the 19th and early 20th centuries waves of Chinese migrants travelled to Siam, where many settled permanently. Some of the wealthiest Chinese families acted as agents in the service of royals, offering daughters to the royal harem and providing vital access to the lucrative Chinese market. Others were able to take on more substantive economic roles because, unlike the local population, they were exempt from corvée duties.[31]

Initially, however, Thai leaders struggled to integrate the Chinese into emergent notions of the Thai nation. First under King Vajiravudh and later Phibun, Chinese communities were labelled foreign and actively discriminated against. Particularly under the first Phibun regime (1938–1944), Chinese dominance over the economy made it easy to portray them as a predatory class intent on exploiting the local population, particularly farmers.[32] Following the war, anti-Chinese sentiment remained, particularly after the fall of China to the Communists in 1949. Throughout the Cold War, local Chinese communities were viewed as a potential source of communist sympathy. At the same time, however, the US-inspired economic agenda of the post-1958 government also privileged the mobilisation of Chinese capital over previous attempts to protect the Thai "national" community. Many of the practical restrictions placed on Sino-Thai capitalists were dropped and the emphasis shifted towards encouraging integration.[33]

The Thai monarchy also re-formed the close alliance with Chinese capital. Beginning in the 1960s, and expanded under Prem, Sino-Thai capitalists were encouraged to support royal projects as a means of gaining access to royal business networks.[34] The former were small development projects, established initially in remote hill-tribe regions deemed at risk of communist insurrection. Throughout the 1980s, such projects were rebranded economic enterprises that supported distant rural communities to maintain a subsistence way of life. From the 1960s on, Chinese capitalists were thus brought more formally into the social and cultural milieu of the Thai elite. Gradually, anti-Chinese sentiment across Thai society declined. During the tiger years, wealthy Chinese were even embraced as model entrepreneurs whose personal stories were marketed through a raft of self-help books and whose access to East Asian culture and enterprise was seen as a benefit to nation.[35]

In the aftermath of the 1997 crisis, anti-American sentiment was particularly noticeable among urban intellectuals and capitalists, many with Chinese heritage. Their anger was understandable. By framing the crisis as the product of crony capitalism, Alan Greenspan and company threatened to resurrect the racism of previous generations. In May 1998, after the crisis spread to Indonesia, the Chinese community was scapegoated and hundreds of them were killed in widespread attacks. In Thailand, as millions contended with a dramatic decline in living standards (per capita income declined by about 20 per cent in rural Thailand), Chinese-Thai communities rallied to demonstrate a commitment to the Thai nation, a love for the king, and anger towards the United States.[36]

Thaksin Shinawatra, himself of ethnic Chinese heritage, was elected prime minister in January 2001 upon a Thai nationalist platform. This included promises of universal health care and village loan schemes, as well as a rejection of the so-called Washington Consensus espoused by the IMF. Once in office, Thaksin embraced the Chinese community, populating over half his first cabinet with its members. He also moved to firm up relations with the PRC and expand economic co-operation. China was now emerging as a global economic power, and a majority of Thais believed China would become their country's most import ally.[37]

The royal family also began to embrace China. By the late 1990s, Princess Sirindhorn was making regular trips, and in 2000 she was awarded a Language, Culture and Freedom Award by the Chinese state. Ultimately, however, the royalist ideology espoused by King Bhumibol struggled to decouple from the American discourse upon which it had piggybacked since the 1950s. In the 1980s, Reaganomics argued that wealth in a free market, low-tax economy trickled down to all. The ideological parallel in Thailand was that unfettered capitalist accumulation and consumer spending could be made virtuous through donations to royal charities and projects: a royally managed "trickle" that bestowed virtue on the patron. In 1992, Bhumibol personally associated himself with the transfer of power from military dictatorship to liberal democracy and free markets – latching on the moment Francis Fukuyama described as the "end of history". Then, following the 1997 crisis, Bhumibol pitched the so-called sufficiency economy. Drawing on Buddhist themes of moderation and simple living, this called for rural Thais to return to village-based subsistence farming. Once again, the philosophy was a thinly veiled reproduction of an American idea in Thai nativist clothing – this time, the austerity economics espoused by the IMF.

As royal power faltered, Thaksin was increasingly viewed as a threat. While of Chinese heritage, Thaksin was from outside the old Chinese-Thai-royal nexus. His connections with China were largely maintained through a network that circumvented the old guard. At the same time, the palace had grown frustrated by his tendency to usurp royal privilege and treat the state as a personal fiefdom. From 2005, the language of late 20th-century civilisation was redeployed against Thaksin by a growing list of enemies. Despite being the most popular prime minister in Thai history, he was deposed in a coup in September 2006. His crime, apparently, was cronyism, corruption and populism. Led by a royalist elite and capitalist class that had cultivated vast fortunes during the Cold War, the putsch has been described as "the last American coup".[38]

Yet by the early 20th century, US interest in Thailand was minimal and the Cold War alliance was a distant memory. Having once supported military government and promoted Bhumibol as a stabilising figure, US framings of Thailand on the world stage now began to shift dramatically. The publication of a critical biography of Bhumibol by an American academic publisher in 2006 (*The King Never Smiles*) was a humiliation. Following the coup, Thailand descended into a decade of political crisis in which frequent battles between Thaksin's supporters and detractors dominated. A violent crackdown on Thaksin-cheering Red Shirts in 2010 resulted in two separate massacres. The American media reported the killings as an attack on democratic government and accused Thailand's so-called liberal elite of hypocrisy. Across social media, anti-Thaksin groups voiced anger at the failure of the Western media to label the Red Shirts terrorists.

A military coup in May 2014, led by military chief General Prayuth Chan-ocha, removed another Thaksin-allied government. Western media outlets once again reported the move as a backward step and US officials openly reprimanded the regime. Coup-makers in 2006 had reassured the world they were committed to civilian rule and elections. In 2014, the promises were much looser. Democratic government would return, but not before a period of significant reform overseen by the military government. Supporters of Prayuth openly rallied against the United States, issuing violent verbal attacks on the Obama government and even protesting outside the US embassy against American "neo-colonialism".

Conclusion

Since the election of Donald Trump in 2016, the United States has shown much less interest in reprimanding the Prayuth regime for its draconian and authoritarian rule. Meanwhile, China has overtaken the United States as Thailand's top trading partner. Some commentators are now suggesting the emergence of a new world order where closer relations with China, combined with a declining West, facilitate a distinctly authoritarian turn across Asia.

Yet, charges of despotism still hurt. An article published in *Time* magazine in 2018 sparked fury within the Thai establishment for suggesting that Thailand was undergoing a "permanent authoritarian regression" and for referring to Prayuth by his erstwhile nickname "Little Sarit", the aforementioned Cold War dictator.[39] Prayuth also took issue with the accusation that increased economic involvement with China indicated a realignment in Thai foreign policy. The protest demonstrated that, despite the changing international environment, sensitivity over Thailand's representation on the world stage continues. It also shows how after two centuries of Great Power politics involving the West, Thailand's leaders remain unsure about where to head next.

Notes

1 See Pavin Chachavalpongpun, *A Plastic Nation: The Curse of Thainess in Thai-Burmese Relations* (Lanham, MD: University Press of America, 2005).

2 George Finlayson Esq., *The Mission to Siam, and Hué the Capital of Cochinchina in the Years 1821–22* (London: John Murray, Albemarle-Street, 1826), p. 167.

3 Kullada Kesboonchoo-Mead, *The Rise and Decline of Thai Absolutism* (London and New York: Routledge, 2006), p. 25.

4 Chris Baker and Pasuk Phongpaichit, *A History of Thailand* (Cambridge: Cambridge University Press, 2005), p. 39.

5 Kullada Kesboonchoo Mead explains that while there was an element of threat placed upon Mongkut to sign the Bowring Treaty, "Those in the Saimese elite who saw potential benefits from the trade treaty welcomed the British demands". It was their hard work that brought about conclusion of it. Kesboonchoo-Mead, *The Rise and Decline of Thai Absolutism*, p. 31. Also, B.J. Terwiel, "The Bowring Treaty: Imperialism and the Indigenous Perspective", *Journal of the Siam Society*, Vol. 79, No. 2 (1991), pp. 40–47.

6 Thant Myint-U, *The Making of Modern Burma* (Cambridge: Cambridge University Press, 2001), p. 150.

7 Major E.B. Sladen to C.U. Attchison Esq., C.S.I., Secretary to the Government of India, Foreign Department. "Visit of King of Siam to India", taken from Sachchidanand Sahai, *India in 1872: As Seen by the Siamese* (New Delhi: B.R. Publishing Corporation, 2002), p. 400.

8 Ernest Young, *The Kingdom of the Yellow Robe: Being Sketches of the Domestic and Religious Rites and Ceremonies of the Siamese* (Edinburgh: A. Constable & Company, 1900), p. 132.
9 Thongchai Winichakul, "The Quest for 'Siwilai': A Geographical Discourse of Civilisational Thinking in the Late Nineteenth and Early Twentieth-Century Siam", *The Journal of Asian Studies*, Vol. 59, No. 3 (2000), pp. 528–549.
10 For more on the culture of the Siamese court during the late 19th and early 20th centuries, see Maurizio Peleggi, *Lord of Things: The Fashioning of the Siamese Monarchy's Modern Image* (Honolulu: University of Hawaii Press, 2002).
11 Christopher Bayly describes the phenomenon of "being modern" as "thinking you are modern", an aspiration to be "up with the times" and a process of emulation and borrowing. See C.A. Bayly, *The Birth of the Modern World 1780–1914: Global Connections and Comparisons* (Oxford: Blackwell Publishing, 2004), p. 10.
12 Thongchai Winichakul, *Siam Mapped: A History of the Geo-Body of a Nation* (Chiang Mai: Silkworm Books, 1995), pp. 95–112. Also see Stanley J. Tambiah, "The Galactic Polity: The Structure of Traditional Kingdoms in Southeast Asia", *Annals of the New York Academy of Sciences*, Vol. 239, No. 1 (1977), pp. 69–97.
13 Shane Strate describes an incident whereby a legal dispute between the British and sovereign of Lan Na, Kawilirot, was settled when Mongkut ordered him to respect the Bowring treaty. The "unique requirements of the Bowring Treaty", he therefore notes, "resulted in a more centralised approach to vassal-overlord relations". In Shane Strate, *The Lost Territories: Thailand's History of National Humiliation* (Honolulu: University of Hawaii Press, 2015), p. 30.
14 Shane Strate, *The Lost Territories*, p. 31.
15 William L. Swan, *Japan's Economic Relations with Thailand: The Rise to "Top Trader" 1875–1942* (Bangkok: White Lotus, 2009), p. 55.
16 Ibid., p. 101.
17 Ibid., pp. 106–107.
18 For more on Siam at the League of Nations, see Stefan Hell, *Siam and the League of Nations: Modernisation, Sovereignty and Multilateral Diplomacy 1920–1940* (Bangkok: River Books, 2010).
19 Thamsook Numnonda, "Phubunsongkhram's Thai Nation-Building Programme during the Japanese Military Presence, 1941–1945", *Journal of Southeast Asian Studies*, Vol. 9, No. 2 (1978), pp. 234–247.
20 Benedict Anderson, "Withdrawal Symptoms: Social and Cultural Aspects of the October 6 Coup", *Bulletin of Concerned Asian Scholars*, Vol. 9, No. 3 (1977), p. 15.
21 Alfred W. McCoy, *The Politics of Heroin: CIA Complicity in the Global Drug Trade* (New York: Lawrence Hill Books, 2003), pp. 191–192.
22 This is described by Christine Gray as a "race for virtue". See Christine Gray, *Thailand: The Soteriological State in the 1970s*, PhD thesis, University of Chicago, 1986.
23 Matthew Phillips, "Ancient Past, Modern Ceremony: Thailand's Royal Barge Procession in Historical Context", in *How the Past Was Used: Historical Cultures c. 750–2000*, edited by Peter Lambert and Bjorn Weiler (Oxford: Published for The British Academy by Oxford University Press, 2017).
24 Matthew Phillips, *Thailand in the Cold War* (Abington: Routledge, 2016).
25 Paul Chambers, "U.S.-Thai Relations after 9/11: A New Era in Cooperation?" *Contemporary Southeast Asia*, Vol. 26, No. 3 (2004), p. 461.
26 Jittipat Poonkham argues that the assertion that Thai leaders historically "bent with wind" in their dealing with foreign powers emerged towards the end of the 1960s in response to détente with the Communist powers. Forthcoming, *A Genealogy of Thai Détente: Discourses, Differences and Decline of Thailand's Triangular Diplomacy (1968–1980)*, a Thesis submitted to Aberystwyth University, 2018.
27 Saitip Sukatipan, "The Evolution of Legitimacy", in *Political Legitimacy in Southeast Asia: The Quest for Moral Authority*, edited by Muthiah Alagappa (Stanford: Stanford University Press, 1995), p. 215.
28 Pasuk Phongpaichit and Chris Baker, *Thailand's Crisis* (Chiang Mai: Silkworm Books, 2000), p. 7.
29 Specifically, Alan Greenspan (then Chairman of the Federal Reserve) and Larry Summers (then Under-Secretary of the US Treasury). See Ajit Singh and Ann Zammit, "Corporate Governance, Crony Capitalism and Economic Crises: Should the US Business Model Replace the Asian Way of 'Doing Business'", *Corporate Governance*, Vol. 14, No. 4 (2006), p. 220.
30 The key critique came from Paul Krugman, who delivered a conference paper in 1998. See Paul Krugman, "Fire-Sale FDI", in *Capital Flows and the Emerging Economies: Theory, Evidence, and Controversies*, edited by Sebastian Edwards (Chicago: University of Chicago Press, 2000), pp. 43–58.

31 Wasana Wongsurawat, "Beyond Jews of the Orient: A New Interpretation of the Problematic Relationship between the Thai State and Its Ethnic Chinese Community", *Positions: East Asia Cultures Critique*, Vol. 24, No. 2 (2016), pp. 564–569.
32 Ibid.
33 Sittithep Eaksittipong, "From Chinese 'in' to Chinese 'of' Thailand: The Politics of Knowledge Production during the Cold War", *Rian Thai*, Vol. 10, No. 1 (2017), p. 99.
34 Christine Gray, "Hegemonic Images: Language and Silence in the Royal Thai Polity", *Man*, New Series, Vol. 26, No. 1 (1991).
35 Craig J. Reynolds, "Globalisers Versus Communitarians: Public Intellectuals Debate Thailand's Futures", *Singapore Journal of Tropical Geography*, Vol. 22, No. 3 (2001): 255.
36 Some of these developments are outlined in: Somsak Jeamteerasakul, "Mass Monarchy", *Yum Yuk Rug Samai: Chalerm Chalong 40 Pi 14 Tula* (Bangkok: Heroes of Democracy Foundation and 14 Tula Committee for Democracy, 2013), pp. 107–118. However, the Chinese community in Thailand is not a singular entity and other work has explored the complex nature of Chinese belonging to the nation, including the tensions that engulfed the community following the Asian Financial Crisis. See, for example, William A. Callahan, "Beyond Cosmopolitanism and Nationalism: Diasporic Chinese and Neo-Nationalism in China and Thailand", *International Organisation*, Vol. 57, No. 3 (2003), pp. 481–517. Also Kasian Tejapira, "The Misbehaving Jeks: The Evolving Regime of Thainess and Sino-Thai Challenges", *Asian Ethnicity*, Vol. 10, No. 3 (2009), pp. 263–283.
37 Benjamin Zawacki, *Thailand: Shifting Ground between the United States and a Rising China* (London: Zed Books, 2017), p. 111.
38 Ibid, p. 180.
39 Charlie Campbell, "Thailand PM Prayuth Chan-ocha on Turning to China over US", *Time Magazine*, 21 June 2018.

References

Baker, Chris and Pasuk Phongpaichit. 2005. *A History of Thailand*. Cambridge: Cambridge University Press.

Bayly, C.A. 2004. *The Birth of the Modern World 1780–1914: Global Connections and Comparisons*. Oxford: Blackwell Publishing.

Mead, Kullada Kesboonchoo. 2006. *The Rise and Decline of Thai Absolutism*. London and New York: Routledge.

Myint-U, Thant. 2001. *The Making of Modern Burma*. Cambridge: Cambridge University Press.

Pavin Chachavalpongpun. 2005. *A Plastic Nation: The Curse of Thainess in Thai-Burmese Relations*. Lanham, MD: University Press of America.

Peleggi, Maurizio. 2002. *Lord of Things: The Fashioning of the Siamese Monarchy's Modern Image*. Honolulu: University of Hawaii Press.

Phillips, Matthew. 2016. *Thailand in the Cold War*. Abington: Routledge.

Swan, William L. 2009. *Japan's Economic Relations with Thailand: The Rise to "Top Trader" 1875–1942*. Bangkok: White Lotus.

Thongchai Winichakul. 1995. *Siam Mapped: A History of the Geo-Body of a Nation*. Chiang Mai: Silkworm Books.

Young, Ernest. 1900. *The Kingdom of the Yellow Robe: Being Sketches of the Domestic and Religious Rites and Ceremonies of the Siamese*. Edinburgh: A. Constable & Company.

Zawacki, Benjamin. 2017. *Thailand: Shifting Ground between the US and a Rising China*. London: Zed Books.

34

THAILAND'S FOREIGN POLICY TOWARDS NEIGHBOURING COUNTRIES AND ASEAN

Pongphisoot Busbarat

This chapter examines the nature of Thailand's foreign relations towards Southeast Asia, particularly its neighbouring countries and the Association of Southeast Asian Nations (ASEAN) as a group. It demonstrates that Thailand's foreign policy posture towards this region exhibits an aspiration to maintain a leading role and status. The motive unveils another aspect of Thai foreign policy; that is, it tends to place mainland Southeast Asia as its core of foreign policy formulation, and ASEAN is deemed to be an extension of this policy. Although ASEAN has become an important platform in the post-Cold War era, Thailand's active role in the group is still largely based on its interest on mainland Southeast Asia. As a result, Thailand's regional role in many occasions bypasses ASEAN.

Viewing Thai foreign policy from this perspective contradicts the well-established generalisation that Thai foreign policy is reactive like "bamboo bending with the wind". This image portrays the policy position that is "always solidly rooted, but flexible enough to bend whichever way the wind blows in order to survive".[1] The bamboo diplomacy reflects the nature of small states in the international system that aims to preserve its survival in an anarchical world. In fact, this label has manifested in Thailand's responses to the changing environments since the late 19th century to maintain its survival. Siam, the former name of Thailand, employed different strategies to minimise impacts from European colonialism. Consequently, it partly helped Siam escape colonial rules. European influence was replaced by the United States after the Second World War when Thailand sought the US security umbrella against the emerging communist threats both within and around Thailand. Thailand, then, could escape the fall into the communist bloc predicted by the domino theory. Amid the US withdrawal from Southeast Asian after the Vietnam War in the 1970s, Bangkok once again adjusted its position. The Sino-Thai relationship developed into a de facto security partnership to circumscribe Hanoi's regional hegemonic desire in mainland Southeast Asia.

Bamboo diplomacy is adequate in explaining Thailand's relations with the Great Powers. However, it may not fit well with Thailand's policy posture within Southeast Asia. Therefore, this chapter argues that a significant characteristic of Thai foreign policy towards this region develops around the desire to maintain a leading regional role. Within this policy posture, Thailand's policy towards its neighbouring countries is the core to consolidate such leadership. To demonstrate this, the first section discusses the nature of small states within the international relations (IR) literature, which is the basic assumption for bamboo diplomacy. The next section shows cases in Thailand's foreign policy towards the region that refuse to comply with small-state predictions. In these cases, Thailand's regional aspiration is an important driving force. It is followed by a discussion of Thailand's foreign policy highlights in the post–Cold War period that share this foreign policy posture despite Thailand's commitment within ASEAN.

Bamboo diplomacy: small-state foreign policy in practice

Most of the Thai foreign policy studies are based on the neo-realist paradigm in IR. It forms a general understanding that security is a primary goal of any states. According to Kenneth Waltz, the international system is anarchical without a single authority to govern or provide security assurance for state survival.[2] Therefore, this anarchical nature of the system reminds states to be distrustful of other states' intentions. As a consequence, states always maximise their own security, resulting in the situation of the security dilemma. Actions taken by a state to maximise its security can lead to reactions with similar measures in other states. This will spirally increase tensions, although no one wants the conflict to happen.

In the anarchical world, small states lack power in quantitative terms, especially in their military and economic attributes.[3] In such a country, the leaders "consider that it can never, acting alone or in a small group, make a significant impact on the system".[4] A small power thus often "feel[s] threatened, to some significant and immediate sense, by the play of Great Power politics".[5] Considering its relatively weak power, a small state generally exhibits a reactive and passive posture rather than act as an agent of change.[6] The maintenance of the status quo in the system is, therefore, an essential goal of this reactive foreign policy. Therefore, a small state prefers a policy of neutrality or nonalignment and finds the way in which its involvement in the conflict among great powers is minimal. However, neutrality is difficult to maintain in reality, ironically due to the country's smallness. Balancing and bandwagoning with great powers thus become more realistic to guarantee its survival.[7] If it chooses inappropriate policies, it may end the possibility of free choice. At this juncture, foreign policy has a significant function to protect this survival interest.

Proponents of this approach view Thailand as a small state; hence, its foreign policy behaviour reflects its size and significance within the international system. As the former Thai Foreign Minister Thanat Khoman puts it, "Thailand is not a big or even a medium power, but a small power . . . therefore we have to be realistic and pursue a policy appropriate to timing and commensurate with our capability".[8] Therefore, the flexibility to select an appropriate foreign policy to practically reflect the surrounding environment at a particular time is deemed necessary by its leaders. Scholars of Thai foreign policy subscribing to this proposition argue that as long as the very nature of Thailand as a marginalised state in the international hierarchy does not change, its foreign policy deems to follow the behavioural pattern of a small state. For Thailand, the analogy of bamboo diplomacy reflects its ultimate goal of maintaining survival and national security. That is, it intends to preserve the status quo and continue the relevance of the "national holy trinity", namely, Nation, Religion (Buddhism), and King.[9] There seemed to be no difficulty for the Thai political elite to suddenly change their foreign policy posture when the situation changed, as long as the change served such stability maintenance.

Several notable historical events about Thailand's position during the major wars reflect this behaviour. In the First World War, Thailand officially declared war against the Central Powers, led by the Austria-Hungary, Germany, and the Ottoman Empire in 1917; Thailand became the only Southeast Asian country in the European theatre. This was somewhat unexpected to the public due to Thailand's cordial relations to these countries with no threat to its territorial integrity.[10] Moreover, Germany and Austria contributed to Thailand's balancing strategy against the influence of Britain and France and played an important role in the modernisation and trade for Thailand.[11] However, partly with his careful observation of the situation in Europe and his pro-British background, King Vajiravudh (Rama VI) was convinced that the Allied Powers were gaining the upper hand in the war. He believed that the Allied Powers would win the war, and siding with them would enable Siam to renegotiate unequal treaties with the European nations on an equal footing.[12] Consequently, Siam could regain its full sovereignty, especially over its judiciary and taxation.

Bamboo diplomacy can also explain Thailand's foreign policy during the Second World War. As Thailand wanted to maintain its independence and the status quo during the war, it adjusted its foreign policy position at least three times, from neutrality to taking sides with both the Axis and the Allied Powers. Thailand's neutrality continued until the Japanese troops came ashore in southern Thailand in December 1941. Despite a brief resistance, Bangkok signed an armistice with Tokyo and later joined a military alliance treaty on 21 December 1941. A month later Thailand declared war against Britain and the United States on 25 January 1942. At the same time, a group of Thai military and civilian elites organised an underground movement under the name Seri Thai (Free Thai) against Japan and the co-belligerent Phibun government. The defeat of the Axis in the war threatened the independence of Thailand. However, Thailand's post-war leaders swiftly renounced the country's war declaration and claimed that it was illegitimately enacted without unanimity in the Council of Regency. Moreover, they argued that Thailand was forced to join the Axis while the majority of Thai people disagreed with that decision, as evidenced by the emergence of Seri Thai.[13] With the earlier refusal of the United States to recognise Thailand's war declaration, Thailand was thereafter treated as an enemy-occupied country and was only required to pay reparation in rice to the British and to relinquish its claim on territories gained during the war.[14] Thailand became among the first Axis countries to join the United Nations in December 1946. Thai sovereignty and independence were therefore preserved.

At this juncture, the United States after World War II became "the principle guarantor of [Thailand]'s independence against hostile regional powers".[15] During the peak of the Cold War's ideological conflict in the region, the security syndrome among the Thai elites called for the need of America to help fight against the communist threat perceived to be threatening Thailand's security.[16] The convergence of both the security and the economic interests of the United States and Thailand brought about a close relationship between the two countries. Thailand was thus a key ally in sustaining American hegemony in Asia during the 1950s and 1960s based on two agreements: the Southeast Asian Collective Defence Treaty in 1954 and the Rusk-Thanat Joint Communique in 1962. This security alliance enabled the United States not only to contain the expansion of communism but also to help supply natural resources and a market for the reconstruction project of post-war Japan.[17] In return, Thailand had received a significant assistance package from the United States since the early 1950s, including access to foreign capital through international financial institutions. Embracing the US presence also shielded Thailand from a possible threat from the communist bloc led by both Soviet Union and China after the Second World War, and subsequently Vietnam. Thai foreign policy throughout the 1960s continued to support the US military presence in the region with the latter's involvement in the Vietnam War to protect South Vietnam and Thailand's buffer zone in Laos and Cambodia from falling into the communist's hands.

In the 1970s, international and regional environments facilitated the adjustment of Thai foreign policy. These important events include the split within the communist bloc between Soviet Union and China, the rapprochement between the United States and China in the early 1970s, the failure of the United States in Vietnam War and its plan for military withdrawals from mainland Southeast Asia, as well as the fall of Saigon, Phnom Penh, and Vientiane, all in 1975. Thailand's foreign policy then turned to seek peaceful coexistence with communist China as well as its Indochinese neighbours. This reflected the reality that Thailand had to live with them regardless of their political differences. The normalisation of Sino-Thai relations in July 1975 was a highlight of this foreign policy adjustment.

Throughout the 1980s, Thai foreign policy courted China as another source of security guarantee following Vietnam's invasion of Cambodia in 1978 and the installation of a pro-Hanoi government there. The fear of the Vietnamese threat and its intention to continue promoting the Federation of Indochina was heightened among Thai policy elites, especially within the right-wing group. This resulted in the return of military regime that led to a break in the foreign policy of peaceful coexistence briefly implemented during 1976–1977. Due to the relaxing international environment, Thailand attempted to diversify support from various major powers and supporters across ideological camps, including not only the United States and the free world but also communist China and even the Soviet Union.[18] While maintaining political support from the United States, China rendered military assistance by supplying weapons to both the Thai armed forces and Cambodian resistance groups through Thailand.[19]

It can be seen from the preceding historical discussion that the notion of Thailand as a small power is well grounded in its experience with the international and regional environments in which external powers are involved. Therefore, Thailand's flexible foreign policy tends to lean towards any great powers that are likely to secure its security and stability. Sometimes this tendency of Thai foreign policy is criticised as opportunistic and lacking principle. However, Thailand's foreign policy posture towards its immediate region takes a different path. The next section will discuss in detail how Bangkok position itself in Southeast Asia – particularly in mainland Southeast Asia – and the regional grouping in general.

Another face of Thai foreign policy: seeking regional leadership

Despite the well-established notion of bamboo diplomacy that reflects the nature of the small state, Thai foreign policy towards its immediate neighbouring countries and Southeast Asia is somewhat different. In general, it can be said that an important characteristic of Thai foreign policy towards this particular geographical area is the pursuance of regional leadership to different degrees.[20] This ranges from building a sphere of influence among mainland Southeast Asian neighbours to seeking a leading role within ASEAN. In many circumstances, the first policy objective can be seen as an immediate priority and drives Thailand's position in ASEAN to a certain extent.

Mainland Southeast Asia has traditionally been Thailand's own backyard, where it gained and attempted to assert its hegemonic power. Historically, the tributary state system was an interstate system in this region, in which weaker kingdoms were subordinate to powerful ones. The latter provided security protection while the former pledged their political loyalty and provided military assistance to the dominant states in wartime. Siam (the old name of Thailand) rose to a dominant kingdom when it conquered the Khmer Empire in the late 14th century. It gained loyalties from other small vassal kingdoms which covered the current area of Cambodia, Laos, parts of southern Myanmar and the northern Malay sultanates. Within this system, however, loyalty was not monopolised by a single dominant power, but vassal states could pay loyalty to multiple nodes of power in order to hedge for their survival. Political loyalty rather than territory was significant in this type of state system. Therefore, becoming a tributary did not equate surrendering full authority or sovereignty to the dominant state.

However, in dealing with colonial powers, especially Britain and France in the late 19th century, Siam embraced a modern state concept which carries the idea of having a single sovereign with a defined territory. Therefore, Siam transmuted the vassal state concept to fit into it and allowed the claim of sovereignty over its vassal states. Some of these claims were overlapping with claims by other dominant powers in mainland Southeast Asia, especially Vietnam and Burma, where European power had advanced their occupation. In order to preserve its authorities over these territories, Siam built a historical narrative that justified its claim over its traditional tributaries. A significant chapter in mainland Southeast Asian history is thus about the interactions between Siam and colonial powers who wanted to clearly draw their and Siamese borders.

The legacy from the adoption of the modern-state concept finds in Thailand having a hegemonic idea over the neighbouring countries as part of its "original territory". The idea was elaborated further with the rise of Thai nationalism by the military elites in the 1930s and 1940s under the *Suvarnabhumi* concept. It maintains that the Thai race is a dominating group in continental Southeast Asia, and Thailand as a representation of the race should be a leader in this area. For instance, Thai foreign policy during the Second World War reflects an attempt by the Thai elites to "regain" its lost territories and to build a Pan-Thai League with the help of Japan. Consequently, Siam became an ally with Japan during the war.

This regional hegemonic idea also has two folds. First, it creates distrust and disputes between Thailand and its neighbours. Thailand has had border issues with all neighbours to the present day because of disagreements over the demarcation under colonial pressures. Negotiations have been slow and some of the border disputes escalated to armed clashes or raised to the international bodies. For example, in 1984 Thailand and Laos engaged in skirmishes over the claim of three villages on the border area between Utaradit Province in Thailand and Xayabuli in Laos. The issue was brought to the UN Security Council but no final solution was accepted. Another Thai-Laos border dispute took place in 1987 over the claim of Ban Romklao village in Thailand's Phitsanulok and Laos' Xayabuli provinces.[21] Armed conflict lasted for ten months before truce and negotiation started. Thai-Cambodia border disputes also erupted occasionally, particularly over the ancient Khmer temple of Preah Vihear. The conflict came to public attention in 1959 when Cambodia brought the issue to the International Court of Justice (ICJ). Although the ICJ ruled in 1962 that the temple and its vicinity belonged to Cambodia, it was unclear what the "vicinity" was. This legal loophole led to another dispute in 2008–2013, when Thailand opposed Cambodia's listing the temple as a UNESCO World Heritage Site. Soldiers exchanged fire across the disputed area, leading to deaths and casualties. After failed attempts to dissolve the conflict through bilateral negotiations and ASEAN, Cambodia requested the ICJ to clarify its 1962 ruling. The court unanimously ruled that Cambodia had sovereignty over the territory of the promontory of Preah Vihear and requested Thailand to withdraw its forces.[22]

In addition, the remaining nationalistic sentiment is connected to the border disputes and in most circumstances stirs up the severity of the conflict. The recent Preah Vihear tension provides a good example. In fact, Thailand in 2008 had supported Cambodia's listing in the first place. However, the issue became problematic in the context of Thailand's domestic conflict. The opposition used this issue to mobilise public support for them to discredit the ruling party that eventually pressured the government to abandon its support to Cambodia. It later led to the border tension and ICJ ruling in 2013. In January 2003, public riot against Thailand also took place in the capital of Cambodia. It started from Cambodia media spreading a fabricated news of a famous Thai movie star's claim that Cambodia stole Angkor Wat from Thailand. The angry rioters invaded and attacked the embassy and Thai businesses with inaction from Cambodian authorities to prevent it. Thailand reduced diplomatic relations with Cambodia as a result.[23]

Analysts suggested that Cambodia's ruling party may have been a beneficiary of the rise in nationalist sentiment for the coming general election in July 2003. Therefore, it can be seen that Thailand's relations with its neighbours are generally complicated with clashes of nationalism as well as misunderstandings and suspicion between them.

Second, the *Suvarnabhumi* concept also influences Thailand's grand strategy towards its neighbouring countries as well as its position in ASEAN since its inception. Thailand's neighbours have remained of utmost interest to Thailand's foreign policy since the Cold War era. On the one hand, Thailand accommodated US-led anti-communist expansion strategy in the region for its security. On the other hand, the establishment of communist Vietnam and its hegemonic expansionary policy in Laos and Cambodia threatened Bangkok's traditional sphere of influence. With both policy backgrounds, the Cold War context may have blurred a fine line between these two policy goals. However, Thailand's support to the United States against communism would eventually help Bangkok maintain its regional position vis-à-vis Vietnam, which was regarded as its traditional rival. This regional rivalry over Laos and Cambodia can be seen in Thailand's support of political groups that opposed communist Vietnam, such as the Lao royalists throughout the 1960s.

However, it is clearer that the Thai state was rather concerned with the expansion of Vietnam's influence than the communist threat itself. After Vietnamese troops invaded Cambodia and installed a pro-Hanoi government in 1978, Thailand's security syndrome rose high. It was the first time that Thailand lost a buffer zone against the Vietnamese and could face them just across the border. Thailand mobilised ASEAN support to lead the international campaign and sanctions on Vietnam and its puppet government in Phnom Penh. It entailed a long process of international peace negotiations on the Cambodian issue that dominated regional and international discussions about Southeast Asia throughout the 1980s. Importantly, Thailand supported the Khmer Rouge fraction to fight against the pro-Vietnam regime with military assistance from China. The special relationship between Bangkok and Beijing was cultivated during this regional episode. The Sino-Thai relationship became a major driving force in Southeast Asia's regional affairs during the same period until the peace agreement in the early 1990s.

Evidently, Thailand's foreign policy towards its mainland Southeast Asian neighbours was occupied by security and strategic concerns. At the international level, the common threat of communist expansion helped converge security interest between Thailand and the Great Powers. Beneath that systemic structure, regional and domestic dynamics also play a role. Thailand's foreign policy towards mainland Southeast Asia was largely influenced by its perception that mainland Southeast Asia was Thailand's pivotal area of interest, especially in Laos and Cambodia. The stronger Vietnam led by the communist party would, therefore, have a regional expansionary goal to challenge Bangkok's leadership.

Recalibrating Thailand's regional leadership in the early post–Cold War era

In the post–Cold War environment, the politics of the balance of power declined in large part due to the absence of traditional security threats. In this situation, Thailand has a larger room to shape its foreign policy with its vision. One of the striking patterns of Thai foreign policy in the post–Cold War period is its persistence to play an active role in the region that will ultimately create a regional setting in which Thailand's national economic and strategic interests are best realised.[24] Once again, Thailand looks at its mainland neighbouring countries as an area in which Thailand can ground this policy activism. Major Thai foreign policy since the end of the Cambodian conflict in the late 1980s can demonstrate this pattern.

A major shift in Thai foreign policy revealed its self-confidence during the final days of the Cold War under the Chatichai Choonhavan government (1988–1991). Thailand reconstituted its rapprochement with Indochinese countries under the slogan of "turning the battlefields into marketplaces". Chatichai's leadership had significant politico-security implications by which it paved the way towards the later policy of "constructive engagement" that promoted peaceful coexistence between neighbours regardless of political systems. This foreign policy was also intended to stimulate economic activities by bridging economic opportunities of both parties initially through border trade. It was later supported by the expansion of transport infrastructure that linked inner regions to seaports on Thailand's eastern seaboard.[25]

The Anand interim government (1991–1992) continued a similar posture towards neighbouring countries. Anand formalised the principle of constructive engagement though Chatichai in practice laid its fundamental rationale through his policy above. Constructive engagement became a major guideline for Thailand's policy towards its neighbours, with a particular focus on Myanmar.[26] This policy was also practically adopted and promoted regionwide by ASEAN.

The focus on neighbouring countries appeared again in the succeeding Chuan government (1992–1995). The highlight of his foreign policy was on the promotion of sub-regional economic cooperation, particularly the Quadrangle Economic Cooperation (QEC). It was proposed to coordinate the development of major transport links among China, Laos, Myanmar, and Thailand – the so-called North-South Economic Corridor. The corridor not only strengthened the region's connectivity with the growing Chinese economy but also placed Thailand at the centre of the dynamic.[27] Thailand played an indispensable role in studying the possibility of transport network development and aligning interests among participating countries. It also acted as a facilitator in brainstorming and mobilising support within the member countries and from external donors such as the Asian Development Bank (ADB), Australia, and Japan.[28] Moreover, it offered technical assistance to Indochinese countries to survey and design many road construction projects, including partially financing their construction.[29] This project was later fully incorporated into the Greater Mekong Sub-region (GMS) project supported by the ADB at the end of the 1990s.

Thailand's leadership aspirations continued in the Chavalit government (1996–1997). This time, Chavalit's foreign policy was clearly inspired by the *Suvarnabhumi* concept widely held among Thai military elites. His Indochina policy aimed to designate Thailand as a centre of mainland Southeast Asia. Chavalit declared that his government was going "to restore Thailand's former international influence which rested mainly on improving relations with neighbouring countries".[30] Importantly, his highlighting of foreign policy revolved around the exercise of Thailand's influence over ASEAN's admission of Cambodia, Myanmar, and Laos, including ASEAN's economic programmes aimed at creating transport links between countries in Indochina and the other ASEAN countries for which Thailand was a major hub.[31] Moreover, Chavalit also pushed forwards the Look West policy that intended to put Thailand as a crucial hook that links Southeast Asia and South Asia under the Bay of Bengal Initiative for Multi-Sectoral Technical and Economic Cooperation (BIMSTEC). It was apparent that Chavalit positioned Thailand as an important piece in the jigsaw puzzle to fill the whole picture of potential economic prosperity with a market size of 1.3 billion people. Thailand's leading role as a bridge between two regions was deemed necessary.

The Asian Financial Crisis that hit Thailand in mid-1997 and later spread throughout the region marked the country's losing competitiveness and seriously affecting its confidence to assert any regional leadership. Although the exercise of leadership was temporarily curtailed due to the major focus on crisis responses, its attempt to play an active role in the region remained intact. Moreover, interestingly, the issue of Thailand's neighbours also constituted a core foreign policy at this time. This can be seen in the flexible engagement proposal during the second Chuan government (1997–2001). This policy was proposed to radically adjust ASEAN's traditional norm of non-interference in each other's domestic affairs. Thai foreign policymakers understood that ASEAN's inability to warn each other of foreseeable dangers that might spill over boundaries was partly responsible for the contagion of the 1997 economic crisis. Thailand viewed that the intramural principle weakened and reduced ASEAN's credibility within the international community. Even though flexible engagement was a timely, creative, and effective improvement in the conduct of ASEAN diplomacy, an important focus was also to address problems from Myanmar rooted in its authoritarian regime. It was, therefore, a breakaway from Thailand's support for Constructive Engagement, emphasising tolerance and gradual change in the regime. Thailand's proposal of flexible engagement in ASEAN in 1998 reflected the idea among Thai foreign policymakers to put the spotlight back on Thailand's leadership in regional affairs.

The pinnacle of Thailand's foreign policy: expanding regional role and consolidating leadership in mainland Southeast Asia

Foreign policy under Prime Minister Thaksin Shinawatra (2001–2006) clearly demonstrated his intention to illuminate Thailand's role in Southeast Asia and beyond. Also, Thailand's neighbouring countries constituted the main part of his policy objective. This foreign policy, so-called forward engagement strategy, revolved around new multilateral initiatives complemented by bilateral economic cooperation.[32] Thaksin viewed that although there were many layers of regional cooperation the link between them was weak and Thailand could bridge that gap. This foreign policy strategy emphasised the creation of a web of cooperation with Thailand's strategic Asian partners. This was pursued notably through creating and renewing Thailand's bilateral and regional initiatives. These include various free trade agreements, the Asia Cooperation Dialogue (ACD), Bay of Bengal Initiative for Multi-Sectoral Technical and Economic Cooperation (BIMSTEC), and the Ayeyarwady-Chao Phraya-Mekong Economic Cooperation Strategy (ACMECS).

Viewing from his major initiatives above, it was clear that Thaksin did not abandon Thailand's role in mainland Southeast Asia. Despite the existing cooperation in the Mekong sub-region under the ADB, this time Thailand initiated its own subregional scheme, ACMECS. That body was not only an apparatus for Thailand to narrow the development gap between old and new ASEAN states but also to reconsolidate its own leadership in this area. However, what differed from previous initiatives for the Mekong Basin was that the Thaksin government was committed to perform a greater leading role by including all mainland Southeast Asian nations and building a more systemic cooperation and aid schemes.

Thailand's leadership under ACMECS can be observed in a number of ways. First, it provided intellectual and entrepreneurial leadership that aimed to help farmers in the region. In this scheme, contract farming played an important link in ACMECS. While it provided stable investment and incomes for neighbouring countries' farmers, Thailand would procure produces at cheaper prices for consumption and intermediate goods production such as biofuels.[33] Thailand could also be a link to supply these products between ACMECS countries and other wider region such as ASEAN and China.[34] Second, Thailand was committed to offer unilateral concession and financial assistance to its ACMECS neighbours to facilitate the operation of supply chain. For instance, Thailand unilaterally reduced tariffs under the ASEAN Integration System of Preference (AISP) for ACMECS countries and set forth to eliminate tariffs on the import of all agricultural and handicraft products under the contract farming projects.[35] Thailand offered initial financial assistance in 2004 for US$2.5 million, and promised to add another US$250 million into ACMECS development projects.[36] It also offered export credit of 3 billion baht to Myanmar and 2 billion baht to Cambodia and Laos.[37]

Moreover, Thailand restructured its aid policy, which signified its attempt to directly strengthen its regional influence in ACMECS and other low-income countries. After successfully repaying the IMF standby arrangement in the end of July 2003, Thaksin declared that Thailand would only solicit aid partnerships on the basis of equality and become a donor country. In 2004, the Thailand International Development Cooperation Agency (TICA) was moved under the Foreign Ministry, which intentionally unified and strengthened Thailand's foreign affairs team with financial tools. This change supported Thailand's proactive policy and practice promoted itself as a centre for technical cooperation in Asia. TICA's new role was to "steer and coordinate Thailand's technical cooperation activities under bilateral, regional, and multilateral frameworks".[38] Thailand's policy assumed an "aid for development" approach, which means it mainly focus technical assistance, based on Thai expertise and experience. In 2006, Thailand contributed 100 million baht for an anti-avian flu programme. It pledged financial support for the construction of a bridge between Thailand and Laos in border province of Nakhon Phanom, and for 50 per cent of the construction cost of a new bridge to Laos, Chiang Khong–Houayxay Bridge.[39]

In addition, Thailand helped mobilise third-party support under the Partnership for Development scheme especially with traditional donor countries, such as Australia, France, Germany, Japan, New Zealand and the ADB. For example, Japan pledged its support in 2004 for the improvement project for Savannakhet Airport in Laos. France and Germany also agreed to support Thailand's development schemes in Cambodia, Laos, Myanmar, and Vietnam.[40]

Thailand's major commitment in its aid policy still focused on mainland Southeast Asian countries. According to TICA's statistics, Cambodia, Laos, Myanmar, and Vietnam benefited most from the TICA assistance programmes.[41] Thus Thailand's aid policy and ACMECS were mutually supportive, enabling Thailand to project its leading role in regional development. As a result of this attempt, Thailand provided overseas assistance totally accounting for 0.6 per cent of gross national income in 2003, and stabilised at around 0.1–0.13 per cent in from 2005 onwards (World Bank 2010). These figures are interesting if compared with those of the United States, which were 0.11 per cent in 2001[42] and 0.17 per cent in 2004.[43]

The attempt during the Thaksin government demonstrated that Thailand wanted to strengthen its regional leadership. Coupled with Thaksin's stability gaining from its control of the majority in parliament, his government could pursue a foreign policy with clearer and elaborated strategies. In Thaksin's vision, mainland Southeast Asia was still clearly Thailand's strong base for its wider regional leadership. Thailand's vision is well captured by the statement of TICA's former deputy director: "Thailand's overseas assistance helps build a pro-Thai perception among new generation leaders especially in CLMV countries in the long run, similar to what the American used to do in East Asia during the Cold War".[44]

ASEAN as an extension of Thailand's leadership aspirations

What about Thailand's position in ASEAN? Certainly ASEAN became an important regional entity and increasingly more integrated and is part of Thailand's foreign policy calculation. Thailand has played an important role in the group in many respects since its start. As alluded to earlier, Thailand's major foreign policy has centred on securing stability and establishing Thailand's leadership around its neighbouring countries. Looking through the idea of neighbouring policy as Thailand's major policy platform, ASEAN fits whenever it serves such policy.

Foremost, Thailand is a founding member of ASEAN. Arguably, ASEAN was established partly because of Thailand's concern of the expansion of communism around itself amid the failure of the earlier multilateral collective defence under SEATO and increasing unreliability of the American security guarantee.[45] Although ASEAN officially emphasised socio-economic cooperation, it is undeniable that ASEAN's major success in the Cold War era was its ability to mobilise regional and international support against Vietnam's invasion of Cambodia throughout the 1980s. Within ASEAN, Thailand was the main stakeholder that played an important role. It is not exaggerating to say that ASEAN's position on the Cambodia issue was largely Thailand's position.

Second, Thailand's role can be seen in its continuous support for ASEAN institutionalisation in the post–Cold War period. In the security area, Thailand helped push ASEAN as a central platform for a multilateral security dialogue under the ASEAN Regional Forum (ARF). Its objective is to encourage confidence-building and preventive diplomacy in Asia Pacific, as well as to foster constructive dialogue and consultation on political and security issues between member states. Thailand hosted the first ARF meeting in July 1994.[46] Another evidence can be seen from the establishment of the ASEAN Troika in 1999. Despite its failed proposal for forward engagement in ASEAN, the idea was modified to accommodate more close consultation among ASEAN members. It acts as a ministerial coordination mechanism between the former, current, and incoming ASEAN chairpersons to ensure the continuity of ASEAN direction and policy. Issues of immediate concern to the region can be raised so that ASEAN could seek ways to handle in a more timely and effective manner.[47]

Thailand was also active in promoting ASEAN integration and its engagement with regional partners. As mentioned above, Thailand was a spearhead in building ASEAN 10 which embraced the membership of Cambodia, Laos, Myanmar, and Vietnam after the Cold War in the organisation. Bangkok also played a significant role in other process of regional integration. For instance, the ASEAN Free Trade Area (AFTA) was proposed by Thailand in 1992, which was a significant step for ASEAN economic integration. Thailand also pioneered a free trade agreement with China by jumpstarting an early harvest agreement on agricultural products in 2003, seven years ahead of the whole group. ASEAN Plus Three (APT/ASEN+3) idea was also formed during the first Asia-Europe Meeting in Bangkok in 1995 as a policy coordinating platform for between ASEAN and Northeast Asian countries, including China, Japan, and South Korea.[48] APT played a crucial role during the Asian Financial Crisis in 1997 when it actively sought solutions for affected countries in East Asia. An agreement to conduct a series of bilateral currency swaps under the Chiang Mai Initiative (CMI) resources was outstanding cooperation under APT.

Therefore, it is clear that ASEAN became another important platform for Thailand's foreign policy in the post-Cold War period. However, Thailand's policy towards ASEAN is not consistent and sometimes became an obstacle for the group to move forward. Critics reflect that despite the fact that some important economic schemes in ASEAN were pushed by Thailand, for example, it still remains one of the most protected markets in ASEAN. Particularly, the Asian Financial Crisis in 1997 raised the fear and suspicion on economic liberalisation and internationalisation among policymakers and local industries in Thailand.[49]

Furthermore, Thailand's domestic politics seriously affects its leading role in ASEAN. The frequent change of government, especially in the 1990s, due to the fragile coalition weakened Thailand's policy continuity. Although each political leaders carried foreign policy vision, it lacked articulation as a long-term strategy. This can be seen from the fact that it is rare to see Thai leaders who attended at least two ASEAN summits.[50] Political struggles since 2005 has also waned Thailand's leading role in ASEAN as Thailand experienced another two military coups in 2006 and 2014. Post-2006 coup politics is largely characterised by political polarisation led to a series of street protests between the anti-Thaksin and pro-Thaksin groups. The political division has become clear and affected the direction of policymaking and Thailand's regional leadership in a number of ways. First, the political conflict damaged political stability engineered by the 1997 constitution. Thailand changed five prime ministers since the coup in 2006 until the recent one in 2014. This change surely disrupts policy continuity and focus. Foreign policy was then subsumed to political survival of the government in power. In many circumstance, "the Foreign Ministry has been so busy with having to explain Thailand's political situation to the world that it has been difficult to plan a long-term strategy in the domain of foreign affairs".[51]

The weak government prior to the coup in 2014 also crippled Thailand's capability to even carry on some important activities in ASEAN. This can be seen in Thailand's ASEAN chairmanship between July 2008 and December 2009. Due to the concerns of street protest against the government, the two important ASEAN summits that year were held separately in different locations. The 14th ASEAN Summit was organised in February 2009 in Hua Hin, while the summit with dialogue partners was in April in Pattaya. This was unusual for ASEAN to have these two major events in separation. However, the ASEAN+3 and ASEAN+6 summits in Pattaya were cancelled due to the anti-Abhisit government demonstration storming part of the venue. Fifteen leaders were evacuated by helicopter.[52] ASEAN member countries and observers started to doubt Thailand's leadership and its ability to participate in the regional affairs in an effective manner amid this political crisis.[53]

Furthermore, the manipulation of nationalism amid the political crisis also affected Thailand's leading role in ASEAN. As mentioned earlier, the re-emergence of border dispute on the area of Pheah Vihear Temple that led to Cambodia's request for ICJ to clarify its 1962 ruling was an important case. Not only did the case show Thailand's fragile relations with its neighbours, but it also suggested how ASEAN has a weight in Thailand's position. ASEAN asserted a mediating role as an observer to monitor to ceasefire. While Cambodia welcomed such move, Thailand was reluctant with it and insisted on bilateral talks rather than involving international actors.[54] As a founding member of ASEAN and long-time active actor in the group, Thailand's unwelcoming attitude towards the role of the organisation in this case demonstrates the severe impact of nationalism on Thailand's foreign affairs and its leading role in ASEAN.

Conclusion

Contrary to the conventional wisdom that postulates Thailand's foreign policy behaviour as a policy of a small state, labelled as "bamboo bending with the wind" diplomacy, this chapter offers another side of the coin. That is, Thai foreign policy towards Southeast Asia, including its immediate neighbours and ASEAN as a whole, exhibits its own regional leadership ambitions. With its mainland neighbours, historical legacies arguably have a significant influence on Thailand's foreign policy directions. Thailand perceives itself as a leading actor in this immediate area and attempts to assert both political and economic influences throughout the modern history. Many major policy initiatives and decision-making of Thailand about its neighbours can refer back to this lingering idea of *Suvarnabhumi*. Therefore, Thailand's neighbouring policy can also be seen as a manifestation of this ideational construct. Certainly, vested interests among key stakeholders including traditional elites and local businesses also partake in policy formulation. This result in different strategies and tactic to attain such business outcomes. However, they have a consensus that Thailand's neighbouring area is the key backyard where Thailand should sustain its leading role as the centre of the regional dynamics.

ASEAN becomes another platform in which Thailand can play an active role. As a founding member of the group, Bangkok has contributed to the organisation not only for its establishment but later both regarding policy direction and institutional development. However, Thailand's active role in ASEAN is not consistent after the end of Cambodia conflict in the early 1990s. Although many initiatives were pushed by Thailand in the post-Cold War era, its proactivity largely depends on Thailand's domestic settings, including political leadership discontinuity, weak business sector, and recently, political crisis.

What is more unswerving in Thailand's policy is that neighbouring countries still constitute the first policy priority whereas ASEAN is rather secondary to that. The cases mentioned in the previous section, including constructive engagement, flexible engagement, the promotion of sub-regional economic cooperation both in the early post–Cold War and during the Thaksin administration, all mainly aim at Thailand's neighbours. This brings us to an interesting observation that many critics on Thailand's foreign policy have raised about the country's ASEAN policy: since the Cambodia conflict, Thailand has not paid much attention to ASEAN in itself as its major policy platform. Many of Thailand's regional initiatives bypassed ASEAN or saw ASEAN as a side-line, not the main supporter.

Looking at Thailand's attempts to sustain its regional leadership, especially in its neighbouring mainland Southeast Asia, offers another window to generate different facets of understanding Thailand's foreign policy. Apart from the small state approach in IR, other concepts can lend different analytical tools to explain nuances of Thailand's policy behaviour. For example, the emerging middle-power concept argues that emerging middle powers play more of a role in the post–Cold War period and are mostly found in developing countries. While traditional middle powers in the advanced Western countries share the policy that aims to maintain the existing international system, the emerging middle powers in developing world tend to use their dominant position in a particular region to reform or restructure the existing rules and structure. They are likely to initiate or lead regional cooperation actively.[55] Therefore, Thailand's active role in Southeast Asia as discussed throughout the chapter can be better explained by this emerging middle-power concept rather than the conventional wisdom of bamboo diplomacy.

Notes

1 Arne Kislenko, "Bending with the Wind: The Continuity and Flexibility of Thai Foreign Policy". *International Journal*, Vol. 57, No. 4 (2002), p. 537.
2 Kenneth Neal Waltz, *Theory of International Politics* (London: Addison-Wesley Pub. Co., 1979), p. 162.
3 Ingebritsen et al., *Small States in International Relations* (Seattle: University of Washington Press, 2006), p. 4.
4 Robert O. Keohane, "Lilliputians' Dilemmas: Small States in International Politics". Edited by Liska, George, Osgood, Robert E., Rothstein, Robert L. and Vital, David, *International Organisation*, Vol. 23, No. 2 (1969), p. 60.
5 Robert L. Rothstein, *Alliances and Small Powers* (New York: Columbia University Press, 1968), p. 4.
6 Jeanne A.K. Hey, *Small States in World Politics: Explaining Foreign Policy Behavior* (Boulder: Lynne Rienner Publishers, 2003), p. 6.
7 Hans Joachim Morgenthau, *Politics Among Nations: The Struggle for Power and Peace* (New York: A.A. Knopf, 1948), p. 133.
8 Thanat Khoman, "Riroem Kotang Samakhom Prachachat Echiatawan-Okchiangtai (Asian) [The Initiative of Establishing the Association of South East Asian Nations (ASEAN)]", in *Ruam Ngan Khian Lae Pathakatha Rueang Kan Tangprathet Khong Thai Chak Aditta Thueng Patchuban* [Collection of Articles and Speeches on Thai Foreign Affairs from the Past to the Present] (Bangkok: Thammasart University Press, 1999), p. 186.

9 Clark Neher, "The Foreign Policy of Thailand", in *The Political Economy of Foreign Policy in Southeast Asia,* edited by Wurfel, David, and Burton, Bruce (London: Macmillan, 1990), p. 201.
10 Clark Neher, "Foreign Relations", in *Thailand: Its People, Its Society, Its Culture,* edited by Frank J. Moore (New Haven, CT: HRAF Press, 1974), p. 311.
11 B.J. Terwiel, *A History of Modern Thailand, 1767–1942* (St. Lucia; New York: University of Queensland Press, 1983), p. 304.
12 David K. Wyatt, *Thailand: A Short History* (New Haven: Yale University Press, 2003), pp. 217–219.
13 Ibid., pp. 250–251.
14 Chris Baker and Pasuk Phongpaichit, *A History of Thailand* (Cambridge: Cambridge University Press, 2005), pp. 135–137.
15 Robert J. Muscat, *Thailand and the United States: Development, Security, and Foreign Aid* (New York: Columbia University Press, 1990).
16 Wiwat Mungkandi, "The Security Syndrome: 1945–1975", in *A Century and a Half of Thai-American Relations,* edited by Wiwat Mungkandi and Warren, William (Bangkok: Chulalongkorn University Press, 1982).
17 Kullada Kesboonchoo-Mead, "A Revisionist History of Thai-US Relations", in *Asian Review 2003: Globalisation and Hegemony* (Bangkok: Institute of Asian Studies, Chulalongkorn University, 2003), pp. 48–49.
18 The then prime minister of Thailand, General Kriangsak Chamanan, made a state visit to Moscow in 1979 after his visit to the United States and China, becoming the first Thai prime minister to visit the Soviet Union.
19 Sukhumbhand Paribatra and Suchit Bunbongkarn, "Thai Politics and Foreign Policy in the 1980s: *Plus Ça Change, Plus C'est La Même Chose*?" in *ASEAN in Regional and Global Context* (Berkeley, CA: Institute of East Asian Studies, University of California, 1986), p. 284.
20 Pongphisoot Busbarat, "A Review of Thailand's Foreign Policy in Mainland Southeast Asia", *European Journal of East Asian Studies,* Vol. 11, No. 1 (2012).
21 Twekiat Janprajak, *Kho Phiphat Khetdaen Thai-Lao* [Thai-Lao Boundary Disputes]. Foundation for the Promotion of Social Sciences and Humanities Textbook Project (Bangkok: Thammasart University Press, 1997).
22 "UN Court Rules for Cambodia in Preah Vihear Temple Dispute with Thailand", *UN News,* 2013 <https://news.un.org/en/story/2013/11/455062-un-court-rules-cambodia-preah-vihear-temple-dispute-thailand>.
23 *The New York Times.* "Cambodia Apologises To Thailand Over Riot", 31 January 2003 <www.nytimes.com/2003/01/31/world/cambodia-apologizes-to-thailand-over-riot.html>.
24 Kusuma Snitwongse, *Thailand's Foreign Policy in the New Millennium,* edited by Shalendra D. Sharma (Berkeley: Institute of East Asian Studies, University of California, 2000).
25 Chuan Leekpai, "Policy Statement of the Council of Ministers of Prime Minister Chuan Leekpai Delivered to Parliament on Wednesday, 21 October 1992". The Secretariat of the House of Representatives. National Assembly Report.
26 Pavin Chachavalpongpun, *A Plastic Nation: The Curse of Thainess in Thai-Burmese Relations* (Lanham, MD: University Press of America, 2005), p. 128.
27 E.C. Chapman and Peter Hinton, "The Emerging Mekong Corridor: A Note on Recent Development (to May 1993)", *Thai-Yunnan Project Newsletter,* No. 21 (June 1993) <www.nectec.or.th/thai-yunnan/21.html#6>.
28 "Rang Naeothang Kan Phattana Khwamruammue Phaitai Khet Siliam Setthakit Rawang Thai-Phama-Lao-Chin (Ton Tai) [Draft of Direction for the Development of Cooperation under the Quadrangle Economic Growth between Thailand, Burma, Laos, and China (Southern)]", Ministry of Foreign Affairs of Thailand. Foreign Ministry Archive, 1994.
29 "Sarup Phon Kan Prachum Siliam Setthakit Thi Mon Thon Yun Nan [Summary of the Quadrangle Economic Growth in Yunnan Province]". Office of the Cabinet, Prime Minister Office, 1994.
30 "Change under Chuan", *Southeast Asian Affairs,* 1998 <https://search.proquest.com/business/docview/216951865/citation/E488890BD81A46A4PQ/1>.
31 John Funston, "Thai Foreign Policy: Seeking Influence", *Southeast Asian Affairs* (1998), p. 304.
32 Thitinan Pongsudhirak, "Thai Foreign Policy under the Thaksin Government: Out of the Box for Whom?" *Thai World Affairs Center (Thai World),* 2004 <www.thaiworld.org/en/thailand_monitor/answer.php?question_id=70>.
33 "Kan Prachum Ratthamontri Acmecs Yang Mai Pen Thangkan [ACMECS Unofficial Ministerial Meeting]", Thai Foreign Ministry, 2004.

34 Anthony M. Zola, "Contract Farming for Exports in ACMECS: Lessons & Policy", Paper presented at the Investment, Trade and Transport Facilitation in ACMECS, Bangkok, 13 March 2007.
35 "Kan Prachum Radap Phunam Acmecs Khrang Thi 2 [Second ACMECS's Leader Summit]", The Secretariat of the Cabinet, 2006.
36 "Kan Prachum Ratthamontri Acmecs Yang Mai Pen Thangkan [ACMECS Unofficial Ministerial Meeting]".
37 Thaksin Shinawatra, "Pathakatha Phiset Khong Nayokratthamontri Huakho 'Mueangthai Nai Anakhot' Wanthi 19 Than Wa Khom 2546 [Special Lecture of the Prime Minister Entitled 'Thailand in the Future', 19 December 2003]", Prime Minister Office.
38 Thailand International Development Cooperation Agency, *Thailand International Cooperation Programme: 2003–2004 Report* (Bangkok: Thai Foreign Ministry, 2005), pp. 6, 8.
39 "Kan Prachum Radap Phunam Acmecs Khrang Thi 2 [Second ACMECS's Leader Summit]".
40 "Kan Prachum Ratthamontri Acmecs Yang Mai Pen Thangkan [ACMECS Unofficial Ministerial Meeting]".
41 Thailand International Development Cooperation Agency, *Thailand International Cooperation Programme: 2003–2004 Report.*
42 United Nations Development Programme, "Millennium Development Goals: A Compact among Nations to End Human Poverty", in *Human Development Report 2003* (New York: United Nations Development Programme, 2003), p. 228.
43 United Nations Development Programme, "Beyond Scarcity: Power and the Global Water Crisis", in *Human Development Report 2006* (New York: United Nations Development Programme, 2006), p. 343.
44 Apinand Phatarathiyanon, Interview, 15 February 2008.
45 Kusuma Snitwongse, "Thailand and ASEAN: Thirty Years On", *Asian Journal of Political Science*, Vol. 5, No. 1 (1997).
46 "About ASEAN Regional Forum".
47 Haacke, *ASEAN's Diplomatic and Security Culture.*
48 Stubbs, "ASEAN Plus Three", pp. 441–442.
49 Stubbs, "Signing on to Liberalisation"; Bowles and MacLean, "Understanding Trade Bloc Formation".
50 Kusuma, "Thailand and ASEAN", p. 96.
51 Pavin, "Thai Foreign Policy: An Unenviable Job to Rebuild Thailand's Credibility", *Bangkok Post*, 19 December.
52 Donald K Emmerson, "Asean's Pattaya Problem", *Asia Time Online*, 2009 <www.atimes.com/atimes/Southeast_Asia/KD18Ae01.html>.
53 Morton Abramowitz and Natalie Parke, "Asean's Continuing Dilemma", Jakarta Post, 1 May 2009 <www.thejakartapost.com/news/2009/05/01/asean039s-continuing-dilemma.html>.
54 *Deutsche Welle.* "Indonesia's Role in the Thai-Cambodia Border Dispute Still Unclear". 25 March 2011 <www.dw.com/en/indonesias-role-in-the-thai-cambodia-border-dispute-still-unclear/a-6483440>.
55 Jordaan Eduard, "The Concept of a Middle Power in International Relations: Distinguishing between Emerging and Traditional Middle Powers", *Politikon*, Vol. 30, No. 1 (2003).

References

"About ASEAN Regional Forum". n.d. ASEAN Regional Forum. <http://aseanregionalforum.asean.org/about.html> (accessed 7 June 2018).

Baker, Chris and Pasuk Phongpaichit. 2005. *A History of Thailand.* Cambridge: Cambridge University Press.

Bangkok Post. 2017. "Time for Thailand to Step in and Lead ASEAS (Again)". 7 August <www.bangkokpost.com/opinion/opinion/1301295/time-for-thailand-to-step-in-and-lead-asean-again>.

Bowles, Paul and MacLean, Brian. 1996. "Understanding Trade Bloc Formation: The Case of the ASEAN Free Trade Area". *Review of International Political Economy*, Vol. 3, No. 2: 319–348.

"Change under Chuan". 1998. *Southeast Asian Affairs.* <https://search.proquest.com/business/docview/216951865/citation/E488890BD81A46A4PQ/1>.

Chapman, E.C. and Hinton, Peter. 1993. "The Emerging Mekong Corridor: A Note on Recent Development (to May 1993)". *Thai-Yunnan Project Newsletter*, No. 21 (June) <www.nectec.or.th/thai-yunnan/21.html#6>.

Chuan Leekpai. 1992. "Policy Statement of the Council of Ministers of Prime Minister Chuan Leekpai Delivered to Parliament on Wednesday, 21 October 1992". The Secretariat of the House of Representatives. National Assembly Report.

Chulanee Thianthai and Thompson, Eric C. 2007. "Thai Perceptions of the ASEAN Region: Southeast Asia as *Prathet Phuean Ban*". *Asian Studies Review*, Vol. 31, No. 1: 41–60.

Deutsche Welle. 2011. "Indonesia's Role in the Thai-Cambodia Border Dispute Still Unclear". 25 March <www.dw.com/en/indonesias-role-in-the-thai-cambodia-border-dispute-still-unclear/a-6483440>.

Funston, John. 1998. "Thai Foreign Policy: Seeking Influence". *Southeast Asian Affairs*, pp. 292–306.

Haacke, Jurgen. 2013. *ASEAN's Diplomatic and Security Culture: Origins, Development and Prospects*. London: Routledge.

Hey, Jeanne A.K. 2003. *Small States in World Politics: Explaining Foreign Policy Behavior*. Boulder: Lynne Rienner Publishers.

Hund, Markus. 2003. "ASEAN Plus Three: Towards a New Age of Pan-East Asian Regionalism? A Skeptic's Appraisal". *The Pacific Review*, Vol. 16, No. 3: 383–417.

Ingebritsen, Christine, Neumann, Iver, Gstohl, Sieglinde and Beyer, Jessica. 2006. *Small States in International Relations*. Seattle: University of Washington Press.

Jordaan, Eduard. 2003. "The Concept of a Middle Power in International Relations: Distinguishing between Emerging and Traditional Middle Powers". *Politikon*, Vol. 30, No. 1: 165–181.

"Kan Prachum Radap Phunam Acmecs Khrang Thi 2 [Second ACMECS's Leader Summit]". 2006. The Secretariat of the Cabinet.

"Kan Prachum Ratthamontri Acmecs Yang Mai Pen Thangkan [ACMECS Unofficial Ministerial Meeting]". 2004. Thai Foreign Ministry.

Keohane, Robert O. 1969. "'Lilliputians' Dilemmas: Small States in International Politics". Edited by Liska, George, Osgood, Robert E., Rothstein, Robert L. and Vital, David. *International Organisation*, Vol. 23, No. 2: 291–310.

Kislenko, Arne. 2002. "Bending with the Wind: The Continuity and Flexibility of Thai Foreign Policy". *International Journal*, Vol. 57, No. 4: 537–561.

Kullada Kesboonchoo-Mead. 2003. "A Revisionist History of Thai-US Relations". In *Asian Review 2003: Globalisation and Hegemony*. Bangkok: Institute of Asian Studies, Chulalongkorn University, pp. 45–67.

Kusuma Snitwongse. 1997. "Thailand and ASEAN: Thirty Years On". *Asian Journal of Political Science*, Vol. 5, No. 1: 87–101.

———. 2000. *Thailand's Foreign Policy in the New Millennium*. Edited by Sharma, Shalendra D. Berkeley: Institute of East Asian Studies, University of California.

Morgenthau, Hans Joachim. 1948. *Politics among Nations: The Struggle for Power and Peace*. New York: A.A. Knopf.

Muscat, Robert J. 1990. *Thailand and the United States: Development, Security, and Foreign Aid*. New York: Columbia University Press.

Neher, Clark. 1974. "Foreign Relations". In *Thailand: Its People, Its Society, Its Culture*. Edited by Moore, Frank J. New Haven, CT: HRAF Press.

———. 1990. "The Foreign Policy of Thailand". In *The Political Economy of Foreign Policy in Southeast Asia*. Edited by Wurfel, David, and Burton, Bruce. London: Macmillan, pp. 177–203.

Nesadurai, Helen E.S. 2003. "Attempting Developmental Regionalism through AFTA : The Domestic Sources of Regional Governance". *Third World Quarterly*, Vol. 24, No. 2: 235–253.

Patience, Allan. 2014. "Imagining Middle Powers". *Australian Journal of International Affairs*, Vol. 68, No. 2: 210–224.

Pavin Chachavalpongpun. 2005. *A Plastic Nation: The Curse of Thainess in Thai-Burmese Relations*. Lanham, MD: University Press of America.

———. 2008. "Thai Foreign Policy: An Unenviable Job to Rebuild Thailand's Credibility". *Bangkok Post*, 19 December.

Pongphisoot Busbarat. 2012. "A Review of Thailand's Foreign Policy in Mainland Southeast Asia: Exploring an Ideational Approach". *European Journal of East Asian Studies*, Vol. 11, No. 1: 127–154.

"Rang Naeothang Kan Phattana Khwamruammue Phaitai Khet Siliam Setthakit Rawang Thai-Phama-Lao-Chin (Ton Tai) [Draft of Direction for the Development of Cooperation under the Quadrangle Economic Growth between Thailand, Burma, Laos, and China (Southern)]". 1994. Ministry of Foreign Affairs of Thailand. Foreign Ministry Archive.

Rothstein, Robert L. 1968. *Alliances and Small Powers*. New York: Columbia University Press.

"Sarup Phon Kan Prachum Siliam Setthakit Thi Mon Thon Yun Nan [Summary of Quadrangle Economic Growth Meeting in Yunnan Province]". 1994. Office of the Cabinet, Prime Minister Office.

Sukhumbhand Paribatra, and Suchit Bunbongkarn. 1986. "Thai Politics and Foreign Policy in the 1980s: Plus Ça Change, Plus C'est La Même Chose?" In *ASEAN in Regional and Global Context*. Berkeley, CA: Institute of East Asian Studies, University of California, pp. 52–76.

Stubbs, Richard. 2000. "Signing on to Liberalisation: AFTA and the Politics of Regional Economic Cooperation". *The Pacific Review*, Vol. 13, No. 2: 297–318.

———. 2002. "ASEAN Plus Three: Emerging East Asian Regionalism?" *Asian Survey*, Vol. 42, No. 3: 440–455.

Terwiel, B.J. 1983. *A History of Modern Thailand, 1767–1942*. St. Lucia; New York: University of Queensland Press.

Thailand International Development Cooperation Agency. 2005. *Thailand International Cooperation Programme: 2003–2004 Report*. Bangkok: Thai Foreign Ministry.

Thaksin Shinawatra. 2003. "Pathakatha Phiset Khong Nayokratthamontri Huakho 'Mueangthai Nai Anakhot' Wanthi 19 Than Wa Khom 2546 [Special Lecture of the Prime Minister Entitled 'Thailand in the Future', 19 December 2003]". Prime Minister Office.

Thanat Khoman. 1999. "Riroem Kotang Samakhom Prachachat Echiatawan-Okchiangtai (Asian) [The initiative of establishing the Association of Southeast Asian Nations (ASEAN)]". In *Ruam Ngan Khian Lae Pathakatha Rueang Kan Tangprathet Khong Thai Chak Aditta Thueng Patchuban [Collection of Articles and Speeches on Thai Foreign Affairs from the Past to the Present]*. Bangkok: Thammasart University Press, pp. 167–187.

Thitinan Pongsudhirak. 2004. "Thai Foreign Policy Under the Thaksin Government: Out of the Box for Whom?" *Thai World Affairs Center (Thai World)* <www.thaiworld.org/en/thailand_monitor/answer.php?question_id=70>.

Twekiat Janprajak. 1997. *Kho Phiphat Khetdaen Thai-Lao [Thai-Lao Boundary Disputes]*. Foundation for the Promotion of Social Sciences and Humanities Textbook Project. Bangkok: Thammasart University Press.

The New York Times. 2003. "Cambodia Apologises to Thailand Over Riot". 31 January <www.nytimes.com/2003/01/31/world/cambodia-apologizes-to-thailand-over-riot.html>.

"UN Court Rules for Cambodia in Preah Vihear Temple Dispute with Thailand". 2013. *UN News*. <https://news.un.org/en/story/2013/11/455062-un-court-rules-cambodia-preah-vihear-temple-dispute-thailand>.

United Nations Development Programme. 2003. "Millennium Development Goals: A Compact among Nations to End Human Poverty". In *Human Development Report 2003*. New York: United Nations Development Programme.

———. 2006. "Beyond Scarcity: Power and the Global Water Crisis". *Human Development Report 2006*. New York: United Nations Development Programme.

Waltz, Kenneth Neal. 1979. *Theory of International Politics*. Reading, MA: Addison-Wesley Pub. Co.

Wiwat Mungkandi. 1982. "The Security Syndrome: 1945–1975". In *A Century and a Half of Thai-American Relations*. Edited by Wiwat Mungkandi and Warren, William. Bangkok: Chulalongkorn University Press, pp. 59–114.

Wyatt, David K. 2003. *Thailand: A Short History*. New Haven: Yale University Press.

Zola, Anthony M. 2007. "Contract Farming for Exports in ACMECS: Lessons & Policy". Paper presented at the Investment, Trade and Transport Facilitation in ACMECS, Bangkok, March 13.

INDEX

Note: *Italicised* page numbers indicate a figure on the corresponding page. Page numbers in **bold** indicate a table on the corresponding page.

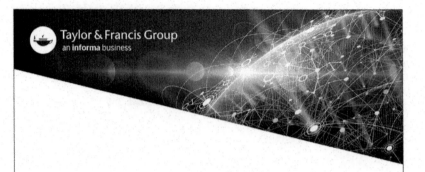